T0329119

HYPOXIA

This is Volume 27 in the

FISH PHYSIOLOGY series
Edited by Anthony P. Farrell and Colin J. Brauner
Honorary Editor: William S. Hoar and David J. Randall

A complete list of books in this series appears at the end of the volume

HYPOXIA

Edited by

Dr. JEFFREY G. RICHARDS
Department of Zoology
The University of British Columbia
Vancouver, British Columbia
Canada

Dr. ANTHONY P. FARRELL
Faculty of Land and Food Systems & Department of Zoology
The University of British Columbia
Vancouver, British Columbia
Canada

Dr. COLIN J. BRAUNER
Department of Zoology
The University of British Columbia
Vancouver, British Columbia
Canada

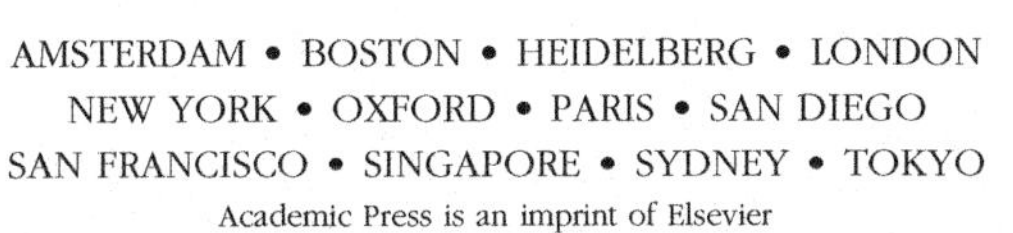

Academic Press is an imprint of Elsevier
84 Theobald's Road, London WC1X 8RR, UK
Radarweg 29, PO Box 211, 1000 AE Amsterdam, The Netherlands
30 Corporate Drive, Suite 400, Burlington, MA 01803, USA
525 B Street, Suite 1900, San Diego, CA 92101-4495, USA

First edition 2009

British Library Cataloging in Publication Data
A catalog record for this book is available from the British Library

Library of Congress Cataloging in Publication Data
A catalog record for this book is available from the Library of Congress

ISBN: 978-0-12-374632-0

For information on all Academic Press publications
visit our website at www.elsevierdirect.com

Printed and bound in

Transferred to Digital Printing, 2011

CONTENTS

4. Oxygen and Capacity Limited Thermal Tolerance
Hans O. Pörtner and Gisela Lannig

5. Oxygen Sensing and The Hypoxic Ventilatory Response
S. F. Perry, M. G. Jonz, and K. M. Gilmour

6. Blood-Gas Transport and Hemoglobin Function: Adaptations for Functional and Environmental Hypoxia
Rufus M. G. Wells

7. Cardiovascular Function and Cardiac Metabolism
A. Kurt Gamperl and W. R. Driedzic

8. The Effects of Hypoxia On Growth and Digestion
Tobias Wang, Sjannie Lefevre, Do thi Thanh Huong, Nguyen Van Cong, and Mark Bayley

9. The Anoxia-Tolerant Crucian Carp (*Carassius Carassius* L.)
Matti Vornanen, Jonathan A.W. Stecyk, and Göran E. Nilsson

10. Metabolic and Molecular Responses of Fish to Hypoxia
Jeffrey G. Richards

CONTRIBUTORS

The numbers in parentheses indicate the pages on which the authors' contributions begin.

MARK BAYLEY *(361), Institute of Biological Sciences Zoophysiology, University of Aarhus, Aarhus C Denmark*

DENISE L. BREITBURG *(1), Senior Scientist, Smithsonian Environmental, Research Center, Edgewater, Maryland*

LAUREN J. CHAPMAN *(25), Department of Biology, McGill University, Montreal, Quebec, Canada*

NGUYEN VAN CONG *(361), College of Environment and Natural Resources, Cantho University, Cantho City, Vietnam*

ROBERT J. DIAZ *(1), School of Marine Science, Virginia Institute of Marine Science, College of William and Mary, Gloucester Point, USA*

W. R. DRIEDZIC *(301), Ocean Sciences Centre, Memorial University of Newfoundland, St. John's, Newfoundland, Canada*

ANTHONY P. FARRELL *(487), Faculty of Land and Food Systems & Department of Zoology, The University of British Columbia, Vancouver, British Columbia, Canada*

A. KURT GAMPERL *(301), Ocean Sciences Centre, Memorial University of Newfoundland, St. John's, Newfoundland, Canada*

K. M. GILMOUR *(193), Department Biology, University of Ottawa, Ottawa, Ontario, Canada*

DO THI THANH HUONG *(361), College of Aquaculture and Fisheries, Cantho University, Cantho City, Vietnam*

M. G. JONZ *(193), Department Biology, University of Ottawa, Ottawa, Ontario, Canada*

GISELA LANNIG *(143), Alfred Wegener Institute, Am Handelshafen 12, Bremerhaven, Germany*

SJANNIE LEFEVRE *(361), Institute of Biological Sciences Zoophysiology, University of Aarhus, Aarhus C Denmark*

DAVID J. MCKENZIE *(25), Institut des Sciences de l'Evolution de Montpellier (UMR 5554 CNRS-Université de Montpellier 2), Station Méditerranéenne de l'Environnement Littoral, 1, quai de la Daurade, France*

GÖRAN E. NILSSON *(397), Division of General Physiology, Department of Biology, University of Oslo, Oslo, Norway*

S. F. PERRY *(193), Department Biology, University of Ottawa, Ottawa, Ontario, Canada*

HANS O. PÖRTNER *(143), Laboratory of Ecophysiology and Ecotoxicology, Alfred Wegener Institute, Am Handelshafen 12, Bremerhaven, Germany*

JEFFREY G. RICHARDS *(443), Department of Zoology, The University of British Columbia, Vancouver, British Columbia, Canada*

JONATHAN A. W. STEYCK *(397), Department of Molecular Biosciences, University of Oslo, Blindern, Oslo, Norway*

MATTI VORNANEN *(397), Professor of Animal Physiology, University of Joensuu, Faculty of Biosciences, Joensuu, Finland*

TOBIAS WANG *(361), Institute of Biological Sciences Zoophysiology, University of Aarhus, Aarhus C Denmark*

RUFUS M. G. WELLS *(255), School of Biological Sciences, The University of Auckland, Auckland, New Zealand*

RUDOLF WU *(79), Department of Biology and Chemistry, City University of Hong Kong, Kowloon Tong, Hong Kong*

PREFACE

Periods of environmental hypoxia are extremely common in aquatic systems due to both natural causes such as diurnal oscillations in algal respiration, seasonal flooding, stratification, ice cover in lakes, and isolation of densely vegetated water bodies, as well as more recent anthropogenic causes (e.g., eutrophication). In view of this, it is perhaps not surprising that among all vertebrates, half of which are fish, fish boast the largest number of hypoxia-tolerant species; hypoxia has clearly played an important role in shaping the evolution of many unique adaptive strategies. These unique adaptive strategies either allow fish to maintain function at low environmental oxygen levels, thus extending hypoxia tolerance limits, or permit them to defend against the metabolic consequences of oxygen levels that fall below a threshold where metabolic functions cannot be maintained.

The past several decades have seen an explosion of research on the responses of fish to hypoxia. The breadth of advances include the evolutionary and ecological consequences of hypoxia exposure in fish in addition to the morphological, behavioral, physiological, biochemical, cellular, and molecular responses that occur in some fish in response to hypoxia exposure. However, with an ever-expanding area of research, the breadth of information available on the responses and adaptations of fish to hypoxia has grown beyond the capacity of a single review article. Fish respond to and survive hypoxia exposure through the integration of numerous adaptive traits, thus a review of the current literature that integrates and synthesizes across levels of biological organization is needed. With this need in mind, we conceived the idea of devoting a single volume of *Fish Physiology* to the responses and adaptations of fish to hypoxia. As a result, the aim of this volume is two-fold. First, this book will review the behavioral, morphological, physiological, biochemical, and molecular strategies used by fish to survive hypoxia exposure and place them within an environmental and ecological context. Second, through the development of a synthesis chapter this book attempts to provide an integrative overview of the responses of fish to hypoxia.

The production of this volume would not have been possible without the contributions of our colleagues. We are truly grateful to all of our colleagues for their thoughtful, knowledgeable, and enthusiastic contributions to this volume. Also, we are grateful to the many reviewers for their constructive comments. Finally, we thank Kristi Gomez and the staff of Elsevier for their support.

Jeffrey G. Richards
Anthony P. Farrell
Colin J. Brauner

1

THE HYPOXIC ENVIRONMENT

ROBERT J. DIAZ
DENISE L. BREITBURG

Low dissolved oxygen environments occur in a wide range of aquatic systems, and vary in temporal frequency, seasonality, and persistence. While there have always been naturally occurring low dissolved oxygen habitats, anthropogenic activities related primarily to organic and nutrient enrichment have led to increases in hypoxia and anoxia in both freshwater and marine systems. Lakes and coastal areas with seasonal stratification tend to be highly sensitive to the consequences of anthropogenic nutrient enrichment. Many systems that are currently hypoxic were not reported to have low dissolved oxygen concentrations when first studied. The rapid rise in the number of coastal hypoxic systems lagged about 20 years behind the increased use of industrial fertilizer. The future status of hypoxia and its consequences for fishes will depend on a combination of climate change (primarily from warming, and altered patterns for wind, currents, and precipitation) and land use change (primarily from expanded agriculture and nutrient loadings). If in the next 50 years humans continue to modify and degrade coastal systems as in previous years, human population pressure will likely be the main driving factor in spreading of coastal dead zones and climate change

Hypoxia: Volume 27
FISH PHYSIOLOGY

DOI: 10.1016/S1546-5098(08)00001-0

factors will be secondary. Climate forcing, however, will tend to make systems more susceptible to development of hypoxia through direct effects on stratification, solubility of oxygen, metabolism, and mineralization rates, particularly in lakes and semienclosed coastal areas.

1. IMPORTANCE OF OXYGEN AND HYPOXIA

Oxygen is necessary to sustain the life of fishes and invertebrates dependent on aerobic respiration. When the supply of oxygen is cut off or consumption exceeds resupply, dissolved oxygen (DO) concentrations can decline below levels required by most animal life. This condition of low DO is known as hypoxia; water devoid of oxygen is referred to as anoxic and can contain lethal concentrations of metabolic products of microbial anaerobic respiration. Thus hypoxia and anoxia differ quantitatively in the availability of oxygen, as well as qualitatively in the presence of toxic compounds such as hydrogen sulfide. While many authors and water quality regulations focus on concentrations of DO below 2–3 mg O_2/L as a threshold value for marine and estuarine environments, and 5–6 mg O_2/L in some freshwater habitats, such arbitrary limits may be unsuitable when examining potential impacts of hypoxia on any one given species or on the way that oxygen concentrations affect interactions among species. Species and life stages differ in their basic oxygen requirements, and oxygen requirements increase as energy-demanding metabolic processes are mobilized.

Depending on temperature and salinity, water contains 20–40 times less oxygen by volume and diffuses about ten thousand times more slowly through water than air (Graham, 1990). This relatively low solubility and diffusion of oxygen in water combined with two principal factors lead to the development of hypoxia and anoxia. These factors are density stratification of the water column that isolates the bottom water from exchange with oxygen-rich surface water and the atmosphere, and decomposition of organic matter in the isolated bottom water that consumes dissolved oxygen. The combination of these factors can allow hypoxia to develop and persist in deeper waters by causing oxygen consumption to exceed resupply. For lakes, factors affecting vertical water mixing such as wind and temperature affect seasonal changes in the DO depth profile and can lower DO in bottom waters (Green *et al.*, 1973). Ice and snow cover on lakes and streams can also block photosynthesis and reaeration, and may lead to hypoxia and "winterkills" (Greenbank, 1945; Magnuson *et al.*, 1985; Graham, 2006). In tropical freshwaters oxygenation is often greater in rainy seasons with more water flow than during dry stagnant water seasons (Val and Almeida-Val, 1995; Graham, 2006).

Because of the low solubility of oxygen in water small changes in the absolute amount of oxygen dissolved in water (resulting from microbial or macrofaunal respiration) lead to large differences in per cent air saturation. Thus what appear to be small changes in DO can have major consequences to animals living in an oxygen-limited milieu. For example, 9.1 mg O_2 will dissolve in a liter of freshwater at 20°C; at this temperature a 1 mg O_2/L drop in oxygen is equivalent to an 11% decline in air saturation (Figure 1.1). Going from freshwater to seawater (35 psu) at the same temperature reduces air saturation to 7.2 mg O_2/L (Benson and Krause, 1984). Some species are particularly sensitive to even small changes in oxygen concentrations. For example, for some salmonids, the limiting factor of DO becomes operative at relatively high values and even air saturation can be limiting at higher temperatures (Fry, 1971).

Hypoxia has been a potent force in evolution. Air breathing and the ultimate evolution of terrestrial vertebrates is thought to have been an evolutionary response to low atmospheric and dissolved oxygen concentrations during the Devonian (Clack, 2007). Within aquatic environments, fishes have developed a wide range of mechanisms to secure more oxygen

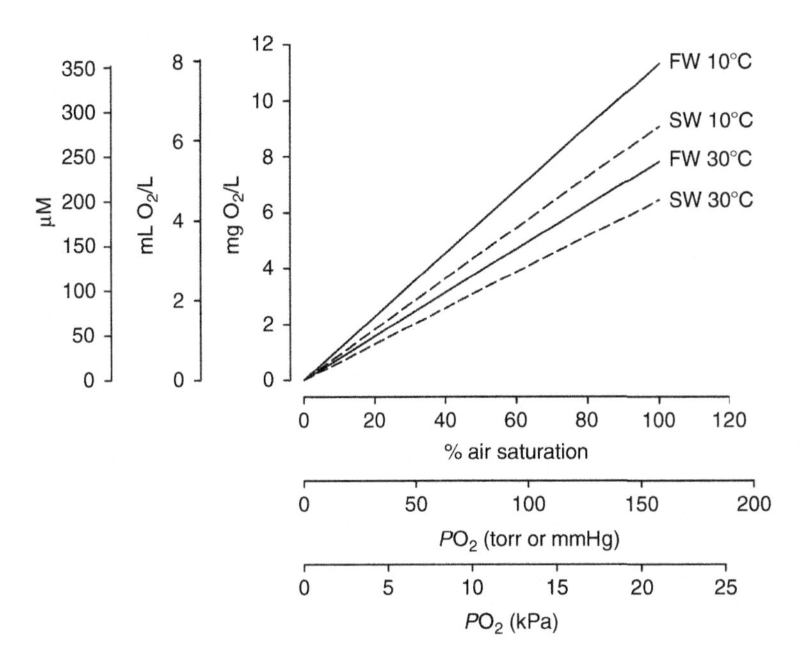

Fig. 1.1. Nomogram for dissolved oxygen in freshwater (FW) and seawater (SW) at 10°C and 30°C. (J. G. Richards, Unpublished data).

from their environment in situations where oxygen availability is critically low (Hoar and Randall, 1984; Brauner *et al.*, 1995; Gonzales *et al.*, 2006). The large number of hypoxia-tolerant aquatic species, and wide variety of anatomical, physiological, and behavioral adaptations to hypoxia, indicate that after the great atmospheric oxygenation event some 2.3 billion years ago (Catling *et al.*, 2001), low DO environments have played an important role in the evolution of many adaptive strategies (Guppy and Withers, 1999; Val, 2000; Bickler and Buck, 2007; see Chapters 2 to 9).

2. HYPOXIA DISTRIBUTION AND CAUSES

2.1. Where Hypoxia Occurs

Oceanic oxygen minimum zones (OMZs) are the largest low DO areas on earth. OMZs form under areas of high surface productivity, which sinks and in the process of microbial metabolism oxygen is consumed (Figure 1.2). They are widespread and stable oceanic features occurring at intermediate depths (typically 400–1000 m), are particularly severe in regions of sluggish circulation, persist for long periods of time (at greater than decadal scales), and are controlled by natural processes and cycles (Wyrtki, 1966; Kamykowski and Zentara, 1990; Helly and Levin, 2004). Where OMZs contact the bottom, globally about a million square kilometers along the continental margins, specialized communities have evolved to survive at DO concentrations as low as 0.1 mg O_2/L (Graham, 1990; Childress and Seibel, 1998; Levin, 2002; Helly and Levin, 2004). Upwelling areas can also develop extensive hypoxia as deep-water nutrients are added to surface waters increasing production that eventually sinks and decomposes. Hypoxia associated with upwelling is not as long-lived and stable as that associated with OMZs.

Hypoxia is a natural component of many freshwater habitats such as swamps and backwaters that circulate poorly, stratify, and have large loads of terrestrial organic matter. Primary productivity, depth, and temperature are the main determinants of the degree of hypolimnetic oxygen depletion in lakes, with both naturally and culturally eutrophic lakes experiencing summer oxygen depletion (Cornett and Rigler, 1980; Wetzel, 2001). In addition, some deep, amictic oligotrophic lakes, like Lake Tanganyika (Coulter, 1967), develop year-round hypoxia and anoxia gradually over time through sinking and decomposition of organic matter. Hypoxia is also common in reservoirs, and lateral variability of hypoxia tends to be greater in these systems than in lakes because of spatial variability in inflow, withdrawal, and loads of particulate organic matter (Thornton *et al.*, 1990). Reservoirs are also more prone to metalimnetic oxygen minima, which are rare in lakes.

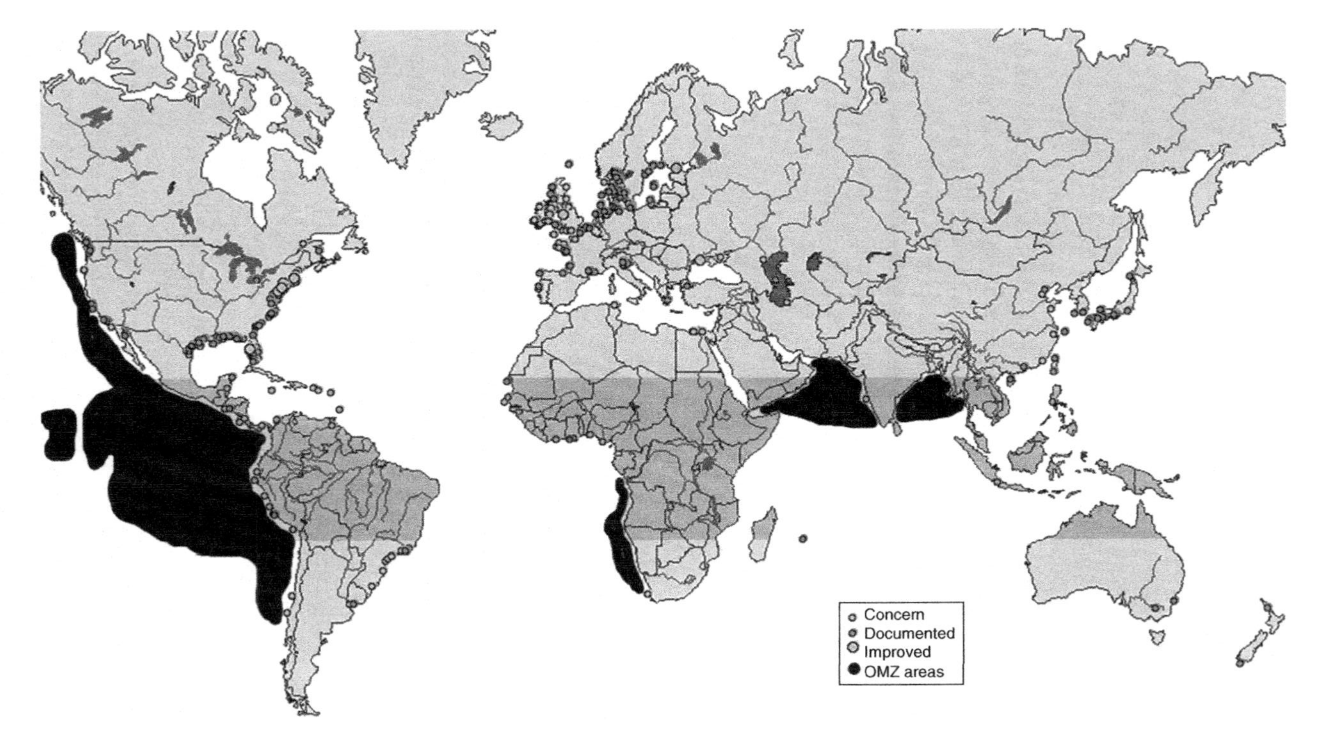

Fig. 1.2. Global distribution of major OMZ areas and coastal hypoxic systems. Systems with documented hypoxia are red circles, areas of concern for being hypoxic are blue circles, and areas that have recovered from hypoxic conditions are yellow circles. Shading indicates the tropical regions (20° north and south of the equator) most likely to experience naturally low dissolved oxygen conditions. [Based on Diaz and Rosenberg (2008), Helly and Levin (2004), and Selman *et al.* (2008).] (See Color Insert.)

Areas of naturally low DO in coastal marine systems are limited to fjord-like systems prone to water column stratification and deep depositional basins, such as Oslofjord, Norway (Karlson *et al.*, 2002) or the central basin of the Black Sea, currently the largest pool of naturally occurring anoxic water on earth (Kideys, 2002). In shallow water, depending on the balance between production and respiration, a natural diel cycling of DO from supersaturation during the day to hypoxic or near anoxic during the night can occur. In highly productive systems, calm weather conditions and extended periods of cloud cover often exacerbate the problem.

Water also becomes hypoxic on floodplains (Townsend and Edwards, 2003; Val *et al.*, 2006), wetlands, and shallow embayments or margins of smaller systems with high productivity and restricted circulation. Among tropical habitats, coral reef crevices can become severely hypoxic at night owing to respiration of coral and associated organisms (Gonzales *et al.*, 2006; Nilsson *et al.*, 2007). During intertidal exposure organisms without adaptations for air-breathing experience hypoxia along with hypercapnia (elevated CO_2) (Warren, 1984; Burnett, 1997).

2.2. Rise of Anthropogenic Influence on Oxygen Budgets

Eutrophication can be defined simply as the production of organic matter in excess of what an ecosystem is normally adapted to processing (Nixon, 1995), however, it is only part of a complex web of stressors that interact to shape and direct ecosystem level processes (Breitburg *et al.*, 1998; Cloern, 2001). The primary driver of eutrophication in both freshwater and marine systems is excess nutrient enrichment, but physical conditions that limit reaeration are also necessary for the development of hypoxia. Thienemann (1926, in Cornett and Rigler, 1980) was one of the first to note that production and morphometry influence oxygen depletion. Phosphorus is generally the limiting nutrient in freshwater (Schindler, 1978) and increases in anthropogenic phosphorus have caused increased algal production and eutrophication in freshwater ecosystems worldwide even where human waste is treated or only a minor contributor to declining water quality (Carpenter *et al.*, 1999; Smith, 2003, 2006). For marine systems the limiting nutrient tends to be nitrogen (Boesch, 2002). This basic difference is related to the physical properties of phosphorus and nitrogen compounds, and their biogeochemical cycling through the freshwater and marine environments.

Eutrophication and associated hypoxia in freshwater systems became widespread in the mid–late 20th century, but effective nutrient management has reversed this trend where it has been rigorously implemented (Jeppesen *et al.*, 2005). In tidal portions of rivers and other water bodies near dense population centers, severe hypoxia and anoxia has been caused by discharge of raw sewage, which is high in both nutrients and organic matter. Areas devoid of fishes were reported at least as early as the late 1800s and persisted until improvements in sewage treatment were implemented (Jones, 2006). Much of the hypoxia and anoxia in shallow coastal marine and estuarine areas is recent in origin (Diaz and Rosenberg, 1995). These areas of hypoxia, commonly called dead zones (Rabalais *et al.*, 2002), tend to be related to a combination of agriculture, human waste, and atmospheric deposition of nitrogen, which has led to a general eutrophication.

Within the last 50 years, dissolved oxygen conditions of many shallow coastal ecosystems around the world have been adversely affected by eutrophication (see Figure 1.2). As more organic matter was produced more oxygen was needed to remineralize the organics, primarily through the microbial loop, and as ecosystems became overloaded DO declined. The declining trend in dissolved oxygen lagged about 20 years behind increased use of chemical fertilizer after World War II (Figure 1.3). For European systems that have historical data from the early 1900s, declines in DO started in the 1950s and 1960s. However, declining dissolved oxygen levels were noted as early as the 1930s in the deep central basin of the Baltic Sea (Fonselius, 1969).

Among marine systems with long-term DO data, benthic hypoxia became a problem in the 1950s in the Baltic Sea proper (Fonselius, 1969), the 1960s in the northern Adriatic (Justić *et al.*, 1987), the 1970s in the Kattegat (Baden *et al.*, 1990), and the 1980s on the Northwest continental shelf of the Black Sea (Mee, 1992). Annual hypoxia does not appear to be a natural condition for marine waters except for those systems previously described. Even in

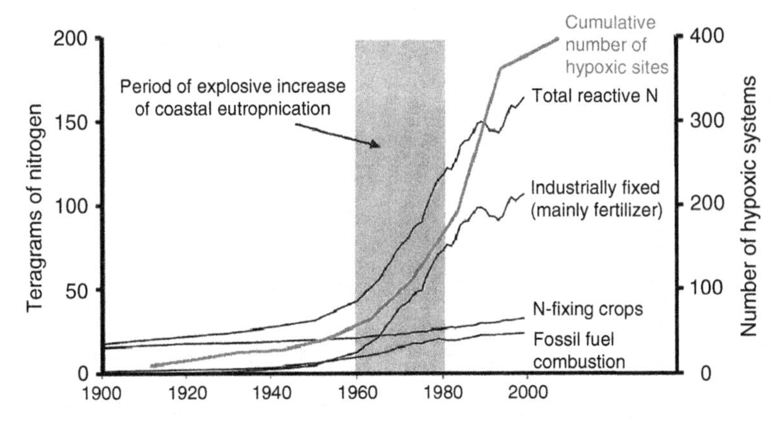

Fig. 1.3. Relationship between fertilizer use and rise of dead zones. [Modified from Boesch (2002) and Diaz and Rosenberg (2008).]

Chesapeake Bay, which had hypoxia when DO measurements were first made in the 1910s in the Potomac River (Sale and Skinner, 1917) and 1930s in the mainstem channel (Officer *et al.*, 1984), the geological record suggests that low DO was not an annual, seasonally persistent feature of the system prior to European colonization (Cooper and Brush, 1991; Zimmerman and Canuel, 2000; Cronin and Vann, 2003). Geochronologies from the hypoxic area on the continental shelf of the northern Gulf of Mexico also indicate that the current seasonal hypoxia, which can cover over 20 000 km^2, did not form annually prior to the 1950s (Sen Gupta *et al.*, 1996). Hypoxia was recorded with the first DO measurement made in the area in the summer of 1973 on the central Louisiana continental shelf (Harper *et al.*, 1981) and has been an annual event ever since. Geochronologies from both of these systems that go back over a 1000 years are at times punctuated by low DO markers that appeared aperiodically and likely marked major discharge events that led to low DO (Osterman *et al.*, 2007).

Recent research and monitoring suggests that once a system develops hypoxia, it can quickly become an annual event and a prominent feature affecting energy flow (Elmgren, 1989; Pearson and Rosenberg, 1992; Baird *et al.*, 2004). From the 1980s to the present, the number of systems reporting hypoxia has increased from <50 in 1960 to about 400 at present (Diaz and Rosenberg, 2008). Only in systems that have experienced intensive regulation of nutrient or carbon inputs have oxygen conditions improved, primarily from initiation of sewage treatment that at first removed organic matter and later from substantial upgrades in treatment level reduced nutrients. Examples include the Hudson River, New York, Delaware River, Pennsylvania-New Jersey, and the Mersey Estuary in England (Patrick, 1988; Brosnan and O'Shea, 1996; Jones, 2006). The northwest shelf of the Black Sea once experienced annual hypoxic events, but is now in a state of recovery largely due to the economic collapse of Eastern Europe in the early 1990s, which greatly reduced the use of fertilizer and subsequent nutrient loading in runoff. Within 3 years, the hypoxic area in the northwest shelf of the Black Sea went from a maximum area of about 40 000 km^2 to none. While no hypoxic events were recorded on the shelf between 1993 and 2001, a full recovery of the Black Sea is far from certain. Climatic conditions caused a large hypoxic area to form during a warmer than average 2001 and expected recovery of farming in Eastern Europe will likely lead to increased nutrient loadings (Mee, 2006).

Temporary improvements have also been seen in systems with changes in hydrology or nutrient inputs. In the northern Gulf of Mexico the size of the hypoxic area responds annually to Mississippi River discharge with low flow years having less hypoxia and high flow years more (Rabalais *et al.*, 2007). Large-scale meteorological events that disrupt stratification are also capable of reducing the area of hypoxia, as in the Gulf of Mexico (Rabalais *et al.*, 2007) and the Gulf of Finland (Karlson *et al.*, 2002).

In most coastal marine systems and in many freshwater habitats, hypoxia appears to be a consequence of general ecosystem eutrophication. As a result, it is difficult to separate the effects of hypoxia from effects of other symptoms of nutrient enrichment or other co-occurring stressors (overfishing, habitat loss, contaminants) on ecosystem functioning (Cloern, 2001; Breitburg, 2002; Breitburg *et al.*, in press). Nutrients are closely linked to a system's secondary productivity and to a point enhance biomass and fisheries yield (Caddy, 1993, 2000; Nixon and Buckley, 2002). The general effect of eutrophication to favor benthic species with opportunistic life histories and eliminate sensitive species leads to higher production of benthic invertebrates (prey resources) during normoxic periods, which can either become available or be lost to higher-level predators depending on the severity and extent of hypoxia (Baird *et al.*, 2004). Another critical point in a system's trajectory of decline is the appearance of anoxia and associated H_2S, which have the potential to produce mass mortality of both benthic and pelagic species. The positive effect of nutrient enrichment on fisheries (i.e., total fisheries' landings, not individual species) may last until as much as 40% of the bottom is affected by hypoxia (Breitburg *et al.*, 2009).

The frequency and duration of hypoxic events vary among systems, over time, and with varying nutrient loads or organic accumulation. Hypoxia ranges from aperiodic events with years to decades between reoccurrences to a persistent year-round feature that can last for years or centuries at a time. Dominant faunal responses differ by type of hypoxia (Figure 1.4).

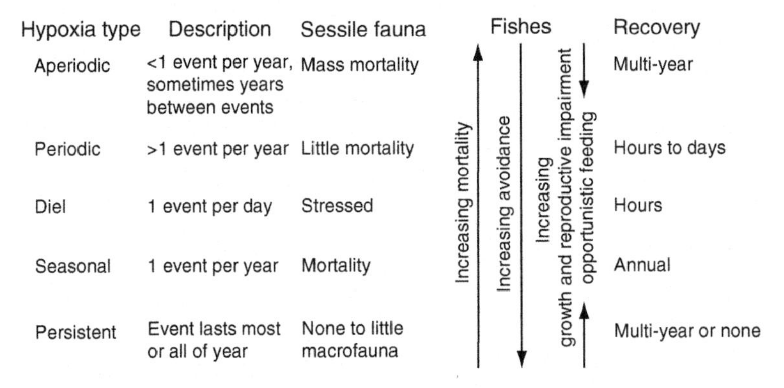

Fig. 1.4. Types of hypoxia and generalized faunal response. Sessile fauna is primarily macrobenthos. Arrows indicate direction of increased impact on fishes. Mortality in fishes is more likely from aperiodic hypoxia, with complete avoidance of persistent hypoxia. Physiological impairment and opportunistic feeding are greatest for periodic and diel hypoxia.

Aperiodic hypoxia, resulting from unusual or uncommon climate conditions, elicits the most dramatic response of mass mortality in sessile and, at times, mobile species. For benthic invertebrates, this dramatic response is due to the large numbers of sensitive species usually present prior to the hypoxic event. For example, the onetime hypoxic event in the New York Bight in 1976 caused mass mortality of many commercial and noncommercial species (Boesch and Rabalais, 1991). Many systems that now experience seasonal hypoxia started out with reports of aperiodic hypoxic events. Wind- or tidal-mixing periodically disrupts stratification and hypoxia in many systems lessening the effects on sessile fauna and at times allowing mobile fanua to return. This form of periodic hypoxia, with several to many events per year, is also common on the edges of seasonal and persistent dead zones as currents or wind-driven upwelling cause hypoxic bottom waters to move. The form of periodic hypoxia with the most frequent recurrences is diel cycling hypoxia, which appears to be common in shallow systems, and is driven by the balance between oxygen production during daylight and respiration at night. Seasonal hypoxia, typically occurring during summer or early autumn, is common and often causes mortality of benthos followed by benthic recolonization, as well as avoidance by mobile species with their return as oxygen concentrations increase. Persistent hypoxia that develops anoxia has the greatest effect on benthic fauna by removing all habitat value of the bottom for extended periods of time.

2.3. Oxygen Budgets and Global Climate Change

If in the next 50 years humans continue to modify and degrade coastal systems as in previous years (Halpern *et al.*, 2008), human population pressure will likely continue to be the main driving factor in the persistence and spreading of coastal dead zones (Figure 1.5). Expanding agriculture for production of crops to be used for food and biofuels will result in increased nutrient loading and expand eutrophication effects (EPA Science Advisory Board, 2007; Rabalais *et al.*, 2007). Climate change, however, may make systems more susceptible to development of hypoxia through direct effects on stratification, solubility of oxygen, metabolism, and mineralization rates. This will likely occur primarily though warming, which will lead to increased water temperatures, decreased oxygen solubility, increased organism metabolism and remineralization rates, and enhanced stratification. Changing temperatures will lead to spatial shifts in habitat suitability for fishes and will favor some species over others in a wide range of habitats. Warming may be particularly important to the development of hypoxia in lakes by leading to expanded periods of thermal stratification and deepening of thermoclines, which can lead to an increase in oxygen demand for aerobic decomposition,

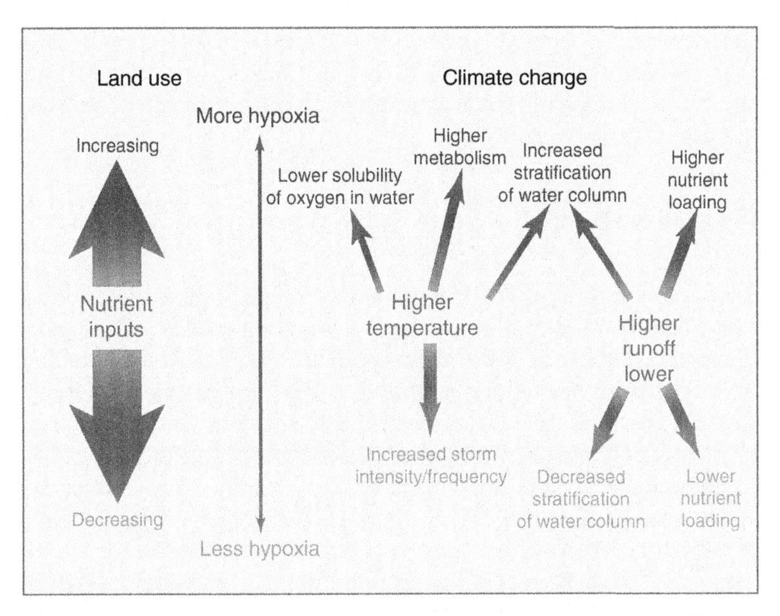

Fig. 1.5. Relative contribution of global climate change and land use to future hypoxia. Thickness of the arrows indicates relative magnitude of effect.

promote an upward flux of phosphorus from sediments, and thereby increase the concentration and amount of phosphorus in the hypolimnion (Magnuson and Destasio, 1996; Komatsu *et al.*, 2007). Climate warming is, however, projected to lessen or eliminate winterkills in some lakes by reducing the period of ice cover in higher latitude lakes (Fang and Stefan, 2000; Fang *et al.*, 2004).

The future pervasiveness of hypoxia in all ecosystems will depend upon a combination of climate change and land management. Climate change will affect water column stratification, organic matter production, nutrient discharges, and rates of oxygen consumption. Land management will also affect the concentrations of nutrients through agriculture. General circulation models predict a decrease in the global oceanic oxygen inventory through increased stratification and warming (Keeling and Garcia, 2002), which may lead to expanding OMZs. Large changes in rainfall patterns are also predicted (IPCC, 2007). If these changes in rainfall lead to increased runoff to estuarine and coastal ecosystems, stratification and nutrient loads are likely to increase and worsen oxygen depletion (Justić *et al.* 2007). Conversely, if stratification decreases due to lower runoff or is disrupted by increased storm activity or intensity, the chances for oxygen depletion should decrease. For the Mississippi River basin associated with the northern Gulf of Mexico seasonal dead zone, climate predictions suggest a 20% increase in river discharge (Miller and Russell, 1992) that would lead to elevated nutrient loading, a 50% increase in primary production, and expansion of the oxygen-depleted area (Justić *et al.*, 1996).

3. HYPOXIA AND FISH

Fishes respond to hypoxia through a wide range of physiological, anatomical, and behavioral adaptations that vary among species, life stages, and habitats (see Chapter 2). At the physiological level, critical adaptations include mechanisms to reduce metabolic rates and increase tolerance of ionic and pH disturbances during exposure to hypoxia, and mechanisms to reduce free-radical damage during reaeration (reviewed in Bickler and Buck, 2007). Anatomical adaptations ranging from highly vascularized buccal cavities to lungs permit the use of atmospheric oxygen by fish residing in waters with low dissolved oxygen concentrations (Brauner *et al.*, 1995; Randall *et al.*, 2004; Soares *et al.*, 2006). Changes in behavior can allow fishes to access more highly oxygenated environments (Kramer, 1987) and reduce the effects of respiration by nearby individuals on local oxygen concentrations (Domenici *et al.*, 2007). The stress of hypoxia can lead to a dramatic decrease in preferred temperature to gain physiological advantages associated with lower temperatures (Crawshaw and O'Connor, 1997). Fish species vary widely in their tolerance to low oxygen, from highly sensitive species to carp (*Carassius* spp.), which can survive months of hypoxia and 2 days of anoxia at low temperatures (reviewed in Nilsson and Renshaw, 2004; see Chapter 9).

3.1. Consequences for Fish

Effects of hypoxia on fishes and adaptations of fishes to low oxygen environments have been longstanding areas of interest in research and management. By 1913, experiments established behavioral avoidance of low dissolved oxygen concentrations (Shelford and Allee, 1913) as well as variation among species and effects of body size in tolerance to low oxygen (Wells, 1913). Studies of adaptations to low oxygen in swamps were already well underway in the 1930s (Carter and Beadle, 1931). Research in the 1940s examined effects of winter hypoxia on fishes in north-temperate lakes (e.g., Greenbank, 1945; Cooper and Washburn, 1949). Jones (1952) recognized that a lack of DO was a major hazard to fishes and that by the 1920s the literature on the effects of oxygen deficiency on fishes was extensive (Gardner, 1926; Black, 1951). Davis (1975) reviewed oxygen requirements of Canadian freshwater and marine species with the intention of determining water quality criteria relative to minimum dissolved oxygen concentrations. These are some of the first works to point to the critical importance of DO, but it was not obvious that DO would become critical in shallow coastal marine systems until the 1970s and 1980s when large areas of low dissolved oxygen started to appear with associated mass mortalities of invertebrates and fishes (Diaz and Rosenberg, 1995). Before the 1950s, there were few reports of mass mortalities of marine animals related to lack of oxygen (Brongersma-Sanders, 1957). However, in the 1940s, hypoxia-driven migrations of mobile organisms to the edge of the water (Jubilee) were reported in Mobile Bay, Alabama (Loesch, 1960). In addition, there were a number of highly urbanized rivers and estuaries that were devoid of fishes by the late 1800s (Araújo *et al.*, 2000; Jones, 2006).

Behavioral responses of fishes to low DO concentrations have been well studied (Duque *et al.*, 1988; Pihl *et al.*, 1991; Plante *et al.*, 1998; see Chapter 2). In general freshwater and marine fishes are capable of actively avoiding low DO water (e.g., Jones, 1952; Schurmann and Steffensen, 1992; Schurmann *et al.*, 1998; Breitburg, 2002). However, the point at which various fishes initiate behavioral response to declining DO and eventually suffocate varies widely among species and habitats. For North American freshwater fishes 5 mg O_2/L appeared to be a lower limit for maintaining a desirable riverine fish fauna (Jones, 1952). But for coldwater salmonids behavioral responses are initiated at 8 mg O_2/L. Marine fishes avoid DO concentrations similar to those that reduce growth (Breitburg, 2002), but growth reductions related to hypoxia occur in the field as a result of imperfect avoidance, the energetic costs of avoiding hypoxia, and density-dependent processes in normoxic parts of systems (Breitburg, 1992; Taylor and Miller, 2001; Perez-Dominguez *et al.*, 2006; Stierhoff *et al.*, 2006).

Consequences of low DO are often sublethal and affect growth (Stewart *et al.*, 1967; Andrews *et al.*, 1973; Pedersen, 1987), immune responses (Thomas *et al.*, 2007), and reproduction (Wu *et al.*, 2003; see Chapter 3). For example, cod from the northern Gulf of St. Lawrence may be less productive than other stocks not only because they live in cold water (Brander, 1995; Dutil *et al.*, 1999), but also because deep waters in the northern Gulf of St. Lawrence are hypoxic and some segments of the cod stock are found in deep waters (D'Amours, 1993; Gilbert *et al.*, 2005). Growth is a determinant of cod surplus production in the northern Gulf of St. Lawrence and factors that affect growth such as DO need to be considered to better forecast stock status (Dutil *et al.*, 1999).

While the global distribution of coastal hypoxic zones is centered on major population centers or closely associated with developed watersheds that deliver large quantities of nutrients (Howarth *et al.*, 1996), these same areas have been major fishing grounds. The direct connection of hypoxia to fisheries' landings at large regional scales is weak because of a number of factors that include confounding effects of overfishing and compensatory mechanisms that alter or mask effects of hypoxia on landings (Breitburg *et al.*, 2009, in press). Both mobile species and fishers can distribute themselves to avoid low DO and utilize prey-enriched areas (Breitburg, 2002; Eby and Crowder, 2004; Craig and Crowder, 2005).

3.2. Consequences for Fish Habitat

When a system becomes hypoxic, fishes have to contend with loss of habitat and, for demersal feeders, loss of prey resources. The interaction of hypoxia with habitat requirements varies by species and life stage. Habitat compression or habitat "squeeze" can occur where hypoxia overlaps with nursery habitat or makes deeper, cooler water unavailable during the summer (Coutant, 1990). Spawning success of cod in the central Baltic is hindered by hypoxic and anoxic water below the halocline (70–80 m) where salinity is high enough to provide buoyancy for cod eggs (Nissling and Vallin, 1996; Cardinale and Modin, 1999).

The elimination of benthic prey and compression of habitat by hypoxia also have profound effects on ecosystem energetics as organisms die and are decomposed by microbes. Under certain circumstances demersal feeding fishes are able to utilize low DO-stressed benthic prey. Over a narrow range of conditions, hypoxia can therefore facilitate the upward trophic transfer as physiologically stressed benthic fauna forced to the sediment surface during hypoxia may be exploited by predators (Pihl *et al.*, 1992; Nestlerode and Diaz, 1998). Aggregation of demersal predators on the edge of dead zones may be a combination of responses that include flight from physiological stress and trophic advantage (Craig and Crowder, 2005). Thus, short-lived and mild hypoxia may not have a net negative effect on trophic transfer as does severe seasonal hypoxia.

This low DO-associated facilitation is most prevalent under diel cycling hypoxic conditions (see Figure 1.4). Diel cycling hypoxia may be a common phenomenon to which fishes respond in an opportunistic manner (D'Avanzo and Kremer, 1994; Layman *et al.*, 2000; Smith and Able, 2003; Tyler and Targett, 2007). In Delaware coastal creek systems, fish emigrated only when DO dropped to very low values and returned quickly with improving DO (Ross, 2003; Tyler and Targett, 2007). Acoustic tagging of juvenile weakfish in Pepper Creek, Delaware, demonstrated that these fish track the dynamics of DO, moving down the creek as DO falls and returning as DO rises (Brady and Targett, unpublished data). While it may be physiologically stressful for juvenile fishes to remain in or near low DO water, the gains in trophic resources and added protection from predation may be greater and diel hypoxic areas can still serve as important nursery habitats. In areas where hypoxia is intermittent and does not cause substantial mortality of the benthos, the behavior of benthos may facilitate energy transfer to oxygen-tolerant bottom feeding fish, which are physiologically capable of withstanding short-term exposure to DO levels to take advantage of weakened benthic prey and receive an energetic gain (Diaz *et al.*, 1992; Pihl *et al.*, 1992; Nestlerode and Diaz, 1998; Seitz *et al.*, 2003).

Upward energy transfer is inhibited in areas where hypoxia is severe as either benthic resources are killed directly and/or predators capable of detecting low DO would avoid the area. In all cases, the increase in the proportion of production transferred to predators is temporary and as mortality of benthos occurs, microbial pathways quickly dominate energy flows (Baird *et al.*, 2004). This energy diversion tends to occur in ecologically important places and at the most inopportune time for predator energy demands, causes overall reduction in ecosystems' functional ability to transfer energy to higher trophic levels, and renders ecosystems less resilient to other stressors.

4. CONCLUSIONS

Hypoxia occurs in a wide range of aquatic systems and varies in temporal frequency, seasonality, and persistence. The oxygen minimum zones of the world's oceans as well as deep basins of permanently stratified lakes and semienclosed seas represent large expanses of severely hypoxic or anoxic environments that exclude all but the most highly specialized fish species and their prey. In these systems, low oxygen is primarily a physically driven process. Persistent stratification prevents reaeration of the lower water column and allows microbial respiration to deplete dissolved oxygen even where anthropogenic enhancement of nutrient loads is low.

In temperate latitudes, bottom waters can remain hypoxic or anoxic for hours to months during summer and autumn, with oxygen concentrations fluctuating with tides, winds, and depth during that period. Hypoxia also develops in shallow waters of organically enriched systems, with oxygen concentrations fluctuating in a diel cycle dependent on the balance between respiration and photosynthesis, and minimum concentrations varying among days depending on cloud cover, wind mixing, and temporal variability in phytoplankton and macrophyte biomass.

A different seasonality characterizes the development of hypoxia in "winterkill" lakes where ice cover prevents reaeration and snow pack reduces light available for photosynthesis and oxygen generation during the coldest parts of the year. Floodplains that develop on a seasonal basis or as a result of storm-related flooding can create large expanses of habitat that physically expand the boundaries of the aquatic realm, but are often characterized by low oxygen concentrations, particularly where they greatly increase loadings of nutrients from crop and livestock agriculture.

There has been a growing appreciation of the importance of low oxygen in microhabitats such as coral reef crevices and burrows in selection for hypoxia tolerance. In addition, fish that utilize intertidal habitats may experience environmental hypoxia in tide pools, and like fishes inhabiting other shallow, chronically hypoxic habitats, many possess adaptations for aerial respiration, or physiological mechanisms to deal with hypoxia and hypercapnia if they remain above the tide line. Finally, under some circumstances fish can deplete oxygen concentrations in embayments and similar habitats with limited circulation.

In general, freshwater systems are more prone to hypoxia and anoxia, with a long history of occurrence. Lakes and coastal areas with seasonal stratification tend to be highly sensitive to anthropogenic nutrient enrichment. Many of the systems that are currently hypoxic were not when first studied. The rapid rise in the number of coastal hypoxic systems lagged about 20 years behind the increase in the use of industrial fertilizer (see Figure 1.3), which led to a general eutrophication of many freshwater and marine systems.

Recovery of a system from a hypoxic event involves two components for fishes: physical habitat value and trophic value. Unless the physical structure of a habitat is biological and perished during hypoxia, the habitat value of an area returns once hypoxia and other toxic compounds, such as H_2S, dissipate. The issue of trophic value is more complex. Generally, hypoxia favors enhanced diversion of energy flows into microbial pathways to the detriment of higher trophic levels. But under certain circumstances hypoxia-enhanced trophic transfer to fishes may occur with periodic hypoxia, diel hypoxia, and along the edges of seasonal hypoxia. For juvenile fishes there may be a benefit to occupying physiologically stressful habitats that are subjected to diel hypoxia as the gains in trophic resources and added protection from predation may be greater and diel hypoxic areas can still serve as important nursery habitats.

The future status of hypoxia and its consequences for fishes will depend on a combination of climate change (primarily from warming, and altered patterns for wind, currents, and precipitation) and land use change (primarily from expanded agriculture and nutrient loadings). If in the next 50 years humans continue to modify and degrade coastal systems as in previous years, human population pressure will be the main driving factor in spreading of coastal dead zones and climate change factors will be secondary (see Figure 1.5). Climate forcing, however, will tend to make systems more susceptible to development of hypoxia through direct effects on stratification, solubility of oxygen, metabolism, and mineralization rates, particularly in lakes and semienclosed coastal areas.

ACKNOWLEDGEMENTS

Supported in part by NOAA, Coastal Hypoxia Research Program grants NA05NOS4781202 to RJD and NA05NOS4781204 to DLB. Contribution 2960 of the Virginia Institute of Marine Science.

REFERENCES

Andrews, J. W., Murai, T., and Gibbons, G. (1973). The influence of dissolved oxygen on the growth of channel catfish. *Trans. Am. Fish. Soc.* **102,** 835–838.

Araújo, F. G., Williams, W. P., and Bailey, R. G. (2000). Fish assemblages as indicators of water quality in the middle Thames Estuary, England (1980–1989). *Estuaries* **23,** 305–317.

Baden, S. P., Loo, L. O., Pihl, L., and Rosenberg, R. (1990). Effects of eutrophication on benthic communities including fish – Swedish west coast. *Ambio.* **19,** 113–122.

Baird, D., Christian, R. R., Peterson, C. H., and Johnson, G. A. (2004). Consequences of hypoxia on estuarine ecosystem function: Energy diversion from consumers to microbes. *Ecol. Appl.* **14,** 805–822.

Benson, B. B., and Krause, D., Jr. (1984). The concentration and isotopic fractionation of oxygen dissolved in freshwater and seawater in equilibrium with the atmosphere. *Limnol. Oceanogr.* **29,** 620–632.

Bickler, P. E., and Buck, L. T. (2007). Hypoxia tolerance in reptiles, amphibians, and fishes: Life with variable oxygen availability. *Ann. Rev. Physiol.* **69,** 145–170.

Black, E. C. (1951). Respiration in fishes. *In* "Some Aspects of the Physiology of Fish. Univ. Toronto Stud. Biol. 59" (Hoar, V. S. B. W. S., and Black, E. C., Eds.), pp. 91–111. Publ. Ont. Fish. Res. Lab. 71.

Boesch, D. F. (2002). Challenges and opportunities for science in reducing nutrient over-enrichment of coastal ecosystems. *Estuaries* **25,** 886–900.

Boesch, D. F., and Rabalais, N. N. (1991). Effects of hypoxia on continental shelf benthos: Comparisons between the New York Bight and the Northern Gulf of Mexico. *In* "Modern and Ancient Continental Shelf Anoxia" (Tyson, R. V., and Pearson, T. H., Eds.), Vol. 58, pp. 27–34. Geological Society.

Brander, K. M. (1995). The effect of temperature on growth of Atlantic cod (*Gadus morhua*). *ICES J. Mar. Sci.* **52,** 1–10.

Brauner, C. J., Ballantyne, C. L., Randall, D. J., and Val, A. L. (1995). Air breathing in the armored catfish (*Hoplosternum littorale*) as an adaptation to hypoxic, acidic, and hydrogen sulfide rich waters. *Can. J. Zool.* **73,** 739–744.

Breitburg, D. L. (1992). Episodic hypoxia in Chesapeake Bay: Interacting effects of recruitment, behavior, and physical disturbance. *Ecol. Monogr.* **62,** 525–546.

Breitburg, D. L. (2002). Effects of hypoxia, and the balance between hypoxia and enrichment, on coastal fishes and fisheries. *Estuaries* **25,** 767–781.

Breitburg, D. L., Baxter, J., Hatfield, C., Howarth, R. W., Jones, C. G., Lovett, G. M., and Wigand, C. (1998). Understanding effects of multiple stresses: Ideas and challenges.

In "Successes, Limitations and Frontiers in Ecosystem Science" (Pace, M, and Groffman, P., Eds.), pp. 416–431. Springer, New York.

Breitburg, D. L., Craig, J. K., Fulford, R. S., Rose, K. A., Boynton, W. R., Brady, D., Ciotti, B. J., Diaz, R. J., Friedland, K. D., Hagy, J. D., III, Hart, D. R., Hines, A. H. *et al.* (in press). Nutrient enrichment and fisheries exploitation: interactive effects on estuarine living resources and their management. *Hydrobiologia* special issue on Eutrophication.

Breitburg, D. L., Hondorp, D. W., Davias, L. W., and Diaz, R. J. (2009). Hypoxia, nitrogen and fisheries: Integrating effects across local and global landscapes. *Ann. Revs. Mar. Sci.* **1,** 329–349.

Brongersma-Sanders, M. (1957). Mass mortality in the sea. *In* "Treatise on Marine Ecology and Paleoecology" (Hedgpeth, J. W., Ed.), Vol. 1, pp. 941–1010. Waverly Press, Baltimore.

Brosnan, T. M., and O'Shea, M. L. (1996). Long-term improvements in water quality due to sewage abatement in the lower Hudson River. *Estuaries* **19,** 890–900.

Burnett, L. E. (1997). The challenges of living in hypoxic and hypercapnic aquatic environments. *Am. Zool.* **37,** 633–640.

Caddy, J. F. (1993). Toward a comparative evaluation of human impacts on fishery ecosystems of enclosed and semi-enclosed seas. *Rev. Fish. Sci.* **1,** 57–95.

Caddy, J. F. (2000). Marine catchment basin effects versus impacts of fisheries on semi-enclosed seas. *ICES J. Mar. Sci.* **57,** 628–640.

Cardinale, M., and Modin, J. (1999). Changes in size-at-maturity of Baltic cod (*Gadus morhua*) during a period of large variations in stock size and environmental conditions. *Fish. Res.* **41,** 285–295.

Carpenter, S. R., Ludwig, D., and Brock, W. A. (1999). Management of eutrophication for lakes subject to potentially irreversible change. *Ecol. Applicat.* **9,** 751–771.

Carter, G. S., and Beadle, L. C. (1931). The fauna of the swamps of the Paraguayan Chaco in relation to its environments II. Respiration in the fishes. *J. Linnaean Soc. London* **37,** 327–368.

Catling, D. C., Zahnle, K. J., and McKay, C. P. (2001). Biogenic methane, hydrogen escape, and the irreversible oxidation of early earth. *Science* **293,** 839–843.

Childress, J. J., and Seibel, B. A. (1998). Life at stable low oxygen levels: Adaptations of animals to oceanic oxygen minimum layers. *J. Exp. Biol.* **201,** 1223–1232.

Clack, J. A. (2007). Devonian climate change, breathing, and the origin of the tetrapod stem group. *Integra. Compar. Biol.* **47,** 510–523.

Cloern, J. E. (2001). Our evolving conceptual model of the coastal eutrophication problem. *Mar. Ecol. Prog. Ser.* **210,** 223–253.

Cooper, G. P., and Washburn, G. N. (1949). Relation of dissolved oxygen to winter mortality of fish in lakes. *Trans. Am. Fish. Soc.* **76,** 23–33.

Cooper, S. R., and Brush, G. S. (1991). Long-term history of Chesapeake Bay anoxia. *Science* **254,** 992–996.

Cornett, R. J., and Rigler, F. H. (1980). The areal hypolimnetic oxygen deficit: An empirical test of the model. *Limnol. Oceanogr.* **25,** 672–679.

Coulter, G. W. (1967). Low apparent oxygen requirements of deep-water fishes in Lake Tanganyika. *Nature* **215,** 317–318.

Coutant, C. C. (1990). Temperature-oxygen habitat for freshwater and coastal striped bass in a changing climate. *Trans. Am. Fish. Soc.* **119,** 240–253.

Craig, J. K., and Crowder, L. B. (2005). Hypoxia-induced habitat shifts and energetic consequences in Atlantic croaker and brown shrimp on the Gulf of Mexico shelf. *Mar. Ecol. Prog. Ser.* **294,** 79–94.

Crawshaw, L. I., and O'Connor, C. S. (1997). Behavioural compensation for long-term thermal change. *In* "Global Warming: Implications for Freshwater and Marine Fish" (Wood, C. M., and McDonald, D. G., Eds.), pp. 351–376. Cambridge University Press, Cambridge, UK.

Cronin, T. M., and Vann, C. D. (2003). The sedimentary record of climatic and anthropogenic influence on the Patuxent estuary and Chesapeake Bay ecosystems. *Estuaries* **26,** 196–209.

D'Amours, D. (1993). The distribution of cod (*Gadus morhua*) in relation to temperature and oxygen level in the Gulf of St. Lawrence. *Fish. Oceanogr.* **2,** 24–29.

D'Avanzo, C., and Kremer, J. N. (1994). Diel oxygen dynamics in an eutrophic estuary of Waquoit Bay, Massachusetts. *Estuaries* **17,** 131–139.

Davis, J. C. (1975). Minimal dissolved oxygen requirements of aquatic life with emphasis on Canadian species: a review. *J. Fish. Res. Bd. Can.* **32,** 2295–2332.

Diaz, R. J., and Rosenberg, R. (1995). Marine benthic hypoxia: A review of its ecological effects and the behavioural responses of benthic macrofauna. *Oceanogr. Mar. Biol. Ann. Rev.* **33,** 245–303.

Diaz, R. J., and Rosenberg, R. (2008). Spreading dead zones and consequences for marine ecosystems. *Science* **321,** 926–929.

Diaz, R. J., Neubauer, R. J., Schaffner, L. C., Pihl, L., and Baden, S. P. (1992). Continuous monitoring of dissolved oxygen in an estuary experiencing periodic hypoxia and the effect of hypoxia on macrobenthos and fish. *Sci. Total Environ.* 1055–1068.

Domenici, P., Lefrançois, C., and Shingles, A. (2007). Hypoxia and the antipredator behaviours of fishes. *Phil. Trans. R. Soc. London* **362,** 2105–2121.

Duque, A. B., Taphorn, D. C., and Winemiller, K. O. (1988). Ecology of the corporo, *Prochilodus mariae* (Characiformes, Prochilodontidae), and status of annual migrations in western Venezuela. *Environ. Biol. Fishes* **53,** 33–46.

Dutil, J.-D., Castonguay, M., Gilbert, D., and Gascon, D. (1999). Growth, condition, and environmental relationships in Atlantic cod (*Gadus morhua*) in the northern Gulf of St. Lawrence and implications for management strategies in the Northwest Atlantic. *Can. J. Fish. Aquat. Sci.* **56,** 1818–1831.

Eby, L. A., and Crowder, L. B. (2004). Effects of hypoxic disturbances on an estuarine nekton assemblage across multiple scales. *Estuaries* **27,** 342–351.

Elmgren, R. (1989). Man's impact on the ecosystem of the Baltic Sea: Energy flows today and at the turn of the century. *Ambio.* **18,** 326–332.

EPA, Environmental Protection Agency Science Advisory Board (2007). Hypoxia in the northern Gulf of Mexico p. 275. U.S. Environmental Protection Agency, Washington D. C., EPA-SAB-08-004.

Fang, X., and Stefan, H. G. (2000). Projected climate change effects on winterkill in shallow lakes in the northern United States. *Environ. Manag.* **25,** 291–304.

Fang, X., Stefan, H. G., Eaton, J. G., McCormick, J. H., and Alam, S. R. (2004). Simulation of thermal/dissolved oxygen habitat for fishes in lakes under different climate scenarios: Part 3. Warm-water fish in the contiguous US. *Ecol. Modeling* **172,** 55–68.

Fonselius, S. H. (1969). Hydrography of the Baltic deep basins III. *In* "Series Hydrography Report No.23," pp. 1–97. Fishery Board of Sweden, Gothenberg.

Fry, F. E. J. (1971). "The Effect of Environmental Factors on the Physiology of Fish." Academic Press, New York.

Gardner, J. A. (1926). Report on the respiratory exchange in freshwater fish, with suggestions as to further investigations. *Fish. Invest., Series I* **3,** 1–17.

Gilbert, D., Sundby, B., Gobeil, C., Mucci, A., and Tremblay, G.-H. (2005). A seventy-two-year record of diminishing deep-water oxygen in the St. Lawrence estuary: The northwest Atlantic connection. *Limnol. Oceanogr.* **50,** 1654–1666.

Gonzales, T. T., Katoh, M., and Ishimatsu, A. (2006). Air breathing of aquatic burrow-dwelling eel goby, *Odontamblyopus lacepedii* (Gobiidae: Amblyopinae). *J. Exp. Biol* **209,** 1085–1092.

Graham, J. B. (1990). Ecological, evolutionary, and physical factors influencing aquatic animal respiration. *Am. Zool.* **30,** 137–146.

Graham, J. B. (2006). Aquatic and aerial respiration. *In* "The Physiology of Fishes" (Evans, D. H., and Claiborne, J. B., Eds.), pp. 85–118. CRC Press, Boca Raton, FL.

Green, J., Corbet, S. A., and Betney, E. (1973). Ecological studies on crater lakes in West Cameroon: The blood of endemic cichlids in Barombi Mbo in relation to stratification and their feeding habits. *J. Zool.* **170,** 299–308.

Greenbank, J. (1945). Limnological conditions in ice-covered lakes, especially as related to winterkill of fish. *Ecol. Monogr.* **15,** 343–392.

Guppy, M., and Withers, P. (1999). Metabolic depression in animals: physiological perspectives and biochemical generalizations. *Biol. Rev.* **74,** 1–40.

Halpern, B. S., Walbridge, S., Selkoe, K. A., Kappel, C. V., Micheli, F., D'Agrosa, C., Bruno, J. F., Casey, K. S., Ebert, C., Fox, H. E., Fujita, R., Heinemann, D., *et al.* (2008). A global map of human impact on marine ecosystems. *Science* **319,** 948–952.

Harper, D. E., McKinney, L. D., Salzer, R. B., and Case, R. J. (1981). The occurrence of hypoxic bottom water off the upper Texas coast and its effects on the benthic biota. *Contrib. Mar. Sci.* **24,** 53–79.

Helly, J. J., and Levin, L. A. (2004). Global distribution of naturally occurring marine hypoxia on continental margins. *Deep-Sea Res.* (*Part I*) **51,** 1159–1168.

Hoar, W. S., and Randall, D. J. (1984). "Fish Physiology, Vol. 10: Gills, Part A: Anatomy, Gas Transfer and Acid-base Regulation." Academic Press, New York.

Howarth, R. W., Billen, G., Swaney, D., Townsend, A., Jaworski, N., Lajtha, K., Downing, J. A., Elmgren, R., Caraco, N., Jordan, T., Berendse, F., Freney, J., *et al.* (1996). Regional nitrogen budgets and riverine N & P fluxes for the drainages to the North Atlantic Ocean: natural and human influences. *Biogeochemistry* **35,** 75–139.

IPCC, International Panel on Climate Change (2007). "Climate Change 2007: The Physical Science Basis." Cambridge University Press, New York.

Jeppesen, E., Søndergaard, M., Jensen, J. P., Havens, K. E., Anneville, O., Carvalho, L., Coveney, M. F., Deneke, R., Dokulil, M. T., Foy, B., Gerdeaux, D., Hampton, S. E., *et al.* (2005). Lake responses to reduced nutrient loading: an analysis of contemporary long-term data from 35 case studies. *Freshwater Biol.* **50,** 1747–1771.

Jones, J. R. E. (1952). The reactions of fish to water of low oxygen concentration. *Journal of Experimental Biology* **29,** 403–415.

Jones, P. D. (2006). Water quality and fisheries in the Mersey estuary, England: A historical perspective. *Mar. Pollut. Bull.* **53,** 144–154.

Justić, D., Legović, T., and Rottini-Sandrini, L. (1987). Trends in oxygen content 1911–1984 and occurrence of benthic mortality in the Northern Adriatic Sea. *Estuar. Coast. Shelf Sci.* **25,** 435–445.

Justić, D., Rabalais, N. N., and Turner, R. E. (1996). Effects of climate change on hypoxia in coastal waters: a doubled CO_2 scenario for the northern Golf of Mexico. *Limnol. Oceanogr.* **41,** 992–1003.

Justić, D., Bierman, Jr., V. J., Scavia, D., and Hetland, R. D. (2007). Forecasting gulf's hypoxia: The next 50 years? *Estuar. Coasts* **30,** 791–801.

Kamykowski, D., and Zentara, S. J. (1990). Hypoxia in the world oceans as recorded in the historical data set. *Deep Sea Res. (Part I)* **37,** 1861–1874.

Karlson, K., Rosenberg, R., and Bonsdorff, E. (2002). Temporal and spatial large-scale effects of eutrophication and oxygen deficiency on benthic fauna in Scandinavian and Baltic waters: A review. *Oceanogr. Mar. Biol. Ann. Rev.* **40,** 427–489.

Keeling, R. F., and Garcia, H. E. (2002). The change in oceanic O_2 inventory associated with recent global warming. *Proc. Nat. Acad. Sci. USA* **99,** 7848–7853.

Kideys, A. E. (2002). Fall and rise of the Black Sea Ecosystem. *Science* **297,** 1482–1484.

Komatsu, E., Fukushima, T., and Harasawa, H. (2007). A modeling approach to forecast the effect of long-term climate change on lake water quality. *Ecol. Model* **209,** 351–366.

Kramer, D. L. (1987). Dissolved oxygen and fish behaviour. *Environ. Biol. Fishes* **18,** 81–92.

Layman, C. A., Smith, D. E., and Herod, J. D. (2000). Seasonally varying importance of abiotic and biotic factors in marsh-pond fish communities. *Mar. Eco. Prog. Ser.* **207,** 155–169.

Levin, L. A. (2002). Deep-ocean life where oxygen is scarce. *Am. Scient.* **90,** 436–444.

Loesch, H. (1960). Sporadic mass shoreward migrations of demersal fish and crustaceans in Mobile Bay, Alabama. *Ecology* **41,** 292–298.

Magnuson, J. J., and Destasio, B. T. (1996). Thermal niche of fishes and global warming. *In* "Global Warming: Implications for Freshwater and Marine Fish" (Wood, C. M., and McDonald, D. G., Eds.), pp. 377–408. Cambridge University Press, Cambridge, UK.

Magnuson, J. J., Beckell, A. L., Mills, K., and Brandt, S. B. (1985). Surviving winter hypoxia: behavioral adaptations of fishes in a northern Wisconsin winterkill lake. *Environ. Biol. Fishes,* **14,** 241–250.

Mee, L. D. (1992). The Black Sea in crisis: A need for concerted international action. *Ambio.* **21,** 278–286.

Mee, L. D. (2006). Reviving dead zones. *Scient. Am.* **295,** 78–85.

Miller, J. R., and Russell, G. L. (1992). The impact of global warming on river runoff. *J. Geophy. Res.* **97,** 2757–2764.

Nestlerode, J. A., and Diaz, R. J. (1998). Effects of periodic environmental hypoxia on predation of a tethered polychaete, *Glycera americana*: Implications for trophic dynamics. *Mar. Ecol. Prog. Ser.* **172,** 185–195.

Nilsson, G. E., and Renshaw, G. M. (2004). Hypoxia survival strategies in two fishes: extreme anoxia tolerance in the North European crucian carp and natural hypoxic preconditioning in a coral-reef shark. *J. Exp. Biol.* **207,** 3131–3139.

Nilsson, G. E., Östlund-Nilsson, S., Penfold, R., and Grutter, A. S. (2007). From record performance to hypoxia tolerance: respiratory transition in damselfish larvae settling on a coral reef. *Proc. R. Soc.* **274,** 79–85.

Nissling, A., and Vallin, L. (1996). The ability of Baltic cod eggs to maintain neutral buoyancy and the opportunity for survival in fluctuating conditions in the Baltic Sea. *J. Fish Biol.* **48,** 217–227.

Nixon, S. W. (1995). Coastal marine eutrophication: A definition, social causes, and future concerns. *Ophelia* **41,** 199–219.

Nixon, S. W., and Buckley, B. A. (2002). "A strikingly rich zone": nutrient enrichment and secondary production in coastal marine ecosystems. *Estuaries* **25,** 782–796.

Officer, C. B., Biggs, R. B., Taft, J. L., Cronin, L. E., Tyler, M. A., and Boynton, W. R. (1984). Chesapeake Bay anoxia; origin, development, and significance. *Science* **223,** 22–27.

Osterman, L. E., Poore, R. Z., and Swarzenski, P. W. (2007). The last 1000 years of natural and anthropogenic low-oxygen bottom water on the Louisiana shelf, Gulf of Mexico. *Mar. Micropaleo.* **66,** 291–303.

Patrick, R. (1988). Changes in the chemical and biological characteristics of the Upper Delaware River estuary in response to environmental laws. (Majumdar, E., Miller, W., and Sage, L. E., Eds.), pp. 332–359. Pennsylvania Academy of Science, Philadelphia, PA.

Pearson, T. H., and Rosenberg, R. (1992). Energy flow through the SE Kattegat: A comparative examination of the eutrophication of a coastal marine ecosystem. *Netherlands J. Sea Res.* **28,** 317–334.

Pedersen, C. L. (1987). Energy budgets for juvenile rainbow trout at various oxygen concentrations. *Aquaculture* **68,** 289–298.

Perez-Dominguez, R., Holt, S. A., and Holt, G. J. (2006). Environmental variability in seagrass meadows: Effects of nursery environment cycles on growth and survival in larval red drum *Sciaenops ocellatus. Mar. Ecol. Prog. Ser.* **321,** 41–53.

Pihl, L., Baden, S. P., and Diaz, R. J. (1991). Effects of periodic hypoxia on distribution of demersal fish and crustaceans. *Mar. Biol* **108,** 349–360.

Pihl, L., Baden, S. P., Diaz, R. J., and Schaffner, L. C. (1992). Hypoxia induces structural changes in the diet of bottom-feeding fish and crustacea. *Mar. Biol.* **112,** 349–361.

Plante, S., Chabot, D., and Dutil, J. (1998). Hypoxia tolerance in Atlantic cod. *J. Fish Biol.* **53,** 1342–1356.

Rabalais, N. N., Turner, R. E., and Wiseman, W. J. (2002). Gulf of Mexico hypoxia, aka The dead zone. *Ann. Rev. Ecol. Systemat.* **33,** 235–263.

Rabalais, N. N., Turner, R. E., Sen Gupta, B. K., Boesch, D. F., Chapman, P., and Murrell, M. C. (2007). Hypoxia in the Northern Gulf of Mexico: Does the science support the plan to reduce, mitigate, and control hypoxia? *Estuar. Coasts* **30,** 753–772.

Randall, D. J., Ip, Y. K., Chew, S. F., and Wilson, J. M. (2004). Air breathing and ammonia excretion in the giant mudskipper, *Periophthalmodon schlosseri. Physiol. Biochem. Zool.* **77,** 783–788.

Ross, S. W. (2003). The relative value of different estuarine nursery areas in North Carolina for transient juvenile marine fishes. *Fish. Bull* **101,** 384–404.

Sale, J. W., and Skinner, W. W. (1917). The vertical distribution of dissolved oxygen and the precipitation by salt water in certain tidal areas. *Journal of the Franklin Institute* **184,** 837–848.

Schindler, D. W. (1978). Factors regulating phytoplankton production and standing crop in the world's freshwaters. *Limnol. Oceanogr.* **23,** 478–486.

Schurmann, H., and Steffensen, J. F. (1992). Lethal oxygen levels at different temperatures and the preferred temperature during hypoxia of the Atlantic cod, *Gadus morhua* L. *J. Fish Biol.* **41,** 927–934.

Schurmann, H., Claireaux, G., and Chartois, H. (1998). Changes in vertical distribution of sea bass (*Dicentrarchus labrax* L.) during a hypoxic episode. *Hydrobiologia* **371/372,** 207–213.

Seitz, R. D., Marshall, L. S., Jr., Hines, A. H., and Clark, K. L. (2003). Effects of hypoxia on predator-prey dynamics of the blue crab *Callinectes sapidus* and the Baltic clam *Macoma balthica* in Chesapeake Bay. *Mar. Ecol. Prog. Ser.* **257,** 179–188.

Selman, M., Greenhalgh, S., Diaz, R., and Sugg, Z. (2008). Eutrophication and hypoxia in coastal areas: A global assessment of the state of knowledge World Resource Institute Policy Note, Water Quality: Eutrophication and Hypoxia, No. 1.

Sen Gupta, B. K., Turner, R. E., and Rabalais, N. N. (1996). Seasonal oxygen depletion in continental-shelf waters of Louisiana: Historical record of benthic foraminifers. *Geology* **24,** 227–230.

Shelford, V. E., and Allee, W. C. (1913). The reactions of fishes to gradients of dissolved atmospheric gases. *J. Exp. Zool.* **14,** 207–266.

Smith, K. J., and Able, K. W. (2003). Dissolved oxygen dynamics in salt marsh pools and its potential impacts on fish assemblages. *Mar. Ecol. Prog. Ser.* **258,** 223–232.

Smith, V. H. (2003). Eutrophication of freshwater and coastal marine ecosystems: A global problem. *Environ. Sci. Pollut. Res.* **10,** 126–139.

Smith, V. H., Joye, S. B., and Howarth, R. W. (2006). Eutrophication of freshwater and marine ecosystems. *Limnol. Oceanogr.* **51,** 351–355.

Soares, M. G. M., Menezes, N. A., and Junk, W. J. (2006). Adaptations of fish species to oxygen depletion in a central Amazonian floodplain lake. *Hydrobiologia* **568,** 353–367.

Stewart, N. E., Shumway, D. L., and Dourdorff, P. (1967). Influence of oxygen concentration on growth of juvenile largemouth bass. *J. Fish. Res. Bd. Can.* **24,** 475–494.

Stierhoff, K. L., Targett, T. E., and Miller, K. L. (2006). Ecophysiological responses of juvenile summer and winter flounder to hypoxia: Experimental and modeling analyses of effects on estuarine nursery quality. *Mar. Ecol. Prog. Ser.* **325,** 255–266.

Taylor, J. C., and Miller, J. M. (2001). Physiological performance of juvenile southern flounder, *Paralichthys lethostigma* (Jordan and Gilbert, 1884), in chronic and episodic hypoxia. *J. Exp. Mar. Biol. Ecol.* **258,** 195–214.
Thienemann, A. (1926). Der Nahrungskreislauf im Wasscr. *Vcrh. Dtsch. Zool. Ges.* **2,** 29–79.
Thomas, P., Rahman, M. S., Khan, I. A., and Kummer, J. A. (2007). Widespread endocrine disruption and reproductive impairment in an estuarine fish population exposed to seasonal hypoxia. *Proc. R. Soc. Biol. Sci.* **274,** 2693–2701.
Thornton, K. W., Kimmel, B. L., and Payne, F. E. (1990). "Reservoir Limnology: Ecological Perspectives." Wiley, New York.
Townsend, S. A., and Edwards, C. A. (2003). A fish kill event, hypoxia and other limnological impacts associated with early wet season flow into a lake on the Mary River floodplain, tropical northern Australia. *Lakes Reservoirs: Res. Manag.* **8,** 169–176.
Tyler, R. M., and Targett, T. E. (2007). Juvenile weakfish *Cynoscion regalis* distribution in relation to diel-cycling dissolved oxygen in an estuarine tributary. *Mar. Ecol. Prog. Ser.* **333,** 257–269.
Val, A. L., Paula da Silva, M. N., and Almeida-Val, V. M. F. (2006). Environmental eutrophication and its effects on fish of the Amazon. *In* "Proceedings of the Ninth International Symposium on Fish Physiology, Toxicology and Water Quality", April 24–28, 2006, Hotel San Michele, Capri, Italy (Brauner C. J., Suvajdzic, K., Nilsson, G., and Randall, D. J., Eds.), pp 65-72, EPS/600/R-7/010. United States Environmental Protection Agency, Environmental Research Laboratory, Athens, Georgia. Val, A. L. and Almeida-Val, V. M. F. (1995). "Fishes of the Amazon and their Environment: Physiological and Biochemical Aspects." Springer-Verlag, Berlin.
Val, V. (2000). Evolutionary features of hypoxia tolerance in fish of the Amazon: from molecular to behavioral aspects. *In* "Evolution of Physiological and Biochemical Traits in Fish. Symposium Proceedings. International Congress on the Biology of Fish" (Val, V., Gonzalez, R., and MacKinlay, D., Eds.). American Fisheries Society.
Warren, L. M. (1984). How intertidal polychaetes survive at low tide. *In* "Proceedings of the First International Polychaete Conference" (Hutchings, P. A., Ed.), pp. 238–253. The Linnaean Society, New South Wales.
Wells, M. M. (1913). The resistance of fishes to different concentrations and combinations of oxygen and carbon dioxide. *Biol. Bull.* **25,** 323–344.
Wetzel, R. G. (2001). "Limnology. Lakes and Rivers Ecosystems." Academic Press, San Diego.
Wu, R. S. S., Zhou, B. S., Randall, D. J., Woo, N. Y. S., and Lam, P. K. S. (2003). Aquatic hypoxia is an endocrine disruptor and impairs fish reproduction. *Environ. Sci. Tech.* **37,** 1137–1141.
Wyrtki, K. (1966). Oceanography of the eastern equatorial Pacific Ocean. *Oceanogr. Mar. Biol. Ann. Rev.* **4,** 33–68.
Zimmerman, A. R., and Canuel, E. A. (2000). A geochemical record of eutrophication and anoxia in Chesapeake Bay sediments: Anthropogenic influence on organic matter composition. *Mar. Chem.* **69,** 117–137.

2

BEHAVIORAL RESPONSES AND ECOLOGICAL CONSEQUENCES

LAUREN J. CHAPMAN
DAVID J. McKENZIE

Fishes employ many and diverse strategies to increase oxygen transfer from the environment to their tissues and/or avoid problems associated with hypoxia. Some of these responses can be activated quickly (e.g., hours, days), whereas others are developmentally plastic and/or genetically fixed. Short-term physiological and biochemical responses provide regulatory mechanisms to deal with variable oxygen in habitats. Behavioral responses can provide additional flexibility to mitigate exposure to hypoxic stress, and many occur at levels of aquatic oxygen availability far higher than lethal levels. Fish may avoid hypoxic areas through movement or they may compensate for hypoxia through air-breathing or aquatic surface respiration (ASR), ventilating their gills with water from the air–water interface. As a result, behavioral responses to hypoxia can influence other critical

Hypoxia: Volume 27
FISH PHYSIOLOGY

DOI: 10.1016/S1546-5098(08)00002-2

components of the behavior of fishes in their environment, including habitat use and selection, predator–prey interactions, competitive interactions, and patterns of aggregation. In this chapter, we review behavioral responses to hypoxia, including aquatic surface respiration, air-breathing, changes in spontaneous activity, and parental care. We then consider the role of hypoxia in modulating ecological interactions, in particular the interaction between predator and prey. Hypoxic alterations to predator–prey interactions can influence other components of the food web and assemblage structure; so predicting whether predator or prey is the beneficiary of hypoxic stress is fundamental to understanding community level impacts of hypoxia, whether natural or anthropogenically induced.

1. INTRODUCTION

The abiotic environment has had a major influence on the ecology and evolution of organisms. For fishes, the availability of dissolved oxygen (DO) is one physicochemical factor that can limit habitat quality, distribution, growth, reproduction, and survival. All fishes require oxygen for long-term survival; however, the physical properties of water (high viscosity, low oxygen content at saturation) make oxygen uptake challenging for fishes even at high DO levels. In addition to these constraints imposed by the physical properties of water, there are many habitats in which dissolved oxygen is depressed below saturation periodically or chronically. Hypoxia occurs naturally in habitats characterized by low mixing or light limitation, such as heavily vegetated swamps, flooded forests, floodplain lakes and ponds, ephemeral pools, spring boils, and the profundal waters of deep lakes; it is particularly widespread in tropical waters where high temperatures elevate rates of organic decomposition and reduce oxygen tensions and contents (Kramer, 1983; Chapman and Liem, 1995; Chapman *et al.*, 1999; see Chapter 1). Hypoxic (and anoxic) environments have existed through geological time, but human environmental degradation is increasing the occurrence of hypoxia, as influxes of municipal wastes and fertilizer runoffs accelerate eutrophication and pollution of water bodies (Diaz, 2001). Increasing hypoxia is now recognized as an environmental issue of global importance to fresh and coastal waters, which can result in changes in species composition, population decline, mass mortalities, and production of extensive "dead zones," such as that in the Gulf of Mexico and Lake Erie that can affect important fisheries (Diaz, 2001; Dybas, 2005; Pollock *et al.*, 2007). Thus, it has become increasingly important to understand the consequences of hypoxic stress on the behavior and ecology of aquatic organisms, and so to predict cascading effects of hypoxia on community function.

Fishes have evolved a variety of solutions to hypoxic stress, including morphological adaptations, physiological adjustments, and biochemical and molecular defenses (see Chapters 5, 6, 7, 9, and 10). Behavioral responses provide fishes with additional flexibility in responding to temporal and spatial variation in DO, through avoidance of hypoxic zones, changes in activity level, use of aquatic surface respiration, and, when possible, changes in air-breathing frequency (Kramer, 1987; Timmerman and Chapman, 2004). These behavioral responses to hypoxia can interact with other physiological and biochemical adjustments to alter crucial components of the interactions between a fish and its environment, in particular predator–prey interactions, schooling behavior, dominance, aggression, and parental care. Given the growing incidence of anthropogenically induced hypoxia in aquatic environments, there is accelerating interest in predicting levels at which hypoxia becomes ecologically active and its role in modifying ecological interactions.

In this chapter, we address behavioral responses to hypoxia and associated changes in spatial and temporal distributions, density, and ecological interactions. We begin by reviewing two major behavioral mechanisms for increasing oxygen uptake (aquatic surface respiration and air-breathing) and the ecological consequences of these behaviors that reflect costs and benefits of surfacing. We then consider effects of hypoxia on a major component of the energy budget of active fishes—swimming. Many fishes exhibit changes in spontaneous swimming activity when exposed to hypoxia; the literature indicates these responses can comprise either reductions or increases in activity, as a function of species and context. Another significant metabolic cost for some fish species is the energy directed to parental care. Costs of parental care may be particularly severe under hypoxia, although high levels of parental care may actually be necessary under hypoxic conditions, to ensure survival of the eggs and young. We explore the degree to which DO is a driver of parental care behavior across a range of strategies, from guarding to viviparity, and the ability of fishes to modify their care behaviors in response to variations in oxygen availability Finally, we integrate behavioral responses to hypoxia and the relative tolerance of species, to develop a picture of how hypoxia influences species interactions, in particular predator–prey relationships, clearly an area of growing concern given the global increase in hypoxia. Whether hypoxia favors the predator or the prey depends, at least in part, on the relative tolerance of the interactants and can ultimately influence other components of the food web and assemblage. Thus predicting whether the prey or the predator is the beneficiary of hypoxic stress is critical for understanding community level impacts of hypoxia, for predicting response to anthropogenically induced hypoxia, and for setting habitat conservation or restoration objectives in the context of rehabilitation or management policy.

2. AQUATIC SURFACE RESPIRATION AND AIR-BREATHING

Aquatic surface respiration (ASR) and air-breathing are two of the most pronounced behavioral responses by bony fishes to aquatic hypoxia. Neither behavior has been described in more plesiomorphic fish groups, the agnathans, elasmobranches, or chondrichthyans; but air-breathing is found in a number of primitive bony fishes and in the ancient crossopterygians (Randall *et al.*, 1981a; Graham, 1997; Reid *et al.*, 2006; McKenzie *et al.*, 2007a).

2.1. Aquatic Surface Respiration

As the name implies, aquatic surface respiration (ASR) involves rising to the surface and ventilating the layer of water in contact with the atmosphere, which is richer in DO than the underlying bulk water (Kramer and McClure, 1982). Many teleost species have evolved this behavioral response, in freshwater and marine environments, both temperate and tropical (Lewis, 1970; Gee *et al.*, 1978; Kramer and McClure, 1982; Kramer, 1983; Gee and Gee, 1995; Nordlie 2006; McNeil and Closs, 2007). A number of species have morphological features, such as upturned mouths and flattened heads, which appear to improve the efficiency of ASR (Lewis, 1970; Cech *et al.*, 1985). In some species, the morphological adaptations are very pronounced such as the dermal lip protuberances in various tropical teleosts (Saint-Paul 1984; Winemiller 1989; Reid *et al.*, 2006). For example, in the Neotropical tambaqui *Colossoma macropomum*, hypoxia causes the lower lip to swell extensively, to form a funnel that skims the surface layer of water into the mouth (Val and Almeida-Val, 1995; Sundin *et al.*, 2000; Figure 2.1). Chapman *et al.* (1994) found evidence to suggest that inverted swimming in the upside-down catfish *Synodontis nigriventris* may also increase the efficiency of ASR. Many species also hold an air bubble (or bubbles) in their mouth when they perform ASR, which may have a dual role of increasing oxygen levels in the bucco-opercular cavity, and maintaining the fish buoyant at the water surface (Burggren, 1982; Gee and Gee, 1991, 1995; Chapman *et al.*, 1995).

Table 2.1 provides a review of species for which ASR responses have been observed and for which hypoxic oxygen partial pressure (PO_2) thresholds for the behavior have been determined. There is a great deal of variability in the thresholds. A higher threshold can be indicative of a lower specific tolerance of hypoxia, because it has been directly associated with a higher threshold for regulation of basal aerobic metabolic rate in hypoxia (the critical oxygen tension PO_2, P_{crit}) in comparative studies over a fairly wide range of species (Chapman *et al.*, 1995, 2002; Rosenberger and Chapman, 2000; Melnychuk and Chapman, 2002; Schofield *et al.*, 2007; Table 2.1). Figure 2.2 shows the

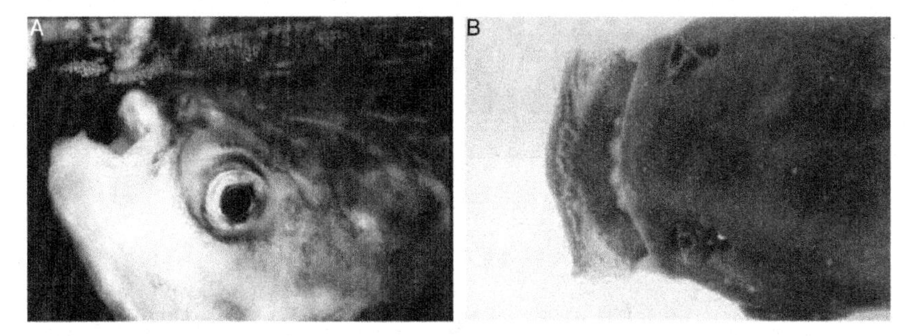

Fig. 2.1. Morphological adaptations for aquatic surface respiration in the tambaqui, *Colossoma macropomum*. (A) Lateral view of the head of a tambaqui approaching the water surface illustrating the initial swelling of the lower lip following exposure to an hypoxic water PO_2 of 2.0 kPa; (B) a dorsal view of the same fish following 3.5 h exposure to the same level of hypoxia, illustrating the forward expansion of the lip near full development. From Sundin *et al.* (2000) with permission from the Company of Biologists.

profile of the ASR response as a function of water PO_2 in three tropical teleosts as described by Kramer and McClure (1982). This response profile, where there is no ASR until quite a discrete threshold at a relatively low PO_2, beyond which it rapidly becomes the dominant behavior, has since been described in the vast majority of species that exhibit the ASR response. Gee *et al.* (1978) demonstrated that the PO_2 threshold at which 50% of fathead minnows, *Pimephales promelas*, performed ASR was directly related to acclimation temperature, rising from approximately 1 kPa (7 mmHg) at 6°C to approximately 6 kPa (28 mmHg) at 31°C. Kramer and Mehegan (1981) also demonstrated that ASR activity increased with increasing water temperature in the guppy *Poecilia reticulata*. Sloman *et al.* (2006; 2008) found that the threshold for ASR increased with body mass in the oscar *Astronotus ocellatus* and the tidepool sculpin *Oligocottus maculosus* (Table 2.1). Prior acclimation to hypoxia (lab-induced or field conditions) has been shown to lower the ASR threshold in a wide variety of species (Kramer and Mehegan, 1981; Olowo and Chapman, 1996; Melnychuk and Chapman, 2002; Chapman *et al.*, 2002; Timmerman and Chapman, 2004; Table 2.1). This is associated with changes to other physiological traits indicative of improved hypoxia tolerance, including increased gill surface area, increased haematocrit, and lower P_{crit} (Olowo and Chapman, 1996; Melnychuk and Chapman, 2002; Chapman *et al.*, 2002; Martinez *et al.*, 2004). ASR thresholds have also been demonstrated to vary with season. Love and Rees (2002) found seasonal differences in ASR thresholds and tolerance of hypoxia in *Fundulis grandis*, whereby thresholds were lower and tolerance greater in summer.

Although ASR is clearly a behavioral response to hypoxia, it is in fact a reflex that is driven by oxygen-sensitive chemoreceptors (Shingles *et al.*, 2005;

Table 2.1

Oxygen partial pressure thresholds (kPa) at which various species of fish perform aquatic surface respiration (ASR), reported as the percentage proportion of individuals performing the behavior, and thresholds for the regulation of routine metabolic rate (P_{crit}, in kPa), as reported by various authors

Species	Mass (g)	Forklength (mm)	Temp (°C)	10% ASR	50% ASR	90% ASR	P_{crit}	Source
Catostomidae								
Catostomus commersoni		63–114	16.5 ± 1		1.79			Gee *et al.* (1978)[a]
Centrarchidae								
Ambloplites rupestris		26–88	16.5 ± 1		2.57			Gee *et al.* (1978)[a]
Lepomis gulosus	4.0 ± 2.6	63 ± 11	22	2.74	2.08	1.65		Schofield *et al.* (2007)
Lepomis marginatus	6.1 ± 1.1	73 ± 5	22	1.41	0.95	0.58		Schofield *et al.* (2007)
Centropomidae								
Lates niloticus	4–28		23	3.33	2.26	1.47	3.51	Chapman *et al.* (2002)
Characidae								
Gymnocorymbus sp.	2.78–3.39		25 ± 1	2.24			2.24	Kramer and McClure (1982)[b]
Hyphessobrycon pulchripinnis	0.53–0.64		25 ± 1	3.55	2.50			Kramer and McClure (1982)[b]
Metynnis sp.	1.69		25 ± 1	3.68	2.76			Kramer and McClure (1982)[b]
Paracheirodon axelrodi	0.51		25 ± 1	2.50	1.97			Kramer and McClure (1982)[b]
Paracheirodon innesi	0.15–0.22		25 ± 1	1.97	1.18			Kramer and McClure (1982)[b]
Roeboides guatemalensis	0.26–0.49		25 ± 1	2.89	2.24			Kramer and McClure (1982)[b]
Cichlidae								
Aequidens coeruleopunctatus	2.37–5.02		25 ± 1	1.97			1.71	Kramer and McClure (1982)[b]

Astatoreochromis alluaudi	3.9 ± 1.0	64 ± 6	23–26	2.64	1.74	1.30		Chapman *et al.* (1995)
Astatotilapia "wrought-iron"	1.9–10.1		18–20	1.97	1.30	0.62	1.78	Melnychuk and Chapman (2002)
Astatotilapia aeneocolor	3.6–7.1		18–20	2.02	1.50	0.98	1.97	Melnychuk and Chapman (2002)
Astatotilapia velifer		42 ± 10	23	3.62	2.55	1.53	1.71	Rosenberger and Chapman (2000)
Astronotus ocellatus	2.05–4.08		25 ± 1	0.92	0.79			Kramer and McClure (1982)[b]
Astronotus ocellatus	16.2 ± 9		28 ± 1	2.93 ± 0.49[e]			~9.00	Sloman *et al.* (2006)
Astronotus ocellatus	230 ± 11		28 ± 1	6.52 ± 1.29[e]			~6.66	Sloman *et al.* (2006)
Cichlasoma biocellatum	2.92–4.60		25 ± 1	1.71	1.45			Kramer and McClure (1982)[b]
Cyprichromis leptosoma	8.8 ± 2.0	96 ± 5	23–26	1.59	1.05			Chapman *et al.* (1995)
Haplochromis "rock kribensis"	3.4 ± 1.4	58 ± 9	23–26	1.92	1.38	1.10		Chapman *et al.* (1995)
Hemichromis bimaculatus	3.59–4.06		25 ± 1	2.23	1.71			Kramer and McClure (1982)[b]
Hemichromis letourneuxi	4.6 ± 2.6	63 ± 11	25	0.80	0.49	0.21		Schofield *et al.* (2007)
Labrochromis ishmaeli	3.9 ± 1.0	68 ± 6	23–26	2.24	1.18	0.20		Chapman *et al.* (1995)
Neochromis nigricans	5.4 ± 1.5	72 ± 6	23–26	1.32	1.00	0.39		Chapman *et al.* (1995)
Neolamprologis tretocephalus	4.1 ± 0.6	60 ± 2	23–26	3.53	2.26			Chapman *et al.* (1995)
Oreochromis esculentis	21.2 ± 5.1	111 ± 9	23–26	0.97	0.80	0.67		Chapman *et al.* (1995)
Oreochromis niloticus	40.1 ± 9.1	137 ± 11	23–26	4.64	0.91	0.59		Chapman *et al.* (1995)
Prognathochromis perrieri	5.8 ± 2.0	79 ± 8	23–26	2.00	1.45	0.92		Chapman *et al.* (1995)

(*continued*)

Table 2.1 *(continued)*

Species	Mass (g)	Forklength (mm)	Temp (°C)	10% ASR	50% ASR	90% ASR	P_{crit}	Source
Prognathochromis venator		75 ± 11	23	2.57	1.83	1.14	1.58	Rosenberger and Chapman (2000)
Pseudocrenilabrus multicolor		59 ± 8	23	1.96	1.11	0.72	1.05	Rosenberger and Chapman (2000)
Pyxichromis orthostoma	1.9 ± 4.4	53 ± 6	23–26	2.24	1.74	0.95		Chapman *et al.* (1995)
Tropheus moorii	4.1 ± 0.6	60 ± 2	23–26	3.88	2.50	1.97		Chapman *et al.* (1995)
Yssichromis argens	2.7 ± 1.6	62 ± 9	23–26	2.72	2.29	2.17		Chapman *et al.* (1995)
Cobitidae								
Botia sidthimunki	0.34–0.37		25 ± 1	3.15	3.15			Kramer and McClure (1982)[b]
Cyprinidae								
Barbus everetti	0.50–0.70		25 ± 1	1.97	1.18			Kramer and McClure (1982)[b]
Barbus neumayeri[c]		68 – 91	20 ± 1	2.94	1.53	0.97		Olowo and Chapman (1996)
Barbus neumayeri[d]		68–91	20 ± 1	1.58	0.68	0.26		Olowo and Chapman (1996)
Barbus nigrofasciatus	0.21–0.40		25 ± 1	3.42	2.89			Kramer and McClure (1982)[b]
Barbus schwanenfeldi	18.66		25 ± 1	2.24	1.97			Kramer and McClure (1982)[b]
Brachydanio albolineatus	0.65–1.04		25 ± 1	3.16	2.50			Kramer and McClure (1982)[b]
Carassius auratus			25	1.99		0.50		McNeil and Closs (2007)
Chrosomus eos		44–51	16.5 ± 1		1.25			Gee *et al.* (1978)[a]
Cyprinus carpio			25	1.99		0.50		McNeil and Closs (2007)

Hybognathus hankinsoni		54–61	16.5 ± 1		1.49		Gee *et al.* (1978)[a]
Labeo bicolor	0.21–0.29		25 ± 1	2.76	2.24		Kramer and McClure (1982)[b]
Nocomis biguttatus		64–112	16.5 ± 1		1.80		Gee *et al.* (1978)[a]
Notropis atherinoides		51–76	16.5 ± 1		2.05		Gee *et al.* (1978)[a]
Notropis cornutus		38–51	16.5 ± 1		2.11		Gee *et al.* (1978)[a]
Notropis hudsonius		63–112	16.5 ± 1		2.26		Gee *et al.* (1978)[a]
Pimephales promelas		40–63	16.5 ± 1		1.80		Gee *et al.* (1978)[a]
Rasboro taeniata	0.28–0.38		25 ± 1	3.95	2.89		Kramer and McClure (1982)[b]
Rhinichthys atratulus		63–76	16.5 ± 1		1.99		Gee *et al.* (1978)[a]
Rhinichthys cataractae		45–78	16.5 ± 1		1.05		Gee *et al.* (1978)[a]
Semotilus atromaculatus		63–89	16.5 ± 1		1.84		Gee *et al.* (1978)[a]
Semotilus margarita		88–115	16.5 ± 1		1.04		Gee *et al.* (1978)[a]
Roeboides guatemalensis	0.26–0.49		25 ± 1	2.89	2.24		Kramer and McClure (1982)[b]
Cyprinodontidae							
Epiplatys dageti	0.70–1.31		25 ± 1	5.39	1.97		Kramer and McClure (1982)[b]
Rivulus hartii	2.91–4.19		25 ± 1	2.89	1.97		Kramer and McClure (1982)[b]
Eleotridae							
Hypseleotris sp.			25	2.74		1.25	McNeil and Closs (2007)
Esocidae							
Esox lucius		89–115	16.5 ± 1		0.8		Gee *et al.* (1978)[a]

(continued)

Table 2.1 *(continued)*

Species	Mass (g)	Forklength (mm)	Temp (°C)	10% ASR	50% ASR	90% ASR	P_{crit}	Source
Galaxidae								
Galaxias rostratus			25	5.49		<2.50		McNeil and Closs (2007)
Culaea inconstans		38–53	16.5 ± 1		2.46			Gee *et al.* (1978)[a]
Pungitus pungitus		31–38	16.5 ± 1		6.57			Gee *et al.* (1978)[a]
Hemiodidae								
Hemiodopis sp.	8.90		25 ± 1	2.89	2.76			Kramer and McClure (1982)[b]
Ictaluridae								
Ictalurus melas		76–115	16.5 ± 1		1.35			Gee *et al.* (1978)[a]
Noturus gyrinus		38–64	16.5 ± 1		0.45			Gee *et al.* (1978)[a]
Mastacembelidae								
Mastacembelus circumcinctus	2.09–2.63		25 ± 1	0.79	0.79			Kramer and McClure (1982)[b]
Mochokidae								
Synodontis nigrita	1.31		25 ± 1	1.97	1.97			Kramer and McClure (1982)[b]
Mormyridae								
Gnathonemus victoriae	1–12		23	1.94	1.43	0.46	1.65	Chapman *et al.* (2002)
Petrocephalus catostoma	1–4		23	1.71	0.40		1.51	Chapman *et al.* (2002)
Mugilidae								
Liza aurata		212 ± 33	20	~3.00[e]				Lefrançois and Domenici (2005)
Mugil cephalus	565 ± 38		24		2.53			Shingles *et al.* (2005)
Mugil cephalus[f]	565 ± 38		24		1.04			Shingles *et al.* (2005)
Percidae								
Etheostoma exile		31–63	16.5 ± 1		1.61			Gee *et al.* (1978)[a]
Etheostoma nigrum		37–50	16.5 ± 1		2.21			Gee *et al.* (1978)[a]

Perca flavescens		50–127	16.5 ± 1		1.54		Gee *et al.* (1978)[a]
Perca fluviatilis			25	5.49		<2.50	McNeil and Closs (2007)
Percina maculata		51–58	16.5 ± 1		1.09		Gee *et al.* (1978)[a]
Percichthyidae							
Nannoperca australis			25	2.3		0.60	McNeil and Closs (2007)
Pimelodidae							
Pimelodella picta	1.38–1.41		25 ± 1	2.24	1.97		Kramer and McClure (1982)[b]
Poeciliidae							
Gambusia holbrooki			25	5.49		1.05	McNeil and Closs (2007)
Poecilia sphenops	0.96–2.05		25 ± 1	4.47	2.76		Kramer and McClure (1982)[b]
Xiphophorus helleri	0.72–1.36		25 ± 1	2.89	2.50		Kramer and McClure (1982)[b]
Retropinnidae							
Retropinna semoni			25	6.23		3.49	McNeil and Closs (2007)
Siluridae							
Krytopterus bicirrhus	0.61–0.62		25 ± 1	4.21	3.68		Kramer and McClure (1982)[b]

[a]Gee *et al.* (1978) found that *Oncorhynchus mykiss*, *Salvelinus alpinus*, *Coregonus clupeaformis* (Salmonidae), and *Stizostedion vitreum* (Percidae) did not perform ASR at any water PO_2.

[b]These values are not thresholds but refer to the PO_2 at which between 1 and 3 fishes of each species performed ASR 10% or 50% of the time (Kramer and McClure, 1982).

[c]Fish collected from relatively normoxic riverine ecotones (Olowo and Chapman, 1996).

[d]Fish collected from hypoxic swamp ecotones (Olowo and Chapman, 1996).

[e]Mean (± SEM) hypoxic PO_2 at which fish broke the surface for the first time.

[f]Same fish as the line above, but in the presence of a model avian predator.

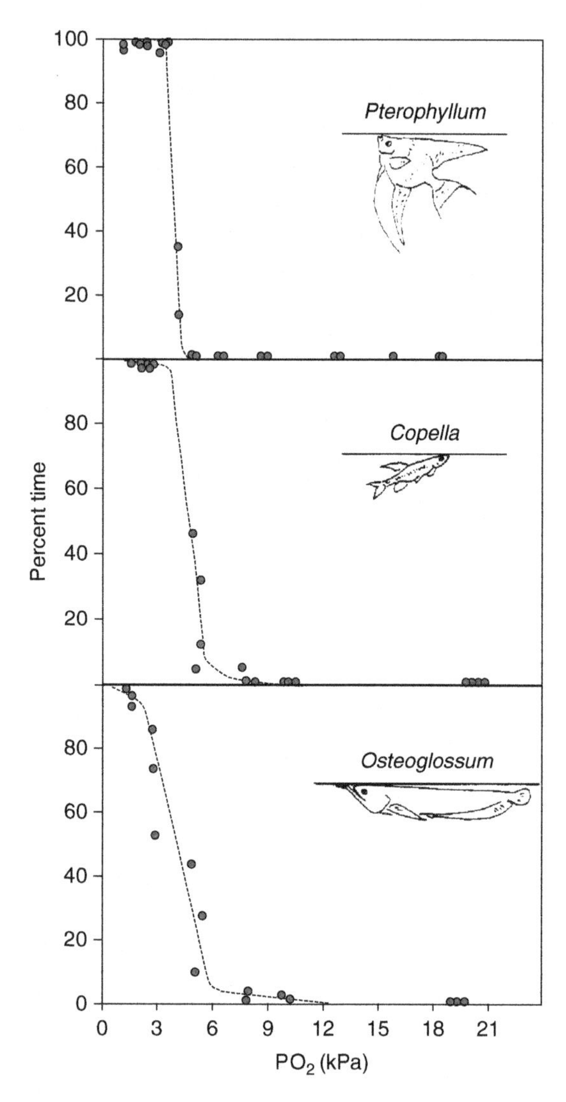

Fig. 2.2. The relationship between bulk water PO_2 and percentage time spent performing aquatic surface respiration (ASR) in three species of tropical freshwater fish. Dotted lines were fitted by eye and drawings show posture adopted by the fish during ASR. Figure redrafted from Kramer and McClure (1982) with permission from Elsevier.

Florindo *et al.*, 2006). That is, when a fish senses that oxygen is low or limiting, they perform the behavior. Shingles *et al.* (2005) used the oxygen-chemoreceptor stimulant sodium cyanide (NaCN) to demonstrate that the ASR response is elicited by chemoreceptors sensitive to oxygen levels in the ventilatory water stream and the blood stream of flathead grey mullet (*Mugil cephalus*). Florindo *et al.* (2006) demonstrated that the chemoreceptors that stimulate ASR in the tambaqui are innervated by cranial nerves that serve the bucco-opercular cavity and gills. These chemoreceptor sensory modalities and innervations would appear to be homologous, therefore, to those that drive reflex gill hyperventilation in all fish groups studied to date (Burleson *et al.*, 1992; Taylor *et al.*, 1999; see Chapter 5), including species that appeared in the fossil record prior to teleosts (McKenzie *et al.*, 1995a; McKendry *et al.*, 2001). Thus, ASR may use the pre-existing sensory arm of such hypoxic ventilatory reflexes, integrating a new motor output that involves rising to the water surface to ventilate the surface layer. Presumably, cessation of this behavior is also driven by information from the same chemoreceptors (Shingles *et al.*, 2005). Gill hyperventilation chemoreflexes in fish are mediated by the medulla, or hind-brain, which contains the central respiratory pattern generator (Taylor *et al.*, 1999). Clearly, ASR is a much more complex chemoreflex with a very large behavioral component, which must involve significant inputs from higher brain centres (Shingles *et al.*, 2005; Sloman *et al.*, 2006, 2008). Teleost fish also exhibit behavioral modulation of gill ventilation patterns (Johnsson *et al.*, 2001), and such higher-order inputs to the respiratory medulla must, presumably, have been a prerequisite for the evolution of the complex ASR motor responses.

The widespread prevalence of the ASR response, and the evolution of specific morphological adaptations for increasing the efficiency of this behavior, must be considered an indicator that ASR provides a physiological advantage to fishes that inhabit periodically or chronically hypoxic habitats (Kramer and McClure, 1982; Kramer, 1987). Nonetheless, there exists no direct quantitative demonstration that ASR increases the ability of fish to regulate their aerobic metabolic rate independent of bulk water oxygen availability in hypoxia. As stated above, the behavior typically shows a sudden increase in prevalence below a relatively discrete threshold water PO_2 (Figure 2.2 and Table 2.1), and this threshold is often quite close to the P_{crit}, which indicates that the behavior only begins when oxygen becomes physiologically limiting in the bulk water (Table 2.1). Rantin *et al.* (1998) found that the threshold for initiation of ASR in the pacu, *Piaractus mesopotamicus*, was slightly below their P_{crit} (Table 2.1). Rosenberger and Chapman (2000), Chapman *et al.* (2002), and Melnychuk and Chapman (2002) have all reported that the threshold for initiation of ASR coincided fairly closely with the P_{crit} in various tropical freshwater species (Table 2.1). Sloman *et al.* (2006) found that the threshold for ASR coincided with the P_{crit} of large oscars, *Astronotus ocellatus*, but was significantly below the P_{crit} of small oscars (Table 2.1), a difference in response that they attributed to greater fear of predation in smaller fish (see Section 5.1 below). On the other hand, Sloman *et al.* (2008) found that the hypoxic threshold for ASR behavior in the tidepool sculpin (Table 2.1) was well above the hypoxic PO_2 at which biochemical disturbances were measurable in fish that were denied access to the response. However, as described in Section 3.2 below, the appearance of metabolic disturbances coincided with the threshold for spontaneous emergence from the water in this species (Sloman *et al.*, 2008). Burggren (1982) demonstrated that permitting goldfish (*Carassius auratus*) to perform ASR and bubble-holding allowed them to maintain arterial oxygen levels higher in deep hypoxia. There are also various demonstrations that ASR improves tolerance of deep hypoxia, measured as survival time (Kramer and Mehegan, 1981; Kramer and McClure, 1982) or as time to loss of equilibrium (Chapman *et al.*, 1995). Rutledge and Beitinger (1989) found that access to ASR in deep hypoxia increased tolerance of acute increases in water temperature (critical thermal maximum) in three species of North American freshwater fish. Lefrançois *et al.* (2005) found that denying access to ASR in deep hypoxia impaired performance of the fast-start escape response in golden grey mullet (*Liza aurata*). Stierhoff *et al.* (2003) found that *Fundulus heteroclitus* exhibited less impairment of growth in deep hypoxia if they were allowed access to the surface to perform ASR. As described in Section 5.1 below, a major ecological benefit to the performance of ASR can be that it allows fish to colonize hypoxic refugia that less tolerant predatory species cannot occupy successfully (Chapman *et al.*, 1995; Rosenberger and Chapman, 1999; Chapman *et al.*, 2002).

Kramer (1987) argues that one major physiological cost to ASR is the increased locomotor activity required for repeated surfacing and skimming. However, as described in Section 5 below, another major ecological cost to ASR relates to predation, in that the behavior places fish at significantly greater risk from aerial predation by birds (Kramer *et al.*, 1983). Perhaps not surprisingly, if fish perceive a risk of predation they can modulate the behavioral component of the ASR chemoreflex (Shingles *et al.*, 2005; Sloman *et al.*, 2006, 2008). Shingles *et al.* (2005) found that exposure of flathead grey mullet to a model avian predator delayed the onset of ASR in hypoxia or in response to direct chemoreceptor stimulation with NaCN. Furthermore, the fish surfaced preferentially under a sheltered area in their experimental chamber or close to the walls (Figure 2.3A). In turbid water, the fish could not see the model predator and it had no effect on the onset of ASR but, in turbidity, all the mullet preferentially surfaced around the walls of their chamber (Figure 2.3B). The tidepool sculpin will release alarm substance from epithelial cells when damaged during a predation event (Hugie *et al.*, 1991). When Sloman *et al.* (2008) added this alarm substance into their water, tidepool sculpins showed a significantly lower threshold for initiation of ASR in response to progressive hypoxia. Thus, the behavioral component of the ASR reflex is plastic; it can be modulated by inputs from higher centers, in particular as a function of perceived risk of predation.

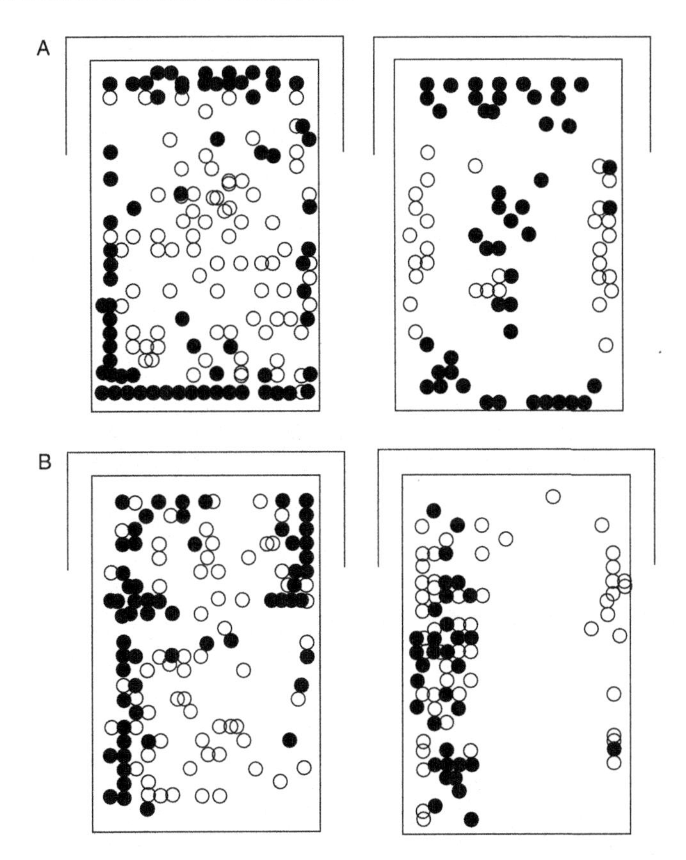

Fig. 2.3. Aerial view of locations of aquatic surface respiration (ASR) events performed by *Mugil cephalus* in response to an external application of 300 mg NaCN into the ventilatory stream via a buccal cannula (A) or when exposed to aquatic hypoxia (B) in clear water (left panel) or in turbid water containing 300 NTU Polsperse 10 kaolin (right panel) at a temperature of 25°C. Open circles represent ASR events in the absence of a model avian predator; solid circles represent events in the presence of the model. The outer lines at the top of the diagrams represent a sheltered area in the aquarium. From Shingles *et al.* (2005) with permission from the University of Chicago Press.

2.2. Air-breathing

A number of bony fishes have evolved bimodal respiration, meaning that they retain functional gills but can also gulp air at the water surface and store this in an air-breathing organ (ABO). The comparative physiology and evolution of air-breathing in fishes have been the subject of extensive reviews and dedicated books (Johansen, 1970; Randall *et al.*, 1981a; Graham, 1997;

Graham and Lee, 2004; Brauner and Val, 2006; Lam *et al.*, 2006; Reid *et al.*, 2006); the reader is referred to these and only a synthetic overview is provided here, with the emphasis upon responses to hypoxia.

In freshwater species, the ABO is typically a highly vascularized diverticulum of the buccal cavity, pharynx, or gut; the primitive air-breathing fishes typically use modified swimbladders whereas the more modern teleosts use the branchial chambers or the gut itself (Graham, 1997). Air-breathing appears to have evolved independently multiple times in freshwater fishes (Randall *et al.*, 1981a; Graham, 1997; Brauner and Val, 2006). The prevailing opinion is that hypoxia was the essential driving force for its evolution, and that it evolved from pre-existing ASR and bubble-holding behaviors (Gee and Gee, 1995; Graham, 1997; Graham and Lee, 2004).

There are a large number of highly derived marine teleosts that occupy the intertidal zone and which are believed to have evolved air-breathing abilities and an amphibious lifestyle independently of the freshwater air-breathers. The selection pressures may have been an ability to tolerate emersion during low tide and to escape extremes of salinity and hypoxia in tidepools (Martin, 1995; Graham, 1997; Graham and Lee, 2004; Sayer, 2005; Lam *et al.*, 2006), although the amphibious lifestyle may also have provided a means of protecting their eggs and young from aquatic predators (Graham, 1997; Shimizu *et al.*, 2006). These species typically use the skin, gills, and branchial chambers as air-breathing organs (Graham, 1997; Graham and Lee, 2004; Sayer, 2005; Lam *et al.*, 2006). It has been suggested that terrestrial vertebrates have evolved from freshwater air-breathing ancestors, rather than from an amphibious marine ancestor (Graham, 1997; Graham and Lee, 2004).

Fishes with bimodal respiration have been classified into two functional groups, either facultative or obligate air-breathers (Johansen, 1970; Graham, 1997). Facultative air-breathers, by definition, supplement oxygen uptake at the surface but can survive by gill ventilation alone if denied access to air in normoxic water. Obligate air-breathers drown if denied access to air-breathing, even in normoxic water, and this is typically because they have very reduced gill surface areas. The reduction in gill surface area is because, in the all bony fishes, oxygenated blood leaving the ABO must traverse the gills to enter the systemic circulation; hence reduced gills will reduce potential loss of oxygen across gills to hypoxic water (reviewed by Graham, 1997). The gills cannot be lost completely because they retain an essential role in excretion of carbon dioxide and ammonia, and for ionic, osmotic, and acid–base balance (Randall *et al.*, 1981a; Graham, 1997; Brauner and Val, 2006). It should be noted that the facultative/obligate classification is not absolute, there are many species that utilize both strategies as a function of their developmental stage or environmental conditions (Graham, 1997; Reid *et al.*, 2006), and animals that have been considered to be obligate air-breathers in laboratory studies, such as species of African lungfish *Protopterus* (Johansen, 1970), have proven to be less so when studied by telemetry in their natural environment (Mlewa *et al.*, 2005).

What is irrefutable is that hypoxia stimulates air-breathing behavior in all fish with bimodal respiration. Figure 2.4 shows the proportion of oxygen uptake that is met by air-breathing as a function of water PO_2 in a number of freshwater species (figure after Graham, 1997). Although the various species show differing degrees of reliance on air-breathing in normoxia, all of them show an increased reliance in hypoxia (Figure 2.4). Much less is known about how hypoxia influences air-breathing behavior in amphibious marine species (Reid *et al.*, 2006; Lam *et al.*, 2006). As described in Section 3.2 intertidal blennies (Blennioidiae) and sculpins (Cottidae) that use their skin as an ABO will spontaneously emerge from hypoxic water into air (Martin, 1995; Yoshiyama *et al.*, 1995; Sloman *et al.*, 2008). The giant mudskipper, *Periophthalmodon schlosseri*, is an obligate air-breathing fish that uses a modified bucco-opercular cavity as an ABO, and it alternates brief periods of gill ventilation using water with longer periods when it holds air in the gill pouch ABO. Aguilar *et al.* (2000) exposed the giant mudskipper to either aquatic or aerial hypoxia, and found that only aerial hypoxia elicited an increase in air-breathing activity, water oxygen levels being without effect. Mudskippers live in j-shaped burrows and create an air-pocket in the end of the burrow by transporting in air in their mouth (Lee *et al.*, 2005). Lee *et al.* (2005) demonstrated that, if oxygen levels in the air are experimentally reduced below a certain PO_2 threshold (about 50% of air saturation) in the burrow of *Scartelaos histophorus*, it will actively expel the air and replace it with fresh normoxic air from the surface. Gonzales *et al.* (2006) found that aquatic hypoxia caused a significant stimulation of air-breathing frequency in the burrow-dwelling eel goby, *Odontamblyopus lacepedii*. This species is sympatric with the mudskippers but is not truly amphibious and remains in its burrow, full of hypoxic water, during low tide (Gonzales *et al.*, 2006).

As was the case for ASR, although air-breathing can be considered a behavioral response to hypoxia, it is a chemoreflex driven by oxygen-sensitive receptors (see Smatresk, 1990; Taylor *et al.*, 1996; Reid *et al.*, 2006, for reviews). The surfacing and gulping response in freshwater air-breathing species can be stimulated by chemoreceptors that sense oxygen levels in either the ventilatory water or the blood stream (Smatresk, 1986; Smatresk *et al.*, 1986; McKenzie *et al.*, 1991), and the gulping element of the motor output is believed to be a modification of pre-existing suction feeding movements (Liem, 1987, reviewed in Taylor *et al.*, 1996; Graham, 1997; Reid *et al.*, 2006). Once again, much less is known about how air-breathing is controlled in marine amphibious species.

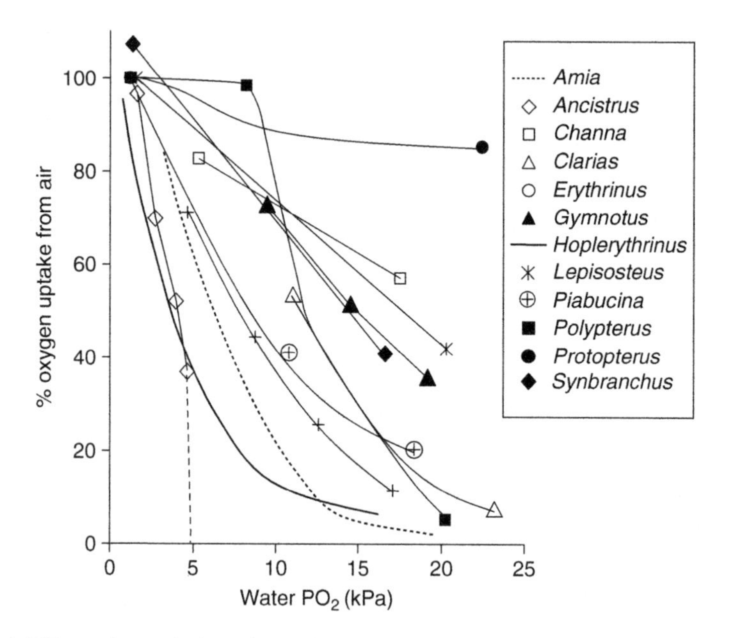

Fig. 2.4. Effects of aquatic hypoxia on the percentage of oxygen uptake that is met by air-breathing in 12 genera of air-breathing fish. The figure is reproduced from Graham (1997), with the addition of new data for two species, *Hoplerythrinus unitaeniatus* and *Synbranchus marmoratus*, taken from McKenzie *et al.* (2007b) and McKenzie *et al.* (submitted). *Amia calva* at 18°C, data estimated from Johansen *et al.* (1970); *Ancistrus chagresi* at 25°C, data from Graham (1983); *Channa argus* at 25°C, data from Itazawa and Ishimatsu (1981); *Clarias lazera* at 28° to 32°C, data from Babiker 1979; *Erythrinus erythrinus* at 27° to 30°C, data from Stevens and Holeton (1978); *Gymnotus carapo* at 29° to 33°C, data from Liem *et al.* (1984); *H. unitaeniatus* at 25°C, data from McKenzie *et al.* (2007b); *Lepisosteus oculatus* at 20°C, data from Smatresk and Cameron (1982); *Piabucina festae* at 25°C, data from Graham *et al.* (1977); *Polypterus senegalus* at 28°C, data from Babiker (1984), *Protopterus annectens* at 28°C, data from Babiker (1984); *S. marmoratus* at 25°C, data from McKenzie *et al.* (submitted), note that the greater than 100% oxygen uptake from air in this species reflects loss of oxygen to the water across the gills in deep hypoxia.

The primary benefit of air-breathing is that it makes oxygen uptake and aerobic metabolism entirely independent of the prevailing water oxygen availability (reviewed by Johansen, 1970; Graham, 1997, see also Perry *et al.*, 2004; Seifert and Chapman, 2006; McKenzie *et al.*, 2007b) such that fish, can, in theory, colonize hypoxic areas successfully. There is some evidence that oxygen taken up from the ABO may be lost to the surrounding water across the gills in very deep hypoxia in some species, but this effect is minor (Randall *et al.*, 1978, 1981b; Smatresk and Cameron, 1982). It has been suggested that the persistence of a few relict species of phylogenetically ancient primitive bony and lobe-finned fishes, such as the polypterids, *Amia*, the gars, and the lungfishes, is due in part to their ability to breathe air (Ilves and Randall, 2007). Air-breathing should, in theory, be energetically more efficient than water-breathing for fishes, because air is so much richer in oxygen and requires much less effort to ventilate (Kramer, 1983, 1987; Graham, 1997). Nonetheless, there are less than 400 known species of air-breathing fish (Graham, 1997), amongst some 25 000 species of bony fish. With the possible exception of the labyrinthine fishes of south-east Asia (Rüber *et al.*, 2006), there is little evidence that the acquisition of a capacity for air-breathing was followed by the extensive adaptive radiation that has occurred in groups of purely water-breathing fish, for example, following the appearance of the hinged jaw apparatus that provides flexibility in feeding strategies (Mabuchi *et al.*, 2007).

Indeed, as reviewed in detail in Section 5 below, there must be significant physiological and ecological costs to air-breathing, which presumably include costs of surfacing and increased risk of predation (Kramer, 1983, 1987; Kramer *et al.*, 1983; Bevan and Kramer, 1986; Randle and Chapman, 2004). As was the case for fish performing ASR, there is evidence that air-breathing patterns and behavior are significantly influenced by perceived risk of predation. Smith and Kramer (1986) reported that exposure of an obligate air-breather, the Florida gar *Lepisosteus platyrhincus*, to a model avian predator resulted in a decrease in air-breathing frequency and an increase in gill ventilation effort. Herbert and Wells (2001) found that fear of predation reduced air-breathing frequency by the blue gourami, *Trichogaster trichopterus*, an obligate air-breather, which compensated by reducing overall rates of activity. Thus, higher processing can influence reflexive air-breathing behaviors, with adaptive responses that would allow the fish to conserve the O_2 stored in their air-breathing organs. There is also evidence that some air-breathing fishes perform the behavior most frequently at night, when the risk of predation might be less (Grigg, 1965; Horn and Riggs, 1973; Babiker, 1979).

It is also possible to speculate about other potential physiological and behavioral costs to air-breathing. The vast majority of air-breathing fishes have reduced relative gill areas compared with closely related and/or sympatric water-breathing species (Graham, 1997). The aerobic metabolic scope necessary for all activities such as sustained swimming, recovery from intense exercise, or digestion of food would, presumably, have to be met to some extent by increased air-breathing activity (Farmer and Jackson, 1998; McKenzie *et al.*, 2007c; Wells *et al.*, 2007). This is an interesting area for future research. It might, presumably, constrain their options in terms of habitat choice (requiring cover from predation) and in terms of their diel rhythms in activity. Air-breathing could also interfere with social interactions if fish were dependent upon constant visits to the surface.

3. EFFECTS OF HYPOXIA ON ACTIVITY

3.1. Spontaneous Swimming Activity

Changes in spontaneous swimming activity have been described in a wide variety of fish groups and species when exposed to hypoxia, including elasmobranchs, chondrosteans, and teleosts. As reviewed in this section, these behavioral responses can comprise either a reduction in activity or an increase in activity, depending upon the species and the context. Table 2.2 lists the species that have been reported to change spontaneous swimming activity in hypoxia and the nature of the response including, when relevant, as a function of the context. Figure 2.5 shows species for which such responses have been quantified. It has been suggested that species that reduce their activity in hypoxia tend to be demersal or bentho/pelagic, with a relatively sedentary lifestyle during which they may often encounter hypoxia in their habitat; whereas species that increase activity tend to be active pelagic schooling fishes (Domenici *et al.*, 2000; Herbert and Steffensen, 2005, 2006).

Swimming is typically considered to represent a major component of the energy budget of active fishes, and high-intensity aerobic swimming can utilize a very significant proportion of a fish's aerobic metabolic scope (Fry, 1971; Claireaux and Lefrançois, 2007). Thus, for those species that reduce levels of spontaneous swimming activity in hypoxia, this has been interpreted as an energy-saving response (Metcalfe and Butler, 1984; Fischer *et al.*, 1992). The Crucian carp, *Carassius carassius*, can tolerate complete anoxia for many days. Nilsson *et al.* (1993) used video-tracking to show that one of the energy-saving strategies used by this species was to reduce their spontaneous activity by 50% (Figure 2.5). Nilsson *et al.* (1993) calculated that the reduced activity would provide a saving of approximately 35% of overall energy requirements in anoxia.

Schurmann and Steffensen (1994) studied effects of gradual stepwise progressive hypoxia on spontaneous swimming activity of Atlantic cod acclimated to three temperatures, 5°C, 10°C, and 15°C. As shown in Figure 2.6, at all temperatures routine normoxic activity was maintained at levels until a threshold degree of hypoxia beyond which it declined. This response pattern is in fact very similar to the pattern for regulation of aerobic metabolic rate by cod (and, indeed, most fish) in hypoxia (Schurmann and Steffensen, 1997). Interestingly, Schurmann and Steffensen observed that the water oxygen threshold at which the decline in activity of the cod occurred was not directly related to temperature but was highest at 10°C and similar at 5°C and 15°C (Figure 2.6). This is somewhat unexpected because the critical PO_2 threshold for regulation of standard metabolic rate (critical PO_2, P_{crit}) is highly temperature-dependent in cod, with cod at 15°C being significantly

Table 2.2

Fish species that have been reported to change their spontaneous locomotor activity in response to hypoxia

Species (Family)	Response	Context	Source
Elasmobranchs			
Carcharhinus acronotus (Carcharhinidae)	Increased swimming speed	Progressive moderate hypoxia	Carlson and Parsons (2001)
Mustelus norrisi (Triakidae)	Decreased swimming speed	Progressive moderate hypoxia	Carlson and Parsons (2001)
Scyliorhinus canicula (Scyliorhinidae)	Decreased activity	Rapid exposure to moderate hypoxia	Metcalfe and Butler (1984)
Sphyrna tiburo (Carcharhinidae)	Increased swimming speed	Progressive moderate hypoxia	Carlson and Parsons (2001)
Chondrosteans			
Acipenser naccarii (Acipenseridae)[a]	Increased swimming speed	Water PO_2 declining to mild hypoxia	McKenzie *et al.* (1995b)
Acipenser naccarii (Acipenseridae)[a]	Reduced swimming speed	Stable mild hypoxia	McKenzie *et al.* (1995b)
Teleosteans			
Ammondytes tobianus (Ammobdytidae)	Reduced swimming speed	Deep hypoxia	Behrens and Steffensen (2007)
Carassius carassius (Cyprinidae)	Reduced swimming speed	Anoxia	Nilsson *et al.* (1993)
Clupea harengus (Clupeidae)	Increased swimming speed	Progressive moderate hypoxia	Domenici *et al.* (2000); Herbert and Steffensen (2006)
Clupea harengus (Clupeidae)	Decreased swimming speed	Deep hypoxia	Domenici *et al.* (2000)
Clupea harengus (Clupeidae)	Increased swimming speed	Water PO_2 declining to moderate and deep hypoxia	Herbert and Steffensen (2006)
Gadus morhua (Gadidae)	Increased swimming speed	Water PO_2 declining to mild hypoxia	Schurmann and Steffensen (1994); Herbert and Steffensen (2005); Johansen *et al.* (2006)
Gadus morhua (Gadidae)	Decreased swimming speed	Stable moderate to deep hypoxia	Schurmann and Steffensen (1994); Herbert and Steffensen (2005); Johansen *et al.* (2006); Skjæraasen *et al.* (2008)

(*continued*)

Table 2.2 (*continued*)

Species (Family)	Response	Context	Source
Katsuwonus pelamis (Scombridae)	Increased swimming speed	Declining water PO_2 in moderate hypoxia	Dizon (1977)
Menidia beryllina (Atherinidae)	Increased swimming speed	Moderate hypoxia; larval fishes	Weltzien *et al.* (1999)
Pomatoschistus minutus (Gobiidae)	Increased "restless" activity	Deep hypoxia	Petersen and Petersen (1990)
Solea solea (Soleidae)[a]	Decreased activity level	Moderate hypoxia	Dalla Via *et al.* (1998)
Thunnus albacares (Scombridae)	Increased swimming speed	Moderate hypoxia	Bushnell and Brill (1991)
Urophycis chuss (Phycidae)	Increased swimming activity	Moderate to deep hypoxia	Bejda *et al.* (1987)
Zoarces viviparous (Zoarcidae)	Progressive decline in activity	Progressive moderate to deep hypoxia	Fischer *et al.* (1992)

[a]Also shows intense agitation in deep hypoxia (Randall *et al.*, 1992; Dalla Via *et al.*, 1998; McKenzie *et al.*, 2008).

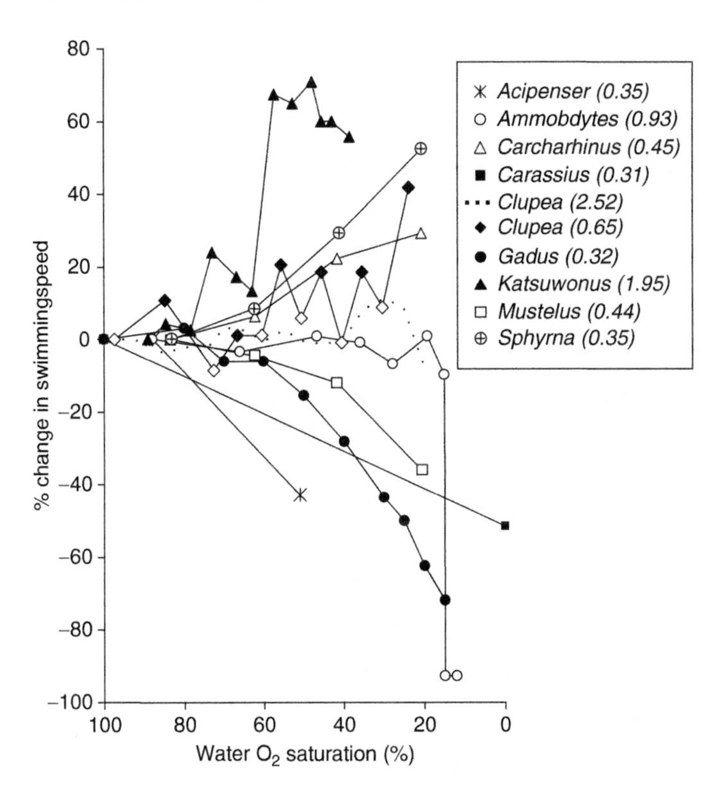

Fig. 2.5. Effects of hypoxia on swimming speed in fishes. Quantitative data are available for nine species, this figure presents mean percentage changes in speed relative to the normoxic control, note also the inverted abscissa. For each species, the number in brackets is the mean normoxic swimming speed in bodylengths sec^{-1}. *Acipenser naccarii* exposed for 3 h to either normoxia or mild hypoxia at 23°C (McKenzie *et al.* 1995b). *Ammobdytes tobianus* exposed to progressive reductions in water PO_2 each 10 min and then maintained for 1 h at a low PO_2 below 20% at 10°C. There was a profound decline in swimming speed when water PO_2 had stabilized below 20% (Behrens and Steffensen, 2007). *Carcarhinus acronotus* exposed to progressive hypoxia in a sealed respirometer at 26°C and speed measured at four PO_2 intervals (Carlson and Parsons, 2001). *Carassius carassius* exposed to anoxia for 1 h at 8°C (Nilsson *et al.*, 1993); *Clupea harengus* dotted line is mean response to progressive reductions in water PO_2 each 10 min at 15°C (Domenici *et al.*, 2000). *Clupea harengus* diamonds show effects of exposure for 30 min to a stepwise series of progressively more hypoxic PO_2s at 10°C (white diamonds) and the response to declining PO_2 between each step (intervening black diamonds). The declining PO_2 caused a significant increase in swimming speed by the herring, which disappeared when PO_2 stabilized (Herbert and Steffensen, 2006). *Gadus morhua* exposed to progressive reductions in water PO_2 each 30 min at 10°C (Schurmann and Steffensen, 1994). Note that Herbert and Steffensen (2005) found that this species increased swimming speed when exposed to declining PO_2, but speed then stabilized at a lower level when the hypoxic PO_2 stabilized (data not shown). *Katsuwonus pelamis* exposed to progressive reduction in PO_2 each 10 min at 24°C (Dizon, 1977). *Mustelus norrisi* and *Sphyrna tiburo* exposed to progressive hypoxia in a sealed respirometer at 26°C and speed measured at four PO_2 intervals (Carlson and Parsons, 2001).

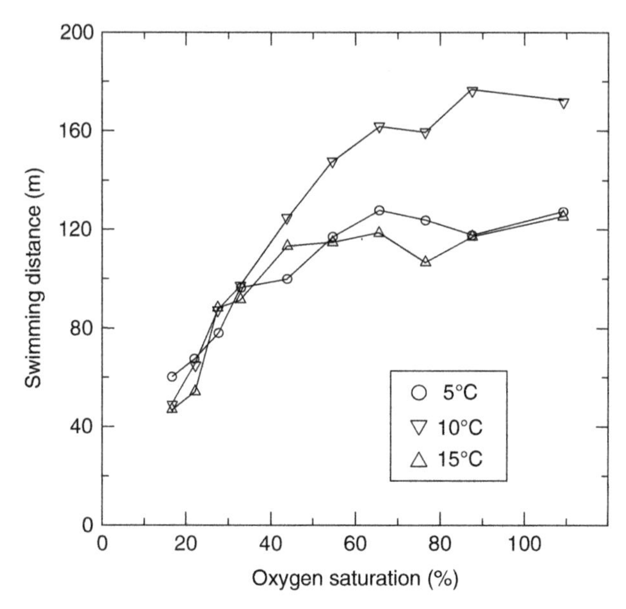

Fig. 2.6. Swimming distance of Atlantic cod *Gadus morhua* at different oxygen saturations, at three temperatures of 5°C, 10°C, and 15°C. Each point represents the mean distance (m) covered by eight cod during 30 min. At all oxygen saturations, the activity levels (swimming distance) at 5°C and 15°C are similar and there was a tendency for cod at 10°C to be more active at higher oxygen saturations. At oxygen saturations below 30%, the activity levels were similar at all temperatures and decreased with decreasing oxygen saturation. From Schurmann and Steffensen (1994) with permission from the Company of Biologists.

less tolerant of hypoxia than those at 5°C (Schurmann and Steffensen, 1997). Skjæraasen *et al.* (2008)) have subsequently, however, demonstrated that cod show greater hypoxia depression of spontaneous activity at 15°C than they do at 5°C, which is consistent with their relative overall hypoxia tolerance at these temperatures (Schurmann and Steffensen, 1997). The thresholds for reduced activity by cod in hypoxia have consistently been higher than the corresponding P_{crit} for cod at that same temperature (Schurmann and Steffensen, 1994, 1997; Herbert and Steffensen, 2005; Skjæraasen *et al.*, 2008), indicating that the reduced swimming activity may have reflected reduced aerobic metabolic scope as hypoxia progressed toward the P_{crit} (Claireaux and Lagardère, 1999).

Behrens and Steffensen (2007) studied effects of progressive hypoxia on spontaneous swimming activity in schools of lesser sandeels, *Ammobdytes tobianus*. These small marine fish bury in the sand for much of the time, but emerge to feed in large schools. Their spontaneous swimming speed was

0.9 BL sec^{-1} in normoxia, and this speed was maintained until a threshold PO_2 of about 15% saturation, beyond which the animals reduced their activity by over 90% and came to rest on the substrate (Figure 2.5). This threshold for reduced activity is below their P_{crit} at the same temperature (~20% saturation), indicating that the reduced activity may have reflected an inescapable metabolic depression, although all of the animals recovered activity when returned to normoxia (Behrens and Steffensen, 2007).

Dalla Via *et al.* (1998) studied the responses of Dover sole, *Solea solea*, to progressive hypoxia. Although swimming activity was not quantified, the authors report that spontaneous activity was reduced at oxygen tensions between 80% and 20% air saturation. Sole of the size used by Dalla Via *et al.* (1998) have a P_{crit} below 20% saturation (van den Thillart *et al.*, 1994; Couturier *et al.*, 2007). Dalla Via *et al.* (1998) found that, as this threshold approached, the fish tended to remain immobile except for a tendency to raise portions of their body off the substrate, possibly to ventilate the blind side (Nonnotte and Kirsch, 1978; McKenzie *et al.*, 2008). At 5% air saturation or lower, the fish exhibited intense agitation followed by a loss of equilibrium and cessation of movements. McKenzie *et al.* (2008) found qualitatively similar patterns of behavioral and metabolic responses to hypoxia in very early life stages of this flatfish species (including in pre-settlement larvae at 9–13 days old), which suggests that the responses arise early in ontogeny (Weltzien *et al.*, 1999). Both Dalla Via *et al.* (1998) and McKenzie *et al.* (2008) found that the agitation response occurred below the corresponding P_{crit} for their life-stages, but the P_{crit} and thresholds for the behaviors were very much higher in the early life stages, being 60% and 56%, respectively, in larvae, and 48% and 29%, respectively, in post-settlement juveniles (15–20 days old). The intense agitation of fish in deep hypoxia below the P_{crit} has been interpreted as an acute escape response (Bejda *et al.*, 1987; Randall *et al.*, 1992; Van Raaij *et al.*, 1996).

For those species that increase their level of spontaneous activity in hypoxia, this has been interpreted as a response to escape the hypoxic area (Dizon, 1977; Bejda *et al.*, 1987; Petersen and Petersen, 1990; Weltzien *et al.*, 1999; Domenici *et al.*, 2000; Herbert and Steffensen, 2005, 2006). Petersen and Petersen (1990) found that oxygen saturations below 40% caused the sand goby *Pomatoschistus minutus* to become restless and perform random activity. Young red hake *Urophycis chuss* moved upwards in the water column if the oxygen concentration fell below about 50% saturation (Bejda *et al.*, 1987). These responses arise very early in life; Weltzien *et al.* (1999) found that larvae of the inland silverside, *Menidia beryllina*, exhibited an avoidance response to hypoxic water within hours of hatching. Larvae were placed in a water column with two salinity layers, when the larvae drifted into a layer that was maintained hypoxic at less than 55% oxygen saturation, this stimulated bursts of swimming activity at up to 6.4 BL sec^{-1} that were directed toward the normoxic layer (Figure 2.5).

Dizon (1977) studied two species of tuna, the skipjack, *Katsuwonus pelamis*, and the yellowfin, *Thunnus albacares*, which swim spontaneously at speeds of about 1.6 and 1.2 BL sec^{-1}, respectively. When these tunas were exposed to a gradual decline in oxygen tension, the skipjack showed an abrupt doubling of swimming speed, up to 2.2 BL sec^{-1}, as oxygen fell below 55% saturation (Figure 2.5), although there was no effect of hypoxia on activity in yellowfin. Dizon (1977) attributed this species difference in behavior to a greater sensitivity to hypoxia in the skipjack, which therefore engendered an escape response. Bushnell and Brill (1991) subsequently reported that the yellowfin did increase swimming speed by about 20% when exposed to hypoxia at 60% of air saturation, and also attributed this to an escape response. Herring school spontaneously at speeds of between 0.8 and 3 BL sec^{-1} and show progressive increases in speed when exposed to stepwise progressive hypoxia (Domenici *et al.*, 2000; Herbert and Steffensen, 2006). Domenici *et al.* (2000) found that the lower the spontaneous speed in normoxia, the greater the increase in speed in hypoxia. Both Domenici *et al.* (2000) and Herbert and Steffensen (2006) found that the most marked increase in activity occurred when water PO_2 fell below about 40% of air saturation; Herbert and Steffensen (2006) found that the most marked increases in swimming speed occurred when water PO_2 was declining rather than at a stable degree of hypoxia (Figure 2.5). These same authors could not detect evidence of a transition to anaerobic metabolism (no increase in plasma lactate) at the hypoxic PO_2s at which the changes in swimming speed occurred, indicating that the herring were above their P_{crit}. This response may help herring to avoid hypoxic areas in their natural environment (Neuenfeldt, 2002; Herbert and Steffensen, 2006). Herring also form vast schools, and the fish at the middle/back of the school may experience hypoxia due to oxygen uptake by those preceding and around them (McFarland and Moss, 1967; Domenici *et al.*, 2002, see Section 5.2). The increased swimming speed would allow the fish to reshuffle and change position, toward better-oxygenated areas in the school (Domenici *et al.*, 2000; Herbert and Steffensen, 2006).

As might be expected, many species can exhibit both types of response (increased or decreased activity), depending on the degree of hypoxia, the rate of change in water oxygen tension, and their degree of activity in normoxia. Thus, as already discussed, Dover sole will reduce activity as they approach their P_{crit} but then exhibit intense agitation prior to loss of equilibrium in deep hypoxia (Dalla Via *et al.*, 1998; McKenzie *et al.*, 2008). Schools of herring exhibit exactly the opposite response, showing increased activity as oxygen levels drop down to about 25% air saturation, followed by a decline in activity and disruption of the school at lower PO_2 (Domenici *et al.*, 2000; Figure 2.5). Schurmann and Steffensen (1994) and Herbert and Steffensen (2005) demonstrated that when oxygen tensions are declining there is a transient increase in activity in Atlantic cod, and it is when hypoxic conditions are stable that activity is down-regulated (Figure 2.5). These authors postulate that the two contrasting responses would be adaptive if the increased activity allowed the fish to escape hypoxia as they encountered it (or as it developed in their surroundings), whereas the reduced activity would allow them to "sit out" hypoxia from which escape was impossible. McKenzie *et al.* (1995b) found that the responses of Adriatic sturgeon, *Acipenser naccarii*, to hypoxia differed as a function of their degree of activity in normoxia. This species shows constant sustained activity at a speed between 0.2 and 0.5 BL sec^{-1}. Animals that were most active in normoxia reduced their activity when exposed to hypoxia, as oxygen levels declined and then stabilized (at 50% of air saturation, Figure 2.5). Animals that were less active in normoxia showed a transient increase in activity as oxygen levels declined toward hypoxia, but then returned to their previous normoxic swimming speed (McKenzie *et al.*, 1995b). This species must swim in order to regulate metabolism in hypoxia, which would argue for a role of ram ventilation (McKenzie *et al.*, 1995b, 2007d) although when animals swam at incremental sustained speeds (McKenzie *et al.*, 2001) there was no evidence of the cessation of gill ventilation that accompanies a transition to ram ventilation in many teleosts (Freadman, 1979; Steffensen, 1985).

Some large open-ocean pelagic species swim constantly and have evolved a dependence upon ram ventilation to generate a flow of water across the gills (Brown and Muir, 1970; Roberts, 1978; Parsons and Carlson, 1998; Carlson and Parsons, 2001). Although tunas rely on ram ventilation (Roberts, 1978), their increased swimming speeds in hypoxia have been attributed to an escape response (Dizon, 1977; Bushnell and Brill, 1991) because a model relating swimming speed and mouth gape to rates of oxygen uptake revealed that observed increases in swimming and gape in hypoxia would not be sufficient to maintain the highly elevated routine normoxic rates of oxygen uptake in these species (Bushnell and Brill, 1991). In a number of large sharks with ram ventilation, however, a spontaneous increase in swimming speed in hypoxia does appear to be an adaptive response to increase rates of gill ventilation (Parsons and Carlson, 1998; Carlson and Parsons, 2001). Carlson and Parsons (2001) compared locomotor and metabolic responses to hypoxia in three species of shark, two large species with ram ventilation (bonnethead, *Sphyrna tiburo*, and blacknose, *Carcharhinus acronotus*) and a dogfish with active gill ventilation (Florida smoothhound shark, *Mustelus norrisi*). All of the species swam spontaneously at a speed of about 0.4 BL sec^{-1} in normoxia; but as the fish were exposed to progressive hypoxia, the bonnethead and blacknose (ram ventilators) increased their swimming speed by up to 50%

and 25%, respectively (Figure 2.5). The dogfish, on the other hand, reduced swimming speed by up to 36% at the lowest degree of hypoxia tested (approximately 40% of air saturation, Figure 2.5). In all species, these changes started at a threshold of about 70% of air saturation; in the ram ventilators they were associated with an increase in oxygen uptake relative to normoxia whereas oxygen uptake was regulated unchanged from normoxic rates in the dogfish (Carlson and Parsons, 2001; Figure 2.5).

In many of these studies, modulation of routine rates of spontaneous activity in hypoxia occurs above the P_{crit} of the fish, hence at levels of hypoxia that are not, presumably, associated with respiratory distress and a transition to anaerobic metabolism (Herbert and Steffensen, 2005, 2006). These changes in routine activity are also, presumably, stimulated by information from the chemoreceptors, which have been demonstrated to monitor oxygen levels in the water and/or the blood of fishes (Burleson *et al.*, 1992; Sundin *et al.*, 2007), although this remains to be confirmed. The routine swimming speeds that have been measured in normoxia all fall within the range of sustainable aerobic activities that are dependent upon the performance of red muscle (Webb, 1993, 1998), hence reducing these activities may relieve hypoxic constraints on aerobic metabolic scope (Claireaux and Lagardère, 1999; Claireaux *et al.*, 2000; Lefrançois and Claireaux, 2003). The locomotor responses to hypoxia arise very early in ontogeny, in particular the avoidance and agitation responses (Weltzien *et al.*, 1999; McKenzie *et al.*, 2008). The agitation responses seen in some species in deep hypoxia below the P_{crit} would seem indicative of the recruitment of white muscle for a transient burst of anaerobic swimming activity (Webb, 1993, 1998).

3.2. Other Locomotor Responses to Hypoxia

There are other locomotor responses to hypoxia that are characteristic of species with particular life histories. Sand eels (Ammodytidae) occupy burrows on sandy sea beds, lying with their snout a few millimeters below the surface. Behrens *et al.* (2007) used novel planar optodes to quantify oxygen levels around buried lesser sandeels, and demonstrated that they breathe by advective transport through the permeable interstice, forming an inverted cone of oxygenated porewater in front of their mouth. Oxygen levels around the buried eel tended, however, to be very low and, every once in a while, the sandeels wriggled in the sand and re-oxygenated their surroundings. It is not known why they do this, because the skin does not appear to play any role in gas exchange (Behrens *et al.*, 2007). If the overlying water is made hypoxic, the sandeel moved toward the surface and its head emerged when PO_2 fell below its P_{crit} (~20% saturation, Behrens and Steffensen, 2007), eventually emerging in extreme hypoxia (5%) to lay on the surface (Behrens *et al.*, 2007).

Another response to aquatic hypoxia is spontaneous emersion, whereby amphibious and semi-amphibious species leave water to go onto land. A number of freshwater and marine species have been reported to venture onto land, a role for aquatic hypoxia in this response has only been studied in a few species (see Graham, 1997; Sayer, 2005, for reviews). The fishes for which the hypoxic emersion response is most well-described are the marine species that inhabit rockpools in the intertidal zone and will emerge from water to maintain gas exchange in air via their skin (Martin, 1995; Graham, 1997; Sayer, 2005). A number of species of blenny (Blennioids) exhibit this response. Graham (1970) found that a tropical blenny, *Mnierpes macrocephalus*, could be stimulated to emerge from hypoxic water in the laboratory. Field observations have reported that other blennies (*Helcogramma medium*, *Blennius pholis*) crawl from hypoxic tidepools (Davenport and Woolmington, 1981; Innes and Wells, 1985). Intertidal sculpins of the family Cottidae inhabit tidepools that can become isolated and extremely hypoxic during low tide, and many of these species will spontaneously emerge from hypoxic water (Davenport and Woolmington, 1981; Yoshiyama *et al.*, 1995; Graham, 1997; Sayer, 2005). Davenport and Woolmington found that *Taurulus (Cottus) bubalis* emerged from hypoxic tidepools when PO_2 fell below 5% saturation. Yoshiyama *et al.* (1995) found that four different species of intertidal sculpin that inhabit rockpools in the intertidal zone of the temperate Pacific (*Oligocottus snyderi*, *O. maculosus*, *Clinocottus globiceps*, and *Ascelichthys rhodorus*) also emerged spontaneously in laboratory experiments. The propensity to perform this behavior was directly related to the capacity of the species to breathe air, being greatest in *C. globiceps* and least in *A. rhodorus* (Yoshiyama *et al.*, 1995); however, the PO_2 thresholds for emergence were not reported. Sloman *et al.* (2008) studied emersion responses by the tidepool sculpin, *O. maculosus*, comparing two size-classes of fish in the laboratory, in artificial tidepool mesocosms, and in their natural habitat. These authors found that both size-classes of fish tended to emerge from the water when oxygen fell below 20% of air saturation, and this was similar in laboratory, mesocosm, and field. This hypoxic threshold coincided with the threshold for the appearance of anaerobic end-products in the tissues of fish that were denied access to emergence. This indicates that in this species, emergence is a last-ditch attempt to avoid hypoxic depression of aerobic metabolism; in air the animals were able to maintain their aerobic metabolism at routine normoxic levels but may suffer increased risk of predation by birds (Yoshiyama *et al.*, 1995; Sloman *et al.*, 2008).

Liem (1987) investigated whether hypoxia stimulated an emersion response in a number of tropical labyrinthine fishes (Anabantoids), air-breathing freshwater species that are known to venture onto land. Hypoxia did not stimulate emersion in the climbing perch (*Anabas testudineus*), the

Siamese fighting fish (*Beta splendens*), or various gourami species (*Macropodus opercularis* and *Trichogaster trichopterus*). In species such as the mangrove killifish *Krytpolebias (Rivulus) marmoratus*, an emergence response can be stimulated by poor water quality (increased water H_2S content, Abel *et al.*, 1987). Similar emersion responses to escape "poor water quality" have been anecdotally reported for many tropical air-breathing fish species such as erythrinids, walking catfish (family Claridae), snakeheads (Channidae), swamp eels (Synbranchidae), and eels of the genus *Anguilla* (Tesch, 1972; Graham, 1997). A potential role of hypoxia in the emergence responses of these species remains, however, to be described.

4. HYPOXIA AND PARENTAL CARE BEHAVIOR

In addition to the large energetic costs associated with swimming and other locomotor responses, another significant metabolic cost for some fish species is the energy directed to parental care. Among fishes that exhibit post-fertilization care, there is an amazing diversity of strategies ranging from simple nest guarding, to mouth brooding, to live bearing. The energetic cost of reproduction increases as more energy is invested into parental care (Fryer and Iles, 1972; Jones and Reynolds, 1999a), and one would predict that alternative oxygen environments would affect the costs and benefits of parental care (Hale *et al.*, 2003). Some oviparous fishes protect their developing young after spawning by selecting suitable nesting sites, as well as nest building and guarding the young, fanning to aerate eggs, and mouthing to clean and remove dead and diseased eggs (Wootton, 1990). In mouth brooders (where the eggs and larvae are held in the mouth of a parent) and other forms of live-bearing fishes (e.g., seahorses or other pouch bearers) the parent can protect the young from predators and other environmental stressors by moving to more suitable environments (Wourms and Lombardi, 1992; Goodwin *et al.*, 2002; Wen-Chi Corrie *et al.*, 2007). Ovoviviparous fish take parental care a step further retaining the embryos in either the ovary or uterus until they reach a more advanced and less vulnerable stage of development. Some live-bearing species are viviparous and have evolved specialized tissues to provide nutrients to developing young (Thibault and Schultz, 1978; Blackburn *et al.*, 1985). The costs to the parent associated with carrying the young either internally or externally include high energetic costs of bearing the young and increased predation risk due to reduced mobility (Thibault and Schultz, 1978; Blackburn *et al.*, 1985; Goodwin *et al.*, 2002; Timmerman and Chapman, 2003). Mouth brooders bear an additional cost as most are prohibited from feeding while brooding.

Costs of parental care in fishes may be particularly severe under hypoxia due to the challenge of providing oxygen to the eggs, the need for physiological and biochemical mechanisms to facilitate oxygen uptake for the parent, and increased predation risk associated with surfacing behavior. However, high levels of parental care may be necessary under hypoxic conditions to ensure survival of the eggs and young. Roberts (1973) noted an interesting link between hypoxic habitats and parental care in his review of the ecology of fishes of the Amazon and Congo basins. He observed that in these large tropical rivers, parental care occurs primarily in fishes in which the adults breed in swamps and other oxygen-deficient habitats. For example, in the African lungfish (*Protopterus* spp.) a nest is constructed and guarded by the male (Johnels and Svensson, 1954; Bouillon, 1961; Greenwood, 1987). In the central Congo basin, *Protopterus dolloi* excavates a burrow that receives air through a chimney-like structure, without which the eggs would be deprived of oxygen. The small anabantid fish *Ctenopoma damasi* and the characoid *Hepsetus odoe* construct floating nests of foam in which the eggs are supported (Berns and Peters, 1969; Roberts, 1973). The most abundant cichlids in the dense interior of East African swamps are mouth-brooding haplochromines and tilapiines (Chapman *et al.*, 2002, 2006a,b) that can move the young to microhabitats with better levels of oxygen.

The effects of low-oxygen stress on parental care of fishes is a topic of growing concern, given the widespread and increasing occurrence of aquatic hypoxia. Literature in this area is still depauperate but, nonetheless, demonstrates that DO is an important driver of parental care behavior across a range of strategies from guarding to viviparity. Many nest-guarding fishes use fanning or other behaviors that increase ambient oxygen levels (Mertz and Barlow, 1966; Wootton, 1976; Zoran and Ward, 1983; Coleman, 1992; Jones and Reynolds, 1999a,b; Takegaki and Nakazono, 1999). Some species have been shown to alter their ventilation behaviors in response to changing levels of DO, including three-spined stickleback (Reebs *et al.*, 1984), the anemonefish, *Amphiprion melanopus* (Green and McCormick, 2004), the sand goby, *Pomatoschistus minutus* (Lissaker *et al.*, 2003), the land-locked goby, *Rhinogobius* sp. (Maruyama *et al.*, 2008), and the common goby, *Pomatoschistus microps* (Jones and Reynolds, 1999a), which supports a role for parental care in oxygen replenishment in fish nests. The importance of male parental care is evident in the sand goby, *P. minutus*, where females were shown consistently to prefer males with elevated levels of parental care under hypoxia (Lindström *et al.*, 2006). However, the response of the Florida flagfish, *Jordanella floridae*, is inconsistent with these earlier studies. Hale *et al.* (2003) found that male flagfish devoted less time to parental care (including fanning) as DO declined. Hale and colleagues hypothesized that the increasing costs of care as DO declined outweighed the benefits for this species, and they noted that earlier studies may not have exposed parental fish to DO levels sufficiently low enough to preclude a benefit of fanning.

More complex parental care has been described for the amphibious mudskippers (Oxudercinae), fish that are well adapted for life on intertidal mudflats, highly productive systems that are characterized by severe environmental challenges. Mudskippers lay their eggs in mud burrows that contain extremely hypoxic water. In their study of the Japanese mudskipper, *Periophthalmus modestus*, Ishimatsu *et al.* (2007) discovered that eggs are deposited on the walls of an air-filled chamber where oxygen is maintained via activity of the guarding male that deposits mouthfuls of fresh air into the chamber at each low tide. After completion of egg development, the male was shown to remove the air from the chamber, flooding the eggs to induce hatching (Ishimatsu *et al.*, 2007).

For mouth-brooding fishes, costs of parental care may be severe under hypoxia because of the increased requirements of oxygenating eggs when the parent cannot eat, and the elevated predation risks associated with any surfacing behavior. Wen-Chi Corrie *et al.* (2007) quantified the behavioral response to progressive hypoxia of the widespread mouth-brooding African cichlid, *Pseudocrenilabrus multicolor victoriae*. This species responded to progressive hypoxia by performing ASR; however, brooding females showed higher ASR thresholds than males, and initiated ASR at a much higher threshold. Non-brooding females did not differ from males for any ASR threshold. A high ASR threshold in brooding females may reflect various costs such as churning behavior, which is used to move the brood inside the mouth, potentially enhancing ventilation and cleaning of the eggs and young (Oppenheimer and Barlow, 1968; Keenleyside, 1991). This may add to the energy expenditure of the female, particularly under hypoxic conditions.

Evidence that brooding affects responses to hypoxia has been reported in other species. In their study of cardinal fishes, paternal mouth brooders, Östlund-Nilsson and Nilsson (2004) found that the critical oxygen tension of brooders was almost twice as high as non-brooders. Similarly, costs of carrying young that affect responses to hypoxia are evident in livebearers. Timmerman and Chapman (2003) found that gestating female sailfin mollies (*Poecilia latipinna*) spent 27% more time at the surface using ASR than non-gestating females. They attributed this to a high mass-specific oxygen requirement in the young, which increased the total oxygen requirement of females (Boehlert and Yoklavich, 1984; Boehlert *et al.*, 1986; Dygert and Gunderson, 1991). The increased time allocated to ASR may directly affect maternal predation risk of mollies in hypoxic waters, as risk of aerial predation has been demonstrated to increase with time spent conducting ASR (Kramer *et al.*, 1983). In the field, one might anticipate gestating females to exhibit behaviors that reduce the risk, such as selection of microhabitats with elevated DO or areas that reduce aerial predation risk, such as vegetated cover.

5. HYPOXIA AND ECOLOGICAL INTERACTIONS

There is a growing body of empirical evidence that hypoxia can influence species interactions, in particular predator–prey relationships, by altering the success rate of the predator and/or the vulnerability of the prey. Low-oxygen conditions may shift the balance of the interaction to favor predator or prey depending, at least in part, on the relative tolerance of the interactants (Domenici *et al.*, 2007). Hypoxia can elicit behaviors such as ASR or air breathing that increase the risk of predation; it can negatively impact fast-start performance of prey, and it can alter the dynamics of schooling behavior. For aquatic water-breathing predators, hypoxia can decrease predation through metabolic depression, lowered appetite, or decreased performance. The outcome of altered predator–prey interactions can ultimately influence other components of the food web and assemblage; therefore, predicting whether the prey or the predator is the beneficiary of hypoxic stress is critical for understanding community level impacts of hypoxia, whether natural or anthropogenically induced.

5.1. Hypoxic Refugia from Piscine Predators

Studies of predator–prey interactions in fishes suggest that hypoxia may be an important modulator; the prey may benefit if high tolerance to hypoxia permits access to refugia from less tolerant predators. For example, for some potential prey in the Lake Victoria basin of East Africa, hypoxic refugia have mitigated impacts of a large introduced piscivore. The explosive speciation of haplochromine cichlid fishes in Lake Victoria is unrivaled among vertebrates; however, over 40% of its endemic fishes disappeared between 1980 and 1986 associated with various anthropogenic perturbations including the upsurge of the invasive predatory Nile perch (*Lates niloticus*) (Kaufman, 1992; Kaufman *et al.*, 1997; Seehausen *et al.*, 1997a,b; Balirwa *et al.*, 2003). A similar pattern of faunal collapse was observed in other lakes in the basin where Nile perch was introduced (Kyoga, Nabugabo). However, some indigenous species persisted with the Nile perch and were resilient to increasing eutrophication and other stressors. Over the years, interest in conservation of this residual fauna has sparked studies directed toward identification of faunal refugia; habitats where native fishes are protected from Nile perch predation. Wetlands in the Lake Victoria basin serve as both structural and low-oxygen refugia for fishes that can tolerate wetland conditions, and function as barriers to dispersal of Nile perch (Chapman *et al.*, 1996a,b; Balirwa, 1998; Schofield and Chapman, 1999; Chapman *et al.*, 2002; Mnaya *et al.*, 2006). Based on a suite of ecophysiological studies on the fishes of the

Lake Nabugabo system (critical oxygen tension, ASR thresholds), Chapman and colleagues found that some cichlids and some native non-cichlids that are relatively tolerant of hypoxia are able to persist in the dense interior of hypoxic swamps (Chapman *et al.*, 1995; Chapman and Chapman, 1998; Rosenberger and Chapman, 2000; Chapman *et al.*, 2002; Rutjes, 2006), while the Nile perch cannot, as indicated by its high threshold for ASR, its high critical oxygen tension, and its distribution (Schofield and Chapman, 2000; Chapman *et al.*, 2002). This permits some fishes to persist in wetlands under reduced predator pressure from Nile perch (Chapman *et al.*, 2002). The ecotone of the wetland/open water is a particularly important refugium because interaction with the main lake waters elevates DO, but structural complexity is still high. Nile perch are present but rare in ecotonal wetlands, and species richness is higher than in the interior swamp (Chapman *et al.*, 1996a,b; Balirwa, 1998; Schofield and Chapman, 1999; Chapman *et al.*, 2002). However, even areas deep within the fringing swamp are important in the maintenance of a subset of the basin fauna (Chapman *et al.*, 1996b, 2002).

There is now a growing body of empirical support, beyond the Lake Victoria basin, for the use of hypoxic refugia by fishes. Nilsson and Östlund-Nilsson (2004) quantified the critical oxygen tension of 31 coral reef fishes in the Great Barrier Reef, Australia and reported a surprisingly high level of tolerance to hypoxia. They suggested that widespread tolerance to aquatic hypoxia in coral reef teleosts may reflect use of hypoxic spaces in the coral as nocturnal refugia from predators or use of isolated tide pools that experience hypoxia (Nilsson and Östlund Nilsson, 2004). In their review of respiratory ecophysiology of coral reef fishes, Nilsson *et al.* (2007) noted that air breathing allows some coral reef gobies to stay in their coral refugia during air exposure at low tides, thereby minimizing predator risk. The crucian carp is another example of a species that may use physiological exclusion to minimize predator risk. This carp is well known for its extreme anoxia tolerance, surviving in shallow ponds in Northern Europe that can become anoxic in the winter and are therefore free of piscine predators (Nilsson and Renshaw, 2004; see Chapter 9). Robb and Abrahams (2003) evaluated hypoxic tolerance of small yellow perch (*Perca flavescens*) and fathead minnows (*Pimephales promelas*), both potential prey of large yellow perch. They found that both within and between species, smaller individuals were the most tolerant of hypoxic environments, and suggested that prey may intentionally seek out low-oxygen habitats under risk of predation. However, Abrahams *et al.* (2007) argued that differences in hypoxia tolerance that fall within the range of physiological acclimation of the predator may only benefit the prey if interactions are emphemeral.

Hypoxia has been implicated as a determinant of fish assemblage structure in other systems where piscine predators are prevalent. For example,

Tonn and Magnuson (1982) demonstrated that in Wisconsin lakes, a centrarchid-*Esox* predator assemblage was dominant if winter oxygen levels were high, while *Umbra*-cyprinid prey assemblages dominated in lakes with low winter oxygen levels. McNeil and Closs (2007) found a generally high level of tolerance to periodic hypoxia in the fishes of the Ovens River floodplain in south-east Australia with the exception of three species, one of which was the predatory introduced redfin perch (*Perca fluviatilis*), again supporting the role of hypoxic habitats as refuge for tolerant prey and emphasizing the implications of species-specific variation in hypoxia-tolerance for community structure. Finally, in their study of the fish community along a DO gradient in a Florida spring, McKinsey and Chapman (1998) found that the mosquitofish, *Gambusia holbrooki*, was the most abundant species at the boil of the spring where DO averaged 0.20 mg L^{-1}. They suggested that the boil area may serve as refugium from predation for *G. holbrooki*; only one large predatory species was observed in the boil region (*Amia calva*, an air breather), whereas other piscivores are found in the main river. Avian predation could compromise the value of the boil refugium; however, *G. holbrooki* resides primarily in heavy vegetation along the boil margins (McKinsey and Chapman, 1998).

In the context of predator–prey interactions under hypoxia, Robb and Abrahams (2003) suggested that there may be an ecological advantage of being small based on studies showing smaller fish to be more hypoxia tolerant than larger fish. Robb and Abrahams reviewed two plausible mechanisms to explain this size sensitivity: the negative allometric relationship for mass-specific gill-surface area (Muir, 1969; Hughes, 1984), and a fractal scaling relationship whereby larger fish may be limited by the fixed size of the red blood cells for gas exchange (West *et al.*, 1997). Recently, Nilsson and Östlund-Nilsson (2008) reviewed the literature on effects of body size on hypoxia tolerance in fishes. They argued that body size *per se* does not influence oxygen uptake ability because the gill respiratory surface has a similar scaling relationship as metabolic rate. In addition, where anaerobic ATP production is required for survival, large fish seem to have an advantage because of their lower mass-specific metabolic rate (Nilsson and Östlund-Nilsson, 2008). Indeed a physiological advantage of small size under hypoxic stress is not always evident, yet hypoxic refugia may still occur. For example, Sloman and colleagues (2006) reported that small Amazonian oscars (*Astronotus ocellatus*) seek out hypoxic habitats as refuge, but found evidence to suggest that the juveniles are not more tolerant than larger conspecifics, but rather, accept a greater physiological compromise to access hypoxic shelter. Although relationships between fish size and hypoxia tolerance are not consistent across studies, there is a growing body of empirical support for the role of hypoxic habitats in modulating piscine predator effects by serving as refugia.

Hypoxia may also decrease vulnerability of fish prey to piscine predators if predator activity is reduced. There are a number of studies demonstrating decreased spontaneous activity (see Section 3 above), feeding rate, metabolism, and/or predator activity under hypoxic stress (e.g., Bejda *et al.*, 1987; Breitburg *et al.*, 1994; Shoji *et al.*, 2005; Ripley and Foran, 2006). In this situation, risk of predation by piscivores may be reduced for fish prey; however, other aquatic predators may still take advantage of the negative effects of hypoxia on the ability of fish to escape (Domenici *et al.*, 2007), thus shifting the balance toward a new player in the game. For example, in a study of the effects of hypoxia on the predation of larval red sea bream (*Pagrus major*) by the jellyfish, *Aurelia aurit*, and the juvenile Spanish mackerel, *Scomberomorous niphonius*, Shoji *et al.* (2005) found that lower DO induced higher predation rates by the tolerant jellyfish and lower predation rates by the juvenile mackerel, which had become physiologically stressed. The authors suggested that the increase in the number of jellyfish and its predation on juvenile sea bream have been driven by high nutrient loading in the Seto Inland Sea, which has accelerated eutrophication and exacerbated oxygen depletion. A similar shift in predator effects was reported by Breitburg *et al.* (1997), who showed that under hypoxia, predation on naked gobies (*Gobiosoma bosc*) by adult piscivores decreased, while predation by the sea nettle (the jellyfish *Chrysaora quniquecirrha*) increased, reflecting both the high tolerance of the sea nettle and impaired fast-start performance (see Section 5.3 below) of the fish prey.

5.2. A Shift in the Beneficiary – Increased Prey Vulnerability under Hypoxia Stress

5.2.1. Risk of Surfacing

Hypoxia may increase vulnerability of fish to predation by affecting their vertical distribution, thereby increasing their potential encounter rate with air-breathing predators. As reviewed in Section 2.1, many fishes respond to hypoxia by using ASR. However, there are clearly costs associated with use of ASR. As DO levels approach zero, many fishes spend most of their time at the surface (Gee *et al.*, 1978; Kramer and McClure, 1982; Chapman *et al.*, 1995; Olowo and Chapman, 1996; Figure 2.2). ASR not only takes time away from other activities, but could place fish at higher risk of predation when they leave shelter and move to the surface where visibility is increased to both avian and other air-breathing or highly tolerant taxa (Kramer *et al.*, 1983; Kramer, 1987; Domenici *et al.*, 2007).

Theoretically, air breathing is energetically more efficient than water breathing for fishes because of air's superior properties as a respiratory medium (Kramer, 1983, 1987). However, as discussed in Section 2.2, the rarity of air-breathing fishes and the richness and diversity of water breathers, even in habitats with low DO, suggest that there must be significant costs to air breathing (Kramer, 1983, 1987). These costs may include increased vulnerability to aerial predation and increased energetic costs of travel to the surface (Kramer, 1983; Kramer *et al.*, 1983; Bevan and Kramer, 1986; Randle and Chapman, 2004). Kramer *et al.* (1983) used a trained heron to evaluate the risk of aerial predation for air-breathing fishes and non-air-breathing fishes that use ASR in response to hypoxic stress. They found that fish using ASR tended to surface at lower DO thresholds than air breathers, though surfacing time was longer. Kramer and colleagues suggested that fish using ASR may incur less risk of avian predation at moderate DO levels, but air breathers seem to have an advantage under extreme hypoxia.

Given the potential costs of surfacing, it is not surprising that fishes show many behaviors to minimize the risk. For example, several air-breathing fishes use some form of synchronous air breathing, where individuals in a group breathe together or in rapid succession. Examples include: *Lepisosteus osseus*, *L. oculatus* (Hill, 1972), *Hoplosternum thoracatum*, *Piabucina festae*, *Trichogaster leeri*, *Ancistrus chagresi* (Kramer and Graham, 1976), and *Clarias liocephalus* (Chapman and Chapman, 1994). It is argued that the selective factor underlying synchrony is predator pressure, with clumped breathing reducing potential for encounter with predators in a manner analogous to schooling (Kramer and Graham, 1976; Gee, 1980; Chapman and Chapman, 1994). Fish have also been shown to reduce the risk of surfacing by selecting less risky habitats (Wolf and Kramer, 1987; Shingles *et al.*, 2005) or at less risky times of the diel cycle (Saint-Paul and Soares, 1987).

5.2.2. Fast-Start

Recent studies have demonstrated negative effects of hypoxia on the escape response in fishes (Domenici *et al.*, 2007). In response to predator attack, fast-starts are a critical evasion strategy in fishes that consist of a sudden acceleration in a direction away from the stimulus. In teleost, this is often characterized by bending the body into a C-shape. The fast-start escape response in fishes is driven anaerobically (Webb, 1993, 1998; Wakeling and Johnston, 1998); however, recent work indicates that hypoxia can still have detrimental effects on the response. In the golden grey mullet (*Liza aurata*), Lefrançois and colleagues (2005) reported that hypoxia affected escape performance by impairing both responsiveness and directionality, suggesting reduced sensitivity of fishes to mechanico-acoustic stimuli. In the golden grey mullet, additional negative effects on locomotor performance were observed when surface access was denied. The golden grey mullet uses ASR in response to hypoxic stress, and by doing so can reduce the negative effects of hypoxia on fast-start performance, but this can increase exposure to avian predation

(Lefrançois *et al.*, 2005). In contrast, the European sea bass, *Dicentrarchus labrax*, does not respond to hypoxia by using ASR. Lefrançois and Domenici (2005) concluded that locomotor variables associated with the fast-start response in the sea bass were not affected by hypoxia exposure; however, similar to the mullet, the main effect was decreased responsiveness and directionality.

At this stage there are an insufficient number of studies to draw generalities with respect to the impact of hypoxia on escape behavior. But, there are clearly species-specific features, and it will be important to broaden the geographical, ecological, and phylogenetic scope of the species studied. In addition, the ecological implications of impaired fast-start performance on vulnerability need to be both studied and interpreted in light of effects on water-breathing predators.

5.2.3. Schooling

Schooling in fishes has been related primarily to predator risk and prey detection (Godin, 1986; Magurran, 1990; Pitcher and Parrish, 1993), and hypoxia can potentially affect the benefits of schooling by influencing spatial structure and velocity of the group. Schooling may actually induce hypoxia along the axis of motion (toward the rear) because of the oxygen consumption of the fish in the front of the school (McFarland and Moss, 1967; Domenici *et al.*, 2007). The response to hypoxia observed in the Atlantic herring, *Clupea harengus* (Domenici *et al.*, 2002), is an increase in the school volume and width (Moss and McFarland, 1970; Domenici *et al.*, 2000, 2002), which may counteract hypoxic stress by increasing the spacing among individuals. However, this could negatively affect the sensory modalities used in school coordination and/or decrease energetic advantages (Domenici *et al.*, 2007). Moss and McFarland (1970) reported an increase in speed under hypoxia in the anchovy, *Engraulis mordax*, but not until near lethal levels were reached. In the Atlantic herring, Herbert and Steffensen (2006) also reported increased swimming speed in response to stepwise progressive hypoxia, which they argued was an adaptive avoidance response.

5.3. When Prey Becomes Predator: Hypoxia and Fish–Invertebrate Interactions

Hypoxia has also been implicated as a factor underlying shifts in trophic interactions between fish and their invertebrate prey. Again, the beneficiary of the interaction often depends on the relative tolerance to hypoxic stress. Invertebrate species that are not tolerant of hypoxia have been observed to avoid seasonal hypolimnetic oxygen depletion by migration up into the water column or to the ice–water interface in the case of winterkill lakes (Nagell, 1977; Magnuson *et al.*, 1985). However, in lakes with fish this can result in very high invertebrate prey mortality (Rahel and Kolar, 1990). Invertebrate species capable of tolerating low DO may use hypoxic benthic areas in stratified lakes as refugia to avoid fish predation (Rahel and Kolar, 1990; Stirling *et al.*, 1990; Kolar and Rahel, 1993). In a series of behavioral experiments, Kolar and Rahel (1993) found that highly mobile taxa with low tolerance to hypoxia (e.g., mayflies and amphipods) moved upward from their benthic refuge in response to oxygen depletion at the substrate–water interface and were preyed upon by fish. Other taxa were less mobile and therefore less vulnerable to the fish. Species vulnerable to both hypoxia and predation altered their behavior to remain longer in the benthic zone in the presence of fish, thus demonstrating a tradeoff between costs of hypoxia and predation risk.

Tolerance to hypoxia by fish predators can limit the effectiveness of benthic refugia. When fish predators are capable of tolerating lethal levels of low oxygen for short periods, they can potentially forage in hypoxic waters. Rahel and Nutzman (1994) examined the foraging behavior of the central mudminnow (*Umbra limi*) in a stratified lake in Wisconsin. Although DO in the benthic zone of the lake was lethal to the mudminnows, they routinely ventured into the environment for short-term foraging bouts. They proposed the conditions promoting foraging in lethal environments represented a tradeoff between food availability in non-lethal waters and the cost of short-term exposure to abiotic stress. McParland and Paszkowski (2006) provided experimental evidence to suggest that small fish that colonize eutrophic, hypoxia-prone prairie potholes in Alberta can reduce aquatic invertebrate densities. By adding brook stickleback (*Culaea inconstans*) and fathead minnow (*Pimephales promelas*) to fishless potholes, they were able to show that these small-bodied hypoxia-tolerant fish reduced invertebrate prey, which altered the foraging behavior of blue-winged teal and other waterbirds.

5.4. Hypoxia and Social Interactions

Given the potential effects of hypoxia exposure on metabolic rate, activity, motivation, and habitat use, it is not surprising to find evidence for changes in social behavior in response to aquatic oxygen availability. Effects of hypoxia on the dynamics and structure of fish schools has been discussed in the context of increased school volume as a mechanism to elevate DO within the school. It has been argued that spacing within schools allows fish to keep track of one another without colliding. Domenici and colleagues (2002) suggested that the fast sound pulses emitted by some fishes, which may assist in synchronous response to predators (Gray and Denton, 1991) may be less effective when school volume is increased under hypoxic stress. Hypoxia may also affect sensory channels involved in fish maneuverability, and thus impair fast antipredator manoeuvres (Domenici *et al.*, 2007).

The use of ASR in response to extreme hypoxia has been demonstrated to affect social behaviors within groups of conspecifics. For example, in their study of ASR in swamp-dwelling and open-water populations of the haplochromine cichlid *Astatotilapia* "wrought-iron," Melnychuk and Chapman (2002) found that the pre-ASR aggression rate was higher in swamp-dwelling "wrought-iron" than in the open-water populations, but the aggression rate dropped in both open-water and swamp-dwelling fish between the pre-ASR and post-ASR periods. The use of ASR may impose both time and energetic constraints that reduce aggression. This could affect the development and maintenance of dominance hierarchies in cichlids and other species with complex social systems. For example, in three-spined stickleback from lotic and lentic sites, Sneddon and Yerbury (2004) found that dominance hierarchies were less stable when fish from the river site were exposed to hypoxia (20% saturation). In addition, fish under hypoxic conditions from both sites showed a decreased frequency of aggressive acts with the exception of the most dominant fish, and the most dominant fish lost mass under hypoxia conditions. Sneddon and Yerbury (2004) hypothesized that the maintenance of aggression under hypoxia in the dominant fish had a significant energetic cost. Hypoxia may also affect aggressive contests among fishes by altering display signals. Abrahams *et al.* (2005) explored the effect of hypoxia on the opercular displays of Siamese fighting fish, *Betta splendens*. The fish reduced their opercular displays under hypoxic conditions, which was interpreted as an honest signal that indicated the physiological condition of the contestant. Although many behavioral studies of physiological stressors employ an acclimation protocol, Marks *et al.* (2005) examined the influence of development under hypoxia in altering aggression in the zebrafish (*Danio rerio*). They found evidence for an effect of the developmental environment and the adult behavioral environment (acclimation). Aggression levels of fish reared in hypoxia were lower than in those reared in normoxia. However, when zebrafish were acclimated to either hypoxia or normoxia, their aggression level was highest in the environment in which they had been reared, providing evidence that hypoxic stress during development affects behavioral responses of this species and emphasizing the importance of understanding both developmental and acclimation responses.

6. SUMMARY

From this review of behavioral responses to hypoxia, it is clear that low DO can become ecologically active at levels far above those that are lethal, and can induce behaviors that alter fish distributions and their inter- and intra-specific interactions. Variation in oxygen availability can influence spatial and temporal patterns of distribution at various scales, from the micro- to the macrohabitat, and alter crucial components of the interactions between a fish and its environment including predator–prey interactions, schooling behavior, dominance, aggression, and parental care.

Aquatic surface respiration and air breathing are two of the most pronounced behavioral responses by bony fishes to aquatic hypoxia. Both behaviors are reflexes driven by oxygen-sensitive chemoreceptors, but can be modulated by inputs from higher centers, in particular as a function of perceived risk of predation and other costs. Many fishes also exhibit changes in spontaneous swimming activity when exposed to hypoxia that can comprise either a reduction in activity or an increase in activity, depending upon the species and the context. A reduction in spontaneous swimming activity has been interpreted as an energy-saving response, while an increase has been interpreted as an avoidance response. As might be expected, many species can exhibit both types of response, depending on the degree of hypoxia, the rate of change in DO, and their degree of activity in normoxia. Another locomotor response that can reduce exposure to hypoxia is spontaneous emersion, whereby amphibious and semi-amphibious species leave water to go onto land; however, the role of hypoxia as a driver has only been studied in a few species, and this remains an area ripe for investigation.

In addition to the large energetic costs associated with locomotor activity, another significant metabolic cost for some fish species is the energy directed to parental care. Costs of parental care in fishes may be particularly severe under hypoxia, due to the challenge of providing oxygen to the eggs. High levels of parental care may, however, be essential under hypoxic conditions, to ensure survival of the eggs and young. Although the literature in this area is still depauperate, DO seems to be an important driver of parental care behavior across a range of strategies from guarding to viviparity.

Given the potential effects of hypoxia exposure on metabolic rate, activity, and motivation, it is not surprising to find evidence for changes in social behavior in response to oxygen availability, including shifts in school volume and synchronous response with schools, as well as levels of aggression that could affect the development and maintenance of dominance hierarchies in cichlids and other species with complex social systems.

An integration of knowledge on behavioral responses to hypoxia and the relative tolerance of species supports the role of hypoxia as a modulator of species interactions, in particular predator–prey relationships where it can alter the success rate of the predator and/or the vulnerability of the prey. Whether hypoxia favors the predator or prey depends, at least in part, on the relative tolerance of the interactants. Hypoxia can elicit behaviors that increase predation risk such as ASR or an increase in the frequency of air breathing; it can negatively impact fast-start performance of prey or alter the dynamics of schooling behavior. For aquatic water-breathing predators, hypoxia can decrease predation through metabolic depression, lowered appetite, or decreased performance. Hypoxia has also been implicated as a factor underlying shifts in trophic interactions between fish and their invertebrate prey. Again, the beneficiary of the interaction often depends on the relative tolerance of the species to hypoxic stress. Tolerance to hypoxia by fish predators can limit the effectiveness of benthic refugia for macroinvertebrates. Clearly, the outcome of altered predator–prey interactions can ultimately influence other components of the food web and assemblage, and thus predicting whether the prey or the predator is the beneficiary of hypoxic stress is critical for understanding community level impacts of hypoxia, whether natural or anthropogenically induced.

ACKNOWLEDGMENTS

The authors wish to thank two anonymous reviewers for comments on an earlier version of this manuscript. Financial support was provided by the Natural Sciences and Engineering Research Council of Canada (LJC) and Canada Research Chair program (LJC).

REFERENCES

Abel, D. C., Koenig, C. C., and Davis, W. P. (1987). Emersion in the mangrove forest fish *Rivulus marmoratus*: A unique response to hydrogen sulfide. *Environ. Biol. Fish.* **18,** 67–72.

Abrahams, M. V., Robb, T. L., and Hare, J. F. (2005). Effect of hypoxia on opercular displays: Evidence for an honest signal? *Anim. Behav.* **70,** 427–432.

Abrahams, M. V., Mangel, M., and Hedges, K. (2007). Predator–prey interactions and changing environments: Who benefits? *Phil. Trans. R. Soc. B* **362,** 2095–2104.

Aguilar, N. M., Ishimatsu, A., Ogawa, K., and Huat, K. K. (2000). Aerial ventilator responses of the mudskipper *Periophthalmodon schlosseri*, to altered aerial and aquatic respiratory gas concentrations. *Comp. Biochem. Physiol.* **127A,** 285–292.

Babiker, M. M. (1979). Respiratory behaviour, oxygen consumption and relative dependence on aerial respiration in the African lungfish (*Protopterus annectens* Owen) and an air-breathing teleost (*Clarias lazera* C.). *Hydrobiologia* **65,** 177–187.

Babiker, M. M. (1984). Development of dependence on aerial respiration in *Polypterus senegalus* (Cuvier). *Hydrobiologia* **110,** 339–349.

Balirwa, J. (1998). Lake Victoria wetlands and the ecology of the Nile tilapia, *Oreochromis niloticus* Linne Ph.D. Dissertation, Wageningen Agricultural University, Wageningen.

Balirwa, J. S., Chapman, C. A., Chapman, L. J., Cowx, I. G., Geheb, K., Kaufman, L., Lowe-McConnell, R. H., Seehausen, O., Wanink, J. H., Welcomme, R. L., and Witte, F. (2003). Biodiversity and fishery sustainability in the Lake Victoria Basin: An unexpected marriage? *Bioscience* **53,** 703–715.

Behrens, J. W., and Steffensen, J. F. (2007). The effect of hypoxia on behavioural and physiological aspects of lesser sandeel, *Ammodytes tobianus* (Linnaeus, 1785). *Mar. Biol.* **150,** 1365–1377.

Behrens, J. W., Stahl, H. J., Steffensen, J. F., and Glud, R. N. (2007). Oxygen dynamics around buried lesser sandeels, *Ammodytes tobianus* (Linnaeus 1758): Mode of ventilation and oxygen requirements. *J. Exp. Biol.* **210,** 1006–1014.

Bejda, A. J., Studholme, A. L., and Olla, B. L. (1987). Behavioral responses of red hake, *Urophycis chuss*, to decreasing concentrations of dissolved oxygen. *Environ. Biol. Fish.* **19,** 261–268.

Berns, S., and Peters, H. M. (1969). On the reproductive behaviour of *Ctenopoma muriei* and *Ctenopoma damasi* (Anabantidae). *Annual Report of the East African Freshwater Fisheries Research Organization* **(1968),** 44–49.

Bevan, D. J., and Kramer, D. L. (1986). The effect of swimming depth on respiratory behavior of the honey gourami. *Colisa chuna* (Pisces, Belontiidae). *Can. J. Zool.* **64,** 1893–1896.

Blackburn, D. G., Evans, H. E., and Vitt, L. J. (1985). The evolution of fetal nutritional adaptations. *In* "Vertebrate Morphology" (Duncker, R., and Fleischer, G. M., Eds.), pp. 437–439. Gustav Fischer Verlag, New York.

Boehlert, G. W., and Yoklavich, M. M. (1984). Reproduction, embryonic energetics, and the maternal–fetal relationship in the viviparous genus *Sebastes* (Pisces: Scorpaenidae). *Biol. Bull.* **167,** 354–370.

Boehlert, G. W., Kusakari, M., Shimizu, M., and Yamada, J. (1986). Energetics during embryonic development in kurosoi, *Sebastes schlegeli*. *J. Exp. Mar. Biol. Ecol.* **101,** 239–256.

Bouillon, J. (1961). The lungfish of Africa. *Nat. Hist.* **70,** 62–71.

Brauner, C. J., and Val, A. L. (2006). Oxygen transfer. *In* "Tropical Fishes" (Val, L. V., Almeida-Val, V. M. F., and Randall, D. J., Eds.), Fish Physiology, Vol. 21, pp. 277–306. Academic Press/Elsevier, San Diego, CA.

Breitburg, D. L., Steinberg, N., DuBeau, S., Cooksey, C., and Houde E. D. (1994). Effects of low dissolved oxygen on predation on estuarine fish larvae. *Mar. Ecol. Prog. Ser.* **104,** 235–246.

Breitburg, D. L., Loher, T., Pacey, C. A., and Gerstein, A. (1997). Varying effects of low dissolved oxygen on trophic interactions in an estuarine food web. *Ecol. Monogr.* **67,** 489–507.

Brown, C. E., and Muir, B. S. (1970). Analysis of ram ventilation of fish gills with application to skipjack tuna (*Katsuwonus pelamis*). *J. Fish. Res. Bd. Can.* **27,** 1637–1652.

Burggren, W. W. (1982). "Air gulping" improves blood oxygen transport during aquatic hypoxia in the goldfish, *Carassius auratus*. *Physiol. Zool.* **55,** 327–334.

Burleson, M. L., Smatresk, N. J., and Milsom, W. K. (1992). Afferent inputs associated with cardioventilatory control in fish. *In* "Fish Physiology" (Hoar, W. S., and Randall, D. J., Eds.), Vol. VIIB, pp. 390–426. Academic Press, New York.

Bushnell, P. G., and Brill, R. W. (1991). Responses of swimming skipjack *Katsuwonus pelamis* and yellowfin *Thunnus albacares* tunas to acute hypoxia, and a model of their cardiovascular function. *Physiol. Zool.* **64,** 787–811.

Carlson, J. K., and Parsons, G. R. (2001). The effects of hypoxia on three sympatric shark species: Physiological and behavioral responses. *Environ. Biol. Fish.* **61,** 427–433.

Cech, J. J., Massingill, M. J., Vondracek, B., and Linden, A. L. (1985). Respiratory metabolism of mosquitofish, *Gambusia affinis*: Effects of temperature, dissolved oxygen, and sex difference. *Environ. Biol. Fish.* **13,** 297–307.

Chapman, L. J., and Chapman, C. A. (1994). Observations on synchronous air breathing in *Clarias liocephalus*. *Copeia.* **1994,** 246–249.

Chapman, L. J., and Chapman, C. A. (1998). Hypoxia tolerance of the mormyrid *Petrocephalus catostoma*: Implications for persistence in swamp refugia. *Copeia.* **1998,** 762–768.

Chapman, L. J., and Liem, K. F. (1995). Papyrus swamps and the respiratory ecology of *Barbus neumayeri*. *Environ. Biol. Fish.* **44,** 183–197.

Chapman, L. J., Kaufman, L. S., and Chapman, C. A. (1994). Why swim upside down?: A comparative study of two mochokid catfishes. *Copeia.* **1994,** 130–135.

Chapman, L. J., Kaufman, L. S., Chapman, C. A., and McKenzie, F. E. (1995). Hypoxia tolerance in twelve species of East African cichlids: Potential for low oxygen refugia in Lake Victoria. *Cons. Biol.* **9,** 1274–1288.

Chapman, L. J., Chapman, C. A., and Chandler, M. (1996a). Wetland ecotones as refugia for endangered fishes. *Biol. Cons.* **78,** 263–270.

Chapman, L. J., Chapman, C. A., Ogutu-Ohwayo, R., Chandler, M., Kaufman, L., and Keiter, A. E. (1996b). Refugia for endangered fishes from an introduced predator in Lake Nabugabo, Uganda. *Cons. Biol.* **10,** 554–561.

Chapman, L. J., Chapman, C. A., Brazeau, D., McGlaughlin, B., and Jordan, M. (1999). Papyrus swamps and faunal diversification: Geographical variation among populations of the African cyprinid *Barbus neumayeri. J. Fish Biol.* **54,** 310–327.

Chapman, L. J., Chapman, C. A., Nordlie, F. G., and Rosenberger, A. E. (2002). Physiological refugia: Swamps, hypoxia tolerance, and maintenance of fish biodiversity in the Lake Victoria Region. *Comp. Biochem. Physiol.* **133,** 421–437.

Claireaux, G., and Lagardère, J.-P. (1999). Influence of temperature, oxygen and salinity on the metabolism of European sea bass. *J. Sea Res* **42,** 157–168.

Claireaux, G., and Lefrançois, C. (2007). Linking environmental variability and fish performance: Integration through the concept of scope for activity. *Phil. Trans. R. Soc. B* **362,** 2031–2042.

Claireaux, G., Webber, D. M., Lagardère, J.-P., and Kerr, S. R. (2000). Influence of water temperature and oxygenation on the aerobic metabolic scope of Atlantic cod (*Gadus morhua*). *J. Sea Res.* **44,** 257–265.

Coleman, R. M. (1992). Reproductive biology and female parental care in the cockscomb prickleback, *Anoplarchus purpurescens* (Pisces, Stichaeidae). *Environ. Biol. Fish.* **35,** 177–186.

Couturier, C., McKenzie, D. J., Galois, R., Joassard, L., and Claireaux, G. (2007). Influence of water viscosity on bioenergetics of the common sole *Solea solea*: Ventilation and metabolism. *Mar. Biol.* **152,** 803–814.

Dalla Via, J., van den Thillart, G., Cattani, O., and Cortesi, P. (1998). Behavioral responses and biochemical correlates in *Solea solea* to gradual hypoxic exposure. *Can. J. Zool.* **76,** 2108–2113.

Davenport, J., and Woolmington, A. D. (1981). Behavioural responses of some rocky shore fish exposed to adverse environmental conditions. *Mar. Behav. Physiol.* **8,** 1–12.

Diaz, R. J. (2001). Overview of hypoxia around the world. *Environ. Qual* **30,** 275–281.

Dizon, A. E. (1977). Effect of dissolved oxygen concentration and salinity on the swimming speed of two species of tuna. *Fish. Bull.* **75,** 649–653.

Domenici, P., Steffensen, J. F., and Batty, R. S. (2000). The effect of progressive hypoxia on swimming activity and schooling in Atlantic herring. *J. Fish. Biol.* **57,** 1526–1538.

Domenici, P., Ferrari, R. S., Steffensen, J. F., and Batty, R. S. (2002). The effects of progressive hypoxia on school structure and dynamics in Atlantic herring *Clupea harengus. Proc. R. Soc. B.* **269,** 2103–2111.

Domenici, P., Lefrançois, C., and Shingles, A. (2007). Hypoxia and the anti-predator behaviour of fishes. *Phil. Trans. R. Soc. B* **362,** 2105–2121.

Dybas, C. L. (2005). Dead zones spreading in world oceans. *BioScience* **55,** 552–557.

Dygert, P. H., and Gunderson, D. R. (1991). Energy utilization by embryos during gestation in viviparous copper rockfish, *Sebastes caurinus. Environ. Biol. Fish.* **30,** 165–171.

Farmer, C. G., and Jackson, D. C. (1998). Air-breathing during activity in the fishes *Amia calva* and *Lepisosteus oculatus. J. Exp. Biol.* **201,** 943–948.

Fischer, P., Rademacher, K., and Kils, K. (1992). *In situ* investigations on the respiration and behaviour of the eelpout *Zoarces viviparus* under short-term hypoxia. *Mar. Ecol. Prog. Ser.* **88,** 181–184.

Florindo, L. H., Leite, C. A. C., Kalinin, A. L., Reid, S. G., Milsom, W. K., and Rantin, F. T. (2006). The role of branchial and orobranchial O_2 chemoreceptors in the control of aquatic surface respiration in the neotropical fish tambaqui (*Colossoma macropomum*): Progressive responses to prolonged hypoxia. *J. Exp. Biol.* **209,** 1709–1715.

Freadman, M. A. (1979). Swimming energetics of striped bass (*Morone saxatilis*) and bluefish (*Pomatomus saltatrix*): Gill ventilation and swimming metabolism. *J. Exp. Biol.* **83,** 217–230.
Fry, F. E. J. (1971). The effect of environmental factors on the physiology of fish. *In* "Fish Physiology" (Hoar, W. S., and Randall, D. J., Eds.), Vol. VI, pp. 1–98. Academic Press, New York.
Fryer, G., and Iles, T.D (1972). "The Cichlid Fishes of the Great Lakes of Africa: Their Biology and Evolution." Oliver and Boyd, London.
Gee, J. H. (1980). Respiratory patterns and antipredator responses in the central mudminnow, *Umbra limi*, a continuous, facultative, air-breathing fish. *Can. J. Zool.* **58,** 819–827.
Gee, J. H., and Gee, P. A. (1991). Reactions of gobioid fishes to hypoxia: Buoyancy control and aquatic surface respiration. *Copeia.* **1991,** 17–28.
Gee, J. H., and Gee, P. A. (1995). Aquatic surface respiration, buoyancy control and the evolution of air-breathing in gobies (*Gobiidae*; Pisces). *J. Exp. Biol.* **198,** 79–89.
Gee, J. H., Tallman, R. F., and Smart, H. J. (1978). Reactions of some Great Plains fishes to progressive hypoxia. *Can. J. Zool.* **56,** 1962–1966.
Godin, J. G. J. (1986). Anti-predator function of shoaling in teleost fishes: A selective review. *Nat. Can.* **113,** 241–250.
Gonzales, T. T., Katoh, M., and Ishimatsu, A. (2006). Air breathing of aquatic burrow-dwelling eel goby, *Odontamblyopus lacepedii* (Gobiidae: Amblyopinae). *J. Exp. Biol.* **209,** 1085–1092.
Goodwin, N. B., Dulvy, N. K., and Reynolds, J. D. (2002). Life-history correlates of the evolution of live bearing in fishes. *Phil. Trans. Roy. Soc. Lond. B.* **357,** 259–267.
Graham, J. B. (1970). Preliminary studies on the biology of the amphibious clinid *Mnierpes macrocephalus*. *Mar. Biol.* **5,** 136–140.
Graham, J. B., Kramer, D. L., and Pineda, E. (1977). Respiration of the air breathing fish *Piabucina festae*. *J. Comp. Physiol.* **122,** 295–310.
Graham, J. B. (1983). The transition to air breathing in fishes: II. Effects of hypoxia acclimation on the bimodal gas exchange of *Ancistrus chagresi* (Loricaridae). *J. Exp. Biol.* **102,** 157–173.
Graham, J. B. (1997). "Air Breathing Fishes: Evolution, Diversity, and Adaptation" Academic Press, San Diego.
Graham, J. B., and Lee, H. J. (2004). Breathing air in air: In what ways might extant amphibious fish biology relate to prevailing concepts about early tetrapods, the evolution of vertebrate air breathing, and the vertebrate land transition? *Physiol. Biochem. Zool.* **77,** 720–731.
Gray, J. A. B., and Denton, E. J. (1991). Fast pressure pulses and communication between fish. *J. Mar. Biol. Assoc.* **71,** UK, 83–106.
Green, B. S., and McCormick, M. I. (2004). O_2 replenishment to fish nests: Males adjust brood care to ambient conditions and brood development. *Beh. Ecol.* **16,** 389–397.
Greenwood, P. H. (1987). The natural history of African lungfishes. *J. Morph. Suppl.* **1,** 163–179 (1986).
Grigg, G. C. (1965). Studies on the Queensland lungfish, *Neoceratodus forsteri* (Kreft). III. Aerial respiration in relation to habits. *Aust. J. Zool.* **13,** 413–421.
Hale, R. E., St Mary, C. M., and Lindstrom, K. (2003). Parental responses to changes in costs and benefits along an environmental gradient. *Environ. Biol. Fish.* **67,** 107–116.
Herbert, N. A., and Steffensen, J. F. (2005). The response of Atlantic cod, *Gadus morhua*, to progressive hypoxia: Fish swimming speed and physiological stress. *Mar. Biol.* **147,** 1403–1412.
Herbert, N. A., and Steffensen, J. F. (2006). Hypoxia increases the behavioural activity of schooling herring: A response to physiological stress or respiratory distress? *Mar. Biol.* **149,** 1217–1225.

Herbert, N. A., and Wells, R. M. G. (2001). The aerobic physiology of the air-breathing blue gourami, *Trichogaster trichopterus*, necessitates behavioural regulation of breath-hold limits during hypoxic stress and predatory challenge. *J. Comp. Physiol. B* **171,** 603–612.

Hill, L. G. (1972). Social aspects of aerial respiration of young gars (*Lepisosteus*). *Southwest. Nat.* **16,** 239–247.

Horn, M. H., and Riggs, C. D. (1973). Effects of temperature and light on the rate of air-breathing of the bowfin, *Amia calva*. *Copeia* **1973,** 653–657.

Hughes, G. M. (1984). Scaling of respiratory areas in relation to oxygen consumption of vertebrates. *Experientia* **41,** 519–652.

Hugie, D. M., Thuringer, P. L., and Smith, R. J. F. (1991). The response of the tidepool sculpin, *Oligocottus maculosus*, to chemical stimuli from injured conspecifics, alarm signalling in the Cottidae (Pisces). *Ethology* **89,** 322–334.

Ilves, K. L., and Randall, D. J. (2007). Why have primitive fishes survived? *In* "Primitive Fishes" (McKenzie, D. J., Brauner, C.J, and Farrell, A. P., Eds.), Fish Physiology, Vol. 26, pp. 516–536. Academic Press/Elsevier, San Diego, CA.

Innes, A. J., and Wells, R. F. G. (1985). Respiration and oxygen transport functions of the blood from an intertidal fish, *Helicogramma medium* (Tripterygiidae). *Environ. Biol. Fish.* **14,** 213–226.

Ishimatsu, A, Yoshida, Y., Itoki, N., Takeda, T., Lee, H. J., and Graham, J.B (2007). Mudskippers brood their eggs in air but submerge them for hatching. *J. Exp. Biol.* **210,** 3946–3954.

Itazawa, Y., and Ishimatsu, A. (1981). Gas exchange in an air breathing fish, the snakehead *Channa argus* in normoxic and hypoxia water and in air. *Bull. Japan Soc. Fish. Sci.* **47,** 829–834.

Johansen, K. (1970). Air-breathing in fishes. *In* "Fish Physiology" (Hoar, W. S., and Randall, D. J., Eds.), Vol. IV, pp. 361–411. Academic Press, New York.

Johnels, A. G., and Svensson, G. S. O. (1954). On the biology of *Protopterus annectens* (Owen). *Ark. Zool.* **7,** 131–164.

Johansen, J. L., Herbert, N. A., and Steffensen, J. F. (2006). The behavioural and physiological response of Atlantic cod *Gadus morhua* L. to short-term acute hypoxia. *J. Fish Biol.* **68,** 1918–1924.

Johansen, K., Hanson, D., and Lenfant, C. (1970). Respiration in a primitive air breather *Amia calva*. *Respir. Physiol.* **9,** 162–174.

Johnsson, J. I., Höjesjo, J., and Flemming, I. A. (2001). Behavioural and heart rate responses to predation risk in wild and domesticated Atlantic salmon. *Can. J. Fish. Aquat. Sci.* **58,** 788–794.

Jones, J. C., and Reynolds, J. D. (1999a). Costs of egg ventilation for male common gobies breeding in conditions of low dissolved oxygen. *Anim. Behav.* **57,** 181–188.

Jones, J. C., and Reynolds, J. D. (1999b). Oxygen and the trade-off between egg ventilation and brood protection in the common goby. *Behaviour* **136,** 819–832.

Kaufman, L. S. (1992). Catastrophic change in species-rich freshwater ecosystems: The lessons of Lake Victoria. *BioScience* **42,** 846–858.

Kaufman, L. S., Chapman, L. J., and Chapman, C. A. (1997). Evolution in fast forward: Haplochromine fishes of the Lake Victoria Region. *Endeavour* **21,** 23–30.

Keenleyside, M. H. A. (1991). Parental care. *In* "Cichlid Fishes: Behaviour, Ecology and Evolution" (Keenleyside, M. H. A., Ed.). Chapman and Hall, New York.

Kolar, C. S., and Rahel, F. J. (1993). Interaction of a biotic factor (predator presence) and an abiotic factor (low oxygen) as an influence on benthic invertebrate communities. *Oecologia* **95,** 210–219.

Kramer, D. L. (1983). The evolutionary ecology of respiratory mode in fishes: An analysis based on the costs of breathing. *Environ. Biol. Fish.* **9,** 145–158.

Kramer, D. L. (1987). Dissolved oxygen and fish behavior. *Environ. Biol. Fish.* **18,** 81–92.

Kramer, D. L., and Graham, J. B. (1976). Synchronous air breathing, a social component of respiration in fishes. *Copeia* **1976,** 689–697.

Kramer, D. L., and McClure, M. (1982). Aquatic surface respiration, a widespread adaptation to hypoxia in tropical freshwater fishes. *Environ. Biol. Fish.* **7,** 47–55.

Kramer, D. L., and Mehegan, J. P. (1981). Aquatic surface respiration, an adaptive response to hypoxia in the guppy, *Poecilia reticulata* (Pisces, Poeciliidae). *Environ. Biol. Fish.* **6,** 299–313.

Kramer, D. L., Manley, D., and Bourgeois, R. (1983). The effects of respiratory mode and oxygen concentration on the risk of aerial predation in fishes. *Can. J. Zool.* **61,** 653–665.

Lam, K., Tsui, T., Nakano, K., and Randall, D. J. (2006). Physiological adaptations of fishes to tropical intertidal environments. *In* "Tropical Fishes" (Val, L. V., Almeida-Val, V. M. F., and Randall, D. J., Eds.), Fish Physiology, Vol. 21, pp. 502–582. Academic Press/Elsevier, San Diego, CA.

Lee, H. J., Martinez, C. A., Hertzberg, K. J., Hamilton, A. L., and Graham, J. B. (2005). Burrow air phase maintenance and respiration by the mudskipper *Scartelaos histophorus* (Gobiidae: Oxudercinae). *J. Exp. Biol.* **208,** 169–177.

Lefrançois, C., and Claireaux, G. (2003). Influence of ambient oxygenation and temperature on metabolic scope and scope for heart rate in the common sole, *Solea solea. Mar. Ecol. Prog. Series* **259,** 273–284.

Lefrançois, C., and Domenici, P. (2005). Locomotor kinematics and behaviour in the escape response of European sea bass, *Dicentrarchus labrax* L., exposed to hypoxia. *Marine Biol.* **149,** 969–977.

Lefrançois, C., Shingles, A., and Domenici, P. (2005). The effect of hypoxia on locomotor performance and behaviour during escape in *Liza aurata. J. Fish. Biol.* **67,** 1711–1729.

Lewis, W. M. (1970). Morphological adaptations of cyprinodontoids for inhabiting oxygen deficient waters. *Copeia.* **1970,** 319–325.

Liem, K. F., Eclancher, B., and Fink, W. L. (1984). Aerial respiration in the banded knifefish *Gymnotus carapo* (Teleostei: Gymnotoidei). *Physiol. Zool.* **57,** 185–195.

Liem, K. F. (1987). Functional design of the air ventilation apparatus and overland excursions by teleosts. *Fieldiana: Zool.* **37,** 1–29.

Lindström, K., St. Mary, C. M., and Pampoulie, C. (2006). Sexual selection for male parental care in the sand goby, *Pomatoschistus minutus. Behav. Ecol. Sociobiol* **60,** 46–51.

Lissaker, M., Kvarnemo, C., and Svensson, O. (2003). Effects of low oxygen environment on parental effort and filian cannibalism in the male sand goby, *Pomatoschistus minutus. Behav. Ecol.* **14,** 374–381.

Love, J. W., and Rees, B. B. (2002). Seasonal differences in hypoxia tolerance in gulf killifish, *Fundulus grandis* (Fundulidae). *Environ. Biol. Fish.* **63,** 103–115.

Mabuchi, K., Miya, M., Azuma, Y., and Nishida, M. (2007). Independent evolution of the specialized pharyngeal jaw apparatus in cichlid and labrid fishes. *BMC Evol. Biol.* **7**, doi: 10.1186/1471-2148-7-10.

Magnuson, J. J., Beckel, A. L., Mills, K., and Brandt, S. B. (1985). Surviving winter hypoxia: Behavioral adaptations of fishes in a northern Wisconsin lake. *Environ. Biol. Fish.* **14,** 242–250.

Magurran, A. E. (1990). The adaptive significance of schooling as an anti-predator defence in fish. *Am. Zool. Fenn.* **27,** 51–66.

Marks, C., West, T. N., Bagatto, G., and Moore, F. B.-G (2005). Developmental environment alters conditional aggression in Zebrafish. *Copeia* **2005**, 901–908.

Martin, K. L. M. (1995). Time and tide wait for no fishes: Intertidal fishes out of water. *Environ. Biol. Fish.* **44,** 165–181.

Martinez, M. S., Chapman, L. J., Grady, J. M., and Rees, B. B. (2004). Interdemic variation in hematocrit and lactate dehydrogenase in the African cyprinid *Barbus neumayeri*. *J. Fish Biol.* **65,** 1056–1069.

Maruyama, A., Onoda, Y., and Yuma, M. (2008). Variation in behavioural response to oxygen stress by egg-tending males of parapatric fluvial and lacustrine populations of a landlocked goby. *J. Fish. Biol.* **72,** 681–692.

McFarland, W. M., and Moss, S. A. (1967). Internal behavior in fish schools. *Science* **156,** 260–262.

McKendry, J. E., Milsom, W. K., and Perry, S. F. (2001). Branchial CO_2 receptors and cardiorespiratory adjustments during hypercarbia in Pacific spiny dogfish (*Squalus acanthias*). *J. Exp. Biol.* **204,** 1519–1527.

McKenzie, D. J., Burleson, M. L., and Randall, D. J. (1991). The effects of branchial denervation and pseudobranch ablation on cardioventilatory control in an air-breathing fish. *J. Exp. Biol.* **161,** 347–365.

McKenzie, D. J., Taylor, E. W., Bronzi, P., and Bolis, L. (1995a). Aspects of cardioventilatory control in the Adriatic sturgeon (*Acipenser naccarii*). *Respir. Physiol.* **100,** 44–52.

McKenzie, D. J., Piraccini, G., Steffensen, J. F., Bolis, C. L., Bronzi, P., and Taylor, E. W. (1995b). Effects of diet on spontaneous locomotor activity and oxygen consumption in Adriatic sturgeon (*Acipenser naccarii*). *Fish. Physiol. Biochem.* **14,** 341–355.

McKenzie, D. J., Cataldi, E., Owen, S., Taylor, E.W, and Bronzi, P. (2001). Effects of acclimation to brackish water on the growth, respiratory metabolism and exercise performance of Adriatic sturgeon (*Acipenser naccarii*). *Can. J. Fish. Aquat. Sci.* **58,** 1104–1112.

McKenzie, D. J., Farrell, A. P., and Brauner, C. J. (Eds.) (2007a). *In* "Primitive Fishes," Fish Physiology Vol. 26. Academic Press/Elsevier, San Diego, CA.

McKenzie, D. J., Campbell, H. A., Taylor, E. W., Andrade, M., Rantin, F. T., and Abe, A. S. (2007b). The autonomic control and functional significance of the changes in heart rate associated with air-breathing in the jeju, *Hoplerythrinus unitaeniatus*. *J. Exp. Biol.* **210,** 4224–4232.

McKenzie, D. J., Hale, M., and Domenici, P. (2007c). Locomotion in primitive fishes. *In* "Primitive Fishes" (McKenzie, D. J., Brauner, C. J., and Farrell, A. P., Eds.), Fish Physiology, Vol. 26, pp. 162–224. Academic Press/Elsevier, San Diego, CA.

McKenzie, D. J., Steffensen, J. F., Korsmeyer, K., Whiteley, N. M., Bronzi, P., and Taylor, E. W. (2007d). Swimming alters responses to hypoxia in the Adriatic sturgeon (*Acipenser naccarii*). *J. Fish Biol.* **70,** 651–658.

McKenzie, D. J., Lund, I., and Pedersen, P. B. (2008). Essential fatty acids influence metabolic rate and tolerance of hypoxia in Dover sole (*Solea solea*) larvae and juveniles. *Mar. Biol.* **154,** 1041–1052.

McKinsey, D. M., and Chapman, L. J. (1998). Dissolved oxygen and fish distribution in a Florida spring. *Environ. Biol. Fish.* **53,** 211–223.

McNeil, D. L., and Closs, G. P. (2007). Behavioural responses of a south-east Australian floodplain fish community to gradual hypoxia. *Freshw. Biol.* **52,** 412–420.

McParland, C. E., and Paszkowski, C. A. (2006). Effects of small-bodied fish on invertebrate prey and foraging patterns of waterbirds in Aspen Parkland wetlands. *Hydrobiologia* **567,** 43–55.

Melnychuk, M. C., and Chapman, L. J. (2002). Hypoxia tolerance of two haplochromine cichlids: Swamp leakage and potential for interlacustrine dispersal. *Environ. Biol. Fish.* **65,** 99–110.

Mertz, J. C., and Barlow, G. W. (1966). On the reproductive behavior of *Jordanella floridae* (Pisces: Cyprinodontidae) with special reference to a quantitative analysis of parental fanning. *Zeit. fur Tierpsychologie* **23,** 537–554.

Metcalfe, J. D., and Butler, P. J. (1984). Changes in activity and ventilation in response to hypoxia in unrestrained, unoperated dogfish, *Scyliorhinus canicula*. *J. Exp. Biol.* **108,** 411–418.

Mlewa, C. M., Green, J. M., and Simms, A. (2005). Movement and habitat use by the marbled lungfish *Protopterus aethiopicus* Heckel 1851 in Lake Baringo, Kenya. *Hydrobiologia* **537,** 229–238.

Mnaya, B., Wolanski, E., and Kiwango, Y. (2006). Papyrus wetlands: A lunar-modulated refuge for aquatic fauna. *Wet. Ecol. Manage.* **14,** 359–363.

Moss, S. A., and McFarland, W. N. (1970). The influence of dissolved oxygen and carbon dioxide on fish schooling behaviour. *Mar. Biol.* **5,** 100–107.

Muir, B. S. (1969). Gill dimensions as a function of size. *J. Fish. Res. Bd. Can.* **26,** 165–170.

Nagell, B. (1977). Phototactic and thermotactic response facilitating survival of *Cloeon dipterum* (Ephemeropotera) larvae under winter anoxia. *Oikos* **29,** 342–347.

Neuenfeldt, S. (2002). The influence of oxygen saturation on the distributional overlap of predator (cod, *Gadus morhua*) and prey (herring, *Clupea harengus*) in the Bornholm Basin of the Baltic Sea. *Fish. Oceangr.* **11,** 11–17.

Nilsson, G. E., and Östlund-Nilsson, S. (2004). Hypoxia in paradise: Widespread hypoxia tolerance in coral reef fishes. *Proc. R. Soc. Lond. B (Suppl.)* **271,** S30–S33.

Nilsson, G. E., and Östlund-Nilsson, S. (2008). Does size matter for hypoxia tolerance in fish? *Biol. Rev.* **83,** 173–189.

Nilsson, G. E., and Renshaw, G. M. C. (2004). Hypoxic survival strategies in two fishes: Extreme anoxia tolerance in the North European crucian carp and natural hypoxic preconditioning in a coral-reef shark. *J. Exp. Biol.* **207,** 3131–3139.

Nilsson, G. E., Rosen, P., and Johansson, D. (1993). Anoxic depression of spontaneous locomotor activity in crucian carp quantified by a computerized imaging technique. *J. Exp. Biol.* **180,** 153–162.

Nilsson, G. E., Hobbs, J.-P. A., and Östlund-Nilsson, S. (2007). Tribute to P. L. Lutz: Respiratory ecophysiology of coral-reef teleosts. *J. Exp. Biol.* **210,** 1673–1686.

Nonnotte, G., and Kirsch, R. (1978). Cutaneous respiration in seven sea-water teleosts. *Respir. Physiol.* **35,** 111–118.

Nordlie, F. G. (2006). Physicochemical environments and tolerances of cyprinodontoid fishes found in estuaries and salt marshes of eastern North America. *Rev. Fish. Biol. Fisheries* **16,** 51–106.

Olowo, J. P., and Chapman, L. J. (1996). Papyrus swamps and variation in the respiratory behaviour of the African fish *Barbus neumayeri*. *Afr. J. Ecol.* **34,** 211–222.

Oppenheimer, J. R., and Barlow, G. W. (1968). Dynamics of parental behavior in the black-chinned mouthbreeder, *Tilapia melanotheron* (Pisces: Cichlidae). *Z. Tierpsychol.* **25,** 889–914.

Östlund-Nilsson, S., and Nilsson, G. E. (2004). Breathing with a mouth full of eggs: Respiratory consequences of mouthbrooding in cardinalfish. *Proc. R. Soc. Lond. B* **271,** 1015–1022.

Parsons, G. R., and Carlson, J. K. (1998). Physiological and behavioural responses to hypoxia in the bonnethead shark, *Sphyrna tiburo*: Routine swimming and respiratory regulation. *Fish Physiol. Biochem.* **19,** 189–196.

Perry, S. F., Reid, S. G., Gilmour, K. M., Boijink, C. L., Lopes, J. M., Milsom, W. K., and Rantin, F. T. (2004). A comparison of adrenergic stress responses in three tropical teleosts exposed to acute hypoxia. *Am. J. Physiol.* **287,** R188–R197.

Petersen, J. K., and Petersen, G. I. (1990). Tolerance, behaviour, and oxygen consumption in the sand goby *Pomatoschistus minutes* (Pallas), exposed to hypoxia. *J. Fish. Biol.* **37,** 921–933.

Pitcher, T. J., and Parrish, J. K. (1993). Function of shoaling behaviour in teleosts. *In* "Behaviour of Teleost Fishes." (Pitcher, T. J., Ed.), pp. 363–439. Chapman and Hall, London, UK.

Pollock, M. S., Clarke, L. M. J., and Dubé, M. G. (2007). The effects of hypoxia on fishes: From ecological relevance to physiological effects. *Environ. Rev.* **15,** 1–14.

Rahel, F. J., and Kolar, C. S. (1990). Trade-offs in the response of mayflies to low oxygen and fish predation. *Oecologia* **84,** 39–44.

Rahel, F. J., and Nutzman, J. W. (1994). Foraging in a lethal environment: Fish predation in hypoxic waters of a stratified lake. *Ecology* **75,** 1246–1253.

Randall, D. J., Farrell, A. P., and Haswell, M. S. (1978). Carbon dioxide excretion in jeju, *Hoplerythrinus unitaeniatus*, a facultative air-breathing teleost. *Can. J. Zool.* **56,** 970–973.

Randall, D. J., Burggren, W. W., Farrell, A. P., and Haswell, M. S. (1981a). "The Evolution of Air-Breathing in Vertebrates." Cambridge University Press, Cambridge, UK.

Randall, D. J., Cameron, J. N., Daxboeck, C., and Smatresk, N. J. (1981b). Aspects of bimodal gas exchange in the bowfin, *Amia calva* (Actinopterygii: Amiiformes). *Respir. Physiol.* **43,** 339–348.

Randall, D. J., McKenzie, D. J., Abrami, G., Bondiolotti, G. P., Natiello, F., Bronzi, P., Bolis, L., and Agradi, E. (1992). Effects of diet on responses to hypoxia in the sturgeon (*Acipenser naccarii*). *J. Exp. Biol.* **170,** 113–125.

Randle, A. R., and Chapman, L. J. (2004). Habitat use by the air-breathing fish *Ctenopoma muriei:* Implications for costs of breathing. *Ecol. Freshwat. Fish.* **13,** 37–45.

Rantin, F. T., Guerra, C. D. R., Kalinin, A. L., and Glass, M. L. (1998). The influence of aquatic surface respiration (ASR) on cardio-respiratory function of the serrasalmid fish *Piaractus mesopotamicus*. *Comp. Biochem. Physiol.* **119A,** 991–997.

Reebs, S. G., Whoriskey, F. G., and Fitzgerald, G. J. (1984). Diel patterns of fanning activity, egg respiration, and the nocturnal behavior of male three-spined sticklebacks, *Gasterosteus aculeatus* L. (F-Trachurus). *Can. J. Zool.* **62,** 329–334.

Reid, S. G., Sundin, L., and Milsom, W. K. (2006). The cardiorespiratory system in tropical fishes: Structure, function, and control. *In* "Tropical Fishes" (Val, L. V., Almeida-Val, V. M. F., and Randall, D. J., Eds.), Fish Physiology, Vol. 21, pp. 225–274. Academic Press/Elsevier, San Diego, CA.

Ripley, J. L., and Foran, C. M. (2006). Influence of estuarine hypoxia on feeding and sound production by two sympatric pipefish species (Syngnathidae). *Mar. Env. Res.* **63,** 350–367.

Robb, T., and Abrahams, M. V. (2003). Variation in tolerance to hypoxia in a predator and prey species: An ecological advantage of being small? *J. Fish. Biol.* **62,** 1067–1081.

Roberts, J. L. (1978). Ram gill ventilation in fishes. *In* "The Physiological Ecology of Tunas" (Sharp, G. D., and Dizon, A. E., Eds.), pp. 83–88. Academic Press, New York.

Roberts, T. R. (1973). Ecology of fishes in the Amazon and Congo Basins. *In* "Tropical Forest Ecosystems in Africa and South America: A Comparative Review" (Meggers, B. J., Ayensu, E. S., and Duckworth, W. D., Eds.), pp. 239–254. Smithsonian Institution Press, Washington, DC.

Rosenberger, A. E., and Chapman, L. J. (1999). Hypoxic wetland tributaries as faunal refugia from an introduced predator. *Ecol. Freshwat. Fish* **8,** 22–34.

Rosenberger, A. E., and Chapman, L. J. (2000). Respiratory characters of three haplochromine cichlids: Implications for persistence in wetland refugia. *J. Fish Biol.* **57,** 483–501.

Rüber, L., Britz, R., and Zardoya, R. (2006). Molecular phylogenetics and evolutionary diversification of labyrinth fishes (Perciformes: Anabantoidei). *Syst. Biol.* **55,** 374–397.

Rutjes, H. A. (2006). Phenotypic responses to lifelong hypoxia in cichlids Ph.D., dissertation, Leiden University.

Rutledge, C. J., and Beitinger, T. L. (1989). The effects of dissolved oxygen and aquatic surface respiration on critical thermal maxima of three intermittent-stream fishes. *Environ. Biol. Fish.* **24,** 137–143.

Saint-Paul, U. (1984). Physiological adaptation to hypoxia of a neotropical characoid fish *Colossoma macropomum*, Serrasalmidae. *Environ. Biol. Fish.* **11,** 53–62.

Saint-Paul, U., and Soares, B. M. (1987). Diurnal distribution and behavioral responses of fishes to extreme hypoxia in an Amazon floodplain lake. *Environ. Biol. Fish.* **20,** 91–104.

Sayer, M. D. J. (2005). Adaptations of amphibious fish for surviving life out of water. *Fish and Fisheries* **6,** 186–211.

Schofield, P. J., and Chapman, L. J. (1999). Interactions between Nile perch, *Lates niloticus*, and other fishes in Lake Nabugabo, Uganda. *Environ. Biol. Fish.* **55,** 343–358.

Schofield, P. J., and Chapman, L. J. (2000). Hypoxia tolerance of introduced Nile perch: Implications for survival of indigenous fishes in the Lake Victoria basin. *Afr. Zool.* **35,** 35–42.

Schofield, P. J., Loftus, W. F., and Brown, M. E. (2007). Hypoxia tolerance of two centrarchid sunfishes and an introduced cichlid from karstic Everglades wetlands of southern Florida, U.S.A. *J. Fish Biol.* **71,** 87–99.

Schurmann, H., and Steffensen, J. F. (1994). Spontaneous swimming activity of Atlantic cod *Gadus morhua* exposed to graded hypoxia at three temperatures. *J. Exp. Biol.* **197,** 29–142.

Schurmann, H., and Steffensen, J. F. (1997). Effects of temperature, hypoxia and activity on the metabolism of juvenile cod. *J. Fish. Biol.* **50,** 1166–1180.

Seehausen, O., van Alphen, J. J. M., and Witte, F. (1997a). Cichlid fish diversity threatened by eutrophication that curbs sexual selection. *Science* **277,** 1808–1811.

Seehausen, O., Witte, F., Katunzi, E. F., Smits, J., and Bouton, N. (1997b). Patterns of the remnant cichlid fauna in southern Lake Victoria. *Cons. Biol.* **11,** 890–904.

Seifert, A. W., and Chapman, L. J. (2006). Respiratory allocation and standard rate of metabolism in the African lungfish, *Protopterus aethiopicus*. *Comp. Biochem. Physiol.* **143A,** 142–148.

Shimizu, N., Sakai, Y., Hashimoto, H., and Gushima, K. (2006). Terrestrial reproduction by the air-breathing fish *Andamia tetradactyla* (Pisces; Blenniidae) on supralittoral reefs. *J. Zool.* **269,** 357–364.

Shingles, A., McKenzie, D. J., Claireaux, G., and Domenici, P. (2005). Reflex cardioventilatory responses to hypoxia in the flathead grey mullet (*Mugil cephalus*) and their behavioural modulation by perceived threat of predation and water turbidity. *Physiol. Biochem. Zool.* **78,** 744–755.

Shoji, J., Masuda, R., Yamashita, Y., and Tanaka, M. (2005). Effect of low dissolved oxygen concentrations on behavior and predation rates on red sea bream *Pagrus major* larvae by the jellyfish *Aurelia aurita* and by juvenile Spanish mackerel *Scomberomorus niphonius*. *Mar. Biol.* **147,** 863–868.

Skjæraasen, J. E., Nilsen, T., Meager, J. J., Herbert, N. A., Moberg, O., Tronci, V., Johansen, T., and Salvanes, A. G. V. (2008). Hypoxic avoidance behaviour in cod (*Gadus morhua* L.): The effect of temperature and haemoglobin genotype. *J. Exp. Mar.Biol. Ecol.* **358,** 70–77.

Sloman, K. A., Wood, C. M., Scott, G. R., Wood, S., Kajiumura, M., Johannsson, O. E., Almeida-Val, V. M. F., and Val, D. (2006). Tribute to R. G. Boutilier: The effect of size on the physiological and behavioural responses of oscar, *Astronotus ocellatus*, to hypoxia. *J. Exp. Biol.* **209,** 1197–1205.

Sloman, K. A., Mandic, M., Todgham, A. E., Fangue, N. A., Subrt, P., and Richards, J. G. (2008). The response of the tidepool sculpin, *Oligocottus maculosus*, to hypoxia in laboratory, mesocosm and field environments. *Comp. Biochem. Physiol.* **149A,** 284–292.

Smatresk, N. J. (1986). Ventilatory and cardiac reflex responses to hypoxia and NaCN in *Lepisosteus osseus*, an air-breathing fish. *Physiol. Zool.* **59,** 385–397.

Smatresk, N. J. (1990). Chemoreceptor modulation of endogenous respiratory rhythms in vertebrates. *Am. J. Physiol.* **259,** R887–R897.

Smatresk, N. J., and Cameron, J. N. (1982). Respiration and acid-base physiology of the spotted gar, a bimodal breather. I. Normal values and the responses to severe hypoxia. *J. Exp. Biol.* **96,** 263–280.

Smatresk, N. J., Burleson, M. L., and Azizi, S. Q. (1986). Chemoreflexive responses to hypoxia and NaCN in longnose gar: Evidence for two chemoreceptive loci. *Am. J. Physiol.* **251,** R116–R125.

Smith, R. S., and Kramer, D. L. (1986). The effect of apparent predation risk on the respiratory behavior of the Florida gar (*Lepisosteus platyrhincus*). *Can. J. Zool.* **64,** 2133–2136.

Sneddon, L. U., and Yerbury, J. (2004). Differences in response to hypoxia in the three-spined stickleback from lotic and lentic localities: Dominance and an anaerobic metabolite. *J. Fish. Biol.* **64,** 799–804.

Steffensen, J. F. (1985). The transition between branchial pumping and ram ventilation in fishes: Energetic consequences and dependence on water oxygen tension. *J. Exp. Biol.* **114,** 141–150.

Stevens, E. D., and Holeton, G. F. (1978). The partitioning of oxygen uptake from air and from water by erythrinids. *Can. J. Zool.* **56,** 965–696.

Stierhoff, K. L., Target, T. E., and Gracey, P. A. (2003). Hypoxia tolerance of the mummichog: The role of access to the water surface. *J. Fish. Biol.* **63,** 580–592.

Stirling, D. G., McQueen, D. J., and Johannes, M. R. S. (1990). Vertical migration in *Daphnia galeata mendotae* (Brooks): Demographic responses to change in planktivore abundance. *Can. J. Fish. Aquat. Sci.* **47,** 395–400.

Sundin, L., Reid, S. G., Rantin, F. T., and Milsom, W. K. (2000). Branchial receptors and cardiovascular reflexes in a neotropical fish, the tambaqui (*Colossoma macropomum*). *J. Exp. Biol.* **203,** 1225–1239.

Sundin, L., Burleson, M. L., Sanchez, A. P., Amin-Naves, J., Kinkead, R., Gargaglioni, L. H., Hartzler, L. K., Wiemann, M., Kumar, P., and Glass, M. L. (2007). Respiratory chemoreceptor function in vertebrates-comparative and evolutionary aspects. *Integr. Comp. Biol.* **47,** 592–600.

Takegaki, T., and Nakazono, A. (1999). Response of the egg-tending gobiid fish *Valenciennea longipinnis* to the fluctuation of dissolved oxygen in the burrow. *Bull. Mar. Sci.* **65,** 815–823.

Taylor, E. W., McKenzie, D. J., Levings, J. J., and Randall, D. J. (1996). Control of ventilation in air-breathing fish. *In* "Physiology and Biochemistry of Fishes of the Amazon" (Val, A. L., Almeida-Val, V. M. F., and Randall, D. J., Eds.), pp. 155–168. INPAS, Manaus.

Taylor, E. W., Jordan, D., and Coote, J. H. (1999). Central control of the cardiovascular and respiratory systems and their interactions in vertebrates. *Physiol. Rev.* **79,** 855–916.

Tesch, F. W. (1972). Experiments on telemetric tracking of spawning migrations of eels (*Anguilla Anguilla*) in North Sea. *Helgolander Wissenschaftliche Meeresuntersuchungen* **23,** 165.

Thibault, R. E., and Schultz, R. J. (1978). Reproductive adaptations among viviparous fishes (Cyprinodontiformes: Poeciliidae). *Evolution* **32,** 320–333.

Timmerman, C. M., and Chapman, L. J. (2003). The effect of gestational state on oxygen consumption and response to hypoxia in the sailfin molly, *Poecilia latipinna. Environ. Biol. Fish.* **68,** 293–299.

Timmerman, C. M., and Chapman, L. J. (2004). Behavioral and physiological compensation for chronic hypoxia in the live-bearing sailfin molly (*Poecilia latipinna*). *Physiol. Biochem. Zool.* **77,** 601–610.

Tonn, W. M., and Magnuson, J. J. (1982). Patterns in the species composition and richness of fish assemblages in northern Wisconsin Lakes. *Ecology* **63,** 1149–1166.

Val, A. L., and Almeida-Val, V. M. F. (1995). "Fishes of the Amazon and their Environment. Physiological and Biochemical Aspects." Springer-Verlag, Heidelberg.

van den Thillart, G., Dalla Via, J., Vitali, G., and Cortesi, P. (1994). Influence of long-term hypoxia exposure on the energy metabolism of *Solea solea*. I. Critical O_2 levels for aerobic and anaerobic metabolism. *Mar. Ecol. Prog. Ser.* **104,** 109–117.

Van Raaij, M. T. M., Pitt, D. S. S., Balm, P. H. M., Steffens, A. B., and van den Thillart, G. (1996). Behavioural strategy and the physiological stress response in rainbow trout exposed to severe hypoxia. *Horm. Behav.* **30,** 85–92.

Wakeling, J. M., and Johnston, I. A. (1998). Muscle power output limits fast-start performance in fish. *J. Exp. Biol.* **201,** 1505–1526.

Webb, P. W. (1993). Swimming. *In* "The Physiology of Fishes" (Evans, D. H., Ed.), pp. 47–73. CRC Press Marine Science Series, Boca Raton, FL.

Webb, P. W. (1998). Swimming. *In* "The Physiology of Fishes" (Evans, D. H., Ed.), pp. 3–24. CRC Press Marine Science Series, Boca Raton, FL.

Wells, R. M. G., Baldwin, J., Seymour, R. S., Christian, K. A., and Farrell, A. P. (2007). Air breathing minimizes post-exercise lactate load in the tropical Pacific tarpon, *Megalops cyprinoides* Broussonet 1782, but oxygen debt is repaid by aquatic breathing. *J. Fish Biol.* **71,** 1649–1661.

Weltzien, F., Døving, K. B., and Carr, W. E. S. (1999). Avoidance reaction of yolk-sac larvae of the inland silverside *Menidia beryllina* (Atherinidae) to hypoxia. *J. Exp. Biol.* **202,** 2869–2876.

Wen-Chi Corrie, L., Chapman, L. J., and Reardon, E. (2007). Brood protection at a cost: Mouthbrooding under hypoxia in an African cichlid. *Environ. Biol. Fish.* **82,** 41–49.

West, G. B., Brown, J. H., and Enquist, B. J. (1997). A general model for the origin of allometric scaling laws in biology. *Science* **276,** 122–126.

Winemiller, K. O. (1989). Development of dermal lip protuberances for aquatic surface respiration in South American characid fishes. *Copeia* **1989,** 382–390.

Wolf, N. G., and Kramer, D. L. (1987). Use of cover and the need to breathe: The effects of hypoxia on vulnerability of dwarf gouramis to predatory snakeheads. *Oecologia* **73,** 127–132.

Wootton, R. J. (1976). "The Biology of Sticklebacks." Academic Press, London.

Wootton, R. J. (1990). "Ecology of Teleost Fishes." Chapman and Hall, New York.

Wourms, J. P., and Lombardi, J. (1992). Reflections on the evolution of piscine viviparity. *Am. Zool.* **21,** 276–293.

Yoshiyama, R. M., Valpey, C. J., Schalk, L. L., Oswald, N. M., Vaness, K. K., Lauritzen, D., and Limm, M. (1995). Differential propensities for aerial emergence in intertidal sculpins (Teleostei; Cottidae). *J. Exp. Mar. Biol. Ecol.* **191,** 195–207.

Zoran, M. J., and Ward, J. A. (1983). Parental egg care behavior and fanning activitiy for the orange chromide, *Etroplus maculatus*. *Environ. Biol. Fish.* **8,** 301–310.

3

EFFECTS OF HYPOXIA ON FISH REPRODUCTION AND DEVELOPMENT

RUDOLF S. S. WU

Hypoxia has a profound effect on fish reproduction and development. Behavioral studies revealed that hypoxia can affect courtship behaviors, mate choice, and reproductive efforts in fish. Both laboratory and field evidence showed that hypoxia can cause major reproductive impairments by inhibiting testicular and ovarian development, affecting production and quality of sperm and egg, reducing fertilization and hatching success, and affecting larval survivorship as well as the quality and fitness of juveniles. Emerging evidence further showed that hypoxia does not impair these key reproductive processes through a general down-regulation of metabolism and reproductive functions, but does so by affecting specific hormones, neurotransmitters, and receptors along the

Hypoxia: Volume 27
FISH PHYSIOLOGY

DOI: 10.1016/S1546-5098(08)00003-4

hypothalamus–pituitary–gonad axis as well as certain enzymes controlling steroidogenesis and vitellogenesis. In zebrafish, hypoxia has been shown to down-regulate CYP19 and alter the ratio of testosterone to estradiol during early sex development, leading to a male-biased F1 generation.

Hypoxia has been shown to delay embryonic development and hatching in many fish species, and embryos in some species may undergo complete developmental arrest under anoxia. In zebrafish embryos, blastomeres were arrested during the S and G2 phases of the cell cycle under anoxia. Fish embryos developed under hypoxia lost their normal synchronization, and abnormalities in spinal and vascular development are commonly observed. Results of both laboratory and field studies showed a higher percentage of malformation in fish developed under hypoxic conditions, possibly through altering their normal apoptosis.

Both *in vitro* and *in vivo* studies demonstrated that expression levels of certain genes directly or indirectly related to cell cycle, cell proliferation, and apoptosis, which underpin some of the fundamental processes related to development, are affected by hypoxia. Whether hypoxic inducible factor is involved in mediating the changes in gene expression and the observed reproductive and development impairments remains unclear.

1. INTRODUCTION

1.2. Occurrence of Hypoxia in the Aquatic Environment

Hypoxia is generally defined as dissolved oxygen less than 2.8 mg O_2/L (equivalent to 2 mL O_2/L or 91.4 mM) (Diaz and Rosenberg, 1995) and anoxia means no oxygen. Hypoxia/anoxia occurs in a variety of marine, estuarine, and freshwater habitats, and can be a natural phenomenon caused by vertical stratification such as formation of haloclines and thermoclines (Rosenberg *et al.*, 1991; Pihl *et al.*, 1992; Hoback and Barnhart, 1996). Globally, the total area of permanently hypoxic continental shelf and bathyal sea floor with dissolved oxygen <0.5 ml O_2/L (minimal oxygen zones) is estimated at more than one million square kilometers (Helly and Levin, 2004). More often, however, the occurrence of hypoxia is due to excessive anthropogenic input of nutrients and organic matters into water bodies with poor circulation (Pihl *et al.*, 1992; Dalla Via *et al.*, 1994; Peckol and Rivers, 1995; Gamenick *et al.*, 1996; Sandberg, 1997; Wu and Lam, 1997; Aarnio *et al.*, 1998; Mason, 1998). Nowadays, hypoxia or anoxia affecting thousands of square kilometers of marine waters has been commonly reported for waters around North and South America, Africa, Europe, India, Southeast Asia, Australia, Japan, and China (Nixon, 1990; Diaz and Rosenberg,

1995; Wu, 1999). Likewise, hypoxia also commonly occurs in freshwater systems in many countries (Sabo *et al.*, 1999; Keister *et al.*, 2000; Fontenot *et al.*, 2001; Breitburg *et al.*, 2003). In China, for example, over 77% of the freshwater ecosystems are now considered under serious threat by hypoxia (Ma and Li, 2002). Indeed, hypoxia caused by eutrophication is now considered to be one of the most serious threats to aquatic ecosystems worldwide. Hypoxia has not only increased in terms of frequency, severity, and areas affected in the last two decades, but is likely to be further exacerbated in the coming years (Diaz and Rosenberg, 1995; Goldberg, 1995; Gray *et al.*, 2002; Wu, 2002). The eminence of the problem is clearly exemplified by the global increase in the number of "dead zones" from 150 in 2004 to 200 in 2006 (UNEP, 2006).

1.2. Global Changes in Fish Populations and Communities

Unlike mammals, which can only tolerate a narrow range of oxygen regimes, fish often have to contend with large fluctuations of oxygen in their natural environment, which sometimes can occur very rapidly (e.g., within a day) or within minutes if they swim through a hypoxic region. Indeed, there is no other environmental parameter, except perhaps temperature, in the aquatic ecosystem that can change so drastically, within such a short time, as dissolved oxygen. Thus, it is not surprising that many fish species have evolved a variety of molecular, biochemical, and physiological adaptations to cope with hypoxia in the course of evolution (see Hochachka and Somero, 2002).

Hypoxia has already led to major changes in fish species composition, alteration of food webs and community structure, decrease in species richness and diversity, population declines and extinction of sensitive species in both marine and freshwater systems in many parts of the world (Wu, 1982; Dauer, 1993; Pihl, 1994; Diaz and Rosenberg, 1995; Alexander *et al.*, 2000; Diaz, 2001; Wanink *et al.*, 2001; Wu, 2002). Massive fish kills over large areas due to hypoxia have been reported in coastal areas all over the world, and sensitive species have been permanently or periodically removed in many places (Wu, 1982; Diaz and Rosenberg, 1995). Massive fish kills in aquaculture due to hypoxia are equally common (Townsend *et al.*, 1992; Grantham *et al.*, 2004; Azanza *et al.*, 2005; Parvez *et al.*, 2006; Bouchet *et al.*, 2007).

Besides causing direct death, hypoxia may also reduce growth, alter behaviors of fishes, and change their food items, thereby reducing their abundance and diversity (Breitburg, 2002). Reductions in the biomass and landing of fish have been reported in many hypoxic areas (Dyer *et al.*, 1983; Rosenberg and Loo, 1988; Baden *et al.*, 1990; Pihl *et al.*, 1991; Breitburg, 1992; Lekve *et al.*, 1999). Petersen and Pihl (1995) demonstrated a significant

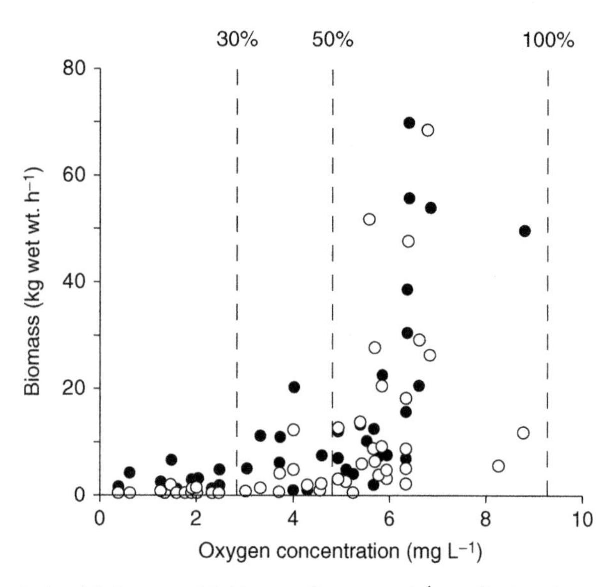

Fig. 3.1. The relationship between fish biomass (kg wet wt. h^{-1} trawling) and oxygen concentration in the bottom water of SE Kattegat. Plaice (<, $p < 0.02$), Dab (•, $p < 0.01$) in the years 1984 to 1990 (After Petersen and Pihl, 1995.)

relationship between biomass (catch per unit effort) of plaice and dab and oxygen concentration in the bottom water of Kattegat, Sweden (Figure 3.1). Hypoxia might also favor the selection of small benthic species with a shorter life cycle, and such long-term changes in prey species, coupled with a lower level of oxygen in bottom waters, have been related to a shift in dominance from demersal to pelagic fish in the Kattegat, Sweden (Pihl, 1994).

The observed decline in natural fish populations may also be caused by reproductive impairments resulting from chronic hypoxia, although it would be difficult to decipher the exact cause, or to attribute the observed population decline and community changes to hypoxia *per se*, since hypoxia in the natural environment is often associated with other confounding factors such as pollution and overfishing.

Arguably, reproductive output, quality of gametes, and survival of larvae and juveniles are the most important factors in determining reproductive success and hence fitness and survival of any species. At the same time, both reproduction and development involve a myriad of intricate processes, making these life stages particularly vulnerable to environmental stresses (Connell *et al.*, 1999), especially since these intricate processes are tightly controlled by hormones that are very sensitive to environmental changes (Bhattacharya, 1999; Seale *et al.*, 2002; Okuzawa *et al.*, 2003). Notably, many coastal areas, which serve as important spawning and nursery grounds, are located in areas where occurrence of hypoxia is common. Surprisingly, the effects of hypoxia on reproduction and development of fish, especially on natural populations, remain poorly understood (Wu, 2002).

A field study in the Atchafalaya River of Louisiana (Fontenot *et al.*, 2001) demonstrated a strong, positive relationship between dissolved oxygen level and the abundance of larval sunfish (*Lepomis* spp.) and shad (*Dorosoma* spp.). The field study of Ingendahl (2001) reported that sea trout (*Salmo trutta*) alevins only emerged from covered redds in tributaries of the Rhine where the mean dissolved oxygen level was above 6.9 mg O_2/L (~56% saturation). More recently, Dumas *et al.* (2007) reported that low oxygen delayed brown trout alevin development and growth in a tributary of the Adour river in south-west France. The above field evidence offers indirect evidence supporting the postulation that hypoxia can affect reproduction and/or larval development in their natural habitats, contributing to the population decline and community changes observed in hypoxic areas worldwide.

2. HYPOXIA AND FISH REPRODUCTION

Hypoxia impairs reproductive success by affecting a number of key reproductive processes, including gametogenesis, the number and quality of sperm and egg, reproductive behaviors, fertilization success, hatching, and, subsequently, larval survivorship and the quality and fitness of juveniles. These impairments may be mediated through disrupting the various hormones and enzymes regulating these key reproductive processes, or by reducing food intake and, hence, the energy available for reproductive investment.

2.1. Control of Reproductive Processes in Fish

2.1.1. The Hypothalamus–Pituitary–Gonad Axis

Despite the fact that reproductive processes in fish are highly diverse and vary among species, the intricate process is very conservative and tightly regulated by the hypothalamus–pituitary–gonad (HPG) axis. The HPG axis controls gametogenesis, reproductive behavior, and reproduction, including the release of gametes and fertilization via positive and/or negative feedback loops (for a detailed review, please see Ankley and Johnson, 2004; Thomas, 2008).

The control of reproductive hormone synthesis and secretion along the HPG axis is schematically shown in Figure 3.2. The basic features and control of the HPG axis in fish closely resembles those in higher vertebrates.

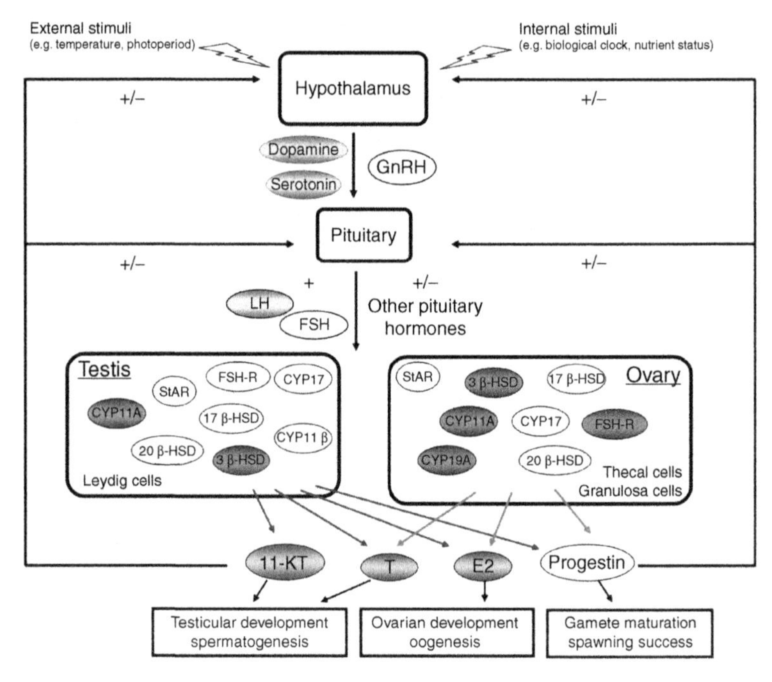

Fig. 3.2. The HPG axis and control of reproductive hormone synthesis and secretion in male and female teleosts. Key regulatory hormones for steroidogenesis are highlighted in yellow, hypoxia-responsive genes are highlighted in green, and steroid hormones involved in gametogenesis are highlighted in purple. (Modified from Weltzien *et al.*, 2004.) (See Color Insert.)

Environmental cues (e.g., temperature, photoperiod, and nutritional changes as well as hypoxia) detected by various sensory systems are relayed to the hypothalamus. The hypothalamus releases various neurotransmitters and neuropeptides, leading to the secretion of gonadotropin-releasing hormone (GnRH) into the intercellular space of the hypophysis of the pituitary through the hypothalamic neurons. The decapeptide GnRH then binds to the specific receptors on the plasma membrane of the gonadotropes from the anterior pituitary and stimulates the production and/or release of two types of glycoprotein gonadotropic hormones (GtHs), follicle-stimulating hormone (FSH) and luteinizing hormone (LH). Both FSH and LH consist of an α subunit (which is common to both GtHs and thyroid-stimulating hormone) and a β subunit, which is hormone specific. These two gonadotropins are then transported to the gonads through blood circulation where they bind to specific G-protein-coupled membrane GtH receptors (GtH-Rs) and activate G-proteins, adenyl cyclase, and Ca^{2+}-dependent second messenger signaling pathways, which subsequently lead to the production and secretion of steroid hormones. FSH primarily induces oogenesis and spermatogenesis, while LH directs maturation and release of gametes. FSH and LH stimulate the thecal and granulosa cells of the ovary to produce the female steroid hormone 17β-estradiol, and FSH induces the enzyme aromatase in the granulosa cells, which converts testosterone to estradiol. 17β-estradiol stimulates oocyte development in the ovary and the synthesis of vitellogenin (the egg yolk precursor protein) in the liver for release into the bloodstream. In the male fish, gonadotropin stimulates Leydig cells to produce the androgens (testosterone and 11-ketotestosterone), which, in turn, activates Sertoli cells to stimulate premitotic spermatogonia to complete spermatogenesis (Nagahama *et al.*, 1994; Thomas, 2008). Breakdown products of sex hormones may act as pheromones directing behavior (Sorensen *et al.*, 2004).

A number of neurotransmitters also play an important role in modulating reproductive processes in fish. For example, the monoamine neurotransmitter serotonin (5-HT) acts on GnRH and potentiates LH secretion. Another neurotransmitter, dopamine (DA), inhibits LH secretion in some species (e.g., carp and catfish) but not in others (e.g., Atlantic croakers). Other neurotransmitters and neuropeptides such as neuropeptide Y, gamma aminobutyric acid (GABA), glutamate, and taurine are also implicated in the neuroendocrine control of GTH release in teleosts (Kah *et al.*, 1993).

2.1.2. Steroidogenesis

Similar to other vertebrates, gametogenesis and sex behaviors in fish are directly controlled by sex steroid hormones. The synthesis of sex steroid hormones (steroidogenesis) mainly takes place in adrenal tissues (zona glomerulosa, zona fasciculate, and zona reticularis) and gonadal tissues (male testes and female ovaries) (Young *et al.*, 2004). A schematic representation of the key steps involved in steroidogenesis in teleosts is shown in Figure 3.3.

Cholesterol is the common precursor for all sex steroid hormones. The first rate-determining step involves the importation of cholesterol into the inner mitochondrial membrane. This step, which initiates steroidogenesis, is regulated by the steroidogenic acute regulatory protein (StAR), and the production of StAR is upregulated by GtH in fish (Bauer *et al.*, 2000; Stocco, 2001; Kusakabe *et al.*, 2002). Subsequent steps of the steroidogenic pathway are controlled by a number of steroidogenic enzymes including cytochrome P450 enzymes and hydroxysteroid dehydrogenases (HSDs) (Miller, 1988; Senthilkumaran *et al.*, 2004; Weltzien *et al.*, 2004; Miller, 2005). Cholesterol is converted to pregnenolone by the P450 enzyme cholesterol side chain cleavage (P450scc or CYP 11A) on the inner membrane of the mitochondria (Takahashi *et al.*, 1993). Pregnenolone is then converted through a series of steps to androgens by 3β-hydrosteroid dehydrogenase (3βHSD),

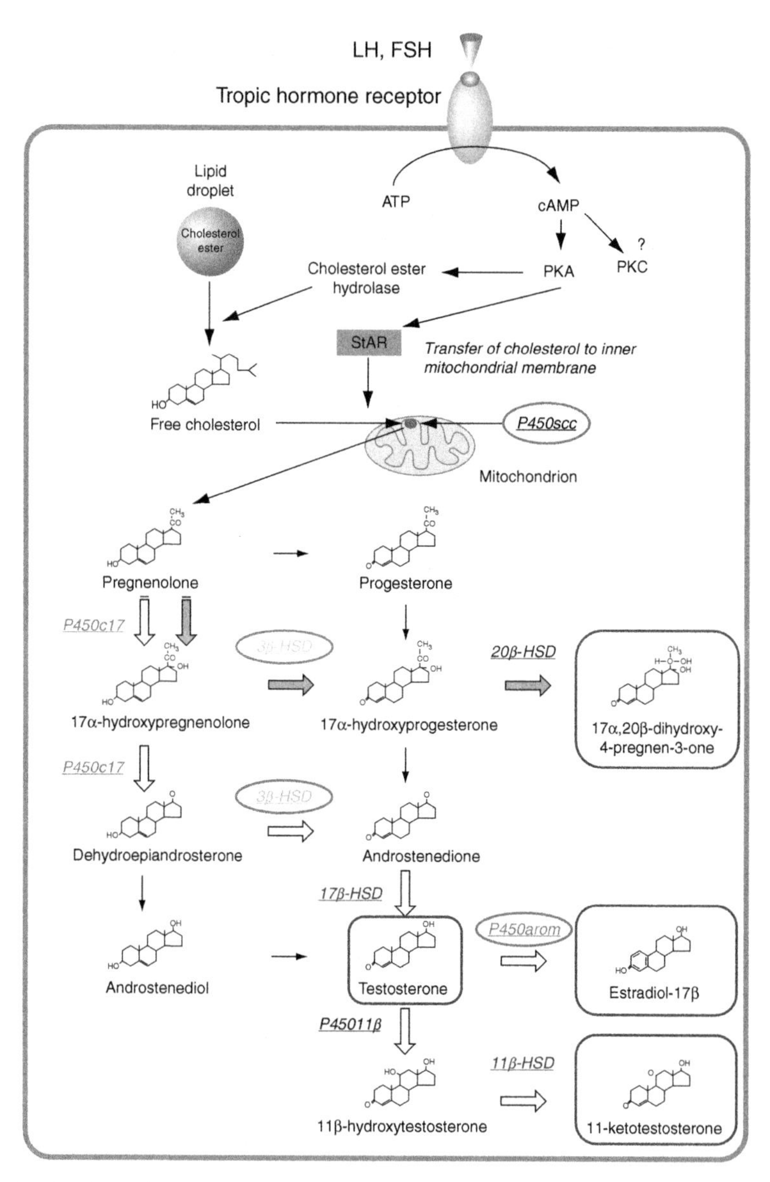

Fig. 3.3. A schematic pathway of steroidogenesis in the gonads of teleost fish. White arrows indicate the proposed androgen synthesis pathway. Gray arrows indicate the proposed progestogen synthesis pathway. Genes known to be inducible by hypoxia are circled in red; hormones known to be affected by hypoxia are framed in blue. (Modified from Young *et al.*, 2004.) (See Color Insert.)

17α-hydroxylase (P450c17), 21-hydroxylase (P450c21), 11β-hydrolase (P450c11), and 20β-dehydroxysteroid dehydrogenase (20β-HSD) to form progestins (Nagahama, 2000) and then 11-ketotestosterone (11-KT). Finally, testosterone is converted to estrogen by aromatase (encoded by P450arom or CYP19).

There is good evidence to show that reproductive processes in fish are regulated by the expression levels of the various steroidogenic enzymes. In the channel catfish (*Ictalurus punctatus*), for example, P450c17, P450scc, and P450arom were up-regulated at the onset of ovarian recrudescence and during early vitellogenic growth of the oocytes, but subsided upon completion of vitellogenesis (Kumar *et al.*, 2000). In salmonids, FSH stimulates the expression and activity of P450arom, and regulates the production of E2 in the ovary (Montserrat *et al.*, 2004). The above evidence shows that GtH may regulate sex steroid levels through regulating various steroidogenic enzymes. It also appears that the genes encoding these three steroidogenic cytochrome P450s have a similar regulatory mechanism (Kumar *et al.*, 2000).

2.1.3. Sex Differentiation and Sex Determination

Unlike mammals, fish exhibit considerable plasticity in sex determination irrespective of their genotypic sex. Many environmental factors (e.g., temperature, photoperiod, and social behavior), chemicals, and sex hormones may influence sex differentiation and determination (Jobling, 1995). In some gonochoristic species such as zebrafish, the gonads will first pass through a juvenile "ovary" phase before differentiating into testes or ovary. A similar sexual developmental pattern is generally found in the protogynous species.

The genetic and molecular mechanisms underlying sex determination and differentiation in fish remain unclear. In particular, the reason why genetic makeup in fish is relatively easy to override by environmental factors as compared to mammals remains unknown, both from a mechanistic and evolutionary points of view. Unlike mammals, sex chromosomes have only been identified in only about 10% of fish (Devlin and Nagahama, 2002). Recently, the *DMY* (Y-specific DM domain) gene has been identified as the sex-determining gene in the Y chromosome of freshwater medaka (*Oryzias latipes*) (Matsuda *et al.*, 2002). Gonadotropin, thyroid hormones, growth hormone, insulin, and insulin-like growth factors have been shown to affect ovarian growth in brown trout (*Salmo trutta*), rainbow trout (*Oncorhynchus mykiss*), and Chinook salmon (*Oncorhynchus tshawytscha*) (Tyler and Sumpter, 1996), and hence sex development. Since the regulation of the HPG axis is a complex and highly intricate process, disruption of the HPG axis is likely to alter sex differentiation and gametogenesis.

The early work of Yamamoto (1961) revealed that complete sex reversal can occur in medaka when steroid hormones (estrogens, androgens, or progestins) were administered during their early developmental stages, resulting in a phenotypic male or female irrespective of genetic sex. Similarly, administration of androgens can alter sex differentiation in Chinook salmon (*Oncorrhychus tshawytcha*), turning genotypic females into males (Piferrer *et al.*, 1993). Subsequent investigations further revealed that inhibition of the cytochrome P450 aromatase complex (by androgens or aromatase inhibitors) during sex differentiation in fish can turn genotypic females into phenotypic males. For example, sex change was reported in the Japanese flounder (*Paralichthys olivaceus*) when treated with aromatase inhibitor and 17α-methyl-testosterone (Kitano *et al.*, 2000); sex change in the goby (*Gobiodon histrio)* was attributable to aromatase activities and levels of 11-KT (Kroon *et al.*, 2003). In the Nile tilapia (*Oreochromis niloticus*), genotypic female fry treated with dietary Fadrozole (an aromatase inhibitor) during sexual differentiation led to an increase in the percentage of males (Kwon *et al.*, 2000). Fenske and Segner (2004) showed that aromatase modulation alters gonad differentiation in zebrafish. The above evidence clearly demonstrated that sex differentiation and sex determination in many fish species are modulated by sex hormones, and factors affecting key enzymes regulating steroidogenesis may alter the balance of sex hormones and hence sex determination. In particular, P450arom, which converts testosterone into estradiol and affects the ratio of androgens to estrogens, could be expected to play a critical role in fish reproduction and sex differentiation. Since the cytochrome P450 enzymes demand oxygen (Nishimura *et al.*, 2006), hypoxia may potentially disrupt normal steriodogenesis and interfere with sex differentiation and sex determination via these enzymes. This, however, may be only one of the many ways in which hypoxia modulates sex differentiation and determination in fish.

2.2. Effects of Hypoxia on the HPG Axis, Steroidogenesis, and Sex Hormones

2.2.1. Gene Expression Profile

A detailed review of the effects of hypoxia on gene expression profile is given in Chapter 10, and only those hypoxia-responsive genes relating to fish reproduction and development will be reviewed here.

Several attempts have been made to map out the global responses of genes to hypoxia, using cDNA microarray technology. Some of the genes responsive to hypoxia, as revealed in these studies, are indirectly or remotely related to neurotransmitters, hormones, cell cycle, cell proliferation, and apoptosis, which underpin some of the fundamental processes related to reproduction and development.

Gracey *et al.* (2001) examined gene expression in liver, brain, skeletal muscle, and heart from adult gobies (*Gillichthys mirabilis* and *G. seta*) exposed to hypoxia (0.8 mg O_2/L, ~10% saturation) for 6 days. Up-regulation of various genes involved in glycolysis, iron metabolism, amino acid metabolism, and growth suppression were found in the liver, and down-regulation of genes involved in protein translation and muscle contraction were found in both skeletal muscle and heart. Induction of MAP kinase phosphatase 1 (KP-1), which stimulates cell growth, was found in all tissues, whereas the anti-proliferation genes transducer, Erb-B2 (Tob) and B-cell translocation gene-1 (BTG-1), were induced in the liver.

Using the same approach, Ton *et al.* (2003) studied the expression patterns of 4512 genes in whole embryos of zebrafish [24 hours post-fertilization (hpf)] exposed to extreme hypoxia (5% saturation, levels at which zebrafish embryos are unable to survive for more than 24 h) for 24 h followed by 5 h of recovery. Hypoxia increased the expression of HIF-1 and certain glycolytic genes, but down-regulated genes involved in oxidative carbohydrate metabolism, muscle contraction, translation, and cell cycle progression. Hypoxia also repressed high-motility group proteins HMG-Y and HMG2a, histone H3, proliferating cell nuclear antigen (PCNA), and cyclin G1 and G2/mitotic-specific cyclin involved in cell division. This is consistent with the observations of Padilla and Roth (2001) that the S and G2 phases of the cell cycle in zebrafish embryonic development (in 4-cell stage embryos) was arrested under hypoxia and anoxia. Down-regulation of intracellular transducers e.g., small GTP-binding protein Rab, which may be related to suppression of cell growth and proliferation under hypoxia, and induction of HSP70, which is known to protect cells against apoptosis, were also found.

Using a microarray containing 8046 medaka (*Oryzias latipes*) genes, Ju *et al.* (2007) reported that 501 genes in the brain, 442 in the gill, and 715 in the liver were differentially expressed in medaka exposed to hypoxia. Among these, two genes relating to neurotransmitter transport in the brain were down-regulated, while two genes in the brain related to response to hormones were up-regulated during hypoxia. The above three microarray screening studies provided evidence that certain genes relating directly or indirectly to reproductive hormones and fundamental processes in development (e.g., apoptosis, cell proliferation, and cell growth) in fish may be affected by hypoxia.

Despite the fact that neither apoptosis nor necrosis was found in the gills of zebrafish subjected to hypoxia (10% saturation for 21 days), a number of genes relating to apoptosis and growth regulation were responsive to hypoxia (Figure 3.4), and the majority of the former are anti-apoptotic genes (van der Meer *et al.*, 2005). HSP70, which is known to protect cells against apoptosis (Höhfeld, 1998), was also up-regulated.

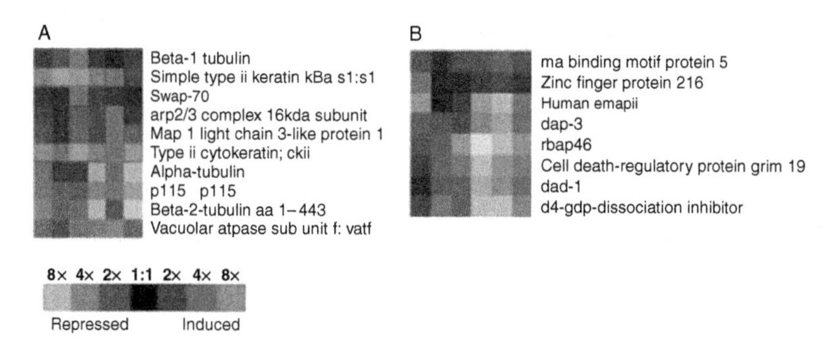

Fig. 3.4. Genes related to (A) apoptosis and (B) growth regulation found to be differentially expressed under hypoxic conditions in the gill of zebrafish. Quantitative changes in gene expression are induced genes (red) and repressed genes (green). (After van de Meer *et al.*, 2005.) (See Color Insert.)

However, many genes specifically related to HPG, steroidogenic enzymes, and neurotransmitters that were found to be affected by hypoxia in zebrafish, medaka, Gulf killifish, and Atlantic croakers in several independent studies (see Sections 2.2.2 and 2.2.4 below) were not revealed in any of these microarray screening studies, suggesting that some of these microarray data should be viewed with great caution, in particular the study by Ton *et al.* (2003) where fish were exposed to an extremely low oxygen level at which death would begin to occur.

2.2.2. GnRH and GtH

It is well known that fish reproduction is regulated by the hypothalamus–pituitary–gonad (HPG) axis; despite this knowledge, studies on the effects of hypoxia on fish reproduction, thus far, have almost exclusively focused on sex steroid hormones in gonads. Only very limited information is available on the effects of hypoxia at the hypothalamus and pituitary levels. Whether hypoxia does affect GnRH, GtHs, and their receptors and the manner in which it does so remain unknown.

Lu *et al.* (2007) reported a significant reduction in mRNA of pituitary FSHβ in female zebrafish after exposure to hypoxia (0.6 mg O_2/L, ~8% saturation) for 3 weeks. Thomas *et al.* (2007) reported that levels of plasma LH were below detection limit in Atlantic croakers exposed to hypoxia (in the saline-injected group), but levels of LH became detectable and showed an inverse relationship to oxygen concentration after GnRH injection. In contrast, Wang *et al.* (2008) found a significant reduction in serum LH level when carp was exposed to long-term hypoxia (1 mg O_2/L, ~11% saturation) for more than 2 months.

2.2.3. Neurotransmitters

Lu *et al.* (2006) showed that 3–4-week-old marine medaka (*Oryzias melastigma*) exposed to hypoxia (1.8 mg O_2/L, ~28% saturation) for 3 months until sexual maturity was reached showed a significant reduction in mRNA of both tryptophan hydroxylase (TPH, the rate limiting enzyme of serotonin synthesis) in the brains and FSH receptor in the ovaries of female fish, while no significant changes could be found in GnRH, GnRH receptors, or FSH and LH in the brain of hypoxic males, suggesting that the responses of neurotransmitters to hypoxia may be sex dependent.

A recent attempt has been made to investigate whether the observed disruption of sex steroid hormones by hypoxia may also affect the neuroendocrine function at the brain and pituitary levels (Thomas *et al.*, 2007). Atlantic croakers (*Micropogonias undulatus*) were injected with GnRHa or saline after exposure to normoxia or hypoxia. LH secretion in response to GnRHa injection was significantly attenuated in croakers exposed to hypoxia, showing a decrease in the responsiveness of the pituitary to GnRHa. The expression of GnRH mRNA was also significantly decreased in the preoptic-anterior hypothalamus. Exposure to hypoxia also caused a decrease in serotonin (5-HT) concentration as well as the activity of tryptophan hydroxylase (the enzyme responsible for synthesis of 5-HT) in the hypothalamus. Artificial restoration of hypothalamic 5-HT levels restored neuroendocrine function, indicating that the stimulatory serotonergic neuroendocrine pathway is a major site of hypoxia-induced inhibition. It was further suggested that inhibition of tryptophan hydroxylase activity could be an adaptive mechanism to down-regulate reproductive activity and survive hypoxia (Thomas *et al.*, 2007).

In vitro mammalian studies showed that hypoxia induced the release of catecholamines, acetylcholine, dopamine, and tyrosine hydroxylase (the enzyme regulating the synthesis of dopamine) in pheochromocytoma 12 cells (Kumar *et al.*, 1998; Kumar *et al.*, 2003; Kim *et al.*, 2004), and offered further evidence to support the notion that neurotransmitters are responsive to hypoxia.

2.2.4. Steroidogenic Enzymes

Steroidogenic enzymes are primarily regulated at the transcriptional level under the control of the pituitary gland (Omura and Morohashi, 1995). Thus, any interference with their transcription may alter the production of sex hormones. Increasing evidence shows that genes regulating steroidogenesis are important target sites for various endocrine disrupting chemicals (Thibaut and Porte, 2004; Sanderson, 2006). The synthesis of sex steroid hormones requires molecular oxygen (Raff & Bruder, 2006). As such, hypoxia may be expected to affect steroidogenesis, and hence the production of sex hormones.

Shang *et al.* (2006) showed that 3β-HSD, CYP11A, and CYP19B in 10 dpf zebrafish were significantly down-regulated by hypoxia (at which time expression of CYP19A was still under the detection limit in both normoxic and hypoxic fish). At 40 dpf, all genes investigated were down-regulated in the hypoxia treatment (Figure 3.5). Expression of β actin (the housekeeping gene for normalization) was not affected by hypoxia, indicating that hypoxic effects on these steroidogenic enzymes were specific, but not due to a general down-regulation of metabolism. *Ex vivo* studies on fish ovarian follicles further showed that FSH can stimulate the expression of CYP19 in brown trout (Montserrat *et al.*, 2004), indicating that the suppression of CYP19A expression in the ovary is under the control of FSHβ in the pituitary.

Both *in vitro* and *in vivo* studies in mammals and mammalian cell lines lend support to the postulation that hypoxia can affect the expression level of steroidogenic enzymes. In rat, CYP11A1 was stimulated while steroidogenic acute regulator (StAR) was inhibited under hypoxia (Bruder *et al.*, 2002,

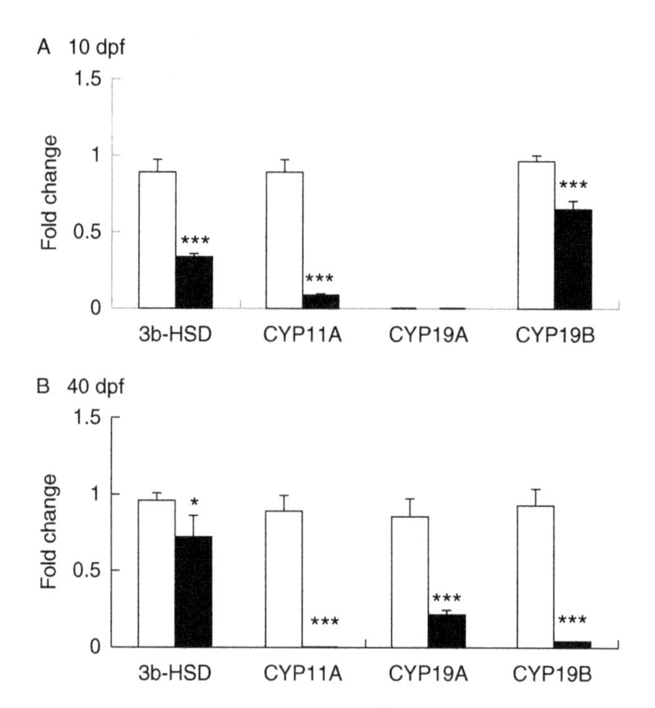

Fig. 3.5. Expression levels of the various sex hormones that control genes in zebrafish at (A) 10 dpf and (B) 40 dpf upon exposure to normoxia (5.8 mg of O_2/L, ~74% saturation) and hypoxia (0.8 mg O_2/L, ~10% saturation) ($n = 4$ replicates, each replicate was pooled from 10 individuals, mean ± SD). Values significantly different from the normoxic control are indicated by asterisks (t-test, * $p < 0.05$, *** $p < 0.001$). (Reproduced from Shang *et al.*, 2006.)

2004, 2005). In sheep, chronic hypoxia represses the expression of CYP11A1 and CYP17 (Myers *et al.*, 2005). *In vitro* studies also showed that hypoxia also reduces the level of CYParom, aldosterone, cortisol, and progesterone receptor in adrenal tissues (Raff *et al.*, 2004).

2.2.5. Sex Hormones

Several studies provided evidence to suggest that hypoxia can disrupt levels of sex hormones, vitellogenin, and triiodothyronine in fish (including carp, zebrafish, Gulf killifish, and Atlantic croakers), and shed light on the underlying mechanisms for the observed reproductive impairments such as retarded gonadal development and a reduction in spawning success, sperm motility, fertilization success, hatching rate, and larval survival.

Serum levels of testosterone (T), estradiol (E2), and triiodothyronine (T_3) were clearly disrupted in carp (*Cyprinus carpio*) upon chronic exposure to hypoxia. A significant increase in T and E2, and a significant decrease in T_3, were clearly evident in male carps exposed to hypoxia for 4 weeks. After 8 weeks of exposure to hypoxia, T and T_3 levels were significantly reduced, but E2 levels increased significantly in male carp. Female carp exposed to hypoxia for 8 weeks showed a significant reduction in serum T, E_2, and T_3 levels (Table 3.1). These hormonal changes were associated with retarded gonadal development in both male and female carp, reduced spawning success, sperm motility, fertilization success, hatching rate, and larval survival (see Figure 3.12), indicating that the adverse effects of hypoxia on reproductive performance resulted from endocrine disruption (Wu *et al.*, 2003).

Chronic exposure of Atlantic croaker (*Micropogonias undulatus*) to hypoxia (1.7 and 2.7 mg O_2/L, ~24 and 38% saturation, respectively) showed dramatic suppression of sex steroid hormones (E2, T, 11-KT), as well as hepatic estrogen receptor and plasma vitellogenin. These hormonal disruptions were clearly related to a decrease in gonadal somatic index, ovarian and testicular development, sperm and egg production, and fecundity (Thomas *et al.*, 2006, 2007). A recent study by Landry *et al.* (2007) also showed similar hormonal disruption and reproductive impairments in the Gulf killifish *Fundulus grandis* exposed to hypoxia (1.34 mg O_2/L, ~19% saturation), in which a 50% reduction in E2 and 11-KT was found in hypoxic females and hypoxic males, respectively. T and VTG, however, remained unchanged in either sex after hypoxic exposure. Female Gulf killifish exposed to hypoxia also produced significantly fewer eggs, and spawning occurred later than in their normoxic counterpart.

A different pattern was observed in the Pacu, *Piaractus brachypomus*, in which plasma T and 11-KT in males, as well as T and E2 in females, were significantly reduced, while 17, 20β-dihydroxy-4-pregnen-3-one (17,20βP) in both sexes remain unchanged when exposed to hypoxia (2.0–4.5 mg O_2/L, ~25–56% saturation) for 3 days. The concentration of spermatozoa, however, was not affected (Dabrowski *et al.*, 2003).

Table 3.1
Hormonal levels (mean ± SEM) in different fish species exposed to hypoxia

Fish species	Sex	Duration of exposure (days)	Mode of study	DO level (mg O_2/L)	% saturation	Temperature (°C)	11-KT (ng/mL)	T (ng/mL)	E2 (ng/mL)	T/E2 ratio	References
Common carp	Immature male	28	Laboratory	7.0	~81	22.5 ± 0.5	N.A.	4.68 ± 1.44	0.04 ± 0.005	117 ± 0.33	Wu *et al.* (2003)
				1.0	~12		N.A.	13.46 ± 2.78 *	0.24 ± 0.035 **	56.08 ± 0.25 ***	
Zebrafish (embryo at blastula stage)	Male	60	Laboratory	5.8	~75	28.5	N.A.	13.4 ± 2.44	10.87 ± 0.79	1.2 ± 0.14	Shang *et al.* (2006)
				0.8	~10		N.A.	3.53 ± 0.3 **	3.27 ± 0.21 ***	1.07 ±0.03	
		120		5.8	~75		N.A.	15.69 ± 5.32	19.42 ± 8.34	1.02 ± 0.17	
				0.8	~10		N.A.	10.75 ± 4.1	13.08 ± 8.09	1.17 ± 0.25	
Zebrafish (embryo at blastula stage)	Female	60	Laboratory	5.8	~75	28.5	N.A.	2.02 ± 0.08	9.4 ± 1.23	0.22 ± 0.02	
				0.8	~10		N.A.	2.37 ± 0.36	5.89 ± 1.1	0.42 ± 0.03 **	
		120		5.8	~75		N.A.	4.34 ± 0.54	9.97 ± 3.33	0.57 ± 0.15	
				0.8	~10		N.A.	6.83 ± 0.68 *	4.33 ± 0.34	1.57 ± 0.05 ***	
Pacu	Mature male	3	Laboratory	5.5–7.5	69–94	26–27.5	37.68 ± 5.48	6.76 ± 0.98	N.A.	N.A.	Dabrowski and Richard (2003)
				2.0–4.5	25–56		5.98 ± 1.61**	0.53 ± 0.1**	N.A.	N.A.	
Pacu	Mature female	3		5.5–7.5	69–94		N.A.	6.1 ± 1.15	7.46 ± 0.32	0.82 ± 0.19	
				2.0–4.5	25–56		N.A.	1.58 ± 0.19**	3.92 ± 0.23**	0.4 ± 0.13	
Gulf killifish	Mature male	30	Laboratory	6.68 ± 2.1	~93	27.1 ± 0.3	0.34 ± 0.076	0.6 ± 0.12	N.A.	N.A.	Landry *et al.* (2007)
				1.34 ± 0.45	~19		0.15 ± 0.03*	0.37 ± 0.03	N.A.	N.A.	
Gulf killifish	Mature female	30		6.68 ± 2.1	~93		N.A.	0.43 ± 0.07	2.99 ± 0.65	0.14 ± 0.27	
				1.34 ± 0.45	~19		N.A.	0.36 ± 0.01	1.27 ± 0.26*	0.28 ± 0.21	
Atlantic croaker	Adult female (1-year-old)	70	Laboratory	5.7	80	23–24	N.A.	N.A.	5.78 ± 0.77	N.A.	Thomas *et al.* (2006)
				2.7	38		N.A.	N.A.	3.45 ± 0.79*	N.A.	
				1.7	24		N.A.	N.A.	0.821 ± 0.22***	N.A.	

Atlantic croaker	Adult male (1-year-old)	70	Laboratory	5.33 ± 0.02	~80	23	5.67 ± 0.33	7.5 ± 0.83	N.A.	N.A.	Thomas *et al.* (2007)
				2.7 ± 0.01	~38		2.67 ± 0.33*	7.92 ± 0.42	N.A.	N.A.	
				1.72 ± 0.01	~24		2.0 ± 0.27*	8.33 ± 0.83	N.A.	N.A.	
Atlantic croaker	Adult male (1-year-old)	N.A.	Field	4.62–5.52	67–80	24.56–25.23 (Oct 2003)	1.38–1.63	1–1.03	N.A.	N.A.	Thomas *et al.* (2007)
				1.2–4.8	18–71		0.88*	0.69	N.A.	N.A.	
				1.32–3.2	19–45		0.63–1.19*	0.5–0.875***	N.A.	N.A.	
				6.65–7.03	97–102	23.59–24.12 (Nov 2003)	1.25–1.5	0.48–0.53	N.A.	N.A.	
				6.68	97		1.25	0.59	N.A.	N.A.	
				2.22–4.72	31–66		0.38–0.63***	0.28–0.45**	N.A.	N.A.	
Atlantic croaker	Adult female (1-year-old)	N.A.	Field	4.62–5.52	67–80	24.56–25.23 (Oct 2003)	N.A.	0.87–1.19	1.63–2.69	0.43–0.52	
				1.2–4.8	18–71		N.A.	0.49*	0.98*	0.5	
				1.32–3.2	19–45		N.A.	0.43–0.76***	0.73–1.47***	0.31–1.31	

Asterisks indicate values in the hypoxic treatments are significantly different from their counterparts in the normoxic control:*, $p < 0.05$; **, $p < 0.01$; ***, $p < 0.001$.

DO, dissolved oxygen; E2, estradiol; 11-KT, 11-ketotestosterone; T, testosterone.

Since hormones in fish also regulate functions other than those involved in reproductive processes, their disruption does more than just impair reproduction and decrease reproductive output and success. It is well known that maternal hormones also play an important role in the development of fish larvae. For example, levels of cortisol (a stress hormone) in female fish can be transferred to the egg yolk and affect larval developmental rates. A field manipulating experiment on damselfish (*Pomacentrus amboinensis*) showed that cortisol levels strongly influenced the yolk size of larvae at hatching, and elevated cortisol levels in the egg reduced larval length. Elevated testosterone also appears to influence yolk utilization rates and increase yolk sac size. Maternally derived cortisol and testosterone have been shown to be important in regulating growth, development, and nutritional reserves of fish embryo and larvae, which may, in turn, affect larval survival and fitness (McCormick, 1998, 1999).

Table 3.1 summarizes the changes in levels of sex steroid hormones (T, E2, and 11-KT), in five species of fish upon exposure to various levels of hypoxia. With a few exceptions, decreases in T, E2, and 11-KT were generally observed in hypoxic fishes, regardless of species and sex, indicating that hypoxia reduces the production of sex steroid hormones, presumably by down-regulating some of the steroidogenic enzymes. A significant increase in the T/E2 ratio was clearly evident in female zebrafish upon exposure to hypoxia (Shang *et al.*, 2006) but not in the other species. A dose–response relationship between hormonal changes and level of hypoxia was also found in the Atlantic croakers (Thomas *et al.*, 2007). Importantly, decreases in the level of sex hormones were associated with reproductive impairments in all these studies.

One of the problems in deciphering the effects of hypoxia on reproduction and development is that hypoxia affects a wide range of physiological and biochemical systems and pathways, and it would be difficult to distinguish between direct and indirect effects of hypoxia. For example, a reduction in metabolism associated with hypoxia will cause many changes as the cell machinery is making adjustments to this new state. These changes are difficult to separate from those hypoxia-induced changes directed at regulating a specific pathway.

2.3. Hypoxia Impairs Fish Reproduction

2.3.1. Reproductive Behaviors

Various laboratory experiments and field studies, to date, have established that many fish species can actively avoid hypoxia (Gray, 1990; Pihl *et al.*, 1991; Wannamaker and Rice, 2000), showing that hypoxia can affect fish behavior. However, only limited studies have shown that hypoxia can also affect reproductive behavior (e.g., mate choice, courtship, reproductive efforts, and investment), which may, in turn, affect reproductive output.

No courtship behavior was observed when male and female carps were reared separately under hypoxia and mixed at the time of spawning, whereas in the normoxic control, male fish followed the females and pushed them with their nose close to the anal papilla about an hour after mixing (Wang *et al.*, 2008), suggesting that normal courtship behavior in fish was affected by chronic hypoxic exposure to hypoxia.

Behavioral studies showed that hypoxia can affect mate choice and reproductive efforts in fish. Female marine gobies (*Pomatoschistus microps*) preferred to spawn with males in nests that already contained eggs that had been spawned earlier by other females. However, this preference was reversed under hypoxia (30% saturation). Under normoxia, females preferred to mate with males with the smallest nest entrance, whereas males exhibiting the signal of willingness to provide parental care would be preferred under hypoxia (Jones and Reynolds, 1999a; Reynolds and Jones, 1999), clearly indicating the adjustment of mate choice in response to selection pressure (in this case, hypoxia) prevailing at different times and in different environments.

Hypoxia can also change reproductive efforts in fish. Under hypoxia (35% saturation), male marine gobies would increase their time and effort in ventilating the eggs, and correspondingly reduced their time in selecting females (Jones and Reynolds, 1999b) (Figure 3.6). During hypoxia, male sand gobies (*Pomatoschistus minutus*) built nests with larger entrances and increased fanning activities to increase oxygen supply to their eggs (Lissåker *et al.*, 2003) (Figure 3.7). In a noncompetitive environment, male mosquito fish (*Gambusia holbrooki*) spent more time following females and increased copulations under hypoxia (15–20% saturation). However, hypoxia had no effect when males were competing for copulations (Carter and Wilson, 2006) (Figure 3.8).

The fact that hypoxia can affect synthesis of sex hormones, while the breakdown products of sex hormones may act as pheromones (Sorensen *et al.*, 2004), suggests the possibility that hypoxia may affect the production of pheromones and hence reproductive behaviors in fish. In goldfish, responses to pheromone is mediated through microvillous olfactory receptor cells (Zippel *et al.*, 1997). In rabbits, it has been shown that the number of olfactory neurons is significantly reduced under hypoxia (Drobyshevsky *et al.*, 2006). Conceivably, hypoxia may also reduce olfactory sensitivity thereby affecting the ability of fish to detect pheromones in the environment. Whether hypoxia can affect fish pheromones or the sensing of pheromone remains unknown, and studies addressing this important topic are required, since even a small change in sex pheromones may lead to reproductive failure in natural fish populations.

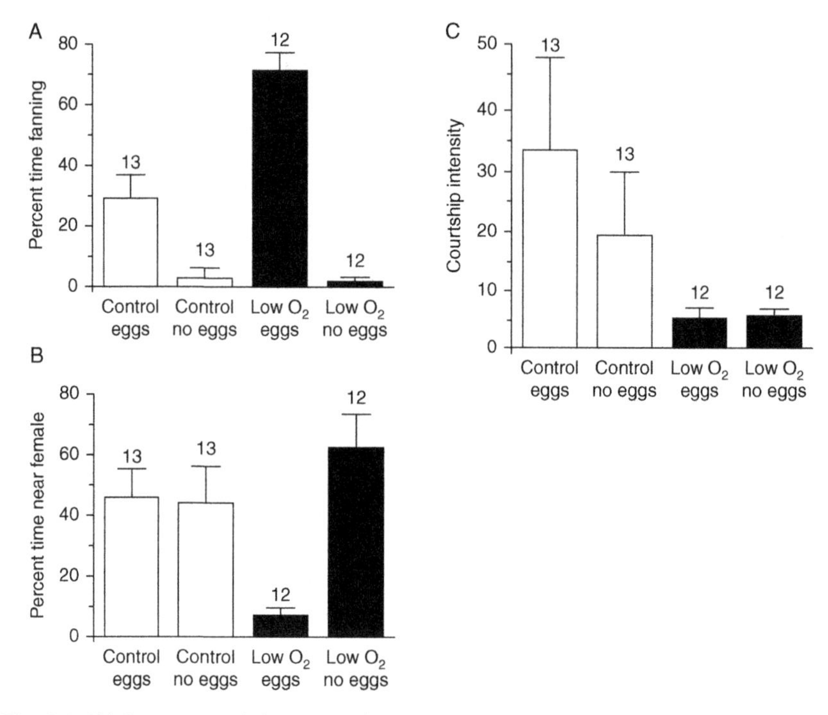

Fig. 3.6. (A) Percentage of time spent fanning by male marine gobies in the presence of a restrained female. (B) Percentage of time spent by males near a restrained female. (C) Intensity of courtship by males during a 20-min observation sessions. Data expressed in mean + SE; numbers above bars are sample sizes. (Reproduced from Reynolds and Jones, 1999.)

2.3.2. Gonad Development and Gametogenesis

Ample laboratory and field evidence shows that chronic exposure to hypoxia can reduce gonad size and retard gametogenesis and gonad development in fish.

The Gonadal Somatic Index (GSI) of adult carp (*Cyprinus carpio*) reared under hypoxia (1 mg O_2/L, ~12% saturation) for 8 weeks was reduced by some 40% and 33% in males and females, respectively (Wu *et al.*, 2003). The GSI of male and female Atlantic croaker (*Micropogonias undulatus*) was reduced by 50% and 75%, respectively, after rearing under hypoxia (1.7 mg O_2/L, ~24% saturation) for 10 weeks (Thomas *et al.*, 2006, 2007). Gulf killifish (*Fundulus grandis*) kept under hypoxia (1.34 mg O_2/L, ~19% saturation) for 1 month had a reduced number of eggs and amount of vitellogenin, and the GSI in females was also significantly lower (Landry *et al.*, 2007). Notably, fish with a higher GSI also produced larvae of larger size with a higher rate of survival (Evans and Geffen, 1998).

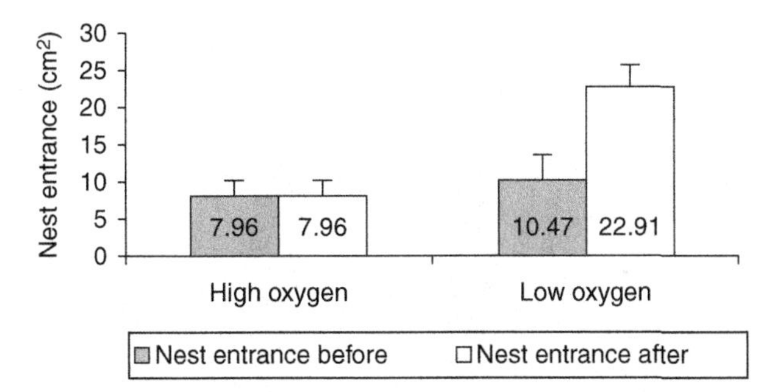

Fig. 3.7. Nest entrance size (mean ± SE) of male sand goby (*Pomatoschistus minutus*) under low (n = 12) and high (n = 15) oxygen regimes. (Reproduced from Lissåker *et al.*, 2003.)

In male carp kept under hypoxia for 8 weeks, a significant reduction in the diameter of testes lobules was observed. Despite the fact that all stages of spermatogenesis can be observed in the testes, the number of spermatocytes (SPC) and spermatids (SPD) was significantly reduced, while a significantly higher number of spermatogonia (SPG) was found, indicating that testicular growth and sperm production was inhibited by hypoxia (Figure 3.9). Similar retardation of gonad development was also found in female carp. The size of gonads was reduced and less yolk deposition was found in each egg. Stage III oocytes were found in 83.3% of all hypoxic females; only 16.7% of hypoxic females carried stage IV oocytes and all hypoxic females failed to produce stage V oocytes. In contrast, eggs were visually observed in all normoxic females, and oocytes in 57.1% of normoxic females reached stage V (Wu *et al.*, 2003).

Retardation of gonad development and gametogenesis were found in zebrafish exposed to hypoxia (0.8 mg/L, ~10% saturation) for 3 months. At 120 dpf, percentages of SPC and SPD were significantly reduced by hypoxia (−46.6% and −36.6%, respectively), while SPG increased by three times in hypoxic males (Figure 3.9). Furthermore, mitosis was commonly observed in normoxic males, but was less common in hypoxic males. In females, oocytes were predominantly in vitellogenic and preovulatory stages in the normoxic fish but were mostly in previtellogenic and vitellogenic stages in hypoxic fish (Figure 3.10). It must be noted that since feeding was also reduced in fish under hypoxia (Zhou *et al.*, 2001), the gonad retardation observed may, in part, be due to reduced feeding.

A marked decrease in mature oocytes and the number of viable eggs was clearly evident when female Atlantic croakers (*Micropogonias undulatus*) were kept under hypoxia (2.7 ppm and 1.7 ppm, ~38% and 24% saturation, respectively) in the laboratory for 10 weeks. Suppression of ovarian and

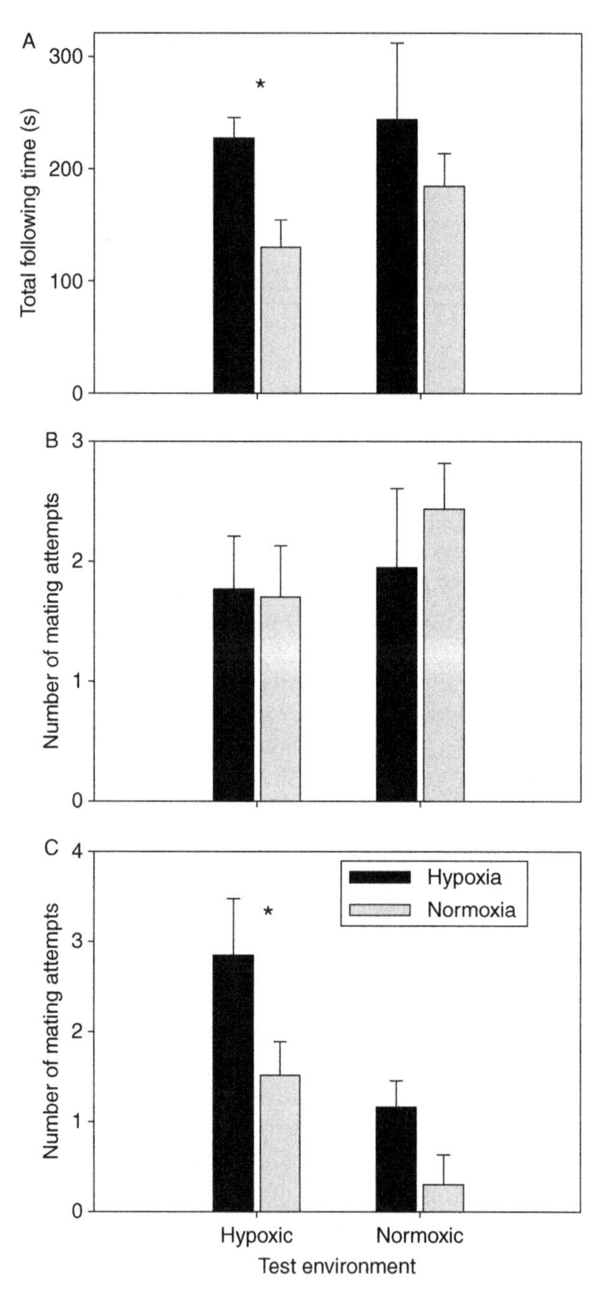

Fig. 3.8. Effect of oxygen on the mating behavior of male mosquito fish, *Gambusia holbrooki*, in a noncompetitive environment. (A) Total time males spent following females, (B) total number of attempted copulations, and (C) total number of successful copulations in 10 min. Data are mean ± SE ($n = 15$). * $p < 0.05$. (Reproduced from Carter and Wilson, 2006.)

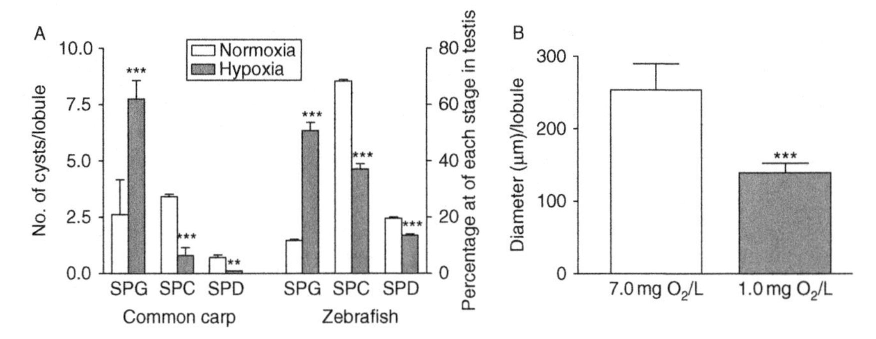

Fig. 3.9. (A) Number of spermatogonia (SPG), spermatocytes (SPC), and spermatids (SPD) in the testis of common carp and zebrafish exposed to hypoxia. Values significantly different from the control are indicated by asterisks (**, $p < 0.01$; ***, $p < 0.001$). (Wu *et al.*, 2003; Shang *et al.*, 2006) (B) Lobule diameter of testes of *C. carpio* upon exposure to 7.0 and 1.0 mg of O_2/L (~81% and 12% saturation, respectively) for 8 weeks (Zhou, 2001). Values significantly different from the control are indicated by asterisks (n = 7–11, mean ± SD) (***, $p < 0.001$).

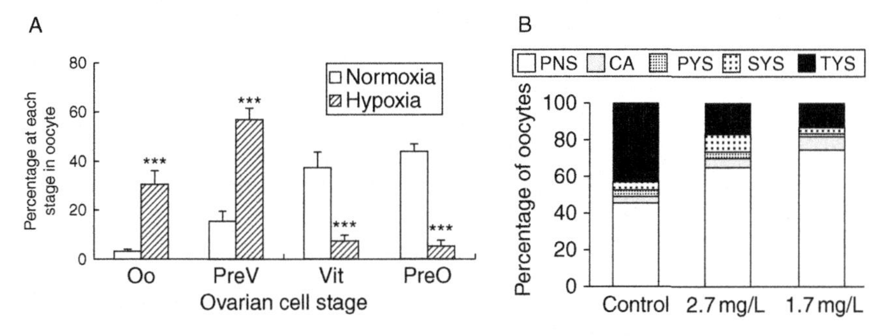

Fig. 3.10. (A) Percentage of oogonia (Oo) and previtellogenic (PreV), vitellogenic (Vit), and preovulatory oocytes (PreO) in female zebrafish after 120 days of development upon exposure to normoxia (5.8 mg of O_2/L, ~75% saturation) and hypoxia (0.8 mg of O_2/L, ~10% saturation), (n = 12–15, mean ± SD). Values significantly different from the normoxic control are indicated by asterisks (t-test, ***, $p < 0.001$). (Reproduced from Shang *et al.*, 2006.) (B) Effects of laboratory hypoxia exposure on ovarian development and endocrine function in female croakers. PNS, peri-nucleolus; CA, cortical alveoli; PYS, primary yolk; SYS, secondary yolk; TYS, tertiary yolk. (Reproduced from Thomas *et al.*, 2007.)

testicular growth was also found in Atlantic croakers collected from hypoxic areas (Thomas *et al.*, 2006, 2007). The study of Landry *et al.* (2007) showed that daily egg production in *F. grandis* was significantly reduced after exposure to hypoxia (1.34 mg O_2/L, ~19% saturation) for 30 days (Figure 3.11).

2.3.3. Quality of Sperm and Eggs

Sperm motility is a reliable predictor for sperm quality and fertilization success (Au *et al.*, 2002). After exposure to hypoxia for 12 weeks, sperm motility (measured by their curvilinear velocity VCL, straight-line velocity VSL, and angular path velocity VAP) was significantly decreased in male carp (Table 3.2), indicating that sperm quality was impaired. All of the normoxic and hypoxic male carp could be induced to spawn using carp pituitary extract; however, the percentage of spawning success in the hypoxic male carp was drastically reduced from 71.4% to 8.3%, clearly demonstrating that the sperm quality produced by males was impaired by hypoxia (Wu *et al.*, 2003).

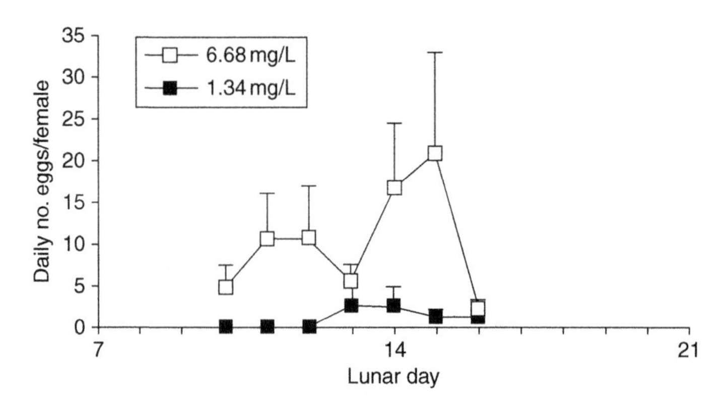

Fig. 3.11. Daily egg production per female *Fundulus grandis* (mean ± SE) exposed to normoxia (6.68 mg O_2/L, ~93% saturation, $n = 7$) and hypoxia (1.34 mg O_2/L, ~19% saturation, $n = 6$) for a 30 days (Reproduced from Landry *et al.*, 2007.)

Table 3.2

Sperm motility of carp after exposure to normoxia (7.0 mg O_2/L, ~81% saturation) and hypoxia (1.0 mg O_2/L, ~12% saturation) for 12 weeks

	7.0 mg O_2/L	1.0 mg O_2/L
VCL	77.42 ± 29.13	46.25 ± 10.83*
VSL	38.83 ± 21.01	10.65 ± 3.89*
VAP	47.69 ± 5.38	21.12 ± 11.41*

Mean ± SD; $n = 6$. The velocity is expressed as micrometers per second.

VCL, mean curvilinear velocity; VSL, mean straight-line velocity; VAP, angular path velocity (Wu *et al.*, 2003).

*Values significantly different from the control (t-test: *, $p < 0.05$).

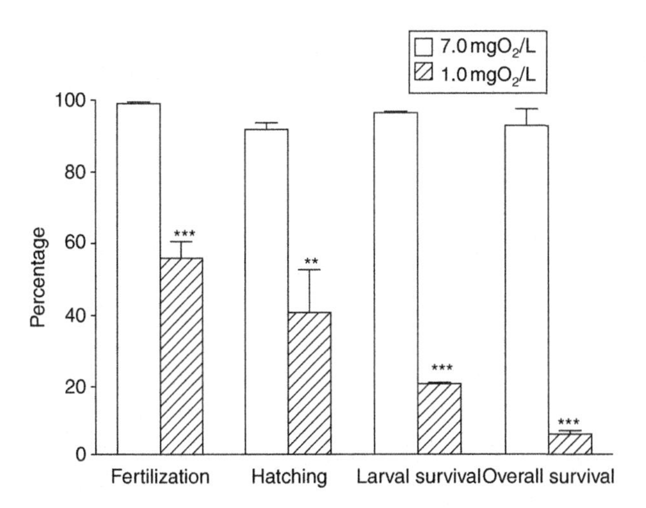

Fig. 3.12. Percentage of fertilization, hatching rate, larval survivorship, and overall survivorship (fertilized egg to 24 h post-hatching) after 12 weeks of normoxic and hypoxic exposure of the parent adult carp (mean ± SD, $n = 6$) (t-test: **, $p < 0.01$; ***, $p < 0.001$). (Reproduced from Wu *et al.*, 2003.)

2.3.4. Spawning, Fertilization Success, and Survival of Larvae

The study of Wu *et al.* (2003) clearly related the decrease in GSI and impaired gametogenesis to sperm quality and subsequently reduced fertilization success and larval survivorship in hypoxic carps. Fertilization success was significantly reduced from 99.4% to 55.5% in hypoxic carps. 98.8% of the fertilized eggs produced by the normoxic carps hatched to larvae, while only 17.2% of fertilized eggs produced by the hypoxic group hatched to larvae. 93.7% of hatched larvae survived in the normoxic group, while larval survival decreased to 46.4% in the hypoxic group 24 h post-hatching. Overall, the survival of fertilized eggs through 24-h-old larvae decreased from 92.3% in the normoxic group to only 4.4% in the hypoxic group (Figure 3.12). Jenkins-Keeran *et al.* (2001) further provided peripheral evidence that oxygen is an important factor in determining sperm motility in fish. Semen from striped bass stored for 48 h under oxygen had a significantly greater percentage of motile sperm (13%) than their counterpart stored under ambient air (9%) or nitrogen (4%).

Hypoxia also significantly delayed the onset of spawning. Female Gulf killifish (*Fundulus grandis*) exposed to hypoxia (1.34 mg/L, ~19% saturation) for 1 month showed a significant delay in their spawning (Landry *et al.*, 2007). Wang *et al.* (2008) showed that although oocytes continued to develop when carp was exposed to long-term hypoxia (1 mg O_2/L, ~11% saturation for more than 2 months), the final oocyte maturation in hypoxic females was significantly retarded, and both ovulation and spawning were inhibited in hypoxic female fish. This was correlated with a significant reduction in serum LH level, indicating that hypoxia may inhibit fish spawning through LH-dependent final oocyte maturation.

3. HYPOXIA AND FISH DEVELOPMENT

It is generally accepted that embryonic and larval development (particularly the gastrula and blastula stages) are the most sensitive stage to stresses in the life cycle of fish (von Westernhagen, 1988; Johnson and Landahl, 1994; Cameron and von Westernhagen, 1997). Normal histogenesis and organogenesis during development rely on a series of intricate, programmed processes in which apoptosis and cell proliferation play a key role (Sanders and Wride, 1995; Jacobson *et al.*, 1997; Vaux and Korsmeyer, 1999). *In vitro* and *in vivo* studies based on mammalian systems provide evidence that hypoxia can induce apoptosis and inhibit cell proliferation (Jung *et al.*, 2001; Saed and Diamond, 2002; Liao *et al.*, 2007; Poon *et al.*, 2007; Lee *et al.*, 2008). Conceivably, hypoxia may also alter cell proliferation and apoptosis in fish, thereby impairing development. In fish, stages of sex differentiation and sex determination have been shown to be particularly sensitive to endocrine disrupting chemicals (Strüssman and Nakamura, 2002). The fact that hypoxia is an endocrine disruptor suggests that hypoxia may also affect sex differentiation and sex determination in fish. Surprisingly, the effects of hypoxia on embryonic and larval development of fish remain largely unknown.

3.1. Regulation of Sex Differentiation, Sex Development, and Sex Determination

Histogenesis and organogenesis during development primarily rely on cell proliferation and apoptosis (reviewed by Vaux and Korsmeyer, 1999; Su, 2000; Lossi *et al.*, 2002). It is widely accepted that intracellular proteins of the Bcl-2 family are involved in the apoptotic signaling pathway; Bcl-2 and Bcl-xL are anti-apoptotic while Bax and Bad are pro-apoptotic (Reed *et al.*, 1996). As such, the ratio of the anti-apoptotic Bcl-2 and the pro-apoptotic Bax is indicative of apoptotic potential (Martin *et al.*, 1995; Misao *et al.*, 1996; Kirshenbaum and de Moissac, 1997; Gross *et al.*, 1998; Saikumar *et al.*, 1998; Cook *et al.*, 1999).

Growth hormone appears to be an important factor in regulating development of teleost fishes. The secretion of growth hormones in fish, in turn, has been shown to be stimulated by neuropeptides, gonadotropin-releasing hormone, growth hormone-releasing hormone, thyrotropin-releasing hormone, neuropeptide Y, serotonin, and pituitary adenylate cyclase-activating polypeptide (Holloway and Leatherland, 1998). Whether hypoxia may affect fish development through directly affecting the secretion of growth hormones or indirectly through affecting their modulating hormones remains completely unknown and warrants further study.

3.2. Hypoxia Impairs Fish Development

3.2.1. Death and Sensitive Window

Early gonad development and reproduction are the two sensitive windows, during which time the HPG axis is particularly susceptible to endocrine disruption caused by chemicals (Ankley and Johnson, 2004). In terms of mortality rate, however, there is no clear evidence to support that a certain life stage would be more susceptible to hypoxia.

Survival times of larval bonefish (*Albula* sp.) in hypoxic sea water (0.68 mg O_2/L, ~10% saturation) decreased from 15 to 5 min over the period of metamorphosis, and this increased sensitivity to hypoxia has been attributed to an increased oxygen demand as metamorphosis advances (Pfeiler, 2001). Susceptibility of small-mouth bass (*Micropterus dolomieui*) to hypoxia also changed with the developmental stage. From the second day to the 10th day after hatching, larvae could not survive for 3 h when exposed to 1 mg O_2/L (~11% saturation), while the majority of the larvae survived under the same conditions after the 11th day (Spoor, 1984). Landman *et al.* (2005), however, found no significant differences in mortality rate between larval and juvenile rainbow trout (*Oncorhynchus mykiss*) and common bully (*Gobiomorphus cotidianus*) when exposed to acute hypoxia (48 h; LC50 values were 1.59–1.62 mg O_2/L, ~16% saturation for rainbow trout parr and fry, and 0.77–0.91 mg O_2/L, ~8% saturation, for bully juvenile and fry).

Dissolved oxygen above 4 mg O_2/L (~38% saturation) did not affect survival of Chinook salmon (*Oncorhynchus tshawytscha*) embryos (Geist *et al.*, 2006). *Scyliorhinus canicula* eggs 13–15 weeks old survived at 50% air saturation and normoxia for 10 weeks. Eggs exposed to 20% air saturation died after 3 weeks, while those exposed to anoxia for 2 h per day died after 10 weeks (Diez and Davenport, 1990). Increased larval mortality rate and reduced hatching success were found for nese (*Chondrostoma nasus*) embryos when exposed to 10% air saturation (Keckeis *et al.*, 1996). Roussel (2007) noted that the survival from fertilization to the end of embryonic development in brown trout (*Salmo trutta*) decreased from 85% to 70% under hypoxia (3.0 mg O_2/L, ~26% saturation). Low oxygen levels (2.0–4.5 mg O_2/L, ~25% to 56% saturation) reduced the survival of embryos of Pacu, *Piaractus brachypomus* (17.3 % in hypoxia as compared with 68.5 % in normoxia) (Dabrowski *et al.*, 2003).

Table 3.3
Viability of different developmental stages of zebrafish in anoxia

Period	Hours after Fertilization*	Percent alive after 24 h of anoxia (N)
Cleavage	2	83.1 (89)
Blastula	4	83.2 (85)
Gastrula	6	97.7 (90)
Segmentation	13	98.8 (85)
Straightening		
Early	25	64.0 (100)
Middle	30	4.4 (91)
Hatching	50	0 (130)

N is total number of embryos.
*hpf when placed in the anoxic environment.
Reproduced from Padilla and Roth (2001).

The study of Padilla and Roth (2001) demonstrated that the susceptibility of zebrafish embryos to anoxia varies considerably with their developmental stage. Most zebrafish embryos before 25 hpf can survive 24 h of anoxia. Tolerance was reduced as embryos developed to the period of straightening (30 hpf), and fish after hatching (beyond 50 hpf) became very sensitive to anoxia (Table 3.3).

A summary of the mortality rate of embryos and larvae of various fish species in response to hypoxia is given in Table 3.4. Clearly, hypoxic tolerance is species specific, which may be related to the ecology and natural habitat of the species. Despite the fact that certain life stages would be more sensitive to hypoxia for a given species, no generalization could be made on life stage specificity across fish species. Hypoxic tolerance also varies considerably according to oxygen levels and duration of exposure. Furthermore, it appears that the hypoxic window for death is very narrow. Above certain oxygen levels, the fish are able to make physiological and biochemical adjustments to survive, but death sets in very rapidly when oxygen levels fall below the threshold beyond which they are incapable of making these adjustments.

3.2.2. Development and Hatching

Earlier studies have shown that hypoxia can retard embryonic development in the Atlantic salmon (*Salmo salar*) and rainbow trout (*Oncorhynus mykiss*), leading to an increase in mortality rate and prematurely hatched embryos through stimulating chorionase secretion (Hamor and Garside, 1976). Subsequent studies have shown that hypoxia has a profound effect on the rate of embryonic development in many fish species (Rombough, 1988).

Table 3.4
Summary of mortality of embryo and larvae of different fish species at different developmental stages under hypoxia

Fish species	DO level	Duration of exposure (days)	Stage	Mortality (%)	Reference
Dogfish (*Scyliorhinus canicula* L.)	50% air saturation	70	13–15 weeks post-fertilization	0	Diez and Davenport (1990)
	20% air saturation	21		100	
	0% for 2h/day	70		100	
Nase (*Chondrostoma nasus*)	10% air saturation	1.5	Fertilized egg to gastrula	7.6 (at hatching)/37.7 (15 dph)	Keckeis *et al.* (1996)
		3.5	Gastrula to eyed stage	8.9 (at hatching)/98.5 (15 dph)	
		5	Fertilized egg to eyed stage	100 (at hatching)	
		8	Eyed stage to hatching	6.2 (at hatching)/100 (15 dph)	
		13	Fertilized egg to hatching	100 (at hatching)	
Pacu (*Piaractus brachypomus*)	2.2 ± 0.5 mg O_2/L (~27% saturation)	N.A.	Embryo	ca. 80	Dabrowski and Richard (2003)
Chinook salmon (*Oncorhynchus tshawytscha*)	4 mg O_2/L (~38% saturation)	40	Embryo	0	Geist *et al.* (2006)
Brown trout (*Salmo trutta*)	3 mg O_2/L (~26% saturation)	110	From fertilization to end of embryonic development	30	Roussel (2007)

DO, dissolved oxygen.

Padilla and Roth (2001) showed that although embryos of zebrafish, *Danio rerio*, can survive anoxia for 24 h, embryos entered into developmental arrest under anoxia, with all movement, cell division, developmental progression, and heart beats ceasing, presumably as adaptive features for energy conservation. No cells were arrested in mitosis, and flow cytometry analysis further revealed that blastomeres were arrested during the S and G2 phases of the cell cycle. Development of zebrafish embryos, however, resumed upon return to normoxia. Shang and Wu (2004) showed that development of zebrafish embryos was clearly delayed when kept under 0.5 mg O_2/L (~6.4% saturation), and took twice as long to develop when compared to the normoxic embryos. Similarly, development of freshwater medaka, *Oryzias latipes*, embryos was retarded upon exposure to hypoxia (0.8 mg O_2/L, ~10% saturation) (Cheung and Wu, 2006).

Berntsen *et al.* (1990) showed that natural hatching in mature salmon eggs was induced by hypoxia, and Jobling (1995) further postulated that hypoxia caused by insufficient diffusion of ambient oxygen across the chorion (to meet oxygen requirements of the developing embryo) may also trigger hatching. When embryos of the nase, *Chondrostoma nasus*, were exposed to hypoxia (10% of air saturation), the hatching period was prolonged from 2.7 days (in the normoxic control) to 4.2–5.3 days, and hatching success was also reduced (Keckeis *et al.*, 1996). Roussel (2007) also reported that hatching in the brown trout, *Salmo trutta*, was delayed from 2–4 days to 5–10 days when the oxygen level was lowered to 3.0 mg O_2/L (26% saturation). Hatching in the Chinook salmon, *Oncorhynchus tshawytscha*, was related to oxygen concentration; fish developed under 4 mg O_2/L (~38% saturation) required 6–10 days longer to hatch, and up to 24 days longer to emerge, when compared with the normoxic control (Geist *et al.*, 2006).

Delays in hatching were often accompanied by impaired development and a lower quality of the offspring. Shang and Wu (2004) showed that the body length of zebrafish hatched under hypoxic conditions was significantly shorter than fish in the normoxic control, and further postulated that a smaller body size may possibly reduce the fitness of adult fish in their natural environment. Eggs of *Syliorhinus canicula* kept under anoxia and hypoxia (0% and 5% saturation) showed retarded growth and reduced proteolytic activities as compared to their normoxic counterpart (Diez and Davenport,1990). Massa *et al.* (1999) reported that not only was hatching delayed, but also a smaller body size of alevins, lower content of water, and lower yolk-sac conversion rate were found when brown trout (*Salmo trutta*) eggs were allowed to develop under low oxygen levels (3 mg O_2/L) for 3 weeks after fertilization. The embryos of *S. trutta* grew more slowly and progressed through delayed hatching under hypoxia; however, both normoxic and hypoxic fish reached similar body sizes when yolk-sac absorption was completed. However, the swimming activity of fish hatched from hypoxic embryos was reduced by 20% and suffered from a 14% higher predation rate compared with normoxic groups (Roussel, 2007) (Figures 3.13 to 3.15).

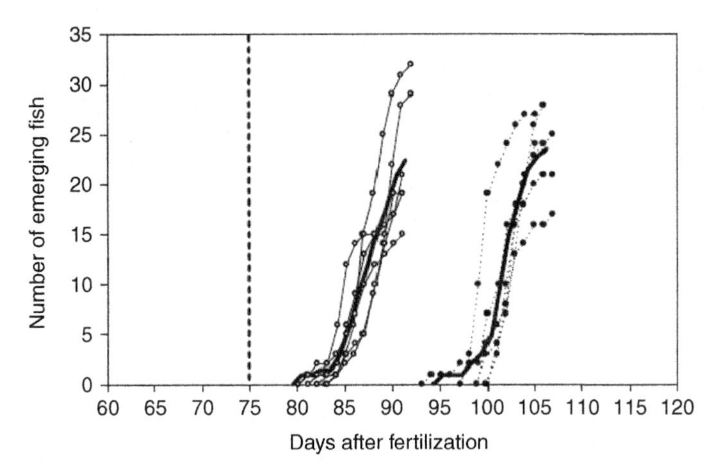

Fig. 3.13. Emergence time (in days after fertilization) of *Salmo trutta* alevins in each channel, for alevins incubated as normoxic embryos (○ and solid lines) or hypoxic embryos (• and broken lines). Bold lines represent average profiles of emergence for each treatment. Vertical broken line indicates the date at which embryos were transferred into channels. (Reproduced from Roussel, 2007.)

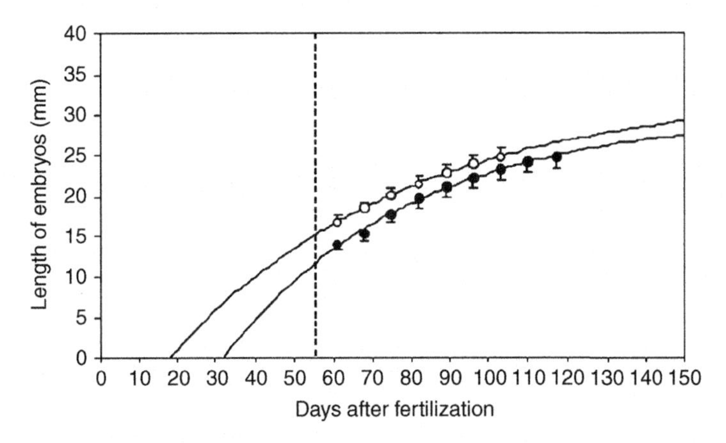

Fig. 3.14. Embryonic growth in brown trout (*Salmo trutta*) from hatching to complete yolk absorption when incubated under normoxia (○) or hypoxia (•). Results are given as mean ± SD, with von Bertalanffy growth models plotted as solid curves. Broken line indicates the beginning of hatching. (Reproduced from Roussel, 2007).

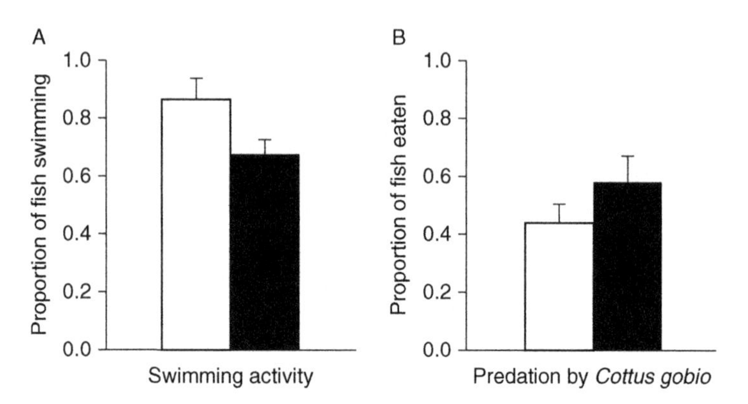

Fig. 3.15. (A) Proportion of alevins of *Salmo trutta* that swim in the water column, and (B) proportion of alevins eaten by the sculpin, *Cottus gobio*, in experimental channels, after being exposed to hypoxia (solid bars) or normoxia (open bars) as embryos (mean ± SD). (Reproduced from Roussel, 2007.)

The above results demonstrated that exposure to hypoxia during development can have carry-over effects on later parts of the life cycle, and may reduce the fitness of adults in the natural habitat, although supporting field evidence is still not available. In many animals, both spawning and hatching is synchronized with environmental factors (e.g., temperature, food availability, density of predators) prevailing in the natural habitats so as to maximize the chance of survival for the juveniles (Smyder and Martin, 2002; Speer-Blank and Martin, 2004; Warkentin, 2007). The ecological consequence of delayed hatching caused by hypoxia is not known.

3.2.3. Malformation

It is well known that hypoxia can cause deformities in fish. 100% of the hatched larvae of *Chondrostoma nasus* were deformed when the fertilized eggs were developed under 10% of air saturation from gastrula to eyed stage and from eyed stage to hatching (Keckeis *et al.*, 1996). A high proportion (20–53%) of female eelpouts (*Zoarces viviparous*) showed developmental defects, including spinal and craniofacial defects, eye lesions or loss of eyes, in broods in Danish fjords receiving domestic and industrial effluents resulting in serious oxygen depletion (Strand *et al.*, 2004). It is also interesting to note that a higher occurrence of malformed fish larvae has been generally reported in polluted areas (Au, 2004), although this increase may not necessarily be attributable to hypoxia because polluted areas are also often contaminated with a variety of chemicals including teratogens and endocrine-disrupting chemicals.

Zebrafish embryos developing under hypoxia lost their normal synchronization, with their tails developing much faster than their heads. External abnormalities such as spinal deformity (predominantly manifesting as altered axial curvature) were also clearly evident (Figure 3.16). Many embryos also failed to develop their vascular systems after several days and died. After 96 h, the percentage of fish with malformations in the hypoxic treatment group was significantly higher than that of the normoxic control (Shang and Wu, 2004) (Figure 3.17).

Most teratogens exert a marked effect during certain stage(s) of embryonic development. Likewise, hypoxia may have different effects when

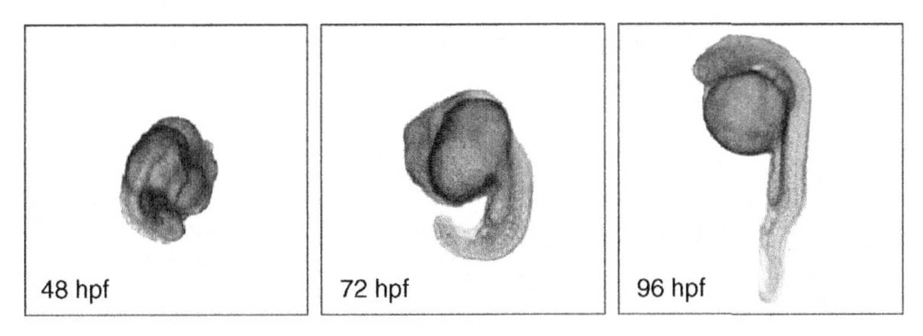

Fig. 3.16. Typical examples of malformation in zebrafish caused by hypoxia (0.8 mg O_2/L, ~10% saturation) at 48 hpf, 72 hpf, and 96 hpf. (Reproduced from Shang, 2005.)

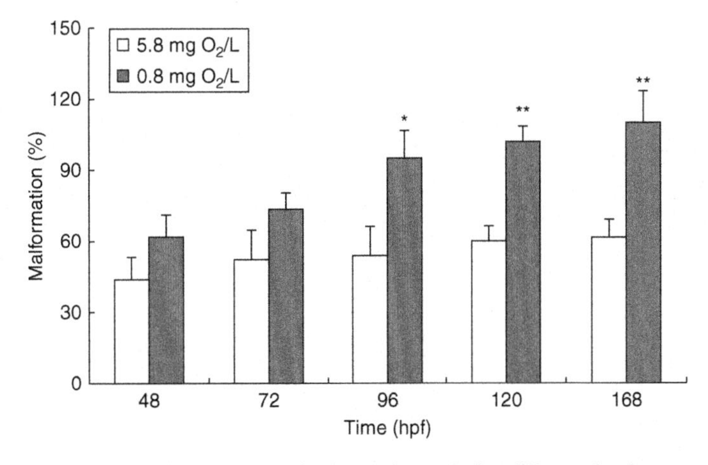

Fig. 3.17. Percentage malformation in zebrafish embryos during different developmental stages (48, 72, 96, 120, and 168 hpf) upon exposure to 5.8 and 0.8 mg O_2/L (~75% and 10% saturation, respectively). Values significantly different from the normoxic control are indicated by asterisks ($n = 100$, mean ± SD) (*, $p < 0.05$; **, $p < 0.01$). (Reproduced from Shang and Wu, 2004.)

administered at different oxygen levels and developmental stages, and different organs may have different "critical windows" during which development is most sensitive and susceptible to hypoxic assault (Burggren, 1999). Conceivably, hypoxia occurring at early developmental stages (organogenesis) may affect fish development more seriously than that occurring during histogenesis at later stages of development. Further study is required to determine the critical window of hypoxia that affects the different stages of gonad development, in order to provide a better understanding of the molecular basis of how hypoxia might affect sex differentiation and determination in fish.

During normal embryonic development, excess cells are commonly removed by apoptosis, and apoptosis is an essential mechanism for normal remodeling and morphogenesis. Hypoxia has been shown to induce *in vitro* apoptosis in a variety of cell types (Schroedl *et al.*, 2002; Wang *et al.*, 2004; Gozal *et al.*, 2005; Lu *et al.*, 2005; Lee *et al.*, 2005; Zhao *et al.*, 2007) and *in vivo* systems (Shin *et al.*, 2004; David and Vert, 2004; Nagai *et al.*, 2007). As such, disruption in apoptosis and change in apoptotic pattern may lead to subsequent malformation in fish. Shang and Wu (2004) demonstrated for the first time that patterns of apoptosis during fish development can be altered by hypoxia. Compared with the normoxic control, apoptotic cells in the tail of hypoxic embryos were significantly reduced (−63.7%). In contrast, a significantly higher percentage (+116%) of apoptotic cells was found in the head region of hypoxic embryos as compared with control embryos (Figure 3.18).

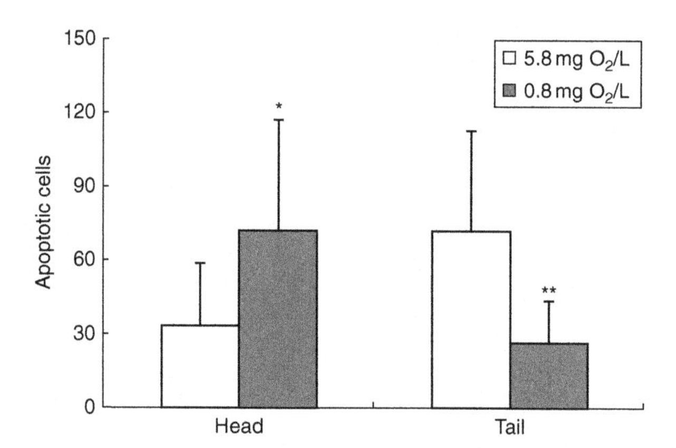

Fig. 3.18. Number of apoptotic cells at 24 hpf in zebrafish embryos upon exposure to 5.8 and 0.8 mg O_2/L (~75% and 10% saturation, respectively). Values significantly different from the control are indicated by asterisks (n = 10, mean ± SD) (t-test: *, $p < 0.05$; **, $p < 0.01$). (Reproduced from Shang and Wu, 2004.)

Concomitantly, a significantly higher ratio of Bax/Bcl-2 was found in the head and a lower ratio of Bax/Bcl-2 in the tail, thus offering further molecular basis to support the observed malformation of hypoxic zebrafish found in the same study (Shang, 2005). The results clearly demonstrated that the apoptotic pattern in zebrafish embryos was altered by hypoxia, thus offering a molecular basis to support the observed malformation in zebrafish caused by hypoxia.

The mechanisms by which hypoxia induces apoptosis are not well understood. Malhotra and Brosius (1999) indicated that hypoxia can trigger apoptosis in different cell types in a way similar to other stresses, and different types of apoptosis (viz., phylogenetic apoptosis, morphogenetic apoptosis, and histogenetic apoptosis) and pathways (e.g., the mitochondrial pathways and the death receptor (Fas-Fasl) pathways) may be involved in histogenesis and organogenesis (Sun *et al.*, 2002; Ribeiro *et al.*, 2003; Adachi-Yamada and O'Connor, 2004; Laurikkala *et al.*, 2006). Normal embryonic development (including brain development and spinal formation, which have been shown to be affected by hypoxia) is regulated by an intricate process of cell proliferation and apoptosis. The fact that apoptosis was affected by hypoxia implies that this intricate process might be disrupted, and provides an explanation of the observed deformities in hypoxic fish. However, in what way hypoxia may affect apoptosis and also the exact relationship between alteration of apoptosis and delayed brain development and spinal deformities remains unclear (Shang and Wu, 2004). In sturgeons (Acipenser shrenckii) that exposed to hypoxia (15% saturation) for 30 min and recovered for 6 h and 30 h, the number of apoptotic cells in the retina, optic tectum, pituitary, and spinal cord showed a significant increase. However, the olfactory lobe, cerebellum, and pons/medulla had relatively few apoptotic cells, showing a differential pattern of apoptosis in response to hypoxia in the central nervous system of fish (Lu *et al.*, 2005). Poon *et al.* (2007), however, found no change in apoptotic rate in liver after carps were exposed to hypoxia (0.5 mg O_2/L, ~6% saturation) for 42 days but extensive DNA damage was found in liver cells. Whether DNA damage resulting from hypoxic exposure may subsequently lead to malformation remains unknown.

Hypoxia (10.3–16.6% saturation) occurring during somitogenesis can cause major vertebral deformity (centrum defect) in the red sea bream (*Pagrus major*), but the 2-cell stage to the blastula stage and gastrula stages were not sensitive to hypoxia (Hattori *et al.*, 2004), thus lending support to the hypothesis that there is a critical window of hypoxic effects on embryonic development. Centrum defects can also occur when eggs are exposed to extremely low oxygen concentrations, even for a brief period of time. For example, somitic disturbances were found in newly hatched larvae of *Pagrus major* upon exposure to anoxia and 10% saturation for only 10 and 120 min, respectively (Sawada *et al.*, 2006) (Figures 3.19 and 3.20).

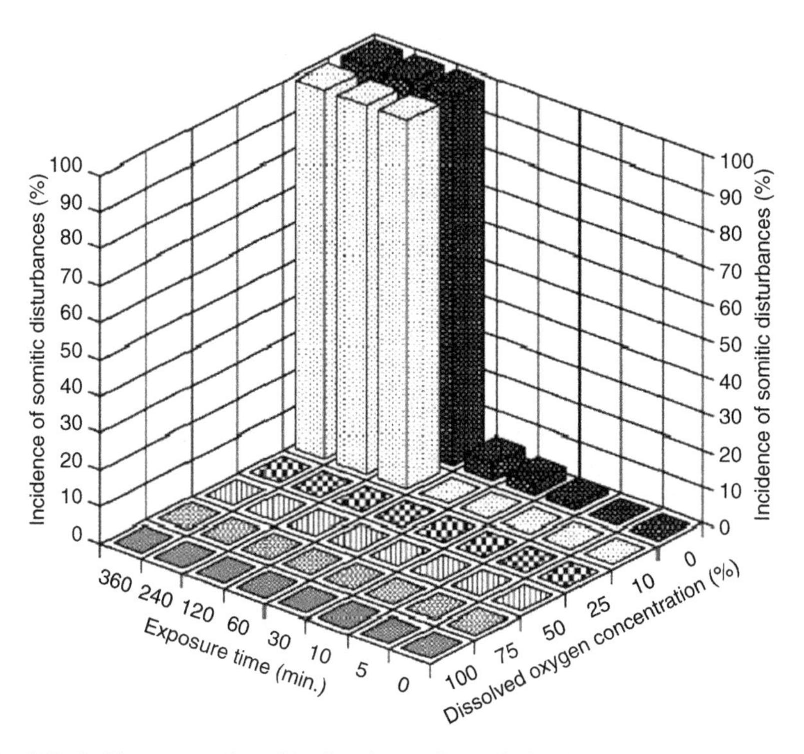

Fig. 3.19. Incidence rate of somitic disturbances in newly hatched larvae of *Pagrus major* induced by exposure to hypoxic conditions. (Reproduced from Sawada *et al.*, 2006.)

Only limited studies have been carried out to decipher the underlying cellular and molecular mechanisms of hypoxia in embryonic development. Kajimura *et al.* (2005) showed that hypoxia strongly induced the expression of insulin-like growth factor binding protein (IGFBP)-1 in zebrafish, but not the expression of insulin-like growth factors (IGFs), IGF receptors, or other IGFBPs, showing that the target of hypoxic effect is specific rather than general. Overexpression of IGFBP-1 resulted in retardation of growth and development under normoxia, while knockdown of IGFBP-1 significantly alleviated the hypoxia-induced growth retardation and developmental delay; the effects were restored by reintroduction of IGFBP-1 to the IGFBP-1 knocked-down embryos. *In vitro* studies using cultured zebrafish embryonic cells showed that IGFBP-1 itself is not mitogenic but can inhibit IGF-1- and IGF-2-stimulated cell proliferation. This inhibitory effect was removed when IGF-1 or IGF-2 was added, suggesting that IGFBP-1 inhibits embryonic development by inhibiting the activities of IGFs.

In zebrafish larvae, cardiac activity was reduced and the formation of blood vessels in various tissues enhanced during early development upon

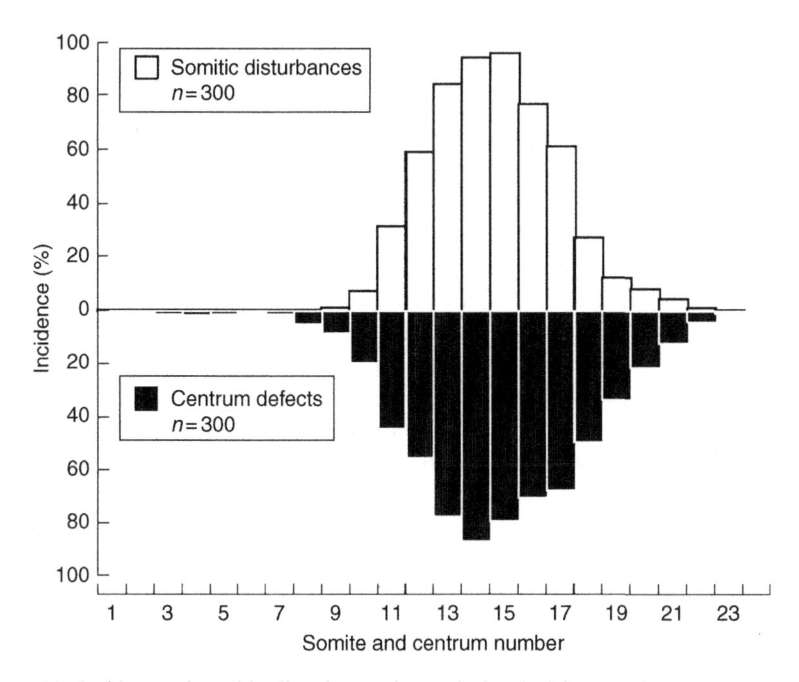

Fig. 3.20. Incidence of somitic disturbances in newly hatched larvae of *Pagrus major* and centrum defects in juveniles exposed to hypoxia (10% saturation) for 240 min during somitogenesis. (Reproduced from Sawada *et al.*, 2006.)

chronic exposure to hypoxia (0.83 mg O_2/L, ~10% saturation) for 7 days (Pelster, 2002), probably due to the up-regulation of vascular endothelial growth factor (VEGF). The reduction in circulation may render the supply of oxygen and nutrient insufficient for the metabolic demand of development, leading to developmental arrest. Shang and Wu (2004) found that the heart rate of zebrafish embryos reared under hypoxia (0.8 mg O_2/L, ~10% saturation) showed an initial increase at 96 hpf and became significantly lower than that of the control embryos at 288 hpf. However, Bagatto (2005) reported that development of zebrafish under severe hypoxia (0.8 mg O_2/L, ~10% saturation) showed a delayed onset of cardiovascular regulation.

3.2.4. Disruption of Hormones

Using zebrafish as a study model, Shang *et al.* (2006) demonstrated that levels of T and E2, as well as the T/E2 ratio, which are critical in modulating developmental processes, can be affected by hypoxia as early as 48 hpf, long before the occurrence of sex differentiation. At 60 dpf, T and E2 were reduced by 73.7% and 69.9%, respectively, in hypoxic males while no significant difference in either hormone was observed between hypoxic and normoxic females. After 120 days, T concentrations increased by 57.4% in hypoxic females, while no change was observable between hypoxic and normoxic males. No significant difference in E2 could be found in either males or females from the normoxic control and the hypoxic treatment after 120 days. At 60 and 120 dpf, significant increases in the T/E2 ratio were clearly evident in hypoxic females (+90.9% and +175.4%, respectively). No change in T/E2 ratio, however, was found in male fish, showing that the hormonal disruption is sex specific (Table 3.1 and Figure 3.21).

Studies on rainbow trout (Tanaka *et al.*, 1992), medaka (Fukada *et al.*, 1996), and tilapia (Chang *et al.*, 1997) have shown that mRNA levels of CYP19 correlated well with aromatase activity, and changes in CYP19 gene expression in the gonad of zebrafish has been shown to associate with alterations of gonadal differentiation (Fenske and Senger, 2004).

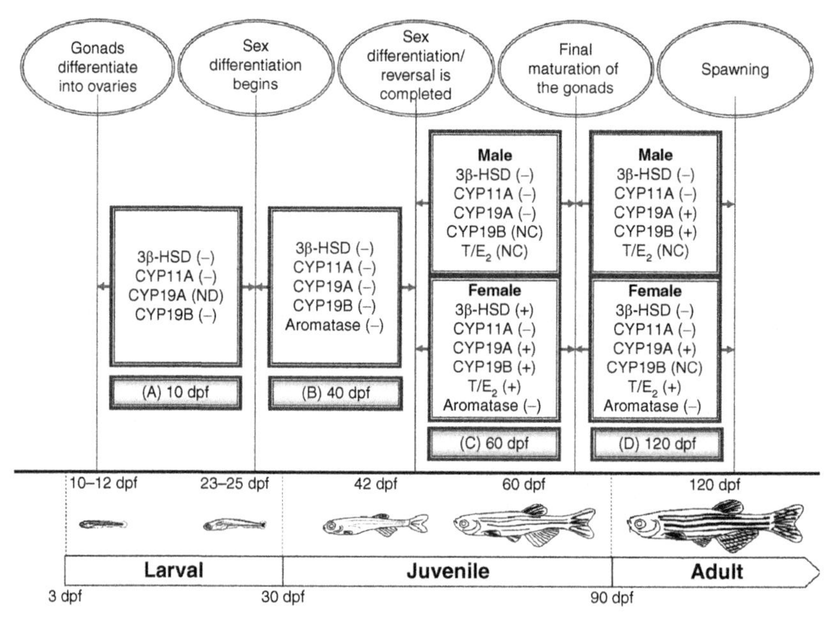

Fig. 3.21. Summary diagram showing changes in expression of various sex hormone control genes, ratio of testosterone/estradiol (T/E2), and CYParom activity with respect to key stages of gonad development in zebrafish exposed to hypoxia (0.8 mg O_2/L, ~10% saturation) and normoxia (5.8 mg O_2/L, ~75% saturation) at 10, 40, 60, and 120 dpf. (+), significant increase in hypoxic treatment with respect to normoxic control; (–), significant decrease in hypoxic treatment with respect to normoxic control; (ND), not detectable; (NC), no significant change between normoxic control and hypoxic treatment. (Modified from Shang *et al.*, 2006.)

Shang *et al.* (2006) set out to test the hypothesis that hypoxia can disrupt genes controlling steroidogenic enzymes and sex hormones, thereby affecting sex differentiation and sex determination in zebrafish. In their study, down-regulation of CYP19B was found in hypoxic fish at 10 dpf, when gonads started to develop into ovaries, and down-regulations of both CYP19A and CYP19B as well as marked reduction of E2 production were found at 40 dpf during sex differentiation/reversal. Both 3β-HSD and CYP11A were significantly down-regulated by hypoxia at 10 and 40 dpf, suggesting a reduction in steroidogenesis during sexual differentiation and before sex determination is completed in the hypoxic group. Decreases in expression of steroidogenic enzymes and production of T and E2 provide a plausible mechanism for the retardation of gametogenesis, which was subsequently observed in both hypoxic males and females. The changes in sex hormone levels, expression levels of various genes controlling steroidogenesis, and aromatase activities with respect to each key stage of gonad development under hypoxia are summarized in Figure 3.21.

Vitellogenin (VTG) production was markedly reduced in both male and female zebrafish developed under hypoxia for 60 and 120 days. In hypoxic females, VTG was markedly reduced by 84.6% at 60 dpf and by 97.6% at 120 dpf. Similarly, VTG was reduced by 78.9% at 60 dpf and 80.6% at 120 dpf in hypoxic males (Figure 3.22). The reduction of VTG correlated well with the retardation of oocyte development and egg production observed upon exposure to hypoxia in the same study (Shang, 2005).

3.2.5. Sex Differentiation, Sex Determination, and Sex Ratio

It is generally believed that sex differentiation in fish is similar to mammalian systems whereby the presence or absence of a testis-determining factor directs male or female differentiation (Jobling, 1995). The balance of sex steroid hormones is important in determining sex differentiation (Kime, 1998), and phenotypic sex of fish may be influenced by various external factors and chemicals regardless of their genotypic sex (Jalabert *et al.*, 2000), especially before gonadal differentiation.

A specific ratio of T/E2 is required for sexual differentiation, and alteration of this ratio can impair gonadal development (Hileman, 1994). Shang *et al.* (2006) designed an experiment to test the hypothesis that hypoxia can alter the balance of sex hormones in fish, which subsequently affects sex differentiation, sex determination, and sex ratio. In their experiment, zebrafish eggs were kept under normoxia (5.8 mg O_2/L, ~75% saturation) and hypoxia (0.8 mg O_2/L, ~10% saturation) for 4 months until they hatched and developed into sexually mature adults. The results showed that chronic exposure to hypoxia can affect sex differentiation during development.

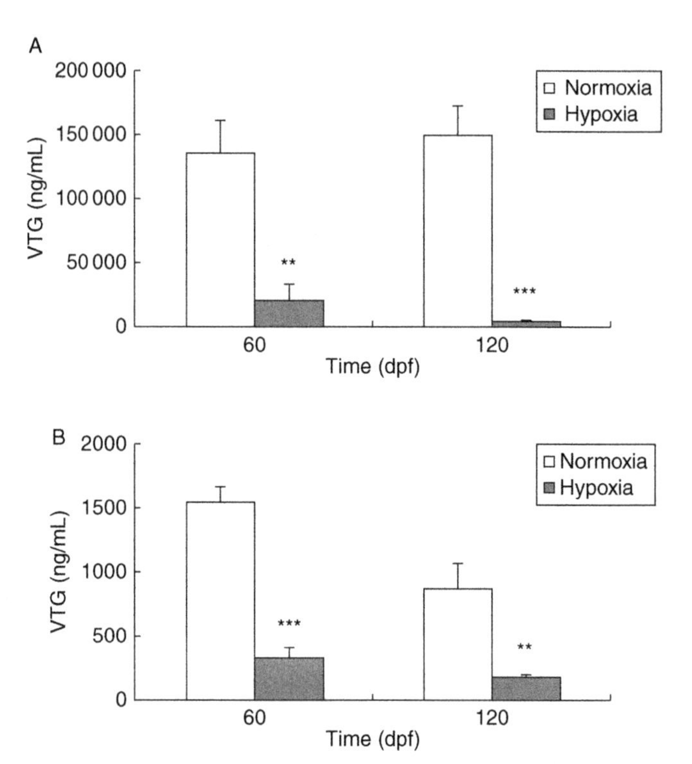

Fig. 3.22. VTG level in (A) female and (B) male zebrafish at 60 dpf and 120 dpf upon exposure to normoxia (5.8 mg O_2/L, ~75% saturation) and hypoxia (0.8 mg O_2/L, ~10% saturation), ($n = 5$, mean ± SE). Values significantly different from the normoxic control are indicated by asterisks (t-test: **, $p < 0.01$; ***, $p < 0.001$). (Reproduced from Shang, 2005.)

Impairment of ovarian development and yolk deposition was found in hypoxic females, which was clearly associated with a decrease in E2 and VTG (Shang *et al.*, 2006). Sex determination was altered, resulting in a male-biased population in the F1 generation (74.4% males in the hypoxic groups versus 61.9% males in the normoxic groups). The fact that no deaths occurred in the hypoxic treatment group after 7 days, long before sex differentiation occurred (10–12 dpf), indicated that the biased sex ratio under hypoxia was not due to differential mortality rates between different sexes. Experimental evidence was further provided to show that the hypoxic effect on sex change was mediated through down-regulations of various genes controlling the synthesis of sex hormones (i.e., 3β-HSD, CYP11A, CYP19A, and CYP19B), leading to changes in levels of T and E2 in female fish at key developmental stages (Figure 3.21). From 60 days onward, the

T/E2 ratio showed a significant increase in hypoxic females but not in hypoxic males, showing that females were more susceptible and that P450arom was inhibited by hypoxia. Taken together with (1) lower E2 and higher T levels found in hypoxic females as compared with normoxic females and (2) reductions in both E2 and T, while the T/E2 ratio remained unchanged in hypoxic males, it appeared that the disruption of the balance between E2 and T could be a major factor contributing to the observed male-biased population in the hypoxic treatment group.

Another possible mechanism leading to a male-biased sex ratio may involve oocyte apoptosis. Both *in vitro* and *in vivo* studies on mammalian systems provide evidence that hypoxia can induce apoptosis (Jung *et al.*, 2001; Saed and Diamond, 2002; Shin *et al.*, 2004). Uchida *et al.* (2002) showed that the disappearance of large numbers of oocytes in male zebrafish during their normal transition from ovary-like tissue to testicular tissue was mediated through apoptosis. Likewise, large numbers of apoptotic early diplotene oocytes and ovarian follicles have been reported in developing male rainbow trout and *Astyanax bimaculatus lacustris* (Janz and Van der Kraak, 1997). The fact that hypoxia could alter the apoptotic pattern of zebrafish as early as 24 hpf (see Section 3.2.5; Figure 3.18) suggests that hypoxia may also alter the scheduled oocyte apoptosis in designated females during sex differentiation and favor the formation of testicular tissues, leading to a male-biased sex ratio in the F1 generation.

In male medaka (*Oryzias latipes*) in which a sex-determining gene, *DMY*, has been found on the Y chromosome (Matsuda *et al.*, 2002), 77% of genotypic XX females reared under hypoxia developed into phenotypic males, while sex change was not found in genotypic females in the normoxic control group (Cheung and Wu, 2006). This is consistent with the findings of Shang *et al.* (2006), who found that hypoxia caused sex change in zebrafish and that hypoxia can also alter sex differentiation and sex determination in species with a sex-determining gene.

The studies of Matsuda (2003) and Hattori *et al.* (2007) showed that sexual development and determination in fish are, in part, determined by germ cell differentiation occurring at early embryonic stages. During sex differentiation, primordial germ cells may differentiate into oogonia or spermatogonia, while the supporting cells may differentiate into granulosa or Sertoli cells in the ovary and testis, respectively (Devlin and Nagahama, 2002). As such, the differentiation of germ cells and their supporting cells during developmental stages may play an important role in gonad differentiation and hence sex determination. Gimeno *et al.* (1996, 1997) showed that male common carp exposed to 4-tert-pentylphenol during the critical period of sex determination (24–51 dph) had a reduced number of primordial germ cells (PGC), which affected gonadal structure, including the induction of oviduct formation. Hypoxia may affect the production and migration of PGC in a similar way, although this has not been demonstrated.

Despite laboratory results showing that hypoxia can lead to a biased F1 generation in two different species, field evidence showing that hypoxia may affect sex determination and sex ratio of fish under naturally occurring hypoxia is not available. Furthermore, so far there has been no attempt to verify the effects of hypoxia on fish development observed in the laboratory in the field.

4. SUPPORTING EVIDENCE FOR THE EFFECTS OF HYPOXIA ON REPRODUCTION AND DEVELOPMENT IN OTHER VERTEBRATES

The scientific evidence provided in the above sections supports the notion that hypoxia is an endocrine disruptor and also a teratogen in fish. Similar to fish, reproductive processes in higher vertebrates are also modulated by sex hormones. The HPG axis and the genes and enzymes controlling steriodogenesis as well as the sex steroid hormones are highly conservative across different vertebrate groups (Ankley and Johnson, 2004). For example, the amino acid sequences of StARs in fish, amphibians, avian species, and mammals are remarkably similar, and non-mammalian StARs share 63–69% sequence identity with human StAR protein (Bauer *et al.*, 2000). The structural organization of the fish receptors (TSHR, FSHR, and LHR) as deduced from the encoding cDNAs is highly homologous to the higher vertebrate receptors (Kumar and Trant, 2001). Thus, the endocrine-disrupting effects of hypoxia found in fish, which subsequently lead to impairment of reproduction and development, may also occur in other vertebrates. Conversely, the effects of hypoxia on reproduction and development revealed in other vertebrate groups may shed light on fish studies, although it must be noted that fish are generally more able to tolerate and adjust to hypoxic conditions than mammals (Ramirez *et al.*, 2007).

4.1. *In Vitro* Evidence

A number of *in vitro* studies provide supporting evidence that expression levels of genes controlling key steriodogenic enzymes and activities of steroidogenic enzymes are reduced in hypoxia. For example, physiologically realistic levels of hypoxia (66–123 torr, ~12%–21% saturation) can specifically inhibit aldosteronogenesis in bovine adrenocortical cells in a dose-dependent manner (Raff and Kohandarvish, 1990). Inhibition of the conversion of corticosterone to aldosterone (the step catalyzed by

P450c11AS) was found in rat adrenal cells exposed to hypoxia (10% O_2, ~48% air saturation) for 3 days. Importantly, change in other cytochrome P450 enzyme activities was not observed (Raff *et al.*, 1996), showing that hypoxic inhibition is specific rather than a general down-regulation of steroidogenesis. A similar conclusion was arrived at in fish studies. Induction of CYP19 expression was found when trophoblast cells isolated from human placenta were maintained under hypoxic conditions (2% O_2, ~10% air saturation), and induction of CYParom mRNA associated with an increase in aromatase activity was clearly evident when hypoxic-treated trophoblasts were returned to normoxia (Jiang *et al.*, 2000).

Kurebayashi *et al.* (2001) found that the expression level of estrogen receptor (ERα) was significantly reduced by hypoxia (1% O_2, ~5% air saturation) in two human breast cancer cell lines (ML-20 and KPL-1). In addition, hypoxia markedly suppressed the induction of progesterone receptor (PgR) mRNA and protein by E2 in both cell lines. *In vitro* studies on steroidogenesis using human adrenal glands with aldosterone-secreting adenomas (Raff and Bruder, 2006) showed that hypoxia (40 mmHg, ~5% air saturation) within the physiological range significantly inhibited cAMP- and ACTH-stimulated cortisol and dehydroepiandrosterone (DHEA) production, showing that steroidogenesis can be affected by hypoxia. VEGF, which is inducible by hypoxia inducible factor-1 (HIF-1), can stimulate proliferation of the mouse TM3 Leydig cells and release of testosterone, while administration of anti-VEGF antibody inhibited the proliferation and release. The results suggest that hypoxia may stimulate cell proliferation and testosterone release in Leydig cells via an increase of VEGF production (Hwang *et al.*, 2007).

4.2. *In Vivo* Evidence

Similar to fish, chronic exposure to hypoxia (3.8 kPa, ~18% saturation) delayed development and hatching in salamanders (*Ambystoma* sp.), and less developed and deformed embryos were produced upon hatching. Development and growth of Australian frog (*Crinia georgiana*) embryos were severely delayed at 2 kPa (~10% saturation) and malformation was observed (Seymonr *et al.*, 2000). In contrast, hypoxia did not affect the developmental rate in the frog (*Rana sp.*), and hypoxic embryos hatched earlier than normoxic embryos, but a higher percentage of less developed embryos was found (Mills and Barnhart, 1999).

The numbers of spermatogenic epithelial cells, Sertoli cells, and Leydig cells in the testicular tissue of male albino rats were significantly reduced after exposure to acute hypobaric hypoxia (Shevantaeva and Kosyuga, 2006). Seven-day-old rats exposed to fetal hypoxia (12% O_2, ~57% air saturation) showed a decrease in plasma aldosterone but no effects on steroidogenic enzyme expression (Raff *et al.*, 2000). In male rats, chronic hypobaric hypoxia has been shown to reduce sperm output and affect spermatogenesis. An increase in FSH and a decrease in LH followed by a decrease in testosterone were found (Farias *et al.*, 2008). The above results support the findings in fish that hypoxia may affect spermatogenesis through hormonal changes along the HPG axis.

Similar to fish, mammalian studies showed that apoptosis, an important process in development, is also affected by hypoxia. Apoptosis in human testicular germ cells was significantly suppressed below 10% oxygen (~48% air saturation) (Erkkila *et al.*, 1999). Using flow cytometry and TUNEL, Liao *et al.* (2007) demonstrated a significant increase in apoptotic germ cells in seminiferous tubules of the hypoxic Wistar rats, especially in spermatogonia and spermatocytes. Both expression level of Bax and the ratio of Bax to Bcl-2 was significantly higher in the hypoxic group after 30 days' exposure, suggesting that chronic hypoxia promotes apoptosis of testicular germ cells in male rats by increasing Bax expression in the rat testis. The results support the findings in fish (Shang and Wu, 2004) that hypoxia can affect apoptosis.

5. THE ROLE OF HYPOXIA-INDUCIBLE FACTORS

Hypoxia-inducible factor 1 (HIF-1) is a heterodimeric transcription factor that is highly conserved and has been found in many species from fish to mammal (Wang and Semenza, 1993; Bunn and Poyton, 1996; Guillemin and Krasnow, 1997; Nikinmaa and Rees, 2005). HIF-1 receives signals from the molecular oxygen sensor through redox reactions and/or phosphorylation (Bunn *et al.*, 1998) and, in turn, regulates the transcription of a number of hypoxia-inducible genes responsible for necessary biochemical and physiological adjustments (Wu, 2002). Since the discovery of HIF-1 (Wang *et al.*, 1995), cumulative evidence has shown that HIFs are the "master regulators" of many molecular responses to hypoxia and, to date, HIF-1 is known to either directly or indirectly regulate the transcriptions of more than 100 genes of diverse functions, including angiogenesis, erythropoiesis, glucose metabolism, vasodilation, cell growth, cell proliferation, transcriptional regulation, differentiation, migration, apoptosis, signaling, and cell fate decisions (Semenza *et al.*, 1994; Arany *et al.*, 1996; Okino *et al.*, 1998; Lisy and Peet, 2008). For a detail review of HIFs, please see Chapter 10.

Some of the above biological processes are fundamental to or indirectly related to reproductive processes as well as embryonic development. Conceivably, HIFs may also regulate certain genes controlling these processes, through which fish reproduction and development are affected, although this has not been clearly demonstrated. Indeed, HIF-1 is implicated in apoptosis (Piret *et al.*, 2002) and reproduction (Park *et al.*, 2007), and has been reported to down-regulate the activity of estrogen receptor (ERα) via the ubiquitin-proteasome degradation pathway in human breast cancer cells (Cho *et al.*, 2005). Deactivation of HIF-1α or HIF-1β in knock-out mice leads to embryonic lethality due to abnormal vascular development (Maltepe *et al.*, 1997; Iyer *et al.*, 1998). These peripheral evidences appear to indicate that some of the observed effects of hypoxia on reproduction and development may be mediated through HIF-1α. It is interesting to note that mRNA expression of both HIF-1α and HIF-2α mRNA in ovaries of Atlantic croaker (*Micropogonias undulatus*) showed a significant increase after exposure to hypoxia (1.7–3.7 mg O_2/L, ~26%–56% saturation for 3 days to 3 weeks), while such up-regulations were not observable in muscle (Rahman and Thomas, 2007). This tissue-specific differential expression appears to suggest that HIFs may be involved in regulating reproductive function in fish, although this has yet to be tested.

HIF-1 consists of two subunits, HIF-1α and HIF-1β, and the latter is the same as the aryl hydrocarbon receptor nuclear translocator (ARNT). The aryl hydrocarbon receptor (AhR) can be ligand-activated to heterodimerize with ARNT, leading to induction of the cytochrome P450 enzymes. Since both the AhR and HIF-1 α compete for ARNT, hypoxia could be expected to decrease the expression of cytochrome P450, which is also involved in steroidogenesis and therefore affects sex hormone production, although there is no clear supporting evidence.

In fish, HIF-1 has been shown to control VEGF and hence angiogenesis (Nikinmaa and Rees, 2005). Vascularization, an important process in fish embryonic development, is controlled by VEGF and many fish embryos reared under hypoxia failed to develop their vascular system and die (Shang and Wu, 2004), suggesting that HIF may also play a role in vascular development via VEGF in the early embryonic stages.

It is not known whether cell proliferation, apoptosis, and development are regulated by HIF. Previous work has established that hypoxia can affect key reproductive and developmental processes in fish through affecting genes controlling steriodogenesis and production of hormones (Dabrowski and Richard, 2003; Shang *et al.*, 2006; Thomas *et al.*, 2006; Landry *et al.*, 2007). Whether these disruptions are mediated through HIF or are independent events remains unknown. *In vitro* transfection studies using the H295R (a human adrenocortical carcinoma) cell line showed down-regulation of 3β-HSD1 and StAR in HIF-1α-overexpressed H295R cells but no change in genes controlling other steroidogenic enzymes, showing that hypoxia can specifically affect certain target genes involved in steroidogenesis

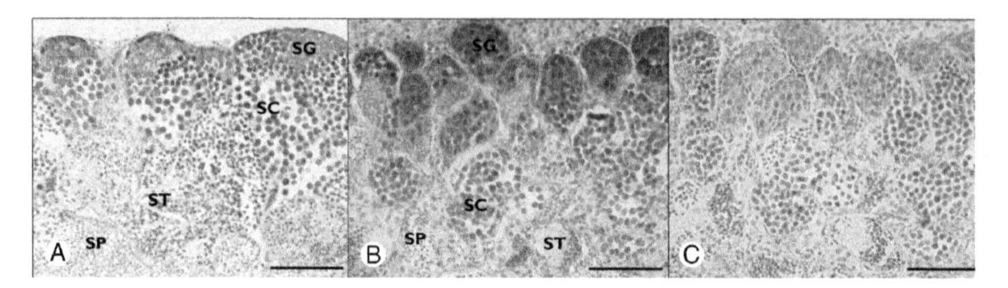

Fig. 3.23. *In situ* localization of *omTERT* mRNA in testes of marine medaka. Scale bars = 50 μm; SG: spermatogonia; SC: spermatocyte; ST: spermatids; SP: spermatozoa. (A) Testis under normoxia (6.4 mg O_2/L, ~96% saturation). Expression of *omTERT* mRNA (blue) is strong in cysts containing spermatogonia and spermatocytes, moderate in differentiating spermatids, and absent in mature spermatozoa (red). (B) Testis exposed to hypoxia (1.8 mg O_2/L, ~27% saturation) for 96 h. Induction of *omTERT* mRNA (blue) is conspicuous in spermatogonia but less prominent in other testicular cells. (C) Adjacent testis section hybridized with an *omTERT* sense riboprobe, serving as a negative control. (Reproduced from Yu *et al.*, 2006.) (See Color Insert.)

(Chu *et al.*, 2006). Whether cell proliferation and apoptosis are regulated by HIF remains unclear. Using *in situ* hybridization, Yu *et al.* (2006) showed that hypoxia can induce telomerase reverse transcriptase (TERT) mRNA in the testis of marine medaka (*Oryzias melastigma*), and results of transfection assays further showed that overexpression of HIF-1α can induce the promoter activity of TERT (Figure 3.23). The results of this study support the notion that hypoxia can up-regulate *TERT* expression via HIF-1 in fish testis *in vivo.* Clearly, a systematic and comprehensive study is required to elucidate the possible role of HIF and its targets of regulation in fish with respect to hypoxic effects on sex hormone production, reproductive impairment, and development.

The functional role of HIF-2 is much less clear, but HIF-2α has been shown to regulate the DNA-damage-inducible alpha protein that induces G2 arrest and apoptosis (Hu *et al.*, 2003). In embryonic stem cells, HIF-2α protects cells against apoptosis during hypoxia (Carmeliet *et al.*, 1998; Brusselmans *et al.*, 2001). As such, HIF-2α may play a role in regulating hypoxic responses in specific cell types (Nangaku and Eckardt, 2007). The possibility may therefore exist that HIF-2α is involved in mediating the effects of hypoxia on cell proliferation and apoptosis during fish development, although this hypothesis has never been tested. Studies have shown that HIF-3α can attenuate HIF-1α-mediated and hypoxia-mediated induction of HRE-driven reporter genes (Hara *et al.*, 2001; Mazure *et al.*, 2002), and may act as an internal repressor of HIF-1α (Makino *et al.*, 2001). Again, whether HIF-3α is involved in mediating the effects of hypoxia on cell proliferation and apoptosis during fish development remains unknown.

6. BIOLOGICAL AND ECOLOGICAL IMPLICATIONS

Given the fact that (1) hormones and genes along the HPG axis, especially those involved in sex steroid synthesis, are highly conservative and (2) there is considerable plasticity in sex determination of fish, the impairment of reproduction and development observed in the few species studied may widely occur in other fish species. Reproductive impairment resulting from hypoxia, as manifested by reduction in reproductive output, quality of sperm and eggs, fertilization success, and larval survival, may have a significant effect on natural fish populations. While reproductive impairment of individuals by hypoxia has been demonstrated in the natural environment, no field data are available to link the observed reproductive impairment to population decline thus far. Understandably, such data would be difficult to collect since population size is often confounded by other factors such as pollution, over fishing, and natural variability prevailing in the same environment, which are difficult to decipher.

Laboratory experiments demonstrated that hypoxia can affect fish development, leading to an increase in embryo mortality rates, delay in hatching, and malformation. It is likely that many of the malformed larvae/juveniles would not be able to survive and contribute to reproduction in the next generation. It is interesting to note that a higher occurrence of malformed fish larvae has been generally reported in polluted areas (Au, 2004), although the observed increase in malformed larvae may not necessarily be attributable to hypoxia because polluted areas are often also contaminated with a variety of chemicals including teratogens and endocrine-disrupting chemicals. Synchronization of the time of hatching with food availability in the natural habitat is important for many species (Eikenaar *et al.*, 2003;Milione and Zeng, 2007), to ensure that natural prey items are available to the newly hatched larvae. Delayed hatching of fish larvae has been reported both in the laboratory and under field conditions (Ingendahl, 2001; Geist *et al.*, 2006; Roussel, 2007); however, the ecological consequence of this occurrence has yet to be elucidated.

Maintaining certain sex ratio is clearly important for ensuring reproductive encounters and hence reproductive success and sustainability of natural populations (Kokko and Brooks, 2003; Le Galliard *et al.*, 2005; Rankin and Kokko, 2006). The laboratory findings that sex differentiation and sex ratio of zebrafish and medaka are affected by hypoxia, resulting in a male-dominated F1 generation (Shang *et al.*, 2006; Cheung and Wu, 2006) is of great environmental concern because this might potentially threaten species survival. A classic, parallel example is imposex in marine whelks (a phenomenon of which females snails exhibiting sexual characteristics of males) caused by tributyltin contamination, which has resulted in a male-biased sex ratio, reproductive failure and extinction of natural populations over large areas worldwide (Bryan *et al.*, 1986). It is important to note that the number of females is the primary limiting factor in determining the reproductive output of a population, and reduction in the number of females may increase competition between mating males and also reduce mating success (Kvarnemo and Ahnesjö, 1996; Jirotkul, 1999). Hypoxia may further reduce both the quantity and quality of gametes and hence reproductive success. The fact that hypoxia affects large areas of aquatic systems worldwide (Diaz and Rosenberg, 1995; Wu, 2002) implies that the ecological consequences caused by hypoxia on natural fish populations could be potentially very serious.

7. CONCLUSIONS

Hypoxia can impair reproduction, alter reproductive behaviors, affect quality of sperm and egg, reduce fertilization success, delay development, reduce hatching success, and increase the incidence of malformation in fish. Results further showed that hypoxia may alter sex differentiation and sex determination. The severity of hypoxic effects, however, depends on the developmental stage and level of oxygen, as well as duration and level of hypoxic exposure.

There is good evidence to suggest that hypoxia impairs fish reproduction by affecting multiple target sites along the HPG axis and enzymes controlling steroidogenesis. Importantly, hypoxia does not cause a general down-regulation of metabolism and reproductive functions, but targets specific hormones, neurotransmitters, and receptors along the HPG axis, as well as certain enzymes controlling steroidogenesis. There is emerging evidence to show that the molecular, hormonal, and behavioral responses of male and female fish to hypoxia may be different.

Hypoxia can arrest or delay fish development by affecting the S and G2 phases of the cell cycle. However, other studies have shown that hypoxia may trigger hatching and development. Although it is well known that hypoxia can cause malformation, the underlying mechanism leading to malformation remains largely unclear. There is good evidence to suggest that this may be mediated through affecting cell proliferation and apoptosis during the various developmental stages.

Laboratory and field studies have shown that hypoxia can alter both the level and balance of androgens and estrogens in several fish species, thereby suppressing ovarian and testicular growth. In particular, hypoxia has been shown to down-regulate CYP19 and alter the ratio of testosterone to estradiol during early development in zebrafish, thereby favoring male development and leading to a male-biased F1 generation. In medaka, a large percentage of genotypic females (with XX chromosomes) showed testicular development and exhibited male phenotypic characteristics when embryos were allowed to develop under hypoxic conditions prior to sex determination.

Scattered evidence appears to indicate that HIF-1 also regulates vascular development as well as some hypoxic responsive genes related to hormonal receptors, cell proliferation, and apoptosis. Whether HIF-1 also plays a role in mediating the observed reproductive and development impairments remain unclear. Transfection assays and knock-down experiments are required to verify the involvement of HIF-1 in mediating the effects of hypoxia on fish reproduction and development.

Since the hormones and regulation of the HPG axis, as well as the enzymes regulating steroidogenesis, are highly conservative in vertebrates, the effects of hypoxia on fish reproduction and development may also occur in other (higher) vertebrates. Indeed, scattered results of *in vitro* and *in vivo* studies in higher vertebrates also lend support to this postulation. More comparative studies between hypoxic responses in different vertebrate groups should be carried out to test this hypothesis and elucidate some common principles and mechanisms.

Reproductive impairment and the adverse effects on development caused by hypoxia revealed in this study suggests that hypoxia poses a significant threat to the sustainability of natural fish populations, especially when considering that hypoxia commonly occurs over very large areas worldwide, and the problem is likely to be exacerbated in the future. For many fish, reproduction is a seasonal event and may involve migration to specific environments, and hypoxia may determine when and where reproduction could occur in these species. Despite this, supporting field evidence is scarce, and the long-term effects of hypoxia on natural fish populations remains virtually unknown.

Finally, it must be cautioned that the vast majority of existing evidence on how hypoxia may affect genes relating to reproduction and development has been based on mRNA transcripts, while the biochemical and physiological responses and adjustments of fish would depend upon the post-translational proteins. Notably, unlike many structural proteins and enzymes, a small change in the amount of sex hormones and neurotransmitters would be sufficient to cause major effects in reproduction and development. Such a small difference may not be detectable by changes in the corresponding mRNA transcripts, especially noting that a two-fold change has been generally employed as a criterion in determining changes in gene profile in microarray studies. Using high-resolution 2-D gel electrophoresis, Bosworth *et al.* (2005) demonstrated that hypoxia did not change the general pattern of protein expression, but only the amounts of six low-abundance proteins in the skeletal muscle of zebrafish. This result contradicts the widespread changes in mRNA levels in hypoxic fish reported in many studies, and the huge difference between the protein and mRNA expression patterns identified calls for a better understanding of proteomic changes in fish during hypoxic exposure.

ACKNOWLEDGEMENTS

I would like to thank Prof. David Randall and Sunny Lu for their comments on a draft of this review. I thank Helen Mok for her technical assistance in data collection and preparation of tables and figures. This work is supported by the Area of Excellence Scheme under the University Grants Committee of the Hong Kong Special Administration Region, China (Project No. AoE/P-04/2004).

REFERENCES

Aarnio, K., Bonsdorff, E., and Norkko, A. (1998). Role of *Halicryptus spinulosus* (Priapulida) in structuring meiofauna and settling macrofauna. *Mar. Ecol. Prog. Ser.* **163,** 145–153.

Adachi-Yamada, T., and O'Connor, M. B. (2004). Mechanisms for removal of developmentally abnormal cells: Cell competition and morphogenetic apoptosis. *J. Biochem.* **136,** 13–17.

Alexander, R. B., Smith, R. A., and Schwarz, G. E. (2000). Effect of stream channel size on the delivery of nitrogen to the Gulf of Mexico. *Nature* **403,** 758–761.

Ankley, G. T., and Johnson, R. D. (2004). Small fish models for identifying and assessing the effects of endocrine-disrupting chemicals. *ILAR J.* **45,** 469–483.

Arany, Z., Huang, L. E., Eckner, R., Bhattacharya, S., Jiang, C., Goldberg, M. A., Bunn, H. F., and Livingston, D. M. (1996). An essential role for p300/CBP in the cellular response to hypoxia. *Proc. Natl. Acad. Sci. USA* **93,** 12969–12973.

Au, D. W. T. (2004). The application of histo-cytopathological biomarkers in marine pollution monitoring: a review. *Mar. Pollut. Bull.* **48,** 817–834.

Au, D. W. T., Chiang, M. W. L., Tang, J. Y. M., Yuen, B. B. H., Wang, Y. L., and Wu, R. S. S. (2002). Impairment of sea urchin sperm quality by UV-B radiation: predicting fertilization success from sperm motility. *Mar. Pollut. Bull.* **44,** 583–589.

Azanza, R. V., Fukuto, Y., Yap, L. G., and Takayama, H. (2005). *Prorocentrum minimum* bloom and its possible link to a massive fish kill in Bolinao, Pangasinan, Northern Philippines. *Harmful Algae* **4,** 519–524.

Baden, S. P., Pihl, L., and Rosenberg, R. (1990). Effects of oxygen depletion on the ecology, blood physiology and fishery of the Norway lobster *Nephrops norvegicus*. *Mar. Ecol. Prog. Ser.* **67,** 141–155.

Bagatto, B. (2005). Ontogeny of cardiovascular control in zebrafish (*Danio rerio*): Effects of developmental environment. *Comp. Biochem. Physiol., Part A Mol. Integr. Physiol.* **141,** 391–400.

Bauer, M. P., Bridgham, J. T., Langenau, D. M., Johnson, A. L., and Goetz, F. W. (2000). Conservation of steroidogenic acute regulatory (StAR) protein structure and expression in vertebrates. *Mol. Cell. Endocrinol.* **168,** 119–125.

Berntsen, O., Bogsnes, A., and Walther, B. T. (1990). The effects of hypoxia, alkalinity and neutrochemicals on hatching of Atlantic salmon (*Salmo salar*) eggs. *Aquaculture* **86,** 417–430.

Bhattacharya, S. (1999). Recent advances in the hormonal regulation of gonadal maturation and spawning in fish. *Curr. Sci.* **76,** 342–349.

Bosworth, C. A., Chou, C. W., Cole, R. B., and Rees, B. B. (2005). Protein expression patterns in zebrafish skeletal muscle: Initial characterization and the effects of hypoxic exposure. *Proteomics* **5,** 1362–1371.

Bouchet, V. M. P., Debenay, J., Sauriau, P., Radford-Knoery, J., and Soletchnik, P. (2007). Effects of short-term environmental disturbances on living benthic foraminifera during the Pacific oyster summer mortality in the Marennes-Oléron Bay (France). *Mar. Environ. Res.* **64,** 358–383.

Breitburg, D. L. (1992). Episodic hypoxia in Chesapeake Bay: Interacting effects of recruitment, behavior, and physical disturbance. *Ecol. Monogr.* **62,** 525–546.

Breitburg, D. (2002). Effects of hypoxia, and the balance between hypoxia and enrichment, on coastal fishes and fisheries. *Estuaries* **25,** 767–781.

Breitburg, D. L., Adamack, A., Rose, K. A., Kolesar, S. E., Decker, M. B., Purcell, J. E., Keister, J. E., and Cowan, J. H. (2003). The pattern and influence of low dissolved oxygen in the Patuxent River, a seasonally hypoxic estuary. *Estuaries* **26,** 280–297.

Bruder, E. D., Nagler, A. K., and Raff, H. (2002). Oxygen-dependence of ACTH-stimulated aldosterone and corticosterone synthesis in the rat adrenal cortex: Developmental aspects. *J. Endocrinol* **172,** 595–604.

Bruder, E. D., Lee, P. C., and Raff, H. (2004). Metabolomic analysis of adrenal lipids during hypoxia in the neonatal rat: Implications in steroidogenesis. *Am. J. Physiol. Endocrinol. Metab.* **286,** E697–E703.

Bruder, E. D., Lee, P. C., and Raff, H. (2005). Lipid and fatty acid profiles in the brain, liver, and stomach contents of neonatal rats: Effects of hypoxia. *Am. J. Physiol. Endocrinol. Metab.* **288,** E314–E320.

Brusselmans, K., Bono, F., Maxwell, P., Dor, Y., Dewerchin, M., Collen, D, Herbert, J. M., and Carmeliet, P. (2001). Hypoxia-inducible factor-2α (HIF-2α) is involved in the apoptotic response to hypoglycaemia, but not to hypoxia. Role of HIF-2α in hypoxia and hypoglycaemia. *J. Biol. Chem.* **276,** 39192–39196.

Bryan, G. W., Gibbs, P. E., Hummerstone, L. G., and Burt, G. R. (1986). The decline of the gastropod *Nucella lapillus* around South-West England: Evidence for the effect of trybutiltin from antifouling paints. *J. Mar. Biolog. Assoc. UK* **66,** 611–640.

Bunn, H. F., and Poyton, R. O. (1996). Oxygen sensing and molecular adaptation to hypoxia. *Physiol. Rev.* **76,** 839–885.

Bunn, H. F., Gu, J., Huang, L. E., Park, J. W., and Zhu, H. (1998). Erythropoietin: A model system for studying oxygen-dependent gene regulation. *J. Exp. Biol.* **201,** 1197–1201.

Burggren, W. (1999). Genetic, environmental and maternal influences on embryonic cardiac rhythms. *Comp. Biochem. Physiol., Part A Mol. Integr. Physiol.* **124,** 423–427.

Cameron, P., and Von Westernhagen, H. (1997). Malformation rates in embryos of North Sea fishes in 1991 and 1992. *Mar. Pollut. Bull.* **34,** 129–134.

Carmeliet, P., Dor, Y., Herbert, J. M., Fukumura, D., Brusselmans, K., Dewerchin, M., Neeman, M., Bono, F., Abramovitch, R., Maxwell, P. H., Koch, C. J., Ratcliffe, P. J., *et al* (1998). Role of HIF-1alpha in hypoxia-mediated apoptosis, cell proliferation and tumor angiogenesis. *Nature* **394,** 485–490.

Carter, A. J., and Wilson, R. S. (2006). Improving sneaky-sex in a low oxygen environment: Reproductive and physiological responses of male mosquito fish to chronic hypoxia. *J. Exp. Biol.* **209,** 4878–4884.

Chang, X. T., Kobayashi, T., Kajiura, H., Nakamura, M., and Nagahama, Y. (1997). Isolation and characterization of the cDNA encoding the tilapia (*Oreochromis niloticus*) cytochrome P450 aromatase (P450arom): Changes in P450arom mRNA, protein and enzyme activity in ovarian follicles during oogenesis. *J. Mol. Endocrinol.* **18,** 57–66.

Cheung, H. Y., and Wu, R. S. S. (2006). Effects of hypoxia on sex determination and differentiation. *In* "SETAC Asia/Pacific 2006: Growth with a Limit: The Integration of Ecosystem Protection for Human Health Benefits" 18–20. September, 2006, Beijing. Abstract B4-2.

Cho, J., Kim, D., Lee, S., and Lee, Y. (2005). Regulation involves hypoxia-inducible factor-1α in MCF-7 human breast cancer cells. *Mol. Endocrinol.* **19,** 1191–1199.

Chu, J. K. Y., Kong, R. Y. C., Giesy, J. P., and Wu, R. S. S. (2006). Reproductive impairments and disruption of steroidogenesis induced by hypoxia: Are these mediated through HIFs? *In* "SETAC Asia/Pacific 2006: Growth with a Limit: The Integration of Ecosystem Protection for Human Health Benefits" 18–20. September, 2006, Beijing. Abstract B4-5.

Connell, D., Lam, P. K. S., Richardson, B., and Wu, R. S. S. (1999). Molecular, biomolecular, physiological and behavioural responses of organisms. *In* "Introduction to Ecotoxicology" pp. 50–76. Blackwell Science, Oxford.

Cook, S. A., Sugden, P. H., and Clerk, A. (1999). Regulation of bcl-2 family proteins during development and in response to oxidative stress in cardiac myocytes: Association with changes in mitochondrial membrane potential. *Circ. Res.* **85,** 940–949.

Dabrowski, K., and Richard, J. (2003). Effect of oxygen saturation in water on reproductive performances of pacu *Piaractus brachypomus*. *J. World Aquacult. Society.* **34,** 441–449.

Dabrowski, K., Rinchard, J., Ottobre, J. S., Alcantara, F., Padilla, P., Ciereszko, A., De Jesus, M. J., and Kohler, C. C. (2003). Effect of oxygen saturation in water on reproductive performances of pacu. *Piaractus brachypomus*. *J. World Aquacult. Soc.* **34,** 441–449.

Dalla Via, J., Vandenthillart, G., Cattani, O., and Dezwaan, A. (1994). Influence of long-term hypoxia exposure on the energy-metabolism of *Solea solea*. 2. Intermediary metabolism in blood, liver and muscle. *Mar. Ecol. Prog. Ser.* **111,** 17–27.

Dauer, D. M. (1993). Biological criteria, environmental health and estuarine macrobenthic community structure. *Mar. Pollut. Bull.* **26,** 249–257.

David, J. L., and Vert, P. (2004). Apoptosis and neurogenesis after transient hypoxia in the developing rat brain. *Semin. Perinatol.* **28,** 257–263.

Devlin, R. H., and Nagahama, Y. (2002). Sex determination and sex differentiation in fish: An overview of genetic, physiological, and environmental influences. *Aquaculture* **208,** 191–364.

Diaz, R. J. (2001). Overview of hypoxia around the world. *J. Environ. Qual.* **30,** 275–281.

Diaz, R. J., and Rosenberg, R. (1995). Marine benthic hypoxia: a review of its ecological effects and the behavioural responses of benthic macrofauna. *Oceanogr. Mar. Biol.* **33,** 245–303.

Diez, J. M., and Davenport, J. (1990). Energy exchange between the yolk and embryo of dogfish (*Scyliorhinus canicula* L.) eggs held under normoxic, hypoxic and transient anoxic conditions. *Comp. Biochem. Physiol.* **96B,** 825–830.

Drobyshevsky, A., Robinson, A. M., Derrick, M., Wyrwicz, A. M., Ji, X., Englof, I., and Tan, S. (2006). Sensory deficits and olfactory system injury detected by novel application of MEMRI in newborn rabbit after antenatal hypoxia-ischemia. *Neuroimage* **32,** 1106–1112.

Dumas, J., Bassenave, J. G., Jarry, M., Barriere, L., and Glise, S. (2007). Effects of fish farm effluents on egg-to-fry development and survival of brown trout in artificial redds. *J. Fish Biol.* **70,** 1734–1758.

Dyer, M. F., Pope, J. G., Fry, F. D., Law, R. J., and Portmann, J. E. (1983). Changes in the fish and benthos catches off the Danish coast in September 1981. *J. Mar. Biolog. Assoc. UK* **63,** 767–775.

Eikenaar, C., Berg, M. L., and Komdeur, J. (2003). Experimental evidence for the influence of food availability on incubation attendance and hatching asynchrony in the Australian reed warbler *Acrocephalus australis*. *J. Avian Biol.* **34,** 419–427.

Erkkila, K., Pentikainen, V., Wikstrom, M., Parvinen, M., and Dunkel, L. (1999). Partial oxygen pressure and mitochondrial permeability transition affect germ cell apoptosis in the human testis. *J. Clin. Endocrinol. Metab.* **84,** 4253–4259.

Evans, J. P., and Geffen, A. J. (1998). Male characteristics, sperm traits, and reproductive success in winter-spawning Celtic Sea Atlantic herring, *Clupea harengus*. *Mar. Biol.* **132,** 179–186.

Farias, J. G., Bustos-Obregón, E., Tapia, P. J., Gutierrez, E., Zepeda, A., Juantok, C., Cruz, G., Soto, G., Benites, J., and Reyes, J. G. (2008). Time course of endocrine changes in the hypophysis–gonad axis induced by hypobaric hypoxia in male rats. *J. Reprod. Dev.* **54,** 18–21.

Fenske, M., and Segner, H. (2004). Aromatase modulation alters gonadal differentiation in developing zebrafish (*Danio rerio*). *Aquat. Toxicol.* **67,** 105–126.

Fontenot, Q. C., Rutherford, D. A., and Kelso, W. E. (2001). Effects of environmental hypoxia associated with the annual flood pulse on the distribution of larval sunfish and shad in the Atchafalaya River Basin, Louisana. *Trans. Am. Fish. Soc.* **130,** 107–116.

Fukada, S., Tanaka, M., Matsuyama, M., Kobayashi, D., and Nagahama, Y. (1996). Isolation, characterization and expression of cDNAs encoding the medaka (*Oryzias latipes*) ovarian follicle cytochrome P450 aromatase. *Mol. Reprod. Dev.* **45,** 285–290.

Gamenick, L., Jahn, A., Vopel, K., and Giere, O. (1996). Hypoxia and sulphide as structuring factors in a macrozoobenthic community on the Baltic Sea shore: Colonization studies and tolerance experiments. *Mar. Ecol. Prog. Ser.* **144,** 73–85.

Geist, D. R., Abernethy, C. S., Hand, K. D., Cullinan, V. I., Chandler, J. A., and Groves, P. A. (2006). Survival, development, and growth of fall Chinook salmon embryos, alevins, and fry exposed to variable thermal and dissolved oxygen regimes. *Trans. Am. Fish. Soc.* **135,** 1462–1477.

Gimeno, S., Gerritsen, A., Bowmer, T., and Komen, H. (1996). Feminization of male carp. *Nature* **384,** 221–222.

Gimeno, S., Komen, H., Venderbosch, P. W. M., and Bowmer, T. (1997). Disruption of sexual differentiation in genetic male common carp (Cyprinus carpio) exposed to an alkyphenol during different life stages. *Environ. Sci. Technol.* **31,** 2884–2890.

Goldberg, E. D. (1995). Emerging problems in the coastal zone for the twenty-first century. *Mar. Pollut. Bull.* **31,** 152–158.

Gozal, E., Sachleben, L. R., Jr., Rane, M. J., Vega, C., and Gozal, D. (2005). Mild sustained and intermittent hypoxia induce apoptosis in PC-12 cells via different mechanisms. *Am. J. Physiol. Cell. Physiol.* **288,** C535–C542.

Gracey, A. Y., Troll, J. V., and Somero, G. N. (2001). Hypoxia-induced gene expression profiling in the euryoxic fish *Gillichthys mirabilis*. *Proc. Natl. Acad. Sci. USA* **98,** 1993–1998.

Grantham, B. A., Chan, F., Nielsen, K. J., Fox, D. S., Barth, J. A., Huyer, A., Lubchenoco, J., and Menge, B. A. (2004). Upwelling-driven nearshore hypoxia signals ecosystem and oceanographic changes in the northeast Pacific. *Nature* **429,** 749–754.

Gray, J. S. (1990). Eutrophication in the sea. *In* "Marine Eutrophication and Population Dynamics" (Colombo, G., Febrari, I, Ceccherelli, V. U., and Rossi, R., Eds.), pp. 3–15. Olsen & Olsen, Fredensborg.

Gray, J. S., Wu, R. S. S., and Or, Y. Y. (2002). Effects of hypoxia and organic enrichment on the coastal marine environment. *Mar. Ecol. Prog. Ser.* **238,** 249–279.

Gross, A., Jockel, J., Wei, M. C., and Korsmeyer, S. J. (1998). Enforced dimerization of BAX results in its translocation, mitochondrial dysfunction and apoptosis. *EMBO J.* **17,** 3878–3885.

Guillemin, K., and Krasnow, M. A. (1997). The hypoxic response: Huffing and HIFing. *Cell* **89,** 9–12.

Hamor, T., and Garside, E. T. (1976). Developmental rates of embryos of Atlantic salmon, *Salmon salar* L., in responses to various levels of temperature, dissolved oxygen, and water exchange. *Can. J. Zool.* **54,** 1912–1917.

Hara, S., Hamada, J., Kobayashi, C., Kondo, Y., and Imura, N. (2001). Expression and characterization of hypoxia-inducible factor (HIF)-3α in human kidney: Suppression of HIF-mediated gene expression by HIF-3α. *Biochem. Biophys. Res. Commun.* **287,** 808–813.

Hattori, M., Sawada, Y., Kurata, M., Yamamoto, S., Kato, K., and Kumai, H. (2004). Oxygen deficiency during somitogenesis causes centrum defects in red sea bream, *Pagrus major* (Temminck et Schlegel). *Aquacult. Res.* **35,** 850–858.

Hattori, R. S., Gould, R. J., Fujioka, T., Saito, T., Kurita, J., Strüssmann, C. A., Yokota, M., and Watanabe, S. (2007). Temperature-dependent sex determination in Hd-rR medaka *Oryzias latipes*: Gender sensitivity, thermal threshold, critical period, and DMRT1 expression profile. *Sex Dev.* **1,** 138–146.

Helly, J. J., and Levin, L. A. (2004). Global distribution of naturally occurring marine hypoxia on continental margins. *Deep Sea Res. Part I Oceanogr. Res. Pap.* **51,** 1159–1168.

Hileman, B. (1994). Environmental estrogens linked to reproductive abnormalities, cancer. *Chem. Eng. News* **72,** 19–23.

Hoback, W. W., and Barnhart, M. C. (1996). Lethal limits and sublethal effects of hypoxia on the amphipod *Gammarus pseudolimnaeus*. *J. North Am. Benthol. Soc.* **15,** 117–126.

Hochachka, P. W., and Somero, G. N. (2002). Cellular metabolism, regulation, and homeostatsis. *In* "Biochemical Adaptation: Mechanism and Process in Physiological Evolution" pp. 20–100. Oxford University Press, New York.

Höhfeld, J. (1998). Regulation of the heat shock cognate Hsc70 in the mammalian cell: The characterization of the anti-apoptotic protein BAG-1 provides novel insights. *Biol. Chem.* **379,** 269–274.

Holloway, A. C., and Leatherland, J. F. (1998). Neuroendocrine regulation of growth hormone secretion in teleost fishes with emphasis on the involvement of gonadal sex steroids. *Rev. Fish Biol. Fish.* **8,** 409–429.

Hu, C. J., Wang, L. Y., Chodosh, L. A., Keith, B., and Simon, M. C. (2003). Differential roles of hypoxia-inducible factor 1alpha (HIF-1alpha) and HIF-2alpha in hypoxic gene regulation. *Mol. Cell Biol.* **23,** 9361–9374.

Hwang, G. S., Wang, S. W., Tseng, W. M., Yu, C. H., and Wang, P. S. (2007). Effect of hypoxia on the release of vascular endothelial growth factor and testosterone in mouse TM3 Leydig cells. *Am. J. Physiol. Endocrinol. Metab.* **292,** E1763–E1769.

Ingendahl, D. (2001). Dissolved oxygen concentration and emergence of sea trout fry from natural redds in tributaries of the River Rhine. *J. Fish Biol.* **58,** 325–341.

Iyer, N. V., Kotch, L. E., Agani, F., Leung, S. W., Laughner, E., Wenger, R. H., Gassmann, M., Gearhart, J. D., Lawler, A. M., Yu, A. Y., and Semenza, G. L. (1998). Cellular and developmental control of O_2 homeostasis by hypoxia-inducible factor 1 alpha. *Genes Dev.* **12,** 149–162.

Jacobson, M. D., Miguel, W., and Martin, C. R. (1997). Programmed cell death in animal development. *Cell* **88,** 347–354.

Jalabert, B., Baroiller, J. F., Breton, B., Fostier, A., Le Gac, F., Guiguen, Y., and Monod, G. (2000). Main neuro-endocrine, endocrine and paracrine regulations of fish reproduction, and vulnerability to xenobiotics. *Ecotoxicology* **9,** 25–40.

Janz, D. M., and Van der Kraak, G. (1997). Suppression of apoptosis by gonadotropin, 17beta-estradiol, and epidermal growth factor in rainbow trout preovulatory ovarian follicles. *Gen. Comp. Endocrinol.* **105,** 186–193.

Jenkins-Keeran, K., Schreuders, P., Edwards, K., and Woods, L. C. (2001). The effects of oxygen on the short-term storage of striped bass semen. *N. Am. J. Aquacult.* **63,** 238–241.

Jiang, B., Kamat, A., and Mendelson, C. R. (2000). Hypoxia prevents induction of aromatase expression in human trophoblast cells in culture: Potential inhibitory role of the hypoxia-inducible transcription factor Mash-2 (mammalian achaete-scute homologous protein-2). *Mol. Endocrinol.* **14,** 1661–1673.

Jirotkul, M. (1999). Operational sex ratio influences female preference and male–male competition in guppies. *Anim. Behav.* **58,** 287–294.

Jobling, M. (1995). Environmental Biology of Fishes, pp. 357–390. Chapman and Hall, London.

Johnson, L. L., and Landahl, J. T. (1994). Chemical contaminants, liver disease, and mortality rates in English sole (*Pleuronectes vetulus*). *Ecol. Appl.* **4,** 59–68.

Jones, J. C., and Reynolds, J. D. (1999a). The influence of oxygen stress on female choice for male nest structure in the common goby. *Anim. Behav.* **57,** 189–196.

Jones, J. C., and Reynolds, J. D. (1999b). Costs of egg ventilation for male common gobies breeding in conditions of low dissolved oxygen. *Anim. Behav.* **57,** 181–188.

Ju, Z., Wells, M. C., Heater, S. J., and Walter, R. B. (2007). Multiple tissue gene expression analyses in Japanese medaka (*Oryzias latipes*) exposed to hypoxia. *Comp. Biochem. Physiol.* **145,** 134–144.

Jung, F., Weiland, U., Johns, R. A., Ihling, C., and Dimmeler, S. (2001). Chronic hypoxia induces apoptosis in cardiac myocytes: A possible role for Bcl-2-like proteins. *Biochem. Biophys. Res. Commun.* **286,** 419–425.

Kah, O., Anglade, I., Lepretre, E., Dubourg, P., and Demonbrison, D. (1993). The reproductive brain in fish. *Fish Physiol. Biochem.* **11,** 85–98.

Kajimura, S., Aida, K., and Duan, C. M. (2005). Insulin-like growth factor-binding protein-1 (IGFBP-1) mediates hypoxia-induced embryonic growth and developmental retardation. *Proc. Natl. Acad. Sci. USA* **102,** 1240–1245.

Keckeis, H., Bauer-Nemeschkal, E., and Kalmer, E. (1996). Effects of reduced oxygen level on the mortality and hatching rate of *Chondrostoma nasus* embryos. *J. Fish Biol.* **49,** 430–440.

Keister, J. E., Houde, E. D., and Breitburg, D. L. (2000). Effects of bottom-layer hypoxia on abundances and depth distributions of organisms in Patuxent River, Chesapeake Bay. *Mar. Ecol. Prog. Ser.* **205,** 43–59.

Kim, D. K., Natarajan, N., Prabhakar, N. R., and Kumar, G.,K. (2004). Facilitation of dopamine and acetylcholine release by intermittent hypoxia in PC12 cells: Involvement of calcium and reactive oxygen species. *J. Appl. Physiol.* **96,** 1206–1215.

Kime, D. E. (1998). Introduction to fish reproduction. *In* Endocrine Disruption in Fish pp. 81–108. Kluwer Academic Publishers, Boston.

Kirshenbaum, L. A., and de Moissac, D. (1997). The bcl-2 gene product prevents programmed cell death of ventricular myocytes. *Circulation* **96,** 1580–1585.

Kitano, T., Takamune, K., Nagahama, Y., and Abe, S. (2000). Aromatase inhibitor and 17 alpha-methyltestosterone cause sex-reversal from genetic females to phenotypic males and suppression of P450 aromatase gene expression in Japanese flounder (*Paralichthys olivaceus*). *Mol. Reprod. Dev.* **56,** 1–5.

Kokko, H., and Brooks, R. (2003). Sexy to die for? Sexual selection and the risk of extinction. *Ann. Zool. Fenn.* **40,** 207–219.

Kroon, F. J., Munday, P. L., and Pankhurst, N. W. (2003). Steroid hormone levels and bi-directional sex change in *Gobiodon histrio*. *J. Fish Biol.* **62,** 153–167.

Kumar, G. K., Overholt, J. L., Bright, G. R., Hui, K. Y., Lu, H. W., Gratzl, M., and Prabhakar, N. R. (1998). Release of dopamine and norepinephrine by hypoxia from PC-12cells. *Am. J. Physiol., Cell Physiol.* **274,** C1592–C1600.

Kumar, G. K., Kim, D. K., Lee, M. S., Ramachandran, R., and Prabhakar, N. R. (2003). Activation of tyrosine hydroxylase by intermittent hypoxia: Involvement of serine phosphorylation. *J. Appl. Physiol.* **95,** 536–544.

Kumar, R. S., and Trant, J. M. (2001). Piscine glycoprotein hormone (gonadotropin and thyrotropin) receptors: A review of recent developments. *Comp. Biochem. Physiol. B, Biochem. Mol. Biol.* **129,** 347–355.

Kumar, R. S., Ijiri, S., and Trant, J. M. (2000). Changes in the expression of genes encoding steroidogenic enzymes in the channel catfish (*Ictalurus punctatus*) ovary throughout a reproductive cycle. *Biol. Reprod.* **63,** 1676–1682.

Kurebayashi, J., Otsuki, T., Moriya, T., and Sonoo, H. (2001). Hypoxia reduces hormone responsiveness of human breast cancer cells. *Jpn. J. Cancer Res.* **92,** 1093–1101.

Kusakabe, M., Todo, T., McQuillan, H. J., Goetz, F. W., and Young, G. (2002). Characterization and expression of steroidogenic acute regulatory protein and MLN64 cDNAs in trout. *Endocrinology* **143,** 2062–2070.

Kvarnemo, C., and Ahnesjö, I. (1996). The dynamics of operational sex ratios and competition for mates. *Trends Ecol. Evol.* **11,** 404–408.

Kwon, J.,Y., Haghpanah, V., Kogson-Hurtado, L.,M, McAndrew, B.,J., and Penman, D. J. (2000). Masculinization of genetic female Nile tilapia (*Oreochromis niloticus)* by dietary administration of an aromatase inhibitor during sexual differentiation. *J. Exp. Zool.* **287,** 46–53.

Landman, M. J., Van Den Heuvel, M. R., and Ling, N. (2005). Relative sensitivities of common freshwater fish and invertebrates to acute hypoxia. *N. Z. J. Mar. Freshwater Res.* **39,** 1061–1067.

Landry, C. A., Steele, S. L., Manning, S., and Cheek, A. O. (2007). Long term hypoxia suppresses reproductive capacity in the estuarine fish, *Fundulus grandis. Comp. Biochem. Physiol.* **148,** 317–323.

Laurikkala, J., Mikkola, M. L., James, M., Tummers, M., Mills, A. A., and Thesleff, I. (2006). p63 regulates multiple signalling pathways required for ectodermal organogenesis and differentiation. *Development* **133,** 1553–1563.

Le Galliard, J. F., Fitze, P. S., Ferriere, R., and Clobert, J. (2005). Sex ratio bias, male aggression, and population collapse in lizards. *Proc. Natl. Acad. Sci. USA* **102,** 18231–18236.

Lee, C. N., Cheng, W. F., Chang, M. C., Su, Y. N., Chen, C. A., and Hsieh, F. J. (2005). Hypoxia-induced apoptosis in endothelial cells and embryonic stem cells. *Apoptosis.* **10,** 887–894.

Lee, S. H., Lee, M. Y., and Han, H. J. (2008). Short period hypoxia increases mouse embryonic stem cell proliferation through cooperation of arachidonic acid and PI3K/Akt signalling pathways. *Cell Prolif.* **41,** 230–247.

Lekve, K., Stenseth, N. C., Gjøsæter, J., Fromentin, J., and Gray, J. S. (1999). Spatio-temporal patterns in diversity of a fish assemblage along the Norwegian Skagerrak coast. *Mar. Ecol. Prog. Ser.* **178,** 17–27.

Liao, W. G., Gao, Y. Q., Cai, M. C., Wu, Y., Huang, J., and Fan, Y. M. (2007). Hypoxia promotes apoptosis of germ cells in rat testes. *Zhonghua Nan Ke Xue* **13,** 487–491.

Lissåker, M., Kvarnemo, C., and Svensson, O. (2003). Effects of a low oxygen environment on parental effort and filial cannibalism in the male sand goby, *Pomatoshistus minutus. Behav. Ecol.* **14,** 374–381.

Lisy, K., and Peet, D. J. (2008). Turn me on: Regulating HIF transcriptional activity. *Cell Death Differ.* **15,** 642–649.

Lossi, L., Coli, A., Giannessi, E., Stornelli, M. R., and Marroni, P. (2002). Cell proliferation and apoptosis during histogenesis of the guinea pig and rabbit cerebellar cortex. *Ital. J. Anat. Embryol.* **107,** 117–125.

Lu, G., Mak, Y. T., Wai, S. M., Kwong, W. H., Fang, M., James, A., Randall, D., and Yew, D. T. (2005). Hypoxia-induced differential apoptosis in the central nervous system of the sturgeon (*Acipenser shrenckii*). *Microsc. Res. Tech.* **68,** 258–263.

Lu, S. X. Y., Yu, R. M. K., Chen, E. X. H., Murphy, M. B., Au, D. W. T., Kong, R. Y. C., and Wu, R. S. S. (2006). Molecular response to hypoxia along the HPG axis of the marine medaka *Oryzias melastigma*. *In* "SETAC Asia/Pacific 2006: Growth with a Limit: The Integration of Ecosystem Protection for Human Health Benefits" 18–20. September, 2006, Beijing. Abstract B4-4.

Lu, S. X. Y., Chen, E. X. H., Yu, R. M. K., Au, D. W. T., and Wu, R. S. S. (2007). Molecular responses to hypoxia along the HPG axis of the zebrafish Danio rerio. *In* "5th Internaltional Conference on Marine Pollution and Ecotoxicology" 3–6. June, 2007, Hong Kong. Abstract O-3.

Makino, Y., Cao, R. H., Svensson, K., Bertilsson, G. R., Asman, M., Tanaka, H., Cao, Y. H., Berkenstam, A., and Poellinger, L. (2001). Inhibitory PAS domain protein is a negative regulator of hypoxia-inducible gene expression. *Nature* **414,** 550–554.

Malhotra, R., and Brosius, F. C. (1999). Glucose uptake and glycolysis reduce hypoxia-induced apoptosis in cultured neonatal rat cardiac myocytes. *J. Biol. Chem.* **274,** 12567–12575.

Maltepe, E., Schmidt, J. V., Baunoch, D., Bradfield, C. A., and Simon, M. C. (1997). Abnormal angiogenesis and responses to glucose and oxygen deprivation in mice lacking the protein ARNT. *Nature* **386,** 403–407.

Ma, J. A., and Li, H. Q. (2002). Preliminary discussion on eutrophication status of lakes, reserviors and rivers in China and overseas. *Res. Environ. Yangtze Basin* **11,** 575–578.

Mason, W. T., Jr. (1998). Macrobenthic monitoring in the lower St. Johns River, Florida. *Environ. Monit. Assess.* **50,** 101–130.

Martin, S. J., Reutelingsperger, C. P., McGahon, A. J., Rader, J. A., van Schie, R. C., LaFace, D. M., and Green, D. R. (1995). Early redistribution of plasma membrane phosphatidylserine is a general feature of apoptosis regardless of the initiating stimulus: Inhibition by overexpression of Bcl-2 and Abl. *J. Exp. Med.* **182,** 1545–1556.

Massa, F., Delorme, C., Bagliniere, J. L., Prunet, P., and Grimaldi, C. (1999). Early life development of brown trout (*Salmo trutta*) eggs under temporary or continuous hypoxial stress: Effects on the gills, yolk sac resorption and morphometric parameters. *Bull. Fr. Peche Prot. Milieux Aquat.* **355,** 421–440.

Matsuda, M. (2003). Sex determination in fish: Lessons from the sex-determining gene of the teleost medaka *Oryzias latipes*. *Dev. Growth Differ.* **45,** 397–403.

Matsuda, M., Nagahama, Y., Shinomiya, A., Sato, T., Matsuda, C., Kobayashi, T., Morrey, C. E., Shibata, N., Asakawa, S., Shimizu, N., Hori, H., Hamaguchi, S., and Sakaizumi, M. (2002). DMY is a Y-specific DM-domain gene required for male development in the medaka fish. *Nature* **417,** 559–563.

Mazure, N. M., Chauvet, C., Bois-Joyeux, B., Bernard, M., Nacer-Chérif, H., and Danan, J. (2002). Repression of α-fetoprotein gene expression under hypoxic conditions in human hepatoma cells – Characterization of a negative hypoxia repsonse element that mediates opposite effects of hypoxia inducible factor-1 and c-Myc. *Cancer Res.* **62,** 1158–1165.

McCormick, M. I. (1998). Behaviorally induced maternal stress in a fish influences progeny quality by a hormonal mechanism. *Ecology* **79,** 1873–1883.

McCormick, M. I. (1999). Experimental test of the effect of maternal hormones on larval quality of a coral reef fish. *Oecologia.* **118,** 412–422.

Milione, M., and Zeng, C. S. (2007). The effects of algal diets on population growth and egg hatching success of the tropical calanoid copepod, *Acartia sinjiensis*. *Tropical Crustacean Aquacult.* **273,** 656–664.

Miller, W. L. (1988). Molecular biology of steroidogenesis. *Endoc. Rev.* **9,** 295–317.

Miller, W. L. (2005). Minireview: Regulation of steroidogenesis by electron transfer. *Endocrinology.* **146,** 2544–2550.

Mills, N. E., and Barnhart, M. C. (1999). Effects of hypoxia on embryonic development in two *Ambystoma* and two *Rana* species. *Physiol. Biochem. Zool.* **72,** 179–188.

Misao, J., Hayakawa, Y., Ohno, M., Kato, S., Fujiwara, T., and Fujiwara, H. (1996). Expression of bcl-2 protein, an inhibitor of apoptosis, and Bax, an accelerator of apoptosis, in ventricular myocytes of human hearts with myocardial infarction. *Circulation* **94,** 1506–1512.

Montserrat, N., Gonzalez, A., Piferrer, F., and Planas, J. V. (2004). Effects of follicle stimulating hormone on estradiol-17 beta production and P-450 aromatase (CYP19) activity and mRNA expression in brown trout vitellogenic ovarian follicle *in vitro*. *Gen. Comp. Endocrinol.* **137,** 123–131.

Myers, D. A., Hyatt, K., Mlynarczyk, M., Bird, I. M., and Ducsay, C. A. (2005). Long-term hypoxia represses the expression of key genes regulating cortisol biosynthesis in the near-term ovine fetus. *Am. J. Physiol. Regul. Integr. Comp. Physiol.* **289,** R1707–1714.

Nagahama, Y. (2000). Gonadal steroid hormones: Major regulators of gonadal sex differentiation and gametogenesis in fish. *In* "Proceedings of the Sixth International Symposium on the Reproductive Physiology of Fish" (Norberg, B., Kjesbu, O. S., Taraanger, G. L., Andersson, E., and Stefansson, S. O., Eds.), pp. 211–232. University of Berge, Berge.

Nagahama, Y., Miura, T., and Kobayshi, T. (1994). The onset of spermatogenesis in fish. *Germline Devel.* **182,** 255–267.

Nagai, S., Asoh, S., Kobayashi, Y., Shidara, Y., Mori, T., Suzuki, M., Moriyama, Y., and Ohta, S. (2007). Protection of hepatic cells from apoptosis induced by ischemia/reperfusion injury by protein therapeutics. *Hepatol. Res.* **37,** 133–142.

Nangaku, M., and Eckardt, K. U. (2007). Hypoxia and the HIF system in kidney disease. *J. Mol. Med.* **85,** 1325–1330.

Nikinmaa, M., and Rees, B. B. (2005). Oxygen-dependent gene expression in fishes. *Am. J. Physiol. Regul. Integr. Comp. Physiol.* **288,** R1079–R1090.

Nishimura, R., Sakumoto, R., Tatsukawa, Y., Acosta, T. J., and Okuda, K. (2006). Oxygen concentration is an important factor for modulating progesterone synthesis in bovine corpus luteum. *Endocrinology.* **147,** 4273–4280.

Nixon, S. (1990). Marine eutrophication: A growing international problem. *Ambio.* **19,** 101.

Okino, S. T., Chichester, C. H., and Whitlock, J. P., Jr. (1998). Hypoxia-inducible mammalian gene expression analyzed *in vivo* at a TATA-driven promoter and at an initiator-driven promoter. *J. Biol. Chem.* **273,** 23837–23843.

Okuzawa, K., Gen, K., Bruysters, M., Bogerd, J., Gothif, Y., and Kagawa, H. (2003). Seasonal variation of the three native gonadotropin-releasing hormone messenger ribonucleic acids levels in the brain of female red seabream. *Gen. Comp. Endocrinol.* **130,** 324–332.

Omura, T., and Morohashi, K. (1995). Gene-regulation of steroidogenesis. *J. Steroid Biochem. Mol. Biol.* **53,** 19–25.

Padilla, P. A., and Roth, M. B. (2001). Oxygen deprivation causes suspended animation in the zebrafish embryo. *Proc. Natl. Acad. Sci. USA* **98,** 7331–7335.

Park, S. E., Park, J. W., Cho, Y. S., Ryu, J. H., Paick, J. S., and Chun, Y. S. (2007). HIF-1 alpha promotes survival of prostate cells at a high zinc environment. *Prostate* **67,** 1514–1523.

Parvez, S., Pandey, S., Ali, M., and Raisuddin, S. (2006). Biomarkers of oxidative stress in *Wallago attu* (Bl. and Sch.) during and after a fish-kill episode at Panipat, India. *Sci. Total Environ.* **368,** 627–363.

Peckol, P., and Rivers, J. S. (1995). Physiological responses of the opportunistic macroalgae *Cladophora vagabunda* (L.) van den Hoek and *Gracilaria tikvahiae* (McLachlan) to environmental disturbances associated with eutrophication. *J. Exp. Mar. Biol. Ecol.* **190,** 1–16.

Pelster, B. (2002). Developmental plasticity in the cardiovascular system of fish, with special reference to the zebrafish. *Comp. Biochem. Physiol.* **133,** 547–553.

Petersen, J. K., and Pihl, L. (1995). Responses to hypoxia of plaice, *Pleuronectes platessa*, and dab, *Limanda limanda*, in the south-east Kattegat: Distribution and growth. *Environ. Biol. Fishes.* **43,** 311–321.

Pfeiler, E. (2001). Changes in hypoxia tolerance during metamorphosis of bonefish leptocephali. *J. Fish Biol.* **59,** 1677–1681.

Piferrer, F., Baker, I. J., and Donaldson, E. M. (1993). Effects of natural, synthetic, aromatizable, and nonaromatizable androgens in inducing male sex-differentiation in genotype female Chinook salmon (*Oncorhynchus tshawytscha*). *Gen. Comp. Endocrinol.* **91,** 59–65.

Pihl, L. (1994). Changes in the diet of demersal fish due to eutrophication-induced hypoxia in the Kattegat, Sweden. *Can. J. Fish. Aquat. Sci.* **51,** 321–336.

Pihl, L., Baden, S. P., and Diaz, R. J. (1991). Effects of periodic hypoxia on distribution of demersal fish and crustaceans. *Mar. Biol.* **108,** 349–360.

Pihl, L., Baden, S. P., Diaz, R. J., and Schaffener, L. C. (1992). Hypoxia-induced structural changes in the diet of bottom-feeding fish and crustaea. *Mar. Biol.* **112,** 349–361.

Piret, J. P., Mottet, D., Raes, M., and Michiels, C. (2002). Is HIF-1 alpha a pro- or an anti-apoptotic protein? *Biochem. Pharmacol.* **65,** 889–892.

Poon, W. L., Hung, C. Y., Nakano, K., and Randall, D. J. (2007). An *in vivo* study of common carp (*Cyprinus carpio* L.) liver during prolonged hypoxia. *Comp. Biochem. Physiol.* **D2,** 295–302.

Raff, H., and Kohandarvish, S. (1990). The effect of oxygen on aldosterone release from bovine adrenocortical cells *in vitro*: PO_2 versus steroidogenesis. *Endocrinology.* **127,** 682–687.

Raff, H., Jankowski, B. M., Engeland, W. C., and Oaks, M. K. (1996). Hypoxia *in vivo* inhibits aldosterone synthesis and aldosterone synthase mRNA in rats. *J. Appl. Physiol.* **81,** 604–610.

Raff, H., Bruder, E. D., Jankowski, B. M., and Engeland, W. C. (2000). The effect of fetal hypoxia on adrenocortical function in the 7-day-old rat. *Endocrine* **13,** 111–116.

Raff, H., Lee, J. J., Widmaier, E. P., Oaks, M. K., and Engeland, W. C. (2004). Basal and adrenocorticotropin-stimulated corticosterone in the neonatal rat exposed to hypoxia from birth: modulation by chemical sympathectomy. *Endocrinology* **145,** 79–86.

Raff, H., Bruder, E. D. and St. Luke's Medical Center Adrenal Tumor Study Group. (2006). Steroidogenesis in human aldosterone-secreting adenomas and adrenal hyperplasia: Effects of hypoxia *in vitro*. *Am. J. Physiol. Endocrinol. Metab.* **290,** E199–E203.

Rahman, M. S., and Thomas, P. (2007). Molecular cloning, characterization and expression of two hypoxia-inducible factor alpha subunits, HIF-1alpha and HIF-2alpha, in a hypoxia-tolerant marine teleost, Atlantic croaker (*Micropogonias undulates*). *Gene* **396,** 273–282.

Ramirez, J., Folkow, L. P., and Blix, A. S. (2007). Hypoxia tolerance in mammals and birds: From the wilderness to the clinic. *Annu. Rev. Physiol.* **69,** 113–143.

Rankin, D. J., and Kokko, H. (2006). Sex, death and tragedy. *Trends Ecol. Evol.* **21,** 225–226.

Reed, J. C., Zha, H., Aime-Sempe, C., Takayama, S., and Wang, H. G. (1996). Structure-function analysis of Bcl-2 family proteins. Regulators of programmed cell death. *Adv. Exp. Med. Biol.* **406,** 99–112.

Reynolds, J. D., and Jones, J. C. (1999). Female preference for preferred males is reversed under low oxygen conditions in the common goby (*Pomatoschistus microps*). *Behav. Ecol.* **10,** 149–154.

Ribeiro, C., Petit, V., and Affolter, M. (2003). Signalling systems, guided cell migration, and organogenesis: Insights from genetic studies in *Drosophila*. *Dev. Biol.* **260,** 1–8.

Rombough, P. J. (1988). Growth, aerobic metabolism, and dissolved-oxygen requirements of embryos and alevins of steelhead, *Salmo gairdneri*. *Can. J. Zool.* **66,** 651–660.

Rosenberg, R., and Loo, L. O. (1988). Marine eutrophication induced oxygen deficiency: Effects on soft bottom fauna, Western Sweden. *Ophelia* **29,** 213–226.

Rosenberg, R., Hellman, B., and Johansson, B. (1991). Hypoxic tolerance of marine benthic fauna. *Mar. Ecol. Prog. Ser.* **79,** 127–131.

Roussel, J. M. (2007). Carry-over effects in brown trout (*Salmo trutta*): Hypoxia on embryos impairs predator avoidance by alevins in experimental channels. *Can. J. Fish. Aquat. Sci.* **64,** 786–792.

Sabo, M. J., Bryan, C. F., Kelso, W. E., and Rutherford, A. (1999). Hydrology and aquatic habitat characteristics of a riverine swamp: II. Hydrology and the occurrence of chronic hypoxia. *Reg. Rivers-Res. Mgt.* **15,** 525–542.

Saed, G. M., and Diamond, M. P. (2002). Apoptosis and proliferation of human peritoneal fibroblasts in response to hypoxia. *Fertil. Steril.* **78,** 137–143.

Saikumar, P., Dong, Z., Weinberg, J. M., and Venkatachalam, M. A. (1998). Mechanisms of cell death in hypoxia/reoxygenation injury. *Oncogene* **17,** 3341–3349.

Sandberg, E. (1997). Does oxygen deficiency modify the functional responses of *Saduria entomon* (Isopoda) to *Bathyporeia pilosa* (Amphopoda)? *Mar. Biol.* **129,** 499–504.

Sanders, E. J., and Wride, M. A. (1995). Programmed cell death in development. *Int. Rev. Cytol.* **163,** 105–173.

Sanderson, J. T. (2006). The steroid hormone biosynthesis pathway as a target for endocrine-disrupting chemicals. *Toxicol. Sci.* **94,** 3–21.

Sawada, Y., Hattori, M., Sudo, N., Kato, K., Takagi, Y., Kazuhiro, U., Kurata, M., Okada, T., and Humai, H. (2006). Hypoxic conditions induce centrum defects in red sea bream *Pagrus major* (Temminck and Schlegel). *Aquacult. Res.* **37,** 805–812.

Schroedl, C., McClintock, D. S., Budinger, G. R. S., and Chandel, N. S. (2002). Hypoxic but not anoxic stabilization of HIF-1α requires mitochondrial reactive oxygen species. *Am. J. Physiol. Lung Cell. Mol. Physiol.* **283,** L922–L931.

Seale, A. P., Riley, L. G., Leedom, T. A., Kajimura, S., Dores, R. M., Hirano, T., and Grau, E. G. (2002). Effects of environmental osomolarity on release of prolactin, growth hormone and ACTH from the tilapia pituitary. *Gen. Comp. Endocrinol.* **128,** 91–101.

Semenza, G. L., Roth, P. H., Fang, H. M., and Wang, G. L. (1994). Transcriptional regulation of genes encoding glycolytic enzymes by hypoxia-inducible factor 1. *J. Biol. Chem.* **269,** 23757–23763.

Senthilkumaran, B., Yoshikuni, M., and Nagahama, Y. (2004). A shift in steroidogenesis occurring in ovarian follicles prior to oocyte maturation. *Mol. Cell. Endocinol.* **215,** 11–18.

Seymonr, R. S., Roberts, J. D., Mitchell, N. J., and Blaylock, A. J. (2000). Influence of environmental oxygen on development and hatching of aquatic eggs of the Australian frog, *Crinia georgiana. Physiol. Biochem. Zool.* **73,** 501–507.

Shang, E. H. H. (2005). Teratogenic and endocrine-disrupting effects of hypoxia on development of zebrafish (*Danio rerio*) PhD. thesis, City University of Hong Kong.

Shang, E. H. H., and Wu, R. S. S. (2004). Aquatic hypoxia is a teratogen and affects fish embryonic development. *Environ. Sci. Technol.* **38,** 4763–4767.

Shang, E. H. H., Yu, R. M. K., and Wu, R. S. S. (2006). Hypoxia affects sex differentiation and development, leading to a male-dominated population in zebrafish (*Danio rerio*). *Environ. Sci. Technol.* **40,** 3118–3122.

Shevantaeva, O. N., and Kosyuga, Y. I. (2006). Effect of acute hypobaric hypoxia on spermatogenesis and lactate concentration in testicular tissue of male albino rats. *Bull. Exp. Biol. Med.* **141,** 20–22.

Shin, D. H., Lee, E., Kim, J., Kwon, B., Jung, M. K., Jee, Y. H., Kim, J., Bae, S., and Chang, Y. P. (2004). Protective effect of growth hormone on neuronal apoptosis after hypoxia-ischemia in the neonatal rat brain. *Neurosci. Lett.* **354,** 64–68.

Smyder, E. A., and Martin, K. L. M. (2002). Temperature effects on egg survival and hatching during the extended incubation period of California grunion, *Leuresthes tenuis*. *Copeia*. **2,** 313–320.

Sorensen, P. W., Murphy, C. A., Loomis, K., Maniak, P., and Thomas, P. (2004). Evidence that 4-pregnen-17, 20β, 21-triol-3-one functions as a maturation-inducing hormone and phermonal precursor in the percid fish, *Gymnocephalys cernuus*. *Gen. Comp. Endocrinol.* **139,** 1–11.

Speer-Blank, T. M., and Martin, K. L. M. (2004). Hatching events in the California grunion, *Leuresthes tenuis*. *Copeia* **1,** 21–27.

Spoor, W. A. (1984). Oxygen requirements of larvae of the smallmouth bass, *Micropterus dolomieui* Lacepede. *J. Fish Biol.* **25,** 587–592.

Stocco, D. M. (2001). StAR protein and the regulation of steroid hormone biosynthesis. *Annu. Rev. Physiol.* **63,** 193–213.

Strand, J., Andersen, L., Dahllöf, I., and Korsgaard, B. (2004). Impaired larval development in broods of eelpout (*Zoarces viviparus*) in Danish coastal waters. *Fish Physiol. Biochem.* **30,** 37–46.

Strüssmann, C. A., and Nakamura, M. (2002). Morphology, endocrinology, and environmental modulation of gonadal sex differentiation in teleost fishes. *Fish Physiol. Biochem.* **26,** 13–29.

Su, T. T. (2000). The regulation of cell growth and proliferation during organogenesis. *In Vivo* **14,** 141–148.

Sun, F., Akazawa, S., Sugahara, K., Kamihira, S., Kawasaki, E., Eguchi, K., and Koji, T. (2002). Apoptosis in normal rat embryo tissues during early organogenesis: The possible involvement of Bax and Bcl-2. *Arch. Histol. Cytol.* **65,** 145–157.

Takahashi, M., Tanaka, M., Sakai, S., Adachi, S., Miller, W. L., and Nagahama, Y. (1993). Rainbow-trout ovarian cholesterol side-chain cleavage cytochrome-P450 (P450SCC) – cDNA cloning and messenger-RNA expression during oogenesis. *FEBS Lett.* **319,** 45–48.

Tanaka, M., Telecky, T. M., Fukada, S., Adachi, S., Chen, S., and Nagahama, Y. (1992). Cloning and sequence analysis of the cDNA encoding P450 aromatase (P450aroma) from a rainbow trout (*Oncorhynchus mykiss*) ovary – relationship between the amount of P450aroma messenger RNA and the production of estradiol-17 beta in the ovary. *J. Mol. Endocrinol.* **8,** 53–61.

Thomas, P. (2008). The endocrine system. *In* "The Toxicology of Fishes" (Di Giulio, R. T., and Hinton, D. E., Eds.), pp. 457–488. CRC Press, Boca Raton.

Thomas, P., Rahman, M. S., Kummer, J. A., and Lawson, S. (2006). Reproductive endocrine dysfunction in Atlantic croaker exposed to hypoxia. *Mar. Environ. Res.* **62,** S249–S252.

Thomas, P., Rahman, M. S., Khan, I. A., and Kummer, J. A. (2007). Widespread endocrine disruption and reproductive impairment in an estuarine fish population exposed to seasonal hypoxia. *Proc. Biol. Sci.* **274,** 2693–2701.

Thibaut, R., and Porte, C. (2004). Effects of endocrine disrupters on sex steroid synthesis and metabolism pathways in fish. *J. Steroid Biochem. Mol. Biol.* **92,** 485–494.

Ton, C., Stamatiou, D., and Liew, C. (2003). Gene expression profile of zebrafish exposed to hypoxia during development. *Physiol. Genomics.* **13,** 97–106.

Townsend, S. A., Boland, K. T., and Wrigley, T. J. (1992). Factors contributing to a fish kill in the Australian wet/dry tropics. *Water Res.* **26,** 1039–1044.

Tyler, C. R., and Sumpter, J. P. (1996). Oocyte growth and development in teleosts. *Rev. Fish Biol. Fish.* **6,** 287–318.

Uchida, D., Yamashita, M., Kitano, T., and Iguchi, T. (2002). Oocyte apoptosis during the transition from ovary-like tissue to testes during sex differentiation of juvenile zebrafish. *J. Exp. Biol.* **205,** 711–718.

UNEP (2006). Further Rise in Number of Marine 'Dead Zones' United Nations Environmental Programme, Nairobi.

Van der Meer, D. L. M., van den Thillart, G. E. E. J. M., Witte, F., de Bakker, M. A. G., Besser, J., Richardson, M. K., Spaink, H. P., Leito, T. D., and Bagowski, C. P. (2005). Gene expression profiling of the long-term adaptive response to hypoxia in the gills of adult zebrafish. *Am. J. Physiol. Regul. Integr. Comp. Physiol.* **289,** R1512–1519.

Vaux, D. L., and Korsmeyer, S. J. (1999). Cell death in development. *Cell* **96,** 245–254.

Von Westernhagen, H. (1988). Sublethal effects of pollutants in fish eggs and larvae. *In* "Fish Physiology, 11-A" (Hoar, W., and Randall, D., Eds.), pp. 253–346. Academic Press, London.

Wang, G. L., and Semenza, G. L. (1993). Characterization of hypoxia-inducible factor 1 and regulation of DNA binding activity by hypoxia. *J. Biol. Chem.* **268,** 21513–21518.

Wang, G. L., Jiang, B. H., Rue, E. A., and Semenza, G. L. (1995). Hypoxia-inducible factor 1 is a basic-helix-loop-helix-PAS heterodimer regulated by cellular O_2 tension. *Proc. Natl. Acad. Sci. USA* **92,** 5510–5514.

Wang, X., Zhou, Y. S., Kim, H. P., Song, R. P., Zarnegar, R., Ryter, S. W., and Choi, A. M. K. (2004). Hepatocyte growth factor protects against hypoxia/ reoxygenation-induced apoptosis in endothelial cells. *J. Biol. Chem.* **279,** 5237–5243.

Wang, S., Yuen, S. S. F., Randall, D. J., Hung, C. Y., Tsui, T. K. N., Poon, W. L., Lai, J. C. C., Zhang, Y., and Lin, H. (2008). Hypoxia inhibits fish spawning via LH-dependent final oocyte maturation. *Comp. Biochem. Physiol. C Toxicol. Pharmacol.* **148,** 363–369.

Wanink, J. H., Kashindye, J. J., Goudswaard, P. C. K., and Witte, F. (2001). Dwelling at the oxycline: Does increased stratification provide a predation refugium for the Lake Victoria sardine *Rastrineobola argentea*? *Freshw. Biol.* **46,** 75–85.

Wannamaker, C. M., and Rice, J. A. (2000). Effects of hypoxia on movements and behavior of selected estuarine organisms from the southeastern United States. *J. Exp. Mar. Biol. Ecol.* **249,** 145–163.

Warkentin, K. M. (2007). Oxygen, gills and embryos behaviour: Mechanisms of adaptive plasticity in hatching. *Comp. Biochem. Physiol., Part A Mol. Integr. Physiol.* **148,** 720–731.

Weltzien, F. A., Andersson, E., Andersen, O., Shalchian-Tabrizi, K., and Norberg, B. (2004). The brain–pituitary–gonad axis in male teleosts, with special emphasis on flatfish (*Pleuroneciformes*). *Comp. Biochem. Physiol., Part A Mol. Integr. Physiol.* **137,** 447–477.

Wu, R. S. S. (1982). Period defaunation and recovery in a sub-tropical epibenthic community in relation to organic pollution. *J. Exp. Mar. Biol. Ecol.* **64,** 253–269.

Wu, R. S. S. (1999). Eutrophication, trace organics and water-borne pathogens: Pressing problems and challenge. *Mar. Pollut. Bull.* **39,** 11–22.

Wu, R. S. S. (2002). Hypoxia: From molecular responses to ecosystem responses. *Mar. Pollut. Bull.* **45,** 35–45.

Wu, R. S. S., and Lam, P. K. S. (1997). Glucose-6-phosphate dehydrogenase and lactate dehydrogenase in the green-lipped mussel (*Perna viridis*): Possible biomarkers for hypoxia in the marine environment. *Water Res.* **31,** 2797–2801.

Wu, R. S. S., Zhou, B. S., Randall, D. J., Woo, N. Y. S., and Lam, P. K. S. (2003). Aqautic hypoxia is an endocrine disruptor and impairs fish reproduction. *Environ. Sci. Technol.* **37,** 1137–1141.

Yamamoto, T. (1961). Progenies of sex-reversal females mated with sex-reversal males in the medaka, *Oryzias latipes*. *J. Exp. Zool.* **146,** 163–179.

Young, G., Kusakabe, M., Nakamura, I., Lokman, P. M., and Goetz, F. W. (2004). Gonadal steroidogenesis in teleost fish. *In* "Hormones and their Receptors in Fish Reproduction" (Melamed, P., and Sherwood, N., Eds.), pp. 155–223. World Scientific, New Jersey.

Yu, R. M. K., Chen, E. X. H., Kong, R. Y. C., Ng, P. K. S., Mok, H. O. L., and Au, D. W. T. (2006). Hypoxia induces telomerase reverse transcriptase (TERT) gene expression in non-tumor fish tissues *in vivo*: The marine medaka (*Oryzias melastigma*) model. *BMC Mol. Biol.* **7,** 27.

Zhao, Y. H., Wang, B., Gao, Y., Zhao, Y. N., Xiao, Z. F., Zhao, W. X., Chen, B., Wang, X., and Dai, J. W. (2007). Olfactory ensheathing cell apoptosis induced by hypoxia and serum deprivation. *Neurosci. Lett.* **421,** 197–202.

Zhou, B. S., Wu, R. S. S., Randall, D. J., and Lam, P. K. S. (2001). Bioenergetics and RNA/DNA ratios in the common carp (*Cyprinus carpio*) under hypoxia. *J. Comp. Physiol. B, Biochem. Syst. Environ. Physiol.* **171,** 49–57.

Zippel, H. P., Sorensen, P. W., and Hansen, A. (1997). High correlation between microvillous olfactory receptor cell abundance and sensitivity to pheromones in olfactory nerve-sectioned goldfish. *J. Comp. Physiol. A.* **180,** 39–52.

4

OXYGEN AND CAPACITY LIMITED THERMAL TOLERANCE

HANS O. PÖRTNER
GISELA LANNIG

Temperature and hypoxia would traditionally be considered as different environmental factors, with specific implications for whole organism functioning. Development of the concept of oxygen and capacity limited thermal tolerance in marine water breathers has revealed how these factors are intertwined. Thermal stress causes systemic hypoxemia and the interaction of temperature and thermally induced hypoxemia will thereby shape acclimation responses at various molecular to whole organism levels. The chapter discusses aspects such as temperature-dependent oxygen supply, width of thermal window and associated energy budget, hypoxemia related stress, and signaling, as well as the cellular mechanisms of thermal adaptation and associated costs including handling and role of calcium. The integration of these responses supports adjustment of metabolic and functional performance at cellular,

Hypoxia : Volume 27
FISH PHYSIOLOGY

DOI: 10.1016/S1546-5098(08)00004-6

tissue, and whole organism levels to within thermal limits. Thereby, processes involved in thermal acclimatization and adaptation counteract thermally induced hypoxemia in fish. Conversely, hypoxia and other stressors will affect thermal tolerance limits and the processes involved in thermal acclimatization and adaptation. As a perspective, the specialization of whole organism functioning on limited temperature ranges emerges as a key element explaining current observations of climate change effects on ecosystems.

1. THERMALLY INDUCED HYPOXEMIA IN FISHES

Studies of temperature-dependent oxygen supply, mode of metabolism, and associated mechanisms of thermal adaptation in marine invertebrates and fishes across latitudes suggested a role of oxygen supply in thermal limitation. Initial evidence came from studies in marine invertebrates (annelids, sipunculids), which showed transition to anaerobic mitochondrial metabolism at both the low and the high end (called critical temperatures) of the thermal tolerance window (Zielinski and Pörtner, 1996; Sommer *et al.*, 1997). These findings stimulated work in bivalves and fishes that demonstrated the onset of anaerobic succinate formation at high temperatures (van Dijk *et al.*, 1999; Pörtner *et al.*, 1999a; Peck *et al.*, 2004). A more recent example confirmed the onset of anaerobic metabolism at low and high temperature extremes in cephalopod mantle tissue (Melzner *et al.*, 2006). The transition to mitochondrial anaerobiosis was shown to result from the development of progressive hypoxemia in arterial haemolymph of a crustacean toward both sides of the thermal window, with an optimum range of maximum body fluid Po_2 in between (Frederich and Pörtner, 2000). Once the critical Po_2 of oxygen diffusion into cells and mitochondria was reached mitochondria started to respire anaerobically. These findings formed the basis of the concept of oxygen and capacity limited thermal tolerance as depicted in Figure 4.1. With it came the conclusion that in a systemic to molecular hierarchy of thermal tolerance the whole organism would experience functional limitations first before biochemical stress events would set in at tissue, cellular, or molecular levels (Pörtner, 2001, 2002; Figure 4.1).

The concept implies that optimized oxygen supply to tissues between lower and upper pejus temperatures combined with the kinetic stimulation of performance rates by warming supports an optimum of performance close to upper pejus temperature. The excess oxygen available above oxygen demand for maintenance fuels the performance capacity of the animal and is reflected in its aerobic scope. Toward both edges of the thermal envelope oxygen supply capacity becomes limiting as oxygen demand of maintenance metabolism progressively exploits all of aerobic scope. This transition reflects

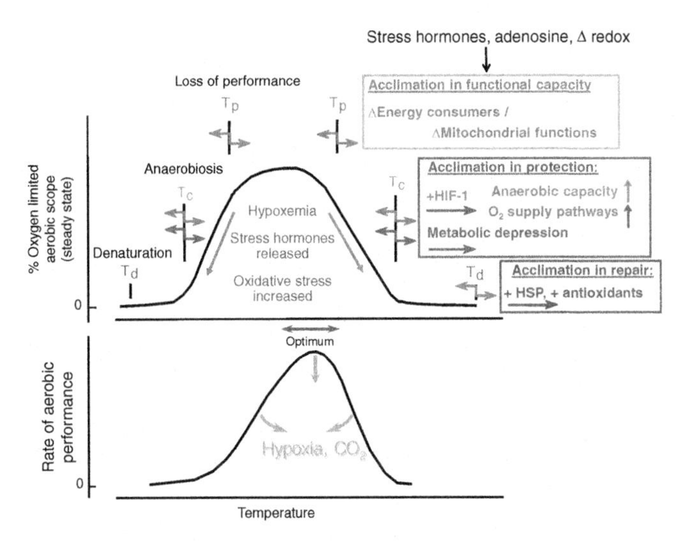

Fig. 4.1. Oxygen and capacity limitation concept of thermal windows indicating the hierarchies (top) of functional limitation (beyond pejus temperatures, T_p), oxygen defiency, anaerobic metabolism and protection through metabolic depression (below and beyond critical temperatures, T_c), and denaturation, as well as repair (beyond denaturation temperatures, T_d). These patterns of thermal limitation lead to a loss in functional capacity and the characteristic right-tilted aerobic performance curve (bottom). Optimized oxygen supply to tissues between low and high pejus temperatures (top) combined with the kinetic stimulation of performance rates by warming supports a performance optimum (i.e., an optimum of aerobic scope) close to upper pejus temperature (bottom). Systemic (e.g., stress hormones, adenosine) and cellular signals (e.g., hypoxia inducible factor HIF-1α, and redox status) associated with temperature-induced hypoxemia contribute to the acclimation response, which leads to a shift in thermal tolerance windows. Ambient hypoxia frequently goes hand in hand with elevated CO_2 levels; both cause a narrowing of thermal windows (Pörtner *et al.*, 2005, Metzger *et al.*, 2007). The graph has been modified and updated from Pörtner and Knust (2007). Note that the figure does not depict details of the signaling pathways involved. (See Color Insert.)

onset of thermal stress and causes an early loss of whole organism functional performance before biochemical stress events take place. A study in the eelpout, *Zoarces viviparus*, in fact demonstrated that the warming-induced decrement in aerobic scope matches the onset of a decrease in growth performance (Pörtner and Knust, 2007). The same thermal threshold is associated with a decrease in abundance of the species in the field, long before anaerobic metabolism sets in due to severe oxygen deficiency and before biochemical stress events take place. These findings clearly indicate that, at the limits of acclimatization capacity, the early onset of a performance decrement is suitable to cause a loss in fitness with the resulting consequences at ecosystem level. Physiological mechanisms setting performance at the whole organism level thus represent the long sought mechanistic link between climate and ecosystem change. Figure 4.1 distinguishes between the temperature range associated with a loss of performance (active range), the subsequent endurance of temperature extremes supported by metabolic depression (passive range), and the range of damage and repair, where protective mechanisms are being used and damaged molecules are accumulating for later removal or repair upon the return of temperatures to control conditions.

To date, evidence of temperature-induced hypoxemia in fish builds on relatively few examples, with the study by Lannig *et al.* (2004) reporting temperature-dependent venous oxygen tensions, the study by van Dijk *et al.* (1999) addressing the transition to anaerobic metabolism. Pörtner and Knust (2007) and Pörtner *et al.* (2001, 2008) integrated these findings with those of temperature-dependent growth. The study by Mark *et al.* (2002) as well as Lannig *et al.* (2004) indicated a limited capacity of cardio-circulation to respond to warming beyond a certain limit, the pejus limit, while temperature-dependent oxygen demand increased upon warming. Excess ambient oxygen improves resistance to warming by shifting pejus limits (Weatherley, 1970; Mark *et al.*, 2002; Figure 4.2). According to these findings hypoxemia results from a mismatch between oxygen supply capacity and oxygen demand, both processes being temperature dependent. Finally, the studies by Pörtner and Knust (2007) as well as Farrell *et al.* (2008) demonstrated the ecological relevance of the oxygen and capacity limitation concept. The study by Pörtner and Knust (2007) showed the link of thermally limited cardio-circulatory performance and aerobic scope to the onset of reduced growth performance and abundance in the natural environment, the German Wadden Sea. Oxygen supply limitations also play a key role in the thermal limitation of muscular exercise of migrating salmon and their inability to reach their spawning grounds in warming rivers (Farrell *et al.*, 2008).

1.1. Temperature-Dependent Oxygen Supply

The temperature-dependent functional capacity of oxygen supply systems (ventilation and cardio-circulation) thus appears crucial in setting whole organism thermal limits. Within the thermal window the capacities of these processes cover the oxygen demand of maintenance and aerobic scope. The level of temperature-dependent oxygen demand in relation to oxygen supply capacity will determine and is thus mirrored in the degree of oxygen saturation of body fluids. Together with blood flow velocity, body fluid oxygenation reflects the scope of oxygen supply to tissues. In the light of the limited number of studies addressing the oxygen limitation concept in fishes, the link between oxygen availability (through cardiac and ventilatory

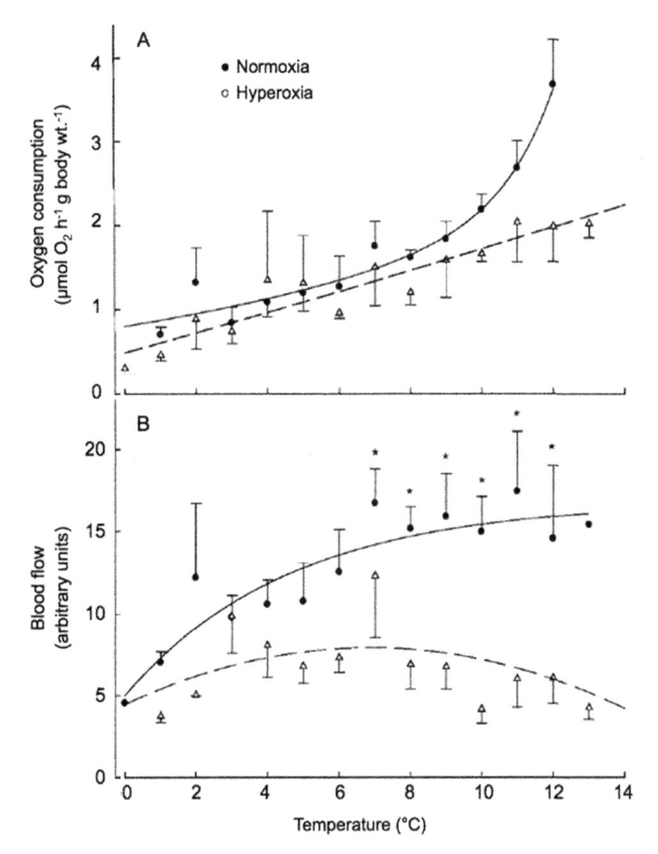

Fig. 4.2. Hyperoxia effects on oxygen consumption (A) and blood flow (B) in the aorta dorsalis of the Antarctic eelpout, *Pachycara brachycephalum*, under normoxia and hyperoxia with rising temperature. Under normoxia, MO_2 showed a large exponential increment, which was eliminated under hyperoxia. At the same time, under normoxia, blood flow increased during warming to 7 °C, and it remained constant and significantly elevated at higher temperatures. In contrast, blood flow remained fairly constant under hyperoxia. (Data by Mark *et al.*, 2002.)

performance) and whole organism aerobic performance still needs to be established on a quantitative basis.

The temperature of optimum oxygen supply and the temperature of maximum energetic efficiency for aerobic performance is supposedly found close to upper pejus temperature (Figure 4.1). However, the maximum of venous Po_2 in Atlantic cod (Lannig *et al.*, 2004) apparently falls below the thermal optimum of growth performance in juvenile cod (Figure 4.3). This apparent discrepancy is alleviated when considering the clear allometry of thermal sensitivity, which

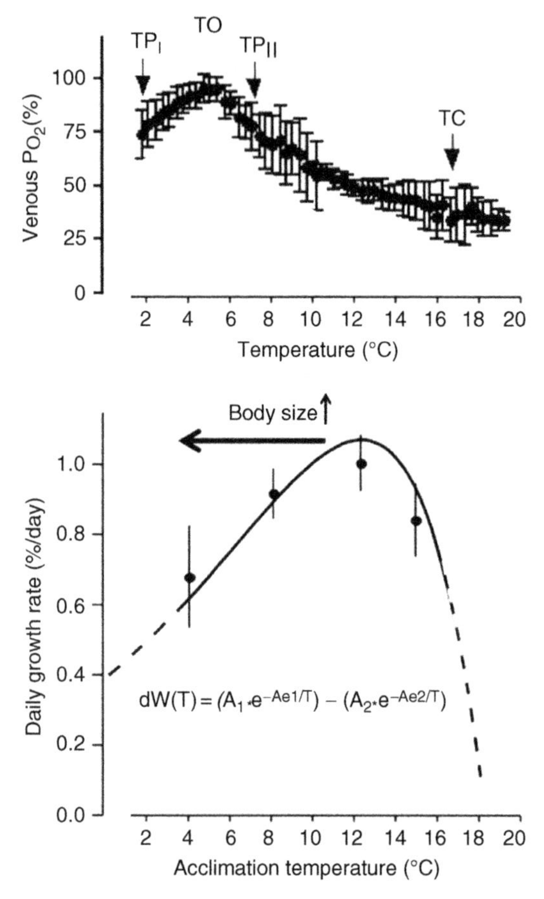

Fig. 4.3. Relative levels (% air saturation) in venous PO_2 in Atlantic cod, *Gadus morhua*, acclimated to 10°C during acutely changing temperature at 1 °C h^{-1} (A, data by Lannig *et al.*, 2004). The acute thermal window of venous blood PO_2 in cod displays a lower "optimum" than expected in relation to acclimation temperature and growth optimum of acclimated juvenile fish (B, data by Fischer, 2003). This apparent discrepancy may be due to various reasons: the figure compares different body sizes such that different optima result. An upward shift of the growth optimum likely occurs upon thermal acclimation. The role of hemoglobin (Hb) oxygen binding in blood oxygen transport in the warmth remains unclear (see text). The progressive increase in the use of Hb for oxygen transport may explain the tailing of venous PO_2 toward warmer temperatures.

causes a shift of growth (and likely, venous PO_2) optima to lower temperatures in cod (Pörtner *et al.*, 2008). This trend would be exacerbated by the 10% and 20% increase in male and female body mass, respectively, during the maturation process and prior to spawning. These trends may in fact explain the sensitivity of the spawning population in the North Sea to warmer winter temperatures above 6°C or 7°C (Perry *et al.*, 2005).

Blood flow regulation and temperature-dependent functioning of hemoglobin also needs to be keyed into the picture of thermally limited oxygen supply (Lannig *et al.*, 2004; Gollock *et al.*, 2006, see Chapter 6). Blood volume is much lower in fish than in invertebrates with an open circulatory system, e.g., crustaceans of similar lifestyles. To compensate and support a higher metabolic rate, the amount of hemoglobin-bound oxygen per unit volume of blood is (more than) one order of magnitude larger in fish than the amount of pigment-bound oxygen in invertebrates. Oxygen transport via the pigment is thus more prominent in fish than in invertebrates. With constantly high arterial Po_2 seen over the whole range of temperatures within the thermal window of cod (Sartoris *et al.*, 2003) the maximum amount of oxygen released from hemoglobin at various temperatures depends on venous Po_2 in relation to temperature-dependent oxygen binding. Cardio-circulation, which appears to comprise a large fraction of the energy budget (see below), will operate at lowest costs in relation to oxygen demand once oxygen release from hemoglobin is maximal. This is likely the case at lower rather than higher maximum venous Po_2 values. Therefore, and in contrast to the respective patterns in invertebrates, the course of venous Po_2 in fishes may not in itself provide the full picture with respect to the optimum of aerobic scope. This conjecture is corroborated by the observation that Pvo_2 did not change with acute warming above the thermal optimum for both resting and swimming sockeye salmon (Steinhausen *et al.*, 2008). All of these considerations remain speculative as long as hemoglobin oxygen binding has not been investigated under *in vivo* conditions, with respect to its role and contribution to the window of thermal tolerance in fishes. The quantitative integration of temperature-dependent oxygen binding of hemoglobin (usually studied *in vitro* and in stripped hemoglobin) with the patterns and cost of circulation within and toward the edges of the thermal tolerance window are thus relevant issues, which, unfortunately, have not yet been explored under the framework of oxygen and capacity limited thermal tolerance.

1.2. Width of Thermal Window and Energy Budget

Aerobic scope available for various "tasks" on various time scales (e.g., long-term: foraging, growth, reproduction, development versus short(er)-term: hunting, migration, escape) emerges as a key parameter shaping fitness. Aerobic scope is not only a matter of oxygen supply capacity but also a matter of energy efficiency, i.e., the reduction of baseline costs to maximize scope and the cost-efficient collection, uptake, and use of available food and substrates. Growth rate, for example, is negatively influenced by the cost of foraging in fish.

With respect to temperature-dependent baseline costs, whole organism oxygen demand is, on the one hand, set by the thermal responses of cellular energy consumers and the resulting level of cellular energy turnover (see Section 3.4). This includes cellular work in ventilatory or circulatory organs, which increases upon rising oxygen demand, even more so when oxygen is in short supply. Evidence for this conclusion comes from the observation that ventilation and circulation represent a significant cost in the energy budget of a fish. The cost to cover oxygen demand rises with the warming-induced increase in baseline oxygen requirements. Vice versa, this cost in itself contributes to whole organism baseline oxygen demand and standard metabolic rate. The contribution of ventilatory and circulatory costs to thermal limitation has been demonstrated through the alleviation of this thermal burden under the effect of hyperoxia (Mark *et al.*, 2002).

On the other hand, functional capacities of cells and tissues co-define the warming-induced increment in the cost of ventilation and circulation and, thus, whole organism oxygen demand. Elevated functional capacity of an organ and of the circulatory system goes hand in hand with elevated baseline energy turnover, partly due to higher densities and maintenance costs of idling mitochondria and transmembrane ion exchange mechanisms, but also due to better capillarization of tissues and volume capacity of blood vessels. At elevated capacity, ventilation and cardio-circulation will find it "easier" to cover the temperature-induced increase in oxygen demand. In consequence, the increment in metabolic cost upon warming will be less and, thus, the onset of thermal stress is alleviated and shifted to higher temperatures. Conversely, baseline costs will be lower at reduced capacity but the cost increment upon warming will be higher at a lower scope, thereby leading to an early limitation. As a corollary, demand, capacity, and supply are intertwined in a way that functional capacity co-defines the width of the thermal window and, thus, the degree of thermal specialization of a species. This is one basic reason for a wider thermal tolerance window in more active species and a narrower window in sessile, hypometabolic species (Pörtner, 2004, 2006).

These relationships are adequately illustrated by the contrasting characteristics of temperate and Antarctic marine fauna. In Antarctic waters, the evolutionary pathways of temperature adaptation can be understood from two points of view: Firstly, animals are exposed to an excess of ambient oxygen at cold temperature, due to high physical solubility of oxygen in ambient water and body fluids. This leads to a larger oxygen reserve than available in warmer waters or body fluids. The expression of intracellular lipid diffusion pathways for oxygen through high mitochondrial densities and networks strengthens this trend even further (Sidell, 1998; Sidell and O'Brien, 2006). This indicates a "relaxed" situation with respect to the effort and energy demand of oxygen transport to tissues, which in turn can be set to lower capacity and cost. This "relaxed" situation is also mirrored in the loss of functional protein in the oxygen transport system and in mitochondria (for review see Sidell and O'Brien, 2006; Pörtner, 2006). The energetic relaxation associated with reduced oxygen supply requirements supports a reduction in whole organism energy turnover and vice versa. Also, the minimized level of cold compensation of molecular, cell, tissue, and whole organism functioning allows minimizing energy turnover even further, such that overall oxygen demand is low, as low as expected from a normal Q_{10} effect on the metabolic rate of a temperate zone fish with similar lifestyles (Clarke and Johnston, 1999). Energy-saving lifestyles in fact typify polar fishes. However, while lower oxygen supply capacity means a lower maintenance cost at cold habitat temperature as a benefit, the trade-off inherent to low capacity is a drastic increment in cost upon mild warming and, as a consequence, an earlier limitation of scope in oxygen supply upon warming. Specialization of polar fishes on high oxygen levels at low ventilatory and circulatory capacities, as well as reduced overall energy turnover, will constrain their capacity to compensate for temperature-induced increments in energy turnover and will thus cause an early hypoxemia and a narrowing of thermal windows.

These considerations give access to understanding the decrease in growth as a consequence of falling aerobic scope. While growth does not fully exploit the scope of aerobic energy turnover it still relies on both excess substrate and energy availability at low baseline costs. Figure 4.4 illustrates this relationship for the Antarctic eelpout by showing that cellular costs are likely minimal at about 4–5°C where growth is maximal. This minimum leaves an excess amount of aerobic energy for maximum growth, at a temperature when not only cellular but also systemic costs are low. However, the cellular minimum is not visible in whole organism oxygen consumption. This may indicate excess energy use for growth. Above this minimum baseline cellular cost rises; at the same time, ventilation and circulation costs rise as well. All of this likely removes excess energy and substrate in competition with long-term aerobic functioning like growth. These considerations indicate that a "relaxed" situation with respect to substrate supply and energy demand is required for the fuelling of long-term aerobic processes and their support on top of standard metabolism. Low cost in oxygen supply, avoidance of hypoxemia, and optimum long-term aerobic performance thus go hand in hand. It remains to be established whether the resulting thermal optima are similar to those seen when the animals actively exploit their aerobic scope for exercise at maximum cost of oxygen supply and at the edge of muscular hypoxemia due to high oxygen demand.

Temperature-induced limitations in aerobic scope for exercise may also bear ecosystem level consequences as is illustrated by available data on Pacific salmon (*Oncorhynchus nerka*) entering the Fraser river, BC, during

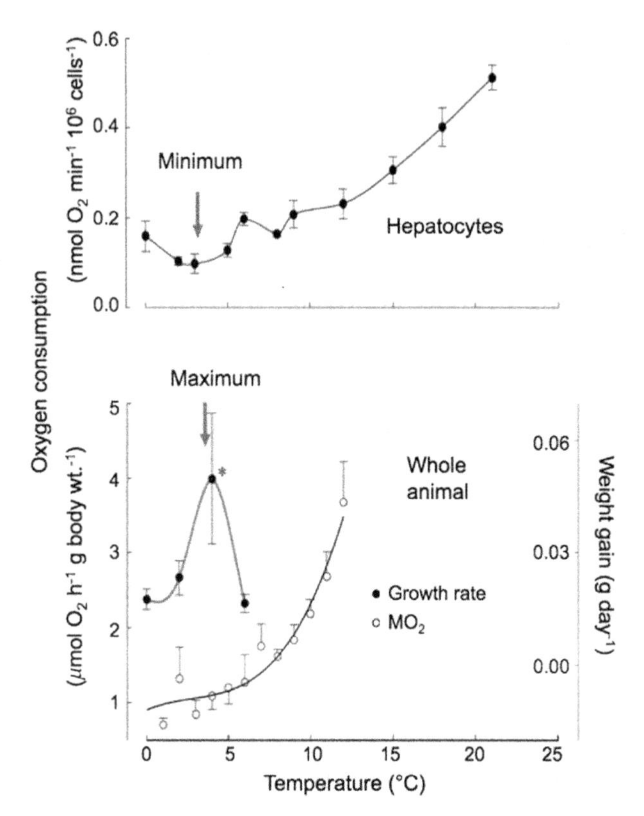

Fig. 4.4. Growth within the thermal window of the Antarctic eelpout, *Pachycara brachycephalum*. As growth determinations occur in thermally acclimated animals, acclimation shifted maximal growth rate to 4°C, a temperature above ambient. Maximum growth occurs at the low end of an acute exponential rise in whole animal oxygen consumption and at lowest cellular costs associated with optimal oxygen supply. (Based on data from Mark *et al.*, 2002, 2005; Lannig *et al.*, 2005; Brodte *et al.*, 2006.)

their spawning migrations (Farrell *et al.*, 2008). Water temperatures in the stream in relation to the temperature window of aerobic scope define whether the salmon will successfully migrate upstream and spawn. This is a special example where a crucial singular event in the life cycle of a species depends upon that maximum capacity for exercise and local climate conditions match. In general, reproduction and early development are processes that rely on supportive climate conditions for any species. Supportive climate conditions are those that reduce the threat of thermally induced hypoxemia and allow keeping baseline costs at a minimum as well as cost-efficient exploitation of aerobic scope for growth or reproduction.

Longer lasting processes like larval development, growth, or foraging activity are equally crucial in maintaining fitness under prevailing climate conditions. In this case, ambient temperature needs to remain below pejus limit for a significant fraction of time to allow these long-term processes to proceed. Such quantitative relationships between fitness levels, widths of the thermal tolerance windows, and ambient temperature variability and sensitivity to thermally induced hypoxemia remain to be established.

2. TEMPERATURE ADAPTATION: ROLE OF HYPOXEMIA

The narrow windows of thermal tolerance that characterize many Antarctic fishes appear as a consequence of their long-term history of permanent cold adaptation over millions of years at minimal risk of exposure to hypoxia or thermally induced hypoxemia. For some zoarcid and notothenioid species, however, a limited but still exploitable capacity to acclimate to warmer conditions (around 5°C) has recently become apparent (Seebacher *et al.*, 2005; Lannig *et al.*, 2005). This matches the finding of inducible heat tolerance in some notothenioid fishes (Podrabsky and Somero, 2006). Warm acclimation capacity is still present in some fishes, in contrast to the situation in many Antarctic invertebrates. The growth optimum of such fauna may lie, in fact, above ambient temperature, as seen in the Antarctic eelpout, *Pachycara brachycephalum*, with an optimum of 4–5°C (Brodte *et al.*, 2006). It needs to be emphasized here that, due to the length of time involved, growth analyses at a specific temperature always comprise effects of prior acclimation if it occurs. From the point of view of acclimation capacity to changing temperatures, some fishes in the Antarctic may currently live in a slow lane, considering their potential to accelerate life functions and live long-term in warmer areas. Antarctic marine invertebrates may not have conserved this apparent capacity (e.g., Pörtner *et al.*, 1999a, b; Peck *et al.*, 2004).

In contrast to Antarctic fauna, temperate fauna and many Arctic or sub-Arctic species or Arctic populations of widely distributed Northern hemisphere species are exposed to more unstable climate and temperature conditions, due to the open nature of Arctic oceans. These species shift thermal windows between seasons and do so for the sake of energy efficiency and savings (Pörtner, 2006). The fact that thermal windows do not match all seasons suggests that the shift of thermal windows may occur in response to or at the verge of thermal stress and associated hypoxemia. Thermal acclimation capacity and thus the capacity to respond to or avoid hypoxemia may vary among species and species populations. According to available data, the degree of cold acclimation capacity is larger in (sub)-Arctic populations of fish than in their Southern-more con-specifics (Lannig *et al.*, 2003; Lucassen *et al.*, 2006; Lurman, 2008). During cold acclimation, an increase in mitochondrial density and capacity contributes to eliminate the capacity limitations of ventilation and circulation; however, the associated rise in metabolic costs enhances sensitivity to warm temperatures as a trade-off, due to the more rapid loss in aerobic scope upon warming. This conjecture could recently be confirmed for Atlantic cod (Z. Zittier, pers. comm.). As a result, the thermal window, i.e., both upper and lower limits of thermal tolerance, are shifted to colder temperatures (Pörtner *et al.*, 2008).

2.1. Systemic Signaling Responses

Seasonal temperature change and latitudinal differences in the temperature regime are well known to be associated with compensation processes at the cellular level including adjustments in the density and functional properties of mitochondria (see Section 3). These changes are associated with the respective shifts in thermal tolerance windows (Pörtner, 2002, 2006). Studies of mitochondrial densities and functional properties in response to temperature change therefore have a long history in the study of thermal adaptation (for review, see Pörtner *et al.*, 2000a; Guderley, 2004). Studying the regulation of these responses provides access to the regulation of shifts in thermal tolerance and temperature-dependent performance optima.

Apart from temperature, exhaustive exercise or hypoxia also induce adjustments of mitochondrial densities and functions (Leary and Moyes, 2000; Hood, 2001). While exercise causes mitochondrial proliferation, hypoxia elicits a decrease of mitochondrial capacities in fish (Johnston and Bernard, 1982; van der Meer *et al.*, 2005). The full range of effector(s) triggering such adjustments still needs to be identified. The key role of whole organism physiology in setting thermal tolerance as well as the concept of oxygen and capacity limited thermal tolerance is suggestive with respect to mechanisms effective in thermal adaptation. At the systemic level central signals would be crucial in coordinating acclimatory responses of individual tissues and cells to temperature.

Eckerle *et al.* (2008) studied the response of hepatocytes isolated from cold or warm acclimated eelpout to subsequent warming or cooling. Warm exposure of hepatocytes from cold acclimated fish led to reduced activities of cytochrome *c* oxidase (COX) and citrate synthase (CS), whereas cold incubation of hepatocytes from warm acclimated fish did not yield any changes in enzyme activities. The observed lack of metabolic cold adaptation of aerobic enzyme capacities at the cellular level *in vitro* might be due to the lack of systemic signaling and oxygen limitation in isolated cells. These observations corroborate that insufficient oxygen supply and associated systemic events as observed in marine ectotherms during acute temperature change might be a key trigger for compensatory adjustments (see Pörtner 2001, 2002). These systemic signals could involve hypoxemia, but also include the endocrine system. According to recent data (Eckerle *et al.*, 2008) eelpout exposed to cooling accumulate adenosine in plasma and tissues. Cold-induced adenosine accumulation in *Z. viviparus* persisted for at least 3 days in plasma and even longer in liver. This period was similar to the time course of cold compensation in liver with respect to changes in the levels of RNA message and activities of mitochondrial enzymes (Lucassen *et al.*, 2003).

Early temperature change is well known to elicit undershoot or overshoot responses in the rate of oxygen consumption as an early "shock" response to cooling or warming, respectively (Cossins and Bowler, 1987). This may occur as a consequence of delays in the functional adjustment of molecules, membranes, cells, and tissues, including oxygen supply systems to temperature, with the potential consequence of early and transiently more severe hypoxemia. In fact, adenosine accumulates in animal tissues in response to hypoxia or anoxia (Lutz and Kabler, 1997; Reipschläger *et al.*, 1997; Renshaw *et al.*, 2002), as a result of a mismatch between ATP production and use. It is released into the extracellular space and can act as a central signal causing metabolic depression (Buck, 2004). At the cellular level adenosine causes various effects, including reduced protein synthesis (Tinton *et al.*, 1995), stimulation of anaerobic glycolysis (Lutz and Nilsson, 1997), and a decrease in oxygen consumption as seen in trout hepatocytes (Krumschnabel *et al.*, 2000). Similar energetic disequilibria would also be involved in situations causing a change in mitochondrial densities or capacities, like during exercise or hypoxia. Associated metabolic signals involved and discussed to elicit or modulate mitochondrial proliferation are Nitric Oxide (NO) (Nisoli *et al.*, 2004) and, most recently, adenosine (Eckerle *et al.*, 2008). Hypoxemia also elicits and exacerbates oxidative stress. In fact hypoxemia developing in the animal toward both ends of the thermal window likely contributes to the pattern of oxidative stress as seen in temperate and Antarctic eelpout (Heise *et al.*, 2007). The level of oxidative stress in response to hypoxemia likely acts as a systemic signal suitable to elicit an adaptive cellular response at the level of most if not all tissues (Section 2.3).

Temperature changes also elicit the release of the classic stress hormones, catecholamines and corticosteroids, which may then influence the process of thermal adaptation (Wendelaar Bonga, 1997). The onset of systemic hypoxemia (cf. Pörtner 2001, 2002) may well be involved in the temperature-induced release of stress hormones. The short-term stress response comprises the rapid release of the catecholamines epinephrine and norepinephrine from their storage site in the chromaffin cells of the head kidney (Reid *et al.*, 1998; Fabbri *et al.*, 1998). They are also rapidly removed from the plasma thereafter. In addition, longer-term accumulation of cortisol occurs from the inter-renal cells of the head kidney (Mommsen *et al.*, 1999). Acute cold exposure caused an accumulation of catecholamines and cortisol in tilapia (Chen *et al.*, 2002), whereas cold acclimation reversed these changes (Perry and Reid, 1994; van Ham *et al.*, 2003; Davis, 2004). The release of cortisol was slightly delayed compared to that of the catecholamines, but the rise in plasma cortisol levels was more prolonged (Chen *et al.*, 2002). Daily infusions of cortisol for 1 week caused an increase in CS activities in the liver, brain, and muscle of catfish (Tripathi and Verma, 2003). Cortisol treatment of isolated eelpout hepatocytes increased the mRNA expression of CS and of the nuclear encoded, but not of the mitochondrial encoded COX subunit. Enzyme activities remained unaffected (Eckerle, 2008). This resembles the early phase of cold acclimation in *Z. viviparus* where enzyme activities also remained unchanged when the message was increased (Lucassen *et al.*, 2003). Cortisol may thus be involved during induction of the cold acclimation process and may be released in response to thermally induced hypoxemia.

Cold acclimation from 20°C to 5°C enhanced the sensitivity of heart and liver in rainbow trout in response to accumulated epinephrine (Keen *et al.*, 1993; McKinley and Hazel, 1993, 2000; Aho and Vornanen, 2001, Shiels *et al.*, 2003). The increase in sensitivity to epinephrine is supported by a higher number of β-adrenoreceptors, as seen in hepatocytes from cold acclimated trout (McKinley and Hazel, 2000). In temperate zone eelpout winter acclimatization of the animals prior to hepatocyte preparation also appeared to enhance sensitivity to epinephrine, arguing again for a seasonal pattern. Epinephrine treatment of hepatocytes isolated from fish in winter caused an increase in the activities of both CS and COX (Eckerle, 2008). Furthermore, thyroid hormones were shown to increase the activities of CS in several tissues of catfish (Tripathi and Verma, 2003) and of COX in mullet (LeRay *et al.*, 1970). It is presently unclear how and where thyroid hormones might fit into the general picture of the regulation of temperature adaptation in poikilotherms. The overall impression is, however, that hypoxemia may play a key role in initiating temperature acclimation mechanisms in fishes.

2.2. Acid-Base and Ion Regulation

While metabolic capacities are adjusted to the prevailing temperature conditions, similar processes would have to adjust cellular and epithelial mechanisms of ion and acid-base regulation (Pörtner *et al.*, 1998). Acid-base regulation is an energy-dependent process since some of the acid-base equivalents are transported by H^+-ATPases or by secondary active processes, for example via the Na^+/H^+ exchanger, which depends upon the Na-gradient established by Na^+/K^+-ATPase. It has recently been suggested that certain species are capable of modulating the cost of acid-base regulation as a means of adjusting the rate of energy turnover to environmental requirements such as temperature change and hypoxia. This can occur acutely, e.g., in response to shifting setpoints of extracellular pH (Pörtner *et al.*, 2000b) or long-term, by modulating the densities and capacities of responsible ion exchange mechanisms (Pörtner *et al.*, 1998).

This requires consideration that the setpoints of pH are temperature dependent. As protein functional capacity is influenced by pH, adequate pH regulation at fluctuating temperature may be required to reduce the risk of performance decrements including temperature-dependent hypoxemia. As a comprehensive concept, Reeves (1972) introduced the imidazole alphastat hypothesis stating that pH regulation in poikilotherms maintains the degree of protonation (α) of imidazole groups in proteins despite changes in body temperature (cf. Burton, 2002). A pH change of around -0.018 pH units $°C^{-1}$ matches a $\Delta pK\ °C^{-1}$ of -0.018 and is expected to support the alphastat pattern and to support pH-dependent protein function at fluctuating temperature. Cameron (1989) proposed a "Z-stat" model where protein net charge Z is maintained rather than α in a protein with diverse histidine groups. This concept takes the variability between $\Delta pK\ °C^{-1}$ into account, which depends upon local charge configurations in the environment of the imidazole group as well as on ionic strength and, therefore, ranges between -0.016 and $-0.024°C^{-1}$ for histidine and free imidazole compounds and between -0.0010 and $-0.051°C^{-1}$ for histidine residues in proteins (Heisler, 1986).

Available data, which are more comprehensive for intracellular than extracellular acid-base status, indicate a trend in line with the concept of alphastat regulation (Ultsch and Jackson, 1996). However, the variability in slopes of $\Delta pH\ °C^{-1}$ is larger than expected from the variability in $\Delta pK\ °C^{-1}$. The alphastat pattern of intracellular pH regulation could be confirmed for marine ectotherms (invertebrates and fish) exposed to various temperatures both depending on the season or in a latitudinal cline suggesting that deviation from the alphastat pattern is involved in or results from metabolic depression (review by Pörtner *et al.*, 1998; Sartoris *et al.*, 2003). Metabolic depression is a typical response to hypoxemia, e.g., at the edges of the thermal window. Consideration of the window of oxygen and capacity limited thermal tolerance in fact revealed that the slope of the pH/temperature line is linear only between critical temperatures (Sommer *et al.*, 1997). Acidotic deviations at temperatures beyond that are caused by the transition to proton-producing anaerobic metabolism (cf. Pörtner, 1987). Concomitantly, a shift in pH regulation may occur during hypoxemia when an acidosis is induced and contributes to metabolic depression. This would be the reason for long-term deviations from the alphastat pattern, as any short-term disturbance would otherwise be compensated for by ion exchange mechanisms. The respective data are scarce in ectotherms if they exist. In this context, the use of pH-stat rather than alphastat conditions is a matter of debate in hypothermic surgery on humans. Compared to the elevation of pH according to alphastat the application of pH-stat perfusate to cooling tissue is equivalent to acidotic exposure and metabolic depression (Ohkura *et al.*, 2004; Li *et al.*, 2004). During surgery under severe hypothermia, further benefits of pH-stat include a relative rightward shift of the hemoglobin oxygen binding curve thereby supporting oxygen delivery. pH-stat causes increased cerebral blood flow and volume, associated with enhanced oxygen availability during circulatory arrest and a greater suppression of cerebral metabolic rate.

In water breathers, the observed changes in pH with temperature are mostly elicited by non-respiratory mechanisms. A passive component is due to proton binding or release from intra- and extracellular buffers owing to the change in dissociation equilibria (pK-values) of the buffer components. In contrast to air breathers, active control of pH by means of ventilatory P_{CO_2} adjustments is minimal; active ion exchange mechanisms predominate. In some species the passive contribution to pH regulation was found to be considerably below the alphastat value seen *in vivo* (van Dijk *et al.*, 1997; Sartoris and Pörtner, 1997). The passive contribution accounted for only 35% of the temperature-induced pH shift in white muscle of the temperate eelpout, whereas it was close to 100% in the Antarctic eelpout. Lower passive pH shifts would lead to more acidic pH values in the cold and leave a larger contribution to ion exchange mechanisms to accomplish alphastat pH regulation. In general, the active component was larger in eurythermal than in stenothermal species (Pörtner and Sartoris, 1999). This indicates that energy savings in Antarctic stenotherms comprise a reduction in the level of active acid-base regulation. At the same time, living at high temperature variability includes the option to use the non-alphastat pH slope. This relative acidification would support metabolic depression as elicited by extreme temperature-induced hypoxemia (see above). In this case, large passive slopes would require active pH regulation to compensate for their effect when more acidic pH values are to be maintained. In contrast, a low passive slope allows flexible adjustments of pH according to metabolic requirements. In the warmth this may be involved in metabolic depression as seen in freshwater burbot, *Lota lota*, during summer (Hardewig *et al.*, 2004; Figure 4.1). On the other side of the temperature spectrum, animals living in large seasonal temperature variations frequently exhibit low pH values at low temperatures in the winter (Thebault and Raffin, 1991; Spicer *et al*, 1994). The shrimp *Palaemon* tends to be inactive at temperatures below 10°C, metabolic depression being reflected by a drop in intracellular pH below the alphastat pattern (Thebault & Raffin, 1991). Acidic pH_i values were also reported by Whiteley *et al.* (1995) for winter crayfish, *Austrapotamobius pallipes*. One might speculate that a capacity for metabolic depression in eurythermal animals is correlated with a reduced contribution of passive mechanisms to pH adjustment during temperature change. As a corollary, a larger active than passive component of alphastat regulation may not only be a prerequisite to colonize shallow coastal waters but may also allow for a variable adjustment of metabolic activity on a seasonal time scale and support flexible response to more frequent exposure of thermally induced hypoxemia under variable temperature conditions.

2.3. Hypoxemia-Related Cellular Stress and Signaling

Temperature-induced metabolic adjustments at the cellular level are key to the maintenance of functioning and cellular energetics and thus survival of the organism. They are most crucial in shifts of the position and width of the thermal tolerance window on the temperature scale. Long-term compensatory adjustments in aerobic metabolism contribute to balance the temperature impact on metabolic processes and to support a new steady state in energy metabolism (see Section 3). For detailed information on the consequences of hypoxia/anoxia see Chapters 9 and 10.

Regulation of thermal adaptation at the cellular level is likely strongly influenced and may even depend upon the response to systemic hypoxemia. The patterns of systemic signaling and oxidative stress in response to temperature-induced hypoxemia indicate a common response following temperature change (Figure 4.1). Both heat and cold exposure, in particular during the recovery phase at control temperatures, led to elevated oxidative stress parameters in hepatic tissue of *Z. viviparus* (Heise *et al.*, 2006a,b), confirming that hypoxemia is suitable to elicit or at least exacerbate onset and effects of oxidative stress. Excess ROS production in marine ectotherms may cause cellular damage (Abele *et al.*, 1998, 2001) and impair cellular functioning (Chabi *et al.*, 2008). Antioxidative defence may thus play an important role in setting passive tolerance to temperature extremes (Pörtner, 2002). Polar ectotherms are thought to be more vulnerable to cellular ROS production during warming than temperate ectotherms as their membranes rich in polyunsaturated fatty acids are easy targets for lipid radical formation (Brand *et al.*, 1991; for review see Abele and Puntarulo, 2004). Strong antioxidative defence, for example, through vitamin E and specific derivatives is thus expressed in Antarctic fish (Dunlap *et al.*, 2002; Heise *et al.*, 2007).

Oxidative stress on both sides of the thermal window may also play a role in shaping the pattern of thermal acclimation. In this context the activation of the hypoxia inducible transcription factor (HIF) as observed during temperature change in fish (Heise *et al.*, 2006a,b, 2007) supports cellular and systemic stress resistance during temperature-induced oxygen shortage

(Figure 4.5). At high temperatures hypoxic signaling and subsequent metabolic reorganization to counterbalance thermal oxygen limitation seems to be effective only in the pejus temperature range, while it appears impaired at critical and even higher temperatures (cf. Figure 4.1). A strong reduction of the cellular redox potential (or a reduced glutathione redox ratio, 2GSSG/GSH) as

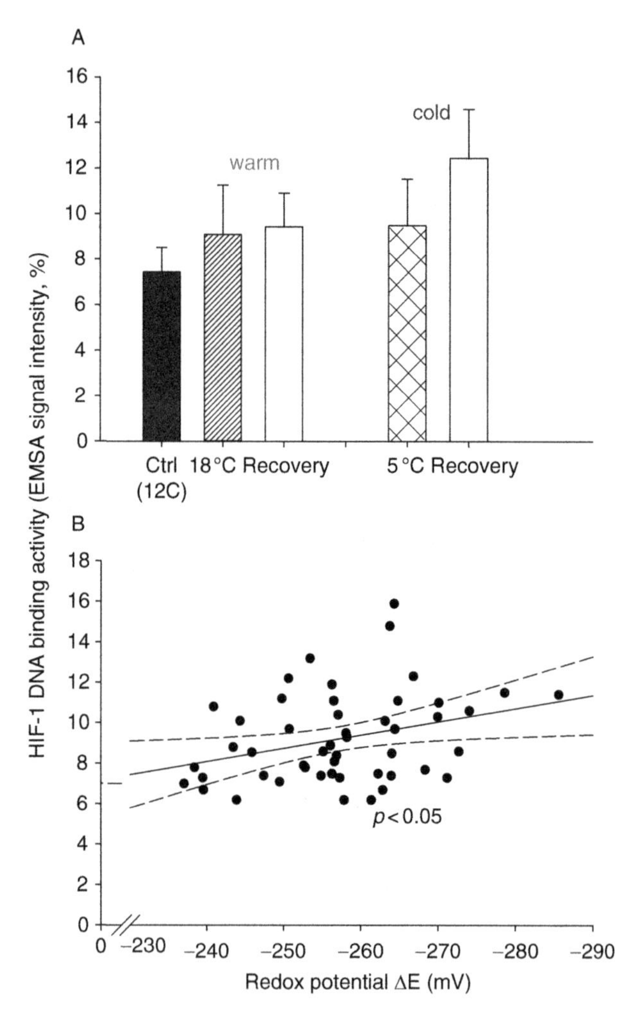

Fig. 4.5. (A) Increased HIF-1 DNA binding in liver at extreme temperatures und during recovery from both cold and heat exposure. (**B**) Linear regression demonstrating increased HIF-1 DNA binding at a more reduced redox potential (i.e., a more reduced glutathione redox ratio in liver samples from *Zoarces viviparus*. (Data from Heise *et al.*, 2006a,b.)

during severe hypoxia supports the binding activity of HIF (Figure 4.5). At present the differentiation between low and high temperatures with respect to the hypoxemia induced involvement of HIF is not clear. However, HIF-1 DNA binding activity was significantly higher at cold temperature, both in the polar and the temperate eelpout species, and when comparing winter and summer in temperate eelpout. In both the Antarctic cold and seasonal cold, the glutathione redox ratio was more oxidized when compared to the warmer condition. This indicates that HIF-1 might stimulate cold compensation mechanisms, but might operate differently in the warmth (Heise *et al.*, 2007). HIF-1 could thus be involved in regulating the adjustment of tissue oxygenation at the border of the thermal envelope of a fish but no longer at more extreme temperatures.

The heat shock response has long been studied in fish and in relation to temperature extremes (for review see Iwama *et al.*, 1998, 2006). Comprehensive evidence on how the heat shock response keys into the general picture of oxygen limited thermal tolerance is not available for fish but comes from a comparative study on two Mediterranean bivalve species, the bearded mussel, *Modiolus barbatus*, and the Mediterranean mussel, *Mytilus galloprovincialis* (Anestis *et al.*, 2007, 2008). *Modiolus barbatus* lives at depths between 8 and 15 m, experiences temperatures up to 21°C and *M. galloprovincialis* lives at depths between 0.5 and 5 m and at temperatures up to 26–28°C. At first sight it is puzzling that mortality sets in at about the same temperature of 26°C but is 20-fold higher in *M. galloprovincialis* at 26°C and more than 80-fold higher at 30°C than in *M. barbatus*. These apparently contradictory findings are resolved when considering that the heat shock response in *M. barbatus* sets in beyond 22°C and in *M. galloprovincialis* only beyond 26°C. Interestingly, *M. barbaratus* also displays a larger capacity to undergo metabolic depression than *M. galloprovincialis*. These findings would suggest an earlier limitation of aerobic scope in *M. barbatus* than in *M. galloprovincialis*. Extreme hypoxemia and anaerobiosis may also set in early and contribute to elicit the earlier onset of the heat shock response (Pörtner, 2002).

However, the distribution of the two species clearly shows that the capacities to undergo metabolic depression and use the heat shock response do not a priori and exclusively define temperature-dependent distribution and heat-induced mortality. Mediterranean mussels (*M. galloprovincialis*), which regularly encounter water temperatures higher than 25°C, live near their incipient lethal temperature. Their extended aerobic range combined with their delayed and limited (compared to *M. barbaratus*) depression of metabolic rate likely support active survival and fitness at warmer temperatures. The earlier onset of the heat shock response in *M. barbatus* than in *M. galloprovincialis* likely mirrors an earlier loss in aerobic scope and onset of hypoxemia in the bearded mussel. This may precondition *M. barbatus* to passively tolerate more extreme temperatures than *M. galloprovincialis,* with the result of better passive survival, i.e., lower mortality of *M. barbatus* at extreme temperatures. At temperatures beyond 23°C the bearded mussel experiences constrained aerobic scope and metabolic depression, which will impair relevant physiological processes such as growth and reproduction and prevent long-term successful settling of shallower, warmer waters by this species.

These data are in line with the systemic to molecular hierarchy of thermal tolerance postulated earlier where the heat shock response is interpreted to shape passive tolerance to thermal extremes, a feature highly relevant in intertidal organisms (Pörtner, 2002). The conclusions drawn for the mussels fully match those drawn for marine fish populations. In fact, abundance of eelpout in the German Wadden Sea begins to decline as soon as aerobic scope for growth is reduced (Pörtner and Knust, 2007; Wang and Overgaard, 2007). This occurs at the upper thermal limits of acclimation capacity. Maintenance of aerobic scope is thus most crucial for long-term survival of extreme temperature conditions in the field (cf. Pörtner and Knust, 2007). This also emphasizes that interpretation of laboratory data on tolerance benefits from consideration of a background of field data.

3. CELLULAR MECHANISMS OF THERMAL ADAPTATION

In the following, we will focus in more detail on how temperature-induced hypoxemia may influence temperature-related compensatory aspects in cell metabolism such as changes in mitochondrial capacities, in membrane structure, and in energy turnover and on the consequences for thermal tolerance. We will discuss temperature-dependent impact on cardiac calcium homeostasis as calcium is of great importance for cardiac performance, and may thus play an important role in setting windows of thermal tolerance in fish. All of these mechanisms will be interpreted in the light of how they might support the organism to overcome the threat of temperature-induced hypoxemia.

3.1. Capacity and Efficiency of Mitochondria

Temperature was shown to greatly affect mitochondrial and enzymatic capacities and thermal plasticity of these parameters/factors was suggested to indicate thermal adaptation (Dahlhoff and Somero, 1993; Weinstein and Somero, 1998; Pörtner *et al.*, 1999a). Temperature adaptation of aerobic scope includes the adjustment of the scope of mitochondrial energy production and of the associated substrate oxidation capacity of mitochondria.

A large number of studies found mitochondrial capacities to fall with decreasing temperature, with the lowest capacities found at high mitochondrial densities in Antarctic fishes. Up-regulation of capacity seems to be restricted to cold exposed eurythermal species (for review see, e.g., Pörtner *et al.*, 2000a, 2005b; Guderley and St-Pierre, 2002; Guderley, 2004; and references therein). "Increasing the volume and surface density of mitochondrial clusters is the primary mechanism" of supporting the aerobic capacity of muscle in cold-adapted Antarctic species (Johnston *et al.*, 1994, 1998). In liver, cold compensation is mainly accomplished by a rise in tissue mass leaving mitochondrial protein per gram liver largely unchanged (Kent *et al.*, 1988; Seddon and Prosser, 1997; Lannig *et al.*, 2003, 2005). Enzyme activities were found increased in cold-acclimated and cold-adapted tissues, depending on enzyme and tissue (e.g., Guderley and Blier, 1988; Crockett and Sidell, 1990; Lannig *et al.*, 2003, 2005; Kawall *et al.*, 2002; Lucassen *et al.*, 2003). Changes in the activities of membrane-bound enzymes are partly induced by changes of membrane structure (see Guderley and St-Pierre, 2002; Pörtner *et al.*, 2005b; and below). The temperature dependence of mitochondrial functions most likely depends more on the integrity of mitochondrial membrane and membrane protein interactions than on protein stability per se (White and Somero, 1982; Guderley *et al.*, 2008; see Section 3.2).

The stable environments of the marine Antarctic supported extreme thermal specialization of its marine inhabitants (for review see Pörtner, 2006 and references therein). Accordingly, Antarctic fishes were regarded as being restricted in their ability to respond to temperature variation. However, Antarctic fish species maintained—at least to some degree—thermal plasticity of metabolic processes and whole animal performance as demonstrated for a zoarcid (Lannig *et al.*, 2005, Brodte *et al.*, 2006) and for a notothenioid (Seebacher *et al.*, 2005). It is presently unclear how their mitochondrial functions are modified to support life in the "warmth." In temperate fish warm acclimation goes hand in hand with reduced mitochondrial capacities (for review see Pörtner *et al.*, 2005b). Warm-acclimated Antarctic eelpout, *P. brachycephalum* (Lannig *et al.*, 2005) displayed unchanged capacities per milligram mitochondrial protein in liver and showed a clear reduction in mitochondrial capacities only at the whole organ level due to decreased liver size after long-term warm acclimation (5°C versus 0°C). Thus, in contrast to the finding of increased standard metabolic rate during acute warming (van Dijk *et al.*, 1999; Mark *et al.*, 2002) warm acclimated *P. brachycephalum* displayed a metabolic rate similar to cold-acclimated specimens when measured at the respective acclimation temperature (0–6°C) (Brodte *et al.*, 2006; Lannig G., unpublished data). Taken together, these findings indicate complete metabolic warm compensation due to a reduction in maintenance costs in the Antarctic eelpout.

The more active Antarctic fish, *Pagothenia borchgrevinki* "displayed astonishing plasticity in cardiovascular response and metabolic control" to maintain locomotory performance at elevated temperatures (Seebacher *et al.*, 2005). After 4–5 weeks of warm acclimation (4°C versus −1°C) activity of muscle lactate dehydrogenase and cytochrome *c* oxidase was significantly elevated indicating an up-regulation of enzymes involved in both anaerobic and aerobic energy production. Interestingly, the activities of other glycolytic and TCA cycle enzymes, like phosphofructokinase and citrate synthase, did not differ in muscle tissue of cold- and warm-acclimated animals. The authors discussed the observed up-regulation of enzyme activities as a compensatory response to meet elevated maintenance costs at increased temperatures (Seebacher *et al.*, 2005). They did not determine standard metabolic rate; however, in contrast to the respective observations in the eelpout, the enzymatic results indicate increased and thus uncompensated standard metabolism following warm acclimation of *P. borchgrevinki*. This assumption appears reasonable as, for comparison, reduced CS and unchanged COX activities in white muscle of North Sea cod, *Gadus morhua*, following warm acclimation (Lannig *et al.*, 2003) were associated with similar standard metabolic rates in warm- and cold-acclimated specimens, measured at the respective acclimation temperatures (Fischer, 2003). As a corollary, the polar zoarcid, *P. brachycephalum*, and the notothenioid, *P. borchgrevinki*, differed in their compensatory response: the former reduced while the latter elevated metabolic rate upon warm acclimation. Both strategies involve successful avoidance of temperature-induced hypoxemia and likely shifted or expanded the thermal window toward higher temperatures. This conjecture, however, warrants further investigation.

Mitochondrial proliferation in the cold compensates for the suppressing effect of low temperature on metabolic and diffusion pathways (Tyler and Sidell, 1984; Egginton and Sidell, 1989; Egginton *et al.*, 2002; O'Brien *et al.*, 2003). In temperate zone species, mitochondrial proliferation causes a rise in cellular energy costs due to increased proton leak rates. The term proton leak describes an inherent proton permeability of the inner mitochondrial membrane. The futile cycle of proton pump and leak occurs without ATP production and can cover 25–50% of standard metabolism in endo- and ectothermal organisms (Brand *et al.*, 1994; Rolfe and Brand, 1996; Brookes *et al.*, 1998). Guderley and St-Pierre (2002) summarized the different mitochondrial strategies of water breathers to cope with cold temperatures and emphasized the key role of proton leak in the regulation of respiratory capacity. In Antarctic species the finding of a high Arrhenius activation energy (E_a) for proton leak rates was thought to minimize dissipative proton flux despite high mitochondrial densities (Hardewig *et al.*, 1999; Pörtner *et al.*, 2000a). A high thermal response of proton leak rates results in Antarctic organisms. In consequence, a mismatch between aerobic ATP production and demand and thus the need for complementary anaerobic ATP production will develop over a narrower temperature range than in eurytherms. The advantage of reduced energy costs for maintenance thereby contributes to an early loss in aerobic scope and thus reduced heat tolerance (Pörtner, 2006). These cellular phenomena complement the whole organism trade-offs in energy budget described above (see Section 1.2).

In this context the roles of uncoupling proteins (UCPs) and their adjustment with cold or warm acclimation and adaptation require consideration. UCPs are located in the inner mitochondrial membrane and uncouple ADP phosphorylation from substrate oxidation by increasing transmembrane proton conductance resulting in the short-circuiting of the redox reaction (Klingenberg *et al.*, 2001). In endothermic mammals UCP1 is used in heat production (Klingenberg and Huang, 1999; Stuart *et al.*, 1999a). UCP homologs (UCP2-5) found in ectothermic organisms suggest further functions unrelated to thermogenesis (Brand *et al.*, 1999; Stuart *et al.*, 1999b; Sokolova and Sokolov, 2005; Mark *et al.*, 2006). To date three mechanisms are discussed: (1) involvement in fatty acid oxidation (Fleury *et al.*, 1997; Samec *et al.*, 1998; Ricquier and Bouillaud, 2000); (2) suppression of oxidative stress (Echtay *et al.*, 2002; Liang *et al.*, 2003; Mark *et al.*, 2006); and (3) facilitation of metabolic flux by futile cycling (Mark *et al.*, 2006). The production of reactive oxygen species (ROS) is significantly increased by mitochondrial electron transport when the proton electrochemical gradient (PEG) across the inner mitochondrial membrane is high. This correlation is explained by a putative feedback loop formed by a PEG-dependent inhibition of further electron flow down the electron transport chain, associated with situations of insufficient ADP availability or reduced activity of ATP synthase.

Mark *et al.* (2006) observed increased UCP2 expression in two zoarcid species, *Z. viviparus* (North Sea) and *P. brachycephalum* (Antarctic) upon cooling or warming, respectively. Nonetheless, mitochondrial proton leak rates remained unchanged (Lannig *et al.*, 2005). Mark *et al.* (2006) suggested that UCP might balance both ATP turnover and ROS formation by controlling the mitochondrial membrane potential. Furthermore, UCP expression paralleled the increased HIF activity upon warming of the Antarctic eelpout and cooling of the temperate eelpout (Heise *et al.*, 2007). Both may thus play a role in controlling tissue oxygenation, metabolic capacity, and oxidative stress. Further work is necessary to evaluate these hypotheses.

3.2. Membrane Structure: Functional Implications and Costs

Membranes play a central role in temperature adaptation (White and Somero, 1982) as the fluidity of the membrane lipid bilayer responds immediately to temperature change and can seriously impair physiological function (Hazel, 1988; Hazel and Williams, 1990). Membrane-mediated processes such as ion and acid-base regulation depend on the maintenance of membrane structure and fluidity. When measured at a common temperature, membrane fluidity was found to decrease with increasing body temperature (see Hazel and Williams, 1990; and references therein). Membrane fluidity was highest in Antarctic fish and decreased in the following order: Antarctic fish (−1°C) > Arctic fish (0°C) > goldfish (5°C) > goldfish (25°C) > pupfish (34°C) > rat (37°C), and is thereby related to thermal windows. When measured at the respective body temperature membrane fluidity was similar among animals, with slightly elevated values in the "warm animals" (*homeoviscous adaptation*, see Sinensky, 1974; Cossins *et al.*, 1981; Hazel and Williams, 1990). In both cold-acclimated and cold-adapted animals cold exposure initiates a rise in the content of unsaturated fatty acids and in the number of double bonds to offset the negative effect of low temperatures on membrane fluidity (Guderley *et al.*, 1997; Bock *et al.*, 2001; for review see, e.g., Wodtke, 1981; Hazel and Williams, 1990; Cossins, 1994). In addition to homeoviscous adaptation McElhaney (1984a, b) introduced the term *homeophasic adaptation*, which refers to alterations in the lipid phase state after thermal compensation of membrane functioning has been observed in the absence of homeoviscous adaptation and vice versa (for more detail see Hazel and Williams 1990; Hazel, 1995).

The level of membrane fluidity and enzyme activity of membrane-associated proteins, such as Na^+/K^+ or Ca^{2+}-ATPases are strongly correlated indicating that the surrounding lipid milieu is of great importance for enzymatic performance (Hazel, 1972; Cossins *et al.*, 1981; Hazel and Williams, 1990; for review see Hoch, 1992). Mitochondrial membranes seem to exhibit stronger temperature-induced modification than membranes of other subcellular compartments (Cossins and Prosser, 1982; Hazel and Williams, 1990). As shown by Guderley *et al.* (2008) temperature-dependent mitochondrial respiration rates are strongly affected by membrane fatty acid composition. Thermal impact on mitochondrial capacities was shown to differ among fish fed with diets of different fatty acid composition. Alterations in membrane composition, particularly in the degree of unsaturation of membrane lipids, might be associated with shifts in temperature-dependent breaks in Arrhenius plots of biochemical processes (known as Arrhenius Break Temperature, ABT). Following cold acclimation lower ABTs were determined e.g., for succinate oxidation by liver mitochondria of carp, *Cyprinus carpio* (Wodtke, 1976) or for Na^+/K^+-ATPase activity in gill of eel, *Anguilla anguilla* (Thomson *et al.*, 1977). Furthermore, ABTs of mitochondrial respiration rates, mainly observed for uncoupled respiration, correlated with the natural habitat temperature of the organism. Lower ABTs, albeit far above habitat temperature, were found in polar species (Weinstein and Somero, 1998).

Accordingly, the temperature-induced impairment and thus thermal sensitivity of cellular functioning is affected by membrane composition. Hypoxemia may play a role or exacerbate these relationships by enhancing the level of oxidative stress and its impact on membrane structure.

As changes in membrane properties occur without central input (Pearson *et al.*, 1999) and in isolated cells (Koban, 1986; Tsugawa and Lagerspetz, 1990), membrane alterations seem, at least in part, to be regulated by cellular temperature alone. Response times differ between warm- and cold-induced metabolic adjustments at least for membrane modification (see below). Short-term adjustments occur between 8 and 24 h and refer to rapid and mostly nonmaintained changes such as alterations in headgroup composition of phospholipids. Long-term adjustments occur more slowly (over days) and are considered responsible for the observed differences in membranes from "cold and warm" animals, e.g., different levels of polyunsaturated fatty acids. The direction of thermal acclimation affects the duration of long-term adjustments from 1 week during warm acclimation to several weeks during cold acclimation (for review see Cossins and Raynard, 1987; Hazel and Williams, 1990). The efficiency of homeoviscous adaptation varies with cell/organelle type and metabolic performance in a way that the degree of fluidity compensation was shown to be highest in mitochondria and decreased in the following order: mitochondria > synaptosomes > myelin (Cossins and Prosser, 1982). The observed heterogeneous response dependent on membrane type as well as the observed slightly lower efficiency of homeoviscous adaptation in cold-acclimated versus cold-adapted organisms led to the suggestion that "either the costs of perfect compensation are too high or the benefits too low to warrant such a pattern of adaptation" (see Hazel and Williams, 1990). Increased ratios of polyunsaturated over monounsaturated fatty acids result in increased molecular activity of membrane proteins. Hulbert and Else (1999) proposed that membrane acyl chain composition can therefore act as a pacemaker for standard metabolism since most of the processes relevant for metabolism such as ion pumps, proton leak, protein synthesis, or oxidative phosphorylation are carried out by membrane-bound systems. In fact, Pernet *et al.* (2008) showed that growth rates are highest with a reduction in standard metabolism and membrane unsaturation index. In the light of minimized proton leak rates in Antarctic animals we propose that this mechanism contributed to enable cold-adapted Antarctic stenotherms to benefit from a highly efficient homeoviscous adaptation without concomitant increments in energy costs. This contributes to lower standard metabolic rates than observed in cold-adapted eurytherms. Higher growth rates seen in Antarctic than in Arctic zoarcids (Brodte *et al.*, 2006) would be in line with these considerations.

3.3. Calcium Homeostasis and Functioning

Calcium (Ca^{2+}) participates in numerous biochemical and physiological processes and adopts a central role in biological systems. It plays a key role in cardiac contraction and relaxation (Bers, 2002), which are critical in thermal tolerance due to their importance in oxygen delivery to tissues. In contrast to mammals where calcium is released from sarcoplasmic reticulum (SR Ca^{2+} release), cardiac function in fish strongly depends on extracellular calcium influx via sarcolemmal L-type calcium channels (SL Ca^{2+} influx) and Na^{+}/Ca^{2+} exchange (Vornanen, 1997; 1999; Hove-Madsen and Tort, 1998; for review see Tibbits *et al.*, 1992; Farrell, 1996; Lillywhite *et al.*, 1999). However, the partitioning between intracellular versus extracellular Ca^{2+} handling for cardiac contraction seems to depend on the fish's lifestyle. In more active fishes such as tuna or trout SR Ca^{2+} stores were shown to be more important (Shiels *et al.*, 1999; Brill and Bushnell, 2001; Landeira-Fernandez *et al.*, 2004). High rates of cardiac SR Ca^{2+}-ATPase activity and of SR Ca^{2+} uptake in tunas indicate "an important evolutionary step for the maintenance of higher heart rates . . .in bluefin tuna" (Castilho *et al.*, 2007). Furthermore, sources of calcium appear to be different between ventricular and atrial myocytes. In ventricular stripes of rainbow trout, *O. mykiss*, Keen *et al.* (1994) observed no contribution of SR Ca^{2+}-release channel in beat-to-beat regulation of cardiac contractility at routine heart rate (>0.6 Hz). In contrast, Aho and Vornanen (1999) observed significant ryanodine sensitivity (ryanodine = a specific and potent inhibitor of SR Ca^{2+} release, Rousseau *et al.*, 1987) in trout atrium at physiological pacing rate.

Inadequate calcium regulation during temperature change would lead to impaired cardiac performance followed by limited oxygen availability at the tissue level finally resulting in hypoxemia (see also Chapter 7). Thus several studies proposed a temperature-dependent alteration in the interplay between SL Ca^{2+} flux and SR Ca^{2+} flux to maintain adequate calcium levels to maintain cardiac performance during temperature change. At high temperatures significant ryanodine-induced impairment in cardiac performance was observed in rainbow trout, *O. mykiss*, at 15, 20, and 22°C (ventricular; Hove-Madsen, 1992; Shiels and Farrell, 1997) and in skipjack tuna, *Katsuwonus pelamis*, at 25°C (atrial; Keen *et al.*, 1992). In contrast, no ryanodine sensitivity was observed at low temperatures in ventricular myocytes of rainbow trout suggesting that cardiac contraction does not depend on SR Ca^{2+} release in the cold (Keen *et al.*, 1994). In general, ryanodine insensitivity in trout ventricular cells was observed at temperatures below 15°C and might be linked to the thermal response of the SR Ca^{2+}-release channels, which tend to remain open at low temperatures (Sitsapesan *et al.*, 1991; Hove-Madsen, 1992; Keen *et al.*, 1992; Gesser, 1996; Shiels and Farrell, 1997). Interestingly, cold acclimation did not change this situation. No effect of ryanodine was observed at low compared to high test temperatures indicating that SR Ca^{2+}-release channels were expressed but played no significant role in ventricular contractility of trout at low temperatures (Keen *et al.*, 1994; Vornanen, 1996).

However, a significant contribution of SR Ca^{2+} flux to contractility following cold acclimation to 4°C was found at low temperatures in trout atrial myocytes (Aho and Vornanen, 1999; see also Gesser, 1996; Shiels *et al.*, 1999). Furthermore, cold-induced compensation of Ca^{2+} handling capacity of the sarcoplasmic reticulum (SR) was also suggested for the ventricle of rainbow trout, *O. mykiss* (Keen *et al.*, 1994; Aho and Vornanen, 1998). When measured at the same temperature Aho and Vornanen (1998) observed higher rates of SR Ca^{2+} uptake in ventricular homogenates of cold- compared to warm-acclimated fish indicating that SR compensated its capacity for calcium load in the cold, while "the Ca^{2+} release channels are not leaky in the cold." In contrast to the suggested complete thermal compensation in ventricular SR Ca^{2+} uptake rate in trout, crucian carp showed reduced SR Ca^{2+} uptake rates at unchanged ryanodine sensitivity following cold acclimation (Vornanen, 1996; Aho and Vornanen, 1998). The authors linked the observed difference in SR Ca^{2+} sequestration to the life styles of the species. Fishes that remain rather active and need adequate cardiac functioning in the cold have higher Ca^{2+}-handling capacity in cardiac (ventricular) SR than less active and cold-dormant species like carp.

Cold temperate species such as the burbot, *Lota lota*, showed significant cold-induced ryanodine sensitivity through a ryanodine-induced reduction of maximum cardiac force by 32 ± 8% in atrial and by 16 ± 3% in ventricular preparations when measured at 1°C (Tiitu and Vornanen, 2002a). This indicates that SR Ca^{2+} release is significantly involved in delivering calcium for cardiac contraction, thereby supporting active performance and associated cold tolerance in this cold-adapted species. SR Ca^{2+} release channels are modified in a way to offset the observed increase in the open probability of the channels with acute cooling (Bers, 1987; Sitsapesan *et al.*, 1991). Furthermore, Ca^{2+}-dependent activation of the ventricular SR Ca^{2+} release channels was found to similar degrees in burbot and rat suggesting that Ca^{2+}-induced Ca^{2+} release (CICR) is involved during excitation-contraction coupling in this cold-adapted fish (Vornanen, 2006). This process, however, is controversial because Shiels and coauthors (Shiels *et al.*, 2006) suggested for the same species that Ca^{2+} is mainly provided via Na^{2+}/Ca^{2+} exchange (NCX) rather than via CICR. Furthermore, sarcolemmal Ca^{2+} flux also displayed cold compensation through elevated surface to volume ratios of smaller cardiac myocytes of burbot compared to larger cells of rainbow trout, resulting in reduced diffusion distances between SL and myofilaments (Tiitu and Vornanen, 2002b).

Shiels *et al.* (2000) found similar thermal responses of SL Ca^{2+} influx and a Q_{10} of around 2 in mammals and fishes, suggesting that SR Ca^{2+} release or another mechanism may support SL Ca^{2+} flux to ensure cardiac functioning during temperature change and compensate for cold exposure. When the authors excluded SR Ca^{2+} flux by adding ryanodine they could show that in physiological situations such as during appropriate action potential waveforms and at the respective test temperatures, SL Ca^{2+} influx did not change with temperature in atrial myocytes of rainbow trout, *O. mykiss*. Temperature-induced modifications in the shape of APs that were proposed to be linked to the temperature-dependent expression of sarcolemmal K^+ channels (Vornanen *et al.*, 2002), may contribute to maintain calcium homeostasis during an acute temperature change and may offset the otherwise Q_{10}-dependent decrease in SL Ca^{2+} influx (Shiels *et al.*, 2000). In a consecutive study, the authors showed that the temperature dependence of SR Ca^{2+} cycling was also mediated by relevant stimulation via shape and frequency of action potentials (Shiels *et al.*, 2002). Shape and frequency of action potentials thus appear to coordinate the role of SL and SR Ca^{2+} fluxes in cardiac contraction.

Furthermore, hormones such as adrenaline, in particular, mediate SL and SR Ca^{2+} flux during temperature change. Several studies revealed the importance of adrenaline for calcium-dependent cardiac performance (Gesser, 1996; Shiels and Farrell, 1997; Rocha *et al.*, 2007). Shiels and coauthors (2003) observed significant temperature-dependent sensitivity to adrenergic stimulation of sarcolemmal Ca^{2+} flux through the L-type Ca^{2+} channel. The response to adrenaline increased at decreasing test temperature. As an acute temperature decrease suppresses SL Ca^{2+} flux (see above) limited cardiac functioning will lead to hypoxemia in the cold *in vivo* if no rapid compensatory response sets in. This may contribute to the cold-induced undershoot phenomenon. Rapid compensation can be achieved through hormonal stimulation and reveals the importance of hormonal signaling to support cardiac performance during acute temperature stress *in vivo* and ameliorate the temperature-dependent alteration in Ca^{2+} flux (see Farrell, 1996). Long-term adjustments in calcium regulation through alterations in gene expression will offset the temperature impact on calcium fluxes and support adequate cardiac performance and thus oxygen supply to tissues in a shifted thermal window. This may, however, involve a shift in the fractional energy demand of calcium signaling and homeostasis. All of these relationships remain largely unexplored.

3.4. Energy Budget, Turnover, and Allocation

In general, the cellular as well as the organismal energy budget (as the sum of all cellular budgets) provides excess energy to growth and other functions only once the energy demand of maintenance and baseline functioning of the organism is met (Wieser, 1994). Reductions in functional scope will occur once the scope for aerobic ATP supply via oxidative phosphorylation is reduced during temperature-induced hypoxemia. Vital cell functions were proposed to display a lower sensitivity to reduced ATP supply than accessory ones (Atkinson, 1977). Shifts in energy allocation to cellular processes might result, which are usually studied using respiration rates of isolated cells under the effect of specific inhibitors. However, this methodological approach bears its risks and may only support qualitative conclusions (Mark *et al.*, 2005). The various O_2-consuming processes influence each other upon inhibition. Respiration was also shown to strongly depend on the concentrations of inhibitors as well as on the previous feeding regime of the experimental animals (see Krumschnabel and Wieser, 1994, Krumschnabel *et al.*, 1997; Wieser and Krumschnabel, 2001). Interpretation of the respective findings and comparisons between studies should be carried out with adequate precaution.

We confine this paragraph mainly to the two most prominent ATP-consumers of cells: Na^+/K^+-ATPase and protein synthesis. Protein turnover and ion-motive ATPases represent key targets of hypoxia causing energy reallocation at the cellular level (Boutilier, 2001). This has implications for temperatures beyond the optimum range when thermally induced hypoxemia sets in. Furthermore, passive ion flux and active K^+ uptake via Na^+/K^+-ATPase display largely different kinetic responses to temperature. Active ion exchange displays a Q_{10} of 2–4 while passive ion flux is rather insensitive to temperature (Ellory and Hall, 1987; Gibbs, 1995). To overcome thermal disturbance of the coupling between active K^+ uptake and passive K^+ efflux two strategies have evolved at the cellular level: compensatory adjustment either of the passive K^+ leaks or of the active Na^+/K^+-ATPase capacities and associated secondary active processes such as $Na^+/K^+/Cl^-$ exchanger that depend on ion gradients, respectively. The underlying mechanisms for the observed changes are still under debate. Temperature-induced changes in K^+ efflux are thought to include: (1) alterations in membrane properties (homeoviscous adaptation, see Section 3.2); (2) down-regulation of ion channels as observed during environmental hypoxia (Péréz-Pinzón *et al.*, 1992); and/or (3) changes in K^+ channel opening through changing concentrations of metabolite ligands (Dunne and Petersen, 1991; Hall and Willis, 1984). In Antarctic fishes with their increased serum osmolality, cost reductions by both the down-regulation of ion channels and reduced ion exchange capacities were suggested to contribute to the observed low rate of standard metabolism (Gonzalez-Cabrera *et al.*, 1995; Pörtner *et al.*, 1998; Guynn *et al.*, 2002). Benefits are increased freezing resistance and decreased energy requirements to maintain the ionic gradient (Somero and DeVries, 1967; Prosser *et al.*, 1970; O'Grady and DeVries, 1982).

This fits the recent comparisons of active and passive pH regulation between stenothermal and eurythermal fish. They revealed that eurythermal fishes mainly use more costly active processes such as carriers dependent on Na^+/K^+-ATPase, whereas cold-stenotherms rather depend on passive processes like nonbicarbonate buffering (Bock *et al.*, 2001; Sartoris *et al.*, 2003, see Section 2.2).

Mark *et al.* (2005) observed a somewhat lower ouabain-sensitive respiration in hepatocytes of high-Antarctic compared to sub-Antarctic notothenids indicating lower capacities for active ion regulation in the former. Following cold acclimation hepatocytes of rainbow trout, *O. mykiss*, showed no compensatory increase in Na^+/K^+-ATPase but achieved a balanced ion regulation at a lower rate through a down-regulation of K^+ efflux (Krumschnabel *et al.*, 1997). In contrast, hepatocytes of roach, *Rutilus rutilus*, showed near complete cold compensation of ion homeostasis following acclimation, through increased Na^+/K^+-ATPase activity and increased $Na^+/K^+/Cl^-$ cotransport activity (Krumschnabel *et al.*, 1997). The latter is insensitive to ouabain and represents a secondary active transport at the expense of Na^+ and Cl^- gradient thereby saving ATP. The observed species-specific differences in ion regulation strategies with temperature acclimation were in line with previous findings by Schwarzbaum *et al.* (1992) and it was concluded that the different strategies might depend on the level of eurythermy (more stenothermal salmonids versus more eurythermal cyprinids) (Krumschnabel *et al.*, 1997).

The different ion regulation strategies correlated with cellular energy expenditure as seen in the study on rainbow trout (Krumschnabel *et al.*, 1997). Similar and thus uncompensated total respiration rates were found between hepatocytes from cold- and warm-acclimated trout when measured at the same temperature. Nonetheless, differences were found in energy allocation. At low test temperatures oxygen consumption accounting for protein synthesis was increased by 10% likely through the benefit of reduced costs for ion regulation in cold- versus warm-acclimated fish (Krumschnabel *et al.*, 1997). Available data on Antarctic fishes (Mark *et al.*, 2005) revealed that a large fraction of about 28% of total cellular respiration (measured at 0°C and 0.1 mM cycloheximide) accounted for protein synthesis. An energy allocation to protein synthesis of between 20% and 25% was observed in hepatocytes of two other Antarctic fish species, *Lepidonotothen kempi* and *P. brachycephalum* (measured at 2°C and 0.03 mM cycloheximide; Langenbuch and Pörtner, 2003). Compensation of protein synthesis capacity for temperature is complete in polar ectotherms (Storch *et al.*, 2003, 2005) and likely supports higher growth efficiency (Heilmayer *et al.*, 2004). However, this point is still somewhat controversial. Conflicting results exist for protein synthesis costs at low temperatures and in polar versus boreal animals, respectively (Pannevis and Houlihan, 1992; Whiteley *et al.*, 1996; Marsh *et al.*, 2001; Storch *et al.*, 2003, 2005; Pace *et al.*, 2004; for review see Fraser and Rogers, 2007). Furthermore, it seems that the costs for protein synthesis are negatively correlated with rates for protein synthesis such that at low synthesis rates the cost is elevated due to a suggested fixed cost component (Pannevis and Houlihan, 1992; Smith and Houlihan, 1995).

To evaluate temperature effects on energy allocation Mark *et al.* (2005) exposed isolated hepatocytes of various Antarctic fish species to an acute temperature rise (up to 15°C). Evidence for a temperature-dependent shift in ATP-consuming processes was minor. There was no effect on energy partitioning between ion regulation (Na^+/K^+-ATPase) or oxidative phosphorylation (Mark *et al.*, 2005). The authors concluded that shifts in energy allocation might become effective during systemic hypoxemia. Pannevis and Houlihan (1992) compared temperature-dependent respiration rates and absolute protein synthesis rates of cells obtained from 10°C acclimated rainbow trout, *O. mykiss*. The authors observed a thermal optimum of protein synthesis rates at intermediate temperatures: 40–50 ng mg protein^{-1} min^{-1} (14–18°C) compared to 15–30 ng mg protein^{-1} min^{-1} (5–10°C and 20°C, respectively). Interestingly, the authors noted no clear evidence for temperature-dependent differences in % cycloheximide inhibition of cellular respiration. The data may also suggest reduced protein synthesis at 20°C, which, however, was not visible when only cycloheximide-sensitive respiration rates were measured.

Furthermore, respiration rates and the fraction of cellular protein synthesis were highest in cells from animals fed ad libitum (Krumschnabel *et al.*, 1997) indicating that the condition of the fish influences the results making comparative approaches more complicated. Overall, the picture of temperature-dependent changes in cellular energy allocation is very unclear and presently does not support the elaboration of unifying principles. Clear control of the acclimation and feeding regime of the fishes as well as monitoring of cellular acclimation processes after isolation may provide a clearer picture. Further investigations are also needed depending on the lifestyle and physiology of the species studied (Wieser and Krumschnabel, 2001). For an appropriate evaluation of temperature-induced changes in cellular energy allocation, measurements should also be performed under simulated *in vivo* conditions, for example, at realistic levels of temperature-dependent tissue or venous Po_2 values rather than at 100% air saturation, which is currently used in most investigations of cellular respiration.

Figure 4.6 lists the various mechanistic aspects covered by the present chapter, with a focus on processes at the cellular level. For only a few of them the interaction between those processes and the interaction of temperature and hypoxemia effects have been adequately explored. To further the study

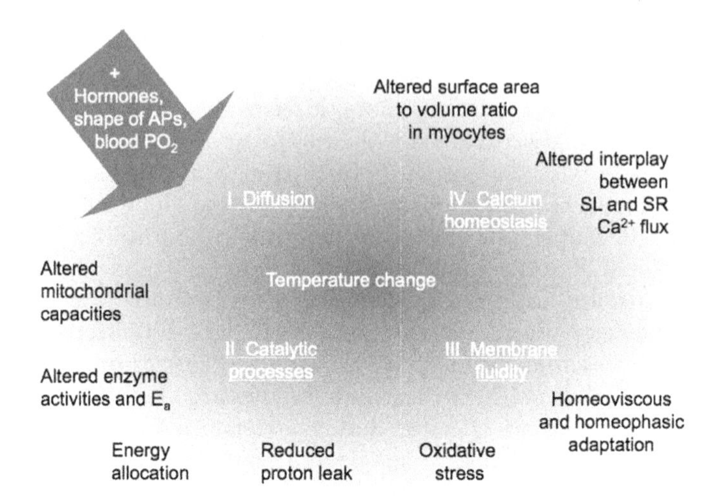

Fig. 4.6. Diagram listing key cellular parameters (I–IV) affected by temperature change and associated responses to overcome the resulting impairment in cellular metabolism. Factors such as hormones, the shape of action potentials (APs), or blood and tissue oxygen levels (Po_2) contribute to modulate the thermal impact on cellular processes. (SL = sacrolemmal and SR = sacroplasmic reticulum).

of such interactions toward an integrative picture of temperature adaptation, inclusion of the role of hypoxemia still needs to be comprehensively developed, as a challenge for the years to come. For example, inadequate calcium regulation during temperature change would lead to impaired cardiac performance followed by impaired oxygen supply finally resulting in hypoxemia. Cardiac failure at critical temperatures may mainly depend on insufficient oxygen supply to myocytes. As many fish species lack a coronary system the heart's oxygen supply relies on venous Po_2 (Farrell, 1993) and specific oxygen thresholds may exist where cardiac arrhythmia sets in (see Farrell and Clutterham, 2003 and references therein). Release of stress hormones, partly as a consequence of hypoxemia, ameliorates calcium regulation during acute temperature stress, thereby improving cardiac functioning and, in turn, alleviating hypoxemia effects.

One defence mechanism in response to hypoxic conditions is to reallocate cellular energy between energy consumers as aerobic ATP supply becomes limited, following a putative priority from less essential to more essential ATP-consuming processes. Furthermore, metabolic depression sets in, which involves a reduction in protein synthesis and in the cost of ion and acid-base regulation. Such cellular energy reallocation is likely more pronounced under temperature stress combined with oxygen limitation than under temperature stress alone. Temperature and hypoxemia also come together in shaping the molecular signaling responses in the cell as exemplified in Figure 4.1.

To overcome the threat of temperature-induced hypoxemia fishes either undergo metabolic depression (Hardewig *et al.*, 2004) or maintain activity levels through metabolic acclimation to the new thermal regime (Seebacher *et al.*, 2005; Lannig *et al.*, 2005). Systemic, cellular, and molecular responses to hypoxemia at both sides of the thermal window need to be elaborated for a comprehensive understanding of thermal adaptation. Available data suggest a role for stress hormones, adenosine and hypoxia inducible factor HIF-1, and redox state in shaping temperature-dependent acclimation.

Mechanisms involved in thermal adaptation include the homeoviscous adaptation of membranes and changes in mitochondrial capacities, as well as altered Arrhenius activation energies to compensate for the temperature-induced alteration in aerobic energy metabolism. These mechanisms also support cardiac performance, in combination with alterations in calcium homeostasis. They include a cold-induced increase in the contribution of intracellular calcium cycling such as SR Ca^{2+} release to calcium homeostasis, which is otherwise mainly determined by sarcolemmal Ca^{2+} flux. Owing to the reliance of thermal tolerance on adequate oxygen supply for aerobic scope, those using mostly sarcolemmal Ca^{2+} flux are less tolerant to acute temperature changes. Further study is needed to qualify and quantify these interdependencies.

4. PERSPECTIVES: HYPOXIA-SENSITIVE THERMAL WINDOWS IN CLIMATE SENSITIVITY

Temperature and hypoxia would traditionally be considered as different environmental factors, each of which has its specific implications for whole organism functioning. Development of the concept of oxygen and capacity limited thermal tolerance has revealed how these factors are intertwined, since thermal stress causes systemic hypoxemia and the interaction of temperature and thermally induced hypoxemia shapes adaptive responses at various molecular to whole organism levels. The integration of these responses supports adjustment and maintenance of metabolic and functional performance at cellular, tissue, and whole organism levels.

These principles play an important role in the context of climate change effects on ecosystems. These are largely due to the current trends of warming in the world's oceans caused by anthropogenic CO_2 accumulation. On larger scales, effects include shifts in geographical distribution such as the observed poleward shifts of phytoplankton, makroalgae, and marine-ectothermal animals along latitudinal clines (Lüning, 1990; Southward *et al.*, 1995; Harrington *et al.*, 1999; Walther *et al.*, 2002; Parmesan and Yohe, 2003; Root *et al.*, 2003; Perry *et al.*, 2005). On smaller scales they also include local decreases in abundances of previously common species with the risk of local extinction of species or even ecosystems like coral reefs (Parmesan and Yohe, 2003; Thomas *et al.*, 2004; Perry *et al.*, 2005; Hoegh-Guldberg *et al.*, 2007; Pörtner and Knust, 2007). In the German Wadden Sea, for example, the falling frequency of colder winters and increased occurrence of warmer summers are key in shaping population structure and community composition (Kröncke *et al.*, 1998; Günther und Niesel, 1999; Pörtner and Knust, 2007). In general, observed ecosystem changes are related more to temperature anomalies (changes in thermal maxima or minima) than to changing temperature means (Stachowicz *et al.*, 2002; Stenseth and Mysterud, 2002), with larger changes and effects at high latitudes (Root *et al.*, 2003). For a long time, the background and relevance of such observations has been obscured by the absence of an understanding of mechanistic cause and effect (Jensen, 2003).

The observations by Perry *et al.* (2005) include the finding that the Northward geographical shifts of various fish species (snake blenny, anglerfish, and cod) occur to various degrees. This may reflect different thermal sensitivities of species coexisting in an ecosystem as a background for changes in community composition. Regime shifts may result like the one between colder years dominated by sardines and warmer years dominated by anchovies on the Pacific coast of Japan (Takasuka *et al.*, 2007). In addition to direct effects of temperature on individual species indirect effects have to be identified at ecosystem level. Here, temperature-dependent changes in species interactions as in the food web may occur and exert their effect at higher levels of the food cascade. As a prominent example, the shift in copepod faunal composition from larger *Calanus finmarchicus* to smaller *Calanus helgolandicus* in the Southern North Sea was seen as a major reason for the decrease in Atlantic cod (*Gadus morhua*) population, due to the reduced size of food particles available for juvenile cod (Beaugrand *et al.*, 2003). More recently, Helaouët and Beaugrand (2007) showed that among the various parameters of the physical environment tested, temperature dominated the distribution of the copepod species, such that the warming trend caused the shift to the smaller species. These results indicate that direct physiological effects of temperature on potential prey can then cause indirect effects of temperature on the predator (cod). The principal understanding of direct temperature effects on individual species is thus key to an understanding of climate-induced changes in species interactions at ecosystem level. This line of thought is in line with findings in terrestrial organisms (higher plants, insects, birds) where climate effects on biogeography and biodiversity are independent from the position of the respective species in the food chain (Huntley *et al.*,

2004). Thermal windows are thus decisive in shaping biogeography, a concept that warrants further analyses in aquatic ecosystems.

Sensitivity to climate results from the specialization of species on climate regimes that include a limited range of temperatures which match a species-specific thermal window. In principle, this thermal window can be expected to match the temperature-dependent range of geographical distribution of a species and thereby defines the response to changing temperatures on both the cold and warm sides of the thermal window. The regime shifts from sardines during colder years to anchovies during warmer years (Takasuka *et al.*, 2007) illustrate such effects. In the case of Atlantic cod the warming-induced reduction of recruitment in the North Sea (Pörtner *et al.*, 2001; Colosimo *et al.*, 2003) and the Northward shift of the species (Perry *et al.*, 2005) are mirrored in the warming-induced increase in recruitment on the cold side of its thermal window, in the Arctic Barents Sea. The direct effects of cold versus warm temperatures on recruitment in various marine provinces strongly suggest that temperature directly influences individual species.

Limited thermal windows of a species relate to limited windows of temperature-dependent growth. These limits are in close association with the temperatures causing climate-induced shifts at ecosystem level (Pörtner and Knust, 2007; Takasuka *et al.*, 2007). In this context it needs to be considered that fluctuating food availability, as an indirect temperature effect mediated through the food chain, may modulate the optimal temperature of growth.

The seasonal timing of events is also being influenced by climate change and contributes to modulate ecosystem dynamics, as exemplified in the North Sea by later development of diatom blooms due to later grazing in previous years (Wiltshire and Manly, 2004), by earlier development of zooplankton (Greve *et al.*, 1996), or by earlier migratory movements of, for example, squid (*Loligo forbesi*) into the North Sea (Sims *et al.*, 2001). During a warming scenario such shifts in timing may be understood as an earlier entry of ambient temperature into species-specific thermal windows of performance during spring (in the case of zooplankton development or squid migration) or their later exit out of this window in the fall (in the case of later diatom grazing by zooplankton).

As a corollary, the physical environment and especially temperature associated with temperature-induced hypoxemia exert large effects on individual member species of a marine ecosystem. Changes in biogeographical distribution result, mirrored in shifts in abundance, species composition, and in species interactions, e.g., via changes in food web composition, at the edges of the thermal window of a species. A mechanistic analysis relies on an understanding of how these environmental conditions exert their direct limiting or supporting effects on individual species and why species specialize on limited environmental windows. Such an understanding of some of the unifying principles of adaptation and limitation has emerged over recent years for effects of temperature (for review see Pörtner *et al.*, 2005a; Pörtner, 2001, 2002) and supported an understanding of the temperature-dependent evolution of animal species and phyla (Pörtner, 2004, 2006).

The present chapter was intended to show that the principles of thermal adaptation and of temperature-dependent oxygen supply through circulation or ventilation are intertwined to an extent that oxygen supply capacity sets the earliest limits of thermal tolerance through the development of hypoxemia. Any factor depressing oxygen supply capacity will thus affect thermal tolerance by narrowing thermal tolerance windows (Pörtner *et al.*, 2005c). Ambient hypoxia reduces oxygen availability and thereby capacity and thermal tolerance. Elevated ambient CO_2 levels frequently parallel aquatic hypoxia and independently but also synergistically cause enhanced sensitivity to thermal extremes (Metzger *et al.*, 2007). However, these relationships remain unexplored in fishes. Altogether, such findings clearly show that cause and effect analyses of past and future climate impacts on ecosystems need to take the synergistic physiological effects of various environmental factors like temperature, carbon dioxide, and hypoxia into account.

ACKNOWLEDGMENTS

Supported by the MarCoPolI program of the Alfred-Wegener-Institute.

REFERENCES

Abele, D., and Puntarulo, S. (2004). Formation of reactive species and induction of antioxidant defence systems in polar and temperate marine invertebrates and fish. *Comp. Biochem. Physiol. A* **138,** 405–415.

Abele, D., Burlango, B., Viarengo, A., and Pörtner, H. O. (1998). Exposure to elevated temperatures and hydrogen peroxide elicits oxidative stress and antioxidant response in the Antarctic intertidal limpet *Nacella concinna*. *Comp. Biochem. Physiol. B* **120,** 425–435.

Abele, D., Tesch, C., Wencke, P., and Pörtner, H. O. (2001). How does oxidative stress relate to thermal tolerance in the Antarctic bivalve *Yoldia eightsi*? *Antarct. Sci.* **13,** 111–118.

Aho, E., and Vornanen, M. (1998). Ca^{2+}-ATPase activity and Ca^{2+} uptake by sarcoplasmic reticulum in fish heart: Effects of thermal acclimation. *J. Exp. Biol.* **201,** 525–532.

Aho, E., and Vornanen, M. (1999). Contractile properties of atrial and ventricular myocardium of the heart of rainbow trout *Oncorhynchus mykiss*: Effects of thermal acclimation. *J. Exp. Biol.* **202,** 2663–2677.

Aho, E., and Vornanen, M. (2001). Cold acclimation increases basal heart rate but decreases its thermal tolerance in rainbow trout (*Oncorhynchus mykiss*). *J. Comp. Physiol. B* **171,** 173–179.

Anestis, A., Lazou, A., Pörtner, H. O., and Michaelidis, B. (2007). Behavioural, metabolic and molecular stress indicators in the marine bivalve *Mytilus galloprovincialis* during long-term acclimation at increasing ambient temperature. *Am. J Physiol.* **293,** R911–921.

Atkinson, D. (1977). "Cellular Metabolism and its Regulation" Academic Press, New York. p. 218.

Beaugrand, G., Brander, K. M., Lindley, J. A., Souissi, S., and Reid, P. C. (2003). Plankton effect on cod recruitment in the North Sea. *Nature* **426,** 661–664.

Bers, D. M. (1987). Ryanodine and the Ca^{2+} content of SR assessed by caffeine and rapid cooling contractures. *Am. J. Physiol.* **253,** C408–C415.

Bers, D. M. (2002). Cardiac excitation–contraction coupling. *Nature* **415,** 198–205.

Bock, C., Sartoris, F. J., Wittig, R. M., and Pörtner, H. O. (2001). Temperature-dependent pH regulation in stenothermal Antarctic and eurythermal temperate eelpout (Zoarcidae): An *in-vivo* NMR study. *Polar Biol.* **24,** 869–874.

Boutilier, R. G. (2001). Mechanisms of cell survival in hypoxia and hypothermia. *J. Exp. Biol.* **204,** 3171–3181.

Brand, M. D., Couture, P., Else, P. L., Withers, K. W., and Hulbert, A. J. (1991). Evolution of energy metabolism. Proton permeability of the inner membrane of liver – mitochondria is greater in a mammal than in a reptile. *Biochem. J.* **275,** 81–86.

Brand, M. D., Chien, L. F., Ainscow, E. K., Rolfe, D. F. S., and Porter, R. K. (1994). The causes and functions of mitochondrial proton leak. *Biochim. Biophys. Acta* **1187,** 132–139.

Brand, M. D., Brindle, K. M., Buckingham, J. A., Harper, J. A., Rolfe, D. F. S., and Stuart, J. A. (1999). The significance and mechanisms of mitochondrial proton conductance. *Int. J. Obes. Relat. Metab. Disord. (Suppl.)* **23,** S4–S11.

Brill, R. W., and Bushnell, P. G. (2001). The cardiovascular system of tunas. *In* "Tunas: Physiology, Ecology and Evolution, Vol. 19" (Block, B. A., and Stevens, E. D., Eds.), pp. 79–120. Academic Press, San Diego.

Brodte, E., Knust, R., and Pörtner, H. O. (2006). Temperature-dependent energy allocation to growth in Antarctic and boreal eelpout (Zoarcidae). *Polar Biol.* **30,** 95–107.

Brookes, P. S., Buckingham, J. A., Tenreiro, A. M., Hulbert, A. J., and Brand, M. D. (1998). The proton permeability of the inner membrane of liver mitochondria from ectothermic and endothermic vertebrates and from obese rats: orrelation with standard metabolic rate and phospholipids fatty acid composition. *Comp. Biochem. Physiol. B* **119,** 325–334.

Buck, L. T. (2004). Adenosine as a signal for ion channel arrest in anoxia-tolerant organisms. *Comp. Biochem. Physiol. B* **139,** 401–414.

Burton, R. F. (2002). Temperature and acid–base balance in ectothermic vertebrates: The imidazole alphastat hypotheses and beyond. *J. Exp. Biol.* **205,** 3587–3600.

Cameron, J. N. (1989). Acid-base homeostasis: Past and present perspectives. *Physiol. Zool.* **62,** 845–865.

Castilho, P. C., Landeira-Fernandez, A. M., Morissette, J., and Block, B. A. (2007). Elevated Ca^{2+}ATPase (SERCA2) activity in tuna hearts: Comparative aspects of temperature dependence. *Comp. Biochem. Physiol. A* **148,** 124–132.

Chabi, B., Ljubicic, V., Menzies, K. J., Huang, J. H., Saleem, A., and Hood, D. A. (2008). Mitochondrial function and apoptotic susceptibility in aging skeletal muscle. *Aging Cell* **7**(1), 2–12.

Chen, W. H., Sun, L. T., Tsai, C. L., Song, Y. L., and Chang, C. F. (2002). Cold-stress induced the modulation of catecholamines, cortisol, immunoglobulin M, and leukocyte phagocytosis in tilapia. *Gen. Comp. Endocrinol.* **126,** 90–100.

Clarke, A., and Johnston, N. M. (1999). Scaling of metabolic rate and temperature in teleost fish. *J. Annu. Ecol.* **68,** 893–905.

Colosimo, A., Giuliani, A., Maranghi, F., Brix, O., Thorkildsen, S., Fischer, T., Knust, R., and Pörtner, H. O. (2003). Physiological and genetical adaptation to temperature in fish populations,. *Cont. Shelf Res.* **23,** 1919–1928.

Cossins, A. R. (1994). Homeoviscous adaptation of biological membranes and its functional significance. *In* "Temperature Adaptation of Biological Membranes" (Cossins, A. R., Ed.), pp. 63–76. Portland Press, London.

Cossins, A. R., and Bowler, K. (1987). "Temperature Biology of Animals." Chapman and Hall, London.

Cossins, A. R., and Prosser, C. L. (1982). Variable homeoviscous responses of different brain membranes of thermally-acclimated goldfish. *Biochim. Biophys. Acta* **687,** 303–309.

Cossins, A. R., and Raynard, R. S. (1987). Adaptive responses of animal cell membranes to temperature. *Sym. Soc. Exp. Biol.* **21,** 95–112.

Cossins, A. R., Prosser, C. L., and Bowler, K. (1981). Homeoviscous adaptation and its effects upon membrane-bound proteins. *J. Therm. Biol.* **6,** 183–187.

Crockett, E. L., and Sidell, B. D. (1990). Some pathways of energy metabolism are cold adapted in Antarctic fishes. *Physiol. Zool.* **63,** 472–488.

Dahlhoff, E., and Somero, G. N. (1993). Effects of temperature on mitochondria from abalone (genus *haliotis*): Adaptive plasticity and its limits. *J. Exp. Biol.* **185,** 151–168.

Davis, K. B. (2004). Temperature affects physiological stress responses to acute confinement in sunshine bass (*Morone chrysops* × *Morone saxatilis*). *Comp. Biochem. Physiol. A* **139,** 433–440.

Dunlap, W. C., Fujisawa, A., Yamamoto, Y., Moylan, T., and Sidell, B. D. (2002). The vitamin E content of notothenioid fish, krill and phytoplankton from Antarctica includes a vitamin E constituent (α- tocomonoenol) functionally associated with cold water adaptation. *Comp. Biochem. Physiol. B* **133,** 299–305.

Dunne, M. J., and Petersen, O. H. (1991). Potassium selective ion channels in insulin-secreting cells: Physiology, pharmacology and their role in stimulus-secretion coupling. *Biochim. Biophys. Acta* **1071,** 67–82.

Echtay, K. S., Roussel, D., St-Pierre, J., Jekabsons, M. B., Cadenas, S., Stuart, J. A., Harper, J. A., Roebuck, S. J., Morrison, A., Pickering, S., Clapham, J. C., and Brand, M. D. (2002). Superoxide activates mitochondrial uncoupling proteins. *Nature* **415,** 96–99.

Eckerle, L. (2008). "Signals and Molecular Mechanisms of Temperature Adaptation of Mitochondrial Functions in Marine Fish." Ph.D. dissertation, Bremen.

Eckerle, L. G., Lucassen, M., Hirse, T., and Pörtner, H. O. (2008). Cold induced changes of adenosine levels in common eelpout (*Zoarces viviparus*): A role in modulating cytochrome-*c*-oxidase expression. *J. Exp. Biol.* **211,** 1262–1269.

Egginton, S., and Sidell, B. D. (1989). Thermal acclimation induces adaptive changes in subcellular structure of fish skeletal muscle. *Am. J. Physiol.* **256,** R1–R9.

Egginton, S., Stilbeck, C., Hoofd, L., Calvo, J., and Johnston, I. A. (2002). Peripheral oxygen transport in skeletal muscle of Antarctic and sub-Antarctic notothenioid fish. *J. Exp. Biol.* **205,** 769–779.

Ellory, J. C., and Hall, A. C. (1987). Temperature effects on red cell membrane transport processes. *Symposia of the Society for Experimental Biology* **21,** 53–66.

Fabbri, E., Capuzzo, A., and Moon, T. W. (1998). The role of circulating catecholamines in the regulation of fish metabolism: An overview. *Comp. Biochem. Physiol. C* **120,** 177–192.

Farrell, A. P. (1993). Cardiac output in fish: Regulation and limitations. *In* "The Vertebrate Gas Transport Cascade: Adaptations to Environment and Mode of Life" (Bicudo, E, Ed.), pp. 208–214. CRC Press, Inc., Boca Raton.

Farrell, A. P. (1996). Effects of temperature on cardiovascular performance. *In* "Global Warming: Implications for Freshwater and Marine Fish" (Wood, C. M., and McDonald, D. G., Eds.), SEB Seminar Series 61, pp. 135–158. Cambridge University Press, Cambridge.

Farrell, A. P., and Clutterham, S. M. (2003). On-line venous oxygen tensions in rainbow trout during graded exercise at two acclimation temperatures. *J. Exp. Biol.* **206,** 487–496.

Farrell, A. P., Hinch, S. G., Cooke, S. J., Patterson, D. A., Crossin, G. T., Lapointe, M., and Mathes, M. T. (2008). Pacific salmon in hot water: Applying aerobic scope models and biotelemetry to predict the success of spawning migrations. *Physiol. Biochem. Zool.* **81,** 697–708.

Fischer, T. (2003). The effects of climate induced temperature changes on cod (*Gadus morhua* L.): Linking ecological and physiological investigations. *Rep. Pol. Mar. Res.* **454,** 1–101.

Fleury, C., Neverova, M., Collins, S., Raimbault, S., Champigny, O., Levi-Meyrueis, C., Bouillaud, F., Seldin, M. F., Surwit, R. S., Ricquier, D., and Warden, C. H. (1997). Uncoupling protein-2: A novel gene linked to obesity and hyperinsulinemia. *Nat. Genet.* **15,** 269–272.

Fraser, K. P. P., and Rogers, A. D. (2007). Protein metabolism in marine animals: The underlying mechanism of growth. *Adv. Mar. Biol.* **52,** 267–363.

Frederich, M., and Pörtner, H. O. (2000). Oxygen limitation of thermal tolerance defined by cardiac and ventilatory performance in the spider crab *Maja squinado*. *Am. J. Physiol.* **279,** R1531–R1538.

Gesser, H. (1996). Cardiac force-interval relationship, adrenaline and sarcoplasmic reticulum in rainbow trout. *J. Comp. Physiol. B* **166,** 278–285.

Gibbs, A. (1995). Temperature, pressure and the sodium pump: The role of homeoviscous adaptation. *In* "Biochemistry and Molecular Biology of Fishes, Vol. 5" (Hochachka, P. W., and Mommsen, T. P., Eds.), pp. 197–212. Elsevier Science B.V., Amsterdam.

Gollock, M. J., Currie, S., Petersen, L. H., and Gamperl, A. K. (2006). Cardiovascular and haematological responses of Atlantic cod (*Gadus morhua*) to acute temperature increase. *J. Exp. Biol.* **209,** 2961–2970.

Gonzalez-Cabrera, P. J., Dowd, F., Pedibhotla, V. K., Rosario, R., Stanley-Samuelson, D., and Petzel, D. (1995). Enhanced hypo-osmoregulation induced by warm-acclimation in antarctic fish is mediated by increased gill and kidney Na^+/K^+-ATPase activities. *J. Exp. Biol.* **198,** 2279–2291.

Greve, W., Reiners, F., and Nast, J. (1996). Biocoenotic changes of the zooplankton in German Bight: The possible effects of eutrophication and climate. *ICES J. Mar. Sci.* **53,** 951–956.

Guderley, H. (2004). Metabolic responses to low temperature in fish muscle. *Biol. Rev.* **79,** 409–427.

Guderley, H., and Blier, P. (1988). Thermal acclimation in fish: Conservative and labile properties of swimming muscle. *Can. J. Zool.* **66,** 1105–1115.

Guderley, H., and St-Pierre, J. (2002). Going with the flow or life in the fast lane: Contrasting mitochondrial responses to thermal change. *J. Exp. Biol.* **205,** 2237–2249.

Guderley, H., St-Pierre, J., Couture, P., and Hulbert, A. J. (1997). Plasticity of the properties of mitochondria from rainbow trout red muscle with seasonal acclimatization. *Fish Physiol. Biochem.* **16,** 531–541.

Guderley, H., Kraffe, E., Bureau, W., and Bureau, D. P. (2008). Dietary fatty acid composition changes mitochondrial phospholipids and oxidative capacities in rainbow trout red muscle. *J. Comp. Physiol. B* **178,** 385–399.

Günther, C. P., and Niesel, V. (1999). Effects of the ice winter 1995/6. *In* "The Wadden Sea Ecosystem. Stability Properties and Mechanisms" (Dittmann, S., Ed.), pp. 193–205. Springer Verlag, Berlin.

Guynn, S., Dowd, F., and Petzel, D. (2002). Characterization of gill Na/K-ATPase activity and ouabain binding in Antarctic and New Zealand nototheniid fishes. *Comp. Biochem. Physiol. A* **131**(2), 363–374.

Hall, A. C., and Willis, J. S. (1984). Differential effects of temperature on three components of passive permeability to potassium in rodent cells. *J. Physiol. (Lond)* **348,** 629–643.

Hardewig, I., Peck, L. S., and Pörtner, H. O. (1999). Thermal sensitivity of mitochondrial function in the Antarctic notothenioid *Lepidonotothen nudifrons*. *J. Comp. Physiol. B* **169,** 597–604.

Hardewig, I., Pörtner, H. O., and van Dijk, P. L. M. (2004). How does the cold stenotherm gadoid *Lota lota* survive high water temperatures during summer? *J. Comp. Physiol. B* **174,** 149–156.

Harrington, R., Woiwod, I., and Sparks, T. (1999). Climate change and trophic interactions. *Trends Ecol. Evol.* **14,** 146–150.

Hazel, J. R. (1972). The effect of temperature acclimation upon succinate dehydrogenase activity from the epaxial muscle of the common goldfish (*Carassius auratus* L.). II. Lipid reactivation of the soluble enzyme. *Comp. Biochem. Physiol. B* **43,** 863–882.

Hazel, J. R. (1988). Homeoviscous adaptation in animal cell membranes. *In* "Advances in Comparative and Environmental Physiology – Physiological Regulation of Membrane Fluidity" (Aloia, R. C., Curtain, C. C., and Gordon, L. M., Eds.), pp. 149–188. A. R. Liss, Inc, New York.

Hazel, J. R. (1995). Thermal adaptation in biological membranes: Is homeoviscous adaptation the explanation? *Annu. Rev. Physiol.* **57,** 19–42.

Hazel, J. R., and Williams, E. E. (1990). The role of alterations in membrane lipid composition in enabling physiological adaptation of organisms to their physical environment. *Prog. Lipid Res.* **29,** 167–227.

Heilmayer, O., Brey, T., and Pörtner, H. O. (2004). Growth efficiency and temperature in scallops: A comparative analysis of species adapted to different temperatures. *Funct. Ecol.* **18,** 641–647.

Heise, K., Puntarulo, S., Nikinmaa, M., Lucassen, M., Pörtner, H. O., and Abele, D. (2006a). Oxidative stress and HIF-1 binding during stressful cold exposure and recovery in the North Sea eelpout (*Zoarces viviparus*). *Comp. Biochem. Physiol. A* **143,** 494–503.

Heise, K., Puntarulo, S., Nikinmaa, M., Abele, D., and Pörtner, H. O. (2006b). Oxidative stress during stressful heat exposure and recovery in the North Sea eelpout *Zoarces viviparus* L. *J. Exp. Biol.* **209,** 353–363.

Heise, K., Estevez, M. S., Puntarulo, S., Galleano, M., Nikinmaa, M., Pörtner, H. O., and Abele, D. (2007). Effects of seasonal and latitudinal cold on oxidative stress parameters and activation of hypoxia inducible factor (HIF-1) in zoarcid fish. *J. Comp. Physiol. B* **177,** 765–777.

Heisler, N. (1986). Comparative aspects of acid-base regulation. *In* "Acid-Base Regulation in Animals" (Heisler, N., Ed.), pp. 397–450. Elsevier North Holland, Amsterdam.

Helaouët, P., and Beaugrand, G. (2007). Macro-ecological study of the niche of *Calanus finmarchicus* and *C. helgolandicus* in the North Atlantic Ocean and adjacent seas. *Mar. Ecol. Progr. Ser.* **345,** 147–165.

Hoch, F. L. (1992). Cardiolipins and biomembrane function. *Biochim. Biophys. Acta* **1113,** 71–133.

Hoegh-Guldberg, O., Mumby, P. J., Hooten, A. J., Steneck, R. S., Greenfield, P., Gomez, E., Harvell, C. D., Sale, P. F., Edwards, A. J., Caldeira, K., Knowlton, N., Eakin, C. M., *et al.*. (2007). Coral reefs under rapid climate change and ocean acidification. *Science* **318,** 1737–1742.

Hood, D. A. (2001). Contractile activity-induced mitochondrial biogenesis in skeletal muscle. *J. Appl. Physiol.* **90,** 1137–1157.

Hove-Madsen, L. (1992). The influence of temperature on ryanodine sensitivity and the force-frequency relationship in the myocardium of rainbow trout. *J. Exp. Biol.* **167,** 47–60.

Hove-Madsen, L., and Tort, L. (1998). L-type Ca^{2+} current and excitation–contraction coupling in single atrial myocytes from rainbow trout. *Am. J. Physiol.* **275,** R2061–R2069.

Hulbert, A. J., and Else, P. L. (1999). Membranes as possible peacemakers of metabolism. *J. Theor. Biol.* **199,** 257–274.

Huntley, B., Green, R. E., Collingham, Y. C., Hill, J. K., Willis, S. G., Bartlein, P. J., Cramer, W., Hagemeijer, W. J. M., and Thomas, C. J. (2004). The performance of models relating species geographical distributions to climate is independent of trophic level. *Ecol. Lett.* **7,** 417–426.

Iwama, G. K., Thomas, P. T., Forsyth, R. H. B., and Vijayan, M. M. (1998). Heat shock protein expression in fish. *Rev. Fish Biol. Fisher.* **8,** 35–56.

Iwama, G. K., Afonso, L. O. B., and Vijayan, M. M. (2006). Stress in fishes. *In* "The Physiology of Fishes" (Evans, D. H., and Claiborne, J. B., Eds.), pp. 319–342. Taylor & Francis, Boca Raton, FL.

Jensen, M. N. (2003). Consensus on ecological impact remains elusive. *Science* **299,** 38.

Johnston, I. A., and Bernard, L. M. (1982). Ultrastructure and metabolism of skeletal muscle fibres in the tench: Effects of long-term acclimation to hypoxia. *Cell Tissue Res.* **227,** 179–199.

Johnston, I. A., Guderley, H., Franklin, C. E., Crockford, T., and Kamunde, C. (1994). Are mitochondria subject to evolutionary temperature adaptation? *J. Exp. Biol.* **195,** 293–306.

Johnston, I. A., Calvo, J., Guderley, H., Fernandez, D., and Palmer, L. (1998). Latitudinal variation in the abundance and oxidative capacities of muscle mitochondria in perciform fishes. *J. Exp. Biol.* **201,** 1–12.

Kawall, H. G., Torres, J. J., Sidell, B. D., and Somero, G. N. (2002). Metabolic cold adaptation in Antarctic fishes: Evidence from enzymatic activities of brain. *Mar. Biol.* **140,** 279–286.

Keen, J. E., Farrell, A. P., Tibbits, G. F., and Brill, R. W. (1992). Cardiac physiology in tunas: II. Effect of ryanodine, calcium and adrenaline on force-frequency relationships in atrial strips from skipjack tuna, *Katsuwonus pelamis*. *Can. J. Zool.* **70,** 1211–1217.

Keen, J. E., Vianzon, D. M., Farrell, A. P., and Tibbits, G. F. (1993). Thermal acclimation alters both adrenergic sensitivity and adrenoceptor density in cardiac tissue of rainbow trout. *J. Exp. Biol.* **181,** 27–47.

Keen, J. E., Vianzon, D. M., Farrell, A. P., and Tibbits, G. F. (1994). Effect of temperature and temperature acclimation on the ryanodine sensitivity of the trout myocardium. *J. Comp. Physiol. B* **164,** 438–443.

Kent, J., Koban, M., and Prosser, C. (1988). Cold-acclimation induced protein hypertrophy in channel catfish and green sunfish. *J. Comp. Physiol. B* **158,** 185–198.

Klingenberg, M., and Huang, S. G. (1999). Structure and function of the uncoupling protein from brown adipose tissue. *Biochim. Biophys. Acta* **1415,** 271–296.

Klingenberg, M., Winkler, E., and Echtay, K. (2001). Uncoupling protein, H^+ transport and regulation. *Biochem. Soc. Trans.* **29,** 806–811.

Koban, M. (1986). Can cultured teleost cells show temperature acclimation? *Am. J. Physiol.* **250,** R211–R220.

Kröncke, I., Dippner, J, W., Heyen, H., and Zeiss, B. (1998). Long-term changes in macrofaunal communities off Norderney (East Frisia, Germany) in relation to climate variability. *Mar. Ecol. Prog. Ser.* **167,** 25–36.

Krumschnabel, G., and Wieser, W. (1994). Inhibition of the sodium pump does not cause stochiometric decrease of ATP-production in energy limited fish hepatocytes. *Experientia* **50,** 483–485.

Krumschnabel, G., Biasi, C., Schwarzbaum, P. J., and Wieser, W. (1997). Acute and chronic effects of temperature, and of nutritional state, on ion homeostasis and energy metabolism in teleost hepatocytes. *J. Comp. Physiol. B* **167,** 280–286.

Krumschnabel, G., Biasi, C., and Wieser, W. (2000). Action of adenosine on energetics, protein synthesis and K(+) homeostasis in teleost hepatocytes. *J. Exp. Biol.* **203,** 2657–2665.

Landeira-Fernandez, A. M., Morrissette, J. M., Blank, J. M., and Block, B. A. (2004). Temperature dependence of the Ca^{2+}-ATPase (SERCA2) in the ventricles of tuna and mackerel. *Am. J. Physiol.* **268,** R398–R404.

Langenbuch, M., and Pörtner, H. O. (2003). Energy budget of hepatocytes from Antarctic fish (*Pachycara brachycephalum* and *Lepidonotothen kempi*) as a function of ambient CO_2: pH-dependent limitations of cellular protein biosynthesis? *J. Exp. Biol.* **206,** 3895–3903.

Lannig, G., Eckerle, L., Serendero, I., Sartoris, F. J., Fischer, T., Knust, R., Johansen, T., and Pörtner, H. O. (2003). Temperature adaptation in eurythermal cod (*Gadus morhua*): A comparison of mitochondrial enzyme capacities in boreal and Arctic populations. *Mar. Biol.* **142,** 589–599.

Lannig, G., Bock, C., Sartoris, F. J., and Pörtner, H. O. (2004). Oxygen limitation of thermal tolerance in cod, *Gadus morhua* L. studied by non-invasive NMR techniques and on-line venous oxygen monitoring. *Am. J. Physiol.* **287,** R902–R910.

Lannig, G., Storch, D., and Pörtner, H. O. (2005). Aerobic mitochondrial capacities in Antarctic and temperate eelpout (Zoarcidae) subjected to warm versus cold acclimation. *Polar Biol.* **28,** 575–584.

Leary, S. C., and Moyes, C. D. (2000). The effects of bioenergetic stress and redox balance on the expression of genes critical to mitochondrial function. *In* "Enviromental Stressors and Gene Responses" (Storey, K. B., and Storey, J., Eds.), pp. 209–229. Elsevier Science, Amsterdam.

LeRay, C., Bonnet, B., Febvre, A., Vallet, F., and Pic, P. (1970). Some peripheral activities of thyroid hormones observed in *Mugil auratus* L. (teleost, Mugilidae). *Ann. Endocrinol.* **31,** 567–572(Paris).

Li, Z. J., Yin, X. M., and Ye, J. (2004). Effects of pH management during deep hypothermic bypass on cerebral oxygenation: alpha-stat versus pH-stat. *J. Zhejiang Univ. SCI* **5,** 1290–1297.

Liang, X. F., Ogata, H. Y., Oku, H., Chen, J., and Hwang, F. (2003). Abundant and constant expression of uncoupling protein 2 in the liver of red sea bream *Pagrus major*. *Comp. Biochem. Physiol. A* **136,** 655–661.

Lillywhite, H. B., Zippel, K. C., and Farrell, A. P. (1999). Resting and maximal heart rates in ectothermic vertebrates. *Comp. Biochem. Physiol. A* **124,** 369–382.

Lucassen, M., Schmidt, A., Eckerle, L. G., and Pörtner, H. O. (2003). Mitochondrial proliferation in the permanent versus temporary cold: enzyme activities and mRNA levels in Antarctic and temperate zoarcid fish. *Am. J. Physiol.* **258,** R1410–R1420.

Lucassen, M., Koschnick, N., Eckerle, L. G., and Pörtner, H. O. (2006). Mitochondrial mechanisms of cold adaptation in cod (*Gadus morhua*) populations from different climatic zones. *J. Exp. Biol.* **209,** 2462–2471.

Lüning, K. (1990). Seaweeds: Their Environment, Biogeography and Ecophysiology John Wiley, New York.

Lurman, G. J. (2008). "Thermal Plasticity and Performance Adaptations in Gadid Muscle: Consequences for Activity and Lifestyle." Ph.D. dissertation, Bremen.

Lutz, P. L., and Kabler, S. (1997). Release of adenosine and ATP in the brain of the freshwater turtle (*Trachemys scripta*) during long-term anoxia. *Brain Res.* **769,** 281–286.

Lutz, P. L., and Nilsson, G. E. (1997). Contrasting strategies for anoxic brain survivalglycolysis up or down. *J. Exp. Biol.* **200,** 411–419.

Mark, F. C., Bock, C., and Pörtner, H. O. (2002). Oxygen limited thermal tolerance in Antarctic fish investigated by magnetic resonance imaging (MRI) and spectroscopy (^{31P}-MRS). *Am. J. Physiol.* **283,** R1254–R1262.

Mark, F. C., Hirse, T., and Pörtner, H. O. (2005). Thermal sensitivity of cellular energy budgets in some Antarctic fish hepatocytes. *Polar Biol.* **28,** 805–814.

Mark, F. C., Lucassen, M., and Pörtner, H. O. (2006). Thermal sensitivity of uncoupling protein expression in polar and temperate fish. *Comp. Biochem. Physiol. D* **1,** 365–374.

Marsh, A. G., Maxson, R. E., and Manahan, D. T. (2001). High macromolecular synthesis with low metabolic cost in Antarctic sea urchin embryos. *Science* **291,** 1950–1952.

McElhaney, R. N. (1984a). The structure and function of the *Acholeplasma laidlawii* plasma membrane. *Biochim. Biophys. Acta* **779,** 1–42.

McElhaney, R. N. (1984b). The relationship between membrane lipid fluidity and phase state and the ability of bacteria and mycoplasmas to grow and survive at various temperatures. *Biomembranes* **12,** 249–276.

McKinley, S. J., and Hazel, J. R. (1993). Epinephrine stimulation of glucose release from perfused trout liver: Effects of assay and acclimation temperature. *J. Exp. Biol.* **177,** 51–62.

McKinley, S. J., and Hazel, J. R. (2000). Does membrane fluidity contribute to thermal compensation of beta-adrenergic signal transduction in isolated trout hepatocytes? *J. Exp. Biol.* **203,** 631–640.

Melzner, F., Bock, C., and Pörtner, H. O. (2006). Critical temperatures in the cephalopod *Sepia officinalis* investigated using *in vivo* ^{31}P NMR spectroscopy. *J. Exp. Biol.* **209,** 891–906.

Metzger, R., Sartoris, F. J., Langenbuch, M., and Pörtner, H. O. (2007). Influence of elevated CO_2 concentrations on thermal tolerance of the edible crab *Cancer pagurus*. *J. Therm. Biol.* **32,** 144–151.

Mommsen, T. P., Vijayan, M. M., and Moon, T. W. (1999). Cortisol in teleosts: dynamics, mechanisms of action, and metabolic regulation. *Rev. Fish Biol. Fisher.* **9,** 211.

Nisoli, E., Clementi, E., Moncada, S., and Carruba, M. O. (2004). Mitochondrial biogenesis as a cellular signaling framework. *Biochem. Pharmacol.* **67,** 1–15.

O'Grady, S. M., and DeVries, A. L. (1982). Osmotic and ion regulation in polar fishes. *J. Exp. Biol. Ecol.* **57,** 219–228.

O'Brien, K. M., Skilbeck, C., Sidell, B. D., and Egginton, S. (2003). Muscle fine structure may maintain the function of oxidative fibers in hemoglobin Antarctic fishes. *J. Exp. Biol.* **206,** 411–421.

Ohkura, K., Kazui, T., Yamamoto, S., Yamashita, K., Terada, H., Washiyama, N., Suzuki, T., Suzuki, K., Fujie, M., and Ohishi, K. (2004). Comparison of pH management during antegrade selective cerebral perfusion in canine models with old cerebral infarction. *J. Thorac. Cardiovasc. Surg.* **128,** 378–385.

Pace, D. A., Maxson, R. E., and Manahan, D. T. (2004). High rates of protein synthesis and rapid ribosomal transit times at low energy cost in Antarctic echinoderm embryos. *Integ. Comp. Biol.* **43,** 1078.

Pannevis, M. C., and Houlihan, D. F. (1992). The energetic cost of protein synthesis in isolated hepatocytes of rainbow trout (*Oncorhynchus mykiss*). *J. Comp. Physiol. B* **162,** 393–400.

Parmesan, C., and Yohe, G. (2003). A globally coherent fingerprint of climate change impacts across natural systems. *Nature* **421,** 37–42.

Pearson, T., Hyde, D., and Bowler, K. (1999). Heterologous acclimation: A novel approach to the study of thermal acclimation in the crab *Cancer pagurus*. *Am. J. Physiol.* **277,** R24–R30.

Peck, L. S., Webb, K. E., and Bailey, D. M. (2004). Extreme sensitivity of biological function to temperature in Antarctic marine species. *Funct. Ecol.* **18,** 625–630.

Péréz-Pinzón, M. A., Rosenthal, M., Sick, T. J., Lutz, P. L., Pablo, J., and Mash, D. (1992). Downregulation of sodium channels during anoxia: A putative survival strategy of turtle brain. *Am. J. Physiol.* **262,** R712–R715.

Pernet, F., Tremblay, R., Redjah, I., Sevigny, J. M., and Gionet, C. (2008). Physiological and biochemical traits correlate with differences in growth rate and temperature adaptation among groups of the eastern oyster *Crassostrea virginica. J. Exp. Biol.* **211,** 969–977.

Perry, S., and Reid, S. (1994). The effects of acclimation temperature on the dynamics of catecholamine release during acute hypoxia in the rainbow trout *Oncorhynchus mykiss. J. Exp. Biol.* **186,** 289–307.

Perry, A. L., Low, P. J., Ellis, J. R., and Reynolds, J. D. (2005). Climate change and distribution shifts in marine fishes. *Science* **308,** 1902–1905.

Podrabsky, J. E., and Somero, G. N. (2006). Inducible heat tolerance in Antarctic notothenioid fishes. *Polar Biol.* **30,** 39–43.

Pörtner, H. O. (1987). Contributions of anaerobic metabolism to pH regulation in animal tissues: theory. *J. Exp. Biol.* **131,** 69–87.

Pörtner, H. O. (2001). Climate change and temperature-dependent biogeography: Oxygen limitation of thermal tolerance in animals. *Naturwissenschaften* **88,** 137–146.

Pörtner, H. O. (2002). Climate variations and the physiological basis of temperature dependent biogeography: Systemic to molecular hierarchy of thermal tolerance in animals. *Comp. Biochem. Physiol. A* **132,** 739–761.

Pörtner, H. O. (2004). Climate variability and the energetic pathways of evolution: the origin of endothermy in mammals and birds. *Physiol. Biochem. Zool.* **77,** 959–981.

Pörtner, H. O. (2006). Climate-dependent evolution of Antarctic ectotherms: An integrative analysis. *Deep-Sea Res. II* **53,** 1071–1104.

Pörtner, H. O., and Knust, R. (2007). Climate change affects marine fishes through the oxygen limitation of thermal tolerance. *Science* **315,** 95–97.

Pörtner, H. O., and Sartoris, F. J. (1999). Invasive studies of intracellular acid-base parameters: environmental and functional aspects. *In* "Regulation of Tissue pH in Plants and Animals: A Reappraisal of Current Techniques" (Taylor, E. W., Egginton, S., and Raven, J. A., Eds.), pp. 69–98. Cambridge University Press, Cambridge.

Pörtner, H. O., Hardewig, I., Sartoris, F. J., and van Dijk, P. L. M. (1998). Energetic aspects of cold adaptation critical temperatures in metabolic, ionic and acid-base regulation? *In* "Cold Ocean Physiology" (Pörtner, H. O., and Playle, R., Eds.), pp. 88–120. Cambridge University Press, Cambridge.

Pörtner, H. O., Hardewig, I., and Peck, L. S. (1999a). Mitochondrial function and critical temperature in the Antarctic bivalve, *Laternula elliptica. Comp. Biochem. Physiol. A* **124,** 179–189.

Pörtner, H. O., Peck, L., Zielinski, S., and Conway, L. Z. (1999b). Intracellular pH and energy metabolism in the highly stenothermal Antarctic bivalve *Limopsis marionensis* as a function of ambient temperature. *Polar Biol.* **22,** 17–30.

Pörtner, H. O., van Dijk, P. L. M., Hardewig, I., and Sommer, A. (2000a). Levels of metabolic cold adaptation: Tradeoffs in eurythermal and stenothermal ectotherms. *In* "Antarctic Ecosystems: Models for Wider Ecological Understanding" (Davison, W., and Williams, C. H., Eds.), pp. 109–122. Caxton Press, Christchurch.

Pörtner, H. O., Bock, C., and Reipschläger, A. (2000b). Modulation of the cost of pHi regulation during metabolic depression: A ^{31}P-NMR study in invertebrate (*Sipunculus nudus*) isolated muscle. *J. Exp. Biol.* **203,** 2417–2428.

Pörtner, H. O., Berdal, B., Blust, R., Brix, O., Colosimo, A., De Wachter, B., Giuliani, A., Johansen, T., Fischer, T., Knust, R., Lannig, G., Naevdal, G., *et al.* (2001). Climate induced temperature effects on growth performance, fecundity and recruitment in marine fish:

Developing a hypothesis for cause and effect relationships in Atlantic cod (*Gadus morhua*) and common eelpout (*Zoarces viviparus*). *Cont. Shelf Res.* **21,** 1975–1997.

Pörtner, H. O., Storch, D., and Heilmayer, O. (2005a). Constraints and trade-offs in climate dependent adaptation: Energy budgets and growth in a latitudinal cline. *Sci. Mar.* **69**(Suppl 2), 271–285.

Pörtner, H. O., Lucassen, M., and Storch, D. (2005b). Metabolic biochemistry: Its role in thermal tolerance and in the capacities of physiological and ecological function. *In* "The Physiology of Polar Fishes:" Fish Physiology Vol. 21 (Steffensen, J. F., Farrell, A. P., Hoar, W. S., Randall, D. R., and Guest, Eds.),pp. 79–154. Elsevier/Academic Press, San Diego.

Pörtner, H. O., Langenbuch, M., and Michaelidis, B. (2005c). Synergistic effects of temperature extremes, hypoxia, and increases in CO_2 on marine animals: From Earth history to global change. *J. Geophys. Res.* **110,** C09S10.

Pörtner, H. O., Bock, C., Knust, R., Lannig, G., Lucassen, M., Mark, F., and Sartoris, F. J. (2008). Cod and climate in a latitudinal cline: Developing a physiological cause and effect understanding of climate effects on marine fishes. *Climate Res.* **37,** 253–270.

Prosser, C. L., Machkay, W., and Kato, K. (1970). Osmotic and ionic concentrations in some Alaska fish and goldfish from different temperatures. *Physiol. Zool.* **43,** 81–89.

Reeves, R. B. (1972). An imidazole alphastat hypothesis for vertebrate acid-base regulation: Tissue carbon dioxide content and body temperature in bullfrogs. *Resp. Physiol.* **14,** 219–236.

Ried, S. G., Bernier, N. J., and Perry, S. F. (1998). The adrenergic stress response in fish: Control of catecholamine storage and release. *Comp. Biochem. Physiol. C* **120,** 1–27.

Reipschläger, A., Nilsson, G. E., and Pörtner, H. O. (1997). A role for adenosine in metabolic depression in the marine invertebrate *Sipunculus nudus*. *Am. J. Physiol.* **272,** R350–R356.

Renshaw, G. M., Kerrisk, C. B., and Nilsson, G. E. (2002). The role of adenosine in the anoxic survival of the epaulette shark, *Hemiscyllium ocellatum*. *Comp Biochem Physiol. B* **131,** 133–141.

Ricquier, D., and Bouillaud, F. (2000). The uncoupling protein homologues: UCP1, UCP2, UCP3, StUCP and AtUCP. *Biochem. J.* **345,** 161–179.

Rocha, M. L., Rantin, F. T., and Kalinin, A. L. (2007). Importance of the sarcoplasmic reticulum and adrenergic stimulation on the cardiac contractility of the neotropical teleost *Synbranchus marmoratus* under different thermal conditions. *J. Comp. Physiol. B* **177,** 713–721.

Rolfe, D. F. S., and Brand, M. D. (1996). Contribution of mitochondrial proton leak to skeletal muscle respiration and to standard metabolic rate. *Am. J. Physiol.* **271,** C1380–C1389.

Root, T. L., Price, J. T., Hall, K. R., Schneider, S. H., Rosenzweig, C., and Pounds, J. A. (2003). Fingerprints of global warming on wild animals and plants. *Nature* **421,** 57–60.

Rousseau, E., Smith, J. S., and Meissner, G. (1987). Ryanodine modifies conductance and gating behaviour of single Ca^{2+} release channels. *Am. J. Physiol.* **253,** C364–C368.

Samec, S., Seydoux, J., and Dulloo, A. G. (1998). Role of UCP homologues in skeletal muscles and brown adipose tissue: Mediators of thermogenesis or regulators of lipids as fuel substrate? *FASEB J.* **12,** 715–724.

Sartoris, F. J., and Pörtner, H. O. (1997). Temperature dependence of ionic and acid-base regulation in boreal and arctic *Crangon crangon* and *Pandalus borealis*. *J. Exp. Mar. Biol. Ecol.* **211,** 69–83.

Sartoris, F. J., Bock, C., and Pörtner, H. O. (2003). Temperature-dependent pH regulation in eurythermal and stenothermal marine fish: An interspecies comparison using ^{31}P-NMR. *J. Therm. Biol.* **28,** 363–371.

Schwarzbaum, P. J., Niederstätter, H., and Wieser, W. (1992). Effects of temperature on the ($Na^+ + K^+$)-ATPase and oxygen consumption in hepatocytes of two species of freshwater fish, roach (*Rutilus rutilus*) and brook trout (*Salvelinus fontinalis*). *Physiol. Zool.* **65,** 699–711.

Seddon, W., and Prosser, C. (1997). Seasonal variations in the temperature acclimation response of the channel catfish, *Ictalurus punctatus*. *Physiol. Zool.* **70,** 33–44.

Seebacher, F., Davison, W., Lowe, C. J., and Franklin, C. E. (2005). A falsification of the thermal specialization paradigm: Compensation for elevated temperatures in Antarctic fishes. *Biol. Lett.* **1,** 151–154.

Shiels, H. A., and Farrell, A. P. (1997). The effects of temperature and adrenaline on the relative importance of the sarcoplasmic reticulum in contributing Ca^{2+} to force development in isolated ventricular trabeculae from rainbow trout. *J. Exp. Biol.* **200,** 1607–1621.

Shiels, H. A., Freund, E. V., Farrell, A. P., and Block, B. A. (1999). The sarcoplasmic reticulum plays a major role in isometric contraction in atrial muscle of yellowfin tuna. *J. Exp. Biol.* **202,** 881–890.

Shiels, H. A., Vornanen, M., and Farrell, A. P. (2000). Temperature-dependence of L-type Ca^{2+} channel current in atrial myocytes from rainbow trout. *J. Exp. Biol.* **203,** 2771–2780.

Shiels, A. H., Vornanen, M., and Farrell, A. P. (2002). Effects of temperature on intracellular [Ca^{2+}] in trout atrial myocytes. *J. Exp. Biol.* **205,** 3641–3650.

Shiels, H. A., Vornanen, M., and Farrell, A. P. (2003). Acute temperature change modulates the response of ICa to adrenergic stimulation in fish cardiomoycytes. *Physiol. Biochem. Zool.* **76,** 816–824.

Shiels, H. A., Paajanen, V., and Vornanen, M. (2006). Sarcolemmal ion currents and sarcoplasmic reticulum Ca^{2+} content in ventricular myocytes from the cold stenothermic fish, the burbot (*Lota lota*). *J. Exp. Biol.* **209,** 3091–3100.

Sidell, B. D. (1998). Intracellular oxygen diffusion: The roles of myoglobin and lipid at cold body temperature. *J. Exp. Biol.* **201,** 1119–1128.

Sidell, B. D., and O'Brien, K. M. (2006). When bad things happen to good fish: The loss of hemoglobin and myoglobin expression in Antarctic fishes. *J. Exp. Biol.* **209,** 1791–1802.

Sims, D. W., Genner, M. J., Southward, A. J., and Hawkins, S. J. (2001). Timing of squid migration reflects North Atlantic climate variability. *Proc. Roy. Soc. Lond. B* **268,** 2607–2611.

Sinensky, M. (1974). Homeoviscous adaptation – A homeostatic process that regulates viscosity of membrane lipids in *Escherichia coli*. *Proc. Natl. Acad. Sci.* USA **71,** 522–525.

Sitsapesan, R., Montgomery, R. A. P., MacLeod, K. T., and Williams, A. J. (1991). Sheep cardiac sarcoplasmic reticulum calcium release channels: Modifications of conductance and gating by temperature. *J. Physiol. (Lond.)* **434,** 469–488.

Smith, R. W., and Houlihan, D. F. (1995). Protein synthesis and oxygen consumption in fish cells. *J. Comp. Physiol. B* **165,** 93–101.

Sokolova, I. M., and Sokolov, E. P. (2005). Evolution of mitochondrial uncoupling proteins: Novel invertebrate UCP homologues suggest early evolutionary divergence of the UCP family. *FEBS Letters* **579,** 313–317.

Somero, G. N., and DeVries, A. L. (1967). Temperature tolerance of some Antarctic fishes. *Science* **156,** 257–258.

Sommer, A., Klein, B., and Pörtner, H. O. (1997). Temperature induced anaerobiosis in two populations of the polychaete worm *Arenicola marina*. *J. Comp. Physiol. B* **167,** 25–35.

Southward, A. J., Hawkins, S. J., and Burrows, M. T. (1995). Seventy years' observations of changes in distribution and abundance of zooplankton and intertidal organisms in the western English Channel in relation to rising sea temperature. *J. Therm. Biol.* **20,** 127–155.

Spicer, J. I., Morritt, D., and Taylor, A. C. (1994). Effect of low temperature on oxygen uptake and haemolymph ions in the sandhopper *Talitrus saltator* (Crustacea: Amphipoda). *J. Mar. Biol. Assoc. UK* **74,** 313–321.

Stachowicz, J. J., Terwin, J. R., Whitlatch, R. B., and Osman, R. W. (2002). Linking climate change and biological invasions: Ocean warming facilitates non-indigenous species invasions. *Proc. Natl. Acad. Sci. USA* **99,** 15497–15500.

Steinhausen M. F., Sandblom E., Eliason E. J., Verhille C., and Farrell A. P. (2008). The effect of acute temperature increases on the cardiorespiratory performance of resting and swimming sockeye salmon (*Oncorhynchus nerka*). *J. Exp. Biol.* **211,** 3915–3926

Stenseth, N. C., and Mysterud, A. (2002). Climate, changing phenology, and other life history traits: Nonlinearity and match-mismatch to the environment. *Proc. Natl. Acad. Sci. USA* **99,** 13379–13381.

Storch, D., Heilmayer, O., Hardewig, I., and Pörtner, H. O. (2003). In vitro protein synthesis capacities in a cold stenothermal and a temperate eurythermal pectinid. J. *Comp. Physiol. B* **173,** 611–620.

Storch, D., Lannig, G., and Pörtner, H. O. (2005). Temperature dependent protein synthesis capacities in Antarctic and temperate (North Sea) fish (Zoarcidae). *J. Exp. Biol.* **208,** 2409–2420.

Stuart, J. A., Brindle, K. M., Harper, J. A., and Brand, M. D. (1999a). Mitochondrial proton leak and the uncoupling proteins. *J. Bioenerg. Biomembr.* **31,** 517–525.

Stuart, J. A., Harper, J. A., Brindle, K. M., and Brand, M. D. (1999b). Uncoupling protein 2 from carp and zebrafish, ectothermic vertebrates. *Biochim. Biophys. Acta* **1413,** 50–54.

Takasuka, A., Oozeki, Y., and Aoki, I. (2007). Optimal growth temperature hypothesis: Why do anchovy flourish and sardine collapse or vice versa under the same ocean regime? *Can. J. Fish. Aqu. Sci.* **64,** 768–776.

Thebault, M. T., and Raffin, J. P. (1991). Seasonal variations in *Palaemon serratus* abdominal muscle metabolism and performance during exercise, as studied by ^{31}P-NMR. *Mar. Ecol. Progr. Ser.* **74,** 175–183.

Thomas, C. D., Cameron, A., Green, R. E., Bakkenes, M., Beaumont, L. J., Collingham, Y. C., Erasmus, B. F. N., Ferreira de Siqueira, M., Grainger, A., Havannah, L., Hughes, L., Huntley, B., *et al.*. (2004). Extinction risk from climate change. *Nature* **427,** 145–148.

Thomson, A. J., Sargent, J. R., and Owen, J. M. (1977). Influence of acclimatization temperature and salinity on (Na^{+}+K^{+})-dependent adenosine triphosphatase and fatty acid composition in the gills of the eel (*Anguilla anguilla*). *Comp. Biochem. Physiol. B* **56,** 223–228.

Tibbits, G. F., Moyes, C. D., and Hove-Madsen, L. (1992). Excitation-contraction coupling in the teleost heart. *In* "Fish Physiology, Vol. 12A" (Hoar, W. S., Randall, D. J., and Farrell, A. P., Eds.), pp. 267–304. Academic Press, San Diego.

Tiitu, V., and Vornanen, M. (2002a). Regulation of cardiac contractility in a stenothermal fish, the burbot (*Lota lota*). *J. Exp. Biol.* **205,** 1597–1606.

Tiitu, V., and Vornanen, M. (2002b). Morphology and fine structure of the heart of the burbot (*Lota lota*), a cold stenothermal fish. *J. Fish Biol.* **61,** 106–121.

Tinton, S. A., Chow, S. C., Buc-Calderon, P. M., Kass, G. E., and Orrenius, S. (1995). Adenosine inhibits protein synthesis in isolated rat hepatocytes. Evidence for a lack of involvement of intracellular calcium in the mechanism of inhibition. *Eur. J. Biochem.* **229,** 419–425.

Tripathi, G., and Verma, P. (2003). Pathway-specific response to cortisol in the metabolism of catfish. *Comp. Biochem. Physiol. B* **136,** 463–471.

Tsugawa, K., and Lagerspetz, K. Y. H. (1990). Direct adaptation of cells to temperature: membrane fluidity of goldfish cells cultured *in vitro* at different temperatures. *Comp. Biochem. Physiol. A* **96,** 57–60.

Tyler, S., and Sidell, B. D. (1984). Changes in mitochondrial distribution and diffusion distances in muscle of goldfish upon acclimation to warm and cold temperatures. *J. Exp. Zool.* **232,** 1–9.

Ultsch, G. R., and Jackson, D. C. (1996). pH and temperature in ectothermic vertebrates. *Bull. Alabama Mus. Nat. Hist.* **18,** 1–41.

ven der Meer, D. L., van den Thillart, G. E., Witte, F., de Bakker, M. A., Besser, J., Richardson, M. K., Spaink, H. P., Leito, J. T., and Bagowski, C. P. (2005). Gene expression profiling of the long-term adaptive response to hypoxia in the gills of adult zebrafish. *Am. J. Physiol.* **289,** R1512–R1519.

van Dijk, P., Hardewig, I., and Pörtner, H. O. (1997). The adjustment of intracellular pH after temperature change in fish: Relative contributions of passive and active processes. *Am. J. Physiol.* **272,** R84–R89.

van Dijk, P. L. M., Tesch, C., Hardewig, I., and Pörtner, H. O. (1999). Physiological disturbances at critically high temperatures: A comparison between stenothermal Antarctic and eurythermal temperate eelpouts (Zoarcidae). *J. Exp. Biol.* **202,** 3611–3621.

van Ham, E. H., van Anholt, R. D., Kruitwagen, G., Imsland, A. K., Foss, A., Sveinsbo, B. O., FitzGerald, R., Parpoura, A. C., Stefansson, S. O., and Bonga, S. E. (2003). Environment affects stress in exercised turbot. *Comp. Biochem. Physiol. A* **136,** 525–538.

Vornanen, M. (1996). Effect of extracellular calcium on the contractility of warm-and cold-acclimated crucian carp heart. *J. Comp. Physiol. B* **165,** 507–517.

Vornanen, M. (1997). Carcolemmal Ca^{2+} influx through L-type Ca channels in ventricular myocytes of a teleost fish. *Am J. Physiol.* **272,** R1432–R1440.

Vornanen, M. (1999). Na^{+}-Ca^{2+} exchange current in ventricular myocytes of fish heart: Contribution to sarcolemmal Ca^{2+} influx. *J. Exp. Biol.* **202,** 1763–1775.

Vornanen, M. (2006). Temperature- and Ca^{2+}-dependence of [^{3}H]ryanodine binding in the burbot (*Lota lota* L.) heart. *Am. J. Physiol.* **290,** R345–R351.

Vornanen, M., Ryokkynen, A., and Nurmi, A. (2002). Temperature dependent expression of sarcolemal K currents in rainbow trout atrial and ventricular myocytes. *Am. J. Physiol.* **282,** R1191–R1199.

Walther, G. R., Post, E., Convey, P., Menzel, A., Parmesan, C., Beebee, T. J. C., Fromentin, J. M., Hoegh-Guldberg, O., and Bairlein, F. (2002). Ecological responses to recent climate change. *Nature* **416,** 389–395.

Wang, T., and Overgaard, J. (2007). The heartbreak of adapting to global warming. *Science* **315,** 49.

Weatherley, A. H. (1970). Effects of superabundant oxygen on thermal tolerance of goldfish. *Biol. Bull.* **139,** 229–238.

Weinstein, R. B., and Somero, G. N. (1998). Effects of temperature on mitochondrial function in the Antarctic fish *Trematomus bernacchii*. *J. Comp. Physiol. B* **168,** 190–196.

Wendelaar Bonga, S. E. (1997). The stress response in fish. *Physiol. Rev.* **77,** 591–625.

White, F. N., and Somero, G. (1982). Acid-base regulation and phospholipid adaptations to temperature: Time courses and physiological significance of modifying the milieu for protein function. *Physiol. Rev.* **62,** 40–90.

Whiteley, N. M., Naylor, J. K., and Taylor, E. W. (1995). Extracellular and intracellular acid-base status in the fresh water crayfish *Austropotamobius pallipes* between 1 and 12°C. *J. Exp. Biol.* **198,** 567–576.

Whiteley, N. M., Taylor, E. W., and el Haj, A. J. (1996). A comparison of the metabolic cost of protein synthesis in stenothermal and eurythermal isopod crustaceans. *Am. J. Physiol.* **271,** R1295–R1303.

Wieser, W. (1994). Cost of growth in cells and organisms: General rules and comparative aspects. *Biol. Rev.* **69,** 1–33.

Wieser, W., and Krumschnabel, G. (2001). Hierarchies of ATP-consuming processes: Direct compared to indirect measurements, and comparative aspects. *Biochem. J.* **355,** 389–395.

Wiltshire, K. H., and Manly, B. F. J. (2004). The warming trend at Helgoland Roads, North Sea: Phytoplankton response. *Helgol. Mar. Res.* **58,** 269–273.

Wodtke, E. (1976). Discontinuities in the arrhenius plots of mitochondrial membrane-bound enzyme systems from a poikilotherm: Acclimation temperature of carp affects transition temperatures. *J. Comp. Physiol. B* **110,** 145–157.

Wodtke, E. (1981). Temperature adaptation of biological membranes. The effects of acclimation temperature on the unsaturation of the main neutral and charged phospholipids in mitochondrial membranes of the carp (*Cyprinus carpio* L.). *Biochim. Biophys. Acta* **640,** 698–709.

Zielinski, S., and Pörtner, H. O. (1996). Energy metabolism and ATP free-energy change of the intertidal worm *Sipunculus nudus* below a critical temperature. *J. Comp. Physiol. B* **166,** 492–500.

5

OXYGEN SENSING AND THE HYPOXIC VENTILATORY RESPONSE

S. F. PERRY
M. G. JONZ
K. M. GILMOUR

The hypoxic ventilatory response is arguably the single most important physiological response accompanying the exposure of fish to lowered ambient Po_2. Increases in ventilation volume, driven by changes in breathing frequency and/or amplitude, serve to raise arterial Po_2 and hence may delay the onset of transition from aerobic to anaerobic metabolism. In air-breathing fish, the hypoxic ventilatory response may consist of increases in aquatic and/or aerial respiration. Ventilatory responses in fish are initiated by O_2 chemoreceptors able to detect changes in water and/or blood PO_2. In zebrafish, the O_2-sensing cells have been identified as neuroepithelial cells of the gill filament. In response to hypoxia, the neuroepithelial cells undergo membrane depolarization owing to an inhibition of outwardly directed

Hypoxia: Volume 27
FISH PHYSIOLOGY

DOI: 10.1016/S1546-5098(08)00005-8

potassium currents that ultimately elcits calcium entry and the release of neurotransmitter(s).

1. INTRODUCTION

The two key factors that determine the rate of gas transfer in fish, and hence metabolic rate, are ventilation and perfusion. Thus, fish must be able to modify these convective processes to respond appropriately to changes in their environment or activity levels. The capacity of fish to mount appropriate cardiorespiratory responses, in turn, requires finely tuned sensory systems able to detect changes in the external and internal environments. During hypoxia, activation of the sensory or afferent pathways initiates a suite of integrated reflexes aimed at promoting homeostasis. In this chapter, we focus on the hypoxic ventilatory response, arguably the single most important physiological reflex occurring during exposure to hypoxia; Chapter 7 in this volume reviews the cardiovascular responses to hypoxia. Obviously, these two chapters are complimentary and together provide a comprehensive overview of hypoxic cardiorespiratory reflexes in fish. In keeping with the tradition of previous volumes of *Fish Physiology*, we provide an extensive historical summary of the literature while also focusing on recent developments including the cellular mechanisms of oxygen sensing.

2. THE HYPOXIC VENTILATORY RESPONSE

As Tables 5.1 and 5.2 demonstrate, fish exposed to hypoxia respond by hyperventilating (see also reviews by Shelton *et al.*, 1986; Perry and Wood, 1989; Glass, 1992; Fritsche and Nilsson, 1993; Graham, 1997; Gilmour, 2001; Perry and Gilmour, 2002; Gilmour and Perry, 2007). This simple statement does not, however, do justice to the diversity of hypoxic ventilatory responses. Among species that are solely or predominately water-breathing, the vast majority respond to aquatic hypoxia by increasing the volume of water ventilated ($\dot{V}_w$) (Table 5.1), while bimodal (water and air) breathers typically increase reliance upon air-breathing (Table 5.2). However, the magnitude of the response, the water O_2 tension (P_wO_2) threshold for hyperventilatory responses, and the mechanism through which ventilation is enhanced all vary greatly among species and, often, among different studies on a single species. Ideally it would be possible to link interspecific differences in the hypoxic ventilatory response to factors such as hypoxia tolerance or ventilatory mechanics, but such comparisons are difficult because experimental protocols have varied widely in the severity of hypoxia imposed, the rate of

Table 5.1
Summary of (gill) ventilatory responses to hypoxia

Species		PO_2 (Torr), time	$\Delta \dot{V}_w$ (%)	ΔV_f (%)	ΔV_{amp} (%)	ΔSV (%)	References
Agnathans							
Eptatretus stoutii	Pacific hagfish	20, 20 min	+149	+274			S. F. Perry, B. Vulesevic, M. Braun and K. M. Gilmour, unpublished
Entosphenus tridentatus	Lamprey	40		+122			Johansen *et al.*, 1973
Lampetra fluviatilis	Lamprey	30, 20 min		+155			Claridge and Potter, 1975
		45, 30 min		+143			Nikinmaa and Weber, 1984
Elasmobranchs							
Scyliorhinus stellaris	Larger spotted dogfish	95, 1 h	+52	ns			Piiper *et al.*, 1970
		54, 1 h	+19	ns			
S. canicula	Dogfish	80, 1 h	+55	ns		+57	Short *et al.*, 1979
Sphyrna tiburo	Bonnethead shark	90, <60 min	+264				Carlson and Parsons, 2003
		70, <60 min	+765				
Squalus acanthias	Spiny dogfish	35, 30 min		+16	+93		Perry and Gilmour, 1996
Torpedo marmorata	Ray ($N = 1$)	33, 20 min		+67	+111		Hughes, 1978
Chondrosteians							
Acipenser baeri	Siberian sturgeon	60, 1 h		+121	+90		Nonnotte *et al.*, 1993
		40, 1 h		+127	+110		

(continued)

Table 5.1 *(continued)*

Species		PO_2 (Torr), time	$\Delta\dot{V}_w$ (%)	ΔV_f (%)	ΔV_{amp} (%)	ΔSV (%)	References
		20, 1 h		+103	+165		
		30, 10 min		+124	+260		Maxime *et al.*, 1995
		10, 30 min		−38	ns		
A. naccarii	Adriatic sturgeon	19, 20 min		+39	+130		McKenzie *et al.*, 1995
		81, 30 min		+28	ns		McKenzie *et al.*, 1997
		50, 30 min		+26	ns		
		35, 20 min		+18	ns		
A. transmontanus	White sturgeon	105, 1 h	−31	ns		−34	Burggren and Randall, 1978
		60, 1 h	−69	−10		−60	
		30, 1 h	−71	−23		−68	
Neopterygiians							
Amia calva	Bowfin	35, 1 h		+67	+66		Hedrick *et al.*, 1991
		47, 15 min		+100	+80		McKenzie *et al.*, 1991
Teleosts							
Anguilla anguilla	European eel	101	+58				Le Moigne *et al.*, 1986
		69	+114				
		39	+150				
		24	+65				
		40	+146	ns		+135	Peyraud-Waitzenegger and Soulier, 1989
		20, 2 h		+5	+43		McKenzie *et al.*, 2000
A. japonica	Japanese eel	40, <2 h	−18	−20		−11	Chan, 1986

Apteronotus leptorhynchus	Brown ghost knifefish	50, 30 min		+10	+30		M. Moorhead, M. Nguyen, J. Lewis, S. F. Perry, and K. M. Gilmour, unpublished data
Callionymus lyra	Dragonet (N = 2)	50, 6 min	+77	−30		+120	Hughes and Umezawa, 1968
Colossoma macropomum	Tambaqui	10, 10 min		+63	+88		Sundin *et al.*, 2000
Cyprinus carpio	Carp	Mild	+287	+218		ns	Itazawa and Takeda, 1978
		40	+475				Lomholt and Johansen, 1979
		110, 1 h	+83	+41		ns	Glass *et al.*, 1990
		75, 1 h	+161	+205		ns	
		100, 30 min	+59	+179		−59	Soncini and Glass, 2000
		90, 30 min	+134	+150		ns	
Danio rerio	Zebrafish	110, 20 min		+30	ns		Vulesevic *et al.*, 2006
		70, 20 min		+55	ns		
		50, 20 min		+73	ns		
		20, 20 min		+63	ns		
		90, 20 min		+49	ns		Vulesevic and Perry, 2006
		50, 20 min		+79	ns		
		20, 20 min		+54	ns		
		80, 5 min		+28	+11		K. Borg, S. Sharam, and W. K. Milsom, unpublished data
		55, 5 min		+39	+50		
		35, 5 min		+48	+61		
Gadus morhua	Atlantic cod	59, 10 min	+39	+35		ns	Kinkead *et al.*, 1991
		46, 25 min	+57	ns		+37	
Hoplerythrinus unitaeniatus	Jeju (no access to air)	70, 1 h	+250	+25		+143	Oliveira *et al.*, 2004

(*continued*)

Table 5.1 *(continued)*

Species		PO_2 (Torr), time	$\Delta \dot{V}_w$ (%)	ΔV_f (%)	ΔV_{amp} (%)	ΔSV (%)	References
		50, 1 h	+575	+35		+371	
		30, 1 h	+107	+37		+700	
		20, 1 h	+102	+35		+643	
	(access to air)	50, 15 min		ns	+55		Perry *et al.*, 2004
		40, 15 min		+24	ns		
		20, 15 min		ns	ns		
Hoplias lacerdae	Trairão	20	+460	+160		+160	Rantin *et al.*, 1992
H. malabaricus	Traira	100, 10 min	+35	+8		+24	Rantin and Johansen, 1984
		50, 10 min	+118	+25		+74	
		25, 10 min	+183	+38		+105	
		15, 10 min	+209	+37		+126	
		20	+470	+40		+375	Rantin *et al.*, 1992
		11, 10 min		+33	+88		Sundin *et al.*, 1999
		20, 30 min	+800	+51		+550	Sakuragui *et al.*, 2003
		20, 15 min		+27	+170		Perry *et al.*, 2004
Hypostomus regani	Ihering	25, 5 h	+778	+20			Mattias *et al.*, 1998
Ictalurus punctatus	Channel catfish	27, < 2 h	+287	+44		+230	Gerald and Cech, 1970
		105, 1 h	+106	ns		+91	Burggren and Cameron, 1980
		65, 1 h	+100	ns		+91	
		46, 10 min		+33	+93		Burleson and Smatresk, 1990a
		73, 5 min		+14	+66		Burleson *et al.*, 2002
		50, 5 min		+16	+154		
		30, 5 min		+33	+100		Sundin *et al.*, 2003
Katsuwonus pelamis	Skipjack tuna	100, 4 min	+25				Bushnell and Brill, 1992
		75, 4 min	+45				

Mugil cephalus	Mullet	68	+371	+58		+149	Cech and Wohlschlag, 1973
		63, 30 min		ns	+111		Shingles *et al.*, 2005
		31, 30 min		ns	+211		
		16, 30 min		ns	+637		
Myoxocephalus scorpius	Shorthorn sculpin	38, 10 min		+39	+57		Turesson and Sundin, 2003
Oncorhynchus mykiss	Rainbow trout	40	+119	+50		+760	Holeton and Randall, 1967a,b
		80, 30 min	+200		+67		Hughes and Saunders, 1970
		60, 30 min	+500		+144		
		40, 30 min	+170		+478		
		60, 5 min	+582	+16		+489	Davis and Cameron, 1971
		75, 30 min	+104	ns		+114	Randall and Jones, 1973
		93, 20 min	+113	ns		+119	Smith and Jones, 1982
		47, 30 min	+142	ns			Aota *et al.*, 1990
		72, 30 min	+192	ns		+179	Kinkead and Perry, 1990
		90, 30 min	+126				Kinkead and Perry, 1991
		70, 25 min		+25	+145		Perry and Thomas, 1991
		50, 20 min		+35	+190		
		40, 20 min		ns	+44		Bindon *et al.*, 1994
		60, 30 min		+18	+21		Gilmour and Perry, 1994
		40, 20 min		+25	+85		Greco *et al.*, 1995
		45, 30 min		ns	+240		Perry and Gilmour, 1996
Ophiodon elongatus	Lingcod	35, 10 min		+107	+390		Farrell and Daxboeck, 1981
Opsanus beta	Gulf toadfish	40, 15 min		ns	+56		S. F. Perry, M. D. McDonald, P. J. Walsh, and K. M. Gilmour, unpublished data
Oreochromis niloticus	Tilapia	80	+50	ns		+50	Fernandes and Rantin, 1989
		60	+100	ns		+125	

(*continued*)

Table 5.1 *(continued)*

Species		PO_2 (Torr), time	$\Delta\dot{V}_w$ (%)	ΔV_f (%)	ΔV_{amp} (%)	ΔSV (%)	References
		20	+550	+18		+400	
		22, 75 min	+300	+33		+333	Kalinin *et al.*, 1999
		9, 75 min	+400	+13		+500	
Orthodon microlepidotus	Sacramento blackfish	90	+50	ns		+38	Campagna and Cech, 1981
		65	+95	ns		+90	
		40	+195	+11		+169	
Parophrys vetulus	English sole	120, 2 h	+69	+8			Boese, 1988
		88, 2 h	+203	+16			
		50, 2 h	+421	+33			
Piaractus mesopotamicus	Pacu	20, 50 min		+158	+300		Leite *et al.*, 2007
		20, 15 min		+103	ns		Perry *et al.*, 2004
Platichthys flesus	Flounder	30, 1 h	+72	ns		+65	Kerstens *et al.*, 1979
		94, 1 h	+63	ns		+58	Steffensen *et al.*, 1982
		62, 1 h	+106	+11		+83	
		39, 1 h	+91	−6		+100	
P. stellatus	Starry flounder	50	+200				Watters and Smith, 1973
Pleuronectes platessa	Plaice	94, 1 h	+11	ns		+14	Steffensen *et al.*, 1982
		62, 1 h	+33	ns		+50	
		39, 1 h	+65	ns		+92	
Prochilodus scrofa	Curimbatá	10	+130	+209		+443	Fernandes *et al.*, 1995
Rhinelepis strigosa	Cascudo preto	26	+900	+23		+557	Takasusuki *et al.*, 1998
Salminus maxillosus	Dourado	70, 30 min	+75	+17		ns	De Salvo-Souza *et al.*, 2001
		40, 30 min	+190	+28		+186	
		20, 30 min	+270	+21		+264	

Scophthalmus maximus	Turbot	60, 1 h		+26	+70		Maxime *et al.*, 2000
Silurus glanis	Sheatfish	81, > 90 min	+100				Forgue *et al.*, 1989
		39, > 90 min	+279				
		23, >90 min	+643				
Thunnus albacares	Yellowfin tuna	100, 4 min	+100				Bushnell and Brill, 1992
		75, 4 min	+75				
		100, 30 min	+37				Bushnell and Brill, 1991
T. obesus	Bigeye tuna	100, 4 min	+30				Bushnell and Brill, 1992
		75, 4 min	+75				
Tinca tinca	Tench	44, 30 min		+44	+250		Randall and Shelton, 1963
		25, 4 h	+500		+68		Eddy, 1974

The change in ventilation volume ($\Delta\dot{V}_w$), ventilation frequency (ΔV_f), ventilation amplitude (ΔV_{amp}), or ventilatory stroke volume (ΔSV) under hypoxic conditions are expressed as a percentage of the normoxic value; thus, a negative value indicates a decrease while a positive value indicates an increase from the normoxic value. Percent changes were calculated from mean data reported in the original studies; changes that were not significant are noted as ns. The level of hypoxia to which the fish was exposed (PO_2 in Torr) and the period of hypoxic exposure (time) are listed; in some studies, the period of hypoxic exposure was not noted.

Table 5.2

Summary of gill ventilation and air-breathing responses to hypoxia in air-breathing fish

Species		PO_2 (Torr), time	P_{crit} (Torr)	Air-breathing threshold (Torr)	Peak gill ventilation (Torr)	f_{AB} (h^{-1})		References
						N	H	
Obligate air-breathers								
Channa argus	Snakehead	35, 3 h	-	NA	-	6.8	9.4	Glass *et al.*, 1986
Protopterus aethiopicus or *P. dolloi*	African lungfish	61, 1 h	-	NA	61	17	40	Jesse *et al.*, 1967
		23, 1 h					54	
P. annectens		30	~100	NA	~70	13.3	29	Babiker, 1979
Lepidosiren paradoxa	South American lungfish (N = 2)	< 80	-	NA	-	12.6	26.1	Johansen and Lenfant, 1967
Facultative air-breathers								
A. calva	Bowfin	30	-	NA	~80	9	30	Johansen *et al.*, 1970
		14	-	NA	-	5.5	12	Randall *et al.*, 1981
		35, 1 h	-	-	-	0	3.2	Hedrick *et al.*, 1991
		47, 15 min	-	NA	NA	~0	5.1	McKenzie *et al.*, 1991
		55	-	NA	-	1.4	3.5	Hedrick and Jones, 1993
		49, 8 h	-	NA	NA	1.2	7.2	Hedrick and Jones, 1999
Amphipnous cuchia	Cuchia eel	15	-	NA	NA	9	16.1	Lomholt and Johansen, 1974
Ancistrus chagresi	Armoured catfish	0, 4 h	40*	33	–	0	14.4	Graham and Baird, 1982; Graham, 1983
Clarias lazera		30	~100	NA	~70	19	63	Babiker, 1979
H. unitaeniatus	Jeju	70, 1 h	40*	64	30*	0	0	Oliveira *et al.*, 2004
		50, 1 h					5.5	

		30, 1 h					14	
		20, 1 h					18.5	
		60, 15 min	-	40	40–50*	0	0	Perry *et al.*, 2004
		40, 15 min					2.5	
		30, 15 min					12	
		20, 15 min					36	
		10, 15 min					36	
		7.5, 3 h	52.5	52.5	NA	2	19.1	McKenzie *et al.*, 2007
Hoplosternum littorale	Armoured catfish	105, 30 min	–	NA	–	2	28	Brauner *et al.*, 1995
		70, 1 h	50*	NA	20	0.5	1	Affonso and Rantin, 2005
		40, 1 h					2.6	
		20, 1 h					3.5	
		10, 1 h					4.5	
Hypostomus plecostomus	Armoured catfish	0, 4 h	–	60	–	0	9.2	Graham and Baird, 1982
H. regani	Ihering	10, 5.5 h	34*	50–65	25*	0	5.8	Mattias *et al.*, 1998
Lepisosteus oculatus	Spotted gar	12, 1 h	–	NA	–	1	8.3	Smatresk and Cameron, 1982
L. osseus	Longnose gar	20, 90 min	–	NA	75	3.5	11.7	Smatresk, 1986
Megalops atlanticus	Tarpon	1, 1–6 h	–	NA	37	1.4	7.7	Geiger *et al.*, 2000
M. cyprinoides	Pacific tarpon	45, 30 min	–	60	60	0	43.8	Seymour *et al.*, 2007
		15, 20 min	–	60	–	1.8	31.8	Clark *et al.*, 2007
Misgurnus anguillicaudatus	Oriental weatherloach	45, 1 h	–	NA	NA	10	20	McMahon and Burggren, 1987
		0, 2 h	–	–	–	0	45.6	McNeil and Closs, 2007
Neoceratodus forsteri	Australian lungfish	22, 60 min	–	75	43	< 1	5	Fritsche *et al.*, 1993
		40, 15 h	–	58	70	0.1	0.3	Kind *et al.*, 2002
Odontamblyopus lacepedii	Eel goby	23, 2 h	–	78	–	0	7.9	Gonzales *et al.*, 2006
		7.5, 2 h					10.8	

(continued)

Table 5.2 *(continued)*

Species		PO_2 (Torr), time	P_{crit} (Torr)	Air-breathing threshold (Torr)	Peak gill ventilation (Torr)	f_{AB} (h^{-1})		References
						N	H	
Piabucina festae		30	70*	NA	–	25	40.4	Graham *et al.*, 1977
Rhinelepis strigosa	Cascudo preto	5	20.5*	22	25*	0	7	Takasusuki *et al.*, 1998
Synbranchus marmoratus	South American swamp eel	< 30	–	30	30	0	4.1	Bicudo and Johansen, 1979
		0, 4 h	150*	33	NA	0	4	Graham and Baird, 1984
Trichogaster trichopterus	Gourami	75, 1 h	-	NA	–	12	20	Burggren, 1979
		37, 1 h					27	
Umbra limi	Mudminnow	20, 90 min	–	45	45	0	7	Gee, 1980

P_{crit}, the partial pressure of O_2 at which the transition from oxygen regulator to oxygen conformer occurs (* indicates that P_{crit} was measured without access to air); Air-breathing threshold, the partial pressure of O_2 at which air-breathing first occurs (species that breathe air under normoxic conditions are indicated with NA); Peak gill ventilation, the partial pressure of O_2 at which ventilation of the gills was maximal (using $\dot{V}_w$ where available, or V_f and/or V_{amp} as appropriate; species in which ventilation continued to increase to the most severe level of hypoxia examined are indicated with NA as it is not clear what impact even more severe hypoxic exposure would have had on ventilation); f_{AB}, air-breathing frequency under normoxic (N) and hypoxic (H) conditions, where the level of hypoxia to which the fish was exposed and the period of hypoxic exposure are listed. Where variables were not measured in a particular study, a dash (–) is used. Additional data on thresholds for air breathing can be found in Table 1 of Takasusuki *et al.* (1998), while Graham (1997) provides comprehensive data both on air-breathing frequencies in normoxia vs. hypoxia (Table 5.2) and air-breathing thresholds (Table 6.1).

change of P_wO_2, and the length of hypoxic exposure, as well as methods of assessing ventilation. Moreover, it is becoming increasingly clear that ventilatory responses to hypoxia within a species may be affected by a multiplicity of conditions, including temperature (e.g., Spitzer *et al.*, 1969; Watters and Smith, 1973; Campagna and Cech, 1981; Berschick *et al.*, 1987; Gehrke and Fielder, 1988; Fernandes and Rantin, 1989; Glass *et al.*, 1990; Fernandes *et al.*, 1995; Geiger *et al.*, 2000; Stecyk and Farrell, 2002, 2008; Cerezo and Garcia, 2004; Valverde *et al.*, 2006), diet (e.g. McKenzie *et al.*, 1997, 2000), perceived predation risk (e.g., Shingles *et al.*, 2005; Randle and Chapman, 2005), developmental plasticity (Vulesevic and Perry, 2006), and prior history of hypoxic exposure (e.g., Kerstens *et al.*, 1979; Lomholt and Johansen, 1979; Kramer and Mehegan, 1981; Graham and Baird, 1982, 1984; Graham, 1983; Nikinmaa and Weber, 1984; Taylor and Miller, 2001; Burleson *et al.*, 2002; Routley *et al.*, 2002; Timmerman and Chapman, 2004a, b; Vulesevic *et al.*, 2006), yet in many cases our understanding of how and why these conditions affect the hypoxic ventilatory response is less than complete. Under these conditions, it becomes difficult to make generalizations about the hypoxic ventilatory response (beyond, of course, the obvious – that fish hyperventilate!) and this point should be kept in mind while considering the generalizations that follow.

Table 5.1 summarizes data on the ventilatory responses to hypoxia of 46 species of fish from several taxonomic groups, focusing on species that are solely or predominantly water-breathing and studies that have measured $\dot{V}_w$ and/or ventilation frequency (V_f) and stroke volume (or its commonly used proxy, ventilation amplitude; V_{amp}). With the exception of lamprey, for which only frequency data appear to have been collected, studies presenting data on V_f alone were not included. Table 5.1 includes four obligate ram ventilators (the bonnethead shark, *Sphyrna tiburo*, and three tuna species), which elevate $\dot{V}_w$ by increasing gape (Figure 5.1) (Bushnell *et al.*, 1990; Carlson and Parsons, 2003), as well as seven species that either lowered ventilation during hypoxia (Japanese eel, *Anguilla japonica*, and white sturgeon, *Acipenser transmontanus*) or for which data sets are incomplete (Pacific hagfish, *Eptatretus stoutii*, lamprey, *Entosphenus tridentatus* and *Lampetra fluviatilis*, starry flounder, *Platichthys stellatus*, and sheatfish, *Silurus glanis*). Of the remaining 35 species, 21 (60%) respond to hypoxia primarily or solely by increasing V_{amp}, 13 species (37%) employ increases in either or both of V_f and V_{amp}, and only one species (carp, *Cyprinus carpio*; 3%) increases V_f in the absence of amplitude or stroke volume adjustments. Included among species that increase V_f during hypoxia are those in which hypoxia promotes a transition from a pattern of episodic to continuous breathing (Figure 5.1) (e.g., Lomholt and Johansen, 1979; Smith *et al.*, 1983; Gehrke and Fielder, 1988; Glass *et al.*, 1990; Nonnotte *et al.*, 1993; Maxime *et al.*, 1995;

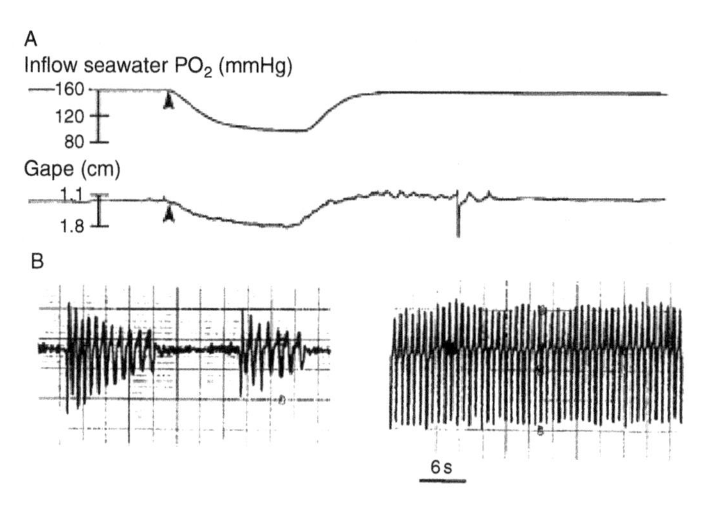

Fig. 5.1. Ventilatory responses to hypoxia. (A) An original recording of the increase in mouth gape that occurs in an obligate ram ventilator, yellowfin tuna (*Thunnus albacares*) as water O_2 tension is lowered. Gape was measured as changes in impedance from electrodes attached to the upper and lower jaws. [Reproduced with permission from Bushnell *et al.* (1990).] (B) Ventilation in the sturgeon *Acipenser baeri* is episodic under normoxic conditions (left-hand panel) but becomes continuous upon exposure to hypoxic water (right-hand panel; PO_2 = 60 Torr). Note the increase in ventilation amplitude that occurs in addition to the increases in frequency. The figure depicts original recordings of pressure changes associated with breathing in the branchial cavity, measured using a water-filled catheter connected to a pressure transducer. [Reproduced with permission from Nonnotte *et al.* (1993).]

Fernandes *et al.*, 1995; Reid *et al.*, 2003; Vulesevic *et al.*, 2006; Leite *et al.*, 2007). This breakdown of responses provides support for the generalization that fish employ the energetically favourable strategy, given the density and viscosity of water as the ventilated medium, of increasing $\dot{V}_w$ through large changes in ventilatory stroke volume coupled to more modest increases of V_f (Shelton *et al.*, 1986; Perry and Wood, 1989; Gilmour, 2001). At the same time, it is clear that strategies for achieving hyperventilation among fish are diverse, at times even within a single species. For example, Vulesevic *et al.* (2006) reported that zebrafish (*Danio rerio*) responded to either hypoxia or hypercapnia (high water CO_2 tension) by hyperventilating, but whereas hypoxia caused V_f to increase in the absence of changes in V_{amp}, the opposite was true of hypercapnia, where breathing amplitude increased but frequency was unaffected. Even more perplexing, the hypoxic hyperventilation in a different group of zebrafish relied largely upon changes in V_{amp} (W. K. Milsom, personal communication; see Table 5.1). The hyperventilatory strategy may also depend upon the level of hypoxia. For example, moderate increases in $\dot{V}_w$ in cod (*Gadus morhua*) were achieved by raising breathing frequency, whereas increases in stroke volume accounted for the greater hyperventilatory response to more severe levels of hypoxia (Kinkead *et al.*, 1991).

Typically, ventilatory responses to hypoxia are initiated very rapidly, i.e., as hypoxic water contacts the gill (Figure 5.2) (Bamford, 1974; Kinkead *et al.*, 1991), and may be sustained for hours to days. Results from several studies that have monitored ventilation for 90 min to 24 h of hypoxia suggest that hyperventilatory responses to hypoxia are independent of exposure period during this time frame (Thomas and Hughes, 1982a,b; Forgue *et al.*, 1989; Glass *et al.*, 1990; Borch *et al.*, 1993; Florindo *et al.*, 2006), except possibly during exposure to near-anoxic conditions (Stecyk and Farrell, 2002). As exposure to hypoxia is prolonged, however, a suite of responses is initiated to enhance O_2 uptake and delivery beyond that achieved by hyperventilation alone. These responses yield increases in hemoglobin–oxygen binding affinity (e.g., Wood and Johansen, 1972; Wood *et al.*, 1975; Tetens and Lykkeboe, 1981; Rutjes *et al.*, 2007) and blood O_2 carrying capacity (e.g. Wood and Johansen, 1973; Lai *et al.*, 2006; Rutjes *et al.*, 2007) that, together with other factors, enhance O_2 transfer (see reviews by Perry and Wood, 1989; Nikinmaa, 2001) and might therefore lower the ventilatory convection requirement. Moreover, prolonged acclimation to hypoxic conditions also may influence the density, size, and morphology of gill neuroepithelial cells (Jonz *et al.*, 2004; Vulesevic *et al.*, 2006), the putative O_2 chemosensors of the fish gill (see Sections 3.2.2 and 3.3), providing a mechanism through which acclimation-associated changes in ventilatory responses could be mediated. Acclimation to hypoxic conditions does appear to influence ventilation parameters. Under normoxic conditions, acclimation

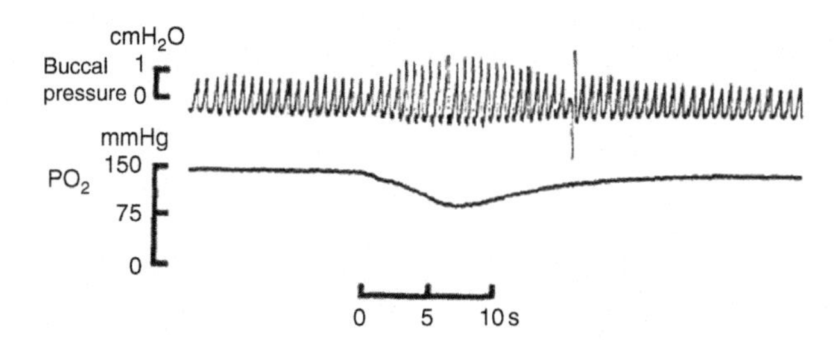

Fig. 5.2. An original recording of ventilation (measured as pressure changes in the buccal cavity associated with breathing using a water-filled catheter connected to a pressure transducer) and water O_2 tension demonstrating that ventilatory responses to hypoxia are initiated within seconds of the hypoxic water contacting the gill. [Reproduced with permission from Bamford (1974).]

to hypoxia eliminates episodic breathing in species that normally exhibit this breathing pattern (Lomholt and Johansen, 1979; Vulesevic *et al.*, 2006), but is otherwise generally without effect (Lomholt and Johansen, 1979; Burleson *et al.*, 2002; Vulesevic *et al.*, 2006). The impact of hypoxic acclimation on the hypoxic ventilatory response is more variable, with increased ventilatory sensitivity to hypoxia (Δventilation/ΔPO_2) reported for some species (Kerstens *et al.*, 1979; Burleson *et al.*, 2002), but no change or reduced ventilation under hypoxic conditions reported for other species (Lomholt and Johansen, 1979; Nikinmaa and Weber, 1984; Vulesevic *et al.*, 2006). There is clearly a need to investigate the time domains of the hypoxic ventilatory response (Powell *et al.*, 1998) in fish in a more systematic fashion, to probe the underlying mechanisms, and to attempt to associate these time domains with the occurrence of other mechanisms that affect branchial O_2 transfer and blood O_2 transport.

Ventilatory responses to hypoxia typically reflect the severity of the hypoxic stimulus, with $\dot{V}_w$ increasing as water PO_2 falls (Table 5.1; Figure 5.3). Estimates of the maximum increase in $\dot{V}_w$ depend not only on the species examined and the severity of hypoxia, but also upon the method used to determine $\dot{V}_w$ and upon acclimation temperature. Notably, use of the Fick method leads to overestimation of $\dot{V}_w$, particularly at high levels of ventilation (Davis and Watters, 1970; Kalinin *et al.*, 1999). Considering only the 26 studies listed in Table 5.1 in which $\dot{V}_w$ was measured directly, maximum increases in $\dot{V}_w$ during hypoxia range from 52% to 581% of the normoxic value (or $\dot{V}_w$ maximally increases 1.5- to 6.9-fold during hypoxia). The maximum increase in $\dot{V}_w$ also depends upon acclimation temperature, with higher ventilation volumes being achieved at higher acclimation temperatures. Although normoxic $\dot{V}_w$ also increases with increasing acclimation temperature, presumably owing to the effect of temperature on metabolic rate, enhanced ventilatory sensitivity to hypoxia (Δventilation/ΔPO_2) in fish acclimated to higher temperatures typically results in greater percentage increases in $\dot{V}_w$ during hypoxia (Table 5.3) (Spitzer *et al.*, 1969; Campagna and Cech, 1981; Fernandes and Rantin, 1989; Glass *et al.*, 1990; Cerezo and Garcia, 2004; Valverde *et al.*, 2006). As with the effects of hypoxic acclimation, the mechanisms underlying the impact of acclimation temperature on the hypoxic ventilatory response remain largely unexplored. The peak ventilatory effort occurs in many studies at the lowest water O_2 tension examined, but it is not uncommon to find that ventilation rises to a maximum and then falls as the level of hypoxia becomes more severe (Figure 5.3A). This fall in ventilation often seems to occur around the critical PO_2 (P_{crit}), i.e., the PO_2 at which the transition from oxyregulator to oxyconformer occurs. Table 5.4 summarizes data for 10 species exhibiting this pattern, where the PO_2 of peak ventilatory effort was significantly correlated with P_{crit} (correlation

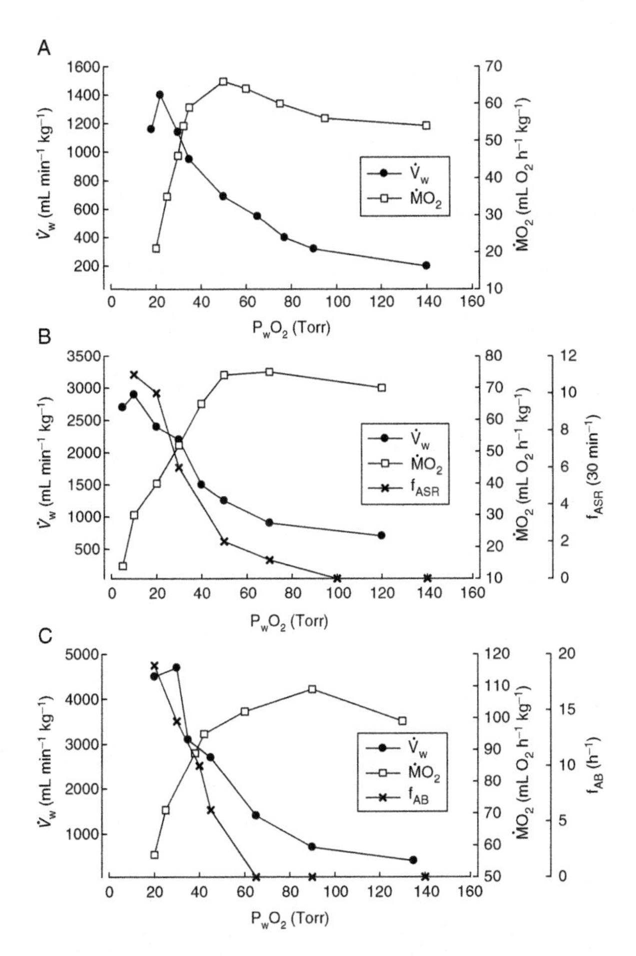

Fig. 5.3. The relationship between O_2 uptake (MO_2) and ventilation during exposure to hypoxia is depicted for three species of fish that make use of different ventilatory strategies. (A) Gill ventilation volume ($\dot{V}_w$) and MO_2 are plotted as a function of water O_2 tension (P_wO_2) for the unimodal water-breather, *Hoplias lacerdae*. $\dot{V}_w$ increases during hypoxia, peaking at a P_wO_2 of about 20 Torr, somewhat below the critical O_2 tension (P_{crit}), which was estimated to be 35 Torr. [Data replotted from Rantin *et al.* (1992).] (B) Data for pacu, *Piaractus mesopotamicus*, a species that utilizes aquatic surface respiration (ASR). In addition to $\dot{V}_w$ and MO_2, the frequency of ascents to the surface to perform ASR (f_{ASR}) is plotted as a function of P_wO_2. ASR was initiated at a P_wO_2 of ~30 Torr, very close to the P_{crit} of 34 Torr (measured without access to the water–air interface), while gill ventilation peaked at a somewhat lower P_wO_2 of 10 Torr. [Modified from Rantin *et al.* (1998).] (C) Data are presented for jeju, *Hoplerythrinus unitaeniatus*, a facultative air-breather. $\dot{V}_w$, MO_2, and the frequency of air-breaths (f_{AB}) are plotted as a function of P_wO_2. Air-breathing was absent until P_wO_2 dropped to 64 Torr, and then increased in frequency as P_wO_2 continued to fall. P_{crit} in the absence of access to air was determined to be 40 Torr, while gill ventilation peaked at ~30 Torr, again illustrating the appearance of an alternative ventilatory strategy at a P_wO_2 when metabolic rate would otherwise fall. [Modified from Oliveira *et al.* (2004).]

Table 5.3
Effects of acclimation temperature on the hypoxic ventilatory response of water-breathing fish

Species		Temp (°C)	Normoxic $\dot{V}_w$	Max $\Delta \dot{V}_w$ (%)	$\Delta \dot{V}_w/\Delta PO_2$	References
C. carpio	Carp	10	50.2	161	1.1	Glass *et al.*, 1990
		20	241	222	7.1	
O. niloticus	Tilapia	20	100	1000	8.3	Fernandes and Rantin, 1989
		25	200			
				1300	21.7	
		30	250	1300	27	
		35	300	1133	34	
O. microlepidotus	Sacramento blackfish	12	60	400	2.7	Campagna and Cech, 1981
		20	200	1950	4.3	
		28	360	3167	12.7	

Normoxic $\dot{V}_w$, ventilation volume under normoxic conditions at the temperature indicated; max $\Delta \dot{V}_w$, the maximum change in ventilation volume expressed as a percentage of the normoxic value; $\Delta \dot{V}_w/\Delta PO_2$, the ventilatory sensitivity to hypoxia, i.e., the change in ventilation volume for a given change in water O_2 tension. Only studies in which $\Delta \dot{V}_w$ was measured were included in the data set.

Table 5.4
A comparison of critical PO_2 (P_{crit}) values with the PO_2 of peak ventilation for water-breathing fish

Species		P_{crit} (Torr)	Peak gill ventilation (Torr)	References
A. anguilla	European eel	70	40	Le Moigne *et al.*, 1986
A. japonica	Japanese eel	100–115	80	Chan, 1986
D. rerio	Zebrafish	20[a]	43	Vulesevic *et al.*, 2006; Vulesevic and Perry, 2006 [a]Barrionuevo and Burggren, 1999
H. lacerdae		35	20	Rantin *et al.*, 1992
H. malabaricus	Traira	20[b]	20–30	Sundin *et al.*, 1999; [b]Rantin *et al.*, 1992
Leiopotherapon unicolor	Spangled perch	27.2	16	Gehrke and Fielder, 1988
O. mykiss	Rainbow trout	22[c]	30	Perry and Gilmour, 1996; [c]Ott *et al.*, 1980
			~49	Greco *et al.*, 1995
			~40	Holeton and Randall, 1967b
P. mesopotamicus	Pacu	34	10	Rantin *et al.*, 1998
			30	Leite *et al.*, 2007
P. flesus	Flounder	60–80	~60	Steffensen *et al.*, 1982
S. maximus	Turbot	20–30	~40	Maxime *et al.*, 2000

P_{crit}, the partial pressure of O_2 at which the transition from oxygen regulator to oxygen conformer occurs; Peak gill ventilation, the partial pressure of O_2 at which ventilation of the gills was maximal (using $\dot{V}_w$ where available, or V_f and/or V_{amp} as appropriate).

coefficient = 0.76; P = 0.01). A tendency for ventilation to fall as water O_2 tension is lowered below the P_{crit} probably explains the two entries in Table 5.1 for species in which ventilation fell with exposure to hypoxia, white sturgeon and Japanese eel. Ventilation parameters for Japanese eel were summarized in Table 5.1 for a P_wO_2 of 40 Torr, but P_{crit} was estimated to be 100–115 Torr (Chan, 1986). White sturgeon appears to be a true oxygen conformer with an unusually high P_{crit}, in that oxygen consumption declined with even very small reductions of water O_2 tension (Burggren and Randall, 1978). Lowering breathing in this species during hypoxia may reduce the cost of ventilation per unit of oxygen uptake, particularly since oxygen extraction was maintained (Burggren and Randall, 1978).

In some species, reductions in gill ventilation at severe levels of hypoxia are associated with the appearance of alternative ventilatory strategies, such as aquatic surface respiration (Table 5.5) or air-breathing (Table 5.2). Aquatic surface respiration (ASR) consists of using the thin zone (<0.5 mm; Burggren, 1982) of relatively oxygen-rich water at the air–water interface to ventilate the gills (e.g., Gee *et al.*, 1978; Kramer and McClure, 1982;

Table 5.5
PO_2 thresholds for aquatic surface respiration (ASR)

Species		PO_2 (Torr), time	P_{crit} (Torr)	ASR threshold (Torr)	Peak gill ventilation (Torr)	References
Astatotilapia aeneocolor		3, 5 h	15*	15	3.4	Melnychuk and Chapman, 2002
A. 'wrought-iron'		3, 5 h	13.5*	15	15	Melnychuk and Chapman, 2002
Astronotus ocellatus	Oscar	20, 2 h	50*	50	–	Sloman *et al.*, 2006
Carassius auratus	Goldfish	0, 2 h	34*[a]	13.7	13.7	McNeil and Closs, 2007; [a]Fry and Hart, 1948
C. carpio	Carp	0, 2 h	20*[b]	13.7	17.1	McNeil and Closs, 2007; [b]Ott *et al.*, 1980
M. cephalus	Mullet	16, 30 min	~44*[c]	32	19.4	Shingles *et al.*, 2005; [c]Nordlie and Lefler, 1975
P. mesopotamicus	Pacu	10, 30 min	34*	30	10	Rantin *et al.*, 1998

P_{crit}, the partial pressure of O_2 at which the transition from oxygen regulator to oxygen conformer occurs (* indicates that P_{crit} was measured without access to the air–water interface); ASR threshold, the partial pressure of O_2 at which ASR first occurs; Peak gill ventilation, the partial pressure of O_2 at which ventilation of the gills was maximal (using $\dot{V}_w$ where available, or V_f and/or V_{amp} as appropriate); the level of hypoxia to which the fish was exposed and the period of hypoxic exposure are also listed. Where variables were not measured in a particular study, a dash (–) is used. See in addition Table 2 in Gee and Gee (1991) for thresholds of ASR in 5 species of Eleotridae and 15 species of Gobiidae, Fig. 3 in Melnychuk and Chapman (2002) for thresholds of ASR and P_{crit} values in 19 species of East African cichlids, Figs. 1 and 2 in McNeil and Closs (2007) for thresholds of ASR and PO_2 of peak gill ventilation in 9 species commonly found in Australian billabongs, Fig. 5 in Soares *et al.* (2006) for thresholds of ASR and PO_2 of peak gill ventilation in 8 Amazonian species, and Table 1 in Gee *et al.* (1978) for thresholds of ASR and PO_2 of peak gill ventilation in 26 species from western Canada.

Kramer, 1983; Gee and Gee, 1991; Melnychuk and Chapman, 2002; Soares *et al.*, 2006; McNeil and Closs, 2007). Fish rise to the surface to skim the surface film of water, adopting a position in which the top of the head lies just at or below the surface of the water. Morphological adaptations such as flattened heads, upturned mouths (Lewis, 1970) and/or the development of lip protuberances during hypoxia (Winemiller, 1989; Sundin *et al.*, 2000; Florindo *et al.*, 2006) may facilitate or increase the effectiveness of ASR. Although the initiation of ASR occurs at much lower O_2 tensions than does hyperventilation, typically appearing at or below the PO_2 of maximum gill ventilation (Table 5.5; Figure 5.3B), in most other respects trends in ASR are similar to those for gill ventilation. For example, ASR effort (measured as time spent in ASR) increases as PO_2 falls (Kramer and Mehegan, 1981; Kramer and McClure, 1982; Rantin *et al.*, 1998), hypoxic acclimation reduces the use of ASR (Kramer and Mehegan, 1981; Timmerman and Chapman, 2004a), and acclimation to higher temperatures raises the threshold at which ASR appears (Gee *et al.*, 1978). Unlike hyperventilation, the use of ASR is associated with an obvious predation risk because of the need to approach the surface of the water. More severe levels of hypoxia are required to elicit ASR in mullet (*Mugil cephalus*) in response to the threat of predation (Shingles *et al.*, 2005). Similarly, small oscar (*Astronotus ocellatus*) surface for ASR at lower O_2 tensions than larger oscar, which are less vulnerable to predation (Sloman *et al.*, 2006). The effect of perceived predation risk on the use of ASR emphasizes a need for investigation of the factors that regulate or modulate O_2 chemosensory reflexes. Moreover, further elucidation of the stimuli that provoke ASR and the neural circuitry underlying this response is needed to confirm that ASR can be classified as an O_2-chemoreflex (or hypoxic reflex) together with alteration of gill ventilation. ASR appears to be mediated by O_2-sensitive chemoreceptors that are located in the orobranchial cavity and innervated by cranial nerve V, since it can be evoked by injection of sodium cyanide (see Section 3.1) into the bloodstream or ventilated water to stimulate internally oriented or externally oriented O_2 chemoreceptors, respectively (Shingles *et al.*, 2005) and it is eliminated by sectioning of the mandibular branches of cranial nerve V (Florindo *et al.*, 2006). However, our knowledge of ASR as a chemosensory reflex is far less detailed than is the case for gill ventilation.

Similar comments apply to air-breathing as an alternative ventilatory strategy during hypoxia, i.e., facultative air-breathing. Whether or to what extent obligate air-breathers respond to aquatic hypoxia is unclear, because several studies have reported effects of aquatic hypoxia on air-breathing frequency (Table 5.2), yet others, often on the same species, have not (Johansen and Lenfant, 1968; Sanchez *et al.*, 2001; Perry *et al.*, 2005). Among facultative air-breathers that do not exhibit air-breathing under normoxic conditions (e.g., armoured catfish, *Ancistrus chagresi*, jeju, *Hoplerythrinus unitaeniatus*, ihering, *Hypostomus regani*, Pacific tarpon, *Megalops atlanticus*, cascudo preto, *Rhinelepis strigosa*; see Table 5.2), the onset of air-breathing coincides (approximately) with the P_{crit} (measured without access to air) and/or the PO_2 of maximum gill ventilation (Table 5.2; Figure 5.3C). Air-breathing frequency in facultative air-breathers in general increases as water PO_2 falls (e.g., Johansen *et al.*, 1970; Graham *et al.*, 1977; Graham and Baird, 1982; Smatresk, 1986; McMahon and Burggren, 1987; Mattias *et al.*, 1998; Takasusuki *et al.*, 1998; Perry *et al.*, 2004; Oliveira *et al.*, 2004; Affonso and Rantin, 2005; Randle and Chapman, 2005), but the effect of aquatic hypoxia on breath (tidal) volume is less clear, with both no effect (Graham, 1983; McMahon and Burggren, 1987) and an increase in tidal volume (Lomholt and Johansen, 1974) having been reported. Few studies have examined the effect of hypoxic acclimation on air-breathing patterns. However, acclimation to hypoxic conditions reduced air-breathing frequency at a given level of hypoxia in two species of armoured catfish (Graham and Baird, 1982), while at the same time increasing the duration and size of each air-breath so as to augment O_2 extraction (Graham, 1983). Hypoxic acclimation was also found to raise the threshold for air-breathing in one study (Bicudo and Johansen, 1979), but not in others (Gee, 1980; Graham and Baird, 1982, 1984). Air-breathing frequency increases with increasing temperature (Johansen *et al.*, 1970; Gee, 1980; Graham and Baird, 1982; McMahon and Burggren, 1987; Geiger *et al.*, 2000) and, like ASR, is affected by perceived predation threat. For example, air-breathing frequency during hypoxia in both gar (*Lepisosteus platyrhincus*; Smith and Kramer, 1986) and gourami (*Colisa lalia*; Wolf and Kramer, 1987) was reduced following exposure to, respectively, an avian or piscine predation threat. Similarly, hypoxic mudminnows (*Umbra limi*) subjected to a disturbance to simulate a predation threat tended to air-breathe in a synchronous fashion (Gee, 1980), while a higher air-breathing frequency under hypoxic conditions was reported for the African anabantoid fish, *Ctenopoma muriei*, when fish were held in groups permitting synchrony of air-breathing behavior (Randle and Chapman, 2005); synchronous air-breathing is thought to reduce vulnerability to aerial predation (Kramer and Graham, 1976). The existence of PO_2 thresholds for air-breathing and the dependence of air-breathing frequency on PO_2 suggest that air-breathing, like ASR and branchial ventilation, can be considered a hypoxic reflex. Some evidence links air-breathing to the activation of branchial O_2-sensitive chemoreceptors. For example, air-breathing is stimulated by injection of cyanide into the bloodstream and/or ventilatory water flow of gar (*L. osseus*) and bowfin (*Amia calva*) (Smatresk *et al.*, 1986; Smatresk, 1986; McKenzie *et al.*, 1991), as well as the bloodstream of the obligate air-breathing African lungfish (*Protopterus aethiopicus*)

(Lahiri *et al.*, 1970). In bowfin, O_2 chemosensors linked to air-breathing reflexes may be located in the pseudobranch, since the combination of branchial denervation together with pseudobranch ablation was necessary to eliminate air-breathing responses to hypoxia (McKenzie *et al.*, 1991; Hedrick and Jones, 1999). Branchial denervation attenuates air-breathing responses to cyanide in lungfish (Lahiri *et al.*, 1970), and preliminary data suggest that the hypoxia-induced air-breathing reflex in gar is also eliminated by branchial denervation (Smatresk, 1994). In general, however, little is known of the O_2 chemoreceptors and afferent pathways that mediate air-breathing reflexes in fish. For a more detailed discussion of air-breathing in fish, Graham's (1997) book on this topic should be consulted.

2.1. The Physiological Significance of the Hypoxic Ventilatory Response

The ventilatory responses to hypoxia, namely gill hyperventilation, ASR and/or air-breathing, represent attempts to maintain O_2 uptake in the face of declining (aquatic) O_2 availability. The rate of O_2 transfer across the gill is governed by diffusive conductance, convection (ventilation and perfusion), and the blood-to-water PO_2 gradient (see reviews by Randall and Daxboeck, 1984; Malte and Weber, 1985; Perry and Wood, 1989; Randall, 1990; Piiper, 1998; Perry and Gilmour, 2002; Evans *et al.*, 2005b). During hyperventilation, increased water flow across the gill decreases the inspired-expired PO_2 difference, raising the mean blood-to-water PO_2 gradient and resulting in an elevation of arterial PO_2. Hyperventilation during hypoxia therefore serves to minimize the extent of the reduction in arterial PO_2 (and hence arterial O_2 content) that is the inevitable consequence of lowering water PO_2 (e.g., Holeton and Randall, 1967b; Wood and Johansen, 1973; Eddy, 1974; Itazawa and Takeda, 1978; Burggren and Cameron, 1980; Forgue *et al.*, 1989; Peyraud-Waitzenegger and Soulier, 1989; Glass *et al.*, 1990; Nonnotte *et al.*, 1993; Bindon *et al.*, 1994; Greco *et al.*, 1995; Maxime *et al.*, 2000; Soncini and Glass, 2000). This benefit of hyperventilation becomes particularly important as arterial PO_2 approaches the P_{50} value of hemoglobin (the PO_2 at which hemoglobin is 50% saturated with O_2), where the steep slope of the O_2 equilibrium curve means that even small differences in arterial PO_2 can have a relatively large impact on arterial O_2 content. In addition to defending arterial PO_2, hyperventilation during hypoxia produces a respiratory alkalosis as arterial PCO_2 is lowered by equilibration of the arterial blood with ventilatory water of lower mean PCO_2 [see Table 1 in Gilmour (2001) for examples]. Elevation of red blood cell pH following from the respiratory alkalosis can, in turn, increase the affinity of hemoglobin for O_2 via the Bohr effect, thereby aiding O_2 uptake (Jensen, 1991; Jensen *et al.*, 1998). The benefit of the hypoxic hyperventilatory response, then, is increased branchial O_2 transfer, which contributes to the maintenance of metabolic rate under hypoxic conditions in oxyregulators. The drawback of the hypoxic hyperventilatory response, on the other hand, is increased energy expenditure on ventilation, the cost of which is high even at rest in water-breathing fish (Cameron and Cech, 1970; Hughes and Saunders, 1970; Edwards, 1971; Jones and Schwarzfeld, 1974; Steffensen, 1985). The maintenance of metabolic rate under these circumstances becomes a battle of diminishing returns, in which the cost of increasing ventilation to maintain O_2 uptake from an environment of reduced O_2 availability eventually exceeds the benefits of the O_2 so obtained. The switch to oxygen conforming, often with a concomitant reduction in $\dot{V}_w$ (e.g., Table 5.4), will lower the energetic expenditure on ventilation.

ASR or air-breathing provides an alternative strategy to augment O_2 uptake under hypoxic conditions. The effectiveness of these alternative hypoxic ventilatory responses as well as their physiological costs and benefits has, however, received little attention to date. Several studies have documented lower mortality rates under hypoxic conditions for fish allowed to perform ASR, indicating that this strategy has survival value (Lewis, 1970; Kramer and Mehegan, 1981; Kramer and McClure, 1982). The impact of ASR on O_2 transfer was assessed by Burggren (1982), who found that goldfish (*Carassius auratus*) permitted to perform ASR under severely hypoxic conditions (water PO_2 = 18 Torr) were able to maintain a significantly higher arterial PO_2 than those denied access to the surface, or those given access to a water–N_2 interface. Although the increase in arterial PO_2 was small (about 1.2 Torr), it was effective in doubling arterial blood O_2 content because it occurred near the P_{50} value of goldfish hemoglobin (Burggren, 1982). Similarly, facultative air-breathers provided with access to air during aquatic hypoxia also exhibit improved blood O_2 status (Perry *et al.*, 2004). In jeju, the ability to defend arterial PO_2 through air-breathing during hypoxia was sufficient to avoid catecholamine mobilization (Perry *et al.*, 2004). Access to air during aquatic hypoxia reduces mortality (Huang *et al.*, 2008) and enables facultative air-breathers both to attenuate the fall in O_2 uptake at a given water PO_2 and to sustain O_2 uptake to more severe levels of hypoxia (see Figure 5.6 in Graham, 1997; as well as Graham *et al.*, 1977). Enhanced pulmonary blood flow (Smatresk and Cameron, 1982; Fritsche *et al.*, 1993), as well as increased pulmonary contribution to total O_2 uptake (Johansen *et al.*, 1967, 1970; Burggren, 1979; Smatresk and Cameron, 1982; Graham and Baird, 1984; see also Table 5.5 and Figure 5.7 in Graham, 1997), under hypoxic conditions also attest to the value of air-breathing as an alternative hypoxic ventilatory response. The above data, while somewhat sparse, lend support to the widely accepted view that ASR or air-breathing is an effective means of augmenting O_2 uptake during severe aquatic hypoxia. However, the costs associated with ASR or air-breathing largely remain to be determined. In both cases, to the cost of ventilation itself (which has yet to be assessed experimentally for air-breathing in fish) must be added the energetic expenditure to access and/or swim at the surface (i.e., costs of locomotion and buoyancy regulation) as well as time lost to other activities (e.g., feeding) and increased predation risk (Kramer, 1983). Theoretical considerations suggest that the travel costs of ASR and air-breathing are significant and will play an important role in determining PO_2 thresholds for use of these alternative strategies (Kramer, 1983).

3. O_2 SENSING AND O_2 SENSORS

The wealth of data demonstrating hypoxic hyperventilation in water-breathing fish (see Tables 5.1 and 5.2), clearly attests to the presence of reliable O_2-sensing mechanisms. However, unlike birds and mammals, which possess a predominant single site of O_2 sensing (the O_2 chemoreceptors of the carotid body), fish (and other lower vertebrates) may exhibit multiple sites of O_2 chemoreception (Milsom and Burleson, 2007). Surprisingly few studies have been conducted to identify the sites of O_2 chemoreceptors in fish, which has made it difficult to formulate general principles, especially considering the marked intraspecific variability that exists in those few species that have been examined. Research has focused on two crucial issues: (1) whether the O_2 chemoreceptors controlling breathing are oriented to sense the external or internal environments (or both); and (2) whether the chemoreceptors are branchial and/or extrabranchial. Elements of these issues have been dealt with in previous reviews (Hughes and Shelton, 1962; Randall, 1982; Shelton *et al.*, 1986; Perry and Wood, 1989; Smatresk, 1990; Burleson *et al.*, 1992; Fritsche and Nilsson, 1993; Burleson, 1995; Milsom *et al.*, 1999; Perry and Gilmour, 1999, 2002; Gilmour, 2001; Gilmour and Perry, 2007; Milsom and Burleson, 2007).

3.1. Internally versus Externally Oriented O_2 Chemoreceptors

Ambient hypoxia causes a lowering of blood PO_2 and, depending on the severity of the hypoxia and the P_{50} of hemoglobin, there may be associated reductions in arterial O_2 content (CaO_2) of variable severity. Thus, the hyperventilation accompanying hypoxia could reflect stimulation of receptors oriented to sense the external environment (so-called external receptors) and/or receptors localized to sense the PO_2 or O_2 concentration of the internal environment (so-called internal receptors). A variety of techniques has been used to assess the relative involvement of external and internal receptors. The most commonly used method is the selective application of respiratory stimuli (hypoxic media or cyanide) to the external and internal compartments by injections into the buccal cavity (to preferentially stimulate external receptors) or pre- or post-branchial blood (to preferentially stimulate internal receptors). Although conflicting results were obtained from trout [notably Eclancher and Dejours (1975) reported an absence of any effect of external cyanide on ventilation in brown trout, *Salmo trutta*], all teleosts that have been studied exhibit hyperventilatory responses to both external and internal cyanide (rainbow trout, *Oncorhynchus mykiss*, Burleson and Milsom, 1995b; Reid and Perry, 2003; channel catfish, *Ictalurus punctatus*, Burleson and Smatresk, 1990b; traira, *Hoplias malabaricus*, Sundin *et al.*, 1999; tambaqui, *Colossoma macropomum*, Sundin *et al.*, 2000). Representative ventilation recordings from gulf toadfish (*Opsanus beta*) before and after administration of external or internal cyanide are depicted in Figure 5.4. Of the non-teleost species that have been examined, the sturgeon (*A. naccarii*, McKenzie *et al.*, 1995) and bowfin (McKenzie *et al.*, 1991) appear to posses both external and internal receptors whereas the gar (*L. osseus)* appears to lack external O_2 receptors linked to hyperventilation (Smatresk *et al.*, 1986). The injection of deoxygenated blood into the ventral aorta of rainbow trout (Bamford, 1974) or dorsal aorta of sea raven (*Hemitripterus americanus*) (Saunders and Sutterlin, 1971) yielded hyperventilatory responses further supporting the presence of internally oriented O_2 chemoreceptors.

Less frequently used techniques to assess the orientation of O_2 chemoreceptors include the use of perfused preparations to selectively manipulate the external and internal compartments (Milsom and Brill, 1986; Burleson and Milsom, 1993) or impairment of blood O_2 transport [induction of anemia (Smith and Jones, 1982) or exposure to carbon monoxide (Holeton, 1971a, 1977)]. Perfused gill preparations (Perry *et al.*, 1984; Perry and Farrell, 1989), while inappropriate for studying most physiological functions (Perry *et al.*, 1985a,b), have been useful in distinguishing external and internal branchial O_2 chemoreceptors. Milsom and Brill (1986) and later Burleson and Milson (1993) recorded single nerve fiber activity originating from O_2 chemoreceptors within the first gill arch of yellowfin tuna (*Thunnus albacares*) and rainbow trout, respectively. These challenging experiments (e.g., only 5% of the 800 fibers tested in trout were actually O_2-sensitive) revealed the presence of three distinct populations of receptors, those exclusively responsive to external or internal hypoxic stimuli and those responsive to both. Interestingly, Burleson and Milsom (1993) reported that receptor discharge frequency declined when PO_2 was lowered below approximately 40 Torr thereby suggesting a depressant effect of severe hypoxia on O_2 chemoreceptor activity. This abrupt switch from chemoreceptor stimulation to inhibition may, in part, underlie the attenuation of the hyperventilatory

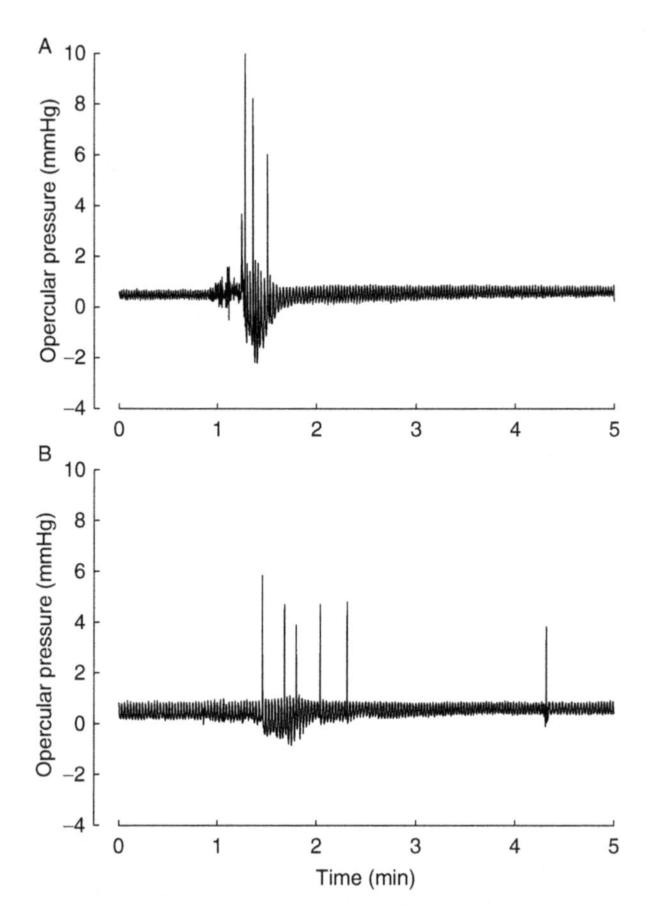

Fig. 5.4. Representative traces of opercular pressure (an index of ventilation amplitude) in Gulf toadfish (*Opsanus beta*) administered bolus injections of the O_2 chemoreceptor stimulant sodium cyanide into (A) the buccal cavity (to preferentially stimulate external receptors) or (B) caudal vein (to preferentially stimulate internal receptors). The large vertical deflections were caused by brief periods of agitation. [S. F. Perry, M. D. McDonald, P. J. Walsh, and K. M. Gilmour, unpublished data.]

response that is sometimes observed once a critical level of hypoxia is achieved (Figures 5.3 and 5.5; see also Table 5.4).

Because rainbow trout hyperventilate in response to acute anemia (Smith and Jones, 1982) or externally applied carbon monoxide (Holeton, 1971b, 1977), it has been suggested that internal O_2 chemoreceptors (or a subset of internal receptors) in fish may respond to changes in blood O_2 content

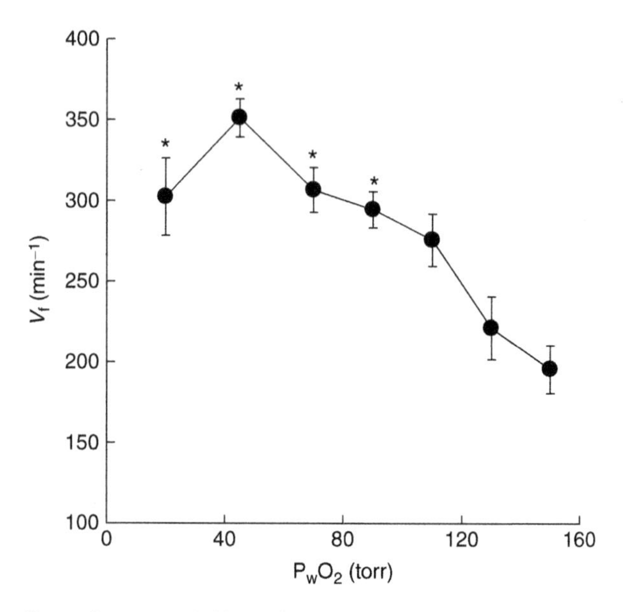

Fig. 5.5. The effects of acute graded hypoxia on ventilation frequency (V_f) in adult zebrafish (*Danio rerio*) demonstrating a linear rise in V_f with increasing hypoxia until a threshold water PO_2 (P_WO_2) is reached at which time V_f decreases. Significant differences from prehypoxia values are indicated by asterisks. [Data from Vulesevic and Perry (2006).]

(see Randall, 1982). While it is difficult to conceive of a cellular mechanism for sensing O_2 content, it is possible that these receptors are responding to the rate of O_2 delivery, which in turn is being influenced by blood O_2 content (Randall, 1982). The increase in O_2 receptor discharge frequency associated with cessation of perfusion in perfused tuna (Milsom and Brill, 1986) and to a lesser extent trout (Burleson and Milsom, 1993) gills is consistent with the notion of an O_2 receptor responding to changes in the rate of O_2 delivery. Alternatively, the receptor cells may simply be responding to a localized decrease in PO_2 owing to continuing O_2 consumption during the period of ischemia. Indeed, given the low O_2 capacitance of the saline used in the perfused gill preparations, it seems unlikely that the stimulation of internal chemoreceptors with a lowering of saline PO_2 could reflect a reduction in O_2 content.

An important caveat to consider when interpreting the results of experiments meant to distinguish between internal and external receptors is that stimuli such as cyanide, if added to the external or internal compartments, may still affect cells within the multilayered epithelium separating these compartments. Thus, as discussed by Gilmour and Perry (2007), the O_2 receptors apparently responding specifically to water- or blood-borne stimuli may indeed be the same receptors (see also Section 3.2.2).

3.2. Branchial versus Extra-Branchial O_2 Chemoreceptors

The gills are predominantly innervated by branches of the glossopharyngeal and vagus nerves (cranial nerves IX and X, respectively), both of which carry sensory nerve fibres (Nilsson, 1983). In teleosts, the first gill arch is innervated by IX and X whereas arches 2–4 are supplied only by branches of X. Thus, bilateral sectioning of these nerves is typically used to reveal the presence of branchial O_2 chemoreceptors and their role in the regulation of breathing. The pseudobranch also is innervated by branches of the glossopharyngeal nerve and has been considered as a possible site of O_2 chemoreception in those species known to possess a pseudobranch (see Section 3.2.1). The invasiveness of the surgery itself may compromise normal breathing and blood gases (e.g., Saunders and Sutterlin, 1971) and it may be for this reason that so few species have yielded useful data on the effects of bilateral gill denervation on the ventilatory responses to hypoxia. Indeed, data from a mere six species have been reported (tench *Tinca tinca*, Hughes and Shelton, 1962; sea raven, Saunders and Sutterlin, 1971; channel catfish, Burleson and Smatresk, 1990a; traira, Sundin *et al.*, 1999; bowfin, McKenzie *et al.*, 1991 and tambaqui, Sundin *et al.*, 2000; Milsom *et al.*, 2002). With the exception of channel catfish, each species retained some capacity to respond normally to hypoxia after bilateral gill denervation, supporting the presence of extra-branchial receptors. In some instances, there was an obvious involvement of both branchial and extra-branchial receptors in promoting the overall response. For example, in the tambaqui, branchial denervation prevented the increase in breathing frequency associated with hypoxia, indicating an exclusive branchial location for these receptors (Sundin *et al.*, 2000). On the other hand, extra-branchial receptors appeared to be responsible for the increase in breathing amplitude during hypoxia.

3.2.1. Extra-branchial Receptors

The pseudobranch (Bridges *et al.*, 1998) has been implicated in O_2 sensing (Laurent and Rouzeau, 1972; Jones and Milsom, 1982) on the basis of its morphology and the responsiveness of *in vitro* perfused preparations to hypoxic perfusate (Laurent and Rouzeau, 1972). Specifically, Laurent and Rouzeau (1972) demonstrated two patterns of neural discharge from afferent vagal branches of the trout pseudobranch; medium amplitude (50–200 μV) impulses of approximately 2 msec duration (termed Type A impulses) and lower amplitude (< 50 μV) impulses of relatively long (> 4 msec) duration (termed Type B impulses). While both discharge activities were sporadic and/or irregular under normoxic conditions, there was a significant increase in Type B activity with increasing levels of hypoxia. Although suggestive of a role for the pseudobranch in O_2 sensing, the response characteristics and sensitivity of the nerve activity to hypoxia differed markedly from nerve activities measured from mammalian carotid body or tuna first gill arch (see Milsom and Brill, 1986). Moreover, neither bilateral sectioning of pseudobranch afferent nerve fibers (Randall and Jones, 1973) nor pseudobranch removal (Bamford, 1974) altered the ventilatory response of trout to external hypoxia. Additionally, many species known to exhibit robust ventilatory responses to hypoxia lack a pseudobranch (e.g., channel catfish). Thus, at present, there is no conclusive evidence to support a role for the pseudobranch as an extra-branchial site of O_2 sensing.

The results of several studies have indirectly implicated the central nervous system (brain) as a site of O_2 chemoreception. Most notably, infusion of hypoxic blood into the dorsal aorta of otherwise normoxic sea raven elicited hyperventilation (Saunders and Sutterlin, 1971) and injections of deoxygenated blood into the ventral aorta caused hyperventilation but only after a significant latency period (Bamford, 1974; Eclancher and Dejours, 1975). The latency period was thought to reflect the time required for the hypoxic blood to travel to central receptors. Subsequent studies have employed a more direct approach to assess the possible role of the brain in O_2 sensing in which the brain is superfused *in situ* with hypoxic saline. The results of these studies performed on tambaqui (Milsom *et al.*, 2002) and bowfin (Hedrick *et al.*, 1991) failed to provide any evidence to support the existence of central O_2 chemoreceptors in fish.

In the absence of any direct data to support a role for the pseudobranch or central nervous system in O_2 sensing, the persistence of ventilatory responses to hypoxia in fish experiencing gill denervation may reflect the presence of extra-branchial receptors within the orobranchial cavity (Milsom *et al.*, 2002). An alternate and not mutually exclusive hypothesis (Randall and Taylor, 1991) is that ventilatory responses attributed to extra-branchial O_2 receptors may in fact arise from the release of catecholamines (adrenaline and noradrenaline) into the circulation. The principle evidence that led to the theory of a supporting role for circulating catecholamines in stimulating breathing is that these hormones are released into the bloodstream during acute hypoxia (Butler *et al.*, 1978; Boutilier *et al.*, 1988; Ristori and Laurent, 1989) coupled with reports that intravascular injections of catecholamines can evoke hyperventilatory responses in European eel, *A. anguilla* (Peyreaud-Waitzenegger, 1979; Peyreaud-Waitzenegger *et al.*, 1980). However, an examination of all available data (Table 5.6), clearly demonstrates that a myriad of responses can be elicited by injections of exogenous

Table 5.6
The effects of exogenous catecholamines on ventilation volume ($\dot{V}_w$), ventilation frequency (V_f), and ventilation amplitude (V_{amp}) or stroke volume (SV) in a variety of teleost and non-teleost species

Species	Injected dose or circulating levels[a]	$\dot{V}_w$	V_f	V_{amp} or SV[b]	Comments	References
O. mykiss	15 nmol l^{-1} NA; 75 nmol l^{-1} A	↓			Injected during moderate hypoxia (90 Torr)	Kinkead and Perry, 1991
	12 nmol l^{-1} NA; 140 nmol l^{-1} A	↓			Injected during moderate hypercapnia (4.5 Torr)	
	5 nmol l^{-1} NA; 75 nmol l^{-1} A	NC			Injected during hyperoxia (640 Torr)	
O. mykiss	1.4 nmol l^{-1} NA; 164 nmol l^{-1} A	↓	NC	↓	30 min infusion of A	Kinkead and Perry, 1990
	200 nmol l^{-1} NA; 5.8 nmol l^{-1} A	↓	↓	↓	30 min infusion of NA	
O. mykiss	3.2 nmol kg^{-1} A	↓			Single injection of A	Playle *et al.*, 1990
	3.2 nmol kg^{-1} NA	↑ → ↓			Single injection of NA	
O. mykiss	2.1 nmol l^{-1} NA; 22 nmol l^{-1} A		↓	NC	Single injection of A	Aota and Randall, 1993
	4.9 nmol l^{-1} NA; 57 nmol l^{-1} A		↓	NC	Single injection of A	
	9.4 nmol l^{-1} NA; 278 nmol l^{-1} A		NC	NC	Single injection of A	
	15 nmol l^{-1} NA; 2.5 nmol l^{-1} A		NC	↑	Single injection of NA	
	71 nmol l^{-1} NA; 4.4 nmol l^{-1} A		↓	↑	Single injection of NA	
	207 nmol l^{-1} NA; 21.8 nmol l^{-1} A		NC	NC	Single injection of NA	

(*continued*)

Table 5.6 *(continued)*

Species	Injected dose or circulating levels[a]	$\dot{V}_w$	V_f	V_{amp} or SV^b	Comments	References
O. mykiss	5 nmol kg^{-1} A		NC	NC	Single injection of A	Burleson and Milsom, 1995b
	100 nmol kg^{-1} A		↑	NC	Single injection of A	
	5 nmol kg^{-1} NA		NC	NC	Single injection of NA	
	100 nmol kg^{-1} NA		↑	NC	Single injection of NA	
O. mykiss	60 nmol kg^{-1} A; 15 nmol kg^{-1} NA		NC	NC	Single injection of catecholamine (A/NA) cocktail during moderate hypoxia	Perry and Gilmour, 1996
A. anguilla	5 nmol kg^{-1} A		↑	↑	Single injection of A during summer	Peyreaud-Waitzenegger, 1979; Peyreaud-Waitzenegger *et al.*, 1980
	5 nmol kg^{-1} NA		↑	↑	Single injection of NA during summer	
	5 nmol kg^{-1} A		↓	↓	Single injection of A during winter	Peyreaud-Waitzenegger *et al.*, 1980
	5 nmol kg^{-1} NA		↓	↓	Single injection of NA during winter	
G. morhua	5.6 nmol kg^{-1} A; 9.4 nmol kg^{-1} NA	↓	↓	↓	Single injection of catecholamine (A/NA) cocktail	Perry *et al.*, 1992
C. macropomum	10 nmol kg^{-1} A		↓	↓	Single injection of A	Milsom *et al.*, 2002
	100 nmol kg^{-1} A		↓	↓	Single injection of A	

A. calva	5 nmol kg^{-1} A	NC[c]	NC[c]	Single infusion of A	McKenzie *et al.*, 1991
	5 nmol kg^{-1} NA	NC	↑	Single infusion of A	
A. nacarii	~30 nmol kg^{-1} NA[d]	↑	↑	Single injection of NA	McKenzie *et al.*, 1995
S. acanthias	38 nmol kg^{-1} A; 38 nmol kg^{-1} NA	↓	NC	Single injection of catecholamine (A/NA) cocktail during moderate hypoxia	Perry and Gilmour, 1996

[a]Wherever possible, we report actual measured levels of circulating catecholamines because of the inherent problems associated with comparing injected doses to levels actually achieved during acute hypoxia.

[b]V_{amp} was estimated in a variety of ways including measurement of opercular or buccal pressures or linear opercular deflections determined from impedance measurements. SV was determined only in those experiments measuring true ventilation volumes and respiratory frequencies.

[c]Data were not statistically significant when analyzed by ANOVA but paired t-tests revealed significant increases at 2.5 min postinfusion.

[d]This dose is our own estimate based on the mean weight of the fish used in the entire study.

NC, no change

catecholamines. In those studies in which ventilation volumes were directly measured, the predominant response to catecholamine injection is hypoventilation (Kinkead and Perry, 1990, 1991; Playle *et al.*, 1990; Perry *et al.*, 1992). Of particular interest are the results of those experiments in which catecholamines were administered under pre-existing conditions of hyperventilation associated with moderate hypoxia (not severe enough, however, to elicit endogenous catecholamine release). In such cases, sudden elevation of circulating catecholamines caused abrupt hypoventilation (rainbow trout; Kinkead and Perry, 1991), a lowering of breathing frequency (spiny dogfish, *Squalus acanthias*; Perry and Gilmour, 1996), or no change (rainbow trout; Perry and Gilmour, 1996). With the exception of two species that show obvious hyperventilatory responses to exogenous catecholamines (European eel in summer months only, Peyreaud-Waitzenegger, 1979; Adriatic sturgeon, *A. nacarii*, McKenzie *et al.*, 1995), most fish that have been examined either exhibit hypoventilation (decreased frequency and/or amplitude) or are unresponsive to injected catecholamines (Table 5.6). Thus, there are no strong data to support a general stimulatory role for circulating catecholamines in the control of breathing during hypoxia, although in certain species it is conceivable that they play a supplementary role at very severe levels of hypoxia when O_2 chemoreceptors may be inhibited (Burleson and Milsom, 1993).

Extra-branchial receptors involved in the mediation of hypoxic ventilatory responses may also be found in the air-breathing organ (ABO) of air-breathing fish. Obligate air-breathers, while probably unresponsive to aquatic hypoxia (see Section 2), typically hyperventilate when exposed to aerial hypoxia (e.g., Johansen and Lenfant, 1968; Burggren, 1979; Glass *et al.*, 1986; Sanchez *et al.*, 2001; Perry *et al.*, 2005, 2008; Alton *et al.*, 2007). This effect may be mediated by internally oriented branchial O_2 chemoreceptors (see Section 3.1), a possibility for which there is some, albeit sparse, experimental evidence (Lahiri *et al.*, 1970; see Section 2). An alternative possibility is that O_2 chemoreceptors are present in the ABO. Physiological evidence supporting the existence of fish ABO chemoreceptors is mixed, with Graham *et al.* (1995) citing the rapidity of a gas-voiding reflex in *Monopterus albus* as evidence for the presence of an ABO chemosensor, but Alton *et al.* (2007) finding little evidence of ABO O_2 chemoreceptors in *Trichogaster leeri*. However, histochemical evidence suggests that pulmonary neuroepithelial cells (NECs) may be present in air-breathing fish that utilize lungs (Zaccone *et al.*, 1989, 1997; Adriaensen and Scheuermann, 1993; Kemp *et al.*, 2003). Additional work is needed to resolve the physiological function of such pulmonary NECs in fish.

3.2.2. Branchial Receptors

The NECs (Figure 5.6) of the gill are considered by many to be suitable candidates for the branchial O_2 chemoreceptors that mediate ventilatory reflexes in fish exposed to hypoxia (Dunel-Erb *et al.*, 1982; Bailly *et al.*, 1992; Sundin *et al.*, 1998; Jonz and Nurse, 2003; Vulesevic *et al.*, 2006; Burleson *et al.*, 2006; Coolidge *et al.*, 2008). The observation that these cells degranulate in response to severe hypoxia (< 10 Torr) likely was the first experimental evidence linking NECs and O_2 sensing in fish (Dunel-Erb *et al.*, 1982). Morphologically, NECs are reminiscent of O_2-chemoreceptive carotid body type I cells and neuroepithelial bodies (NEBs) in mammals (González *et al.*, 1994; Cutz and Jackson, 1999), and are considered to be phylogenetic precursors of these cells. In addition, NECs are highly conserved in fish, as they have been identified in the gills of every species in which these cells have been investigated (see Table 5.7). NECs contain neurotransmitters (see Section 4.3), particularly serotonin (5-hydroxytryptamine, 5-HT), and cytoplasmic synaptic vesicles in which these chemicals are stored (Dunel-Erb *et al.*, 1982; Jonz and Nurse, 2003; Saltys *et al.*, 2006). Furthermore, these vesicles are concentrated at the plasma membrane near adjacent nerve fibres. Nervous innervation of gill NECs has been documented at the ultrastructural level (Dunel-Erb *et al.*, 1982; Bailly *et al.*, 1992), and the association of entire populations of NECs with nerve fibres was observed in whole-mount gills of zebrafish using confocal microscopy, allowing visualization of complex innervation patterns in this tissue (Jonz and Nurse, 2003; Saltys *et al.*, 2006). It is apparent from ultrastructural and histological studies that NECs of the filaments receive multiple sources of innervation (Dunel-Erb *et al.*, 1982; Bailly *et al.*, 1992; Sundin *et al.*, 1998; Jonz and Nurse, 2003; Zaccone *et al.*, 2006) and a component of this innervation includes sensory nerve fibres. In addition, during zebrafish development, innervation of NECs in filament primordia coincided with a significant increase in the hyperventilatory response to environmental hypoxia in lightly anaesthetized larvae, suggesting that these nerve fibres were sensory (Jonz and Nurse, 2005). NECs, some of which contain 5-HT, have also been identified in the secondary lamellae of the gill, but their innervation appears to be species specific (Jonz and Nurse, 2003; Zaccone *et al.*, 2006; Saltys *et al.*, 2006; Coolidge *et al.*, 2008). The role of innervated lamellar NECs in sensing hypoxia is questionable, at least in zebrafish, because lamellae are not even present during early larval stages, when O_2-sensory pathways of the gill filaments are established and the larvae can respond to hypoxia (Jonz and Nurse, 2005). Any potential role in the hypoxic response played by noninnervated NECs of the lamellae may be that of paracrine release of stored chemicals and subsequent effects on surrounding tissue (Coolidge *et al.*, 2008).

Table 5.7
Neurochemicals (or other associated markers) identified in gill neuroepithelial cells of fish by immunohistochemistry

	Species		References
Neurotransmitter			
Serotonin	*A. calva*	Bowfin	Goniakowska-Witalinska *et al.*, 1995
	A. anguilla	Eel	Zaccone *et al.*, 1992
	Blennius sanguinolentus	Blenny	Zaccone *et al.*, 1992
	C. auratus	Goldfish	Saltys *et al.*, 2006; Coolidge *et al.*, 2008
	D. rerio	Zebrafish	Jonz and Nurse, 2003; Jonz and Nurse, 2005; Saltys *et al.*, 2006
	Dicentrarchus labrax	Sea perch	Bailly *et al.*, 1992
	G. morhua	Atlantic cod	Sundin *et al.*, 1998
	Heteropneustes fossilis	Indian catfish	Zaccone *et al.*, 1992
	H. lacerdae	Trairão	Coolidge *et al.*, 2008
	H. malabaricus	Traira	Coolidge *et al.*, 2008
	I. melas	Black bullhead	Bailly *et al.*, 1992
	I. nebulosus	Bullhead	Zaccone *et al.*, 1992
	I. punctatus	Channel catfish	Burleson *et al.*, 2006
	L. osseus	Gar	As cited by Zaccone *et al.*, 1997
	Micropterus dolomieui	Black bass	Bailly *et al.*, 1992
	O. mykiss	Rainbow trout	Bailly *et al.*, 1992; Saltys *et al.*, 2006; Coolidge *et al.*, 2008
	O. massambica	Tilapia	Bailly *et al.*, 1992
	Perca fluviatilis	Perch	Bailly *et al.*, 1992
	P. annectens	African lungfish	Zaccone *et al.*, 1992
	Salmo trutta	Brown trout	Zaccone *et al.*, 1992
	S. stellaris	Dogfish	As cited by Zaccone *et al.*, 1997
Neuropeptide			
Endothelin	*A. calva*	Bowfin	Goniakowska-Witalinska *et al.*, 1995; Zaccone *et al.*, 1996
	Conger conger	Sea eel	Zaccone *et al.*, 1996
	H. fossilis	Indian catfish	Zaccone *et al.*, 1996
	S. canicula	Dogfish	Zaccone *et al.*, 1996

(*continued*)

Table 5.7 (*continued*)

	Species		References
	S. trutta	Brown trout	Zaccone *et al.*, 1996; Mauceri *et al.*, 1999
	T. marmorata	Electric ray	Zaccone *et al.*, 1996
Enkephalins	*A. calva*	Bowfin	Goniakowska-Witalinska *et al.*, 1995
	A. anguilla	Eel	Zaccone *et al.*, 1992
	B. sanguinolentus	Blenny	Zaccone *et al.*, 1992
	H. fossilis	Indian catfish	Zaccone *et al.*, 1992
	I. nebulosus	Bullhead	Zaccone *et al.*, 1992
	Lampetra japonica	Lamprey	As cited by Zaccone *et al.*, 1992
	L. osseus	Gar	As cited by Zaccone *et al.*, 1997
	P. annectens	African lungfish	Zaccone *et al.*, 1992
	S. trutta	Brown trout	Zaccone *et al.*, 1992
Neuropeptide Y	*O. mossambica*	Tilapia	As cited by Zaccone *et al.*, 1994
PACAP 27 and 38[a]	*H. fossilis*	Indian catfish	As cited by Zaccone *et al.*, 2006
	Pangasus hypothalamus	Vietnamese catfish	As cited by Zaccone *et al.*, 2006
Vasoactive intestinal polypeptide	*H. fossilis*	Indian catfish	Zaccone *et al.*, 2003
	S. pavo	Blenny	As cited by Zaccone *et al.*, 2006
Biosynthetic enzyme			
Endothelial nitric oxide synthase[b]	*H. fossilis*	Indian catfish	Zaccone *et al.*, 2006
	L. osseus	Gar	As cited by Zaccone *et al.*, 1997
Neuronal nitric oxide synthase[c]	*H. fossilis*	Indian catfish	Mauceri *et al.*, 1999; Zaccone *et al.*, 2003
Tyrosine hydroxylase[c]	*I. punctatus*	Channel catfish	Burleson *et al.*, 2006

[a]Pituitary adenylate cyclase-activating polypeptide.
[b]Biosynthetic enzyme involved in production of nitric oxide.
[c]Biosynthetic enzyme involved in production of catecholamines.
Note that in some studies NECs were not serotonergic. Parts of this table were compiled from Zaccone *et al.*, 1994 and 1997.

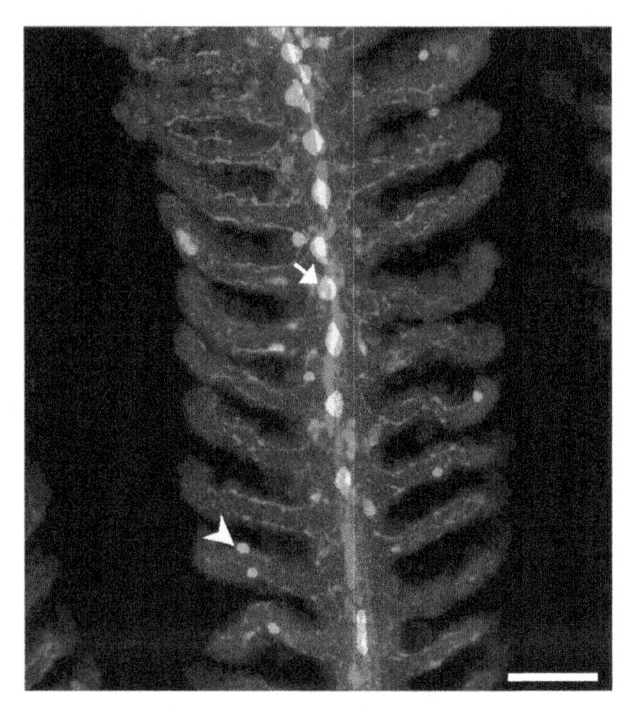

Fig. 5.6. Confocal image of neuroepithelial cells (NECs; green) of the gill filaments (arrow) and lamellae (arrowheads) from zebrafish labelled with antibodies against the neurotransmitter, serotonin (5-HT). Nerve fibers (red) are labeled with antibodies against a neuron-specific antigen. Scale bar 50 μm. [Modified from Jonz and Nurse (2003).] (See Color Insert.)

Notably, 5-HT has been described to have a paracrine role in the rat carotid body (Nurse, 2005). The O_2-sensitivity of innervated or noninnervated lamellar NECs, however, has not been tested.

3.3. Chemoreceptor Plasticity

If gill NECs are indeed the O_2 chemosensors responsible for triggering cardiorespiratory responses to hypoxia, then it should be possible to link changes in chemoreceptor number, i.e., chemoreceptor plasticity, to differences in the hypoxic ventilatory response (plasticity of respiratory control), and vice versa. This possibility has been explored both during early development and in adult fish using zebrafish. During zebrafish development, a hyperventilatory response to hypoxia can be detected as early as 2 days post-fertilization (dpf), but increases significantly at the point (7 dpf) where innervation of NECs of the filament primordia occurs (Jonz and Nurse, 2005), a coincidence of events that supports a critical role for filament NECs in initiating O_2 ventilatory reflexes. Furthermore, chemoreceptor plasticity in adult zebrafish appears to be associated with alteration of ventilatory responses. For example, adult zebrafish acclimated to hyperoxic water for 28 days exhibited a significant reduction in the density of gill filament NECs together with a diminished ventilation frequency response to either hypoxia or cyanide (Vulesevic *et al.*, 2006). Acclimation of zebrafish to hypoxic conditions for 28–60 days did not affect the density of 5-HT-containing NECs (Jonz *et al.*, 2004; Vulesevic and Perry, 2006), nor were ventilatory responses to cyanide or acute hypoxia altered (Vulesevic and Perry, 2006). Interestingly, chronic hypoxia elicited proliferation of non-5-HT-containing NECs and induced morphological changes of 5-HT-containing NECs, including increases in cell size and the growth of neuron-like processes (Jonz *et al.*, 2004), but the functional significance of these changes remains to be determined. Plasticity of respiratory control has also been investigated in zebrafish, by exposing fish during the first week of development to hypoxia or hyperoxia, then assessing the hypoxic ventilatory response of these fish as adults (Vulesevic and Perry, 2006). Although early rearing under hypoxic conditions did not affect adult ventilatory responses, zebrafish exposed to hyperoxia for the first week of development exhibited blunted ventilatory responses to both hypoxia and cyanide (Vulesevic and Perry, 2006). The study of Vulesevic and Perry (2006) did not examine NECs, but clearly future studies should attempt to link such developmental plasticity of ventilatory responses to changes in chemoreceptor number, morphology, and/or function.

4. CELLULAR MECHANISMS OF O_2 SENSING

4.1. Cellular Models of O_2 Sensing and Hypoxic Chemotransduction

It is well established that both prokaryotic and eurkaryotic cells are sensitive to changes in oxygen (Bunn and Poyton, 1996). However, O_2 chemoreceptors in vertebrates are specialized cells of the periphery that respond to acute changes in O_2 tension and initiate appropriate cardiorespiratory responses. Much of what is known about the cellular mechanisms of O_2 sensing has come from decades of research on O_2-sensitive cells from mammalian systems. These include type I (or glomus) cells of the carotid body, neuroepithelial bodies (NEBs) of the lung epithelium, adrenal medullary chromaffin cells (AMCs) of neonates, and vascular smooth muscle cells. Prior to discussing a model for O_2 chemoreception in the fish gill (see Section 4.2), the current working hypotheses of cellular mechanisms of O_2 sensing in vertebrates will be briefly summarized, i.e., how O_2 is actually "sensed" within the cell and how the hypoxic stimulus results in a cellular response, such as membrane depolarization and neurotransmitter release. For a more detailed account of O_2 sensors and mechanisms of chemotransduction, beyond the scope of this chapter, the reader is referred to other recent review articles (López-Barneo *et al.*, 2001, 2004; Weir *et al.*, 2005; Lahiri *et al.*, 2006; Kemp, 2006; Prabhakar, 2006; Buckler, 2007; Dinger *et al.*, 2007; López-López and Pérez-García, 2007; Peers and Wyatt, 2007; Wyatt and Evans, 2007).

At present, a "universal" O_2 sensor common to all O_2-chemosensory cells has not been identified. Moreover, the molecular identity of a sensor in the carotid body, the primary model for O_2 sensing, still remains a controversial issue. Two hypotheses have been proposed to explain how O_2 is sensed by chemoreceptors: the "membrane hypothesis," originally proposed for the carotid body (López-Barneo *et al.*, 1988), predicts that O_2 is sensed at the plasma membrane (i.e., membrane-delimited); whereas in the "mitochondrial hypothesis," detection of hypoxia is linked to changes in oxidative phosphorylation and/or levels of reactive oxygen species (ROS). There is a general consensus in the mammalian literature that chemotransduction of the hypoxic stimulus is primarily mediated by inhibition of plasma membrane K^+ channels and consequent depolarization (see above referenced reviews). This depolarization is believed to result in an increase in cytosolic Ca^{2+} levels, due to activation of voltage-gated Ca^{2+} channels, and the subsequent release of neurotransmitters (Weir *et al.*, 2005). The type of K^+ channel involved in the hypoxic response, however, is specific to species and cell type, and there may even be more than one type of O_2-sensitive K^+ channel within a single chemoreceptor. For example, chemotransduction channels include background (or leak) K^+ channels (K_B) that are O_2-sensitive in type I cells of the rat carotid body (Buckler, 1997), Ca^{2+}-dependent K^+ channels (K_{Ca}) that are believed to play a role in hypoxic chemotransduction in rat type I cells (Peers, 1990), rat AMCs (Thompson and Nurse, 1998) and NEBs of rabbit (Fu *et al.*, 1999), and O_2-sensitive voltage-dependent K_v channel subtypes that are present in the carotid bodies of other mammals (López-López and Pérez-García, 2007). K_B channels, as found in carotid body chemoreceptors, are particularly interesting because they function independently of changes in membrane potential and conduct K^+ ions at resting levels, when other K^+ channels may be closed. These channels make a significant contribution to setting the membrane potential and excitability (i.e., input resistance) of the cell (Lesage and Lazdunski, 2000; Goldstein *et al.*, 2001; Lesage, 2003). O_2 sensitivity of K_B channels may then be advantageous in chemoreceptors because K_B inhibition by hypoxia, which presumably occurs when the cell is at resting membrane potential, is not dependent on the channel first being activated at a particular voltage.

Evidence supporting the membrane hypothesis of O_2 sensing in carotid body type I cells was recently presented. Hemoxygenase-2 (HO-2) was identified within the O_2-sensitive K_{Ca} channel complex in rat type I cells and was reported to act as an O_2 sensor that mediated K_{Ca} channel activity through production of carbon monoxide (CO) (Williams *et al.*, 2004), which itself activates these channels (Wang and Wu, 1997; Riesco-Fagundo *et al.*, 2001). However, in HO-2 knockout mice, O_2 sensitivity in carotid body type I cells and AMCs was unaffected (Ortega-Sáenz *et al.*, 2006), suggesting species specificity of this putative sensing mechanism. An alternative model argues that sensing of hypoxia may occur through production of ROS, such as hydrogen peroxide (H_2O_2), by a membrane-bound NADPH oxidase or the mitochondrion. In NEBs of the rabbit lung, O_2-sensitive K^+ currents are potentiated by H_2O_2 and NADPH oxidase activation (Wang *et al.*, 1996; Fu *et al.*, 2000), and O_2 sensitivity of similar channels is abolished in transgenic mice deficient in the NADPH oxidase subunit, $gp91^{phox}$ (Fu *et al.*, 2000). However, in carotid body type I cells NADPH appears instead to mediate repolarization (He *et al.*, 2002; Dinger *et al.*, 2007), and O_2 sensing in oxidase-deficient mice remains unaltered in pulmonary arterial myocytes (Archer *et al.*, 1999) and neonatal AMCs (Thompson *et al.*, 2002). In the latter case, evidence suggests that O_2 sensing in neonatal AMCs occurs at a rotenone-sensitive site (possibly complex I of the mitochondrial electron transport chain) that couples to decreased ROS production during hypoxia and inhibition of O_2-sensitive K^+ channels at the plasma membrane (Thompson *et al.*, 2007; Buttigieg *et al.*, 2008). However, controversy surrounds ROS as a mediator of chemotransduction, and whether acute hypoxia increases or decreases ROS (Weir *et al.*, 2005).

Two models predict how mitochondrial O_2 sensing, via a decrease in oxidative phosphorylation during hypoxia, may be coupled to regulation of plasma membrane K^+ channels in carotid body type I cells. The first proposes that since ATP enhances the activity of a specific type of K_B ("TASK-like") channel (Williams and Buckler, 2004), during hypoxia a decrease in ATP would lead to its inhibition (Wyatt and Buckler 2004; Varas *et al.*, 2007). The second model proposes that a fall in ATP production during hypoxia leads to an increase in the cytosolic AMP/ATP ratio, followed by subsequent activation of AMP-activated protein kinase and inhibition of O_2-sensitive K^+ channels, such as K_B and K_{Ca}, by phosphorylation (Evans *et al.*, 2005a; Wyatt *et al.*, 2007; Wyatt and Evans, 2007).

A recent study proposed that hydrogen sulphide (H_2S) may act as a sensor or transducer of O_2 and mediate responses to hypoxia in vascular smooth muscle of vertebrates, including fish and mammals (Olson *et al.*, 2006). In this model, the concentration of intracellular H_2S, an endogenous signaling molecule (Wang, 2002), is controlled by H_2S production and its oxidation by available O_2. The model proposes that decreased availability of O_2 during hypoxia leads to reduced oxidation of H_2S and its subsequent accumulation. Since H_2S has been shown to have effects on membrane potential similar to those of hypoxia, this may lead to appropriate vasoactive responses, which may be vasoconstriction or vasodilation depending on the tissue and species (Olson *et al.*, 2006). While a definitive link between H_2S and changes in membrane potential under conditions of hypoxia has not been established, H_2S has been shown to activate K^+ channels in vascular smooth muscle cells and induce hyperpolarization (Zhao *et al.*, 2001; Cheng *et al.*, 2004).

To summarize, while there is little doubt that ion channels play a major role in O_2 chemotransduction at the plasma membrane, a diversity of explanations as to how hypoxia is actually sensed by O_2 chemoreceptors currently exists. Rather than viewing these many possibilities as contradictory, the "chemosome hypothesis" (Prabhakar, 2006) proposes that multiple O_2 sensors are involved in the cellular response to hypoxia, thus allowing for responses across a broad range of PO_2 levels with temporal flexibility.

4.2. O_2 Chemotransduction in Fish Gill NECs

Little is known of the process of hypoxic chemotransduction in NECs of the fish gill. However, recent studies of the response of membrane ion channels to hypoxia in NECs (Jonz *et al.*, 2004) appear to match data from mammalian studies (Buckler, 1997). In NECs isolated from zebrafish gills, a plasma membrane K^+ current was recorded that was reversibly inhibited (mean of 16% at 25 Torr) by a decrease in extracellular PO_2 (Figure 5.7). Pharmacological characterization of this current demonstrated that it was resistant to traditional blockers of voltage-dependent K^+ channels (tetraethylammonium, TEA; and 4-aminopyridine, 4-AP) but inhibited by quinidine, which blocks several membrane conductances, including those of voltage-independent K_B channels. Sensitivity of an ion conductance to quinidine but not TEA and 4-AP has previously been used to identify K_B channels in other O_2-sensitive cells (Buckler, 1997; O'Kelly *et al.*, 1999; Campanucci *et al.*, 2003). Modeling of the current-voltage relationship of the O_2-sensitive current itself produced results in agreement with pharmacological data: that the O_2-sensitive K^+ current was carried by K_B channels (Jonz *et al.*, 2004). Importantly, this inhibition of K_B by hypoxia in zebrafish NECs led to a membrane depolarization of about 6 mV. An O_2-sensitive (presumably K^+) current was also reported in isolated NECs of channel catfish (Burleson *et al.*, 2006). However, in this study hypoxia produced either inhibition or potentiation of an O_2-sensitive current,

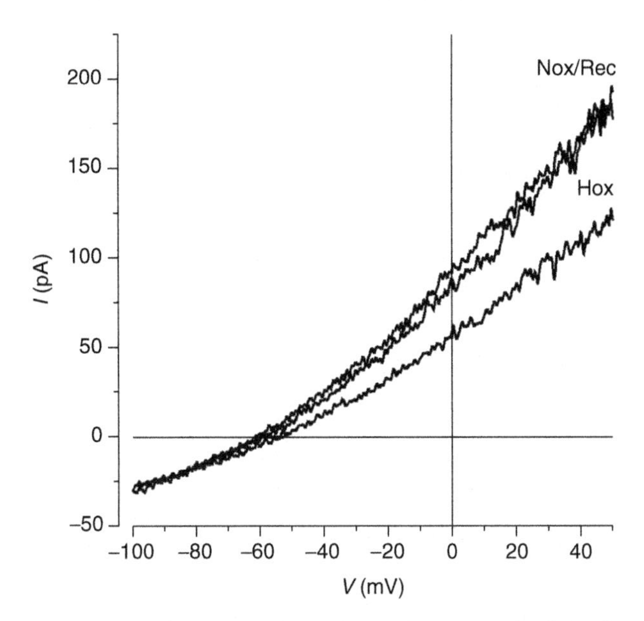

Fig. 5.7. Whole-cell voltage-clamp recording of an isolated neuroepithelial cell (NEC) from a zebrafish (*Danio rerio*) gill filament showing reversible inhibition of an outward K^+ current during acute reduction of PO_2 from normoxia (Nox, 150 Torr) to hypoxia (Hox, 25 Torr). During the recording, the membrane potential was progressively changed from −100 mV to 50 mV to induce the current. Rec, recovery. [Modified from Jonz *et al.* (2004).]

and this may have been due to recording from separate cell populations (Burleson *et al.*, 2006).

Generally, a chemotransduction mechanism similar to that of mammalian O_2 chemoreceptors (López-Barneo *et al.*, 2001), such as carotid body type I cells, can be postulated for gill NECs: a decrease in tissue PO_2, from either environmental or arterial hypoxia, causes membrane depolarization followed by Ca^{2+}-dependent neurotransmitter release and activation of sensory nerve fibres. This proposed model is summarized in Figure 5.8. A number of steps of this putative pathway remain to be identified in gill NECs. Specifically, the presence of plasma membrane Ca^{2+} channels necessary for Ca^{2+}-dependent neurosecretion has not been investigated, and there is no convincing evidence (but see Section 3.2.2 for anecdotal reports) that depolarization following acute hypoxia does, in fact, cause release of 5-HT or any other neurotransmitter. Nevertheless, it would appear that regulation of K_B channels by hypoxia is a fundamental mechanism that has been relatively conserved and may have appeared early in vertebrate evolution.

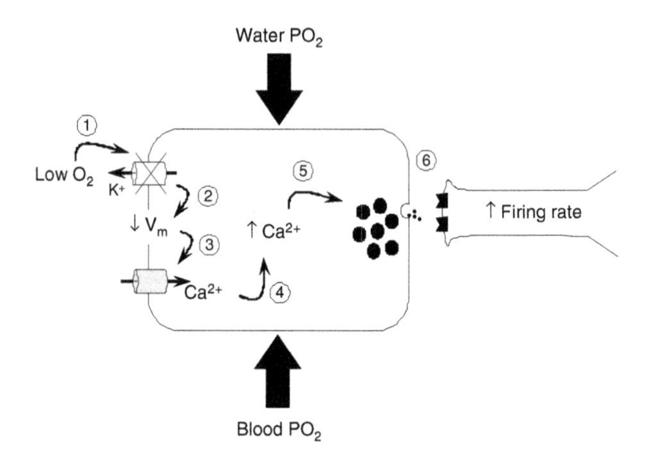

Fig. 5.8. Proposed model for oxygen sensing by gill neuroepithelial cells in fish. A schematic diagram illustrates that gill neuroepithelial cells (NECs) may respond to a decrease in internal (blood) or external (water) PO_2. A decrease in PO_2 is sensed by the NEC and leads to (1) inhibition of membrane-bound K^+ channels. This causes (2) a reduction in the membrane potential (V_m) and (3) subsequent activation of voltage-gated Ca^{2+} channels. An influx of Ca^{2+} across the membrane will (4) increase intracellular Ca^{2+} levels, which will (5) induce secretion of neurotransmitter(s) from cytoplasmic synaptic vesicles into the extracellular (synaptic) space. Neurosecretion from the presynaptic NEC will (6) cause activation of receptors on a postsynaptic (sensory) neuron leading to an increase in firing rate. See text for further details and Table 5.7 for putative neurotransmitters.

A putative O_2 sensor for NECs of the gill filaments has not yet emerged. Hypoxic sensitivity of the O_2-sensitive K_B current over a range of membrane potentials in NECs (Jonz *et al.*, 2004) appears to resemble that of rat carotid body type I cells (Buckler, 1997). It is, therefore, tempting to speculate that K_B channels in NECs may also be susceptible to regulation by cytoplasmic components, or linked to the mitochondrion, as proposed for mammalian chemoreceptors. However, in whole-cell patch-clamp recordings in zebrafish, NECs were dialyzed with an intracellular recording solution, suggesting that inhibition of K_B channels occurred in the absence of native cytoplasmic modulators (Jonz *et al.*, 2004), thus favoring a membrane-delimited mechanism (membrane hypothesis). Thus, as is the case for mammalian models of O_2 sensing, determination of whether O_2 sensing in NECs of the fish gill occurs via a membrane-delimited or mitochondrial mechanism (or both) awaits further experimentation. Such studies may point to an O_2 sensor that is present in both anamniotes and amniotes, that has been conserved throughout vertebrate phylogenesis, and that may be universal among all O_2 chemoreceptors.

4.3. Neurotransmitters

Most studies characterizing the morphology and distribution of gill NECs have conveniently exploited the expression of 5-HT in these cells for their identification (see Table 5.7). However, not all gill NECs contain 5-HT. A relatively small proportion of NECs have been described in the gill filaments and respiratory lamellae of zebrafish, goldfish, and trout that are not serotonergic but contain neurosecretory synaptic vesicles (Jonz and Nurse, 2003; Saltys *et al.*, 2006). In addition, 5-HT-negative NECs have been reported in the gills of other fish species (Zaccone *et al.*, 1994). It seems prudent to suggest that in these studies, 5-HT-negative NECs store an unidentified chemical substance, perhaps a neurotransmitter, and that such chemicals may play a role in neurotransmission between O_2 chemoreceptor and sensory nerve fibre, or in paracrine pathways, within the gill. The role of nonserotonergic NECs as O_2 chemoreceptors has not been confirmed, however, since only NECs containing monoamines (e.g., 5-HT) could be identified for patch-clamp recording and tested for O_2 sensitivity (Jonz *et al.*, 2004).

Many histochemical studies have identified the presence of neurochemicals other than or in addition to 5-HT in gill NECs that may potentially contribute to chemical neurotransmission from O_2 chemoreceptors, or paracrine effects on surrounding tissue, during the hypoxic response. These are summarized in Table 5.7 and include neuropeptides (Zaccone *et al.*, 1992, 1994, 1996, 1997, 2006; Goniakowska-Witalinksa *et al.*, 1995; Mauceri *et al.*, 1999) and biosynthetic enzymes (Zaccone *et al.*, 1997, 2003, 2006; Mauceri *et al.*, 1999; Burleson *et al.*, 2006) involved in the synthesis of neurotransmitters, such as nitric oxide and catecholamines. It is evident that there is a wide variety of neurochemicals found in gill NECs, even within a single species. NECs of the Indian catfish (*Heteropneustes fossilis*), for example, contain 5-HT, endothelin, enkephalins, and both endothelial and neuronal nitric oxide synthase (Zaccone *et al.*, 1992, 1996; Mauceri *et al.*, 1999). However, information regarding the colocalization of specific neurochemicals to NECs, and whether these cells play a role in afferent signaling (via innervation) or paracrine regulation during hypoxia, is incomplete.

There is, unfortunately, very little available evidence in support of neurochemical transmission between NECs and sensory nerve fibres in the fish gill, and so it is difficult to propose what events occur after NECs are depolarized by hypoxic stimulation and what signals are received by sensory nerves. Evidence from the mammalian carotid body suggests that a variety of neurochemicals, such as acetylcholine (ACh), ATP, catecholamines (dopamine, norepinephrine), 5-HT, GABA, and neuropeptides, play excitatory, inhibitory, and modulatory roles in O_2 sensing (González *et al.*, 1994;

Nurse, 2005; Prabhakar, 2006; Lahiri *et al.*, 2006). In addition, secretion of 5-HT from pulmonary NEBs stimulated by hypoxia has been demonstrated (Fu *et al.*, 2002). It is then reasonable to predict that similar chemical signals may be utilized in the fish gill. Tyrosine hydroxylase (TH), which is involved in the biosynthetic pathway of catecholamines, has been localized to NECs of the channel catfish (Burleson *et al.*, 2006). In addition, preliminary evidence supports the presence of cells in the gills of trout and goldfish that contain TH and VAChT, the vesicular transporter that mediates loading of ACh into synaptic vesicles (C. S. Ciuhandu and W. K. Milsom, personal communication). Interestingly, exogenous application of ACh, 5-HT, and dopamine to isolated perfused gill arch preparations in rainbow trout produced an increase in discharge frequency of afferent glossopharyngeal fibres, suggesting activation of hypoxia-sensing pathways (Burleson and Milsom, 1995a). Although not yet supported by data at the cellular level, histochemical and physiological evidence suggests that the neurochemical basis of O_2 chemoreception in the gill may involve multiple populations of NECs (i.e., 5-HT-positive and -negative NECs of the filament and lamellae), multiple neurotransmitters or neuropeptides, and perhaps a diversity of excitatory, inhibitory and modulatory mechanisms.

5. CONCLUSIONS AND PERSPECTIVES

The hypoxic ventilatory response of fish has been studied for more than half a century. Yet despite the wealth of data that has been accumulated on how fish respond to hypoxia (see, e.g., Tables 5.1 and 5.2), a surprising number of questions remains. It is clear that hyperventilation is the dominant response to aquatic hypoxia, at least in unimodal water-breathers and facultative air-breathers. However, the devil is in the details, and understanding the diversity of hypoxic ventilatory responses both within andamong species will require a more systematic approach to assessing the responses and a concerted effort to attribute differences in thresholds, magnitude, sensitivity, and timing of ventilatory responses to underlying differences in branchial gas transfer and blood gas transport. The hypoxic ventilatory reflex is clearly linked to O_2-sensitive chemoreceptors with the gill being the predominant site of O_2 chemosensing. Again, however, greater insight will come from pinning down details of the specific location(s) and orientation of chemosensory cells, and by making explicit links between phenomena such as chemoreceptor plasticity and modulation of hypoxic ventilatory reflexes. Moreover, little is known of the central pathways through which information from peripheral O_2 sensors is integrated to elicit ventilatory chemoreflexes, nor have the neurotransmitters and receptors involved in the afferent, central, or efferent pathways been described in any detail (see reviews by Gilmour and Perry, 2007; Sundin *et al.*, 2007). The chemonsensory pathways underlying ASR and air-breathing are in particular need of elucidation. Finally, while recent years have witnessed significant advances in our understanding of the cellular basis of O_2 chemosensing in fish, much work remains to fully characterize the stimulus transduction mechanisms present in fish O_2-sensitive chemoreceptor cells. Arguably, elucidation of these mechanisms could shed light on the current uncertainty and often conflicting views concerning the mechanisms of chemotransduction present in mammalian cells.

ACKNOWLEDGMENTS

Original research of the authors reported above was supported by NSERC of Canada Discovery and Research Tools and Instruments grants. Thanks are extended to W. K. Milsom for access to unpublished data.

REFERENCES

Adriaensen, D., and Scheuermann, D. W. (1993). Neuroendocrine cells and nerves of the lung. *Anat. Rec.* **236,** 70–85.

Affonso, E. G., and Rantin, F. T. (2005). Respiratory responses of the air-breathing fish *Hoplosternum littorale* to hypoxia and hydrogen sulfide. *Comp. Biochem. Physiol. C* **141,** 275–280.

Alton, L. A., White, C. R., and Seymour, R. S. (2007). Effect of aerial O_2 partial pressure on bimodal gas exchange and air-breathing behaviour in *Trichogaster leeri*. *J. Exp. Biol.* **210,** 2311–2319.

Aota, S., Holmgren, K. D., Gallaugher, P., and Randall, D. J. (1990). A possible role for catecholamines in the ventilatory responses associated with internal acidosis or external hypoxia in rainbow trout *Oncorhynchus mykiss*. *J. Exp. Biol.* **151,** 57–70.

Aota, S., and Randall, D. J. (1993). The effect of exogenous catecholamines on the ventilatory and cardiac responses of normoxic and hyperoxic rainbow trout, *Oncorhynchus mykiss*. *J. Comp. Physiol. B* **163,** 138–146.

Archer, S. L., Reeve, H. L., Michelakis, E., Puttagunta, L., Waite, R., Nelson, D. P., Dinauer, M. C., and Weir, E. K. (1999). O_2 sensing is preserved in mice lacking the gp91 phox subunit of NADPH oxidase. *Proc. Natl. Acad. Sci. USA* **96,** 7944–7949.

Babiker, M. M. (1979). Respiratory behaviour, oxygen consumption and relative dependence on aerial respiration in the African lungfish (*Protopterus annectens*, Owen) and an air-breathing teleost (*Clarias lazera*, C.). *Hydrobiologia* **65,** 177–187.

Bailly, Y., Dunel-Erb, S., and Laurent, P. (1992). The neuroepithelial cells of the fish gill filament: Indolamine-immunocytochemistry and innervation. *Anat. Rec.* **233,** 143–161.

Bamford, O. S. (1974). Oxygen reception in the rainbow trout (*Salmo gairdneri*). *Comp. Biochem. Physiol.* **48A,** 69–76.

Barrionuevo, W. R., and Burggren, W. W. (1999). O_2 consumption and heart rate in developing zebrafish (*Danio rerio*): Influence of temperature and ambient O_2. *Am. J. Physiol.* **276,** R505–R513.

Berschick, P., Bridges, C. R., and Grieshaber, M. K. (1987). The influence of hyperoxia, hypoxia and temperature on the respiratory physiology of the intertidal rockpool fish *Gobius cobitis* Pallas. *J. Exp. Biol.* **130,** 369–387.

Bicudo, J. E. P. W., and Johansen, K. (1979). Respiratory gas exchange in the airbreathing fish, *Synbranchus marmoratus. Env. Biol. Fish.* **4,** 55–64.

Bindon, S. D., Gilmour, K. M., Fenwick, J. C., and Perry, S. F. (1994). The effects of branchial chloride cell proliferation on respiratory function in the rainbow trout *Oncorhynchus mykiss. J. Exp. Biol.* **197,** 47–63.

Boese, B. L. (1988). Hypoxia-induced respiratory changes in English sole (*Parophrys vetulus* Girard). *Comp. Biochem. Physiol.* **89A,** 257–260.

Borch, K., Jensen, F. B., and Andersen, B. B. (1993). Cardiac activity, ventilation rate and acid-base regulation in rainbow trout exposed to hypoxia and combined hypoxia and hypercapnia. *Fish Physiol. Biochem.* **12,** 101–110.

Boutilier, R. G., Dobson, G. P., Hoeger, U., and Randall, D. J. (1988). Acute exposure to graded levels of hypoxia in rainbow trout (*Salmo gairdneri*): Metabolic and respiratory adaptations. *Respir. Physiol.* **71,** 69–82.

Brauner, C. J., Ballantyne, C. L., Randall, D. J., and Val, A. L. (1995). Air breathing in the armoured catfish (*Hoplosternum littorale*) as an adaptation to hypoxic, acidic, and hydrogen sulphide rich waters. *Can. J. Zool.* **73,** 739–744.

Bridges, C. R., Berenbrink, M., Muller, R., and Waser, W. (1998). Physiology and biochemistry of the pseudobranch: An unanswered question? *Comp. Biochem. Physiol. A* **119,** 67–77.

Buckler, K. J. (1997). A novel oxygen-sensitive potassium current in rat carotid body type I cells. *J. Physiol.* **498,** 649–662.

Buckler, K. J. (2007). TASK-like potassium channels and oxygen sensing in the carotid body. *Resp. Physiol. NeuroBiol.* **157,** 55–64.

Bunn, H. F., and Poyton, R. O. (1996). Oxygen sensing and molecular adaptation to hypoxia. *Physiol. Rev.* **76,** 839–885.

Burggren, W. W. (1979). Bimodal gas exchange during variation in environmental oxygen and carbon dioxide in the air breathing fish *Trichogaster trichopterus. J. Exp. Biol.* **82,** 197–213.

Burggren, W. W. (1982). "Air gulping" improves blood oxygen transport during aquatic hypoxia in the goldfish *Carassius auratus. Physiol. Zool.* **55,** 327–334.

Burggren, W. W., and Cameron, J. N. (1980). Anaerobic metabolism, gas exchange, and acid-base balance during hypoxic exposure in the channel catfish, *Ictalurus punctatus. J. Exp. Zool.* **213,** 405–416.

Burggren, W. W., and Randall, D. J. (1978). Oxygen uptake and transport during hypoxic exposure in the sturgeon *Acipenser transmontanus. Respir. Physiol.* **34,** 171–183.

Burleson, M. L. (1995). Oxygen availability: Sensory systems. *In* "Biochemistry and Molecular Biology of Fishes", (Hochachka, P.W, and Mommsen, T.P, Eds.), pp. 1–18. Elsevier, Amsterdam.

Burleson, M. L., Carlton, A. L., and Silva, P. E. (2002). Cardioventilatory effects of acclimatization to aquatic hypoxia in channel catfish. *Respir. Physiol. NeuroBiol.* **131,** 223–232.

Burleson, M. L., Mercer, S. E., and Wilk-Blaszczak, M. A. (2006). Isolation and characterization of putative O_2 chemoreceptor cells from the gills of channel catfish (*Ictalurus punctatus*). *Brain Res.* **1092,** 100–107.

Burleson, M. L., and Milsom, W. K. (1993). Sensory receptors in the first gill arch of rainbow trout. *Respir. Physiol.* **93,** 97–110.

Burleson, M. L., and Milsom, W. K. (1995a). Cardio-ventilatory control in rainbow trout: I. Pharmacology of branchial, oxygen-sensitive chemoreceptors. *Respir. Physiol.* **100,** 231–238.

Burleson, M. L., and Milsom, W. K. (1995b). Cardio-ventilatory control in rainbow trout: II. Reflex effects of exogenous neurochemicals. *Respir. Physiol.* **101,** 289–299.
Burleson, M. L., and Smatresk, N. J. (1990a). Effects of sectioning cranial nerves IX and X on cardiovascular and ventilatory reflex responses to hypoxia and NaCN in channel catfish. *J. Exp. Biol.* **154,** 407–420.
Burleson, M. L., and Smatresk, N. J. (1990b). Evidence for two oxygen-sensitive chemoreceptor loci in channel catfish, *Ictalurus punctatus*. *Physiol. Zool.* **63,** 208–221.
Burleson, M. L., Smatresk, N. J., and Milsom, W. K. (1992). Afferent inputs associated with cardioventilatory control in fish. *In* "The Cardiovascular System" (Hoar, W. S., Randall, D. J., and Farrell, A. P., Eds.), pp. 389–423. Academic Press, San Diego.
Bushnell, P. G., and Brill, R. W. (1991). Responses of swimming skipjack (*Katsuwonus pelamis*) and yellowfin (*Thunnus albacares*) tunas to acute hypoxia, and a model of their cardiorespiratory function. *Physiol. Zool.* **64,** 787–811.
Bushnell, P. G., and Brill, R. W. (1992). Oxygen transport and cardiovascular responses in skipjack tuna (*Katsuwonus pelamis*) and yellowfin tuna (*Thunnus albacares*) exposed to acute hypoxia. *J. Comp. Physiol. B* **162,** 131–143.
Bushnell, P. G., Brill, R. W., and Bourke, R. E. (1990). Cardiorespiratory responses of skipjack tuna (*Katsuwonus pelamis*), yellowfin tuna (*Thunnus albacares*), and bigeye tuna (*Thunnus obesus*) to acute reductions of ambient oxygen. *Can. J. Zool.* **68,** 1857–1865.
Butler, P. J., Taylor, E. W., Capra, M. F., and Davison, W. (1978). The effect of hypoxia on the levels of circulating catecholamines in the dogfish *Scyliorhinus canicula*. *J. Comp. Physiol.* **127,** 325–330.
Buttigieg, J., Brown, S. T., Lowe, M., Zhang, M, and Nurse, C. A. (2008). Functional mitochondria are required for O_2 but not CO_2 sensing in immortalized adrenomedullary chromaffin cells. *Am. J. Physiol.* **294,** C945–C956.
Cameron, J. N., and Cech, J. J. (1970). Notes on the energy cost of gill ventilation in teleosts. *Comp. Biochem. Physiol.* **34,** 447–455.
Campagna, C. G., and Cech, J. J. (1981). Gill ventilation and respiratory efficiency of Sacramento blackfish, *Orthodon microlepidotus* Ayres, in hypoxic environments. *J. Fish Biol.* **19,** 581–591.
Campanucci, V. A., Fearon, I. M., and Nurse, C. A. (2003). A novel O_2-sensing mechanism in rat glossopharyngeal neurones mediated by a halothane-inhibitable background K^+ conductance. *J. Physiol.* **548,** 731–743.
Carlson, J. K., and Parsons, G. R. (2003). Respiratory and hematological responses of the bonnethead shark, *Sphyrna tiburo*, to acute changes in dissolved oxygen. *J. Exp. Mar. Biol. Ecol.* **294,** 15–26.
Cech, J. J., and Wohlschlag, D. E. (1973). Respiratory responses of the striped mullet, *Mugil cephalus* (L.) to hypoxic conditions. *J. Fish Biol.* **5,** 421–428.
Cerezo, J., and Garcia, B. G. (2004). The effects of oxygen levels on oxygen consumption, survival and ventilatory frequency of sharpsnout sea bream (*Diplodus puntazzo* Gmelin, 1789) at different conditions of temperature and fish weight. *J. Appl. Ichthyol.* **20,** 488–492.
Chan, D. K. O. (1986). Cardiovascular, respiratory, and blood adjustments to hypoxia in the Japanese eel, *Anguilla japonica*. *Fish Physiol. Biochem.* **2,** 179–193.
Cheng, Y., Ndisang, J. F., Tang, G., Cao, K., and Wang, R. (2004). Hydrogen sulfide-induced relaxation of resistance mesenteric artery beds of rats. *Am. J. Physiol.* **287,** H2316–H2323.
Claridge, P. N., and Potter, I. C. (1975). Oxygen consumption, ventilatory frequency and heart rate of lampreys (*Lampetra fluviatilis*) during their spawning run. *J. Exp. Biol.* **63,** 193–206.
Clark, T. D., Seymour, R. S., Christian, K., Wells, R. M. G., Baldwin, J., and Farrell, A. P. (2007). Changes in cardiac output during swimming and aquatic hypoxia in the air-breathing Pacific tarpon. *Comp. Biochem. Physiol. A* **148,** 562–571.

Coolidge, E. H., Ciuhandu, C. S., and Milsom, W. K. (2008). A comparative analysis of putative oxygen-sensing cells in the fish gill. *J. Exp. Biol.* **211,** 1231–1242.

Cutz, E., and Jackson, A. (1999). Neuroepithelial bodies as airway oxygen sensors. *Respir. Physiol.* **115,** 201–214.

Davis, J. C., and Cameron, J. N. (1971). Water flow and gas exchange at the gills of rainbow trout, *Salmo gairdneri. J. Exp. Biol.* **54,** 1–18.

Davis, J. C., and Watters, K. W. (1970). Evaluation of opercular catheterisation as a method of sampling expired water from fish. *J. Fish. Res. Bd. Canada* **27,** 1627–1635.

De Salvo-Souza, R. H., Soncini, R., Glass, M. L., Sanches, J. R., and Rantin, F. T. (2001). Ventilation, gill perfusion and blood gases in dourado, *Salminus maxillosus* Valenciennes (Teleostei, Characidae), exposed to graded hypoxia. *J. Comp. Physiol. B* **171,** 483–489.

Dinger, B., He, L., Chen, J., Liu, X., González, C., Obeso, A., Sanders, K., Hoidal, J., Stensaas, L., and Fidone, S. (2007). The role of NADPH oxidase in carotid body arterial chemoreceptors. *Respir. Physiol. NeuroBiol.* **157,** 45–54.

Dunel-Erb, S., Bailly, Y., and Laurent, P. (1982). Neuroepithelial cells in fish gill primary lamellae. *J. Appl. Physiol.* **53,** 1342–1353.

Eclancher, B., and Dejours, P. (1975). Contrôle de la respiration chez les poissons téléostéens: Existence de chémorécepteurs physiologiquement analogues aux chémorécepteurs des vertébrés supérieurs. *C. R. Acad. Sci. Ser. D.* **280,** 451–453.

Eddy, F. B. (1974). Blood gases of the tench (*Tinca tinca*) in well aerated and oxygen-deficient waters. *J. Exp. Biol.* **60,** 71–83.

Edwards, R. R. C. (1971). An assessment of the energy cost of gill ventilation in the plaice (*Pleuronectes platessa* L.). *Comp. Biochem. Physiol. A* **40,** 391–398.

Evans, A. M., Mustard, K. J., Wyatt, C. N., Peers, C., Dipp, M., Kumar, P., Kinnear, N.P, and Hardie, D. G. (2005a). Does AMP-activated protein kinase couple inhibition of mitochondrial oxidative phosphorylation by hypoxia to calcium signalling in O_2-sensing cells? *J. Biol. Chem.* **280,** 41504–41511.

Evans, D. H., Piermarini, P. M., and Choe, K. P. (2005b). The multifunctional fish gill: Dominant site of gas exchange, osmoregulation, acid-base regulation, and excretion of nitrogenous waste. *Physiol. Rev.* **85,** 97–177.

Farrell, A. P., and Daxboeck, C. (1981). Oxygen uptake in the lingcod, *Ophiodon elongatus*, during progressive hypoxia. *Can. J. Zool.* **59,** 1272–1275.

Fernandes, M. N., Barrionuevo, W. R. and Rantin, F. T. (1995). Effects of thermal stress on respiratory responses to hypoxia of a South American Prochilodontid fish, *Prochilodus scrofa. J. Fish Biol.* **46,** 123–133.

Fernandes, M. N., and Rantin, F. T. (1989). Respiratory responses of *Oreochromis niloticus* (Pisces, Cichlidae) to environmental hypoxia under different thermal conditions. *J. Fish Biol.* **35,** 509–519.

Florindo, L. H., Leite, C. A. C., Kalinin, A. L., Reid, S. G., Milsom, W. K., and Rantin, F. T. (2006). The role of branchial and orobranchial O_2 chemoreceptors in the control of aquatic surface respiration in the neotropical fish tambaqui (*Colossoma macropomum*): Progressive responses to prolonged hypoxia. *J. Exp. Biol.* **209,** 1709–1715.

Forgue, J., Burtin, B., and Massabuau, J.-C. (1989). Maintenance of oxygen consumption in resting *Silurus glanis* at different levels of ambient oxygenation. *J. Exp. Biol.* **143,** 305–319.

Fritsche, R., Axelsson, M., Franklin, C. E., Grigg, G. G., Holmgren, S., and Nilsson, S. (1993). Respiratory and cardiovascular responses to hypoxia in the Australian lungfish. *Respir. Physiol.* **94,** 173–187.

Fritsche, R., and Nilsson, S. (1993). Cardiovascular and ventilatory control during hypoxia. *In* "Fish Ecophysiology" (Rankin, J.C, and Jensen, F.B, Eds.), pp. 180–206. Chapman & Hall, London.

Fry, F. E. J., and Hart, J. S. (1948). The relation of temperature to oxygen consumption in the goldfish. *Biol. Bull.* **94,** 66–77.

Fu, X. W., Nurse, C. A., Wang, Y. T., and Cutz, E. (1999). Selective modulation of membrane currents by hypoxia in intact airway chemoreceptors from neonatal rabbit. *J. Physiol.* **514,** 139–150.

Fu, X. W., Wang, D., Nurse, C. A., Dinauer, M. C., and Cutz, E. (2000). NADPH oxidase is an O_2 sensor in airway chemoreceptors: Evidence from K^+ current modulation in wild-type and oxidase-deficient mice. *Proc. Natl. Acad. Sci. USA* **97,** 4374–4379.

Fu, X. W., Nurse, C. A., Wong, V., and Cutz, E. (2002). Hypoxia-induced secretion of serotonin from intact pulmonary neuroepithelial bodies in neonatal rabbit. *J. Physiol.* **539,** 503–510.

Gee, J. H. (1980). Respiratory patterns and antipredator responses in the central mudminnow, *Umbra limi*, a continuous, facultative, air-breathing fish. *Can. J. Zool.* **58,** 819–827.

Gee, J. H., and Gee, P. A. (1991). Reactions of Gobioid fishes to hypoxia: Buoyancy control and aquatic surface respiration. *Copeia* 17–28**1991**.

Gee, J. H., Tallman, R. F., and Smart, H. J. (1978). Reactions of some great plains fishes to progressive hypoxia. *Can. J. Zool.* **56,** 1962–1966.

Gehrke, P. C., and Fielder, D. R. (1988). Effects of temperature and dissolved oxygen on heart rate, ventilation rate and oxygen consumption of spangled perch, *Leiopotherapon unicolor* (Günther 1859), (Percoidei, Teraponidae). *J. Comp. Physiol. B* **157,** 771–782.

Geiger, S. P., Torres, J. J., and Crabtree, R. E. (2000). Air breathing and gill ventilation frequencies in juvenile tarpon, *Megalops atlanticus*: Responses to changes in dissolved oxygen, temperature, hydrogen sulfide, and pH. *Environ. Biol. Fish.* **59,** 181–190.

Gerald, J. W., and Cech, J. J. (1970). Respiratory responses of juvenile catfish (*Ictalurus punctatus*) to hypoxic conditions. *Physiol. Zool.* **43,** 47–54.

Gilmour, K. M. (2001). The CO_2/pH ventilatory drive in fish. *Comp. Biochem. Physiol. A* **130,** 219–240.

Gilmour, K. M., and Perry, S. F. (1994). The effects of hypoxia, hyperoxia or hypercapnia on the acid-base disequilibrium in the arterial blood of rainbow trout. *J. Exp. Biol.* **192,** 269–284.

Gilmour, K. M., and Perry, S. F. (2007). Branchial chemoreceptor regulation of cardiorespiratory function. *In* "Fish Physiology Sensory Systems Neuroscience" Vol. 25, (Zielinski, B., and Hara, T. J., Eds.), pp. 97–151. Academic Press, San Diego.

Glass, M. L. (1992). Ventilatory responses to hypoxia in ectothermic vertebrates. *In* "Physiological Adaptations in Vertebrates, Respiration, Circulation, and Metabolism" (Wood, S.C, Weber, R. E., Hargens, A. R., and Millard, R.W, Eds.), pp. 97–118. Marcel Dekker, Inc, New York.

Glass, M. L., Ishimatsu, A., and Johansen, K. (1986). Responses of aerial ventilation to hypoxia and hypercapnia in *Channa argus*, an air-breathing fish. *J. Comp. Physiol. B* **156,** 425–430.

Glass, M. L., Andersen, N. A., Kruhøffer, M., Williams, E. M., and Heisler, N. (1990). Combined effects of environmental PO_2 and temperature on ventilation and blood gases in the carp *Cyprinus carpio* L. *J. Exp. Biol.* **148,** 1–17.

Goldstein, S. A. N., Backenhauer, D., O'Kelly, I., and Zilberberg, N. (2001). Potassium leak channels and the KCNK family of two-P-domain subunits. *Nat. Rev. Neurosci.* **2,** 1–11.

Goniakowska-Witalinska, L., Zaccone, G., Fasula, S., Mauceri, A., Licata, A., and Youson, J. (1995). Neuroendocrine cells in the gills of the bowfin *Amia calva*. An ultrastructural and immunocytochemical study. *Fol. Histochem. CytoBiol.* **33,** 171–177.

Gonzales, T. T., Katoh, M., and Ishimatsu, A. (2006). Air breathing of aquatic burrow-dwelling eel goby, *Odontamblyopus lacepedii* (Gobiidae: Amblyopinae). *J. Exp. Biol.* **209,** 1085–1092.

González, C., Almaraz, L., Obeso, A., and Rigual, R. (1994). Carotid body chemoreceptors: From natural stimuli to sensory discharges. *Physiol. Rev.* **74,** 829–898.

Graham, J. B. (1983). The transition to air breathing in fishes II. Effects of hypoxia acclimation of the bimodal gas exchange of *Ancistrus chagresi* (Loricariidae). *J. Exp. Biol.* **102,** 157–173.
Graham, J. B. (1997). "Air-Breathing Fishes." Academic Press, San Diego.
Graham, J. B., and Baird, T. A. (1982). The transition to air breathing in fishes I. Environmental effects of the facultative air breathing of *Ancistrus chagresi* and *Hypostomus plecostomus* (Loricariidae). *J. Exp. Biol.* **96,** 53–67.
Graham, J. B., and Baird, T. A. (1984). The transition to air breathing in fishes III. Effects of body size and aquatic hypoxia on the aerial gas exchange of the swamp eel *Synbranchus marmoratus. J. Exp. Biol.* **108,** 357–375.
Graham, J. B., Kramer, D. L., and Pineda, E. (1977). Respiration of the air breathing fish *Piabucina festae. J. Comp. Physiol. B* **122,** 295–310.
Graham, J. B., Lai, N. C., Chiller, D., and Roberts, J. L. (1995). The transition to air breathing in fishes V. Comparative aspects of cardiorespiratory regulation in *Synbranchus marmoratus* and *Monopterus albus* (Synbranchidae). *J. Exp. Biol.* **198,** 1455–1467.
Greco, A. M., Gilmour, K. M., Fenwick, J. C., and Perry, S. F. (1995). The effects of softwater acclimation on respiratory gas transfer in the rainbow trout *Oncorhynchus mykiss. J. Exp. Biol.* **198,** 2557–2567.
He, L., Chen, J., Dinger, B., Sanders, K., Sundar, K., Hoidal, J., and Fidone, S. (2002). Characteristics of carotid body chemosensitivity in NADPH oxidase-deficient mice. *Am. J. Physiol.* **282,** C27–C33.
Hedrick, M. S., and Jones, D. R. (1993). The effects of altered aquatic and aerial respiratory gas concentrations on air-breathing patterns in a primitive fish (*Amia calva*). *J. Exp. Biol.* **181,** 81–94.
Hedrick, M. S., and Jones, D. R. (1999). Control of gill ventilation and air-breathing in the bowfin *Amia calva. J. Exp. Biol.* **202,** 87–94.
Hedrick, M. S., Burleson, M. L., Jones, D. R., and Milsom, W. K. (1991). An examination of central chemosensitivity in an air-breathing fish (*Amia calva*). *J. Exp. Biol.* **155,** 165–174.
Holeton, G. F. (1971a). Oxygen uptake and transport by the rainbow trout during exposure to carbon monoxide. *J. Exp. Biol.* **54,** 239–254.
Holeton, G. F. (1971b). Respiratory and circulatory responses of rainbow trout larvae to carbon monoxide and to hypoxia. *J. Exp. Biol.* **55,** 683–694.
Holeton, G. F. (1977). Constancy of arterial blood pH during CO-induced hypoxia in the rainbow trout. *Can. J. Zool.* **55,** 1010–1013.
Holeton, G. F., and Randall, D. J. (1967a). Changes in blood pressure in the rainbow trout during hypoxia. *J. Exp. Biol.* **46,** 297–305.
Holeton, G. F., and Randall, D. J. (1967b). The effect of hypoxia upon the partial pressure of gases in the blood and water afferent and efferent to the gills of rainbow trout. *J. Exp. Biol.* **46,** 317–327.
Huang, C.-Y, Lee, W, and Lin, H.-C. (2008). Functional differentiation in the anterior gills of the aquatic air-breathing fish, *Trichogaster leeri. J. Comp. Physiol. B* **178,** 111–121.
Hughes, G. M. (1978). On the respiration of *Torpedo marmorata. J. Exp. Biol.* **73,** 85–105.
Hughes, G. M., and Saunders, R. L. (1970). Responses of the respiratory pumps to hypoxia in the rainbow trout (*Salmo gairdneri*). *J. Exp. Biol.* **53,** 529–545.
Hughes, G. M., and Shelton, G. (1962). Respiratory mechanisms and their nervous control in fish. *Adv. Comp Physiol Biochem.* **1,** 275–364.
Hughes, G. M., and Umezawa, S.-I. (1968). On respiration in the dragonet *Callionymus lyra* L. *J. Exp. Biol.* **49,** 565–582.
Itazawa, Y., and Takeda, T. (1978). Gas exchange in the carp gills in normoxic and hypoxic conditions. *Respir. Physiol.* **35,** 263–269.

Jensen, F. B. (1991). Multiple strategies in oxygen and carbon dioxide transport by haemoglobin. *In* "Physiological Strategies for Gas Exchange and Metabolism" (Woakes, A. J., Grieshaber, M. K., and Bridges, C. R., Eds.), pp. 55–78. Cambridge University Press, Cambridge.

Jensen, F. B., Fago, A., and Weber, R. E. (1998). Hemoglobin structure and function. *In* "Fish Respiration" (Perry, S. F., and Tufts, B. L., Eds.), pp. 1–40. Academic Press, San Diego.

Jesse, M. J., Shub, C., and Fishman, A. P. (1967). Lung and gill ventilation of the African lung fish. *Respir. Physiol.* **3,** 267–287.

Johansen, K., and Lenfant, C. (1967). Respiratory function in the South American lungfish, *Lepidosiren paradoxa* (Fitz). *J. Exp. Biol.* **46,** 205–218.

Johansen, K., and Lenfant, C. (1968). Respiration in the African lungfish *Protopterus aethiopicus* II. Control of breathing. *J. Exp. Biol.* **49,** 453–468.

Johansen, K., Lenfant, C., and Grigg, G. C. (1967). Respiratory control in the lungfish, *Neoceratodus forsteri* (Krefft). *Comp. Biochem. Physiol.* **20,** 835–854.

Johansen, K., Hanson, D., and Lenfant, C. (1970). Respiration in a primitive air breather, *Amia calva*. *Respir. Physiol.* **9,** 162–174.

Johansen, K., Lenfant, C., and Hanson, D. (1973). Gas exchange in the lamprey, *Entosphenus tridentatus*. *Comp. Biochem. Physiol.* **44A,** 107–119.

Jones, D. R., and Milsom, W. K. (1982). Peripheral receptors affecting breathing and cardiovascular function in non-mammalian vertebrates. *J. Exp. Biol.* **100,** 59–91.

Jones, D. R., and Schwarzfeld, T. (1974). The oxygen cost to the metabolism and efficiency of breathing in trout (*Salmo gairdneri*). *Respir. Physiol.* **21,** 241–254.

Jonz, M. G., and Nurse, C. A. (2003). Neuroepithelial cells and associated innervation of the zebrafish gill: A confocal immunofluorescence study. *J. Comp. Neurol.* **461,** 1–17.

Jonz, M. G., and Nurse, C. A. (2005). Development of oxygen sensing in the gills of zebrafish. *J. Exp. Biol.* **208,** 1537–1549.

Jonz, M. G., Fearon, I. M., and Nurse, C. A. (2004). Neuroepithelial oxygen chemoreceptors of the zebrafish gill. *J. Physiol.* **560,** 737–752.

Kalinin, A. L., Glass, M. L., and Rantin, F. T. (1999). A comparison of directly measured and estimated gill ventilation in the Nile tilapia, *Oreochromis niloticus*. *Comp. Biochem. Physiol. A* **122,** 207–211.

Kemp, P. J. (2006). Detecting acute changes in oxygen: Will the real sensor please stand up? *Exp. Physiol.* **91,** 829–834.

Kemp, P. J., Searle, G. J., Hartness, M. E., Lewis, A., Miller, P., Williams, S., Wootton, P., Adriaensen, D., and Peers, C. (2003). Acute oxygen sensing in cellular models: Relevance to the physiology of pulmonary neuroepithelial and carotid bodies. *Anat. Rec.* **270A,** 41–50.

Kerstens, A., Lomholt, J. P., and Johansen, K. (1979). The ventilation, extraction and uptake of oxygen in undisturbed flounders, *Platichthys flesus*: Responses to hypoxia acclimation. *J. Exp. Biol.* **83,** 169–179.

Kind, P. K., Grigg, G. C., and Booth, D. T. (2002). Physiological responses to prolonged aquatic hypoxia in the Queensland lungfish *Neoceratodus forsteri*. *Respir. Physiol. NeuroBiol.* **132,** 179–190.

Kinkead, R., and Perry, S. F. (1990). An investigation of the role of circulating catecholamines in the control of ventilation during acute moderate hypoxia in rainbow trout (*Oncorhynchus mykiss*). *J. Comp. Physiol.* **160B,** 441–448.

Kinkead, R., and Perry, S. F. (1991). The effects of catecholamines on ventilation in rainbow trout during external hypoxia or hypercapnia. *Respir. Physiol.* **84,** 77–92.

Kinkead, R., Fritsche, R., Perry, S. F., and Nilsson, S. (1991). The role of circulating catecholamines in the ventilatory and hypertensive responses to hypoxia in the Atlantic cod (*Gadus morhua*). *Physiol. Zool.* **64,** 1087–1109.

Kramer, D. L. (1983). The evolutionary ecology of respiratory mode in fishes: An analysis based on the costs of breathing. *Environ. Biol. Fish.* **9,** 145–158.

Kramer, D. L., and Graham, J. B. (1976). Synchronous air breathing, a social component of respiration in fishes. *Copeia* 689–697**1976**.

Kramer, D. L., and McClure, M. (1982). Aquatic surface respiration, A widespread adaptation to hypoxia in tropical freshwater fishes. *Environ. Biol. Fish.* **7,** 47–55.

Kramer, D. L., and Mehegan, J. P. (1981). Aquatic surface respiration, an adaptive response to hypoxia in the guppy, *Poecilia reticulata* (Pisces, Poeciliidae). *Environ. Biol. Fish.* **6,** 299–313.

Lahiri, S., Szidon, J. P., and Fishman, A. P. (1970). Potential respiratory and circulatory adjustments to hypoxia in the African lungfish. *Fed. Proc.* **29,** 1141–1148.

Lahiri, S., Roy, A., Baby, S. M., Hoshi, T., Semenza, G. L., and Prabhakar, N. R (2006). Oxygen sensing in the body. *Prog. Biophys. Mol. Biol.* **91,** 249–286.

Lai, J. C. C., Kakuta, I., Mok, H. O. L., Rummer, J. L., and Randall, D. J. (2006). Effects of moderate and substantial hypoxia on erythropoietin levels in rainbow trout kidney and spleen. *J. Exp. Biol.* **209,** 2734–2738.

Laurent, P., and Rouzeau, J.-D. (1972). Afferent neural activity from pseudobranch of the teleosts. Effects of PO_2, pH, osmotic pressure and Na^+ ions. *Respir. Physiol.* **14,** 307–331.

Le Moigne, J., Soulier, P., Peyraud-Waitzenegger, M., and Peyraud, C. (1986). Cutaneous and gill O_2 uptake in the European eel (*Anguilla anguilla* L.) in relation to ambient PO_2: 10–400 Torr. *Respir. Physiol.* **66,** 341–354.

Leite, C. A. C., Florindo, L. H., Kalinin, A. L., Milsom, W. K., and Rantin, F. T. (2007). Gill chemoreceptors and cardio-respiratory reflexes in the neotropical teleost pacu, *Piaractus mesopotamicus*. *J. Comp. Physiol. A* **193,** 1001–1011.

Lesage, F. (2003). Pharmacology of neuronal background potassium channels. *Neuropharmacology* **44,** 1–7.

Lesage, F., and Lazdunski, M. (2000). Molecular and functional properties of two-pore-domain potassium channels. *Am. J. Physiol.* **279,** F793–F801.

Lewis, W. M. (1970). Morphological adaptations of Cyprinodontoids for inhabiting oxygen deficient waters. *Copeia* 319–326**1970**.

Lomholt, J. P., and Johansen, K. (1974). Control of breathing in *Amphipnous cuchia*, an amphibious fish. *Respir. Physiol.* **21,** 325–340.

Lomholt, J. P., and Johansen, K. (1979). Hypoxia acclimation in carp – How it affects O_2 uptake, ventilation, and O_2 extraction from water. *Physiol. Zool.* **52,** 38–49.

López-Barneo, J., López-López, J. R., Ureña, J., and González, C. (1988). Chemotransduction in the carotid body: K^+ current modulated by PO_2 in type I chemoreceptor cells. *Science* **241,** 580–582.

López-Barneo, J., Pardal, R., and Ortega-Sáenz, P. (2001). Cellular mechanisms of oxygen sensing. *Annu. Rev. Physiol.* **63,** 259–287.

López-Barneo, J., del Toro, R., Levitsky, K. L., Chiara, M. D., and Ortega-Sáenz, P. (2004). Regulation of oxygen sensing by ion channels. *J. Appl. Physiol.* **96,** 1187–1195.

López-López, J. R., and Pérez-García, M. T. (2007). Oxygen sensitive Kv channels in the carotid body. *Respir. Physiol. NeuroBiol.* **157,** 65–74.

Malte, H., and Weber, R. E. (1985). A mathematical model for gas exchange in the fish gill based on non-linear blood gas equilibrium curves. *Respir. Physiol.* **62,** 359–374.

Mattias, A. T., Rantin, F. T., and Fernandes, M. N. (1998). Gill respiratory parameters during progressive hypoxia in the facultative air-breathing fish, *Hypostomus regani* (Loricariidae). *Comp. Biochem. Physiol. A* **120,** 311–315.

Mauceri, A., Fasulo, S., Ainis, L., Licata, A., Lauriano, E. R., Martinez, A., Mayer, B., and Zaccone, G. (1999). Neuronal nitric oxide synthase (nNOS) expression in the epithelial neuroendocrine cell system and nerve fibers in the gill of the catfish, *Heteropneustes fossilis*. *Acta Histochem* **101,** 437–448.

Maxime, V., Nonnotte, G., Peyraud, C., Williot, P., and Truchot, J.-P. (1995). Circulatory and respiratory effects of an hypoxic stress in the Siberian sturgeon. *Respir. Physiol.* **100,** 203–212.

Maxime, V., Pichavant, K., Boeuf, G., and Nonnotte, G. (2000). Effects of hypoxia on respiratory physiology of turbot, *Scophthalmus maximus*. *Fish Physiol. Biochem.* **22,** 51–59.

McKenzie, D. J., Burleson, M. L., and Randall, D. J. (1991). The effects of branchial denervation and pseudobranch ablation on cardioventilatory control in an air-breathing fish. *J. Exp. Biol.* **161,** 347–365.

McKenzie, D. J., Taylor, E. W., Bronzi, P., and Bolis, C. L. (1995). Aspects of cardioventilatory control in the adriatic sturgeon (*Acipenser naccarii*). *Respir. Physiol.* **100,** 45–53.

McKenzie, D. J., Piraccini, G., Papini, N., Galli, C., Bronzi, P., Bolis, L., and Taylor, E. W. (1997). Oxygen consumption and ventilatory reflex responses are influenced by dietary lipids in sturgeon. *Fish Physiol. Biochem.* **16,** 365–379.

McKenzie, D. J., Piraccini, G., Piccolella, M., Steffensen, J. F., Bolis, L., and Taylor, E. W. (2000). Effects of dietary fatty acid composition on metabolic rate and responses to hypoxia in the European eel (*Anguilla anguilla*). *Fish Physiol. Biochem.* **22,** 281–296.

McKenzie, D. J., Campbell, H. A., Taylor, E. W., Micheli, M., Rantin, F. T., and Abe, A. S. (2007). The autonomic control and functional significance of the changes in heart rate associated with air breathing in the jeju, *Hoplerythrinus unitaeniatus*. *J. Exp. Biol.* **210,** 4224–4232.

McMahon, B. R., and Burggren, W. W. (1987). Respiratory physiology of intestinal air breathing in the teleost fish *Misgurnus anguillicaudatus*. *J. Exp. Biol.* **133,** 371–393.

McNeil, D. G., and Closs, G. P. (2007). Behavioural responses of a south-east Australian floodplain fish community to gradual hypoxia. *Freshw. Biol.* **52,** 412–420.

Melnychuk, M. C., and Chapman, L. J. (2002). Hypoxia tolerance of two haplochromine cichlids: Swamp leakage and potential for interlacustrine dispersal. *Environ. Biol. Fish.* **65,** 99–110.

Milsom, W. K., and Brill, R. W. (1986). Oxygen sensitive afferent information arising from the first gill arch of yellowfin tuna. *Respir. Physiol.* **66,** 193–203.

Milsom, W. K., and Burleson, M. L. (2007). Peripheral arterial chemoreceptors and the evolution of the carotid body. *Respir. Physiol. NeuroBiol.* **157,** 4–11.

Milsom, W. K., Sundin, L., Reid, S., Kalinin, A., and Rantin, F. T. (1999). Chemoreceptor control of cardiovascular reflexes. *In* "Biology of Tropical Fishes" (Val, A. L., and Almeida-Val, V. M. F., Eds.), pp. 363–374. INPA, Manaus.

Milsom, W. K., Reid, S. G., Rantin, F. T., and Sundin, L. (2002). Extrabranchial chemoreceptors involved in respiratory reflexes in the neotropical fish *Colossoma macropomum* (the tambaqui). *J. Exp. Biol.* **205,** 1765–1774.

Nikinmaa, M. (2001). Haemoglobin function in vertebrates: Evolutionary changes in cellular regulation in hypoxia. *Respir. Physiol.* **128,** 317–329.

Nikinmaa, M., and Weber, R. E. (1984). Hypoxic acclimation in the lamprey, *Lampetra fluviatilis*: Organismic and erythrocytic responses. *J. Exp. Biol.* **109,** 109–119.

Nilsson, S. (1983). "Autonomic Nerve Function in the Vertebrates." Springer Verlag, Berlin.

Nonnotte, G., Maxime, V., Truchot, J.-P., Williot, P., and Peyraud, C. (1993). Respiratory responses to progressive ambient hypoxia in the sturgeon, *Acipenser baeri*. *Respir. Physiol.* **91,** 71–82.

Nordlie, F. G., and Lefler, C. W. (1975). Ionic regulation and the energetics of osmoregulation in *Mugil cephalus* Lin. *Comp. Biochem. Physiol.* **51A,** 125–131.

Nurse, C. A. (2005). Neurotransmission and neuromodulation in the chemosensory carotid body. *Autonom. Neurosci.* **120,** 1–9.

O'Kelly, I., Stephens, R. H., Peers, C., and Kemp, P. J. (1999). Potential identification of the O_2-sensitive K^+ current in a human neuroepithelial body-derived cell line. *Am. J. Physiol.* **276,** L96–L104.

Oliveira, R. D., Lopes, J. M., Sanches, J. R., Kalinin, A. L., Glass, M. L., and Rantin, F. T. (2004). Cardiorespiratory responses of the facultative air-breathing fish jeju, *Hoplerythrinus unitaeniatus* (Teleostei, Erythrinidae), exposed to graded ambient hypoxia. *Comp. Biochem. Physiol. A* **139,** 479–485.

Olson, K. R., Dombkowski, R. A., Russell, M. J., Doellman, M.M, Head, S. K., Whitfield, N. L., and Madden, J. A. (2006). Hydrogen sulfide as an oxygen sensor/transducer in vertebrate hypoxic vasoconstriction and hypoxic vasodilation. *J. Exp. Biol.* **209,** 4011–4023.

Ortega-Sáenz, P., Pascual, A., Gómez-Díaz, R., and López-Barneo, J. (2006). Acute oxygen sensing in heme oxygenase-2 null mice. *J. Gen. Physiol.* **128,** 405–411.

Ott, M. E., Heisler, N., and Ultsch, G. R. (1980). A re-evaluation of the relationship between temperature and the critical oxygen tension in freshwater fishes. *Comp. Biochem. Physiol. A* **67,** 337–340.

Peers, C. (1990). Hypoxic suppression of K^+ currents in type-I carotid-body cells – selective effect on the Ca^{2+}-activated K^+ current. *Neurosci. Lett.* **119,** 253–256.

Peers, C., and Wyatt, C. N. (2007). The role of maxiK channels in carotid body chemotransduction. *Respir. Physiol. NeuroBiol.* **157,** 75–82.

Perry, S. F., and Farrell, A. P. (1989). Perfused preparations in comparative respiratory physiology. *In* "Techniques in Comparative Respiratory Physiology." Society for Experimental Biology Seminar Series, Vol. 27 (Bridges, C. R., and Butler, P. J., Eds.), pp. 224–257. Cambridge University Press, Cambridge.

Perry, S. F., and Gilmour, K. M. (1996). Consequences of catecholamine release on ventilation and blood oxygen transport during hypoxia and hypercapnia in an elasmobranch (*Squalus acanthias*) and a teleost (*Oncorhynchys mykiss*). *J. Exp. Biol.* **199,** 2105–2118.

Perry, S. F., and Gilmour, K. M. (1999). Respiratory and cardiovascular systems. *In* "Stress Physiology" (Balm, P. H. M., Ed.), pp. 52–107. Sheffield Academic Press, Sheffield.

Perry, S. F., and Gilmour, K. M. (2002). Sensing and transfer of respiratory gases at the fish gill. *J. Exp. Zool.* **293,** 249–263.

Perry, S. F., and Thomas, S. (1991). The effects of endogenous or exogenous catecholamines on blood respiratory status during acute hypoxia in rainbow trout (*Oncorhynchus mykiss*). *J. Comp. Physiol. B* **161,** 489–497.

Perry, S. F., and Wood, C. M. (1989). Control and coordination of gas transfer in fishes. *Can. J. Zool.* **67,** 2961–2970.

Perry, S. F., Davie, P. S., Daxboeck, C., Ellis, A. G., and Smith, D. G. (1984). Perfusion methods for the study of gill physiology. *In* "Fish Physiology" Vol. 10B, (Hoar, W. S., and Randall, D. J., Eds.), pp. 325–388. Academic Press, New York.

Perry, S. F., Booth, C. E., and McDonald, D. G. (1985a). Isolated perfused head of rainbow trout I. Gas transfer, acid-base balance, and hemodynamics. *Am. J. Physiol.* **249,** R246–R254.

Perry, S. F., Booth, C. E., and McDonald, D. G. (1985b). Isolated perfused head of rainbow trout II. Ionic fluxes. *Am. J. Physiol.* **249,** R255–R261.

Perry, S. F., Kinkead, R., and Fritsche, R. (1992). Are circulating catecholamines involved in the control of breathing by fishes? *Rev. Fish Biol. Fish.* **2,** 65–83.

Perry, S. F., Reid, S. G., Gilmour, K. M., Boijink, C. L., Lopes, J. M., Milsom, W. K., and Rantin, F. T. (2004). A comparison of adrenergic stress responses in three tropical teleosts exposed to acute hypoxia. *Am. J. Physiol.* **287,** R188–R197.

Perry, S. F., Gilmour, K. M., Vulesevic, B., McNeil, B., Chew, S. F., and Ip, Y. K. (2005). Circulating catecholamines and cardiorespiratory responses in hypoxic lungfish (*Protopterus dolloi*): A comparison of aquatic and aerial hypoxia. *Physiol. Biochem. Zool.* **78,** 325–334.

Perry, S. F., Euverman, R. M., Wang, T., Loong, A. M., Chew, S. F., Ip, Y. K., and Gilmour, K. M. (2008). Control of breathing in African lungfish (*Protopterus dolloi*): A comparison of aquatic and cocooned (terrestrialized) animals. *Respir. Physiol. NeuroBiol.* **160,** 8–17.

Peyreaud-Waitzenegger, M. (1979). Simultaneous modifications of ventilation and arterial PO_2 by catecholamines in the eel, *Anguilla anguilla* L.: Participation of alpha and beta effects. *J. Comp. Physiol. B* **129,** 343–354.

Peyreaud-Waitzenegger, M., and Soulier, P. (1989). Ventilatory and circulatory adjustments in the European eel (*Anguilla anguilla* L.) exposed to short term hypoxia. *Exp. Biol.* **48,** 107–122.

Peyreaud-Waitzenegger, M., Barthelemy, L., and Peyreaud, C. (1980). Cardiovascular and ventilatory effects of catecholamines in unrestrained eels (*Anguilla anguilla* L.). *J. Comp. Physiol. B* **138,** 367–375.

Piiper, J. (1998). Branchial gas transfer models. *Comp. Biochem. Physiol.* **119A,** 125–130.

Piiper, J., Baumgarten, D., and Meyer, M. (1970). Effects of hypoxia upon respiration and circulation in the dogfish *Scyliorhinus stellaris*. *Comp. Biochem. Physiol.* **36,** 513–520.

Playle, R. C., Munger, R. S., and Wood, C. M. (1990). Effects of catecholamines on gas exchange and ventilation in rainbow trout (*Salmo gairdneri*). *J. Exp. Biol.* **152,** 353–367.

Powell, F. L., Milsom, W. K., and Mitchell, G. S. (1998). Time domains of the hypoxic ventilatory response. *Respir. Physiol.* **112,** 123–134.

Prabhakar, N. R. (2006). O_2 sensing at the mammalian carotid body: Why multiple O_2 sensors and multiple transmitters? *Exp. Physiol.* **91,** 17–23.

Randall, D. J. (1982). The control of respiration and circulation in fish during exercise and hypoxia. *J. Exp. Biol.* **100,** 275–288.

Randall, D. J. (1990). Control and co-ordination of gas exchange in water breathers. *In* "Advances in Comparative and Environmental Physiology" (Boutilier, R. G., Ed.), pp. 253–278. Springer-Verlag, Berlin.

Randall, D. J., and Daxboeck, C. (1984). Oxygen and carbon dioxide transfer across fish gills. *In* "Fish Physiology" (Hoar, W. S., and Randall, D. J., Eds.), pp. 263–314. Academic Press, London.

Randall, D. J., and Jones, D. R. (1973). The effect of deafferentation of the pseudobranch on the respiratory response to hypoxia and hyperoxia in the trout (*Salmo gairdneri*). *Respir. Physiol.* **17,** 291–301.

Randall, D. J., and Shelton, G. (1963). The effects of changes in environmental gas concentrations on the breathing and heart rate of a teleost fish. *Comp. Biochem. Physiol.* **9,** 229–239.

Randall, D. J., and Taylor, E. W. (1991). Evidence of a role for catecholamines in the control of breathing in fish. *Rev. Fish Biol. Fish.* **1,** 139–157.

Randall, D. J., Cameron, J. N., Daxboeck, C., and Smatresk, N. J. (1981). Aspects of bimodal gas exchange in the bowfin, *Amia calva* L. (Actinopterygii: Amiiformes). *Respir. Physiol.* **43,** 339–348.

Randle, A. M., and Chapman, L. J. (2005). Air-breathing behaviour of the African anabantoid fish *Ctenopoma muriei*. *J. Fish Biol.* **67,** 292–298.

Rantin, F. T., and Johansen, K. (1984). Responses of the teleost *Hoplias malabaricus* to hypoxia. *Environ. Biol. Fish.* **11,** 221–228.

Rantin, F. T., Guerra, C. D. R., Kalinin, A. L., and Glass, M. L. (1998). The influence of aquatic surface respiration (ASR) on cardio-respiratory function of the serrasalmid fish *Piaractus mesopotamicus*. *Comp. Biochem. Physiol. A* **119,** 991–997.

Rantin, F. T., Kalinin, A. L., Glass, M. L., and Fernandes, M. N. (1992). Respiratory responses to hypoxia in relation to mode of life of two erythrinid species (*Hoplias malabaricus* and *Hoplias lacerdae*). *J. Fish Biol.* **41,** 805–812.

Reid, S. G., and Perry, S. F. (2003). Peripheral O_2 chemoreceptors mediate humoral catecholamine secretion from fish chromaffin cells. *Am. J. Physiol.* **284,** R990–R999.

Reid, S. G., Sundin, L., Florindo, L. H., Rantin, F. T., and Milsom, W. K. (2003). Effects of afferent input on the breathing pattern continuum in the tambaqui (*Colosoma macropomum*). *Respir. Physiol. NeuroBiol.* **136,** 39–53.

Riesco-Fagundo, A. M., Pérez-García, M. T., Gonzalez, C., and López-López, J. R. (2001). O_2 modulates large-conductance Ca^{2+}-dependent K^+ channels of rat chemoreceptor cells by a membrane-restricted and CO-sensitive mechanism. *Circ. Res.* **89,** 430–436.

Ristori, M. T., and Laurent, P. (1989). Plasma catecholamines in rainbow trout (*Salmo gairdneri*) during hypoxia. *Exp. Biol.* **48,** 285–290.

Routley, M. H., Nilsson, G. E., and Renshaw, G. M. C. (2002). Exposure to hypoxia primes the respiratory and metabolic responses of the epaulette shark to progressive hypoxia. *Comp. Biochem. Physiol. A* **131,** 313–321.

Rutjes, H. A., Nieveen, M. C., Weber, R. E., Witte, F., and van den Thillart, G. E. E. J. M. (2007). Multiple strategies of Lake Victoria cichlids to cope with lifelong hypoxia include hemoglobin switching. *Am. J. Physiol.* **293,** R1376–R1383.

Sakuragui, M. M., Sanches, J. R., and Fernandes, M. N. (2003). Gill chloride cell proliferation and respiratory responses to hypoxia of the neotropical erythrinid fish *Hoplias malabaricus*. *J. Comp. Physiol. B* **173,** 309–317.

Saltys, H. A., Jonz, M. G., and Nurse, C. A. (2006). Comparative study of gill neuroepithelial cells and their innervation in teleosts and *Xenopus* tadpoles. *Cell Tissue Res.* **323,** 1–10.

Sanchez, A. P., Soncini, R., Wang, T., Koldkjær, P., Taylor, E. W., and Glass, M. L. (2001). The differential cardio-respiratory responses to ambient hypoxia and systemic hypoxaemia in the South American lungfish, *Lepidosiren paradoxa*. *Comp. Biochem. Physiol. A* **130,** 677–687.

Saunders, R. L., and Sutterlin, A. M. (1971). Cardiac and respiratory responses to hypoxia in the sea raven, *Hemitripterus americanus*, and an investigation of possible control mechanisms. *J. Fish Res. Bd. Canada* **28,** 491–503.

Seymour, R. S., Farrell, A. P., Christian, K., Clark, T. D., Bennett, M. B., Wells, R. M. G., and Baldwin, J. (2007). Continuous measurement of oxygen tensions in the air-breathing organ of Pacific tarpon (*Megalops cyprinoides*) in relation to aquatic hypoxia and exercise. *J. Comp. Physiol. B* **177,** 579–587.

Shelton, G., Jones, D. R., and Milsom, W. K. (1986). Control of breathing in ectothermic vertebrates. *In* "Handbook of Physiology, Section 3: The Respiratory System. Volume II Control of Breathing" (Cherniak, N. S., and Widdicombe, J. G., Eds.), pp. 857–909. American Physiological Society, Bethesda, Maryland.

Shingles, A., McKenzie, D. J., Claireaux, G., and Domenici, P. (2005). Reflex cardioventilatory responses to hypoxia in the flathead grey mullet (*Mugil cephalus*) and their behavioural modulation by perceived threat of predation and water turbidity. *Physiol. Biochem. Zool.* **78,** 744–755.

Short, S., Taylor, E. W., and Butler, P. J. (1979). The effectiveness of oxygen transfer during normoxia and hypoxia in the dogfish (*Scyliorhinus canicula* L.) before and after cardiac vagotomy. *J. Comp. Physiol. B* **132,** 289–295.

Sloman, K. A., Wood, C. M., Scott, G. R., Wood, S., Kajimura, M., Johannsson, O. E., Almeida-Val, V. M. F., and Val, A. L. (2006). Tribute to R. G. Boutilier: The effects of size on the physiological and behavioural responses of oscar, *Astronotus ocellatus*, to hypoxia. *J. Exp. Biol.* **209,** 1197–1205.

Smatresk, N. J. (1986). Ventilatory and cardiac reflex responses to hypoxia and NaCN in *Lepisosteus osseus*, an air-breathing fish. *Physiol. Zool.* **59,** 385–397.

Smatresk, N. J. (1990). Chemoreceptor modulation of endogenous respiratory rhythms in vertebrates. *Am. J. Physiol.* **259,** R887–R897.

Smatresk, N. J. (1994). Respiratory control in the transition from water to air breathing in vertebrates. *Amer. Zool.* **34,** 264–279.

Smatresk, N. J., and Cameron, J. N. (1982). Respiration and acid-base physiology of the spotted gar, a bimodal breather I. Normal values, and the response to severe hypoxia. *J. Exp. Biol.* **96,** 263–280.

Smatresk, N. J., Burleson, M. L., and Azizi, S. Q. (1986). Chemoreflexive responses to hypoxia and NaCN in longnose gar: Evidence for two chemoreceptor loci. *Am. J. Physiol.* **251,** R116–R125.

Smith, D. G., Duiker, W., and Cooke, I. R. C. (1983). Sustained branchial apnea in the Australian short-finned eel, *Anguilla australis*. *J. Exp. Zool.* **226,** 37–43.

Smith, F. M., and Jones, D. R. (1982). The effect of changes in blood oxygen carrying capacity on ventilation volume in the rainbow trout (*Salmo gairdneri*). *J. Exp. Biol.* **97,** 325–334.

Smith, R. S., and Kramer, D. L. (1986). The effect of apparent predation risk on the respiratory behavior of the Florida gar (*Lepisosteus platyrhincus*). *Can. J. Zool.* **64,** 2133–2136.

Soares, M. G. M., Menezes, N. A., and Junk, W. J. (2006). Adaptations of fish species to oxygen depletion in a central Amazonian floodplain lake. *Hydrobiologia* **568,** 353–367.

Soncini, R., and Glass, M. L. (2000). Oxygen and acid-base status related drives to gill ventilation in carp. *J. Fish Biol.* **56,** 528–541.

Spitzer, K. W., Marvin, D. E., and Heath, A. G. (1969). The effect of temperature on the respiratory and cardiac response of the bluegill sunfish to hypoxia. *Comp. Biochem. Physiol.* **30,** 83–90.

Stecyk, J. A. W., and Farrell, A. P. (2002). Cardiorespiratory responses of the common carp (*Cyprinus carpio*) to severe hypoxia at three acclimation temperatures. *J. Exp. Biol.* **205,** 759–768.

Stecyk, J. A. W., and Farrell, A. P. (2008). Regulation of the cardiorespiratory system of common carp (*Cyprinus carpio*) during severe hypoxia at three seasonal acclimation temperatures. *Physiol. Biochem. Zool.* **79,** 614–627.

Steffensen, J. F. (1985). The transition between branchial pumping and ram ventilation in fishes: Energetic consequences and dependence on water oxygen tension. *J. Exp. Biol.* **114,** 141–150.

Steffensen, J. F., Lomholt, J. P., and Johansen, K. (1982). Gill ventilation and O_2 extraction during graded hypoxia in two ecologically distinct species of flatfish, the flounder (*Platichthys flesus*) and the plaice (*Pleuronectes platessa*). *Environ. Biol. Fish.* **7,** 157–163.

Sundin, L., Holmgren, S., and Nilsson, S. (1998). The oxygen receptor of the teleost gill? *Acta. Zool.* **79,** 207–214.

Sundin, L., Reid, S. G., Kalinin, A. L., Rantin, F. T., and Milsom, W. K. (1999). Cardiovascular and respiratory reflexes: The tropical fish, traira (*Hoplia malabaricus*) O_2 chemoresponses. *Respir. Physiol.* **116,** 181–199.

Sundin, L., Reid, S. G., Rantin, F. T., and Milsom, W. K. (2000). Branchial receptors and cardiorespiratory reflexes in the neotropical fish, Tambaqui (*Colossoma macropomum*). *J. Exp. Biol.* **203,** 1225–1239.

Sundin, L., Turesson, J., and Burleson, M. L. (2003). Identification of central mechanisms vital for breathing in the channel catfish, *Ictalurus punctatus*. *Respir. Physiol. NeuroBiol.* **138,** 77–86.

Sundin, L., Burleson, M. L., Sanchez, A. P., Amin-Naves, J., Kinkead, R., Gargaglioni, L. H., Hartzler, L. K., Wiemann, M., Kumar, P., and Glass, M. L. (2007). Respiratory chemoreceptor function in vertebrates – comparative and evolutionary aspects. *Integ. Comp. Biol.* **47,** 592–600.

Takasusuki, J., Fernandes, M. N., and Severi, W. (1998). The occurrence of aerial respiration in *Rhinelepis strigosa* during progressive hypoxia. *J. Fish Biol.* **52,** 369–379.

Taylor, J. C., and Miller, J. M. (2001). Physiological performance of juvenile southern flounder, *Paralichthys lethostigma* (Jordan and Gilbert, 1884), in chronic and episodic hypoxia. *J. Exp. Mar. Biol. Ecol.* **258,** 195–214.

Tetens, V., and Lykkeboe, G. (1981). Blood respiratory properties of rainbow trout, *Salmo gairdneri*: Responses to hypoxia acclimation and anoxic incubation of blood *in vitro*. *J. Comp. Physiol.* **145,** 117–125.

Thomas, S., and Hughes, G. M. (1982a). A study of the effects of hypoxia on acid-base status of rainbow trout blood using an extracorporeal blood circulation. *Respir. Physiol.* **49,** 371–382.

Thomas, S., and Hughes, G. M. (1982b). Effects of hypoxia on blood gas and acid-base parameters of sea bass. *J. Appl. Physiol.* **53,** 1336–1341.

Thompson, R. J., and Nurse, C. A. (1998). Anoxia differentially modulates multiple K^+ currents and depolarizes neonatal rat adrenal chromaffin cells. *J. Physiol.* **512,** 421–434.

Thompson, R. J., Farragher, S. M., Cutz, E., and Nurse, C. A. (2002). Developmental regulation of O_2 sensing in neonatal adrenal chromaffin cells from wild-type and NADPH-oxidase-deficient mice. *Pflugers Arch* **444,** 539–548.

Thompson, R. J., Buttigieg, J., Zhang, M., and Nurse, C. A. (2007). A rotenone-sensitive site and H_2O_2 are key components of hypoxia-sensing in neonatal rat adrenomedullary chromaffin cells. *Neuroscience* **145,** 130–141.

Timmerman, C. M., and Chapman, L. J. (2004a). Behavioral and physiological compensation for chronic hypoxia in the sailfin molly (*Poecilia latipinna*). *Physiol. Biochem. Zool.* **77,** 601–610.

Timmerman, C. M., and Chapman, L. J. (2004b). Hypoxia and interdemic variation in *Poecilia latipinna*. *J. Fish Biol.* **65,** 635–650.

Turesson, J., and Sundin, L. (2003). *N*-methyl-D-aspartate receptors mediate chemoreflexes in the shorthorn sculpin *Myoxocephalus scorpius*. *J. Exp. Biol.* **206,** 1251–1259.

Valverde, J. C., Lopez, F.-J. M., and Garcia, B. G. (2006). Oxygen consumption and ventilatory frequency responses to gradual hypoxia in common dentex (*Dentex dentex*): Basis for suitable oxygen level estimations. *Aquaculture* **256,** 542–551.

Varas, R., Wyatt, C. N., and Buckler, K. J. (2007). Modulation of TASK-like background potassium channels in rat arterial chemoreceptor cells by intracellular ATP and other nucleotides. *J. Physiol.* **583,** 521–536.

Vulesevic, B., and Perry, S. F. (2006). Developmental plasticity of ventilatory control in zebrafish, *Danio rerio*. *Respir. Physiol. NeuroBiol.* **154,** 396–405.

Vulesevic, B., McNeill, B., and Perry, S. F. (2006). Chemoreceptor plasticity and respiratory acclimation in the zebrafish, *Danio rerio*. *J. Exp. Biol.* **209,** 1261–1273.

Wang, R. (2002). Two's company, three's a crowd: Can H_2S be the third endogenous gaseous transmitter? *EMBO J.* **16,** 1792–1798.

Wang, R., and Wu, L. (1997). The chemical modification of K_{Ca} channels by carbon monoxide in vascular smooth muscle cells. *J. Biol. Chem.* **272,** 8222–8226.

Wang, R., Youngson, C., Wong, V., Yeger, H., Dinauer, M. C., Vega-Saenz de Miera, E., Rudy, B., and Cutz, E. (1996). NADPH-oxidase and a hydrogen peroxide-sensitive K^+ channel may function as an oxygen sensor complex in airway chemoreceptors and small cell lung carcinoma cell lines. *Proc. Natl. Acad. Sci. USA* **93,** 13182–13187.

Watters, K. W., and Smith, L. S. (1973). Respiratory dynamics of the starry flounder *Platichthys stellatus* in response to low oxygen and high temperature. *Mar. Biol.* **19,** 133–148.
Weir, E. K., López-Barneo, J., Buckler, K. J., and Archer, S. L. (2005). Acute oxygen-sensing mechanisms. *N. Eng. J. Med.* **353,** 2042–2055.
Williams, B. A., and Buckler, K. J. (2004). Biophysical properties and metabolic regulation of a TASK-like potassium channel in rat carotid body type I cells. *Am. J. Physiol.* **286,** L221–L230.
Williams, S. E., Wootton, P., Mason, H. S., Bould, J., Iles, D. E., Riccardi, D., Peers, C., and Kemp, P. J. (2004). Hemoxygenase-2 is an oxygen sensor for a calcium-sensitive potassium channel. *Science* **306,** 2093–2097.
Winemiller, K. O. (1989). Development of dermal lip protuberances for aquatic surface respiration in South American Characid fishes. *Copeia* 382–390**1989**.
Wolf, N. G., and Kramer, D. L. (1987). Use of cover and the need to breathe: The effects of hypoxia on vulnerability of dwarf gouramis to predatory snakeheads. *Oecologia* **73,** 127–132.
Wood, S. C., and Johansen, K. (1972). Adaptation to hypoxia by increased HbO_2 affinity and decreased red cell ATP concentration. *Nature* **237,** 278–279.
Wood, S. C., and Johansen, K. (1973). Blood oxygen transport and acid-base balance in eels during hypoxia. *Am. J. Physiol.* **225,** 849–851.
Wood, S. C., Johansen, K., and Weber, R. E. (1975). Effects of ambient PO_2 on hemoglobin-oxygen affinity and red cell ATP concentrations in a benthic fish, *Pleuronectes platessa. Respir. Physiol.* **25,** 259–267.
Wyatt, C. N., and Buckler, K. J. (2004). The effect of mitochondrial inhibitors on membrane currents in isolated neonatal rat carotid body type I cells. *J. Physiol.* **556,** 175–191.
Wyatt, C. N., and Evans, A. M. (2007). AMP-activated protein kinase and chemotransduction in the carotid body. *Respir. Physiol. NeuroBiol.* **157,** 22–29.
Wyatt, C. N., Mustard, K. J., Pearson, S. A., Dallas, M. L., Atkinson, L., Kumar, P., Peers, C., Hardie, D. G., and Evans, A. M. (2007). AMP-activated protein kinase mediates carotid body excitation by hypoxia. *J. Biol. Chem.* **282,** 8092–8098.
Zaccone, G., Tagliafierro, G., Goniakowska-Witalinska, L., Fasulo, S., Ainis, L., and Mauceri, A. (1989). Serotonin-like immunoreactive cells in the pulmonary epithelium of ancient fish species. *Histochemistry* **92,** 61–63.
Zaccone, G., Lauweryns, S., Fasulo, S., Tagliafierro, G., Ainis, L., and Licata, A. (1992). Immunocytochemical localization of serotonin and neuropeptides in the neuroendocrine paraneurons of teleost and lungfish gills. *Acta Zool.* **73,** 177–183.
Zaccone, G., Fasulo, S., and Ainis, L. (1994). Distribution patterns of the paraneuronal endocrine cells in the skin, gills and the airways of fishes as determined by immunohistochemical and histological methods. *Histochem. J.* **26,** 609–629.
Zaccone, G., Mauceri, A., Fasulo, S., Ainis, L., Lo Cascio, P., and Ricca, M. B. (1996). Localization of immunoreactive endothelin in the neuroendocrine cells of fish gill. *Neuropeptides* **30,** 53–57.
Zaccone, G., Fasulo, S., Ainis, L., and Licata, A. (1997). Paraneurons in the gills and airways of fishes. *Microsc. Res. Tech.* **37,** 4–12.
Zaccone, G., Ainis, L., Mauceri, A., Lo Cascio, P., Lo Guidice, F., and Fasulo, S. (2003). NANC nerves in the respiratory air sac and branchial vasculature of the Indian catfish, *Heteropneustes fossilis. Acta Histochem* **105,** 151–163.
Zaccone, G., Mauceri, A., and Fasulo, S. (2006). Neuropeptides and nitric oxide synthase in the gill and the air-breathing organs of fishes. *J. Exp. Zool.* **305A,** 428–439.
Zhao, W., Zhang, J., Lu, Y., and Wang, R. (2001). The vasorelaxant effect of H_2S as a novel endogenous gaseous K_{ATP} channel opener. *EMBO J* **20,** 6008–6016.

6

BLOOD-GAS TRANSPORT AND HEMOGLOBIN FUNCTION: ADAPTATIONS FOR FUNCTIONAL AND ENVIRONMENTAL HYPOXIA

RUFUS M. G. WELLS

The physiological diversity of blood oxygen transport traits in fishes appears designed to maintain tissue oxygenation under challenges from both metabolic demand and environmental oxygen supply. Brief episodes of functional hypoxia may occur during strenuous exercise when aerobic

Hypoxia: Volume 27
FISH PHYSIOLOGY

DOI: 10.1016/S1546-5098(08)00006-X

metabolism cannot be maintained and, when ambient oxygen tensions are low, more persistent environmental hypoxia may result. Hypoxic responses may be acclimatory (phenotypic plasticity) or adaptational (evolutionary plasticity). Fish adapted for an athletic lifestyle do not generally thrive under environmental hypoxia. Highly active species tend to have high O_2-carrying capacities, relatively low blood O_2-affinities, sigmoidal binding curves, marked Bohr and Root effects, and O_2-affinity is modulated by adenosine triphosphate (ATP). Fish living in habitats that are periodically low in oxygen may also have high oxygen-carrying capacity, but generally have high blood O_2-affinities, low Hill coefficients, and hemoglobin (Hb) function is modulated by both guanosine triphosphate (GTP) and ATP. Hb function is further regulated by erythrocyte surface adrenoceptors when present. Multiple Hb components are functionally differentiated in some species, but not in others, and are not generally altered by acclimation. Low heterogeneity in Antarctic fish does not appear to be an adaptation for environmental stability. The O_2-binding properties of purified Hbs are difficult to interpret ecologically and consideration of the erythrocyte environment is critical to sensible interpretation of physiological traits. The Bohr effect depends on an ateriovenous pH gradient sustained by respiratory acidosis (CO_2), whereas the Root effect is activated by fixed acid (lactate) and, unless localized in specific retial tissues, may seriously compromise effective oxygen transport in hypoxic situations. Reduced temperature sensitivity of Hb–O_2 binding occurs in endothermic fishes that encounter thermal shifts at the gill exchange surface. Recent progress has been made in understanding the environmental thresholds for expression of factors compensating for hypoxia. These include globin synthesis, a role for Hb in regulation of the paracrine vasodilator NO, and changes in gene expression of HIF targets. Responses to hypoxia may be species specific, and comparisons become more difficult to interpret with increasing phylogenetic distance. The challenge for the future is to place research findings in the context of physiological ecology and behavior.

1. INTRODUCTION

The most successful group of vertebrates, both in terms of species diversity and habitat distribution, are the bony fishes. The diversity of modern fishes represents a long evolutionary history and complex phylogeny and any attempt to interpret adaptive features of the oxygen transport system requires not only an understanding of the inter-relationship of extant fishes, but consideration of historical selection pressures. On the basis of spiracular anatomy from Devonian fossil fishes, Clack (2007) proposed that lungs were present in most of the early bony fishes, whether of freshwater, estuarine, or marginal marine origin. While air-gulping supports facultative air breathing and presented evolutionary opportunities that led to the tetrapods (Graham, 1997), the hypoxic condition of aquatic habitats in the mid–late Devonian indicates widespread environmental hypoxia (Berner, 2006). The subsequent evolution of the swim bladder in teleost fishes was a major innovation permitting vertical exploitation of the water column, but presented new challenges for obligate aquatic breathers in coping with hypoxia.

The evolutionary success of teleost fishes is largely due to an oxygen secretion mechanism involving special hemoglobins (Hbs) that are unique to this group. These Hbs, called Root effect Hbs, are present in most teleosts and enable smart vision and buoyancy control (Pelster and Randall, 1998; Berenbrink *et al.*, 2005). Oxygen secretion into the avascular eye is important for high-performance visual discrimination (Herbert *et al.*, 2002), and for adjusting gas volumes in the swim bladder without having to visit the air–water interface to gulp air (Graham, 1997).

Along with the critical importance of gas secretion, an efficient oxygen transport system is required for aerobic performance. It is worth noting that the most athletic fish easily outperform any mammal both in terms of maximum sustained and sprint speeds, and distances travelled during migrations. This is all the more remarkable considering the density of the aquatic medium, and the much lower oxygen content of water. Fish therefore have the potential to suffer oxygen deprivation both as a result of strenuous exercise (functional hypoxia), and when living in water of low and variable oxygen content (environmental hypoxia). How then, do fish cope with hypoxia?

Several reviews emphasize different aspects of this question and are recommended for further reading. Molecular approaches to structure–function relationships in fish Hbs have been reviewed by Jensen *et al.* (1998), Weber and Fago (2004), and de Souza and Bonilla-Rodriguez (2007). The unique properties of fish Hbs manifested through the Root effect with an emphasis on molecular interpretation have been reviewed by Brittain (2005, 2008). Nikinmaa (2006) has reviewed the role of erythrocyte membrane exchangers and pH regulation of blood-gas transport, and the effect of temperature on oxygen transport has been reviewed by Jensen *et al.*, (1993). Brauner and Val (2006) reviewed oxygen transport in tropical fishes and included adaptations of the gas exchange organs. Adaptive mechanisms contributing to the final step in the oxygen cascade from erythrocyte to mitochondrion have been reviewed by Wilhelm Filho (2007).

The present chapter attempts to deal with the question of how fish cope with hypoxia in the context of the blood oxygen transport system, both in terms of adaptations and phenotypic adjustments (acclimation). The evolutionary plasticity of the Hb molecule and its functional interactions within the complex environment of the erythrocyte allows for adaptation to both environmental and functional hypoxia. In addition, so-called phenotypic plasticity permits individual scope for habitat exploitation under variable oxygen conditions. In earlier reviews, Wells (1990, 1999) commented on the difficulty of comparing physiological adaptations in divergent species, and on the absence of information on environmental thresholds for expression of factors compensating for hypoxia. Considerable progress has been made in the last decade. In addition to the central role of Hb, further consideration is given to other members of the globin family, and to recently discovered special roles of Hb in hypoxia protection.

2. THE Hb SYSTEM

2.1. Concepts of Oxygen Transport

The design of the blood oxygen transport system in fishes is expected to have sufficient resilience to maintain an adequate oxygen supply to tissues both in the face of short-term functional hypoxia, where internal oxygen pressures fall as a result of strenuous exercise, and under transient or permanent levels of environmental hypoxia. The protein Hb is critical in meeting these expectations, because it enables oxygen to diffuse across the gas exchange membrane against its concentration gradient. Approximately 1.35 mL oxygen can be bound by 1 g Hb, and so the higher the Hb concentration, the higher the oxygen-carrying capacity of blood becomes, although optimum capacity is limited by the viscosity of blood at high hematocrit (Wells and Baldwin, 1990). Measurements of blood Hb concentration, however, do not tell the full story. The concentration of Hb within the erythrocyte appears to vary considerably among fish species and is quantified by mean cell Hb concentration (MCHC) calculated by dividing [Hb] by the hematocrit (the fraction of blood volume occupied by erythrocytes). Although there does not appear to be a systematic review of MCHC in fish, the parameter seems to correlate with both activity levels and environmental temperature (see Wells, 2005).

The adjustments of the Hb system required to compensate for hypoxia are more subtle than simply increasing the erythrocyte mass. It is the oxygen-binding properties of Hb that impart the functional diversity of this remarkable protein to match oxygen supply and demand. These properties are conventionally described by the oxygen equilibrium curve (OEC), which describes the relationship between the partial pressure of oxygen (PO_2) and the fraction of Hb-bound oxygen. OECs range in shape from essentially hyperbolic through to sigmoidal and the shape is quantified by Hill's coefficient, *n*, which will have values from 1.0 where the OEC is hyperbolic, to approximately 3 for a strongly sigmoidal OEC (Figure 6.1A). The diagnostic parameter is the P_{50} or PO_2 at which half of the Hb content is oxygenated. The P_{50}, however, is increased by both temperature and protons (where reduced pH expresses the Bohr effect), and by phosphate compounds in the erythrocytes (Figure 6.1B). The principal erythrocyte organic phosphate compounds in fish are adenosine and guanosine triphosphates (ATP, GTP), which bind to Hb and decrease its affinity for oxygen (Weber and Wells, 1989; Val, 2000). Fish blood differs from that of other vertebrates in that it may show an extreme reaction to low pH such that full saturation in air is impossible. This phenomenon is known as the Root effect (Figure 6.1C). These relationships are potentially confusing because the OEC is an *in vitro* determination and not a characteristic of a living fish. Nonetheless, there are strong reasons for believing that features of the OEC have physiological significance when considering how fish adapt to hypoxia.

An example of how the OEC appears adaptive to either environmental hypoxia or to functional hypoxia is shown in Figure 6.2. Here, the catfish, which is adapted to a low oxygen environment, has a relatively high blood-oxygen affinity, low sigmoidal coefficient (Hill's *n*-value), and blood that is relatively insensitive to pH; these features depend on the ability of tissues to function at low internal PO_2, and favor the loading of oxygen in the gills. The trout in contrast, has a relatively low blood-oxygen affinity, strongly sigmoidal OEC, and sensitivity of the P_{50} to pH; these features favor tissue unloading and the maximum oxygen loading and unloading occurs over the steep part of the OEC thus delaying the onset of functional hypoxia during exercise.

2.2. Oxygen-Carrying Capacity Responses

The correlation between Hb concentration and the potential for functional hypoxia in fast-swimming teleost fishes is well established. For example, when pelagic and benthic tropical reef fishes are compared, the most active fishes had approximately twice the Hb content of inactive species (Wells and Baldwin, 1990). Hb concentration in elasmobranchs does not, however, correlate with their propensity for functional hypoxia (Baldwin and Wells, 1990). Fish living at sub-zero temperatures in the Antarctic seas have extremely low metabolic rates and have either very low Hb contents or none at all (Wells *et al.*, 1980). Athletic fish, however, do not maintain a permanently high oxygen-carrying capacity, but during exercise release additional erythrocytes from the adrenergicallystimulated spleen into the circulation (Wells and Weber, 1990). Big gamefish such as tuna and marlin have

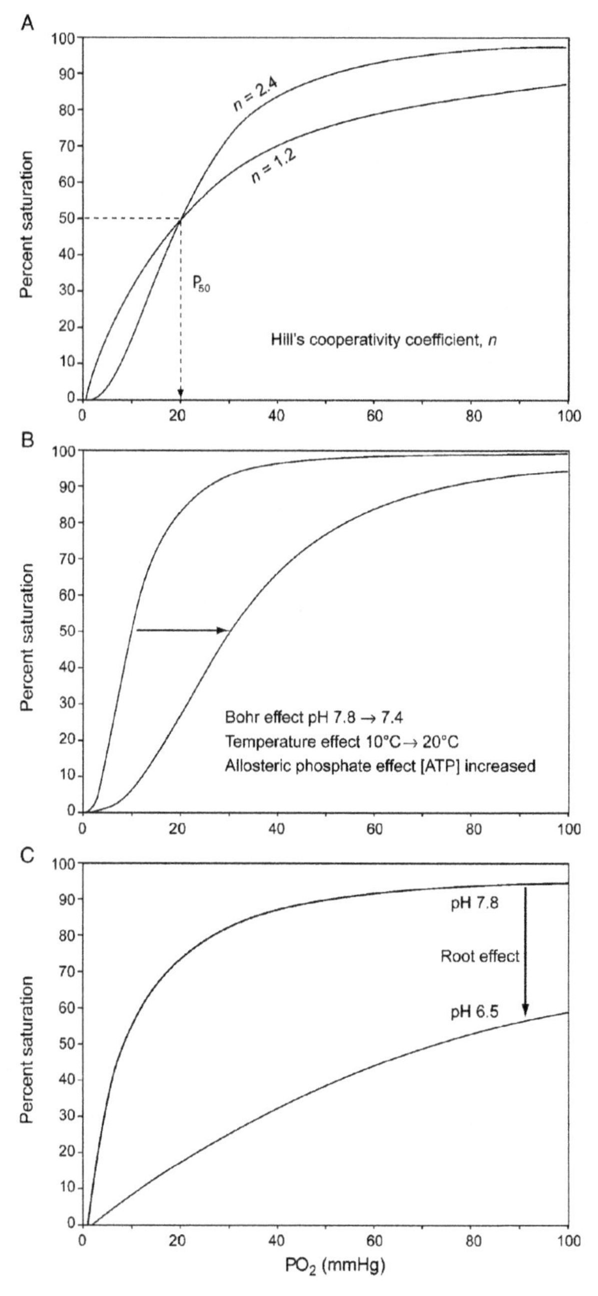

Fig. 6.1. Variations in the shape and position of theoretical blood oxygen equilibrium curves (OEC). (A) Comparison of two OECs with similar half-saturation values (P_{50}), but contrasting

extensive Hb reserves that may result in transient hematocrit values that exceed 70% during extreme activity (Wells *et al.*, 1986). Exercise training in rainbow trout, *Oncorhynchus mykiss*, however, did not lead to an increase in either oxygen-carrying capacity or P_{50}, and optimization of O_2 transport occurred through improved microcirculation (Davie *et al.*, 1986).

The question of whether exposure to environmental hypoxia leads to similar acclimatory responses is more complex. Initial exposure to environmental hypoxia in rainbow trout, *O. mykiss*, resulted in an increase in Hb concentration through release of erythrocytes from the spleen, but under persistent hypoxia, an erythropoietin-mediated synthesis of new erythrocytes increased the oxygen-carrying capacity of the blood (Lai *et al.*, 2006). Oxygen-carrying capacity may also be increased following brief exposure to environmental hypoxia (about 30% saturation) via hemoconcentration and in this case a reduction in plasma volume occurred (Tervonen *et al.*, 2006). Chronic exposure to extreme hypoxia in the sailfin molly, *Poecilia latipinna*, initially resulted in aquatic surface breathing behavior that diminished with time, though increased Hb concentration was maintained throughout the 6-week period of hypoxic exposure (Timmerman and Chapman, 2004).

It is not clear whether the capacity responses of trout are typical of teleosts. Acclimation for 40 days under three levels of chronic hypoxia did not result in physiologically significant changes to O_2-carrying capacity in turbot, *Scophthalamus maximus*, or sea bass, *Dicentrachus labrax*, although capacity in the more active sea bass under normoxic conditions was already twice that of the inactive benthic species (Pichavant *et al.*, 2003).

The tambaqui, *Colossoma macropomum*, is an inhabitant of the Amazonian floodplain and subjected seasonally not only to hypoxia, but also to hypercapnia and elevated levels of sulfide to which it seems tolerant. The initial response to hypoxia was a temporary increase in Hb concentration, and levels were readjusted toward normal carrying capacity within a few days as respiratory and metabolic processes compensated to maintain oxygen delivery (Affonso *et al.*, 2002). Carter and Wilson (2006) demonstrated

---

cooperativity coefficients (Hill's *n*-value) showing high cooperativity ($n = 2.4$) where oxygen may be loaded or unloaded over a narrow range of PO_2 on the steep part of the curve, and low cooperativity ($n = 1.2$) where oxygen may be effectively loaded or unloaded over a broader range of PO_2. (B) The rightward shift of the OEC (increased P_{50}) resulting from either a fall in pH (Bohr factor $\Phi = \Delta \log P_{50}/\Delta pH$), an increase in temperature, or a high erythrocyte ATP content. These three factors improve oxygen delivery to tissues, although saturation in the gills may be compromised if environmental PO_2 falls much below 80 mmHg. (C) The exaggerated rightward shift of the OEC in response to proton load (low pH) in some fish may result in failure to saturate the blood even at high PO_2, and indicates a Root effect.

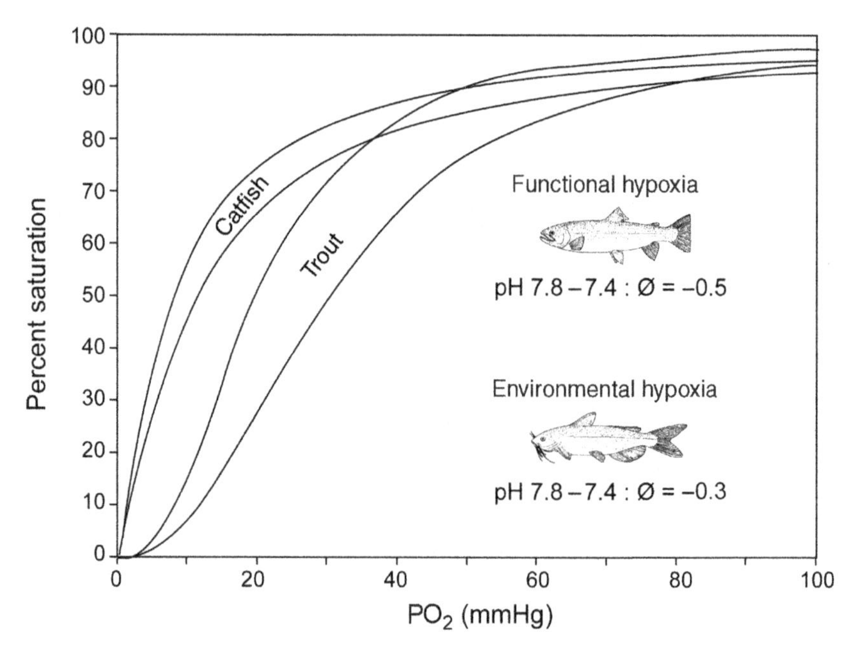

Fig. 6.2. A comparison of the main operational features of whole blood OEC in a fish adapted to environmental hypoxia (e.g., catfish; Grigg 1969), and an athletic fish adapted for functional hypoxia (e.g., trout; Eddy *et al.*, 1977). The former is characterized by a relatively high O_2 affinity, low cooperativity, and small Bohr factor thus favoring O_2 uptake in the gills. The latter is characterized by lower O_2 affinity, strongly cooperative O_2 binding, and a larger Bohr factor favoring O_2 unloading to tissues.

improved sexual fitness of hypoxia-acclimated male mosquito fish, *Gambusia holbrooki*, largely as a result of elevated oxygen-carrying capacity. It therefore seems that increasing Hb content is a useful short-term acclimatory strategy to cope with transient environmental hypoxia, but that persistent exposure requires responses that do not compromise the substantial increase in viscosity of blood expected from an excess of erythrocytes in the circulation. These responses involve regulation of Hb-oxygen affinity and are discussed in Section 8.1.

Fish living permanently in low-oxygen environments appear to have comparatively high Hb contents. Chapman *et al.* (2002) introduced the concept of physiological refugia in populations of aquatic-breathing indigenous cichlid fishes in African lakes. These species had high Hb contents ($>$100 g l^{-1}) and their populations were maintained in deep, hypoxic swamp refugia despite impacts from invasive species not so well adapted to environmental hypoxia.

3. PROTON LOAD MAY IMPROVE OXYGEN DELIVERY: BOHR AND ROOT EFFECTS

3.1. Responses to Respiratory and Metabolic Acidosis

When the demand for oxygen exceeds supply, the pH of the internal environment falls as a result of both accumulated dissolved carbon dioxide from respiration, and the activation of anaerobic energy production leading to lactic acid. Under these conditions of environmental or functional hypoxia, the excess of protons arising from the hydration reaction of CO_2 and dissociation of lactic acid is largely buffered by the histidine groups of muscle proteins (Abe *et al.*, 1985) and fish that are more athletic tend to have higher buffering capacities than less active species (Dickson and Somero, 1987; Wells *et al.*, 1988). The buffering power of surface histidine components of fish Hbs, however, is rather poor with the consequence that pH effects on Hb function may be quite dramatic (Jensen, 2004). Protons that bind to Hb signal a conformational change in the protein that results in a reduced oxygen affinity, such that more oxygen is released in response to higher proton load. This response is called the Bohr effect and may be quantified by $\Phi = \Delta\log P_{50}/\Delta pH$ (see Figure 6.1B). Values are typically negative in the physiological pH range because the blood-oxygen affinity parameter, P_{50}, increases with *decreasing* pH. Values of Φ are themselves pH-dependent and at very low pH, oxygen affinity starts to increase, signaling the positive acid-Bohr effect (Pelster and Weber, 1990; Weber, 2000). Negative and positive Bohr effects are sometimes referred to as alkaline and acid Bohr effects, respectively. The positive (acid) Bohr effect may be of interest to protein chemists, but is probably without physiological significance. Nevertheless, teleost fish have some of the biggest negative Bohr effects in the animal kingdom (Jensen, 1989) and the molecular structures responsible for this strong pH response appear different from those in mammals or in primitive fish groups such as the agnathans (Qiu *et al.*, 2000; Nikinmaa, 2004).

Many fish show a decrease in Hb-oxygen binding capacity at low pH, even when blood PO_2 is high. This phenomenon, the Root effect, is exclusive to fishes and saturation may not even be possible under an atmosphere of pure oxygen. Counter-current multipliers in the choroid rete of the eye, and in the gas gland of the swim bladder, secrete Hb-bound oxygen and assist vision and buoyancy (Pelster and Randall, 1998). There is a characteristic loss of Hb-cooperativity at low pH and *n*-values may on occasion be <1.0. This feature, together with an absence of an acid-Bohr effect, may be used to confirm the presence of a Root effect Hb (Brittain, 2008). The effect is in part due to an extreme conformational shift in the Hb molecule to the low affinity state, and in part due to modification of the quaternary architecture of Hb by allosteric effectors in the presence of protons (Bonaventura *et al.*, 2004; Brittain, 2005). Further complexities arise from the interactions of anion exchangers and carbonic anhydrase in modulating the Bohr effect in the complex intra-erythrocytic environment whereby carbonic anhydrase catalyzes the fast hydration of CO_2 in the capillaries, thereby activating the Bohr effect (Jensen, 2004). We should therefore view the Bohr effect as operational under an arterial–venous pH difference generated by respiratory acidosis, and the Root effect as operational under a localized circulatory metabolic acidosis. This distinction is obviously important when comparing environmental and functional hypoxias.

Physiological interpretation of the adaptive significance of Bohr and Root effects at the level of protein function, however, are fraught with difficulty because the Bohr effect can only be measured by reference to full saturation (P_{100}) and is quantified from the pH-induced change in P_{50}. In practice, P_{100} is generally assumed when blood or Hb is in equilibrium with air at a specified pH (and possibly PCO_2) and temperature. By contrast, the Root effect can only be evaluated at a particular pH by reference to P_{100}, or oxygen-carrying capacity determined under *different* experimental conditions, and there is no agreed parameter for its quantification. For example, Berenbrink *et al.* (2005) quantified the Root effect as the reduction in saturation of an air-equilibrated hemolysate at pH > 8.0 when the pH was dropped to pH $= 5.5$. Accordingly, some physiologists consider the Root effect to be a distinct characteristic of fish blood in which oxygen bound to Hb can be released by protons at constant high PO_2, whereas others see it as an exaggerated Bohr effect (Wells, 1999; Weber and Fago, 2004; Brittain, 2005). Although from a physiological viewpoint, it is difficult to know how large a Bohr effect needs to be before it becomes a Root effect, the magnitude of both effects across species is highly correlated, and it has been suggested that the Root effect originally evolved as an extension of the Bohr effect by varying the number of surface histidine substitutions (Berenbrink *et al.*, 2005). There appears no simple molecular explanation at the protein structural level to explain the Root effect. The failure to transpose a Root effect through suspected key residues into recombinant mutant human HbA (Nagai *et al.*, 1985), and the production of chimaeric Hbs by site-directed mutagenesis (Unzai *et al.*, 2009) have thus far failed to demonstrate the expected structure–function correlation. The present view is that the Root effect arises from several evolutionary pathways, each producing species-specific synergistic clusters of many residues that contribute quantitatively to saturation inhibition (Bonaventura *et al.*, 2004; Brittain, 2008).

A further difficulty in the physiological interpretation of Bohr and Root effects arises from a consideration of the origin of the proton burden when a fish becomes hypoxic. When a respiratory acidosis generates protons from the hydration of CO_2 and subsequent dissociation of carbonic acid (as determined from the Henderson-Hasselbalch relationship), the reaction is reversed at the gas exchange surface when the lower external PCO_2 enables release of the gas to the surrounding medium, and the postbranchial efferent blood pH rises. This is important because an operational CO_2-Bohr effect depends on the arterial–venous pH difference from respiratory acidosis. However, in the case of a metabolic acidosis, the acid is "fixed" in the sense that the quantity of protons from dissociated lactic acid cycles between venous and arterial networks without the necessary pH change required to elicit a functional Bohr effect – in other words, blood-oxygen affinity theoretically remains approximately the same at both the uptake and delivery sites, and continues until lactate is recycled through the gluconeogenic pathway. The proton burden from functional hypoxia is likely to be vastly in excess of the proton load from aerobic respiration. Hb function may, however, be largely insulated from this excess through the action of the erythrocyte surface Na^+/H^+ exchangers that occur in most teleost fishes (see Section 9).

As a result of these conceptual difficulties, and because some investigators include the Root effect component within measurements of the Bohr effect, a comparison of the pH sensitivity of blood oxygen transport systems among species with different athletic traits and with fish from oxygen-labile habitats does not yield a consistent picture (see Jensen *et. al.*, 1998; Pelster and Randall, 1998; Jensen, 2004; Pelster and Weber, 2004). Acclimation of the cyprinid fish, *Tinca tinca*, to environmental hypoxia resulted in a marked increase in blood-oxygen affinity brought about by a sharp decrease in erythrocyte GTP (Jensen and Weber, 1982). That there was no change in the Bohr factor remains puzzling because erythrocyte phosphates should enhance the Bohr factor. We can expect, however, that aerobically active fishes should have both marked Bohr and Root effects, and fishes that generate significant metabolic acidosis are unlikely to thrive in hypoxic environments unless they have recourse to air-breathing.

3.2. Evidence for Visual Impairment in Functional Hypoxia

The Root effect correlates with the presence of a dense capillary network in the fish eye known as the choroid rete. This structure supports retinal oxygen flux through secretion of lactic acid in an essentially closed system, thus liberating oxygen via the Root effect to the highly aerobic retinal cells. Berenbrink *et al.* (2005) surveyed the Root effect in diverse groups of fishes and found that a fall in pH from 8.5 to 5.5 reduced Hb saturation by around 10% in sharks and lungfish, but most teleost species showed a depression of around 40% with values ranging from 2% in the catfish *Silurus* sp. that are benthic dwelling and perhaps have limited need for visual acuity to 80% in the cod, *Gadus morhua*, which are mesopelagic species.

Anecdotal evidence suggests that captured and highly agitated fish suffer visual dysfunction and often appear unable to navigate solid objects. A likely explanation is that the high fixed acid load in the circulation generated by anaerobic burst activity switches the Root Hbs into the extreme low-affinity state, thus impeding oxygen supply to the eye. There is some experimental evidence that visual discrimination is altered in functionally hypoxic fish possessing Root effect Hbs (Herbert and Wells, 2002; Herbert *et al.*, 2002). Vision is likely to be one of the first systems affected during exposure to hypoxia.

4. ENVIRONMENTAL TEMPERATURE: OXYGEN SUPPLY AND DEMAND

There is a close relationship between aquatic hypoxia and temperature, for as temperature increases, the obligatory decrease in oxygen solubility results in less dissolved oxygen available for respiration. Compounding the reduced availability of oxygen, metabolism—and hence biological oxygen demand—increases at warmer temperatures creating the potential for substantial hypoxic effects. Fish living in shallow, freshwater, and estuarine regions are likely to be most affected. The oxygen transport systems of fishes appear to be geared for hypoxic acclimation resulting from elevated temperature due to the exothermic nature of Hb-O_2 binding, where the blood-oxygen affinity decreases as temperature rises. This has the consequence of releasing more oxygen to tissues as metabolic demand increases in tandem with temperatures. The effect may be quantified by the van't Hoff relationship:

$$\Delta H \alpha \left((T_1 T_2)/(T_2 - T_1) \cdot \log\left(\mathrm{P}^1_{50}/\mathrm{P}^2_{50}\right) \right)$$

where ΔH is the heat of oxygenation and T_1 and T_2 are the lower and higher temperatures in K, and P^1_{50} and P^2_{50} are the Hb-oxygen affinity coefficients at T_1 and T_2, respectively. Values close to zero indicate relative temperature-independence of oxygen binding (Weber and Wells, 1989). Accordingly, fish living in thermally stable or seasonally changing habitats tend to show normal temperature sensitivity with negative values for ΔH (Weber and Wells, 1989) although ΔH may itself be temperature-dependent (Fago *et al.*, 1997).

The cyprinid fishes, *Carassius carassius* and *C. auratus*, showed a decrease in blood-oxygen affinity following exposure to higher temperatures, and a concomitant increase in gill surface area, thus ensuring that oxygen turnover to tissues was maintained (Sollid *et al.*, 2005). Thermal acclimation to warmer temperature partly reverses the change in oxygen affinity so that oxygen uptake in the gills can be improved (Albers *et al.*, 1983). In an extreme stenothermal example, whole blood oxygen equilibrium studies with Antarctic fish acclimated to −1.5°C and 4.5°C revealed ΔH values comparable to those expected from seasonal adjustments in temperate fishes, though with low P_{50} values (Tetens *et al.*, 1984). A seasonal decrease in blood-oxygen affinity is not due solely to the entropy of Hb–O_2 binding. The allosteric modulator GTP also increased in summer-acclimated eels, *Anguilla anguilla*, further assisting oxygen turnover to actively metabolizing tissues (Andersen *et al.*, 1985). Powers *et al.* (1979) compared the temperature effects on oxygen binding of Hbs and whole blood from a range of tropical and temperate teleosts, and, interestingly, differences in the thermal sensitivity of O_2 binding could not be shown for purified Hb, but only in the presence of phosphate cofactors. This study emphasizes the need for care in ecological interpretation of data from purified Hb.

A high value for ΔH in the cool temperate teleost, *Odax pullus* was hypothesized to compromise the warmer limits of its geographic range (Brix *et al.*, 1998a). By contrast, intertidal triplefin fishes subjected to rapid fluctuations in temperature showed low values of ΔH at low pH, suggesting that oxygen transport is maintained despite rapid thermal shifts (Brix *et al.*, 1999).

5. ENDOTHERMIC FISHES: STABILIZING INTERNAL OXYGEN TENSIONS

A different thermal problem arises for the several groups of fishes that have developed partial endothermy. Tunas, swordfishes, and lamnid sharks generate heat in the red, mitochondrial-rich swimming musculature that is exchanged via counter-current multiplication in order to generate more thermodynamically efficient processes in swimming, and sensory information processing (Block and Carey, 1985; Altringham and Block, 1997; Block *et al.*, 2001; Fritsches *et al.*, 2005). When cool blood with a negative ΔH (as is usually the case) meets warm blood in the gills, there is a potential for loss of oxygen from the arterial to venous circuit. The Hbs of fish with internal temperature gradients of 10–20°C appear to show temperature-insensitive oxygen binding as indicated by values of ΔH that are close to zero. For example, the albacore tuna, *Thunnus alalunga*, had a reversed temperature effect (ΔH = +1.72) from 10°C to 30°C, closely matching the maximal thermal gradient from ambient water to core body temperature (Cech *et al.*, 1984). Interestingly, a strong Bohr factor ($\Phi = -1.17$) was found, yet negligible Root and cooperativity effects (Hill's $n = 1.1$) were also found. These results were contested by Jones *et al.* (1986) who reported marked sigmoidal whole blood binding curves (Hill's $n = 1.72$) and a smaller Bohr effect ($\Phi = -0.59$) for the kawakawa, *Euthynnus affinis*. The discrepancy could be explained either if full saturation was not experimentally achieved in *T. alalunga* or if the much smaller species *E. affinis* is not significantly endothermic. The reverse temperature effect in Hb component I isolated from bluefin tuna, *Thunnus thynnus*, is explained by a large Bohr factor in which the protons bind endothermically resulting in a positive value for ΔH (Ikeda-Saito *et al.*, 1983). The temperature effects on whole blood from *T. thynnus* were similar to those seen in Hb solutions, suggesting that erythrocyte ATP or other cofactors were not responsible for mediating the observed temperature insensitivity (Brill and Bushnell, 2006).

Recent work has furthermore demonstrated that ΔH itself is temperature-dependent in tunas. Clark *et al.* (2008) working with southern bluefin tuna, *Thunnus maccoyii*, showed that the temperature effect was reversed at low temperatures, but whole blood oxygen binding was essentially independent of temperature at warmer temperatures. The investigators proposed that this was to avoid premature O_2 off-loading around the heat exchanger.

The bigeye tuna, *Thunnus obesus*, faces a unique hypoxic conflict. In addition to extreme depth excursions exposing the gills to a temperature shift from as high as 28°C down to 7°C, the fish spends considerable periods in the hypoxic oxygen minimum zone and thus faces both functional and environmental hypoxia (Lowe *et al.*, 2000). The authors found that unlike other tunas, *T. obesus* blood has in addition to temperature insensitivity an exceptionally high affinity for oxygen, coupled with a large Bohr factor, thus appearing to optimize both uptake in the gills and release to warmer tissues.

Close examination of the functional properties of Hb in the porbeagle shark, *Lamna nasus*, has revealed that the temperature effect is saturation dependent, with a normal temperature effect at low saturation (deoxygenated, venous blood), and a reverse temperature effect at high saturation (reflecting arterial blood), thus oxygen transport is protected despite the thermal gradient (Larsen *et al.*, 2003). This special feature is not, however, solely mediated via the intrinsic properties of Hb, but induced by ATP–Hb binding. A comparable mechanism seems to operate in other endothermic fishes such as the striped marlin, *Tetrapterus audax* (Weber and Jensen, 1988).

6. EXPRESSION AND SIGNIFICANCE OF MULTIPLE Hb COMPONENTS

Nearly all animals have more than one kind of Hb present in their erythrocytes. Multiple forms of Hb are particularly common in ectothermic animals, especially in fish that are expected to cope with fluctuating conditions of environmental oxygen and temperature (Weber, 1990; Wells, 1999). Our knowledge of the functional significance of multiple Hbs is largely due to the continuing investigations of R. E. Weber and colleagues, and the search for examples of Hb heterogeneity that might suggest adaptation to hypoxia in fish has attracted considerable interest (for reviews see Weber and Jensen, 1988; Weber and Wells, 1989; Weber, 1990, 1996, 2000; Weber and Fago, 2004).

Two groups of freshwater fishes have received special attention. The anguilliform eels are likely to be found in water low in dissolved oxygen and subjected to environmental hypoxia. Hb components isolated from the European eel, *Anguilla anguilla*, can be broadly resolved by electrophoresis into anodic Hbs that have low oxygen affinities accompanied by marked Bohr and Root effects, and cathodic Hbs that lack significant pH effects on either oxygen affinity or cooperativity. The cathodic Hbs are assumed to confer protection against hypoxia and acidosis (Fago *et al.*, 1995; Tamburrini *et al.*, 2001). Hemoglobins isolated from other eels, the common moray, *Muraena helena* (Pellegrini *et al.*, 1995), brown moray, *Gymnothorax unicolor* (Tamburrini *et al.*, 2001), and the conger eel, *Conger conger* (Pellegrini *et al.*, 2003), show a similar pattern of functional heterogeneity. The deep-sea eel, *Symenchelis parasitica*, while showing similar heterogeneity, possesses very little of the cathodic component and this feature is consistent with a more stable environment (Weber *et al.*, 2003).

Salmonid fishes, and in particular rainbow trout, *Oncorhynchus mykiss* (formerly *Salmo gairdneri*), have also been well researched. In contrast to eels, salmonids tend to inhabit well-aerated habitats and are athletic fish. Accordingly, trout are expected to develop some degree of functional hypoxia during strenuous exercise (McKenzie *et al.*, 2004). As in eels, trout possess multiple Hbs that resolve into anodal and cathodal components whereby the latter are largely insensitive to pH and allosteric effectors (Weber *et al.*, 1976a). Though the Hb system of rainbow trout, *O. mykiss*, has been studied for decades, only recently have further functionally heterogeneous components been detected, bringing the current total from four to nine (Fago *et al.*, 2002). A comparison of the Hb system in eels and trout does not suggest that Hb multiplicity is a particular characteristic of fish living in oxygen-deficient habitats. Plaice, *Platessa platessa*, and flounder, *Platchthys flesus*, contain 8 Hbs apiece, though there is little functional differentiation and their predominantly anodic components appear adapted for hypoxia through low sensitivity to phosphates and pH (Weber and de Wilde, 1976). Cathodic Hbs are not therefore a prerequisite for hypoxia adaptation. Hb multiplicity also occurs in cartilaginous fishes (Dafre and Reischl, 1997; Galderisi *et al.*, 1996) and in primitive teleosts such as the sturgeons (Luk'yanenko, 1978).

Hb multiplicity was compared in three phylogenetically distant teleosts inhabiting the same oxygen-deficient tropical billabongs: the Hb system in the osteoglossiform saratoga, *Scleropages jardinii*, was characterized by a single Hb component, the elopiform tarpon, *Megalops cyprinoides* (a facultative air breather) by one major and one minor component, and the perciform barramundi, *Lates calcarifer*, by seven components (Wells *et al.*, 1997). The latter species is a more advanced teleost. GTP was the principal modulator of Hb function in the water breathers, and ATP in the active, air-breathing tarpon.

Stronger evidence for the adaptive value of functionally heterogeneous Hbs is to be found among closely related species occupying different ecological niches. A functional analysis of Hb components isolated by isoelectric focusing from congeneric species of triplefin fishes (Family Tripterygiidae) found that Hbs less sensitive to pH and temperature occurred in species found in the oxy-labile and thermally unstable habitats of rock pools (Brix *et al.*, 1999). Further, the functional attributes of Hb components from marine kyphosid fishes appeared to support the distinction in temperature sensitivities of oxygen binding in relation to the thermal habitats of the different species investigated (Brix *et al.*, 1998). In another example, the African cichlid fishes of Lake Victoria have rapidly evolved into specialized niches, and both Hb heterogeneity and the Hb–O_2 binding characteristics of hemolysates appeared to correlate with hypoxia (Verheyen *et al.*, 1986), although cichlids generally lacked the pH-insensitive cathodic components (Weber, 1990). The eel pouts (Family Zoarcidae) are inactive, benthic, and generally deep-sea species occurring at all latitudes. In an extreme example, *Thermarces cerberus* is associated with deep sea hydrothermal vents where temperatures are high and oxygen content is very low; its Hbs are not functionally differentiated and showed pH and phosphate sensitivity, but much higher affinities than other zoarcids (Weber *et al.*, 2003).

By contrast, nototheniid fishes from the cold, thermally stable Antarctic seas where oxygen content is generally high showed far less Hb heterogeneity than temperate and tropical fishes, and their Hbs showed marked phosphate and pH-sensitive oxygen binding (reviewed by di Prisco *et al.*, 1998; Wells, 2005). Five components have been isolated from the cryopelagic species, *Pagothenia borchgrevinki* (Cocca *et al.*, 2000), and three from *Pleuragramma antarcticum* (Tamburrini *et al.*, 1997). The Hbs from *P. borchgrevinki* are functionally distinct and are presumed to support the more active behavior of this species (Riccio *et al.*, 2000). Other benthic notothenioids have either a single Hb component (Kunzmann, 1991), or an additional minor component comprising less that 5% total Hb (di Prisco *et al.*, 1991). Interestingly, the nototheniid, *Notothenia angustata*, occurs at much lower latitudes in southern New Zealand and also shares this pattern of low Hb diversity (Fago *et al.*, 1992).The minor fraction is expressed in greater amounts in *Gobionotothen gibberifrons* and has been correlated with hypoxic adaptation (Marinakis *et al.*, 2003). The Antarctic species *Anotopterus pharao* and *Macrourus holotrachys* are neither endemic nor members of the Notothenioidei, and have four to five Hb components (Kunzmann, 1991).

The number of Hb components present in Arctic fishes is generally higher than that of Antarctic fishes (di Prisco and Tamburrini, 1992; D'Avino and di Prisco, 1997). Two hypotheses have been advanced to explain these observations. Arctic fishes are frequently distributed in a latitudinal cline covering significant thermal variation whereas Antarctic fishes are thermally isolated by the circumpolar seas and thus face less environmental perturbation in oxygen supply and demand. Alternatively, the Arctic fishes are pleisiomorphic, and contrast with the monophyletic origins of the dominant nototheniods of Antarctica leading to lower diversity of Hb components (see reviews by Wells, 2005; Verde *et al.*, 2006). A comparison of Hb multiplicity in several groups of fish is shown in Figure 6.3 and does not suggest any correlation between athletic fish groups likely to experience functional hypoxia, those likely to experience environmental hypoxia, or phylogenetic relationships.

At the population level, there is scant evidence for Hb polymorphisms that may be linked to environmental oxygen tensions, though polymorphisms appear common in salmonids and sturgeons (Giles, 1991; Soldatov, 2002). Studies with Arctic and cold temperate fishes have revealed Hb polymorphisms in Atlantic cod, *Gadus morhua* (Fyhn *et al.*, 1994; Brix *et al.*, 1998b), and turbot, *Scophthalamus maximus* (Imsland *et al.*, 2000), but links

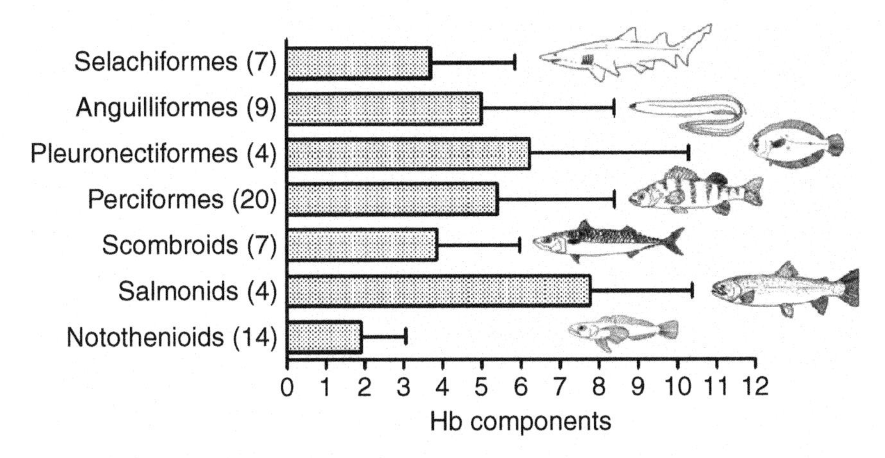

Fig. 6.3. Number of hemoglobin (Hb) components in representative species from distinct phylogenies (mean ± S.D.). The taxonomic hierarchies are not equivalent and the number of species represented within each group is in parentheses.

to environmental oxygen are circumstantial. A predominance of low pH-sensitive Hb components may be linked with more variable environmental temperatures in the turbot (Samuelsen *et al.*, 1999). Hb multiplicity in the Atlantic croaker, *Micropogon undulates*, revealed a complex polymorphism with functional phosphate-mediated differences in O_2 binding, but no clear ecological interpretation emerged (Shelly and Mangum, 1997).

On the basis of a "division of labor" into pH- and phosphate-insensitive high-affinity cathodic components, and generally lower affinity anodic components with marked phosphate and Bohr effects, a loose categorization of fish into Type I species having only anodic Hb and Type II with both anodic and cathodic Hb (incorporating an oxygen reserve for hypoxic episodes via cathodic Hb) has been proposed (Weber, 1990, 2000; Weber *et al.*, 2003). The mudfish, *Labeo capensis*, is an exception, in which both anodal and cathodal components are phosphate-sensitive (Frey *et al.*, 1998). Despite the large number of studies documenting the presence of multiple Hbs in fish, however, caution should be exercised before accepting that any isolated component reflects the *in vivo* condition, for Hbs are fragile molecules easily modified through the sometimes harsh purification treatments. Characterization of Hb components is most often dependent on electrophoretic mobility corresponding with differences in molecular surface charges; neither neutral substitutions nor internal structural differences may be revealed. In addition, the formation of hybrid globin complexes might not represent the Hb species in the intact erythrocyte (see Soldatov, 2002).

Physiological interpretation based on the functional properties of isolated, purified Hb components should be made with caution. Giardina *et al.* (2004) pointed out that *in vivo* functionality may be markedly different where the pH range under which Bohr and Root effects operate are modulated by red cell organic phosphates, and the operational range is shifted to a higher pH range. Unusual whole blood effects may also arise from the interactions of different Hb isoforms with functionally distinct binding characteristics to produce cooperativity coefficients with values less than unity (Deker and Nadja, 2007). At present, there is insufficient evidence for adaptive differences in fishes from different oxygen environments with respect to interspecies variance in ratios of cathodal to anodal components. There remain, however, convincing ecological associations with the functional attributes of the dominant Hb component.

A key question is, of course, whether the pattern of Hb components alters when a fish is exposed to chronic hypoxia. Surprisingly, there is very little evidence for acclimatory adjustments in Hb isoforms that might compensate hypoxic exposure. Marinsky *et al.* (1990) chemically rendered rainbow trout, *O. mykiss*, anaemic and then observed differences in the Hb pattern of fish recovering in controlled normoxic and hypoxic environments. Acclimation of goldfish, *Carassius auratus*, to temperature cycling also resulted in a shift in the Hb isomorph profile (Houston and Gingras-Bedard, 1994). The authors considered a rearrangement of the globins rather than *de novo* synthesis of new Hbs, which might be adaptive to the altered oxygen demands. Exposure to hypoxia or increased temperature of the mudfish, *Labeo capensis*, resulted in raised Hb concentration, but affected neither the pattern of Hb multiplicity nor the intrinsic Hb oxygen-binding properties, indicating that the principal acclimatory mechanism of the Hb system to hypoxia is elicited through phosphate–Hb interactions (Frey *et al.*, 1998). More recently, several species of African cichlid fishes were raised from an early stage in development under hypoxic conditions, and differences in the Hb systems contrasted with normoxic controls (Rutjes *et al.*, 2007). The authors demonstrated the expected increases in oxygen-carrying capacity, and phosphate shifts increasing Hb-O_2 affinity, but have critically shown that the synthesis of high-affinity Hb isomorphs occurs and a clear case of adaptive response seems justified.

The most recent study of changes in the Hb system following hypoxic exposure has been conducted by Campo *et al.* (2008). The gilthead sea bream, *Sparus auratus*, has two Hb components thatshare a common β-globin gene. Under normoxic conditions (7 mg L^{-1} oxygen), the components are present in approximately equal proportions, but these change progressively as fish are rendered hypoxic (2.5 mg L^{-1} oxygen). Campo *et al.* (2008) were not, however, able to demonstrate functional differences between the two components, either in intrinsic oxygen affinity, ATP modulation, Bohr and Root effects, or cooperativity, and hence the acclimatory value of the component ratio is unknown.

7. ROLE OF OTHER GLOBINS IN HYPOXIA

7.1. Myoglobin: Intracellular Oxygen Transfer

It is evident from the appearance of fish fillets in the supermarket that more athletic fishes have both more red muscle and intense colour, indicating higher myoglobin (Mb) content. The concentration of monomeric heme protein, Mb, correlates with mitochondrial oxygen demand in a variety of muscle tissues and in the oxy-form diffuses at about 1/20th the rate of diatomic oxygen, thereby contributing usefully to the oxygen cascade (Wittenberg, 2007). The presence of Mb in both locomotory and ventricular muscles suggests a role in maintaining aerobic metabolism when activity increases. The swimming muscles in fish are differentiated into fast-contracting fibers deployed during anaerobic burst activity, and slow-contracting fibers rich in mitochondria and Mb that are used predominantly during sustained aerobic swimming. Direct experimental evidence obtained using optical fiber sensors points to a role for Mb in providing improved intramuscular oxygen tensions during strenuous swimming in free-swimming rainbow trout, *Oncorhynchus mykiss* (McKenzie *et al.*, 2004). Moreover, the authors showed that the PO_2 in the differentiated fish muscle was significantly higher than that typically found in other vertebrates.

The presence of myoglobin in the fish ventricle appears to be an insurance against myocardial hypoxia. Functional studies on perfused hearts of the myoglobin-rich sea raven, *Hemitripterus americanus*, and the myoglobin-poor ocean pout, *Macrozoarces americanus*, showed that the former species was better able to maintain oxygen consumption under hypoxia, but failed to do so following chemical blocking of Mb (Bailey and Driedzic, 1986). However, electrically paced ventricular strips from the sculpin, *Myoxocephalus octodecimspinosus*, suggested that Mb did not play a critical role in maintaining performance under normoxia (Canty and Driedzic, 1987). Assumptions about tissue hypoxia based on dissolved oxygen levels in a perfusate are also problematic in *in vitro* studies of this kind. Further, Mb is not expressed in a range of unrelated species of sedentary fish, suggesting that Mb plays a role in protection against functional rather than environmental hypoxia (Grove and Sidell, 2002). Further research is needed in order to see whether these observations can be extrapolated to include elite swimmers such as carangids or gamefish. However, Mb concentration increased with cardiac growth in tuna (*Thunnus thynnus*), and showed a breakpoint increase at the phase of development coinciding with high-performance swimming (Poupa *et al.*, 1981).

Current research using molecular tools has focused on questions about the adaptive significance of globin gene expression. Fraser *et al.* (2006) demonstrated that hypoxia-tolerant common carp, *Cyprinus carpio*, increased expression of Mb under chronic hypoxia. Roesner *et al.* (2008) have extended these observations using real-time PCR and report that only Mb, but not α- or β-globin, neuroglobin or cytoglobin, expression is altered under hypoxic challenge in the carp. Enhanced expression of Mb was also observed in zebrafish, *Danio reria*, acclimated to severe hypoxic conditions for 3 weeks (van der Meer *et al.*, 2005). The plasticity of Mb expression in Atlantic cod, *Gadus morhua*, was also evident during temperature acclimation when both demand for oxygen and its solubility are altered, and in addition to modulating oxygen diffusivity, the Mb appeared also to play a role in scavenging nitric oxide (Lurman *et al.*, 2007; see Section 10.1).

In the chronically cold environment of the Antarctic seas, the loss of both Mb and Hb expression in icefishes (Family Channichthyidae) has occurred, but there are associated costs of reduced cardiac performance and a requirement to re-engineer the circulatory system (Sidell and O'Brien, 2006).

Gracey *et al.* (2001) measured the critical oxygen tension for the hypoxia-tolerant and burrow-dwelling goby, *Gillichthys mirabilis*, then determined the transcriptional response to hypoxia using cDNA microarray technology. The authors noted differential patterns of gene expression in various tissues directed toward closing down a number of major energy-requiring pathways and changes to heme metabolism. Further studies of gene expression under appropriate physiological conditions are required for other species of fish that regularly experience either functional or environmental hypoxia.

Functional studies on the oxygen-binding characteristics of fish myoglobins are sparse. Nichols and Weber (1989) correlated Mb-oxygen binding affinities with demand for oxygen in fish from different habitats. Mb extracted from the red swimming muscles of various fish species point to adaptive differences according to the activity of the species, such that inactive species living at low environmental temperatures (e.g., buffalo sculpin, *Enophrys bison*) had comparatively lower affinities when compared to more active species (e.g., yellowfin tuna, *Thunnus albacares*) living at warmer temperatures. This diversity of function is supported by Marcinek *et al.* (2001) who compared highly athletic endothermic and ectothermic fishes and related functional differences to the requirement to transfer oxygen at normal muscle operating temperatures. Optimization of Mb-O_2 affinity to tissue temperatures suggests that most animals are capable of adaptive adjustments to oxygen availability (Wittenberg, 2007). The implication of these studies on Mb is that hypoxia has a major effect on the PO_2 of the mitochondrial environment and that Mb may serve to buffer against such changes.

7.2. Neuroglobin: Protection of Neural Tissues

The presence of novel heme compounds in the neural tissues of vertebrates has recently been described (Burmester *et al.*, 2000) and has led to questions about their role in protecting neurons against either too much, or too little oxygen. These neuroglobins (Ngb), together with Hb and Mb, appear to be expressed in zebrafish (*Danio rerio*) in response to activation of hypoxia-inducible genes (Roesner *et al.*, 2006). Using real-time PCR, the authors noted marked increases in Ngb mRNA in brain but not retinal tissues, and Mb mRNA in cardiac tissue when fish were exposed to hypoxic insult. These observations are consistent with the differential effects of hypoxia whereby the anoxic crucian carp brain (*Carassius carassius*) maintained functionality (Nilsson, 2001), but the fish suffered severe visual impairment (Johansson *et al.*, 1997).

Whether Ngb function parallels the role of Mb in myotome or is involved in detoxification of reactive oxygen species and prevention of apoptosis is not known. However, Hundahl *et al.*, (2006) described the oxygenational characteristics of Ngb and reported non-cooperative binding and the absence of a Bohr effect. These characteristics allow for a role in either oxygen storage or enhanced diffusivity of oxygen. On this basis the authors proposed that neural excitability was extended under hypoxia. Ngb from zebrafish has an exceptionally high affinity for oxygen ($P_{50} = 1$ mmHg) and is co-located with mitochondria (Fuchs *et al.*, 2004). Up-regulation of Ngb after extended hypoxia in mice suggests a universal mechanism of neural protection from hypoxic injury (Hundahl *et al.*, 2005).

Recently discovered Globin-X, occurring only in fishes and amphibians, shares a common origin with Ngb and appears to be an ancient globin arising at the divergence of bony fishes and tetrapods (Roesner *et al.*, 2005). The functions of Globin-X, together with ubiquitous cytoglobins (Burmester *et al.*, 2002), remain unclear in relation to their expression under hypoxic challenge.

8. ERYTHROCYTE RESPONSES TO HYPOXIA

8.1. Phosphate Regulation of Hb-Oxygen Affinity

The erythrocyte phosphates are a main line of both acclimatory and adaptational defence in hypoxia protection. ATP is present in all fish erythrocytes, but fish regularly exposed to aquatic hypoxia also tend to have high proportions of GTP; an alternative strategy is present in air-breathing fish that have 2,3-diphosphoglyceric acid (DPG) (e.g., armoured catfish, *Pteroglopichthys* spp.) or lungfish, *Protopterus* spp. that have inositol phosphates (IP) (Val, 2000). These phosphates bind to specific sites in the central cavity of the Hb tetramer and stabilize the structure in the low-affinity, deoxy conformation. Known as allosteric regulation, this feature of Hb provides for adjustments to oxygen affinity via changes in the molar ratio of phosphate cofactor: Hb. A decrease in phosphates results in increased Hb oxygen affinity, thus favoring full saturation in the gills. The mechanism was first described for the eel (Wood and Johansen, 1972) with the cofactor GTP exerting the main influence (Weber *et al.*, 1976b). Subsequently, many fish living in hypoxic waters have been shown to use a similar survival strategy (Weber, 1996; Val, 2000).

The distribution of phosphate cofactors across species suggests a correlation with environmental hypoxia. GTP appears to predominate in species such as eels that are regularly subjected to environmental hypoxia, whereas ATP predominates in fish such as trout that are more likely to experience functional hypoxia. The distinction between ATP and GTP on the one hand

and DPG and IP on the other is important because their synthesis proceeds according to aerobic and anaerobic pathways, respectively. Trinucleotide production is linked to oxidative phosphorylation, whereas DPG and IP production is linked to the glycolytic pathway. Thus, under hypoxic challenge, shifting from aerobic to anaerobic metabolism with accompanying acidosis favors the sharp reduction of ATP and GTP, or, if present, an increase in DPG. This in turn results either in an increase in blood-oxygen affinity (reduced ATP, GTP), favoring oxygen uptake in the gills, or a decrease in blood-oxygen affinity (increased DPG, IP), promoting efficient oxygen unloading to tissues. One further role for NTP regulation deserves mention. Associated with the mass-specific reduction in oxygen requirements with increasing size in fish is a concomitant increase in oxygen affinity. In the piranha, *Serrasalmus rhombius*, the growth-correlated reduction in P_{50} is due entirely to reduced erythrocyte GTP rather than changes to the Hb isoforms (Wood *et al.*, 1979). It is not known whether larger individuals have less capacity for hypoxia acclimation than do smaller individuals. Weber (1996) has reviewed the distribution of phosphate compounds in the erythrocytes of some Amazonian fish. However, the control mechanisms signaling regulation of erythrocyte organic phosphates remain unclear (Nikinmaa, 2002).

In addition to the direct allosteric control of Hb-O_2 affinity by phosphates, both ATP and GTP modulate the Bohr effect in, for example, the anodic Hb of the eel, *Gymnothorax unicolour* (Tamburrini *et al.*, 2001). The enhanced binding of protons to the anodic Hb component of the eel, *Anguilla anguilla*, in the presence of GTP reveals the obverse of the Bohr effect—i.e., the Haldane effect (Brauner and Weber, 1998). The binding of ATP or GTP to the anodic components of several species of fish is essential for expression of the Root effect; reduction in the NTP:Hb ratio during hypoxic acclimation, however, is not sufficient to diminish the strength of the Root effect (Pelster and Weber, 1990).

The presence of high concentrations of the cofactor DPG in erythrocytes is typically associated with mammals, but also occurs in the air-breathing Amazonia catfish, *Hoplosternum littorale* (Weber *et al.*, 2000). In addition, lungfish erythrocytes contain the potent modulator inositol polyphosphate (IP; Val, 2000). Both DPG and the typical fish cofactors ATP and GTP are present in the erythrocytes of the catfish. Weber *et al.* (2000) reported pronounced but differentiated phosphate effects on the cathodal Hb, DPG<ATP<GTP, and concluded that the catfish Hb system appeared to impart no selective advantage for DPG binding given the lower sensitivity to DPG. An alternative interpretation considers the strategies for regulation of red cell organic phosphates under hypoxic pressure, whereby the modified pathway of red cell glycolysis leads to an accumulation of DPG, and via suppression of oxidative phosphorylation, a marked reduction of ATP and

other trinucleotides (see Val, 2000). Accordingly, it may be predicted that the typical air-breathing strategy of elevated DPG or IP modulation under hypoxia thus decreases Hb-oxygen affinity thereby securing adequate tissue oxygen delivery, and contrasts with the hypoxic reduction of ATP/GTP content alone that would increase Hb-oxygen affinity, thus ensuring maximal oxygen uptake at the exchange surface. Clearly, the catfish is a worthy candidate for experiments on acclimation to hypoxia under appropriate physiological conditions.

In the primitive erythrocytes of jawless agnathan fishes (lampreys and hagfish) the response to hypoxia is cell swelling. This results in an increased Hb-O_2 affinity through favoring the dissociation of the monomer–oligomer Hb complex (Nikinmaa, 2001). Cell swelling also occurs in teleosts under hypoxia, but the mechanism is entirely different. Here, the regulatory role of the organic phosphate–Hb complex is involved and the equilibrium favors the free unbound state resulting in increased oxygen affinity with erythrocyte swelling (see Nikinmaa, 2001 for review).

8.2. Allosteric Effects of Chloride and Water

Whereas red cell organic phosphates have been clearly shown to play a part in hypoxia regulation of Hb oxygen transport, the reaction of Hb with numerous other cytosolic and protoplasmic factors complicates an adaptive interpretation of allosteric regulation at the subcellular level (Weber and Voelter, 2004). Chloride ions have been shown to play an allosteric role in the eel, *Gymnothorax unicolour*; Cl^- increases cooperativity thereby maximizing oxygen uptake and delivery over a precisely delineated change in PO_2 (Tamburrini *et al.*, 2001). The activity of water molecules in oxygen-linked allosteric regulation of Hb function is also widespread, and a characteristic of the anodic Hb components of the eel, *Anguilla anguilla*, and rainbow trout, *O. mykiss* (Hundahl *et al.*, 2003). Water decreases oxygen affinity by a different mechanism in the agnathan hagfish, *Myxine glutinosa*, by stabilizing the oligomeric state of Hb in the erythrocytes (Müller *et al.*, 2003). Both water and chloride ions are ubiquitous and hence unlikely to play an adaptive role in modulating oxygen affinity in response to hypoxic challenge.

8.3. Integrative Functions of the Erythrocyte

The fish erythrocyte is a functional unit of oxygen transport. Studies on whole blood or intact erythrocytes in relation to hypoxia are likely to be of higher ecophysiological relevance than those of the reduced elements of the oxygen transport system because the Hb is operating under realistic *in vivo* conditions. This is not to say, however, that the extensive literature on Hb structure-function is not critical to our interpretation of how the integrated system works. Nonetheless, it has proved far easier to collect and freeze blood samples from fish with interesting phylogenies living in interesting habitats, and to study the purified Hbs at a time convenient for the investigator. Recent whole blood studies are few and far between because the intact erythrocytes cannot be stored for more than a few hours without substantial degradation of erythrocyte metabolites and Hb integrity. Caldwell *et al.* (2006) found that erythrocytes from *O. mykiss* can be stored for up to 96 h with almost no effect on the magnitude of the erythrocyte β-adrenergic response.

Comparing the oxygen transport functions of purified lysates with whole blood from three phylogenetically distant fish living in the hypoxic billabongs of northern Australia revealed discrepancies that led to the conclusion that an adaptive interpretation of isolated Hbs was unwarranted (Wells *et al.*, 1997). The athletic, facultative air-breathing tarpon, *Megalops cyprinoides*, had the lowest whole blood-oxygen affinity and largest Bohr factor, whereas the water-breathing saratoga, *Scleropages jardinii*, had the highest affinity and smallest Bohr factor. However, in purified lysate, barramundi, *Lates calcarifer*, showed the highest affinity and saratoga the strongest Bohr effect. The respective erythrocyte environments therefore modify the effective properties of Hb that adapt these fish for hypoxic habitats.

In an integrated study, Yang *et al.* (1992) compared the suite of metabolic and respiratory adaptations in two scorpaenid fishes that differed in their depth distribution. *Scorpaena guttata* lives in relatively shallow water on the upper continental shelf, whereas *Sebastolobus alascanus* occurs at depths in excess of 1000 m in the oxygen minimum zone. The authors found that the species adapted to the hypoxic zone had a higher whole blood-oxygen affinity compared to its shallow water counterpart. Both species showed cooperative oxygen binding with similar Hill coefficients, and similar hematocrit values suggested that oxygen-carrying capacities were similar. These features suggest that the principal adaptive feature of the oxygen transport system is the P_{50}, and that this serves in *S. alascanus* to maintain full saturation in the gills when the fish is exposed to low environmental oxygen. Whether these functional differences are due to the intrinsic properties of the Hb or the erythrocyte ATP/GTP modulators has not yet been determined.

While regulation of blood-oxygen affinity in the face of hypoxia seems a sensible adaptation for fish living in habitats that become periodically depleted in oxygen, one would not expect to find similar hypoxic responses in inactive species living in a well-oxygenated environment. Curiously, the Antarctic fish, *Pagothenia borchgrevinki*, showed a robust acclimatory response to hypoxia that included increased oxygen-carrying capacity and increased whole blood-oxygen affinity arising from down-regulation of erythrocyte ATP production (Wells *et al.*, 1989). The authors suggested that the hypoxic response represents a generalized, phenotypic plasticity that is likely to be present in most fish species, rather than a specific adaptation to environmental hypoxia.

The facultative air-breathing tarpon, *Megalops cyprinoides*, and the salmon catfish, *Arius leptaspis*, both inhabit oxygen-poor billabongs. Comparison of whole blood oxygen binding in the two species showed that the former had a low affinity blood with marked Bohr effect and cooperativity, and allosteric modulation by ATP; by contrast, the catfish had high affinity blood, smaller Bohr effect and cooperativity, with regulation by GTP (Wells *et al.*, 2005). Thus, high affinity, reduced Bohr and Hill coefficients, and GTP modulation of Hb function seem adaptive for water-breathers living under chronic hypoxia.

Acclimation studies in the obligate air-breathing lungfish, *Protopterus amphibius*, showed a similar response. The oxygen affinity of whole blood increased sharply during aestivation in response to low oxygen in the near environment of the mud cocoon and regulation of affinity was effected by reduced erythrocyte GTP (Johansen *et al.*, 1976).

Initial expectations that air-breathing fishes would have higher blood oxygen affinities than water-breathers and in view of the frequently hypercapnic conditions of many oxygen-deficient habitats, reduced Bohr and Root effects have not been confirmed in a survey of the blood oxygen transport properties of many air- and water-breathing species (reviewed by Graham, 1997). There was some evidence to support the hypothesis when comparing closely related osteoglossid genera, but the correlation disappeared with phylogenetic distance. Nonetheless, a high oxygen-carrying capacity appears to be a feature of most air-breathing fishes (Graham, 1997).

Acclimation of the sedentary benthic turbot, *Scophthalamus maximus*, and the more active sea bass, *Dicentrarchus labrax*, to several levels of environmental hypoxia did not result in changes to blood-O_2 affinity; the species differences in O_2-binding properties under normoxia revealed a smaller Bohr factor and lower Hb concentration in the turbot, and these characteristics appear sufficient to deal with reduced O_2 loading in the gills while maintaining efficient transfer to tissues (Pichavant *et al.*, 2003).

Studying the mechanisms of hypoxia tolerance in rainbow trout, *O. mykiss*, Boutilier *et al.* (1988) showed that whereas the ATP:Hb ratio decreased upon hypoxic exposure, so too did the pH gradient across the erythrocyte membrane, so that the integrated effect was no change in blood-oxygen affinity. This study calls into question the assumed adaptive significance of allosteric compensation for hypoxia, since the presumed Bohr effect offsets the ATP effect. It also calls into question the relevance of interpreting physiological measurements on the basis of plasma pH when the erythrocytic proton–Hb relationship is unknown. Subsequent investigation into the transfer of oxygen from blood to myoglobin-rich red swimming muscle in *O. mykiss* supports the hypothesis that red muscle in teleost fishes leads to higher intramuscular PO_2 as a result of the sigmoidal oxygen equilibrium curve (McKenzie *et al.*, 2004). The authors came to this conclusion by exposing trout to mild environmental hypoxia (50% water saturation) and monitoring red muscle PO_2 using O_2-sensitive optical fiber sensors while exercising the fish. They further discounted a role for the Root effect Hb in oxygenating the muscle.

A recent study found unexpected differences in blood-oxygen affinity between four unrelated cold-temperate marine teleost fishes that appeared unrelated to the likelihood of functional hypoxia or to the predicted responses of each species to environmental hypoxia (Herbert *et al.*, 2006). The authors advised against overestimating the adaptive functional properties of Hb when comparing unrelated species.

Sturgeons are primitive, chondrostean fishes (Family Acipenseridae) that differ from teleost fishes in their hypoxic responses. Under the acute hypoxic challenge of either moderate (50% saturation) or severe (20%) environmental hypoxia, the typical teleostean response of increased oxygen-carrying capacity did not occur; hematocrits in this group under normoxic conditions were not notably high and compensation was respiratory and metabolic rather than hematological (Baker *et al.*, 2005). Sturgeons, however, despite their benthic cruising behavior, have much lower whole blood oxygen affinities than are typical of teleosts with robust hypoxic tolerances (Crocker and Cech, 1998), although the oxygen equilibrium properties of the anadromous green sturgeon, *Acipenser medirostris*, seem typical of fish adapted to environmental hypoxia: modest Bohr effect, low Hill coefficient (1.4–1.5), and GTP modulation (Kauffman *et al.*, 2007). Furthermore, oxygen binding is temperature sensitive with values of ΔH tending to become more positive with increasing temperature ($\Delta H = -34.2$ kJ mol^{-1} at 11°C; −6.7 at 24°C).

9. ROLE OF β-ADRENERGIC RECEPTORS IN ERYTHROCYTE OXYGEN TRANSFER

Exposure to low environmental oxygen or exercise challenge triggers the release into the circulation of a hormonal flush of catecholamines and corticosteroids (Randall and Perry, 1992; Gamperl *et al.*, 1994; Lowe and Wells, 1996). Although this is a typical vertebrate response to stress, teleost fishes are unusual in having erythrocyte surface receptors that bind adrenaline and noradrenaline (Nikinmaa and Heustis, 1984). These receptors are of ancient evolutionary origin and are thought to predate the fish–tetrapod division (Nikinmaa, 2003). In teleost fishes, β_3-adrenergic receptors control the function of a sodium-proton pump in the erythrocyte membrane so that intracellular protons are exchanged for extracellular sodium ions (Nickerson *et al.*, 2003). The effect of the Na^+/H^+ exchanger is twofold: first, the reduction of proton activity inside the erythrocyte raises the pH of the erythrocyte and thereby increases Hb-O_2 affinity via the Bohr effect; and second, the inward flow of sodium is balanced by a parallel water flux with the result that the erythrocyte swells (Nikinmaa, 2002). The low buffering capacity of lamprey and teleost fish Hbs, coupled with large Bohr and Haldane effects, means that the Na^+/H^+ exchanger plays a critical role in regulating blood-oxygen affinity in response to acute hypoxia (Nikinmaa, 1997, 2001). These reactions were first noted by Nikinmaa (1982, 1983) in erythrocytes from rainbow trout (*Salmo gairdneri*, now known as *Oncorhynchus mykiss*). Evidence for an adrenergic effect during functional or environmental hypoxia is most obviously seen from a sharp rise in hematocrit that cannot be explained by an increase in the numbers of erythrocytes released under concomitant adrenergic stimulation of the spleen (Wells and Weber, 1990). This has the additional effect of an increase in Hb-O_2 affinity due to dilution of the cell contents and dissociation of the phosphate–Hb complex. These processes can result in a rapid increase in blood-oxygen affinity, thus securing adequate oxygen-binding in the gills when the fish is subjected to hypoxia. Perry and Reid (1992) suggested that a decrease in arterial PO_2 below the P_{50} was required to trigger the catecholamine response.

Specific examples are the response to environmental hypoxia by rainbow trout, *O. mykiss* (Tetens and Christensen, 1987), and exercise-induced functional hypoxia in *O. mykiss* (Primmett *et al.*, 1986) and striped bass, *Morone saxatilis* (Nikinmaa *et al.*, 1984). This hypoxic response is also influenced by temperature because of the direct effect on metabolism, and the reduced availability of oxygen at warmer environmental temperatures. The seasonal response in Arctic charr, *Salvelinus alpinus*, is an up-regulation of the erythrocyte adrenergic system at low seasonal temperatures (Lecklin and Nikinmaa, 1999).

Jensen (2001) hypothesized that the low buffer values for Hb in teleosts might be a necessary prerequisite for the regulation of erythrocyte pH via the Na^+/H^+ exchanger. The observation that three species of tuna all showed low Hb-specific buffer values, despite the remarkable metabolic acidosis that develops during burst swimming, supported the hypothesis (Jensen, 2001). With plasma pH reduced by as much as 0.4 units, it might be supposed that the erythrocyte response to catecholamines would be greater than that in other less active species. Lowe *et al.* (1998) found, however, that the responses of two tuna species were similar to those in less active teleosts, and that adaptations to extreme functional hypoxia do not occur at the level of erythrocyte function.

Until recently, there was little evidence for activation of a Na^+/H^+ exchanger in response to functional hypoxia in sharks or other cartilaginous fishes. Brill *et al.* (2008), however, have shown that anaerobic exercise in the sandbar shark, *Carcharhinus plumbeus*, is not accompanied by a strongly decreased blood-O_2 affinity because the metabolic acidosis is compensated by alkalinization of the erythrocytes following activation of the Na^+/H^+ exchanger. Moreover, the significant increase in hematocrit reflected not only erythrocyte swelling, but a real increase in oxygen-carrying capacity.

The Na^+/H^+ exchanger is absent from the Osteoglossomorpha, representing primitive teleost fishes. β-Adrenergic stimulation of erythrocytes from the primitive agnathan fish, *Lampetra fluviatilis*, resulted in volume change via chloride channels rather than the Na^+/H^+ exchanger, which appears to be absent in this group (Nikinmaa *et al.*, 2001).

The swelling of erythrocytes under hypoxia might be expected to impede the flow of blood through the capillary circulation. This does not appear to be the case, however, and β-adrenergic stimulation of trout (*O. mykiss*) erythrocytes actually decreased the shear-dependence of blood viscosity (Wells *et al.*, 1991). That the adrenergic mechanism really does improve oxygen transport under functional hypoxia was demonstrated experimentally by testing visual function in the β-blocked trout, *O. mykiss* (Herbert and Wells, 2002). Retinal function is highly oxygen dependent, and in the absence of the adrenergic mechanism, severe visual impairment is likely when fish become hypoxic.

10. NOVEL MOLECULAR MECHANISMS FOR HYPOXIA PROTECTION

10.1. Putative Role of Hb-Nitric Oxide Binding

Much interest has followed the discovery that nitric oxide (NO) released from vascular endothelium may act as a paracrine vasodilator to relax vascular smooth muscle, allowing for improved blood flow (Jaffrey and Snyder, 1995). How widespread across the vertebrate groups this effect is has not yet been determined. Since NO is both short-lived and neurotoxic in excess, its production and distribution must be closely regulated. While NO-synthase plays an important role in its production, the observation that NO binds to the thiol groups of proteins (Jaffrey and Snyder, 1995) has invited speculation concerning a new role for Hb (Reischl *et al.*, 2007). Accordingly, Weber and Fago (2004) considered whether oxygenation-linked NO binding might form the basis of NO transport and release to induce more widespread vasodilation under conditions of tissue hypoxia.

This speculation assumes further significance for fish living in hypoxic freshwater habitats where nitrite levels are often high. Jensen (2003) suggested that Hb may play a role in nitrate reduction and production of NO thereby improving circulation under conditions of low environmental oxygen. Under hypoxia, the higher proportion of deoxyhaemoglobin promotes greater nitrite reductase activity, and hence NO production (Jensen, 2008). In fishes, the NO-mediated regulation of vascular dilation appears well developed in the branchial vasculatures of the eel, *Anguilla anguilla* (Pellegrino *et al.*, 2002), and the Atlantic salmon, *Salmo salar* (Ebbesson *et al.*, 2005). A role for Hb in NO regulation has also been suggested for the endothelial lining of the swimbladder, and in lungs of air-breathing fishes (Zaccone *et al.*, 2006). Surprisingly, the Antarctic icefish, *Chionodraco hamatus*, has a well-developed NO-synthase mechanism, but lacks Hb altogether (Pellegrino *et al.*, 2004; Amelio *et al.*, 2006). Sidell and O'Brien (2006) suggested that icefish have larger blood vessels as a result of NO not being scavenged by Hb. Thus, NO has both acute acclimatory and adaptational possibilities for regulation of oxygen transport.

Few physiological experiments have been undertaken to resolve the role of NO-Hb in hypoxic fishes. Swenson *et al.* (2005) noted a rapid up-regulation of NO production in hypoxic dogfish, *Squalus acanthias*, that appeared to promote vasodilation. The authors favored the view that Hb plays a role as a NO scavenger since the physiological effects of NO were most marked in Hb-free preparations. These observations in sharks are supported by those from a teleost fish; channel catfish, *Ictalurus puncatus*, when subjected to severe hypoxia showed an increase in nitrergic nerve fibres in the branchial region (Zaccone *et al.*, 2006).

It seems reasonable to conclude that hypoxia will up-regulate NO-synthase expression in fishes, allowing for NO-mediated vasodilation. In the meantime, Hb–NO interactions remain poorly understood (Fago *et al.*, 2003) and although current opinion supports the role of Hbs from fish and other animals in both releasing NO upon deoxygenation and in thiolated removal of NO, the results of new research in this field are eagerly awaited.

10.2. Post-Translational Modification of Hb Function

The possibility of nongenetic modifications to Hb function in fishes seems not to have been considered. Yet, there is growing evidence that this is a common phenomenon in avian and mammalian Hbs where glycosylation, glutathionylation, and deamidation of β-globins may result in altered functionality (Di Simplico *et al.*, 1996; Dafre and Reischl, 1998; Niwa *et al.*, 2000;

Henty *et al.*, 2007). Glutathionylation appears to occur in the scalloped hammerhead shark, *Sphyrna lewini* (Dafre and Reischl, 1997) but there is no evidence yet that these mechanisms could provide phenotypic compensation for hypoxia.

11. HYPOXIA INDUCIBLE FACTOR HIF-1α: EVIDENCE FOR ROLE IN HYPOXIC RESISTANCE

The search for adaptive responses to hypoxia through changes in gene expression has gathered considerable momentum in recent years. Oxygen sensing via hypoxia inducible factors (HIFs) and molecular responses to hypoxic challenge has been demonstrated throughout the animal kingdom (Hoogewijs *et al.*, 2007). Fish are at the forefront of this research because of comparative species differences in responses to hypoxia from fish inhabiting diverse aquatic habitats. The HIF-1α protein in fish is an important transcription factor that mediates a range of responses to hypoxia through the expression of genes controlling the oxygen transport system, and possible HIF targets are erythropoietin, globin synthesis, angiogenesis, and gill surface area (reviewed by Nikinmaa and Rees, 2005). Suppression of apoptosis and metabolic arrest in the hypoxia-resistant crucian carp, *Carassius carassius*, resulted in a significant increase in gill surface area when the fish was chronically exposed to hypoxia (Sollid *et al.*, 2006).

Both seasonal and latitudinal differences in water temperature have a significant effect on oxygen solubility with oxygen content being much reduced in warmer waters. The organismal response to increased temperature is generally a higher metabolic demand for oxygen. These two opposing factors interact to exacerbate internal hypoxia. Rissanen *et al.* (2006) have shown that temperature has a marked effect on HIF-1α expression in *C. carassius* suggesting possible adaptation to temperature-induced hypoxia. Recent studies with the polar zoarcids *Zoarces viviparous* and *Pachycara brachycephalum* provide further evidence for a transcriptional control mechanism for oxygen transport in temperature acclimation of extreme poikilotherms (Heise *et al.*, 2006, 2007). Oxidative defence mechanisms in both cold-adapted and acclimated fishes therefore appear to be mediated by HIF-1α expression.

A critical question remains as to whether sequence variation in the HIF-1α gene can be linked to intraspecific differences in hypoxia defence. Current research by Scandinavian investigators is attempting to address this question by comparing both phylogenetically similar species with different hypoxic tolerances, and unrelated species with similar oxygen requirements (Rytkönen *et al.*, 2007). The investigators are thus potentially able to distinguish adaptive from evolutionarily neutral changes. Adaptive responses of HIF-1α expression allow fish to delay the onset of metabolically inefficient anaerobiosis. It is not yet clear, however, at which point in the HIF-1α pathway species-specific differences occur.

12. CONCLUSIONS AND COMMENTARY

Environmental hypoxia is common in aquatic habitats at all latitudes, and under natural conditions is associated with increased carbon dioxide, ammonia, hydrogen sulfide, and reduced pH. Apart from the oxygen-minimum zone and areas around deep sea thermal vents, the marine environment is not generally associated with hypoxic habitats, yet even coral reef fish show significant mechanisms of hypoxia tolerance (Nilsson and Renshaw, 2004; Nilsson and Östlund-Nilsson, 2006). It is not often enough emphasized that the first responses of fish to environmental hypoxia are generally behavioral and include reduced locomotion, feeding, and reproductive activities, and, where possible, an attempt to seek out cooler water (Randall *et al.*, 2006). Species adapted to persistent environmental hypoxia often seem to have high Hb concentrations, as do populations compensating for reduced oxygen availability. To a large extent, this increased oxygen-carrying capacity compensates for the reduced turnover of oxygen to tissues caused by apparently adaptive increases in blood-oxygen affinity under hypoxia. Brauner and Wang (1997) calculated that increased oxygen-carrying capacity was significantly more beneficial to tissue oxygen delivery during environmental hypoxia than were changes in blood-oxygen affinity. There is, therefore, some uncertainty concerning the adaptive value of acclimatory shifts in Hb-O_2 affinity, particularly in the absence of more complete information on compensatory adjustments in the cardiovasular system and metabolic processes. In the meantime, it may be useful to evaluate "benefit" in the sense that an increase in Hb concentration in response to environmental hypoxia and strenuous exercise benefits perfusion limited situations, whereas allosteric effects modulating Hb function, and perhaps myoglobin, benefit diffusion limited gas exchange.

Adaptations to functional hypoxia are more obviously manifested through oxygen transport characteristics. The role of the spleen as a reservoir of erythrocytes in raising oxygen-carrying capacity during exercise is well understood, as are the advantages of low blood-O_2 affinity favoring unloading to tissues, a robust Bohr factor permitting oxygen turnover in response to demand, and highly cooperative Hb allowing for rapid loading and unloading of oxygen and carbon dioxide over comparatively narrow pressure gradients.

Oxygen transport systems appear geared toward maintaining the oxygen cascade from the gas exchangers facing the environment, to the cytochrome oxidase enzymes in the mitochondria of working tissues where oxygen pressures may be <1 mmHg. Accordingly, much of our adaptive interpretation of blood oxygen transport has been toward optimizing this gradient. It seems counterintuitive then that the evolution of an oxygen transport system may not solely be directed toward maximizing oxygen supply. There is growing evidence that a major constraint is the requirement to protect internal tissues against reactive oxygen species, especially in high performance fish where mitochondrial densities in the red swimming muscles are among the highest in vertebrates, and antioxidant protection is essential to prevent oxidative damage (Wilhelm Filho, 2007). Furthermore, the cascade should be interpreted against a historical background of extensive fluctuations in atmospheric oxygen levels over the Phanaerozoic era, when bimodal breathing became common (Berner *et al.*, 2007; Flück *et al.*, 2007). Fish have a long evolutionary history in coping with fluctuations in oxygen availability, and have emerged as the most functionally diverse vertebrate group. In present day habitats, hypoxia is very common and an important determinant of species distribution.

Recent research has reinterpreted the role of Hb as a cellular oxygen sensor inducing a cascade of changes in response to hypoxia (Wu, 2002). These include transcription of factors such as HIF-1α, glycolytic and phosphorylation pathways, and globin synthesis. The role of Hb as an oxygen-sensing mechanism linking K^+ flux in erythrocytes of trout, *O. mykiss*, has also been proposed (Berenbrink *et al.*, 2000). The interaction of deoxygenated Hb with the cytoplasmic domains of Band 3 erythrocyte membrane proteins suggests further modulation of several metabolic functions, including glycolysis, the pentose phosphate pathway, and ion exchanges (Weber *et al.*, 2004). The importance of the Band 3 protein in fish is not well characterized, given the lesser role of glycolysis in nucleated fish erythrocytes, but could play a part in hypoxia acclimation in air-breathing fish such as armoured catfish, *Pteroglopichthys* spp., and lungfish, *Protopterus* spp., which possess the glycolytic O_2 affinity modulators DPG and IP, respectively (Val, 2000).

Much of what we know about functional adaptations of the oxygen transport system to hypoxia in fishes has not been the result of acclimation or field experiments, but has been inferred from laboratory-based studies on isolated Hbs and erythrocytes. Despite our detailed understanding of the molecular basis of functional adaptations in fish Hbs (see Weber and Fago, 2004), the physiological significance of Hb function at the whole organism level remains less well understood. Considerable plasticity has been shown in the respiratory system in response to changes in oxygen supply (environmental hypoxia) and demand (functional hypoxia). As pointed out by Bavis *et al.* (2007) it is difficult to predict the responses of an animal based on the plasticity of a single system. This lesson is also important in interpreting the oxygen transport system since that in turn is comprised of several different elements. Nonetheless, it seems that contributing to respiratory plasticity are: adjustments to oxygen-carrying capacity, Hb isoforms, modulation of Hb-oxygen affinity, regulation of the internal erythrocyte environment, and expression of hypoxia inducible factors. Results from expression profiling in the long-jaw mudsucker, *Gillichthys mirabilis*, are likely to be typical of most fish that are routinely exposed to hypoxia (see Nikinmaa and Rees, 2005), and showed that a large number of genes are both induced and suppressed during hypoxic exposure, revealing the different roles of specific tissues during hypoxia (Gracey *et al.*, 2001). Changes in hypoxic response during development are less well understood. Exposure of zebrafish larvae (*Danio rerio*) to a low oxygen environment during development resulted in the stimulation of convective oxygen transport (Jacob *et al.*, 2002), but the capacity for the Hb system to adapt is unknown.

Although continued whole-organism research into how fish cope with demand for oxygen under restricted supply is likely to remain a productive area, differences in habitat and activity level need to be related to hypoxia-related gene expression in order to more fully understand the molecular basis for adaptation to oxygen fluctuations. Emerging research on hypoxia inducible factors, the role of NO, and erythrocyte surface receptors emphasizes the integrative approach to understanding hypoxic responses at the organismal level.

ACKNOWLEDGMENTS

The author wishes to thank Thomas Brittain for useful discussion.

REFERENCES

Abe, H., Dobson, G. P., Hoeger, U., and Parkhouse, W. S. (1985). Role of histidine-related compounds to intracellular buffering in fish skeletal muscle. *Am. J. Physiol.* **249,** R449–R454.

Affonso, E. G., Polez, V. L. P., Corrêa, C. F., Mazon, A. F., Araujo, M. R. R., Moraes, G., and Rantin, F. T. (2002). Blood parameters and metabolites in the teleost fish *Colossoma macropomum* exposed to sulfide or hypoxia. *Comp. Biochem. Physiol.* **133C,** 375–382.

Albers, C., Manz, R., Muster, D., and Hughes, G. M. (1983). Effect of acclimation temperature on oxygen transport in the blood of the carp, *Cyprinus carpio*. *Resp. Physiol.* **52,** 165–179.

Altringham, J. D., and Block, B. A. (1997). Why do tuna maintain elevated slow muscle temperatures? Power output of muscle isolated from endothermic and ectothermic fish. *J. Exp. Biol.* **200,** 2617–2627.

Amelio, D., Garofalo, F., Pellegrino, D., Giordano, F., Tota, B., and Cerra, M. C. (2006). Cardiac expression and distribution of nitric oxide synthases in the ventricle of the cold-adapted Antarctic teleosts, the hemoglobinless *Chionodraca hamatus* and the red-blooded *Trematomus bernacchii*. *Nitric Oxide* **15,** 190–198.

Andersen, N. A., Laursen, J. S., and Lykkeboe, G. (1985). Seasonal variations in hematocrit, red cell hemoglobin and nucleoside triphosphate concentrations, in the European eel *Anguilla anguilla*. *Comp. Biochem. Physiol.* **81A,** 87–92.

Bailey, J. R., and Driedzic, W. R. (1986). Function of myoglobin in oxygen consumption by isolated perfused fish hearts. *Am. J. Physiol. Regul. Itegr. Comp. Physiol.* **251,** R1144–R1150.

Baker, D. W., Wood, A. M., and Kieffer, J. D. (2005). Juvenile Atlantic and shortnose sturgeons (Family: Acipenseridae) have different hematological responses to acute environmental hypoxia. *Physiol. Biochem. Zool.* **78,** 916–925.

Baldwin, J., and Wells, R. M. G. (1990). Oxygen transport potential in tropical elasmobranchs from the Great Barrier Reef: Relationship between haematology and blood viscosity. *J. Exp. Mar. Biol. Ecol.* **144,** 145–155.

Bavis, R. W., Powell, F. L., Bradford, A., Hsia, C. C. W., Peltonen, J. E., Soliz, J., Zeis, B., Fergusson, E. K., Fu, Z., Gassmann, M., Kim, C. B., Maurer, J., *et al.* (2007). Respiratory plasticity in response to changes in oxygen supply and demand. *Integr. Comp. Biol.* **47,** 532–551.

Berenbrink, M., Völkel, S., Heisler, N., and Nikinmaa, M. (2000). O_2-dependent K^+ fluxes in trout red blood cells: The nature of O_2 sensing revealed by the O_2 affinity, cooperativity and pH dependence of transport. *J. Physiol.* **526,** 69–80.

Berenbrink, M., Koldkjær, P., Kepp, O., and Cousins, A. R. (2005). Evolution of oxygen secretion in fishes and the emergence of a complex physiological system. *Science* **307,** 1752–1757.

Berner, R. A. (2006). GEOCARBSULF: A combined model for Phanaerozoic atmospheric O_2 and CO_2. *Geochim. Cosmochim. Ac.* **70,** 5653–5664.

Berner, R. A., VandenBrooks, J. M., and Ward, P. D. (2007). Oxygen and evolution. *Science* **316,** 557–558.

Block, B. A., and Carey, F. G. (1985). Regulation of brain and eye temperatures of the bluefin tuna. *Comp. Biochem. Physiol.* **43A,** 425–433.

Block, B. A., Dewar, H., Blackwell, S. B., Williams, T. D., Prince, E. D., Farwell, C. J., Boustany, E., Teo, S. L. H., Seitz, A., Walli, A., and Fudge, D. (2001). Migratory movements, depth preferences, and thermal biology of Atlantic bluefin tuna. *Science* **293,** 1310–1314.

Bonaventura, C., Crumbliss, A. L., and Weber, R. E. (2004). New insights into the proton-dependent oxygen affinity of Root effect haemoglobins. *Acta Physiol. Scand.* **182,** 245–258.

Boutilier, R. G., Dobson, G., Hoeger, U., and Randall, D. J. (1988). Acute exposure to graded levels of hypoxia in rainbow trout (*Salmo gairdneri*): Metabolic and respiratory adaptations. *Resp. Physiol.* **71,** 69–82.

Brauner, C. J., and Wang, T. (1997). The optimal oxygen equilibrium curve: A comparison between environmental hypoxia and anaemia. *Amer. Zool.* **37,** 101–108.

Brauner, C. J., and Weber, R. E. (1998). Hydrogen ion titrations of the anodic and cathodic haemoglobin components of the European eel *Anguilla anguilla*. *J. Exp. Biol.* **201,** 2507–2514.

Brauner, C. J., and Val, A. L. (2006). Oxygen transfer. *In* "The Physiology of Tropical Fishes" (Val, A. L., de Almeida-Val, V. M. F., and Randall, D. J., Eds.), "Fish Physiology" Vol. 21 (Hoar, W. S., Randall, D. J., and Farrell, A. P., series Eds.). pp. 277–306. Academic Press, London.

Brill, R. W., and Bushnell, P. G. (2006). Effects of open- and closed-system temperature changes on blood O_2-binding characteristics of Atlantic bluefin tuna (*Thunnus thynnus*). *Fish Physiol. Biochem.* **32,** 283–294.

Brill, R., Bushnell, P., Schroff, S., Seifert, M., and Galvin, M. (2008). Effects of anaerobic exercise accompanying catch-and-release fishing on blood-oxygen affinity of the sandbar shark (*Carcharhinus plumbeus*, Nardo). *J. Exp. Mar. Biol. Ecol.* **354,** 132–143.

Brittain, T. (2005). Root effect hemoglobins. *J. Inorg. Biochem.* **99,** 120–129.

Brittain, T. (2008). Extreme pH sensitivity in the binding of oxygen to some fish hemoglobins in the Root effect. *In* "The Smallest Biomolecules. Diatomics and their Interactions with Heme Proteins" (Ghosh, A, Ed.), pp. 219–234. Elsevier, UK.

Brix, O., Clements, K. D., and Wells, R. M. G. (1998a). An ecophysiological interpretation of hemoglobin multiplicity in three herbivorous marine teleost species from New Zealand. *Comp. Biochem. Physiol.* **121A,** 189–195.

Brix, O., Forås, E., and Strand, I. (1998b). Genetic variation and functional properties of Atlantic cod haemoglobins: Introducing a modified tonometric method for studying fragile haemoglobins. *Comp. Biochem. Physiol.* **119A,** 575–583.

Brix, O., Clements, K. D., and Wells, R. M. G. (1999). Haemoglobin components and oxygen transport in relation to habitat distribution in triplefin fishes (Tripterygiidae). *J. Comp. Physiol.* **169B,** 329–334.

Burmester, T., Weich, B., Reinhardt, S., and Hankeln, T. (2000). A vertebrate globin expressed in the brain. *Nature* **407,** 520–523.

Burmester, T., Ebner, B., Weich, B., and Hankeln, T. (2002). Cytoglobin: A novel globin type ubiquitously expressed in vertebrate tissues. *Mol. Biol. E* **19,** 416–421.

Caldwell, S., Rummer, J. L., and Brauner, C. J. (2006). Blood sampling techniques and storage duration: Effects on the presence and magnitude of the red blood cell β-adrenergic response in rainbow trout (*Oncorhynchus mykiss*). *Comp. Biochem. Physiol.* **144A,** 188–195.

Campo, S., Nastasi, G., D'Ascola, A., Campo, G. M., Avenoso, A., Traina, P., Calatroni, A., Burrascano, E., Ferlazzo, A., Lupidi, G., Gabbianelli, R., and Falcioni, G. (2008). Hemoglobin system of *Sparus aurata*: Changes in fishes farmed under extreme conditions. *Sci. Tot. Environ.* **403,** 148–153.

Canty, A. A., and Driedzic, W. R. (1987). Evidence that myoglobin does not support heart performance at maximal levels of oxygen demand. *J. Exp. Biol.* **128,** 469–473.

Carter, A. J., and Wilson, R. S. (2006). Improving sneaky-sex in a low oxygen environment: Reproductive and physiological responses of male mosquito fish to chronic hypoxia. *J. Exp. Biol.* **209,** 4878–4884.

Cech, J. J., Laurs, R. M., and Graham, J. B. (1984). Temperature-induced changes in blood gas equilibria in the albacore, *Thunnus alalunga*, a warm-bodied tuna. *J. Exp. Biol.* **109,** 21–34.

Chapman, L. J., Chapman, C. A., Nordlie, F. G., and Rosenberger, A. E. (2002). Physiological refugia: Hypoxia tolerance and maintenance of fish diversity in the Lake Victoria region. *Comp. Biochem. Physiol.* **133A,** 421–437.

Clack, J. A. (2007). Devonian climate change, breathing, and the origin of the tetrapod stem group. *Integ. Comp. Biol.* **47,** 510–523.

Clark, T. D., Seymour, R. S., Wells, R. M. G., and Frappell, P. B. (2008). Thermal effects on the blood respiratory properties of southern bluefin tuna, *Thunnus maccoyii. Comp. Biochem. Physiol.* **150A,** 239–246.

Cocca, E., Detrich, H. W., Parker, S. K., and di Prisco, G. (2000). A cluster of four globin genes from the Antarctic fish *Notothenia coriiceps*. *J. Fish Biol.* **57**(Suppl. A), 33–50.

Crocker, C. E., and Cech, J. J. (1998). Effects of hypercapnia on blood-gas and acid-base states in the white sturgeon, *Acipenser transmontanus*. *J. Comp. Physiol.* **198B,** 50–60.

D'Avino, R., and di Prisco, G. (1997). The hemoglobin system of Antarctic and non-Antarctic notothenioid fishes. *Comp. Biochem. Physiol.* **118A,** 1045–1049.

Dafre, A. L., and Reischl, E. (1997). Asymmetric hemoglobins, their thiol content, and blood glutathione of the scalloped hammerhead shark, *Sphyrna lewini. Comp. Biochem. Physiol.* **116B,** 323–331.

Dafre, A. L., and Reischl, E. (1998). Oxidative stress causes intracellular reversible *S*-thiolation of chicken haemoglobin under diamide and xanthine oxidase treatment. *Arch. Biochem. Biophys.* **358,** 291–296.

Davie, P. S., Wells, R. M. G., and Tetens, V. (1986). Effects of sustained swimming on rainbow trout muscle structure, blood oxygen transport, and lactate dehydrogenase isozymes: Evidence for increased aerobic capacity of white muscle. *J. Exp. Zool.* **237,** 159–171.

Deker, H., and Nadja, H. (2007). Negative cooperativity in Root effect hemoglobins: Role of heterogeneity. *Integr. Comp. Biol.* **47,** 656–661.

de Souza, P. C., and Bonilla-Rodriguez, G. O. (2007). Fish hemoglobins. *Braz. J. Med. Biol. Res.* **40,** 769–778.

Dickson, K. A., and Somero, G. N. (1987). Partial characterization of the buffering components of the red and white myotomal muscle of marine teleosts, with special emphasis on scombrid fishes. *Physiol. Zool.* **60,** 699–706.

di Prisco, G., and Tamburrini, M. (1992). The hemoglobins of marine and freshwater fish: The search for correlations with physiological adaptation. *Comp. Biochem. Physiol.* **102B,** 661–671.

di Prisco, G., D'Avino, R., Caruso, C., Tamburrini, M., Carmadella, L., Rutigliano, B., Carratore, V., and Romano, M. (1991). The biochemistry of oxygen transport in red-blooded Antarctic fish. *In* "Biology of Antarctic Fish" (di Prisco, G., Maresca, B., and Tota, B., Eds.), pp. 263–281. Springer-Verlag, Berlin.

di Prisco, G., Tamburrini, M., and D'Avino, R. (1998). Oxygen transport systems in extreme environments: Multiplicity and structure-function relationship in haemoglobins of Antarctic fish. *In* "Cold Ocean Physiology" (Pörtner, H. O., and Playle, R. C., Eds.). Society for Experimental Biology, Series 66. pp. 143–165. Cambridge University Press, Cambridge.

Di Simplico, P., Lupis, E., and Rossi, R. (1996). Different mechanisms of glutathione-protein mixed disulphides of diamide and *tert*-butyl hydroperoxide in rat blood. *Biochim. Biophys. Acta* **1289,** 252–260.

Ebbesson, L. O. E., Tipsmark, C. K., Holmqvist, B., Nilsen, T., Andersson, E., Stefansson, S. O., and Madsen, S. S. (2005). Nitric oxide synthase in the gill of Atlantic salmon: Colocalization with and inhibition of Na^+, K^+-ATPase. *J. Exp. Biol.* **208,** 1011–1017.

Eddy, F. B., Lomholt, J. P., Weber, R. E., and Johansen, K. (1977). Blood respiratory properties of rainbow trout (*Salmo gairdneri*) kept in water of high CO_2 tension. *J. Exp. Biol.* **67,** 37–47.

Fago, A., D'Avino, R., and di Prisco, G. (1992). The hemoglobins of *Notothenia angustata,* a temperate fish belonging to a family largely endemic to the Antarctic Ocean. *Eur. J. Biochem.* **210,** 963–970.

Fago, A., Carratore, V., di Prisco, G., Feurlein, R. J., Sottrup-Jensen, L., and Weber, R. E. (1995). The cathodic hemoglobin of *Anguilla anguilla.* Amino acid sequence and oxygen equilibria of a reverse Bohr effect hemoglobin with high oxygen affinity and high phosphate sensitivity. *J. Biol. Chem.* **270,** 18897–18902.

Fago, A., Wells, R. M. G., and Weber, R. E. (1997). Temperature-dependent enthalpy of oxygenation in Antarctic fish hemoglobins. *Comp. Biochem. Physiol.* **118B,** 319–326.

Fago, A., Forest, E., and Weber, R. E. (2002). Hemoglobin and subunit multiplicity in the rainbow trout (*Oncorhynchus mykiss*) haemoglobin system. *Fish Physiol. Biochem.* **24,** 335–342.

Fago, A., Crumbliss, A. L., Peterson, J., Pearce, L. L., and Bonaventura, C. (2003). The case of the missing NO-hemoglobin: Spectral changes suggestive of heme redox reactions reflect changes in NO-heme geometry. *Proc. Natl Acad. Sci. USA* **100,** 12087–12092.

Flück, M., Webster, K. D., Graham, J., Giomi, F., Gerlach, F., and Schmitz, A. (2007). Coping with cyclic oxygen availability: Evolutionary aspects. *Integr. Comp. Biol.* **47,** 524–531.

Fraser, J., de Mello, L. V., Ward, D., Rees, H. H., Williams, D. R., Fang, Y., Brass, A., Gracey, A. Y., and Cossins, A. R. (2006). Hypoxia-inducible myoglobin expression in nonmuscle tissues. *PNAS* **103,** 2977–2981.

Frey, B. J., Weber, R. E., van Aardt, W. J., and Fago, A. (1998). The haemoglobin system of the mudfish, *Labeo capensis*: Adaptations to temperature and hypoxia. *Comp. Biochem. Physiol.* **120B,** 735–742.

Fritsches, K. A., Brill, R. W., and Warrant, E. J. (2005). Warm eyes provide superior vision in swordfishes. *Current Biol.* **15,** 55–58.

Fuchs, C., Heib, V., Kiger, L., Haberkamp, M., Roesner, A., Schmidt, M., Hamdane, D., Marden, M. C., Hankeln, T., and Burmester, T. (2004). Zebrafish reveals different and conserved features of vertebrate neuroglobin gene structure, expression pattern, and ligand binding. *J. Biol. Chem.* **279,** 24110–24112.

Fyhn, U. E. H., Brix, O., Nævdal, G., and Johansen, T. (1994). New variants of the haemoglobins of Atlantic cod: A tool for discriminating between coastal and Arctic cod populations. *ICES Mar. Sci. Symp.* **198,** 666–670.

Galderisi, U., Fucci, L., and Geraci, G. (1996). Multiple hemoglobins in the electric ray: *Torpedo marmorata*. *Comp. Biochem. Physiol.* **113B,** 645–651.

Gamperl, A. K., Vijayan, M. M., and Boutilier, R. G. (1994). Experimental control of stress hormone levels in fish: Techniques and applications. *Rev. Fish Biol. Fisheries* **4,** 215–255.

Giardina, B., Mosca, D., and De Rosa, M. C. (2004). The Bohr effect of haemoglobin in vertebrates: An example of molecular adaptation to different physiological requirements. *Acta Physiol. Scand.* **182,** 229–244.

Giles, M. A. (1991). Strain differences in hemoglobin polymorphism, oxygen consumption, and blood oxygen equilibria in three hatchery broodstocks of Arctic charr, *Salvelinus alpinus*. *Fish Physiol. Biochem.* **9,** 291–301.

Gracey, A. Y., Troll, J. V., and Somero, G. N. (2001). Hypoxia-induced gene expression profiling in the euryoxic fish *Gillichthys mirabilis*. *PNAS* **98,** 1993–1998.

Graham, J. B. (1997). "Air-Breathing Fishes: Evolution, Diversity, and Adaptation." Academic Press, London.

Grigg, G. C. (1969). Temperature-induced changes in the oxygen equilibrium curve of the blood of the brown bullhead, *Ictalurus nebulosus*. *Comp. Biochem. Physiol.* **28,** 1203–1223.

Grove, T. J., and Sidell, B. D. (2002). Myoglobin deficiency in the hearts of phylogenetically diverse temperate-zone fish species. *Can. J. Zool.* **80,** 893–901.

Heise, K., Puntarulo, S., Nikinmaa, M., Lucassen, M., Pörtner, H.-O., and Abele, D. (2006). Oxidative stress and HIF-1 DNA binding during stressful cold exposure and recovery in the North Sea eelpout (*Zoarces viviparus*). *Comp. Biochem. Physiol.* **143A,** 494–503.

Heise, K., Estevez, M. S., Puntarulo, S., Galleano, M., Nikinmaa, M., Pörtner, H. O., and Abele, D. (2007). Effects of seasonal and latitudinal cold on oxidative stress parameters and activation of hypoxia inducible factor (HIF-1) in zoarcid fish. *J. Comp. Physiol.* **177B,** 765–777.

Henty, K., Wells, R. M. G., and Brittain, T. (2007). Characterization of the hemoglobins of the adult brushtailed possum, *Trichosurus vulpecula* (Kerr) reveals non-genetic heterogeneity. *Comp. Biochem. Physiol.* **148A,** 498–503.

Herbert, N. A., and Wells, R. M. G. (2002). The effect of strenuous exercise and β-adrenergic blockade on the visual performance of juvenile rainbow trout, *Oncorhynchus mykiss*. *J. Comp. Physiol.* **172B,** 725–731.

Herbert, N. A., Wells, R. M. G., and Baldwin, J. (2002). Correlates of choroid rete development with the metabolic potential of various tropical reef fish and the effect of strenuous exercise on visual performance. *J. Exp. Mar. Biol. Ecol.* **275,** 31–46.

Herbert, N. A., Skov, P. V., Wells, R. M. G., and Steffensen, J. F. (2006). Whole blood-oxygen binding properties of four cold-temperate marine fishes: Blood affinity is independent of pH-dependent binding, routine swimming performance, and environmental hypoxia. *Physiol. Biochem. Zool.* **79,** 909–918.

Hoogewijs, D., Terwilliger, N. B., Webster, K. A., Powell-Coffman, J. A., Tokishita, S., Yamagata, H., Hankeln, T., Burmester, T., Rytkönen, K. T., Nikinmaa, M., Abele, D., Heise, K., Lucassen, M., Fandrey, J., Maxwell, P. H., Påhlman, S., and Gorr, T. A. (2007). From critters to cancers: Bridging comparative and clinical research on oxygen sensing, HIF signalling, and adaptations towards hypoxia. *Integr. Comp. Biol.* **47,** 552–577.

Houston, A. H., and Gingras-Bedard, J. H. (1994). Variable versus constant temperature acclimation regimes: Effects on hemoglobin isomorph profile in goldfish, *Carassius auratus*. *Fish Physiol. Biochem.* **13,** 445–450.

Hundahl, C., Fago, A., Malte, H., and Weber, R. E. (2003). Allosteric effect of water in fish and human hemoglobins. *J. Biol. Chem.* **278,** 42769–42773.

Hundahl, C., Fago, A., Dewilde, S., Moens, L., Hankeln, T., Burmester, T., and Weber, R. E. (2006). Oxygen binding properties of non-mammalian nerve globins. *FEBS Journal* **273,** 1323–1329.

Hundahl, C., Stoltenberg, M., Fago, A., Weber, R. E., Dewilde, S., Fordel, E., and Danscher, G. (2005). Effects of short-term hypoxia on neuroglobin levels and localization in mouse brain tissues. *Neuropathol. Appl. Neurobiol.* **31,** 610–617.

Ikeda-Saito, M., Yonetani, T., and Gibson, Q. H. (1983). Oxygen equilibrium studies on hemoglobin from the bluefin tuna (*Thunnus thynnus*). *J. Mol. Biol.* **168,** 673–686.

Imsland, A. K., Foss, A., Stefansson, S. O., and Nævdal, G. (2000). Hemoglobin genotypes of turbot (*Scophthalamus maximus*): Consequences for growth and variations in optimal temperature for growth. *Fish Physiol. Biochem.* **23,** 75–81.

Jacob, E., Drexel, M., Schwerte, T., and Pelster, B. (2002). Influence of hypoxia and hypoxemia on the development of cardiac activity in zebrafish larvae. *Amer. J. Physiol.* **283,** R911–R917.

Jaffrey, S. R., and Snyder, S. H. (1995). Nitric oxide: A neural messenger. *Annu. Rev. Cell Dev. Biol.* **11,** 417–440.

Jensen, F. B. (1989). Hydrogen ion equilibria in fish haemoglobins. *J. Exp. Biol.* **143,** 225–234.

Jensen, F. B. (2001). Hydrogen ion binding properties of tuna haemoglobins. *Comp. Biochem. Physiol.* **129A,** 511–517.

Jensen, F. B. (2003). Nitrite disrupts multiple physiological functions in aquatic animals. *Comp. Biochem. Physiol.* **135A,** 9–24.

Jensen, F. B. (2004). Red blood cell pH, the Bohr effect, and other oxygenation-linked phenomena in blood O_2 and CO_2 transport. *Acta Physiol. Scand.* **182,** 215–227.

Jensen, F. B. (2008). Nitric oxide formation from the reaction of nitrite with carp and rabbit haemoglobin at intermediate oxygen saturations. *FEBS Journal* **275,** 3375–3387.

Jensen, F. B., and Weber, R. E. (1982). Respiratory properties of tench blood and hemoglobin adaptations to hypoxic-hypercapnic water. *Mol. Physiol.* **2,** 235–250.

Jensen, F. B., Nikinmaa, M., and Weber, R. E. (1993). Environmental perturbations of oxygen transport in teleost fishes: Causes, consequences, and compensations. *In* "Fish Ecophysiology" (Rankin, J. C., and Jensen, F. B., Eds.), pp. 161–179. Chapman and Hall, London.

Jensen, F. B., Fago, A., and Weber, R. E. (1998). Hemoglobin structure and function. *In* "Fish Physiology Vol. 17" (Perry, S. F., and Tufts, B. L., Eds.), pp. 1–40. Academic Press, San Diego.

Johansen, K., Lykkeboe, G., Weber, R. E., and Maloiy, G. M. O. (1976). Respiratory properties of blood in awake and estivating lungfish, *Protopterus amphibius*. *Resp. Physiol.* **27,** 335–345.

Johansson, D., Nilsson, G. E., and Døvink, B. (1997). Anoxic depression of light-evoked potentials in retina and optic tectum of crucian carp. *Neurosci. Letts* **237,** 73–76.

Jones, D. R., Brill, R. W., and Mense, D. C. (1986). The influence of blood gas properties on gas tensions and pH of ventral and dorsal aortic blood in free-swimming tuna, *Euthynnus affinis*. *J. Exp. Biol.* **120,** 201–213.

Kauffman, R. C., Houck, A. G., and Cech, J. J. (2007). Effects of temperature and carbon dioxide on green sturgeon blood-oxygen equilibrium. *Environ. Biol. Fish* **79,** 201–210.

Kunzmann, A. (1991). Blood physiology and ecological consequences in Weddell Sea fishes (Antarctica). *Berichte zur Polarforschung* **91,** 1–79.

Lai, J. C. C., Kakuta, I., Mok, H. O. L., Rummer, J. L., and Randall, D. (2006). Effects of moderate and substantial hypoxia on erythropoietin levels in rainbow trout kidney and spleen. *J. Exp. Biol.* **209,** 2734–2738.

Larsen, C., Malte, H., and Weber, R. E. (2003). ATP-induced reverse temperature effect in isohemoglobins from the endothermic porbeagle shark (*Lamna nasus*). *J. Biol. Chem.* **278,** 30741–30747.

Lecklin, T., and Nikinmaa, M. (1999). Seasonal and temperature effects on the adrenergic responses of Arctic charr (*Salvelinus alpinus*) erythrocytes. *J. Exp. Biol.* **202,** 2233–2238.

Lowe, T. E., and Wells, R. M. G. (1996). Primary and secondary stress responses to line capture in the blue mao mao. *J. Fish Biol.* **49,** 287–300.

Lowe, T. E., Brill, R. W., and Cousins, K. L. (1998). Responses of the red blood cells from two high-energy-demand teleosts, yellowfin tuna (*Thunnus albacares*) and skipjack tuna (*Katsuwonus pelamis*), to catecholamines. *J. Comp. Physiol.* **168B,** 405–418.

Lowe, T. E., Brill, R. W., and Cousins, K. L. (2000). Blood oxygen-binding characteristics of bigeye tuna (*Thunnus obesus*), a high-energy-demand teleost that is tolerant of low oxygen. *Mar. Biol.* **136,** 1087–1098.

Luk'yanenko, V. I. (1978). Ecological peculiarities of hemoglobinograms of three species of sturgeons. *Zh. Evol. Biokhim. Fiziol.* **14,** 347–350.

Lurman, G. J., Koschnick, N., Pörtner, H-O., and Lucassen, M. (2007). Molecular characterisation and expression of Atlantic cod (*Gadus morhua*) myoglobin from two populations held at two different acclimation temperatures. *Comp. Biochem. Physiol.* **148A,** 681–689.

Marcinek, B. J., Bonaventura, J., Wittenberg, J. B., and Block, B. A. (2001). Oxygen affinity and amino acid sequence of myoglobins from endothermic and ectothermic fish. *Am. J. Physiol.* **280,** R1123–R1133.

Marinakis, P., Tamburrini, M., Carratore, V., and di Prisco, G. (2003). Unique features of the haemoglobin system of the Antarctic notothenioid fish *Gobionotothen gibberifrons*. *Eur. J. Biochem.* **270,** 3981–3987.

Marinsky, C. A., Houston, A. H., and Murad, A. (1990). Effect of hypoxia on haemoglobin isomorph abundances in rainbow trout, *Salmo gairdneri*. *Can. J. Zool.* **68,** 884–888.

McKenzie, D. J., Wong, S., Randall, D. J., Egginton, S., Taylor, E. W., and Farrell, A. P. (2004). The effects of sustained exercise and hypoxia upon oxygen tensions in the red muscle of rainbow trout. *J. Exp. Biol.* **207,** 3629–3637.

Müller, G., Fago, A., and Weber, R. E. (2003). Water regulates oxygen binding in hagfish (*Myxine glutinosa*) haemoglobin. *J. Exp. Biol.* **206,** 1389–1395.

Nagai, K., Perutz, M. F., and Poyart, C. (1985). Oxygen binding properties of human mutant hemoglobins synthesized in *Escherichia coli*. *Proc. Natl Acad. Sci.* USA **82,** 7252–7255.

Nichols, J. W., and Weber, L. J. (1989). Comparative oxygen affinity of fish and mammalian myoglobins. *J. Comp. Physiol.* **159B,** 205–209.

Nickerson, J. G., Dugan, S. G., Drouin, G., Perry, S. F., and Moon, T. M. (2003). Activity of the unique β-adrenergic Na^+/H^+ exchanger in trout erythrocytes is controlled by a novel β_3-AR subtype. *Am. J. Physiol.* **285,** R562–R535.

Nikinmaa, M. (1982). Effects of adrenaline on red cell volume and concentration gradient of protons across the red cell membrane in the rainbow trout, *Salmo gardneri*. *Mol. Physiol.* **2,** 287–297.

Nikinmaa, M. (1983). Adrenergic regulation of haemoglobin oxygen affinity in rainbow trout cells. *J. Comp. Physiol.* **152B,** 67–72.

Nikinmaa, M. (1997). Oxygen and carbon dioxide transport in vertebrate erythrocytes: An evolutionary change in the role of membrane transport. *J. Exp. Biol.* **200,** 369–380.

Nikinmaa, M. (2001). Haemoglobin function in vertebrates: Evolutionary changes in cellular regulation in hypoxia. *Respir. Physiol.* **128,** 317–329.

Nikinmaa, M. (2002). Oxygen-dependent cellular functions – why fish and their aquatic environment are a prime choice of study. *Comp. Biochem. Physiol.* **133A,** 1–16.

Nikinmaa, M. (2003). β_3-Adrenergic receptors – studies on rainbow trout reveal ancient evolutionary origins and functions distinct from the thermogenic response. *Am. J. Physiol.* **285,** R515–R516.

Nikinmaa, M. (2004). The Bohr effect – a discovery 100 years ago, with intensive studies about the effect of protons on haemoglobins still going on. *Acta Physiol. Scand.* **182,** 213–214.

Nikinmaa, M. (2006). Gas transport. *In* "The Physiology of Fishes" (Evans, D. H., and Claiborne, J. B., Eds.), 3rd edn, pp. 153–174. CRC Press, Taylor and Francis, Boca Raton, FL.

Nikinmaa, M., and Heustis, W. H. (1984). Adrenergic swelling in nucleated erythrocytes: Cellular mechanisms in a bird, domestic goose, and two teleosts, striped bass and rainbow trout. *J. Exp. Biol.* **113,** 215–224.

Nikinmaa, M., and Rees, R. B. (2005). Oxygen-dependent gene expression in fishes. *Amer. J. Physiol.* **288,** R1079–R1090.

Nikinmaa, M., Cech, J. J., and McEnroe, M. (1984). Blood oxygen transport in stressed striped bass (*Morone saxatilis*): Role of β-adrenergic responses. *J. Comp. Physiol.* **154B,** 365–369.

Nikinmaa, M., Salama, A., Bogdanova, A., and Virkki, L. V. (2001). β-Adrenergic stimulation of volume-sensitive chloride transport in lamprey erythrocytes. *Physiol. Biochem. Zool.* **74,** 45–51.

Nilsson, G. E. (2001). Surviving anoxia with the brain turned on. *News Physiol. Sci.* **16,** 217–221.

Nilsson, G. E., and Östlund-Nilsson, S. (2006). Hypoxia tolerance in coral reef fishes. *In* "The Physiology of Tropical Fishes" (Val, A. L., de Almeida-Val, V. M. F., and Randall, D. J., Eds.), "Fish Physiology" Vol. 21 (Hoar, W. S., Randall, D. J., and Farrell, A. P., series Eds.), pp. 583–596. Academic Press, London.

Nilsson, G. E., and Renshaw, G. M. C. (2004). Hypoxic survival strategies in two fishes: Extreme anoxia tolerance in the North European crucian carp and natural hypoxic preconditioning in a coral-reef shark. *J. Exp. Biol.* **207,** 3131–3139.

Niwa, T., Naito, C., Mawjood, A. H., and Imai, K. (2000). Increased glutathionyl haemoglobin in diabetes mellitus and hyperlipidemia demonstrated by liquid chromatography/electrospray ionization-mass spectrometry. *Clin. Chem.* **46,** 399–419.

Pellegrini, M., Giardina, B., Olianas, A., Sanna, M. T., Deiana, A. M., Salvadori, S., di Prisco, G., Tamburrini, M., and Corda, M. (1995). Structure/function relationships in the hemoglobin components from moray (*Muraena helena*). *Eur. J. Biochem.* **234,** 431–436.

Pellegrini, M., Giardina, B., Verde, C., Carratore, V., Olianas, A., Sollai, L., Sanna, M. T., Castagnola, M., and di Prisco, G. (2003). Structural-functional characterization of the cathodal hemoglobin of the conger eel *Conger conger*: Molecular modelling study of an additional phosphate-binding site. *Biochem. J.* **372,** 679–686.

Pellegrino, D., Sprovieri, E., Mazza, R., Randall, D. J., and Tota, B. (2002). Nitric oxide-cGMP-mediated vasoconstriction and effects of acetylcholine in the branchial circulation of the eel. *Comp. Biochem. Physiol.* **132A,** 447–457.

Pellegrino, D., Palmerini, C. A., and Tota, B. (2004). No hemoglobin but NO: The icefish (*Chionodraco hamatus*) heart as a paradigm. *J. Exp. Biol.* **207,** 3855–3864.

Pelster, B., and Randall, D. (1998). The physiology of the Root effect. *In* "Fish Physiology Vol. 17" (Perry, S. F., and Tufts, B. L., Eds.), pp. 113–139. Academic Press, London.

Pelster, B., and Weber, R. E. (1990). Influence of organic phosphates on the Root effect of multiple fish haemoglobins. *J. Exp. Biol.* **149,** 425–437.

Pelster, B., and Weber, R. E. (2004). The physiology of the Root effect. *Adv. Comp. Environ. Physiol.* **8,** 51–77.

Perry, S. F., and Reid, S. D. (1992). Relationship between blood O_2 content and catecholamine levels during hypoxia in rainbow trout and American eel. *Am. J. Physiol.* **263,** R240–R249.

Pichavant, K., Maxime, V., Soulier, P., Boeuf, G., and Nonnotte, G. (2003). A comparative study of blood oxygen transport in turbot and sea bass: Effect of chronic hypoxia. *J. Fish Biol.* **62,** 928–937.

Poupa, O., Lindström, L., Maresca, A., and Tota, B. (1981). Cardiac growth, myoglobin proteins and DNA in developing tuna (*Thunnus thynnus thynnus* L.). *Comp. Biochem. Physiol.* **70A,** 217–222.

Powers, D. A., Martin, J. P., Garlick, R. L., Fyhn, H. J., and Fyhn, U. E. H. (1979). The effect of temperature on the oxygen equilibria of fish hemoglobins in relation to environmental thermal variability. *Comp. Biochem. Physiol.* **62A,** 87–94.

Primmett, D. R. N., Randall, D. J., Mazeaud, M., and Boutilier, R. G. (1986). The role of catecholamines in erythrocyte pH regulation and oxygen transport in rainbow trout (*Salmo gairdneri*) during exercise. *J. Exp. Biol.* **122,** 139–148.

Qiu, Y., Maillett, D. H., Knapp, J., Olsen, J. S., and Riggs, A. F. (2000). Lamprey hemoglobin – structural basis of the Bohr effect. *J. Biol. Chem.* **275,** 13517–13528.

Randall, D. J., and Perry, S. F. (1992). Catecholamines. *In* "Fish Physiology Vol. 12B" (Hoar, W. S., Randall, D. J., and Farrell, A. P., Eds.), pp. 255–300. Academic Press, London.

Randall, D. J., Hung, C. Y., and Poon, W. L. (2006). Response of aquatic vertebrates to hypoxia. *In* "Fish Physiology, Toxicology, and Water Quality" (Randall, D. J., and Mandy, D. Y. M., Eds.), pp. 1–10. Ecosystems Research Division, Athens, GA.

Reischl, E., Dafre, A. L., Franco, F. L., and Wilhelm Filho, D. (2007). Distribution, adaptation and physiological meaning of thiols from vertebrate hemoglobins. *Comp. Biochem. Physiol.* **146C,** 22–53.

Riccio, A., Tamburrini, M., Carratore, V., and di Prisco, G. (2000). Functionally distinct haemoglobins of the cryopelagic Antarctic teleost *Pagothenia borchgrevinki*. *J. Fish Biol.* **57**(Suppl. A), 20–32.

Rissanen, E., Tranberg, H. K., Sollid, J., Nilsson, G. E., and Nikinmaa, M. (2006). Temperature regulates hypoxia-inducible factor-1 (HIF-1) in a poikilothermic vertebrate, crucian carp (*Carassius carassius*). *J. Exp. Biol.* **209,** 994–1003.

Roesner, A., Fuchs, C., Hankeln, T., and Burmester, T. (2005). A globin gene of ancient evolutionary origin in lower vertebrates: Evidence for two distinct globin families in animals. *Mol. Biol. Evol.* **22,** 12–22.

Roesner, A., Hankeln, T., and Burmester, T. (2006). Hypoxia induces a complex response of globin expression in zebrafish (*Danio rerio*). *J. Exp. Biol.* **209,** 2129–2137.
Roesner, A., Mitz, S. A., Hankeln, T., and Burmester, T. (2008). Globins and hypoxia in the goldfish, *Carassius auratus. FEBS Journal* **275,** 3633–3643.
Rutjes, H. A., Nieveen, M. C., Weber, R. E., Witte, F., and Van den Thillart, G. E. E. J. M. (2007). Multiple strategies of Lake Victoria cichlids to cope with lifelong hypoxia include hemoglobin switching. *Am. J. Physiol.* **293,** R1376–R1383.
Rytkönen, K. T., Vuori, K. A. M., Primmer, C. R., and Nikinmaa, M. (2007). Comparison of hypoxia-inducible factor-1 alpha in hypoxia-sensitive and hypoxia-tolerant fish species. *Comp. Biochem. Physiol.* **2D,** 177–186.
Samuelsen, E. N., Imsland, A. K., and Brix, O. (1999). Oxygen binding properties of three different hemoglobin genotypes in turbot (*Scophthalamus maximus* Rafinesque): Effect of temperature and pH. *Fish Physiol. Biochem.* **20,** 135–141.
Shelly, D. A., and Mangum, C. P. (1997). Hemoglobin polymorphism in the Atlantic croaker, *Micropogon undulatus. Comp. Biochem. Physiol.* **118B,** 1419–1428.
Sidell, B. D., and O'Brien, K. M. (2006). When bad things happen to good fish: The loss of haemoglobin and myoglobin expression in Antarctic icefishes. *J. Exp. Biol.* **209,** 1791–1802.
Soldatov, A. A. (2002). Peculiarities of structure, polymorphism, and resistance to oxidation of fish hemoglobins. *J. Evol. Biochem. Physiol.* **38,** 392–400.
Sollid, J., Weber, R. E., and Nilsson, G. E. (2005). Temperature alters the respiratory surface area of crucian carp *Carassius carassius* and goldfish *Carassius auratus. J. Exp. Biol.* **208,** 1109–1116.
Sollid, J., Rissanen, E., Tranberg, H. K., Thorstensen, T., Vuori, K. A. M., Nikinmaa, M., and Nilsson, G. E. (2006). HIF-1α and iNOS levels in crucian carp gills during hypoxia-induced transformation. *J. Comp. Physiol* **176B,** 359–369.
Swenson, K. E., Eveland, R. L., Gladwin, M. T., and Swenson, E. R. (2005). Nitric oxide (NO) in normal and hypoxic vascular regulation of the spiny dogfish, *Squalus acanthias. J. Exp. Zool.* **303A,** 154–160.
Tamburrini, M., D'Avino, R., Carratore, V., Kunzmann, A., and di Prisco, G. (1997). The hemoglobin system of *Pleuragramma antarcticum* – correlation of hematological and biochemical adaptations with lifestyle. *Comp. Biochem. Physiol.* **118A,** 1037–1044.
Tamburrini, M., Verde, C., Olianas, A., Giardina, B., Corda, M., Sanna, M. T., Fais, A., Deiana, A. M., di Prisco, G., and Pellegrini, M. (2001). The hemoglobin system of the brown moray *Gymnothorax unicolor*: Structure/function relationships. *Eur. J. Biochem.* **268,** 4104–4111.
Tervonen, V., Vuolteenaho, O., and Nikinmaa, M. (2006). Haemoconcentration via diuresis in short-term hypoxia: Apossible role for cardiac natriuretic peptide in rainbow trout. *Comp. Biochem. Physiol.* **144A,** 86–92.
Tetens, V., and Christensen, N. J. (1987). Beta-adrenergic control of blood oxygen affinity in acutely hypoxic exposed rainbow trout. *J. Comp. Physiol.* **157B,** 667–675.
Tetens, V., Wells, R. M. G., and DeVries, A. L. (1984). Antarctic fish blood: Respiratory properties and the effects of thermal acclimation. *J. Exp. Biol.* **109,** 265–279.
Timmerman, C. M., and Chapman, L. J. (2004). Behavioral and physiological compensation for chronic hypoxia in the sailfin molly (*Poecilia latipinna*). *Physiol. Biochem.* Zool. **77,** 601–610.
Unzai, S., Park, S-Y., Nagai, K., Brittain, T., and Tame, J. R. H. (2009). Mutagenic studies on the origins of the Root effect. *In* "Dioxygen Binding and Sensing Proteins" (Bolognesi, M., Di Prisco, G., and Verde, C., Eds.). Springer-Verlag, Italy.
Val, A. L. (2000). Organic phosphates in the red blood cells of fish. *Comp. Biochem. Physiol.* **125,** 417–435.

van der Meer, D. L. M., van den Thillart, G. E. E. J. M., Witte, F., de Bakker, M. A. G., Besser, J., Richardson, M. K., Spaink, H. P., Leito, J. T. D., and Bagowski, C. P. (2005). Gene expression profiling of the long-term adaptive response to hypoxia in the gills of adult zebrafish. *Am. J. Physiol.* **289,** R1512–R1519.

Verde, C., Giordano, D., and di Prisco, G. (2006). Molecular evolution of haemoglobins of polar fishes. *Rev. Envir. Sci. Biotech.* **5,** 297–308.

Verheyen, E., Blust, R., and Decleir, W. (1986). Hemoglobin heterogeneity and the oxygen affinity of the hemolysate of some Victorian cichlids. *Comp. Biochem. Physiol.* **84A,** 315–318.

Weber, R. E. (1990). Functional significance and structural basis of multiple hemoglobins with special reference to ectothermic vertebrates. *In* "Comparative Physiology; Animal Nutrition and Transport Processes, 2. Transport, Respiration and Excretion: Comparative and Environmental Aspects. II. Blood Oxygen Transport: Adjustment to Physiological and Environmental Conditions" (Truchot, J. P., and Lahlou, B., Eds.), Vol. 6, pp. 58–75. Karger, Basel.

Weber, R. E. (1996). Hemoglobin adaptations in Amazonian and temperate fish with special reference to hypoxia, allosteric effectors and functional heterogeneity. *In* "Physiology and Biochemistry of the Fishes of the Amazon" (Val, A. L., Almeida-Val, V. M. F., and Randall, D. J., Eds.), pp. 75–90. INPA, Brazil.

Weber, R. E. (2000). Adaptations for oxygen transport: Lessons from fish hemoglobins. *In* "Hemoglobin Function in Vertebrates. Molecular Adaptation in Extreme and Temperate Environments" (Di Prisco, G., Giardina, B., and Weber, R. E., Eds.), pp. 23–37. Springer-Verlag, Italy.

Weber, R. E., and de Wilde, J. A. M. (1976). Multiple haemoglobins in plaice and flounder and their functional properties. *Comp. Biochem. Physiol.* **54B,** 433–437.

Weber, R. E., and Fago, A. (2004). Functional adaptation and its molecular basis in vertebrate hemoglobins, neuroglobins and cytoglobins. *Resp. Physiol. Neurobiol.* **144,** 141–159.

Weber, R. E., and Jensen, F. B. (1988). Functional adaptations in hemoglobins from ectothermic vertebrates. *Annu. Rev. Physiol.* **50,** 161–179.

Weber, R. E., and Voelter, W. (2004). "Novel" factors that regulate oxygen binding in vertebrate hemoglobins. *Micron* **35,** 45–46.

Weber, R. E., and Wells, R. M. G. (1989). Haemoglobin structure and function. *In* "Comparative Pulmonary Physiology" (Wood, S. C., Ed.), pp. 279–310. Marcell Dekker, New York.

Weber, R. E., Wood, S. C., and Lomholt, J. P. (1976a). Temperature acclimation and oxygen-binding properties of blood and multiple haemoglobins from rainbow trout. *J. Exp. Biol.* **65,** 333–345.

Weber, R. E., Lykkeboe, G., and Johansen, K. (1976b). Physiological properties of eel haemoglobin: Hypoxic acclimation, phosphate effects and multiplicity. *J. Exp. Biol.* **64,** 75–88.

Weber, R. E., Fago, A., Val, A. L., Bang, A., Van Hauwaert, M.-L., Dewilde, S., Zal, F., and Moens, L. (2000). Isohemoglobin differentiation in the bimodal-breathing Amazon catfish *Hoplosternum littorale*. *J. Biol. Chem.* **275,** 17297–17305.

Weber, R. E., Hourdez, S., Knowles, F., and Lallier, F. (2003). Hemoglobin function in deep-sea and hydrothermal-vent endemic fish: *Symenchelis parasitica* (Anguillidae) and *Thermarces cerberus* (Zoarcidae). *J. Exp. Biol.* **206,** 2693–2702.

Weber, R. E., Voelter, W., Fago, A., Echner, H., Campanella, E., and Low, P. S. (2004). Modulation of red cell glycolysis: Interactions between vertebrate hemoglobins and cytoplasmic domains of band 3 red cell membrane proteins. *Am. J. Physiol.* **287,** R454–R464.

Wells, R. M. G. (1990). Hemoglobin physiology in vertebrate animals: A cautionary approach to adaptationist thinking. *In* "Advances in Comparative and Environmental Physiology" (Boutilier, R. G., Ed.), pp. 143–161. Springer, Heidelberg.

Wells, R. M. G. (1999). Haemoglobin function in aquatic animals: Molecular adaptations to environmental challenge. *Mar. Freshw. Res.* **50,** 933–939.

Wells, R. M. G. (2005). Blood-gas transport and haemoglobin function in polar fishes: Does low temperature explain physiological characters? *In* "The Physiology of Polar Fishes" (Steffensen, J. F., and Farrell, A. P., Eds.), Fish Physiology Vol. 22 (Hoar, W. S., Randall, D. R., and Farrell, A. P., series Eds.), pp. 281–316. Academic Press, New York.

Wells, R. M. G., and Baldwin, J. (1990). Oxygen transport potential in tropical reef fish with special reference to blood viscosity and haematocrit. *J. Exp. Mar. Biol. Ecol.* **141,** 131–143.

Wells, R. M. G., and Weber, R. E. (1990). The spleen in hypoxic and exercised rainbow trout. *J. Exp. Biol.* **150,** 461–466.

Wells, R. M. G., Ashby, M. D., Duncan, S. J., and Macdonald, J. A. (1980). Comparative study of the erythrocytes and haemoglobins in nototheniid fishes from Antarctica. *J. Fish Biol.* **17,** 517–527.

Wells, R. M. G., McIntyre, R. H., Morgan, A. K., and Davie, P. S. (1986). Physiological stress responses in big gamefish after capture: Observations on plasma chemistry and blood factors. *Comp. Biochem. Physiol.* **84A,** 565–571.

Wells, R. M. G., Summers, G., Beard, L. A., and Grigg, G. C. (1988). Ecological and behavioural correlates of intracellular buffering capacity in the muscles of Antarctic fishes. *Polar Biol.* **8,** 321–326.

Wells, R. M. G., Grigg, G. C., Beard, L. A., and Summers, G. (1989). Hypoxic responses in a fish from a stable environment: Blood oxygen transport in the Antarctic fish *Pagothenia borchgrevinki*. *J. Exp. Biol.* **141,** 97–111.

Wells, R. M. G., Davie, P. S., and Weber, R. E. (1991). The effect of β-adrenergic stimulation of trout erythrocytes on blood viscosity. *Comp. Biochem. Physiol.* **100C,** 653–655.

Wells, R. M. G., Baldwin, J., Seymour, R. S., and Weber, R. E. (1997). Blood oxygen transport and hemoglobin function in three tropical fish species from northern Australian freshwater billabongs. *Fish Physiol. Biochem.* **16,** 247–258.

Wells, R. M. G., Baldwin, J., Seymour, R. S., Christian, K., and Brittain, T. (2005). Red blood cell function and haematology in two freshwater fishes from Australia. *Comp. Biochem. Physiol.* **141A,** 87–93.

Wilhelm Filho, D. (2007). Reactive oxygen species, antioxidants and fish mitochondria. *Frontiers in Bioscience* **12,** 1229–1237.

Wittenberg, J. B. (2007). On optima: The case of myoglobin-facilitated oxygen diffusion. *Gene* **398,** 156–161.

Wood, S. C., and Johansen, K. (1972). Adaptation to hypoxia by increased HbO_2 affinity and decreased red cell ATP concentration. *Nature, New Biol.* **237,** 278–279.

Wood, S. C., Weber, R. E., and Powers, D. A. (1979). Respiratory properties of blood and hemoglobin solutions from the piranha. *Comp. Biochem. Physiol.* **62A,** 163–167.

Wu, R. S. S. (2002). Hypoxia: From molecular responses to ecosystem responses. *Mar. Poll. Bull.* **45,** 35–45.

Yang, T.-H., Lai, N. C., and Somero, G. N. (1992). Respiratory, blood, and heart enzymatic adaptations of *Sebastolobus alascanus* (Scorpaenidae; Teleostei) to the oxygen minimum zone: A comparative study. *Biol. Bull.* **183,** 490–499.

Zaccone, G., Mauceri, A., and Fasulo, S. (2006). Neuropeptides and nitric oxide synthase in the gill and air-breathing organs of fishes. *J. Exp. Zool.* **305A,** 428–439.

7

CARDIOVASCULAR FUNCTION AND CARDIAC METABOLISM

A. KURT GAMPERL
W. R. DRIEDZIC

Next to extremes in temperature, hypoxia is arguably the most significant environmental challenge faced by fishes. This is because of the disruptions/consequences it has for the fish's physiology, reproduction, and survival, and the fact that hypoxia is an increasing problem in aquatic systems worldwide. The cardiovascular system is critical for the effective uptake of oxygen from the environment and the distribution/transport of oxygen and nutrients to the tissues, and its proper functioning is paramount to activities such as locomotion and digestion, and to the capacity to deal with environmental pertubations. In this chapter, an overview of cardiovascular responses to hypoxia in fishes, of some of the mechanisms that influence/mediate the effects of hypoxia on the fish's cardiovascular system, and of how myocardial energy metabolism is regulated under hypoxia (this aspect is critical to the continued functioning of the heart during periods of oxygen shortage) is

Hypoxia: Volume 27
FISH PHYSIOLOGY

DOI: 10.1016/S1546-5098(08)00007-1

provided. Information is reviewed on the effects of both acute and chronic hypoxia and on interspecific variation in the magnitude and timing of responses, and covers life stages from embryo to adult and levels of biological organization from gene expression to the whole animal. Further, where possible, recent advances in our understanding of the influence of hypoxia on fish cardiovascular function are highlighted, and unresolved issues are identified. It is expected that this chapter will become a valued resource for those interested in the interplay between hypoxia and cardiovascular function, and will stimulate research in this interesting area of fish physiology.

1. INTRODUCTION

Aquatic habitats are subject to many environmental variations and one of the most important parameters affecting non air-breathing vertebrates is dissolved oxygen. Hypoxia, or oxygen depletion, is a phenomenon that occurs in a wide variety of aquatic environments, from the Amazon drainage basin, to iced-over shallow water bodies in winter, and ever increasingly, to coastal marine areas around the world (including the Black and Baltic Seas, the Gulf of Mexico, and the Gulf of St. Lawrence). The response of fish to hypoxic environments includes complex behavioral changes such as decreased locomotion and predator avoidance capacity (Dalla Via *et al.*, 1998; Lefrançois and Domenici, 2006; Behrens and Steffensen, 2007) or movement away from/avoidance of areas of low dissolved oxygen (Pihl *et al.*, 1991; Claireaux *et al.*, 1995) (see Chapter 2). When escape from the hypoxic stress is not possible, a variety of physiological adjustments may be invoked to compensate for low oxygen availability, thus allowing fish to withstand short-term (acute) hypoxic exposure (Jensen *et al.*, 1993; *Val et al.*, 1998) or to eventually allow for the restoration of essential activities such as feeding, reproduction, and escape from predators (Jensen *et al.*, 1993) (see Chapter 10). Given the critical role that cardiovascular function plays in blood oxygen transport and substrate delivery, and that fish heart function is solely or partially (for those with coronary arteries, or with lungs or accessory breathing organs) dependent on whatever oxygen is left in the venous blood after it has traversed the fish's other tissues, it is not surprising that fish cardiovascular function during hypoxia has been an active area of research for over five decades.

In this chapter we provide an overview of cardiovascular responses to hypoxia in fishes (with the caveat that only minimal reference is made to the crucian carp, which is the focus of Chapter 9), some of the mechanisms that influence/mediate the effects of hypoxia on the cardiovascular system, and how myocardial energy metabolism is regulated under hypoxic conditions.

Further, where possible, we emphasize specific areas where recent advances in our understanding of the influence of hypoxia on fish cardiovascular function have been realized, and where unresolved issues remain.

This chapter was challenging to put together given the wide range of hypoxia tolerance exhibited amongst fish species, and the large variations in the severity and duration of hypoxia, and temperature, used in experimental protocols. For example, the common carp (*Cyprinus carpio*) can withstand rapidly induced anoxia for 2–24 h (depending on temperature; Stecyk and Farrell, 2007), while the Atlantic cod (*Gadus morhua*) only tolerates exposure to a water PO_2 of 10 mmHg (at 10°C) for a matter of minutes (Petersen and Gamperl, unpublished data). Water O_2 content varies inversely with temperature, while metabolic demand generally goes up by a factor of 2–3 with each 10°C increase in temperature, and thus hypoxia-tolerance for a given species is temperature dependent. Finally, many authors have examined the effects of hypoxia on fish cardiovascular function by rapidly exposing fish to 8–12 min of anoxia/severe hypoxia, whereas others have used protocols where water oxygen levels were gradually reduced over several hours, fish were rapidly exposed to severe hypoxia and held at this level of oxygenation for extended periods (h), or fish were maintained at moderate levels of hypoxia (i.e., 35–60 mmHg) for several weeks. Given the methodological complexity of the available literature, we divided the information contained in this chapter into two broad categories for ease of presentation (although not all research falls easily within either category; e.g., see data from Stecyk and Farrell, 2007 in section 4.1): acute hypoxia referring to research that exposed fish to minutes to hours of lowered water O_2; and chronic hypoxia indicating reductions in water O_2 levels that lasted days or weeks, i.e., long enough to have changes in gene and protein expression. Further, we used oxygen partial pressure (mmHg) as our unit of water O_2 measurement throughout the chapter. This was largely done to facilitate multispecies comparisons (i.e., the reader can easily judge the severity of hypoxia for a species at a given temperature based on oxygen availability; fully saturated water having a PO_2 of approximately 155 mmHg at sea level), and because arterial blood PO_2, which generally reflects water PO_2, has a number of potential implications for hypoxia tolerance. For example, hemoglobin–oxygen binding affinity is expressed in mmHg as the P_{50} (PO_2 at which hemoglobin is 50% saturated with oxygen), and there is evidence that the release of catecholamines from the chromaffin tissue, which stimulates a number of physiological alterations that would improve hypoxia tolerance [e.g., oxygen uptake, blood oxygen transport, and cardiac function; e.g., see Farrell and Jones (1992) and Randall and Perry (1992)], appears to occur near the fish's P_{50} value (Reid and Perry, 1994).

2. HYPOXIC EFFECTS ON *IN VIVO* CARDIOVASCULAR FUNCTION

2.1. Acute Hypoxia

2.1.1. Heart Rate

The vast majority of research on the effects of hypoxia on fish cardiovascular function has investigated the effects of a few minutes or hours of exposure to reduced oxygen levels. In fishes, the most common cardiac response to hypoxia is reflex bradycardia (a decrease in heart rate, f_H), and a recent review by Farrell (2007) provides a comprehensive overview of the data in this area and proposes several direct benefits of hypoxic bradycardia to the fish heart. These benefits include: (1) improved cardiac contractility through the negative force-frequency effect; (2) enhanced oxygenation of the myocardium due to an increase in the diastolic residence time of blood in the lumen of the heart (i.e., increased time for oxygen diffusion), and stretching of the myocardium (i.e., decreased diffusion distance), in species that respond to hypoxia with a concomitant increase in stroke volume (SV, see below); (3) a reduction in myocardial oxygen demand due to a decrease in the rate of ventricular pressure development (dP/dt); and (4) an increase in coronary blood flow, and thus a diminished reliance on the oxygen content and partial pressure of venous blood, due to an extended diastolic period (diastole the portion of the cardiac cycle where the majority of coronary blood flow occurs; ~75–85%; Davie and Franklin, 1993; Gamperl *et al.*, 1995). Thus, in this chapter, we only provide a brief summary of the effect of acute hypoxia on heart rate in fishes, of how temperature and hypoxia tolerance influence the onset of bradycardia, and what control mechanisms may account for species and other differences.

First, there are some taxonomic groups that do not exhibit changes in heart rate when exposed to hypoxia or where no clear response pattern to hypoxia has been established. For example, since reflex bradycardia is primarily mediated by vagal cardioinhibitory tone, and hagfishes lack autonomic cardiac innervation (Nilsson, 1983), it is not surprising that heart rate (f_H) in this taxa remains unchanged in response to severe hypoxia (Axelsson *et al.*, 1990). Bradycardia is absent in all three genera of lungfish when exposed to aquatic hypoxia (*Neoceratodus*, Fritsche *et al.*, 1993; *Leptidosiren*, Sanchez *et al.*, 2001; *Protopterus*, Perry *et al.*, 2005). This finding is likely related to the absence of external gill O_2 chemoreceptors or external O_2 chemoreceptors that are unresponsive/very insensitive to changes in water O_2 levels in this taxa (Perry *et al.*, 2005), and suggests that the loss of hypoxic bradycardia may have coincided with the evolution of air-breathing in fishes. However, the picture is not so clear when the f_H response of other air-breathing fishes to aquatic hypoxia is examined. For example, the jeju (*Hoploerythrinus unitaeniatus*) developed hypoxic bradycardia when hypoxia began to compromise oxygen consumption (approx. 40 mmHg; Oliveira *et al.*, 2004), f_H in two species of facultative air-breathing Amazonian armoured catfish (*Liposarcus pardalis* and *Glyptoperichthyes gibbceps*) (MacCormack *et al.*, 2003a) did not change significantly (although average f_H decreased by ~15 and 40 beats min^{-1}, respectively) at dissolved oxygen levels down to ~1 mg l^{-1}, and the garfish (*Lepisosteus oculatus*; Smatresk and Cameron, 1982) and *Synbranchus marmoratus* (Skals *et al.*, 2006) showed modest tachycardia during exposure to aquatic hypoxia (approx. 12 mmHg and 50 mmHg, respectively). Further, the interpretation of these latter data is complicated because: (1) the garfish and *S. marmoratus* were allowed access to air, and the increase in f_H with aquatic hypoxia (even during periods of aquatic ventilation) may have resulted because inflation of their accessory breathing organs overrode the drive for hypoxic bradycardia initiated by O_2 receptors in the gills (e.g., see Graham, 1997; Skals *et al.*, 2006); and (2) experiments with two water-breathing Amazonian fish species have shown that internally oriented gill O_2 chemoreceptors (i.e., those sensing changes in blood oxygen) can exclusively (in *Hoplias malabaricus*; Sundin *et al.*, 1999b) or in combination with externally oriented O_2 receptors (in *Colossoma macropodum*; Sundin *et al.*, 2000) elicit bradycardia. Clearly, in the absence of other measurements of changes in f_H with exposure to aquatic hypoxia, it is not possible to determine to what extent air-breathing fishes have lost/retained the capacity for hypoxia-induced bradycardia, or whether the loss of capacity for bradycardia in at least some air-breathing fishes is related to the concurrent absence of external O_2 receptors.

The final group in which there is no clear picture with regards to the presence/absence of hypoxic bradycardia is Antarctic fishes. Despite the fact that these species live in a cold stenothermal environment with very stable water oxygen levels, the response to acute hypoxia varies among species, studies, and individuals ranging from no effect, to slight tachycardia, to a clear hypoxic bradycardia. The variable results for red-blooded Antarctic species (i.e., *Trematomus bernachii* and *P. borchgrvinki*) may be due to a high and variable cholinergic tone on the heart (where only individuals with low cholinergic tone show substantial decreases in heart rate) and those for the icefish (*Chaenocephalus aceratus*) may be due to differences in experimental protocols or hypoxic thresholds (Axelsson, 2005). However, as with the air-breathing fishes, clarification of the f_H response of Antarctic taxa to environmental hypoxia requires careful study, where the rate of hypoxic initiation is consistent (i.e., gradual or abrupt) and where oxygen levels in the water are lowered to such an extent as to preclude differential f_H responses due to varied responsiveness of the O_2 chemoreceptors that trigger hypoxic bradycardia.

With regards to the majority of water-breathing fishes, it is clear that diminished water oxygen levels lead to hypoxic bradycardia, and that hypoxia tolerance and temperature influence the water oxygen level (PO_2) at which the reduction in heart rate is initiated. To illustrate how PO_2 influences the onset of bradycardia we have plotted the f_H–water PO_2 relationships for 10 species of water-breathing fishes that were acclimated to either 22–25°C or 8–12°C and exposed to an experimental protocol that involved the gradual reduction of water oxygen levels; this latter criteria was used because rapid versus gradual decreases in water O_2 levels can affect the f_H response to hypoxia (e.g., see Butler and Taylor, 1971; Figure 7.1). What can be seen at both water temperatures is that there is a large range of PO_2 values at which f_H becomes noticeably reduced; f_H reductions occurring at PO_2 levels as high as 110mm Hg in the dourado (*Salminus maxillosus*) and 70 mmHg in the rainbow trout (*Oncorhynchus mykiss*) and Japanese eel (*Anguilla japonica*), to as low as 25 mmHg for *Hoplias lacerdae*, 35 mmHg for the cod (*Gadus morhua*), and <40 mmHg for the tench (*Tinca tinca* L.). What factors determine the PO_2 at which bradycardia is initiated has not been extensively studied but data from a number of investigations suggest that it is related to a fish's lifestyle and hypoxia tolerance. For example, Rantin *et al.* (1993) performed a direct comparison of *H. lacerdae* (an Amazonian species that inhabits well-oxygenated rivers) and *H. malabaricus* (considered to be well adapted to hypoxic conditions), and the critical PO_2 (the PO_2 at which routine oxygen consumption can no longer be maintained) and PO_2 at which bradycardia was initiated were approximately 35 and 20 mmHg in the two species, respectively. Furimsky *et al.* (2003) showed that largemouth bass (*Micropterus salmoides*), which prefer shallow/weedy areas (i.e., a habitat prone to large fluctuations in dissolved oxygen), initiate bradycardia at least 45 mmHg later than the smallmouth bass (*M. dolomieu*), which inhabits deeper and colder waters. They also showed that this delayed onset of bradycardia was associated with several physiological variables (e.g., a lower P_{50} value for hemoglobin–oxygen binding) that would have allowed for enhanced hypoxia tolerance in the former species (although our analysis of the available data failed to reveal a significant relationship between the PO_2 at which bradycardia was initiated during graded hypoxic exposure and literature values for a species' P_{50} value for hemoglobin–oxygen binding). Finally, the dourado and rainbow trout, which are very active species that normally inhabit well-oxygenated waters, have thresholds for the induction of bradycardia of >70 mmHg while those for the hypoxia-tolerant carp (*C. carpio*) and tench are <40 mmHg.

With respect to temperature, although it appears from Figure 7.1 that this parameter does not constrain the range of PO_2 values over which hypoxia-tolerant and hypoxia-sensitive species initiate bradycardia, there are several studies which show that the hypoxic threshold for bradycardia increases with

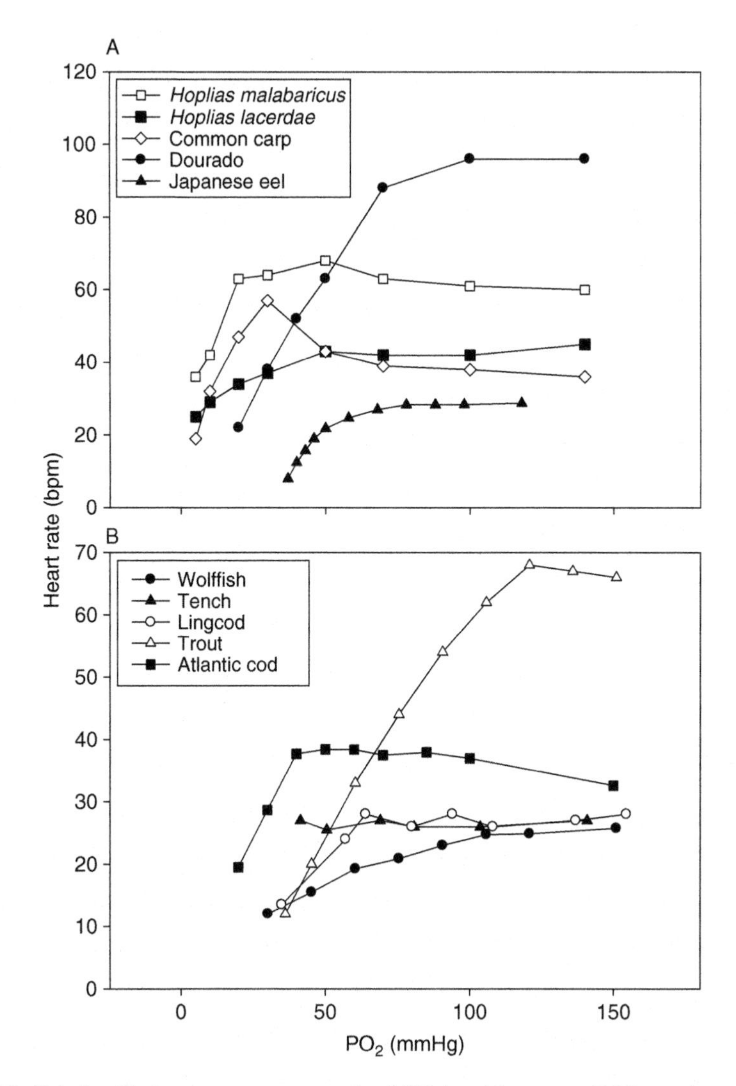

Fig. 7.1. Relationship between water oxygen level (PO_2) and heart rate (f_H) for various teleost species acclimated to temperatures of 22–25°C (A) and 8–12°C (B). Data for *Hoplias malabaricus*, *Hoplias lacerdae*, and common carp (*Cyprinus carpio*) from Rantin *et al.* (1993); Dourado (*Salminus maxillosus*) from De Salvo Souza *et al.* (2001); Japanese eel (*Anguilla japonica*) from Chan (1986); Tench (*Tinca tinca*) and trout (*Oncorhynchus mykiss*) from Marvin and Heath (1968); Lingcod (*Ophiodon elongates*) from Farrell (1982); Atlantic cod (*Gadus morhua*) from Petersen and Gamperl (unpublished data); wolffish (*Anarhichas lupus*) from Joaquim and Gamperl (unpublished data).

temperature; the PO_2 at which bradycardia was initiated increased from ~20 to ~60 mmHg in spangled perch (*Leiopotherapon unicolor*) acclimated to 10 and 30°C (Gehrke and Fielder, 1988) and from <40 to >120 mmHg in dogfish (*Syliorhinus canicula*) at seasonal temperatures of 7 and 17°C (Figure 7.2A) (Butler and Taylor, 1975). While this appears to be the "typical" relationship between temperature and the water PO_2 at which bradycardia occurs, and makes sense given the similar relationship between water temperature and *in vivo* hemoglobin–oxygen affinity (e.g., see Perry and Reid, 1994), recent experiments by Mendonca and Gamperl (unpublished data) indicate that this relationship does not apply to all teleost species. These authors acclimated winter flounder (*Pleuronectes americanus*) to 8 and 15°C and then exposed them to a gradual hypoxic challenge by decreasing water O_2 levels by 10% air saturation (approximately 15 mmHg) per hour. Surprisingly, while the onset of bradycardia was ~90 mmHg in flounder acclimated to 8°C, f_H remained constant in fish acclimated to 15°C down to a PO_2 of at least 30 mmHg (i.e., the response was opposite to that observed in the spangled perch and dogfish)(Figure 7.2B). Further, this result does not appear to be peculiar to the particular experimental conditions utilized by Mendonca *et al.* (unpublished data) as Cech *et al.* (1977) showed that exposure of this species to water of 45% air saturation (i.e., a PO_2 of approx. 65 mmHg) had no effect on f_H at 10°C.

Flatfish lack adrenergic cardiac innervation (Santer, 1972; Donald and Campbell, 1982) and cholinergic tone on the heart increases (see Sureau *et al.*, 1989), not decreases, with temperature as has been shown for other teleosts such as the rainbow trout (e.g., see Wood *et al.*, 1979). Thus, one could speculate that these differences in cardiac nervous control are responsible for the temperature-dependent differences in the response of the flounder versus the dogfish and spangled perch to graded hypoxia; the hypothesis being that a higher cholinergic tone in flounder at 15°C precluded increases in vagal tone from mediating a decrease in f_H in response to lowered water oxygen levels. This explanation is unlikely, however, as elasmobranchs also lack cardiac adrenergic innervation, and Taylor *et al.* (1977) showed that cholinergic tone on the heart increases in the dogfish with temperature. At present we have no physiological mechanism to explain why there is no bradycardia in 15°C-acclimated winter flounder down to water oxygen levels of 30 mmHg. However, we cannot preclude the possibility that f_H did not decrease in flounder at this higher temperature due to some, as yet unexplained, ability to avoid myocardial dysfunction. For example, MacCormack and Driedzic (2002) demonstrated that ventricle strips of the yellowtail flounder (*Limanda ferruginea*) show a transient increase in force development when subjected to anoxia (i.e., N_2 gassing). Sundin *et al.* (2000) showed that hypoxia still induced bradycardia in tambaqui (*C. macropomum*) after

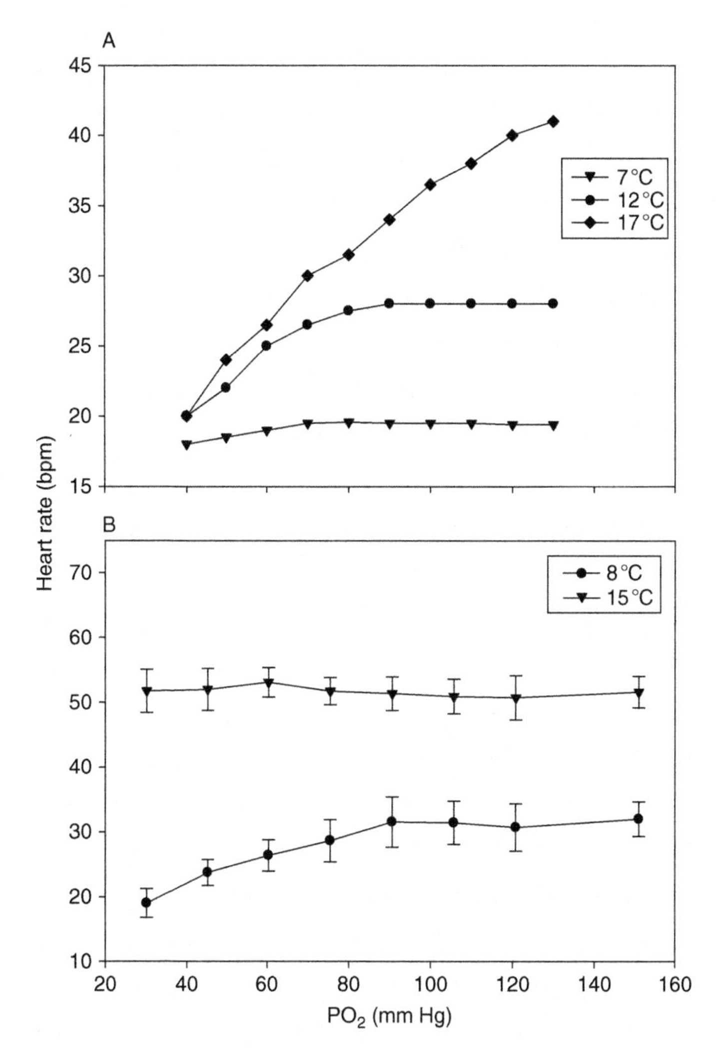

Fig. 7.2. The effect of acclimation temperature on the relationship between water PO_2 and heart rate for (A) the dogfish (*Scyliorhinus canicula*; Butler and Taylor, 1975) and (B) the winter flounder (*Pleuronectes americanus*; Mendonca and Gamperl, unpublished data) Values for the flounder are means ± S.E.

sectioning of cranial nerves IX and X to the gill arches and pretreatment with atropine, although the bradycardia was only approximately 30% of that seen in "intact" animals; i.e., there is a non-neural component to hypoxia-induced bradycardia in some fish species. Finally, Rantin *et al.* (1995) showed that *C. carpio* has an unusual pattern of f_H changes in response to hypoxia (i.e., f_H increasing by approx. 25 beats min^{-1} prior to the onset of bradycardia at 35 mmHg; see Figure 7.1A) and that changes in the electrocardiogram (ECG) of this species were decidedly different as compared to three other tropical fish species examined. In *C. carpio*, the direction of the ECG reversed from + to – with the onset of severe hypoxia, as compared to no change (in *Piaractus mesopotamicus*) or a – to + transition in *H. malabaricus* and *H. lacerdae*, and there was only minimal change in the amplitude of the T-wave of *C. carpio* with graded hypoxia as opposed to a 1.8- to 4-fold increase in this parameter in the other three species.

To this point, we have confined our discussion of the effects of acute hypoxia on f_H to adult fishes. However, there are several papers that have now examined the ontogeny of f_H control in fishes under hypoxic conditions. From these studies it is clear that hypoxia-induced reductions in f_H and cardiac activity during early development are a direct effect of severe oxygen shortage on the cardiac myocytes, and that the exact nature of f_H responses to acute hypoxia depends on developmental stage and the degree of hypoxia. While the lack of a hypoxia-induced bradycardia in early larval stages has been noted for the rainbow trout (Holeton, 1971) and Arctic charr (*Salvelinus alpinus*)(McDonald and McMahon, 1977), it is recent work on the zebrafish (*Zebra danio*) that is the primary basis for these conclusions. For example, although Padilla and Roth (2001) showed that 4 h of anoxic exposure reduced the f_H of zebrafish by 40% at 29 hours post-fertilization, Jonz and Nurse (2005) reported that gill neuroepithelial cells (NEC, considered to be the gill's O_2 chemoreceptors) are not expressed until 5 days post-fertilization (dpf) and not innervated until 7 dpf, and Schwerte *et al.* (2006) indicated that nervous cholinergic tone on the heart is not established until approximately 12 dpf. Further, bradycardia is absent before 20 dpf at rearing temperatures between 25 and 31°C when PO_2 does not fall below 10 mmHg during a graded hypoxic challenge (see Figure 7.3; Barrionuevo and Burggren, 1999). With regards to temperature and developmental effects on the degree and onset of hypoxia-induced bradycardia in zebrafish, the results of Barrionuevo and Burggren (1999) are difficult to interpret as temperature also affects developmental rate. However, it appears that f_H is more susceptible to hypoxia-induced reductions at warmer rearing temperatures at any given developmental stage (the examined range 0–100 dpf) (e.g., see Figure 7.3).

While these studies have greatly advanced our understanding of heart rate development in teleosts, they also raise one intriguing question. If both NEC and the heart are innervated by 12 dpf in zebrafish, and cholinergic receptors are functional in the zebrafish heart by 5 dpf (Schwerte *et al.*, 2006), why does bradycardia not develop until >20 dpf? The answer to this question will require further study, but the most logical interpretation is that sensitivity of the NEC to reductions in water oxygenation, or of myocardial

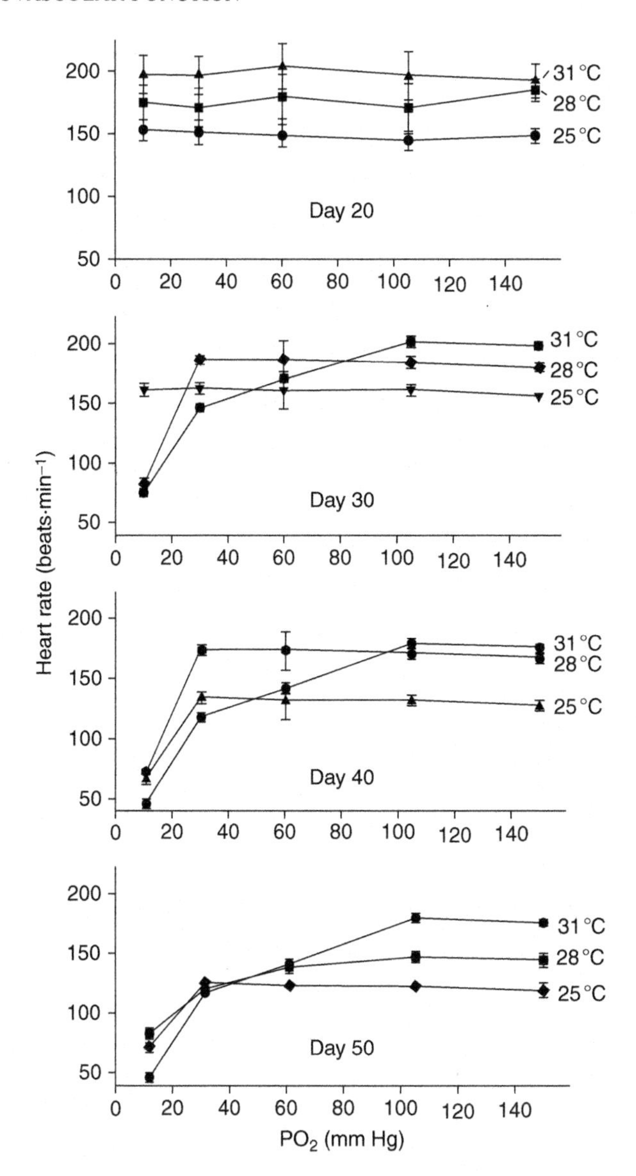

Fig. 7.3. Influence of acute hypoxic exposure on heart rate in zebrafish (*D. rerio*) larvae (day 20 post-fertilization) and juveniles (days 30, 40, and 50 post-fertilization) reared at various temperatures. Mean values ± SE are plotted; $n = 10$ for each plotted developmental stage at all three temperatures. [Modified from Barrionuevo and Burggren (1999) with the permission of the *American Journal of Physiology*.]

cholinergic receptors to nervous stimulation, is low during early life-history stages and increases with development.

2.1.2. Cardiac Output, Stroke Volume, and Venous Tone

With the exception of hagfish (Axelsson *et al.*, 1990), which have an extremely low cardiac output (Q) and power output (PO) at rest (Axelsson *et al.*, 1990; Forster *et al.*, 1991), and do not exhibit hypoxia-induced bradycardia, all fish examined to date show an increase in stroke volume (SV) when hypoxia-induced bradycardia develops. The difference among species, however, lies in the level of hypoxia at which they initiate increases in SV and the extent that increases in SV compensate for the effect of hypoxic-induced bradycardia on cardiac output (Q). In general, there are three patterns that are exhibited by fishes, and these are illustrated in Figure 7.4 using the Atlantic cod, Atlantic wolffish (*Anarhichas lupus*), and winter flounder as examples. In the first response pattern, as seen in the Atlantic cod and rainbow trout (also see Wood and Shelton, 1980; Sandblom and Axelsson, 2005), SV starts to increase prior to hypoxic bradycardia leading to an initial increase in Q, and SV is initially able to compensate for hypoxia-induced decreases in f_H before Q eventually falls. In the second pattern, increases in SV are either concomitant with the onset of bradycardia (e.g., dourado—de Salvo Souza *et al.*, 2001; sea bass, *Dicentrarchus labrax*—Axelsson *et al.*, 2002; lingcod—Farrell., 1982; Japanese eel—Chan, 1986) or begin after the bradycardia is initiated (often near the limit of hypoxia tolerance, e.g., in wolfish; smallmouth bass; Furimsky *et al.*, 2003). However, these increases in SV are inadequate to compensate for decreases in f_H and Q falls almost continuously (albeit slower than f_H) with the severity of hypoxia. Finally, in some fishes, for example the winter flounder (Figure 7.4), dogfish shark (*Scyliorhinus canicula*; Butler and Taylor, 1971), and sturgeon (*Acipenser naccarii*; Agnisola *et al.*, 1999) (at least down to a PO_2 of ~35 mmHg), increases in SV initiated during hypoxia are able to fully compensate for the drop in f_H with hypoxia, such that Q is maintained. These latter two patterns, if one utilizes the same terminology as applied to the oxygen consumption–water PO_2 relationship, are characterized as representing conformers and regulators, respectively, with regards to their Q responses.

Given the limited number of species on which direct measurements of Q and SV have been performed under well-controlled experimental conditions, it is not possible to determine to what extent interspecific differences in the responses of these parameters to hypoxia are related to differences in activity, lifestyle, and hypoxia tolerance. For example, although data on the flounder and sturgeon suggest that hypoxia-tolerant species can maintain Q through increases in SV until low oxygen levels, data on the eel (*Anguilla anguilla*) indicate that this hypoxia-tolerant species (critical water O_2 tension 25 mmHg

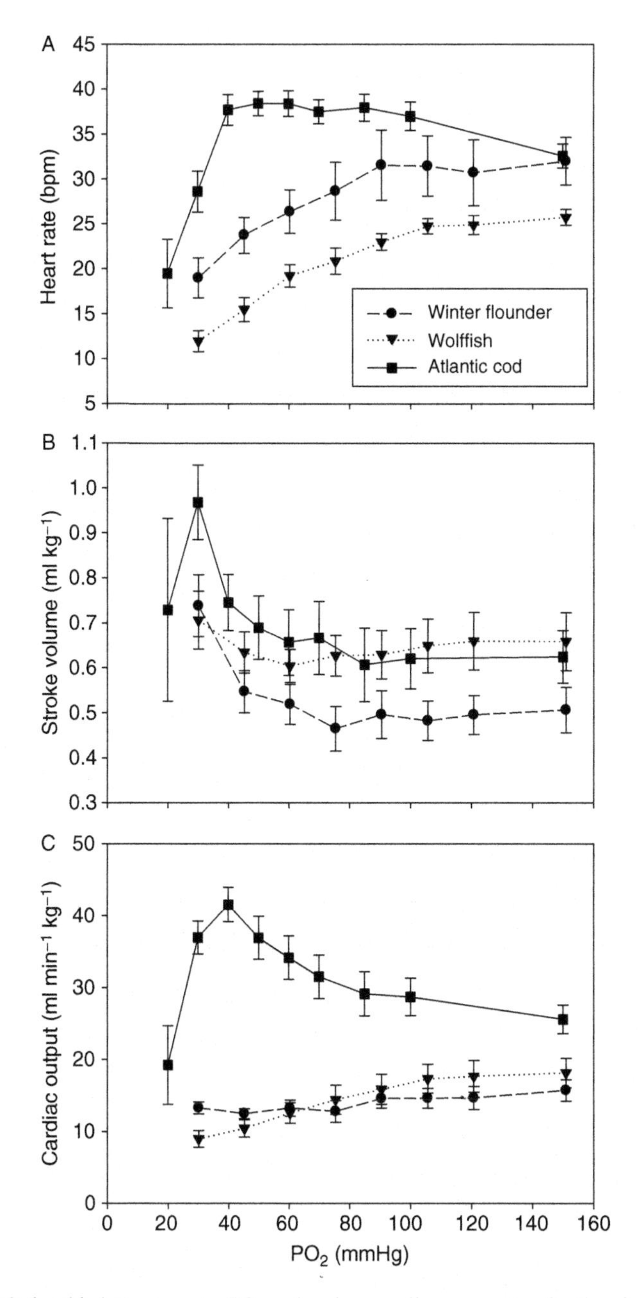

Fig. 7.4. Relationship between water PO_2 and various cardiac parameters for the winter flounder (*Pleuronectes americanus*; Mendonca and Gamperl, unpublished data), wolffish (*Anarhichas lupus*; Joaquim and Gamperl, unpublished data), and Atlantic cod (*G. morhua*; Petersen and Gamperl, unpublished data) at temperatures of 8–10°C. Values are means ± S.E.

at 25°C; Cruz-Neto and Steffensen, 1997) does not elevate SV in response to bradycardia at 40 mmHg O_2 (Peyraud-Waitzenegger and Soulier, 1989). Further, we have a less than complete picture of what mechanisms, other than the increase in filling time and filling pressure (resulting from the pooling of central venous blood) that are concomitant with bradycardia (Farrell, 1991; Altimiras and Axelsson, 2004), enable fish to elevate SV in response to aquatic hypoxia. However, evidence has accumulated over the past decade that the active regulation of venous tone and cardiac filling are equally important for regulating SV in fishes, including during hypoxia. For example, Sandblom and Axelsson (2005) showed that venous pressure (P_{ven}) and SV increase in rainbow trout at water oxygen levels that do not elicit bradycardia (Figure 7.5). Sandblom and Axelsson (2006) showed, using venous capacitance curves, that some of the venous blood volume is actively shifted into the stressed vascular compartment by an increase in venous smooth muscle tonus during hypoxia, and that this results in an elevated mean circulatory filling pressure (Figure 7.6). Sandblom and Axelsson (2006) showed that hypoxia-induced changes in trout venous capacitance are primarily under α-adrenergic control, and that this regulation has both neural and hormonal components [based on the differential effects of the α-adrenergic agonist prazosin and the neuronal blocking agent bretylium on the mean circulatory filling pressure (MCFP); see Figure 7.6]. Finally, Skals *et al.* (2006) showed the venous system plays an important regulatory role with regards to cardiac filling and SV in the air-breathing swamp eel (*Synbranhus marmoratus*) during hypoxia, and that this control is dependent upon both α- and β-adrenergic mechanisms. This latter study is important because it shows that venous tone may be controlled by similar mechanisms across a range of teleost species.

Although venous tone appears to be a major factor controlling cardiac filling and SV/Q during hypoxia, several other mechanisms may be involved. These include: (1) diminished myocardial force development; (2) hypoxia-mediated changes in gill vascular resistance, potentially leading to alterations in cardiac afterload and end-systolic volume; and (3) local (regional) alterations in vascular tone resulting in reduced systemic vascular resistance (R_{sys}) and a decreased arteriovenous pressure gradient (Sandblom and Axelsson, 2005). Although we will examine the effect of anoxia/severe hypoxia on myocardial function/contractility later in the chapter, the next two sections discuss the effect of hypoxia on branchial and systemic resistance, and evaluate their capacity to contribute to changes in arterial blood pressure.

2.1.3. Branchial Vascular Responses to Hypoxia

Most teleosts and elasmobranchs respond to severe hypoxia with an increase in branchial vascular resistance (R_{gill})(Butler and Taylor, 1975; Farrell, 1982; Pettersson and Johansen, 1982; Sundin and Nilsson, 1997;

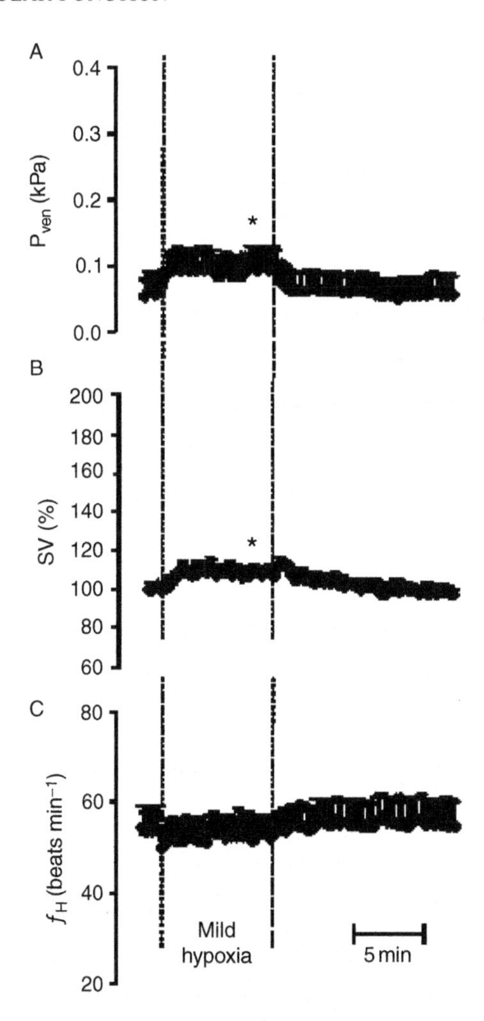

Fig. 7.5. Effect of 8 min of mild hypoxia (PO_2 ~85 mmHg; section located between dotted lines) on central venous pressure (P_{ven}), stroke volume (SV), and heart rate (f_H) in rainbow trout (*Oncorhynchus mykiss*). Statistically significant difference between the average value for the normoxic period and the average value of the last 2 min of the hypoxic period. [Modified from Sandblom and Axelsson (2005).]

Stensløkken *et al.*, 2004), and the mechanisms mediating changes in pressure and flow within the gill have been studied using a number of techniques, including epi-illumination microscopy. Once blood reaches the teleost gill from the heart it flows through the afferent filament arteries (AFA), is oxygenated in the secondary lamellae (SL), and then reaches the efferent

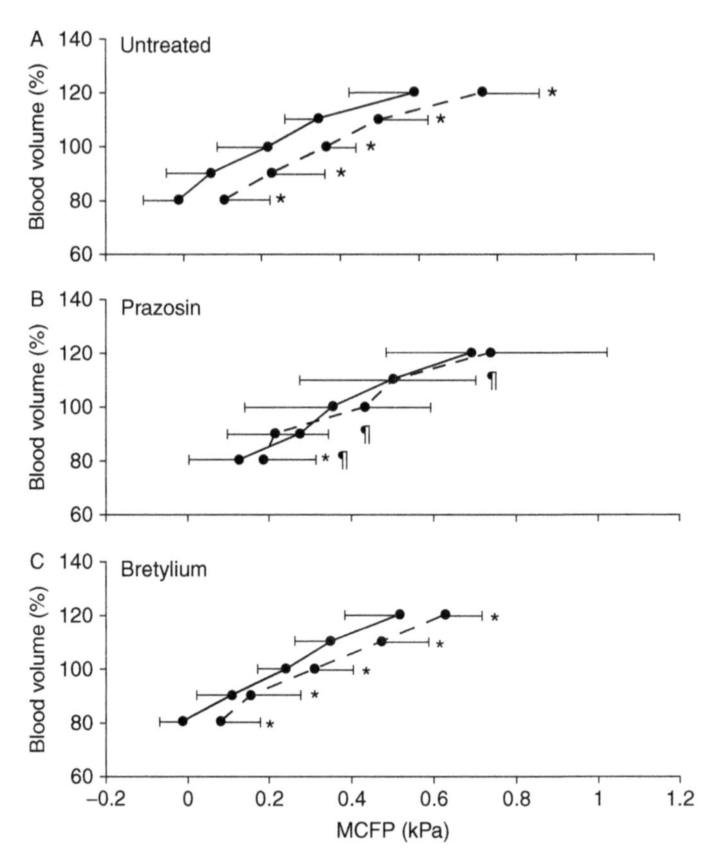

Fig. 7.6. Mean values (± S.D.) of mean circulatory filling pressure (MCFP) at 80–120% of total blood volume in untreated ($n = 11$) (A), prazosin-treated (1 mg/kg; $n = 9$) (B), and bretylium-treated (10 mg/kg; $n = 9$) (C) rainbow trout (*O. mykiss*). Solid lines represent normoxia and broken lines represent hypoxia (~70 mmHg). *Statistical difference between normoxia and hypoxia; and ¶ statistical difference between prazosin- and bretylium-treated normoxic values compared with corresponding untreated normoxic values ($P < 0.05$). [Reproduced from Sandblom and Axelsson (2006) with permission from the *American Journal of Physiology*.]

filamentous artery (EFA). At this point, however, it has two pathways that it can follow, and control of gill blood flow during hypoxia dramatically alters the distribution between these two pathways (Figure 7.7). Under resting normoxic conditions, the majority of blood leaving the SL enters the EFA, with only 5–30% of the blood flow entering the arteriovenous anastomoses (AVA) and nutritive vasculature (NV) where it is returned to the venous system through the branchial vein (BV) (Hughes *et al.*, 1982; Ishimatsu *et al.*, 1988; Sundin and Nilsson, 1992). In contrast, during severe hypoxia,

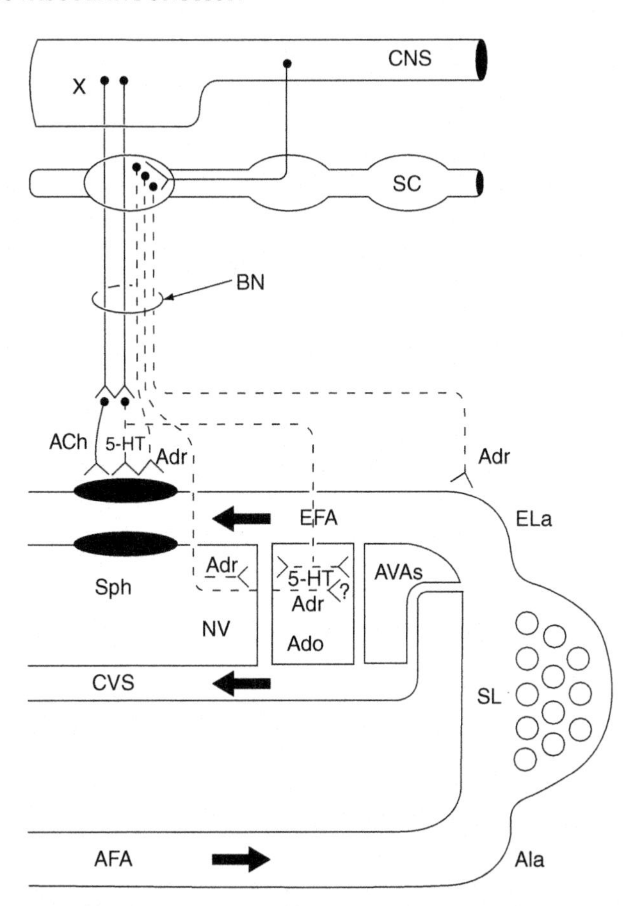

Fig. 7.7. Working model for autonomic control of teleost gill vasculature, demonstrating potential sites for control of vascular resistance and thereby blood flow distribution between the arterioarterial and arteriovenous pathways. Sphincter (represented by dark oblong shapes) at base of the efferent filamental arteries (EFA) is a key site affecting blood pressure and flow in the gill vasculature. Cholinergic and serotonergic innervation of the sphincter produce constriction, whereas adrenergic innervation may cause dilation, acting through β-adrenoceptors. Other potential sites for blood distribution control are the nutritive vasculature (NV) and arteriovenous anastomoses (AVAs). Adrenergic innervation of NV produces vasoconstriction via α-adrenoceptors, and this may also apply to AVAs, whereas serotinergic innervation and locally released adenosine (Ado) have been shown to cause dilation. AFA, afferent filamental artery; ALa, afferent lamellar arteriole; SL, (secondary) lamella; AVAs, anastomoses between EFA, efferent lamellar arteriole, and central venous system (CVS); ELa, efferent lamellar arteriole; BN, branchial nerve; ACh, Adr, and 5-HT putative nerve types (cholinergic, adrenergic, and serotonergic, respectively); CNS, central nervous system; SC, sympathetic chain; Sph, sphincter at base of EFA; X, vagus nerve. [Modified from Sundin and Nilsson (1997), with permission of the *American Journal of Physiology*.]

blood flow in the BV increases considerably (by about 1.5-fold in cod; Sundin, 1995), and a greater proportion of oxygenated blood is returned to the venous circulation. This redistribution of blood flow potentially plays a critical role in hypoxia tolerance by facilitating the energy-demanding work of ion-transporting cells that are located in the filamental epithelium, and of the heart by raising the oxygen content and partial pressure of the venous blood that supplies the myocardium (at least the spongy myocardium).

Control of flow and resistance in the teleost and elasmobranch gill vasculature has been studied using a number of pharmacological agonists and antagonists (both in normoxia and hypoxia)(Pettersson and Nilsson, 1979; Nilsson, 1984; Sundin, 1995; Sundin and Nilsson, 1996; Sundin and Nilsson, 1997; Smith *et al.*, 2001; Stensløkken *et al.*, 2004), and although abrupt (rapid) hypoxic exposures have been used exclusively in these investigations (and thus extension of this knowledge to the effects of moderate or gradual hypoxic exposures is unclear), we have a fairly comprehensive understanding of the control of gill vascular tone and blood flow under these conditions. First, although vasoactive compounds such as acetylcholine (Sundin and Nilsson, 1997) and adenosine (Sundin and Nilsson, 1996) can constrict the distal parts of the EFA and/or AFA, no constriction of these vessels has been observed during hypoxia using *in vivo* epi-illumination microscopy (Sundin and Nilsson, 1997; Stensløkken *et al.*, 2004). This suggests that the net effect of all mechanisms that mediate the hypoxic response does not involve vasoconstriction of the distal vasculature on either the afferent or efferent side. Second, alterations in neurohormonal control of AVA/NV and EFA resistance work in concert to re-direct flow into the teleost gill's venous circulation during hypoxia (Sundin, 1995; Sundin and Nilsson, 1997). For example, it appears that during normoxia (and possibly moderate hypoxia) β-adrenergic mediated dilation of the sphincters located at the base of the EFA (and distal to the AVA opening), and α-adrenergic mediated constriction of the AVA/NV predominate, and thus the majority of blood flow enters the arterioarterial system. Whereas, during severe hypoxia, cholinergic and serotinergic nerves cause constriction of the sphincter, while adenosine (mediated though A1 receptors) and increases in seritonergic nerve activity dilate the AVA and/or NV: the result of these neurohormonal changes is an increase in R_{gill} and enhanced flow through the AVA and NV (i.e., arteriovenous system). Third, although hypoxia can directly constrict the arterioarterial pathway (i.e., the EFA in teleosts; Sundin *et al.*, 1995; Smith *et al.*, 2001), it appears that hypoxia-induced vasoconstriction of the EFA is normally balanced by the dilatory effects of norepinephrine released from adrenergic nerves (Sundin *et al.*, 1995). Finally, although the branchial vascular anatomy of elasmobranchs differs significantly from that of teleosts, it appears from the study of Stensløkken *et al.* (2004) that cholinergic-mediated constriction of the EFA sphincter and adenosine-induced dilation of gill longitudinal vessels (the functional equivalent of the AVA in teleosts) also play a role in gill vasomotor responses to hypoxia in this taxa.

2.1.4. Systemic Vascular Resistance and Changes in Arterial Pressures

When examining the literature relating to these parameters, it is difficult to make even the most basic generalizations about how they are affected by hypoxia. There are several reasons for this. First, the majority of experiments have involved the rapid exposure of fish to hypoxia of limited duration (i.e., 8–12 min), and many of the cardiovascular responses to this type of experimental protocol are transitory or inconsistent. For example, Axelsson and Fritsche (1991), Sundin (1995), and Fritsche and Nilsson (1989) all exposed Atlantic cod to a water PO_2 of ~40–50 mmHg for 8–10 min and report that while R_{sys} increased initially by 20–50%, R_{sys} returned to prehypoxic levels within the first 5–6 min of hypoxia (i.e., the response of R_{sys} was likely an acute stress response due to the protocol and not associated with hypoxia itself; e.g., see Ristori and Laurent, 1989). There does not appear to be a clear relationship between the severity of hypoxia and R_{sys}, as the R_{sys} of rainbow trout increases by approximately 10% at a PO_2 of 85 mmHg, decreases by approximately 30% at a PO_2 of ~50 mmHg, but is essentially unchanged by exposure to severe hypoxia (PO_2 <10 mmHg) (Sundin and Nilsson, 1997; Sandblom and Axelsson, 2005). Finally, it is difficult to predict how dorsal aortic pressure (PDA) and ventral aortic pressure (PVA) will change when exposed to rapid hypoxia, because of the three major responses of Q. Also, Q can vary considerably even at the same level of hypoxia within a species [e.g., the Q of Atlantic cod did not change in Fritsche and Nilsson (1989) but increased by ~50% in Axelsson and Fritsche (1991 and Sundin (1995)]. Second, although isolated vessels of the rainbow trout respond to hypoxia, whether the vessels are refractory, constrict, or dilate depends on the type of vessel and the nature of pre-existing stimulation, and other factors (e.g., season or other environmental factors) appear to play a modulatory role in conditioning the response of the vessels (Smith *et al.*, 2001). Third, the response of PDA and PVA to graded hypoxia is species-dependent among teleosts; the rainbow trout showing significant increases in both parameters starting at approximately 70–80 mmHg (Holeton and Randall, 1967), tuna showing no change in either parameter down to 50 mmHg (Bushnell and Brill, 1992), whereas both the Japanese eel (*Anguilla japonica*; Chan, 1986) and lingcod (Farrell, 1982) become hypotensive somewhere between 75 and 35 mmHg.

It is obvious from the preceding discussion that our knowledge of the control of systemic vascular resistance and blood pressure in fishes is extremely limited, and that carefully designed experiments using appropriate acclimation times/conditions and varied species will be required before our understanding is significantly enhanced. However, there are a few points that are worth making at this time. Based on the limited species examined to date, it appears that elasmobranchs and sturgeons regulate systemic vascular resistance in a different way from teleosts. Systemic vascular resistance decreases in the dogfish (Butler and Taylor, 1971) and sturgeon (Agnisola *et al.*, 1999) at PO_2s less than 60 mmHg, while it increases in those teleosts examined to date, including the rainbow trout (Holeton and Randall, 1967), lingcod (Farrell, 1982), and Japanese eel (Chan, 1986). Two studies have investigated the effects of hypoxia on gastrointestinal (GI) blood flow (Axelsson and Fritsche, 1991; Axelsson *et al.*, 2002), and provide important insights into the regulation of R_{sys} and blood flow distribution during hypoxia (e.g., see Figure 7.8). (1) Resistance in vessels supplying the GI

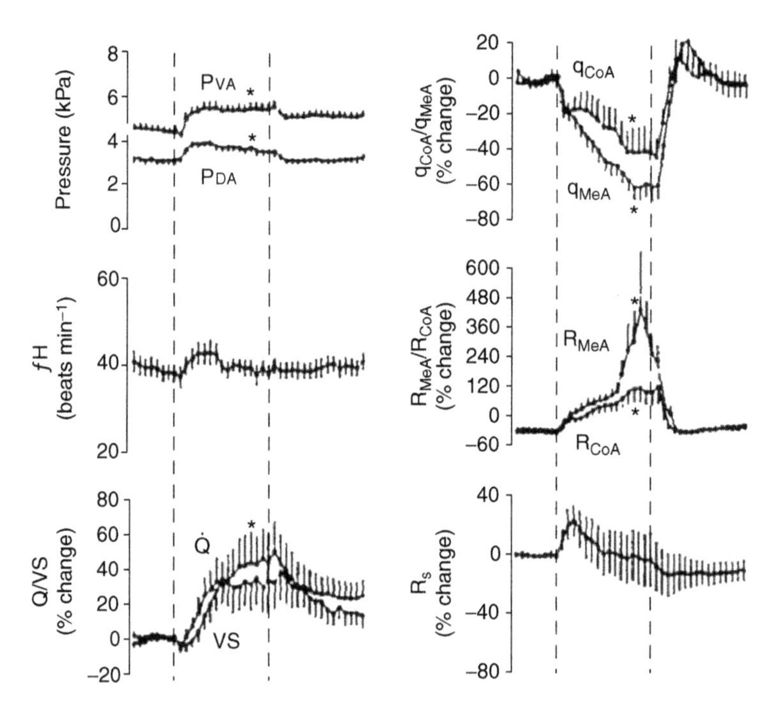

Fig. 7.8. The response of Atlantic cod (*Gadus morhua*) cardiovascular parameters to 8 min of hypoxic exposure (PO_2 ~30–40 mmHg) ($N = 8$–12). Data are means ± S.E. Asterisks indicate statistically significant ($P < 0.05$) differences compared with normoxic values. PVA, ventral aortic pressure; PDA, dorsal aortic pressure; f_H, heart rate; Q, cardiac output; VS, stroke volume; q_{Coa}, flow in coeliac artery; q_{MeA}, flow in mesenteric artery; R_{CoA}, resistance in coeliac artery; R_{MeA}, resistance in mesenteric artery; R_{sys}, systemic vascular resistance. [Modified from Axelsson and Fritsche (1991).]

vessels of cod increases by ~150–320% when exposed to severe hypoxia (30–40 mmHg), and this vasoconstriction is dependent on both nervous and humoral adrenergic mechanisms; these two mechanisms playing varied, but important, roles in controlling overall systemic vascular resistance in cod and rainbow trout at rest and during exercise and hypoxic exposure (Randall and Daxboeck, 1982; Smith *et al.*, 1985; Axelsson and Nilsson, 1986; Fritsche and Nilsson, 1990). (2) The relative proportion of Q reaching the gut was reduced by approximately 50% and 20% in unfed versus fed seabass, respectively, when exposed to hypoxia (Axelsson *et al.*, 2002). This study shows that locally released vasoactive substances can offset the effects of hypoxia on GI vascular resistance and blood flow. (3) The observation that these large increases in resistance of the GI circulation did not lead to increases in R_{sys}, and that GI blood flow as a proportion of cardiac output fell by 50% during hypoxia indicates that the somatic vasculature is significantly dilated at these water oxygen levels and that there is a redistribution of blood flow from the GI circulation to the somatic circulation. These direct measurements of blood flow contradict earlier measurements using microspheres, which suggested that the proportion of Q directed to the muscles and visceral organs does not change during hypoxia (Cameron, 1975). Finally, it has been recently reported that concentrations of H_2S (hydrogen sulphide) in rainbow trout plasma reach levels that produce vasoactive effects in isolated vessels (Dombkowski *et al.*, 2004), that H_2S is constitutively synthesized by vascular smooth muscle and cellular concentrations are determined by a simple balance between H_2S production and the amount of O_2 available for H_2S (hydrogen sulphide) oxidation (Olson, 2008), and that hypoxia and H_2S evoke the same response in vertebrate isolated blood vessels, irrespective of whether the response is a contraction, relaxation, or is multiphasic (see Fig. 1 in Olson, 2008). Given the extremely varied response of isolated trout vessels to hypoxia (Smith *et al.*, 2001) and the data on H_2S to date, it appears that H_2S acts as an "oxygen sensing molecule" in smooth muscle cells (e.g., see Olson *et al.*, 2006). Further research efforts in this area will significantly enhance our understanding of the regulation of blood flow and vascular resistance in fishes during both hypoxia and hyperoxia.

2.2. Chronic Hypoxia

In contrast to the large body of literature that exists on the effects of acute hypoxia on fish cardiovascular function, the effects of chronic (days to weeks) hypoxia has been largely overlooked; with the exception of the crucian carp (see Chapter 9). However, there are now a few studies that have looked at the effects of chronic hypoxia on cardiovascular function in developing zebrafish and adult fish.

2.2.1. Hypoxic Effects on Zebrafish Cardiovascular Development/Function

To date, four studies have investigated the effect of chronic hypoxia on cardiovascular development, and although the use of different temperatures, levels of hypoxia, and strains complicates some interpretations, these studies reveal novel insights into how chronic hypoxia modifies cardiovascular function and development in the early life history stages of this species. In the first study on the effects of chronic hypoxia on zebrafish cardiac function, Jacob *et al.* (2002) placed embryos into moderately hypoxic (approx. 75 mmHg O_2) water and measured heart function using videomicroscopy from 1 to 12 dpf at temperatures ranging from 25 to 31°C. This study revealed some temperature-dependent differences in the response of cardiac function to hypoxia (i.e, when dpf changes appeared), but clearly showed that chronic hypoxia increases Q as early as 3–4 dpf (at 25 and 28°C) and that this elevated Q was due to both increases in f_H and end-diastolic volume (Figure 7.9). This is a very interesting finding, as adrenergic receptors do not appear in the normoxic zebrafish heart until 5–6 dpf (Bagatto, 2005; Schwerte *et al.*, 2006), exposure to severe hypoxia (15 mmHg) apparently delays the appearance of adrenergic receptors in the heart by 2 days at 25°C (Bagatto, 2005), and vagal tone on the heart is not established until 12 dpf at 28°C (Schwerte *et al.*, 2006). Further, the direct effect of hypoxia on vertebrate cardiomyocytes is normally a decrease in activity, and chronic exposure of zebrafish embryos/larvae to more severe hypoxia results in a sustained decrease, not increase, in f_H (Figure 7.10, Bagatto, 2005; Moore *et al.*, 2006). Collectively, these results suggest that cardiac function was increased in the study by Jacob *et al.* (2002) because convective oxygen transport becomes important in fish larvae when the gradient for bulk oxygen diffusion (i.e., from the environment to the tissues) is reduced, that the afferent nervous system is capable of sensing hypoxic conditions very early in life, and that central control units are active and can indirectly stimulate the heart via some presently unidentified hormone. While the increase in f_H was most likely due to stimulation of the cardiac pacemaker or the cardiomyoctyes, Jacob *et al.* (2002) do not provide any explanation for the increase in end-diastolic volume with hypoxia. However, this could be due to an increase in ventricle size (a parameter not measured in any of the studies), ventricular remodeling (but see Marques *et al.*, 2008), or an increase in venous tone/pressure and thus cardiac filling (this homeostatic mechanism clearly established for adult fish; see Sandblom and Axelsson, 2005, 2006). With regards to the effects of chronic hypoxia on the zebrafish's vasculature, a number of effects were also noted. These included increases and decreases in blood flow distribution to the muscle (by approx. 350%) and GI tract (by approx. 45%), respectively,

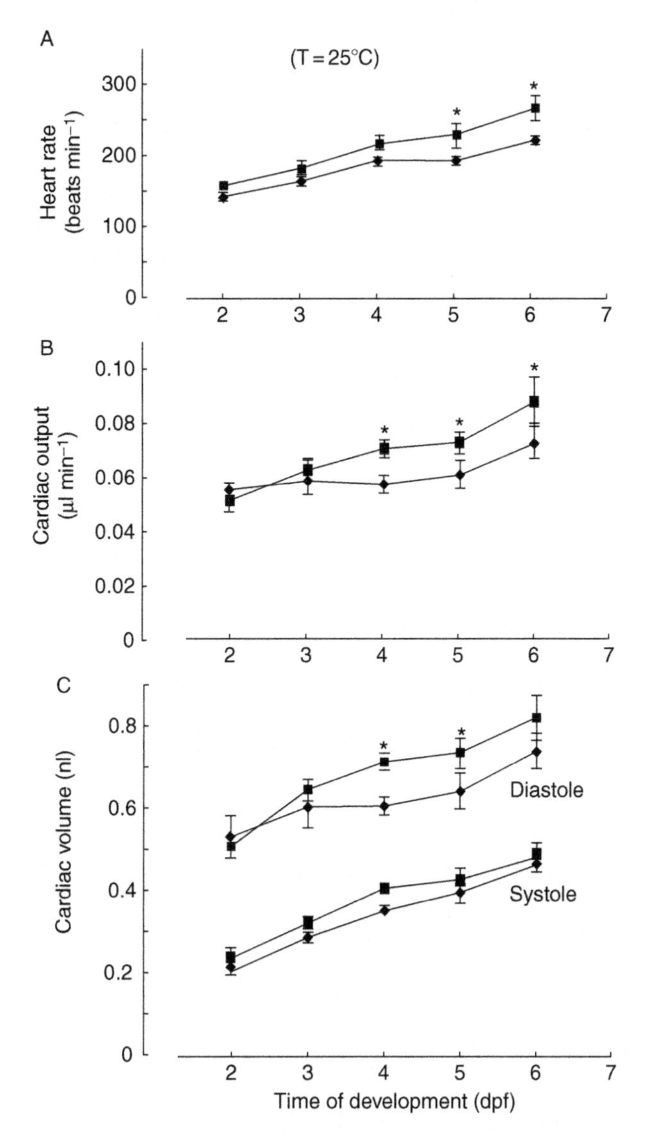

Fig. 7.9. Cardiac activity [heart rate (A); cardiac output (B); systolic and diastolic ventricular volumes (C)] in zebrafish (*D. rerio*) larvae raised under normoxia ($PO_2 = 150$ mmHg) and under chronic hypoxia ($PO_2 = 75$ mmHg) at a temperature of 25°C ($n = 10$). dpf = days post-fertilization.*Significantly different from controls ($P < 0.05$). [Modified from Jacob *et al.* (2002), with permission of the *American Journal of Physiology*.]

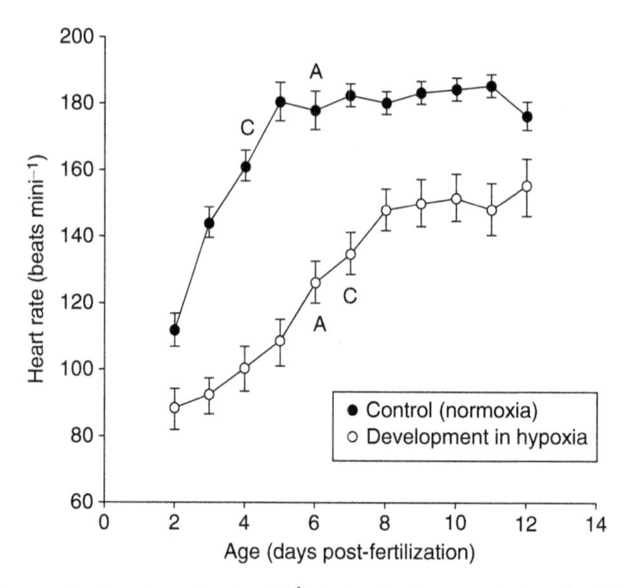

Fig. 7.10. Mean resting heart rate (beats min^{-1}) during development in the zebrafish (*D. rerio*) at 25 °C when reared under normoxia and hypoxia (PO_2 <10 mmHg). The letter C indicates the first significant negative chronotropic response to cholinergic agonists during development, and the letter A indicates the first significant positive chronotropic response to adrenergic agonists. All data points are significantly different from the corresponding controls. Data shown are means ± S.E. [Modified from Bagatto (2005).]

of 12–15 dpf zebrafish (Schwerte *et al.*, 2003), and changes in the size of major arteries and veins (although these latter results were study and age dependent: Bagatto, 2005; Moore *et al.*, 2006). An increase in blood flow to the muscle (especially superficial red muscle) would distribute blood from the core to the surface of the larvae, and thus potentially enhance cutaneous oxygen uptake and internal oxygen convection during hypoxic conditions. While the mechanism(s) facilitating the redistribution of blood flow in hypoxic zebrafish larvae are unknown, α-adrenergically controlled precapillary sphincters are present in the intersegmental muscle tissue of zebrafish by 8 dpf (Schwerte and Pelster, 2000) and Fritsche *et al.* (2000) showed that nitric oxide can cause vasodilation in the zebrafish as early as 5 dpf. However, the redistribution of blood flow is clearly not due to changes in angiogenesis (increased/decreased vascularization) as Schwerte *et al.* (2003) failed to show any effect of chronic hypoxia (PO_2 65 mmHg) on vascularization of the tail muscle or the gut.

While the above results are novel, and informative, the work of Moore *et al.* (2006) may be the most important as it sets the stage for potentially transformative discoveries relative to how genetics and environmental challenges during the embryonic/larval period influence adult cardiovascular function and morphology. This study tested for family-specific differences in the response of an integrated set of cardiovascular traits to severe hypoxia (approx. 15 mmHg) and reported that considerable variation in the degree of familial response to hypoxia exists in cardiovascular traits that relate to Q (Figure 7.11). While these authors only measured these traits at 4 dpf, and thus it is not known whether differences in traits at this stage of development translate into differences in the juvenile or adult, there is indirect evidence

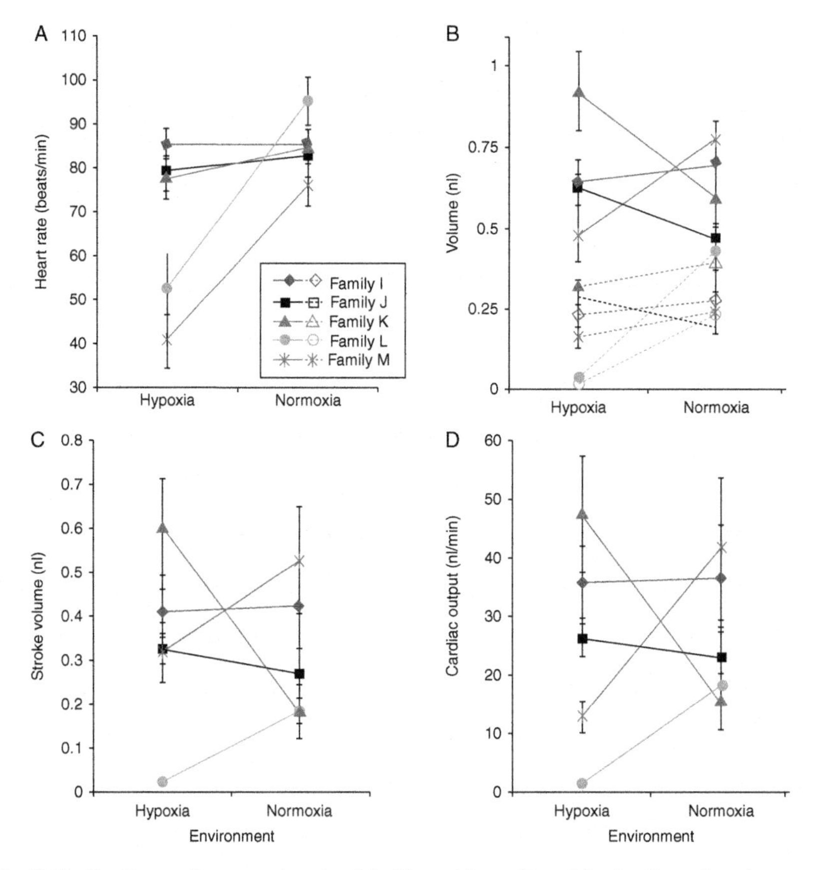

Fig. 7.11. Cardiac performance in zebrafish (*D. rerio*) as affected by family and environment (hypoxia, <10 mmHg; normoxia). In B, solid lines and symbols represent means for end diastolic volume (EDV) while intermittent lines and open symbols represent end systolic volume (ESV). Values are means ± 1 S.E. Family and family × environment interactions were significant sources of variation in all five traits, while a direct effect of environment was only significant in heart rate and ESV. [Reproduced from Moore *et al.* (2006).]

that genetic variation in response plasticity may provide the basic ingredient for adaptation to variable environments. For example the work of Gamperl *et al.* (Gamperl *et al.*, 2004; Faust *et al.*, 2004; Overgaard *et al.*, 2004b) shows that there is considerable variation in the inherent myocardial hypoxia tolerance of rainbow trout from different hatcheries, and that this influences their capacity to be preconditioned (see below).

2.2.2. Effects on Adult Cardiovascular Function

Surprisingly, to date, there is only one publication on the effects of chronic hypoxia on *in vivo* cardiovascular function. Burleson *et al.* (2002) acclimated channel catfish to normoxia and hypoxia (PO_2 75 mmHg) for 1 week, and reported that f_H in hypoxia-acclimated fish was 15–20% higher as compared with normoxia-acclimated individuals when tested under both normoxia and hypoxia. However, Petersen and Gamperl (unpublished data) recently made the first measurements of fish cardiorespiratory function during exercise (from rest to critical swimming speed) under hypoxia (PO_2 approx. 60 mmHg), and of how acclimation of cod to this same level of hypoxia for 6–12 weeks influenced resting and exercise-induced cardiac function under both hypoxic and normoxic conditions. This study confirmed the findings of Burleson *et al.* (2002) with regards to chronic hypoxia increasing resting f_H (see below; also Figure 7.12), and suggests that this is a regulated response. This is because Petersen and Gamperl (unpublished data) report that hypoxia-acclimated cod had significantly lower values for resting and maximum SV and Q in both swim tests, and a significantly lower scope for SV when swum under hypoxic conditions, as compared with the normoxia-acclimated group (Figure 7.12). There are at least three potential explanations for the poor pumping capacity of hearts from hypoxia-acclimated fish. First, it is possible that the cod myocardium was damaged by constant exposure to low oxygen conditions. Such a conclusion would be consistent with the findings of Lennard and Huddart (1992) who reported that cardiomyocytes in flounder (*Platichthys flesus*) subjected to 3 weeks of hypoxia (water PO_2 ~35 mmHg) showed striking changes in mitochondrial morphology (decreased size, budding, and necrosis) and evidence of myofibril degeneration. However, the level of hypoxia utilized in the Petersen and Gamperl (unpublished data) (approx. 60 mmHg) was not nearly as severe as that used by Lennard and Huddart (1992), and several studies have shown that, at least in the trout heart, acute (<30 min) exposure to severe anoxia (perfusate PO_2 <1 kPa) does not result in myocardial necrosis or a disruption in myocardial energetic and enzymatic status (Faust *et al.*, 2004; Overgaard *et al.*, 2004a,b). These data, thus, question whether myocardial damage/necrosis was experienced by the hypoxia-acclimated cod. Second, it is possible that the hearts of hypoxia-acclimated cod were merely "stunned" (i.e., experiencing mechanical dysfunction related to a decrease in myocardial

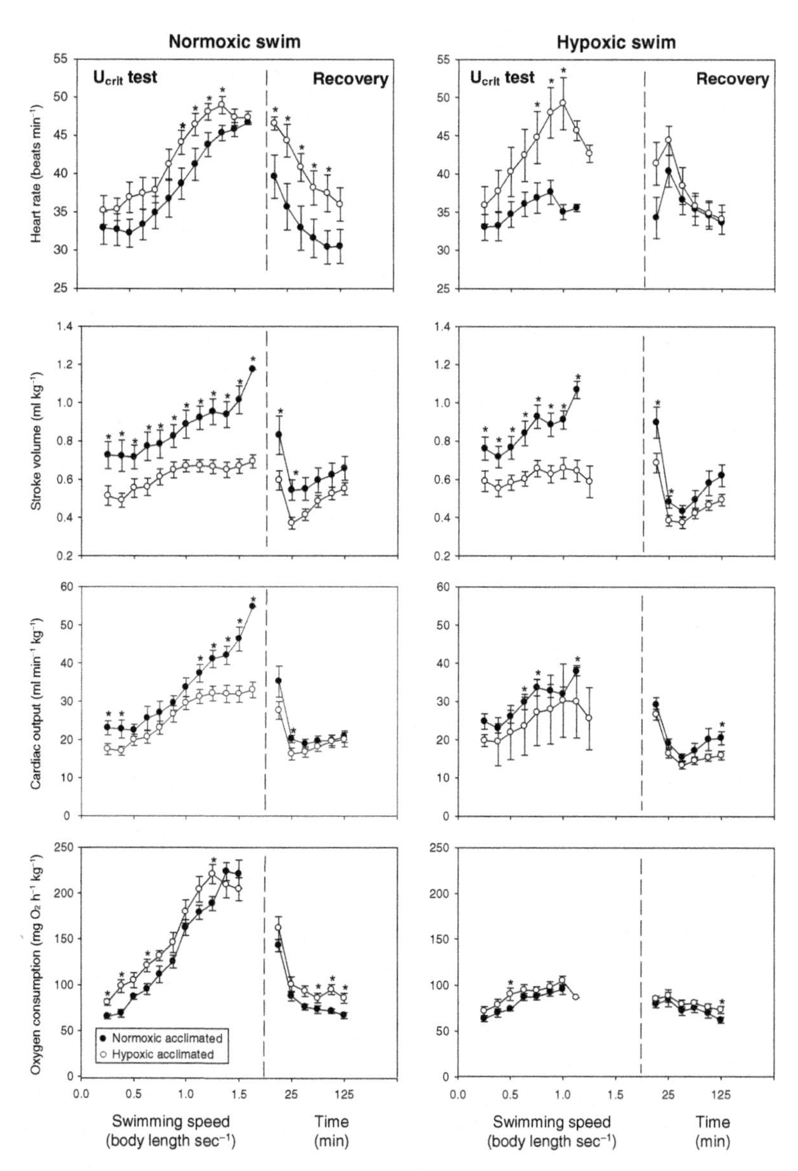

Fig. 7.12. Cardiac parameters and oxygen consumption in normoxia- ($N = 10$) and hypoxia-acclimated (PO_2 60 mmHg, $N = 12$) Atlantic cod (*Gadus morhua*) during critical swimming speed (U_{crit}) tests, and during postexercise recovery. All fish were swum in normoxic water on day 1 and hypoxic water on day 2, but recovery was performed in normoxic water for both swims. *Indicates a significant difference ($P < 0.05$) between the normoxia- and hypoxia-acclimated groups at a particular swimming speed (Petersen and Gamperl, unpublished data).

calcium sensitivity; see Bolli and Marban, 1999). However, this seems unlikely as Driedzic *et al.* (1985) showed that 4–6 weeks of hypoxic acclimation (at 45 mmHg) enhanced the contractility of normoxic myocardial strips from *Zoarces vivparous* under conditions of elevated calcium. Further, Gamperl and Petersen (unpublished data) have shown that while the maximum *in situ* cardiac performance of hearts from hypoxia-acclimated cod is reduced by a similar amount to that measured *in vivo* under normoxic conditions (by approx. 25%; see Figure 7.12), they can maintain maximum cardiac function under severe hypoxia (PO_2 5–10 mmHg) longer than hearts from normoxia-acclimated cod, and show enhanced post-hypoxia recovery of maximum cardiac function as compared to their normoxia-acclimated counterparts. Finally, it is possible there was no myocardial damage or significant dysfunction in the hearts of hypoxia-acclimated cod, and that hypoxia-induced myocardial remodeling reduced the maximum SV of the heart. Although, the similar relative ventricular mass (RVM) in hypoxia- and normoxia-acclimated cod (in contrast to the increase in RVM after trout are made hypoxemic by repeated injections of phenylhydrazine; Simonot and Farrell, 2007) provides some evidence against extensive cardiac remodeling in chronically hypoxic cod, Marques *et al.* (2008) showed that acclimation of zebrafish and the cichlid (*Haplochromis piceatus*) to 75 mmHg O_2 increased cardiac myocyte density (presumably though hyperplasia) and that this resulted in a smaller ventricular outflow tract and reductions in the size of the central ventricular cavity and lacunae (Figure 7.13). Such a decrease in the capacity of the ventricle to fill with blood would certainly explain why maximum *in vivo* SV was reduced by 28% in the hypoxia-acclimated cod, and raises the possibility that cardiac remodeling caused by hypoxic acclimation aims to reduce the wall tension required to eject blood from the ventricle (i.e., see Law of LaPlace) and the workload of individual cardiomyocytes. Clearly, more research needs to be conducted before the mechanism(s) mediating the diminished pumping capacity of hearts from hypoxia-acclimated fishes can be understood.

Interestingly, despite the diminished SV and elevated resting f_H, hypoxia-acclimated cod were able to increase f_H by a similar amount in the normoxic swim (i.e., the scope for f_H was not altered by hypoxia acclimation), and they were able to elevate f_H during the hypoxic swim to levels measured during normoxia (Figure 7.12). This latter result allowed them to have a significantly greater scope for f_H (12.6 vs. 5.8 beats min^{-1} in normoxia-acclimated fish) when tested under hypoxic conditions, and to achieve the same maximum Q as compared to normoxia-acclimated fish when swum at 60 mmHg O_2. The mechanism(s) resulting in the differential regulation of f_H in the two groups when swum under hypoxic conditions cannot be ascertained from the work of Petersen and Gamperl (unpublished data) or the literature. However, as the f_H of *in situ* hearts from hypoxia-acclimated cod was similar to that of

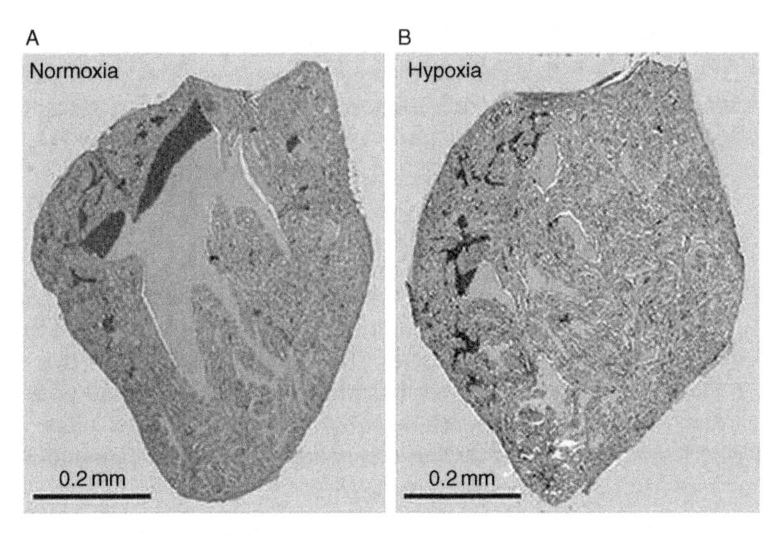

Fig. 7.13. Histological changes in cichlid (*Haplochromis piceatus*) hearts after exposure to chronic constant hypoxia (PO_2 = ~15 mmHg for 3 weeks). (A) Heart of normoxia-acclimated individual. (B) Heart of hypoxia-acclimated individual. Hearts were sectioned and stained with hematoxylin-eosin. [Modified from Marques *et al.* (2008).] (See Color Insert.)

normoxia-acclimated hearts at resting or maximal Q (Petersen and Gamperl, unpublished data), it is clear that the enhanced capacity of hearts from hypoxia-acclimated fish to elevate f_H is under neural and/or hormonal control. These results, in combination with recent data showing that rainbow trout at 24°C can maintain Q even when f_H is cut in half using the pharmacological agent zetabradine (Gamperl *et al.*, 2008), highlight the tremendous plasticity in how fish cardiorespiratory physiology responds to environmental challenges and that our understanding of control mechanisms that mediate myocardial function and adaptation in fishes is far from complete.

3. CARDIAC ENERGY METABOLISM

Earlier reviews summarize the control of energy metabolism in fish hearts (Driedzic, 1992; Driedzic and Gesser, 1994). In addition, Farrell and Stecyk (2007) provide a more recent discussion of rates of ATP turnover under normoxia and hypoxia and how they relate to the heart's power output requirements, especially in hagfish, carps, and the rainbow trout. They point out that there are two general strategies in the cardiac response to hypoxia. The first is to meet routine power output (ATP requirement) with a much enhanced anaerobic ATP production (i.e., glycolysis); the second is to reduce power output (ATP demand). In either situation, however, energy generation could be supported by an enhanced oxygen extraction. This section builds upon these contributions and addresses more recent issues of how metabolism under hypoxic conditions is regulated and serves to extend survival time. Earlier papers are referred to only in the development of key points.

3.1. Creatine Phosphate and ATP Levels

The primary function of energy metabolism in the heart is to maintain ATP levels to support the ATPases of the contractile apparatus and of ion pumping. Studies with whole animals and with *in vitro* preparations reveal that total heart ATP levels are generally well defended under hypoxic conditions and that cardiac failure ensues before any substantive drop in total tissue ATP (see Driedzic and Gesser, 1994). This is especially true if ATP demand is low (Arthur *et al.*, 1992; Driedzic and Gesser, 1994; Overgaard and Gesser, 2004). In contrast, Creatine Phosphate (CP) levels often fall to extremely low levels under hypoxic conditions, and although rarely measured, this would result in concomitant increases in the free phosphate pool (an increase in phosphate reportedly one of the mechanisms responsible for contractile failure; Allen *et al.*, 1985; Godt and Nosek, 1989). For instance, in the perfused rainbow trout heart performing basal levels of work, CP decreased by 80% when exposed to anoxia while ATP levels remained constant (Arthur *et al.*, 1992). Interestingly, microarray experiments reveal 3- to 4-fold decreases in heart creatine kinase mRNA levels as a result of hypoxic acclimation. This was shown for *Gillichthys mirabilis* held for 6 days (Gracey *et al.*, 2001) and for zebrafish held for 21 days, both at 20 mmHg oxygen (Marques *et al.*, 2008). The functional significance of this remains to be ascertained, but it could result in slower rates of CP discharge (and thus phosphate accumulation) in hypoxia-adapted fish when faced with an acute hypoxic challenge or slower rates of CP replenishment following hypoxic episodes.

Ventricle preparations functioning at reduced levels of work under hypoxia can often increase force development in response to Ca^{2+} application (Driedzic and Gesser, 1994; Bailey *et al.*, 2000). This is an important finding since it suggests that ATP production mechanisms need not be operating at the maximum rates under some hypoxic conditions, and therefore that the rate of ATP production is not necessarily the limiting factor to performance.

3.2. Decreasing ATP Demand

A hallmark of anoxic tolerance is the ability to decrease ATP demand. As discussed earlier in this review, the hypoxia-induced bradycardia shown by some species will decrease demands on myofibrillar ATPase. Further,

Vornanen *et al.* (see Chapter 9) discuss myosin isoforms that change from summer to winter in crucian carp, and energy conservation under hypoxia occurs through other processes as presented below.

Hypoxia is associated with decreased demands of ion pumping mechanisms. The reader is again referred to Chapter 9 for a discussion of the relationship between electrical activity and Ca^{2+} management. Crucian carp (*C. carassius*) acclimated to <6 mmHg oxygen showed a 30% decrease in maximal *in vitro* Na^+–K^+–ATPase activity (Paajanen and Vornanen, 2003). The authors propose that this is related to the decreased number of action potentials associated with slower heart rates, and that this alone could contribute to reduced ATP usage. Surprisingly, there was no change in properties of the inwardly rectifying K^+ current, a major leak and repolarizing pathway. However, it was stated (although not documented) that in a few recordings sarcolemmal K_{ATP} (sarcK_{ATP}) channel activity was observed and this occurred more frequently in hypoxic than normoxic myocytes. SarcK_{ATP} channels have been identified in crucian carp and rainbow trout myocytes, and these channels increase their open probability *in vitro* in response to lack of ATP or complete metabolic inhibition (oxygen stripping and glycoytic poisoning) (Paajanen and Vornanen, 2002).

Total ATP levels generally remain relatively stable under hypoxic conditions; however, it is likely that there are subcellular microenvironments in which the ATP/ADP ratio decreases. This could in turn open sarcK_{ATP} channels resulting in K^+ efflux that serves to shorten the duration of the action potential, limiting Ca^{2+} influx, and consequently reducing contractility. There is compelling evidence that the opening of these channels is important in some species. Application of anoxia (NaCN plus N_2) to rainbow trout ventricle strips resulted in a rapid loss of force associated with a transient decrease in action potential duration (Gesser and Høglund, 1988). In the goldfish heart, hypoxia results in a decrease in action potential duration in association with the opening of sarcK_{ATP} channels and the opening of these channels improves hypoxic cell viability. Furthermore, the opening of sarcK_{ATP} channels appears to be mediated by nitric oxide activation of guanylyl cyclase (Cameron *et al.*, 2003; Chen *et al.*, 2005). This is presumably followed by phosphorylation of the channel protein via protein kinase G (Han *et al.*, 2001). Regardless of the mechanism, the opening of sarcK_{ATP} channels under hypoxia could serve to decrease demands on Na^+–K^+–ATPase, Ca^{2+}–ATPases and indirectly on myofibrillar ATPases. In effect, there is an elegant feedback mechanism at the metabolic level whereby a small decrease in ATP could by itself reduce further ATP demand.

Mitochondrial K_{ATP} (mitK_{ATP}) channels may also be involved in the hypoxia defense mechanism, both through decreasing contractility and maintaining mitochondrial integrity. Treatment of hypoxic ventricle strips from yellowtail flounder (*Limanda ferruginea*) with the drug diazoxide, a $mitK_{ATP}$ channel opener, exacerbated the loss of twitch force presumably conserving ATP (MacCormack and Driedzic, 2002) and extending the hypoxic viability of isolated heart cells from goldfish (Cameron *et al.*, 2003). In contrast, force development was increased under hypoxia in ventricle strips from the Amazonian armoured catfish, *Liposarcus pardalis*, when treated with 5-hydroxydecanoic acid (5HD), a $mitK_{ATP}$ channel blocker (MacCormack *et al.*, 2003b), and consistent with this finding 5HD decreased the protective effect of channel opening in hypoxic goldfish hearts (Chen *et al.*, 2005). How the opening of $mitK_{ATP}$ channels during hypoxia can alter force development in not known, but activation of $mitK_{ATP}$ channels results in depolarization of the mitochondrial membrane and altered mitochondrial Ca^{2+} uptake and release (Holmuhamedov *et al.*, 1998). Since force development is intimately associated with calcium levels, any alteration in calcium cycling could affect twitch force development. Regardless, the physiological implications of $mitK_{ATP}$ regulation appear to be important although the mechanisms of action remain to be resolved. Aside from contractile aspects, dos Santos *et al.* (2002) argue that in the ischemic rat heart, open $mitK_{ATP}$ channels maintain mitochondrial volume and the tight structure of the intermembrane space, and that this prevents ATP hydrolysis by mitochondria under oxygen limitation. If this occurs in fish hearts, it would be a further mechanism for extending energy reserves.

Adenosine is well recognized as an agent that protects the mammalian heart under oxygen limitation through a variety of actions including reduced cardiac performance (for extensive reference to the mammalian literature see MacCormack and Driedzic, 2007; Stecyk *et al.*, 2007). Adenosine is formed from the breakdown of the adenylate pool, and serves as a signaling molecule that links ATP supply with ATP demand. For example, the injection of adenosine resulted in a decrease in heart rate in the normoxic Antarctic notothenioid *Pagotehenia borchgrevinki* (Sundin *et al.*, 1999a) and the epaulette shark (*Hemiscyllium ocellatum*) (Stensløkken *et al.*, 2004). Also, under normoxic conditions adenosine caused a decrease in heart rate in isolated, whole heart preparations and a decrease in force developed by electrically paced ventricle strips from rainbow trout (Aho and Vornanen, 2002). However, adenosine control does not appear to be generally important for the hypoxic myocardium. Short-horned sculpin subjected to hypoxia for up to 6 h showed marked bradycardia but no change in heart adenosine levels, despite the finding that adenosine levels increased following 30 min of reoxygenation (MacCormack and Driedzic, 2004). Atropine treatment resulted in a release of bradycardia and an increase in Q under hypoxia, but there was no further effect of an adenosine blocker suggesting that the heart of the hypoxic short-horned sculpin is not under adenosine control

(MacCormack and Driedzic, 2007). In *L. pardalis*, a species that does not exhibit bradycardia (MacCormack *et al.*, 2003a), there was no change in heart adenosine content as a function of hypoxia; however, adenosine levels were significantly higher in fish maintained in laboratory aquaria than in fish sampled directly from a pond (MacCormack *et al.*, 2006). The studies with short-horned sculpin and *L. pardalis*, two species that show quite different cardiac responses to hypoxia, are important in revealing that heart adenosine concentrations can vary under different conditions, but at least in these fish, adenosine does not play a role in protecting the heart under hypoxia. Similar conclusions were reached in experiments with crucian carp where injection of the adenosine receptor antagonist aminophylline did not release hypoxia-induced bradycardia or change Q. Furthermore, the application of aminophylline had no impact on contractile failure of ventricle strips caused by NaCN (Vornanen and Toumennoro, 1999; Stecyk *et al.*, 2007). Finally, injection of aminophylline did not alter hypoxia-induced bradycardia in the epualette shark (Stensløkken *et al.*, 2004). The only study that suggests adenosine control plays a role under hypoxia is with common carp (*Cyprius carpio*). Heart rate in normoxic water at 5°C was about 8 bpm and Q about 4 mL min^{-1} kg^{-1}, and these values decreased to about 4 bpm and 2.3 mL min^{-1} kg^{-1} under hypoxia. Adenosine receptor blockade with aminophylline increased these values to only about 5 bpm and 3.3 mL min^{-1} kg^{-1} (Stecyk *et al.*, 2007). Collectively, this body of work eliminates the appealing conjecture that adenosine control plays an important role in reducing ATP demand by the hypoxic fish heart, although the possibility of a modest contribution in the case of the common carp cannot be ruled out.

Next to contraction and ion pumping, protein synthesis accounts for the greatest ATP demand in the fish heart. Under hypoxia, protein synthesis is decreased by about 50% in both crucian carp (*C. carassius*) and oscar (*Astronotus ocellatus*) (Smith *et al.*, 1996; Lewis *et al.*, 2007). These studies followed the incorporation of radiolabeled phenylalanine into the protein pool and thus reflect decreases in rates of translation. How this decrease in protein synthesis is achieved is unknown but a decrease in pH is a likely possibility.

3.3. The Potential of Enhanced Oxygen Utilization Under Hypoxia

One potential metabolic strategy to cope with hypoxia would be to extend the lower limit at which oxygen extraction from the extracellular space is possible. It is well established across different species that the presence of myoglobin (Mb), at least in acutely challenged isolated hearts, results in a better maintenance of ATP levels, oxygen consumption, and performance under hypoxia (Driedzic and Gesser, 1994; Acierno *et al.*, 1997). As such, an increase in heart Mb content is a potential adaptive response to hypoxia.

Indeed, Mb protein levels increased by 20% in hearts of zebrafish maintained for 48 h at 22% oxygen (Roesner *et al.*, 2006). However, an increase in Mb content does not appear to be a common response. There was no change in Mb content in hearts of *Zoarces viviparous* held under hypoxia for 4–6 weeks (Driedzic *et al.*, 1985). Changes in Mb expression were also not observed in microarray studies on *G. mirabilis* and zebrafish (Gracey *et al.*, 2001; Marques *et al.*, 2008), yet given the conservative nature of the protein it should have appeared in these experiments. More recently, Hall *et al.* (unpublished data) showed no change in Mb mRNA levels in hearts of Atlantic cod held at 40% oxygen for 3 or 6 days (Figure 7.14). In contrast, a low temperature challenge does result in an increase in both protein and Mb transcript level in hearts of Atlantic cod (Lurman *et al.*, 2007), showing that the modulation of the content of this protein can occur. If Mb is so powerful in allowing enhanced oxygen extraction, we question why

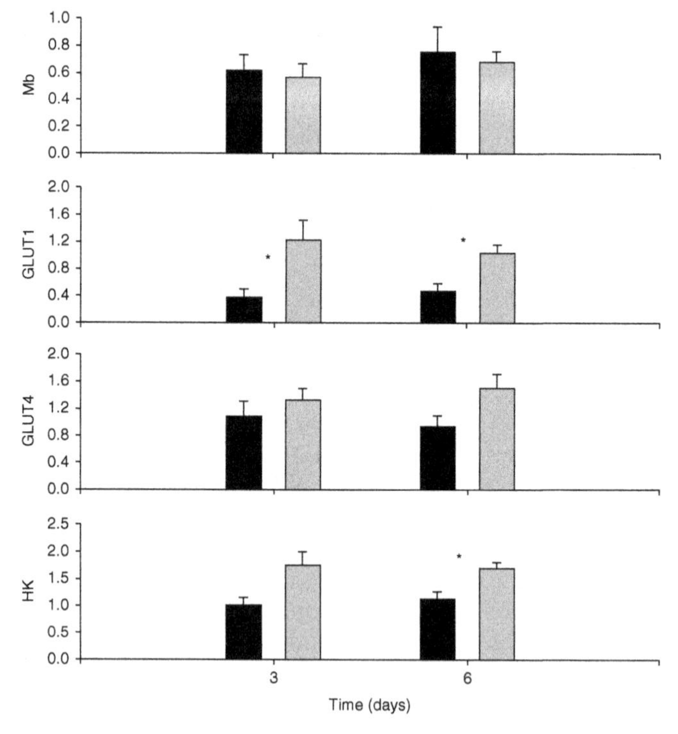

Fig. 7.14. Relative gene expression levels, determined by qPCR, in the heart of Atlantic cod (*G. morhua*) held under normoxic (black bars) or hypoxic (grey bars) conditions (~60 mmHg oxygen) (J. R. Hall, unpublished data).

substantially increased levels of this protein are not observed in the hypoxic heart. The answer may reside in the penetrating commentary of Sidell and O'Brien (2006). These authors point out that Mb is a nitric oxide-oxygenase, and as such, any increase in Mb content could result in a decrease in nitric oxide (NO). In some fish hearts, NO has a negative inotropic effect (Tota *et al.*, 2005). In the context of the hypoxic fish heart, an elevation of Mb could decrease NO levels and in turn result in increased energy demand. Thus, there may be a complex compromise between the benefits of increased oxygen extraction through high Mb levels and the trade off of increased ATP demand due to loss of the inhibitory actions of NO. This conjecture remains to be tested.

It is theoretically possible that increased mitochondrial surface area could allow for more effective uptake of oxygen at low concentration gradients. For example, flounder (*Platitchthys flesus*) subjected to hypoxic conditions for 3 weeks showed microscopic evidence of mitochondrial necrosis, but in addition, many mitochondria were greater in length and the number of cristae in each mitochondrion appeared to have increased (Lennard and Huddart, 1992). However, three separate studies involving hypoxic acclimation have failed to show any increase in marker mitochondrial enzymes including citrate synthase, malate dehydrogenase, and cytochrome C oxidase (*Z. viviparous*, Driedzic *et al.*, 1985; *Hoplias microlepsis*, Dickson and Graham, 1986; *Astronotus crassipinnis*, Chippari-Gomes *et al.*, 2005). It is therefore unlikely that modification of mitochondrial properties occurs with respect to effective utilization of available oxygen and metabolic fuels under hypoxia. However, as discussed under mitK_{ATP} channels and hexokinase (HK), there may be other changes to mitochondrial function under hypoxia that are critical.

3.4. Anaerobic Energy Metabolism

Under hypoxic conditions heart glycogen stores and blood-borne glucose are called upon as anaerobic energy sources with lactate accumulating as the end product (e.g., Bailey *et al.*, 2000; Vornanen and Paajanen, 2004; Overgaard and Gesser, 2004; MacCormack *et al.*, 2006).

3.4.1. Acute Response

3.4.1.1. Glycogen Breakdown. Heart glycogen is mobilized under oxygen limiting conditions, and although the availability of extracellular glucose curtails glycogen utilization in American eel (*Anguilla rostrata*) and Atlantic cod heart preparations, it does not prevent glycogen breakdown (Bailey *et al.*, 2000; Clow *et al.*, 2004). Glycogen metabolism is generally a well understood phenomenon related to activation of the glycogen phosphorylase cascade. Contrary to what would be anticipated, however, goldfish subjected to 24 h of anoxia showed a decrease in % phosphorylase-*a* (Storey, 1987).

As presented in Driedzic and Gesser (1994), this may be related to exhaustion of heart glycogen at that time. But, this issue remains to be resolved.

3.4.1.2. Glucose is Essential for Heart Performance Under Hypoxia. The necessity for glucose to support heart performance is well established. For example, isolated, perfused heart preparations from American eel and Atlantic cod could sustain 50% of normoxic power development for 2 h under hypoxia if glucose was available in the medium, whereas hearts from both species failed within 40 min without glucose (Bailey *et al.*, 2000; Clow *et al.*, 2004). Ventricle strips from rainbow trout subjected to anoxia for 30 min showed better force development and maintenance of resting tension when glucose was available in the medium than without glucose when glycogen was partially depleted by prior challenge (Gesser, 2002). Finally, isolated, perfused rainbow trout hearts can sustain sub-basal levels of performance under hypoxia with glucose in the medium, but with reduced levels of ATP turnover based on oxygen consumption and lactate production measurements (Arthur *et al.*, 1992; Farrell and Stecyk, 2007). Interestingly, the use of glucose under hypoxia may involve features beyond the provision of fuel for total ATP production, including a direct interplay between glucose metabolism and ion balance. This issue is addressed in the section on hexokinase (HK) below. Regardless, it is clear that glucose utilization is required for ATP production to support the contractile apparatus under oxygen limitation, especially if glycogen stores are compromised. How this is achieved is addressed in the following sections.

3.4.1.3. Glucose Concentration Gradient Increases Under hypoxia. Glucose enters cells by facilitated diffusion, and uptake is determined by the glucose concentration gradient and the abundance of GLUTs (glucose transporter proteins). Hypoxia places increased demands on the glucose transport system as anaerobic metabolism is highly activated (unless the tissue is entering a severe hypometabolic state), and often results in an increase in blood glucose (Table 7.1) that may help support the concentration gradient from the extra- to the intracellular space [e.g. flounder (Jørgensen and Mustafa, 1980); goldfish (Shoubridge and Hochachka, 1983); Atlantic cod (Claireaux and Dutil, 1992); rainbow trout (Haman *et al.*, 1997); *A. crassipinnis* (Chippari-Gomes *et al.*, 2005); Amazonian armoured catfish (*L. pardalis*) (MacCormack *et al.*, 2006), and many other species of Amazonian fishes (see Table 10.2 in Val *et al.*, 2006)]. In rainbow trout an increase in plasma glucose under the initial stages of hypoxia is associated with a transient increase in the rate of glucose appearance, presumably from liver glycogen, without a change in the rate of whole animal glucose disappearance (Haman *et al.*, 1997).

For glucose uptake though, it is the glucose gradient that is critical and not blood glucose levels per se. Table 7.1 presents values for blood/plasma

Table 7.1
Blood and heart glucose levels and heart lactate levels under normoxic and hypoxic conditions

Species	Glucose normoxia			Glucose hypoxia			Lactate		Conditions
	Blood/plasma	Heart	Gradient	Blood/plasma	Heart	Gradient	Normoxic	Hypoxic	
Flounder[a]	1.33	2.14	–0.81	2.91	1.58	1.33	6.2	10.2	29 h; 15 mmHg; 10°C
Goldfish[b]	1.39	1.94	–0.55	7.21	4.43	2.78	0.42	11.35	60 h; anoxia; 4°C
Atlantic cod[c]	8.40	4.48	3.92	11	3.84	7.16	1.84	17.55	6 h; 30 mmHg; 5°C
African lungfish[d]	0.23	0.96	–0.73	2	1	1	1.9	5.2	12 h; 22°C
Rainbow trout[e]	12.50	8.80	3.70	7.3	5.8	1.5	0.71	6.2	3 h; 20 mmHg; 4°C
Armoured catfish[f]	2.20	2.50	–0.30	5	2.9	2.1	0.05	20.2	3 h; <45 mmHg; 26°C
Short-horned sculpin[g]	0.23	0.35	–0.12	0.6	0.7	–0.1	8	14	5 h; <33 mmHg; 8°C

[a] *Platichthys flesus*; Jorgensen and Mustafa (1980).
[b] *C. carrassius*; Shourbridge and Hochachka (1983).
[c] *Gadus morhua*; Claireaux and Dutil (1992).
[d] *Protopterus aethiopicus*; Dunn *et al.* (1983).
[e] *O. mykiss*; Dunn and Hochachka (1986).
[f] *Liposarcus pardalis*; MacCormack *et al.* (2006).
[g] *Myxocephalus scorpius*; MacCormack *et al.* (2006).

Air breathing lungfish and armoured catfish were denied access to air. Glucose values expressed as μmol glucose/mL blood or plasma. Glucose and lactate in heart is expressed as μmol/g wet weight. Glucose gradient was calculated as the blood/plasma value minus the heart value. No correction was made for extracellular space. All values are taken directly from published papers.

glucose and heart glucose. No attempt has been made to calculate intra- and extracellular glucose so this analysis must be viewed as a first approximation only. In each of the seven cases, fish were subjected to a hypoxic challenge sufficient to at least result in elevated average levels of heart lactate. Under normoxic conditions five species showed higher levels of heart glucose than blood glucose, which may be an artifact of the lack of precise values for intra- and extracellular glucose. However, an alternate explanation is that heart glucose values are high due to mobilization of glycogen during the sampling period and/or gluconeogenesis, both of which would require an active glucose 6-phosphatase. These contentions remain to be tested and are beyond the scope of this review. More importantly, in the context of this analysis, is that under hypoxia all of the species show a positive gradient for glucose diffusion from the extra- to the intracellular space with the exception of shorthorn sculpin, where there is essentially no difference between plasma and cellular levels. Further, in most cases, the hypoxic challenge was associated with a more favorable inward glucose gradient. An increase in the inward diffusion gradient in association with hypoxia suggests that the removal of glucose (i.e., by phosphorylation) is elevated to a greater extent than glucose entry into the cell. This may be a cue to one of the features of maintaining anaerobic metabolism.

3.4.1.4. Glucose Uptake is Increased Under Hypoxia via Enhancement of Facilitated Diffusion. Injection of anoxic goldfish with C^{14}-labeled glucose resulted in C^{14} glucose-specific activity in the heart that was equivalent to that measured in the blood after 3 h. Lactate was also labeled, and although the data are limited, it appears that the specific activity in heart was higher than in blood (Shoubridge and Hochachka, 1983). This is an important study in that it provides evidence for glucose uptake, equilibration of glucose between the blood and intracellular space, and the production of lactate. Thereafter, it was shown in isolated, perfused rainbow trout hearts that glucose uptake, as assessed by C^{14} 2-deoxyglucose, was stimulated 10-fold in NaCN-treated preparations relative to normoxic hearts performing low levels of work (West *et al.*, 1993). The stimulation of glucose uptake under hypoxia was subsequently confirmed in American eel ventricle strips (45% increase) (Rodnick *et al.*, 1997) and in isolated, perfused Atlantic cod hearts (approx. 3-fold increase) (Clow *et al.*, 2004). In American eel ventricle strips, cytochalasin B, a general inhibitor of GLUTs, prevented both anoxia-stimulated and contraction-stimulated increases in glucose uptake (Rodnick *et al.*, 1997). In isolated, perfused hearts of Atlantic cod, the inclusion of cytochalasin B in the medium, during hypoxia, resulted in a significant decrease in glucose uptake. This was associated with a consistent trend of lower levels of performance, lower levels of tissue glucose, and increased glycogen breakdown (Figure. 7.15)

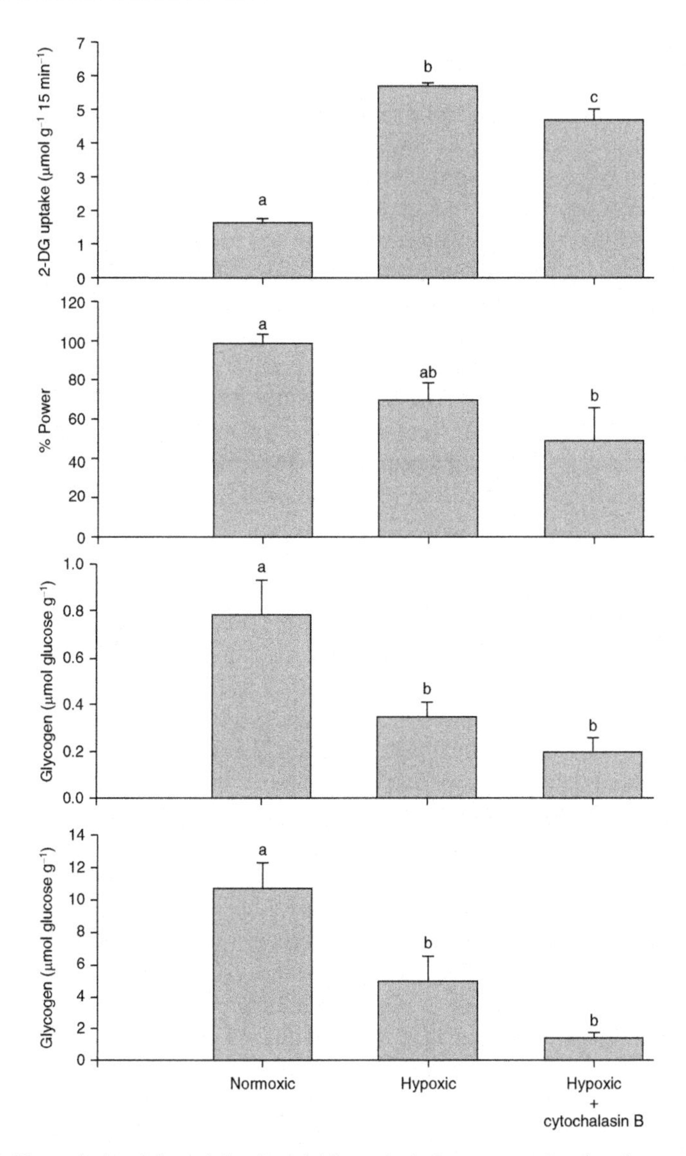

Fig. 7.15. Hearts isolated from Atlantic cod (*G. morhua*) that were perfused under normoxic or hypoxic conditions with media containing cytochalasin B to inhibit glucose transporter proteins. In all cases the media contained 5 mM glucose. 2-Deoxyglucose (2-DG) uptake was determined following 15 min of perfusion. % Power output shows values following 30 min of perfusion when at least 7 of 8 preparations were viable. Glucose and glycogen levels were assessed following 120 min of perfusion or immediately after heart failure. Values that do not share a common letter are significantly different. [Data are taken from Clow *et al.* (2004).]

(Clow *et al.*, 2004). All of these features are consistent with an inhibition of hypoxia-induced enhancement of facilitated glucose diffusion.

There is only one study dealing with the impact of hypoxia on heart glucose uptake at the whole animal level (MacCormack *et al.*, 2007). Shorthorn sculpin were subjected to approximately 30 mmHg oxygen for up to 4 h, and these authors report a 30% reduction in heart rate (with no change in stroke volume), and that the hypoxic challenge did not result in an increase in glucose uptake despite an increase in plasma glucose (Note, however, that levels of glucose in this species are extraordinarily low at <0.6 mM.) The simplest explanation for this finding is that aerobic metabolism, even at reduced oxygen availability, was able to support ATP demand. Consistent with this interpretation is the high Mb level in hearts of shorthorn sculpin (Driedzic and Stewart, 1982). In addition, treatment with atropine under hypoxia resulted in increases in f_H, Q, and glucose uptake (MacCormack *et al.*, 2007), again implying that the maximal rate of glycolysis associated with ATP production is not necessarily the limiting factor in performance of the hypoxic myocardium.

3.4.1.5. Glucose Transporters. In mammals, facilitated glucose transport is achieved primarily via four Na-independent proteins (Wood and Trayhurn, 2003). Homologs of mammalian GLUTs 1, 2, 3, and 4 have been characterized in Atlantic cod (Hall *et al.*, 2004, 2005, 2006). Similar to well-studied mammalian systems GLUT1 is found in most tissues, GLUT2 is the predominant liver isoform, GLUT3 is in kidney/spleen, and GLUT4 is in heart and muscle. Rainbow trout hearts shows high expression levels of GLUT1 (Teerijoki *et al.*, 2000), while the tilapia (*O. nilotica*) heart shows high amounts of both transcript and protein (Wright *et al.*, 1998). GLUT4 is abundant in red and white skeletal muscle of brown trout (*Salmo truta*) but only at very low levels in heart (Planas *et al.*, 2000). In the rat heart, ischemia results in the movement of GLUT4 protein from intracellular vesicles to the T-tubular membrane and sacrolemma, and provides a beneficial effect at high glucose levels (Ramasamy *et al.*, 2001; Davey *et al.*, 2007). We are not aware of any investigations of this nature in the fish heart but this certainly should be assessed as a potential mechanism for defense against acute hypoxia. The low level of GLUT4 mRNA in the heart of brown trout is particularly provocative as it might be associated with the relative hypoxia sensitivity of salmonid hearts.

3.4.1.6. Hexokinase Activity. Hexokinase (HK) is a regulated enzyme that catalyzes the first step in the use of glucose according to the following reaction: glucose + ATP $\rightarrow$ G6P + ADP.

HK may be critical for the maximum utilization of glucose through its direct catalytic activity and its role in maintaining low intracellular glucose, and thus maximizing the glucose diffusion gradient. In addition, HK binds to mitochondria where it may have functions in addition to simple catalysis.

A number of studies suggest that HK catalytic activity per se in heart may be important under hypoxia. The calculated anaerobic ATP yield based on *in vitro* HK activity matches total ATPase activity for many ectothermic and endothermic vertebrates (Driedzic and Gesser, 1994). There is a linear relationship between total HK and LDH in hearts of ectothermic vertebrates (Driedzic and Gesser, 1994). In three species of Amazonian fish (but not north temperate species), force development of ventricle strips under NaCN treatment rank orders with maximal HK activity (Bailey *et al.*, 1999; West *et al.*, 1999). The maximal activity of HK in the fish heart is high by mammalian and avian standards (Driedzic and Gesser, 1994), and although there may be a correlation between HK activity and rates of ATP production, it should be appreciated that the maximal rate of glucose utilization in heart preparations, as estimated from glucose uptake, is only a small fraction of the total maximal HK activity assessed with *in vitro* assays. For example, following correction for assay temperatures, heart glucose uptake amounts to only 1.3, 2.8, and 8.2% of maximal HK activities for American eel, rainbow trout, and Atlantic cod, respectively (West *et al.*, 1993; Driedzic and Gesser, 1994; Rodnick *et al.*, 1997; Clow *et al.*, 2004). Similarly, maximal *in vitro* rates of HK activity are much higher than rates of lactate production in hypoxic heart preparations from rainbow trout or *L. pardalis* (Overgaard and Gesser, 2004; Overgaard *et al.*, 2004; Treberg *et al.*, 2007). We continue to query why there are such high levels of HK in fish hearts.

The binding of HK to the outer mitochondrial membrane may be a specific control feature of glucose utilization. In the rat heart there are two major isoforms of HK, HKI and HKII. A proportion of both isoforms is bound to mitochondria under normoxic conditions and the level of binding increases under ischemia (Zuurbier *et al.*, 2005; Southworth *et al.*, 2007). The binding of HK to the particulate fraction was assessed in hypoxia-resistant ventricle strips of the fish *L. pardalis*. Heart preparations were subjected to 2 h of anoxia, which was sufficient to result in an increase in lactate from 2.5 to 12.7 μmol g^{-1}. Simultaneous measurements of HK and citrate synthase activities in cell fractions revealed that a much higher proportion (>4 times) of HK is associated with the mitochondrial pellet in *L. pardalis* than in rat hearts. Following hypoxia HK binding to mitochondria tended to increase ($P = 0.08$) on the basis of HK/CS ratio, and using an alternative approach of assessing generalized binding to a particulate fraction via sucrose dilution there was a substantial and significant increase in enzyme binding (Treberg *et al.*, 2007). Similar to *L. pardalis*, the proportion of HK activity in the particulate fraction of goldfish heart, and in the mitochondrial enriched pellet of eel heart, is high by mammalian standards (Duncan and Storey, 1991; Rodnick *et al.*, 1997). This might explain why hypoxia did not increase HK binding in either of these two species. On balance it appears that in fish hearts, even under normoxic conditions, there is a high proportion of HK bound to the mitochondrial membrane and in some species this may increase with a hypoxic challenge. As data are available for only three species of hypoxia-tolerant fishes, it would be interesting to determine the level of enzyme binding in hearts from hypoxia-sensitive species to assess the generality of this feature.

Enzyme binding may be a response to defend against ischemic/hypoxic insults on kinetic grounds. HK is generally inhibited by glucose-6-phosphate (G6P) as has been shown for American eel heart (Rodnick *et al.*, 1997), and in rat brain mitochondrial inhibition of HK by G6P is decreased when the enzyme is bound (Wilson, 2003). In addition, HK binds to voltage-dependent anion channels (VDACs), and mice lacking VDACs have a reduced capacity to metabolize glucose (Anflous-Pharayra *et al.*, 2007). The function of enzyme binding though is probably much more complex than a relationship to simple enzyme activity per se. For instance, VDACs serve as conduits for metabolite movement, including adenylates, across the outer mitochondrial membrane. In rat liver mitochondria the binding of HK closes the transition pore, reducing Ca^{2+} release and possibly the release of cytochrome *c* that leads to apoptotic cell death (Azoulay-Zohar *et al.*, 2004). We suggest that HK binding and subsequent activity in association with VDACs could translate into a localized decrease in the ATP/ADP ratio within the intermembrane space, and in turn, open mitK_{ATP} channels. If this is the case, it would link two of the key elements in the hypoxia defense process.

3.4.1.7. Phosphofructokinase. Phosphofructokinase (PFK) is activated during the early stages of oxygen limitation as evidenced by cross-over analysis of metabolite levels. Fish heart PFK, similar to other systems, is inhibited by low pH, ATP, and citrate while activators include AMP, Pi, and fructose-2,6,-diphosphate (F-2,6-P_2). F-2,6-P_2 levels increase in the heart of anoxic goldfish and may function as a potent activator (Storey, 1987; Driedzic and Gesser, 1994). Subcellular binding of PFK may also be important in glycolytic control. Following 21 h of anoxia the percentage of PFK bound to the particulate fraction of goldfish heart increased from 35% to 48% (Duncan and Storey, 1991). Similar increases from 20% to 45% bound enzyme were noted for isolated ventricle sheet preparations of *L. pardalis* subjected to 2 h of severe hypoxia (Treberg *et al.*, 2007). Although not shown directly in these studies, it is most likely that PFK is binding to myofibrils. In the bound configuration inhibitors are less effective and activators are more effective, thus catalytic activity is enhanced (Brooks and Storey, 1995). The binding of PFK to the particulate fraction under normoxia is higher in goldfish and *L. pardalis* than rat and mice; moreover, under hypoxia the percentage of bound PFK approaches 50% in the fish species but only 25% in rat heart (see Treberg *et al.*, 2007 for details). As such, this may also represent a key feature in the hypoxia defense mechanism of the fish heart.

3.4.1.8. Pyruvate Kinase and Lactate Dehydrogenase. There is no increase in binding of pyruvate kinase (PyK) or lactate dehydrogenase (LDH) to the particulate fraction as there is for HK and PFK in hypoxic ventricle sheets (Treberg *et al.*, 2007). PyK is likely activated by increases in the activator F-1,6-P_2 under oxygen limitation (see Driedzic and Gesser, 1994). However, there are no obvious correlations between the tolerance of isolated preparations to anoxia and either maximal *in vitro* activity of PyK or LDH, or in LDH kinetics (e.g., isozyme assembledge, K_m for pyruvate, activity ratios) (see Driedzic and Gesser, 1994; Bailey *et al.*, 1999; West *et al.*, 1999).

3.4.1.9. General Model for Activation of Glycolysis Under Acute Hypoxia. A substantial body of literature exists on the impact of hypoxia on heart metabolism, as detailed above, and there is now sufficient information to propose a generalized model that results in anaerobic energy generation in response to acute hypoxia in the fish heart. Foremost, CP is utilized to maintain cellular ATP levels; a metabolic corollary is that P_i levels increase. The impact of the rise in Pi on contractility and its possible role in controlled down-regulation of heart performance, however, warrants further investigation. Although total ATP levels remain relatively constant we cannot rule out the possibility of localized decreases in ATP/ADP ratio that would open sarcK_{ATP} channels leading to reduced ATP demand. Hypoxia leads to glycogen mobilization in both liver and heart. Glucose released from the liver results in an increase in blood glucose that creates a more favorable gradient for glucose diffusion into the myocytes. G6P produced in the heart from glycogen can enter glycolysis directly. Glucose transport is activated, and this might be due to movement of the GLUT4 isoform to the sarcolemmal membrane (untested hypothesis). PFK binds to contractile fibrils and is activated through increases in P_i and F-2,6-P_2. Activation of PFK would serve to decrease G6P levels. HK binds to the mitochondrial membrane and becomes less susceptible to inhibition by G6P. Binding of HK may also play a role in mitK_{ATP} channel regulation. An activated HK would decrease intracellular glucose, and thus, improve glucose entry. The detailed aspects of the sequelae of events is yet to be proven, especially the causes of enzyme binding as our picture is drawn from many parts of a puzzle. The challenge is to intellectually connect altered gene expression and subsequent alterations in protein levels to changes in Ca^{2+} levels that trigger contraction and ATP levels that provide energy for contraction, and to do it in a living animal as opposed to reductionist preparations.

3.4.2. Chronic Response

This section deals with changes in processes of anaerobic metabolism that may occur after exposure to hypoxia for days to weeks, that is, a time frame long enough to result in changes in gene expression and protein levels. Our understanding of the events at this level, however, is poor and rests primarily on a few papers dealing with heart performance, enzyme activity levels, and gene expression.

Z. viviparous were acclimated to hypoxia (~35 mmHg) for 4–6 weeks. Thereafter, ventricle strips were challenged under anoxia with and without glucose in the medium. All preparations failed to a similar extent under anoxia, and after 30 min developed about 65% of their initial force. Then, an increase in extracellular Ca^{2+} (from 1 to 5 mM) was used to assess maximal force development, and this resulted in a substantial and sustainable increase in force development only with glucose in the medium, and only in hearts from hypoxia-adapted animals (Driedzic *et al.*, 1985; see Driedzic and Bailey, 1999 for further discussion). Interestingly, however, there was no change in the maximal activity levels of PFK or PyK following hypoxia acclimation. This experiment suggests that there is an adaptable feature(s) in the hypoxic response that may be related to glucose utilization perhaps at the level of glucose entry or HK. In this context, although no change was noted in GLUT1 expression in hearts of Atlantic cod held at 40–45% O_2 (approx. 60 mmHg) after either 6 h (Hall *et al.*, 2004) or 24 h (Hall *et al.*, 2005), more recent studies show an increase following either 3 or 6 days (Figure 7.14; Hall *et al.*, unpublished data). Further, the mean values for GLUT4 mRNA and HK mRNA increased, with the increase in HK expression after 6 days of hypoxia reaching significance. It would be of interest to determine if these changes result in increased rates of glucose uptake.

In killifish (*Fundulus grandis*) following 4 weeks at 30 mmHg oxygen, the maximal activity of total homogenate HK, triose phosphate isomerase, and PK increased by 27, 18, and 30%, respectively (Martínez *et al.*, 2006). In contrast, there was no change in the activity of eight other glycolytic enzymes including PFK and LDH, and Dickson and Graham (1986) showed no significant change in PK or LDH in *Hoplias microlepis* held under hypoxic conditions for 16–25 days. The extremely hypoxia-tolerantAmazonian cichlid (*A. crassipinnis*) presents an interesting case study. At 6% oxygen (approx. 10 mmHg) these fish can maintain an MO_2 of about 60% of that measured during normoxia, and a stepwise decrease to 1% oxygen (1.5 mmHg) over 2 days results in a decrease in PyK and an increase in LDH maximal *in vitro* activity (Chippari-Gomes *et al.*, 2005). Microarray studies also provide us with contrasting results. For example, Gracey *et al.* (2001) exposed *G. mirabilis* to 15 mmHg O_2 for 6 days and showed a down-regulation in heart levels of transcripts for two glycolytic enzymes, enolase and glyceraldehyde-3-phosphate dehydrogenase, with no change in LDH-A mRNA. In contrast, in zebrafish held under hypoxia for 21 days there was an up-regulation of heart PK and aldolase (Marques *et al.*, 2008).

The results from the enzyme activity and gene microarray studies do not present a consistent or coherent response to chronic hypoxia. What is missing in the field at present are comprehensive and integrative studies that relate the elements of gene expression, enzyme activity, glucose transport, lactate production, myocardial performance, etc. Hopefully these will be performed soon, and will add significantly to our understanding of the biochemical and metabolic responses of the fish heart to chronic hypoxia.

4. ADDITIONAL INSIGHTS

Thus far, we have attempted to synthesize the available literature on the effects of chronic and acute hypoxia on *in vivo* cardiovascular function, and on aspects of biochemistry and metabolism related to cardiac hypoxia tolerance and performance under hypoxia. Although we cannot cover all topics related to hypoxia and cardiovascular physiology/biochemistry in this chapter, there are several important/interesting aspects that still need to be addressed.

4.1. Interactive Effects of Temperature and Hypoxia

While it is clear that hypoxia suppresses cardiac function in fishes, the degree to which cardiac function is affected is related to both the severity of hypoxia and water temperature. With regards to the interactive effects of temperature and severe hypoxia (anoxia) on aspects of cardiac function, there are several studies that provide particularly relevant information in addition to that presented in Section 2.1.

First, Overgaard *et al.* (2004a) acclimated rainbow trout to 10°C and evaluated the capacity of their hearts to maintain basal *in situ* cardiac function during severe hypoxia/anoxia (PO_2 ~5 mmHg), and to recover maximal cardiac function when returned to normoxia, when tested at 5, 10, 15, or 18°C. This study showed that, although hearts at 5°C could maintain cardiac performance throughout 20 min of severe hypoxia and maximal cardiac performance recovered fully after the severe hypoxic period, there was a significant increase in functional impairment during anoxia and recovery from anoxia as temperature increased (i.e., *in situ* heart performance during severe hypoxia and upon recovery was inversely related to temperature) (Figure 7.16). Further, they showed that this functional impairment (of both SV and f_H) at elevated temperatures occurred even though cardiac glycolytic enzyme activity and the rate of lactate production were increased proportionally with temperature, and there was no evidence of myocardial necrosis or differences in biochemical and energetic parameters between groups. These results lead to two important conclusions:

(1) That while a decrease in cardiac performance with severe hypoxia at any particular temperature results from insufficient anaerobic energy production to meet demand (e.g., see Arthur *et al.*, 1992), the inverse

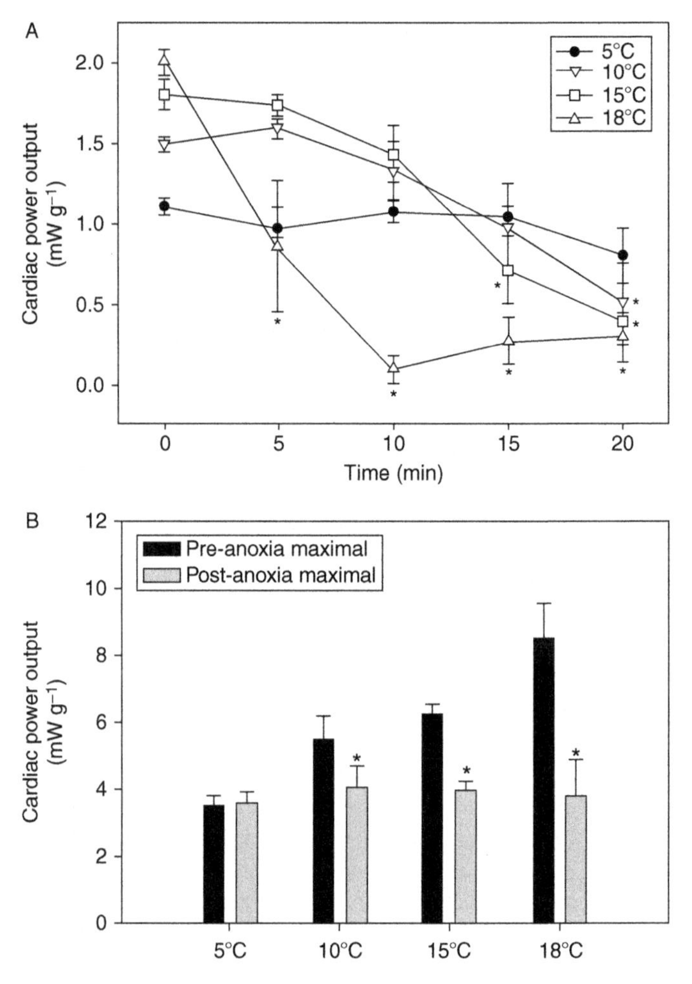

Fig. 7.16. Effect of test temperature on changes in trout (*O. mykiss*) basal cardiac power output during 20 min of anoxic perfusion (A), and on maximum power output after recovery from 20 min of anoxic perfusion as compared with preanoxic (normoxic) values (B). In both A and B an asterisk (*) indicates a value different from preanoxic values at a given temperature. Values are means ± S.E.M; N = 6–7. Trout were all acclimated at 10°C. [Modified from Farrell and Stecyk (2007); original data from Overgaard *et al.* (2004a).]

relationship between cardiac performance during severe hypoxia and temperature was due to a faster accumulation of waste products, in particular, intracellular phosphate and protons. Indeed, cardiac failure during hypoxia/anoxia appears to be caused by increased levels of intracellular inorganic phosphates and reduced intracellular pH

(Turner and Driedzic, 1980; Allen *et al.*, 1985; Godt and Nosek, 1989; Arthur *et al.*, 1992).

(2) That decreased postanoxic performance of the trout heart is due to myocardial stunning (mechanical dysfunction that persists after reoxygenation/reperfusion due to a reduction in Ca^{2+} responsiveness caused by damage of the contractile apparatus by oxygen radicals and/or Ca^{2+} overload; see Bolli and Marban, 1999), and that this loss of Ca^{2+} responsiveness is related to temperature and the duration of flow deprivation.

Second, while the above *in situ* study on rainbow trout suggests that reductions in cardiac function with anoxic exposure are gradual, with the rate of decrease depending on temperature, Stecyk *et al.* (2007) show that the *in vivo* cardiovascular response of the common carp (*C. carpio*) to anoxia is triphasic, and that not all changes are directly related to temperature. For example, these authors showed that the rate of loss of cardiac function (mainly as a result of changes in f_H, as opposed to both f_H and SV) during phase 1 (acute phase) was slower at 10°C than at 5°C (with 15°C being intermediate), that in the middle (prolonged) phase the heart achieved minimal levels of cardiac activity that were temperature independent ($Q_{10} = 1.2$), and that in phase 3 (expiratory phase) cardiac activity temporarily increased at all temperatures before the carp approached death. These data suggest that complex cardiorespiratory control mechanisms are utilized by the common carp to survive anoxia (these were revealed in a later study; Stecyk and Farrell, 2006), and that elevations in cardiac function to meet the needs of the whole animal may ultimately lead to cardiac damage or failure. For example, Stecyk *et al.* (2007) suggest that the increase in cardiac activity during phase 3 of anoxia was related to the heart's role in transporting nutrients to, and wastes from, the tissues.

Third, it is well established that aquatic hypoxia is not the only situation where the heart's oxygen supply may be limited (e.g., severe anemia, exhaustive exercise), and that adrenergic nervous tone and circulating catecholamines stimulate positive inotropic and chronotropic responses that allow the fish heart to maintain or elevate its performance under conditions that would normally compromise myocardial function (Gesser *et al.*, 1982; Farrell, 1984; Stecyk and Farrell, 2006; Hanson *et al.*, 2006). Given the importance of circulating catecholamines and adrenergic nervous tone (e.g., see Stecyk and Farrell, 2006) to cardiac function under conditions that result in hypoxia or hypoxemia, and that adrenergic sensitivity of the trout myocardium is decreased at elevated temperatures due in large part to a decrease in cell-surface β-adrenoreceptor density (Keen *et al.*, 1993), Hanson and Farrell (2007) assessed the hypoxic threshold for maximum *in situ* cardiac performance at 18°C under conditions that simulated venous blood composition during exhaustive exercise [e.g., acidosis, pH 7.5; hyperkalemia, K^+, 5 mM; and maximal adrenergic stimulation, 500 nM adrenaline) and compared it to previous data collected at 10°C (Hanson *et al.*, 2007). Hanson and Farrell (2007) found that complete cardiac failure occurred at a perfusate PO_2 of 38 mmHg at 18°C, an oxygen tension that far exceeded the hypoxic threshold at 10°C (15 mmHg for diminished cardiac function and 7.5 mmHg for heart failure; Hanson and Farrell, 2006) and venous oxygen partial pressures measured in salmonids swimming maximally at temperatures between 6 and 16°C (PO_2 7–28 mmHg; Steffensen and Farrell, 1998; Farrell and Clutterham, 2003). Collectively, this research suggests that the capacity of adrenergic mechanisms to support cardiac function is diminished at high temperatures, and that this may limit myocardial hypoxia tolerance *in vivo* under conditions requiring elevated cardiac performance.

4.2. Preconditioning

Thus far we have discussed the influence of environmental (hypoxic)–genetic interactions during development/rearing on cardiac function (see Section 2.2.1), and shown that reductions in SV caused by chronic hypoxia are compensated/partially compensated for by an enhanced capacity to elevate f_H (see Section 2.2.2). While these studies highlight the capacity of the myocardium to respond/adapt to prolonged low oxygen conditions, it is clear from studies of preconditioning that even short-term exposure to hypoxia (i.e., minutes) can have profound implications for the fish heart's capacity to deal with oxygen deprivation. Preconditioning has been studied extensively in the mammalian heart, and is a phenomenon whereby prior exposure of the mammalian heart to a physiological insult (ischemia, hypoxia, acidosis, stretch, rapid pacing, etc.) or biologically active molecules (adenosine, adrenaline, bradykinin, etc.) protects the myocardium from damage or loss of function resulting from a subsequent hypoxic/ischemic episode (e.g., see reviews by Downey *et al.*, 2007; Gross and Gross, 2007). Gamperl *et al.* (2001) provided the first evidence (using hypoxia-sensitive trout) that preconditioning exists in fishes, and thus that preconditioning is a mechanism of cardioprotection that appeared early in the evolution of vertebrates (see Figure 7.17A). Further, research in this area to date has shown that: (1) increased anaerobic glycolysis, fueled by exogenous glucose, is associated with preconditioning (Gamperl *et al.*, 2001); (2) trout hearts with inherent myocardial hypoxia tolerance cannot be preconditioned (Figure 7.17B, Gamperl *et al.*, 2004; Overgaard *et al.*, 2004b); (3) unlike mammalian cardiac cells, fish myocardial cells are not irreversibly damaged (i.e., do not die) following exposure to periods of oxygen deprivation $\leq$ 30 min. (Gamperl *et al.*, 2001; Overgaard *et al.*, 2004b); and (4) preconditioning is not limited to myocardium that normally receives

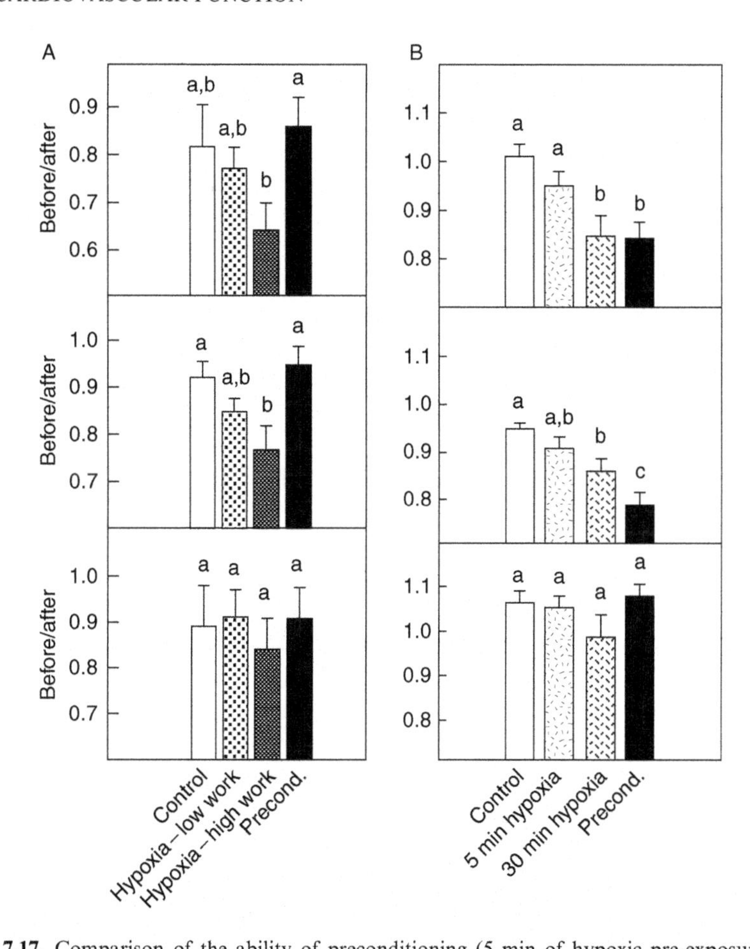

Fig. 7.17. Comparison of the ability of preconditioning (5 min of hypoxic pre-exposure) to protect (A) hypoxia-sensitive (Gamperl *et al.*, 2001) and (B) hypoxia-tolerant (Gamperl *et al.*, 2004) trout (*O. mykiss*) hearts from the myocardial dysfunction that follows more prolonged exposure to hypoxia. In A, 5 min of hypoxic pre-exposure completely eliminated the loss of myocardial function that normally followed the "Hypoxia-high workload" protocol. In B, preconditioning with 5 min of hypoxia either did not affect, or increased, the amount of myocardial dysfunction following exposure to "30 min of hypoxia." Top panels, maximum cardiac output; middle panels, maximum stroke volume; bottom panels, heart rate. Note that the hypoxia-tolerant trout hearts in B required twice the duration of hypoxia (15 vs. 30 min), and 6 times the workload, as compared with hypoxia-sensitive hearts (A) to achieve a comparable (15–20%) decrease in posthypoxic myocardial function. Values were obtained by comparing maximum *in situ* cardiac function before and after the treatment protocols. All values are means ± S.E.M; $N = 7$–9. Dissimilar letters indicate a significant difference at $P<0.05$, as determined by one-way ANOVA. Hypoxia in these experiments was defined as perfusate $PO_2 = 5$–10 mmHg. Control hearts were only exposed to oxygenated saline. [Reproduced from Gamperl and Farrell (2004), with permission from the *Journal of Experimental Biology*.]

highly oxygenated blood from the coronary circulation (Gamperl and Genge, unpublished data; see Fig. 6 in Gamperl and Farrell, 2004). These studies provide important insights into fish myocardial hypoxia tolerance, and provide indirect evidence that the cellular mechanisms/signaling pathways involved in providing protection to the myocardium following short-term (acute) and long-term (chronic) oxygen deprivation are similar. This hypothesis is consistent with the mammalian literature (e.g., see Kolář and Ostadal, 2004), though it is not known which of the multitude of pathways/mechanisms that have been identified in the mammalian heart to have cardioprotective effects are involved in fishes.

5. CONCLUDING REMARKS

In this chapter we have shown that our understanding of some aspects of fish cardiovascular responses to hypoxia (e.g., heart rate responses to acute hypoxia, control of branchial resistance/blood flow) is fairly advanced. However, it is also obvious from the information provided that our knowledge of many, even basic, aspects of cardiovascular function/control under hypoxic conditions is extremely limited. For example, information on the control of systemic vascular resistance during hypoxia and on the effects of chronic hypoxia is less than extensive. Unlike the effects of temperature (e.g., see Shiels *et al.*, 2002), we have little idea of whether excitation-contraction coupling and calcium dynamics are affected by myocardial hypoxia in fishes (with the exception of the crucian carp; Vornanen and Paajanen, 2004). Finally, we have only begun to understand how complex changes in gene expression, protein levels/function/subcellular localization, signaling cascades, and the control of oxidative and anaerobic metabolism result in intra- and interspecific differences in myocardial hypoxia tolerance or mediate the phenomenon of preconditioning. However, this is an extremely active field of scientific investigation which, along with continued advances in *in vivo* physiological measurement techniques, cellular imaging, molecular cloning, and functional genomics, will reveal many novel insights into myocardial plasticity and adaptation in fishes (vertebrates), and the molecular and biochemical pathways that protect the heart from environmental insults that might normally lead to cardiac dysfunction, myocardial damage, and eventually death.

ACKNOWLEDGMENTS

We would like to thank Marc Bolli, Juan Perez-Casanova, and Connie Short for their assistance in putting this body of work together. This contribution and the research programs of the authors are supported by grants from the Natural Sciences and Engineering Research

Council of Canada (NSERC), the Canadian Institutes of Health Research (CIHR), and funds made available through the Canada Foundation for Innovation. WRD holds the Canada Research Chair in Marine Bioscience.

REFERENCES

Acierno, R., Agnisola, C., Tota, B., and Sidell, B. D. (1997). Myoglobin enhances cardiac performance in Antarctic icefish species that express the protein. *Am. J. Physiol.* **273,** R100–R106.

Agnisola, C., McKenzie, D. J., Pellegrino, D., Bronzi, P., Tota, B., and Taylor, E. W. (1999). Cardiovascular responses to hypoxia in the Adriatic sturgeon (*Acipenser naccarii*). *J. Appl. Ichthyol.* **15,** 67–72.

Aho, E., and Vornanen, M. (2002). Effects of adenosine on the contractility of normoxic rainbow trout heart. *J. Comp. Physiol.* **172,** 217–225.

Allen, D. G., Morris, P. G., Orchard, C. H., and Pirolo, J. S. (1985). A nuclear magnetic resonance study of metabolism in the ferret heart during hypoxia and inhibition of glycolysis. *J. Physiol.* **361,** 185–204.

Almeida-Val, V. M. F., Chippari Gomes, A. R., and Lopes, N. P. (2006). Metabolic and physiological adjustments to low oxygen and high temperature in fishes of the Amazon. *In* "Fish Physiology Vol. 21, The Physiology of Tropical Fishes" (Val, A. L., Almeida-Val, V. M. F., and Randall, D., Eds.), pp. 443–500. Elsevier Inc., The Netherlands.

Altimiras, J., and Axelsson, M. (2004). Intrinsic autoregulation of cardiac output in rainbow trout (*Oncorhynchus mykiss*) at different heart rates. *J. Exp. Biol.* **207,** 195–201.

Anflous-Pharayra, K., Cai, Z., and Craigen, W. J. (2007). VDAC1 serves as a mitochondrial binding site for hexokinase in oxidative muscles. *Biochim. Biophys. Acta* **1767,** 136–142.

Arthur, P. G., Keen, J. E., Hochachka, P. W., and Farrell, A. P. (1992). Metabolic state of the *in situ* perfused trout heart during severe hypoxia. *Am. J. Physiol.* **263,** R798–R804.

Axelsson, M. (2005). The circulatory system and its control. *In* "Fish Physiology Vol. 22, The Physiology of Polar Fishes" (Farrell, A. P., and Steffensen, J. F., Eds.), pp. 239–280. Elsevier Inc, California.

Axelsson, M., and Fritsche, R. (1991). Effects of exercise, hypoxia and feeding on the gastrointestinal blood flow in the Atlantic cod *Gadus morhua*. *J. Exp. Biol.* **158,** 181–198.

Axelsson, M., and Nilsson, S. (1986). Blood pressure regulation during exercise in the Atlantic cod, *Gadus morhua*. *J. Exp. Biol.* **126,** 225–236.

Axelsson, M., Farrell, A. P., and Nilsson, S. (1990). Effects of hypoxia and drugs on cardiovascular dynamics of the Atlantic hagfish *Myxine glutinosa*. *J. Exp. Biol.* **151,** 297–316.

Axelsson, M., Altimiras, J., and Claireaux, G. (2002). Post-prandial blood flow to the gastrointestinal tract is not compromised during hypoxia in the sea bass *Dicentrarchus labrax*. *J. Exp. Biol.* **205,** 2891–2896.

Azoulay-Zohar, H., Israelson, A., Abu-Hamad, S., and Shoshan-Barmatz, V. (2004). In self-defense: Hexokinase promotes voltage-dependent anion channel closure and prevents mitochondria-mediated apoptotic cell death. *Biochem. J.* **377,** 347–355.

Bagatto, B (2005). Ontogeny of cardiovascular control in zebrafish (*Danio rerio*): Effects of developmental environment. *Comp. Biochem. Physiol.* **141A,** 391–400.

Bailey, J. R., Val, A. L., Almeida-Val, V. M. F., and Driedzic, W. R. (1999). Anoxic cardiac performance in Amazon and north-temperate teleosts. *Can. J. Zool.* **77,** 683–689.

Bailey, J. R., Rodnick, K. J., MacDougall, R., Clowe, S., and Driedzic, W. R. (2000). Anoxic performance of the American eel (*Anguilla rostrata* L.) heart requires extracellular glucose. *J. Exp. Zool.* **286,** 699–706.

Barrionuevo, W. R., and Burggren, W. W. (1999). O_2 consumption and heart rate in developing zebrafish (*Danio rerio*): Influence of temperature and ambient O_2. *Am. J. Physiol.* **276,** R505–R513.

Behrens, J. W., and Steffensen, J. F. (2007). The effect of hypoxia on behavioural and physiological aspects of lesser sandeel, *Ammodytes tobianus* (Linnaeus, 1785). *Mar. Biol.* **150,** 1365–1377.

Bolli, R., and Marban, E. (1999). Molecular and cellular mechanisms of myocardial stunning. *Physiol. Rev.* **79,** 609–634.

Brooks, S. P. J., and Storey, K. B. (1995). Is glycolytic rate controlled by the reversible binding of enzymes to subcellular structures? *In* "Biochemistry and Molecular Biology of Fishes" (Hochachka, P. W., and Mommsen, T. P., Eds.), Vol. 4, pp. 291–307. Elsevier, Amsterdam.

Burleson, M. L., Carlton, A. L., and Silva, P. E. (2002). Cardioventilatory effects of acclimatization to aquatic hypoxia in channel catfish. *Res. Physiol. Neurobiol.* **131,** 223–232.

Bushnell, P. G., and Brill, R. W. (1992). Oxygen transport and cardiovascular responses in skipjack tuna (*Katsuwonus pelamis*) and yellowfin tuna (*Thunnus albacares*). *J. Comp. Physiol.* **162B,** 131–143.

Butler, P. J., and Taylor, E.W (1971). Responses of the dogfish (*Scyliorhinus canicula* L.) to slowly induced and rapidly induced hypoxia. *Comp. Biochem. Physiol.* **39A,** 307–323.

Butler, P. J., and Taylor, E. W. (1975). The effect of progressive hypoxia on respiration in the dogfish (*Scyliorhinus canicula*) at different seasonal temperatures. *J. Exp. Biol.* **63,** 117–130.

Cameron, J. N. (1975). Blood tracer distribution as indicated by tracer microspheres in resting and hypoxic Arctic grayling (*Thymalus arcticus*). *Comp. Biochem. Physiol.* **52A,** 441–445.

Cameron, J. S., Hoffman, K. E., Zia, C., Hemmett, H. M., Kronsteiner, A., and Lee, C. M. (2003). A role for nitric oxide in hypoxia-induced activation of cardiac K_{ATP} channels in goldfish (*Carassius auratus*). *J. Exp. Biol.* **206,** 4057–4065.

Cech, J. J., Jr., Rowell, D. M., and Glasgow, J. S. (1977). Cardiovascular responses of the winter flounder *Pseudopleuronectes americanus* to hypoxia. *Comp. Biochem. Physiol.* **57A,** 123–125.

Chan, D. K. O. (1986). Cardiovascular, respiratory, and blood adjustments to hypoxia in the Japanese eel, *Anguilla japonica. Fish Physiol. Biochem.* **2,** 179–193.

Chen, J., Zhu, J. X., Wilson, I., and Cameron, J. S. (2005). Cardioprotective effects of K_{ATP} channel activation during hypoxia in goldfish *Carassius auratus. J. Exp. Biol.* **208,** 2765–2772.

Chippari-Gomes, A. R., Gomes, L. C., Lopes, N. P., Val, A. L., and Almeida-Val, V. M. F. (2005). Metabolic adjustments in two Amazonian cichlids exposed to hypoxia and anoxia. *Comp. Biochem. Physiol.* **141B,** 347–355.

Claireaux, G., and Dutil, J. D. (1992). Physiological response of the Atlantic cod (*Gadus morhua*) to hypoxia at various environmental salinities. *J. Exp. Biol.* **163,** 97–118.

Claireaux, G., Webber, D. M., Kerr, S. R., and Boutilier, R. G. (1995). Physiology and behaviour of free swimming Atlantic cod (*Gadus morhua*) facing fluctuating salinity and oxygenation. *J. Exp. Biol.* **198,** 61–69.

Clow, K. A., Rodnick, K. J., MacCormack, T. J., and Driedzic, W. R. (2004). The regulation and importance of glucose uptake in the isolated Atlantic cod heart: Rate-limiting steps and effects of hypoxia. *J. Exp. Biol.* **207,** 1865–1874.

Cruz-Neto, A. P., and Steffensen, J. F. (1997). The effects of acute hypoxia and hypercapnia on oxygen consumption in the freshwater European eel. *J. Fish Biol.* **50,** 759–769.

Dalla Via, J., Van den Thillart, G., Cattani, O., and Cortesi, P. (1998). Behavioural responses and biochemical correlates in *Solea solea* to gradual hypoxic exposure. *Can. J. Zool.* **76,** 2108–2113.

Davey, K. A. B., Garlick, P. B., Warley, A., and Southworth, R. (2007). Immunogold labeling study of the distribution of GLUT-1 and GLUT-4 in cardiac tissue following stimulation by insulin or ischemia. *Am. J. Physiol.* **292,** H2009–H2019.

Davie, P. S., and Franklin, C. E. (1993). Preliminary observations on blood flow in the coronary arteries of two school sharks (*Galeorhinus australis*). *Can. J. Zool.* **71,** 1238–1241.

De Salvo Souza, R. H., Soncini, R., Glass, M. L., Sanches, J. R., and Rantin, F. T. (2001). Ventilation, gill perfusion and blood gases in dourado *Salminus maxillosus* Valenciennes (Teleostei, Characidae), exposed to graded hypoxia. *J. Comp. Phyiol.* **171B,** 483–489.

Dickson, K. A., and Graham, J. B. (1986). Adaptations to hypoxic environments in the erythrinid fish *Hoplias microlepsis*. *Environ. Biol. Fishes* **15,** 301–308.

Dombkowski, R. A., Rusell, M. J., and Olson, K. R. (2004). Hydrogen sulfide as an endogenous regulator of vascular muscle tone in trout. *Am. J. Physiol.* **286,** R678–R685.

Donald, J., and Campbell, G. (1982). A comparative study of the adrenergic innervation of the teleost heart. *J. Appl. Physiol.* **147,** 85–91.

dos Santos, P., Kowaltowski, A. J., Laclau, M. N., Seetharaman, S., Paucek, P., Boudina, S., Thambo, J.-B., Taroisse, L., and Garlid, K. D. (2002). Mechanisms by which opening the mitochondrial ATP-sensitive K^+ channel protects the ischemic heart. *Am. J. Physiol.* **283,** H284–H295.

Downey, J. M., Davies, A. M., and Cohen, M. V. (2007). Signaling pathways in ischemic preconditioning. *Heart Fail. Rev.* **12,** 181–188.

Driedzic, W. R. (1992). Cardiac energy metabolism. *In* "Fish Physiology Vol. 12A" (Hoar, W. S., Randall, D. J., and Farrell, A. P., Eds.), pp. 219–266. Academic Press Inc., New York.

Driedzic, W. R., and Bailey, J. R. (1999). Anoxia cardiac tolerance in Amazonian and North temperate teleosts is related to the potential to utilize extracellular glucose. *In* "Biology of Tropical Fishes" (Val, A. L., and Almeida-Val, V. M. F., Eds.), pp. 217–227. INPA, Manaus.

Driedzic, W. R., and Gesser, H. (1994). Energy metabolism and contractility in ectothermic vertebrate hearts: Hypoxia, acidosis and low temperature. *Physiol. Rev.* **74,** 221–258.

Driedzic, W. R., and Stewart, J. M. (1982). Myoglobin content and the activities of enzymes of energy metabolism in red and white fish hearts. *J. Comp. Physiol.* **149,** 67–74.

Driedzic, W. R., Gesser, H., and Johansen, K. (1985). Effects of hypoxic adaptation on myocardial performance and metabolism of *Zoarces viviparous*. *Can. J. Zool.* **63,** 821–823.

Duncan, J. A., and Storey, K. B. (1991). Role of enzyme binding in muscle metabolism of the goldfish. *Can. J. Zool.* **69,** 1571–1576.

Farrell, A. P. (1982). Cardiovascular changes in the unanaesthetized lingcod (*Ophiodon elongates*) during short-term, progressive hypoxia and spontaneous activity. *Can. J. Zool.* **60,** 933–941.

Farrell, A. P. (1984). A review of cardiac performance in the teleost heart: Intrinsic and humoral regulation. *Can. J. Zool.* **62,** 523–536.

Farrell, A. P. (1991). From hagfish to tuna: A perspective on cardiac function in fish. *Physiol. Zool.* **64,** 1137–1164.

Farrell, A. P. (2007). Tribute to Peter Lutz: A message from the heart – why hypoxic bradycardia in fishes? *J. Exp. Biol.* **210,** 1715–1725.

Farrell, A. P., and Clutterham, S. M. (2003). On-line venous oxygen tensions in rainbow trout during graded exercise at two acclimation temperatures. *J. Exp. Biol.* **206,** 487–496.

Farrell, A. P., and Jones, D. R. (1992). The heart. *In* "Fish Physiology Vol. 12A, The Cardiovascular System" (Hoar, W. S., Randall, D. J., and Farrell, A. P., Eds.), pp. 1–88. Academic Press, San Diego.

Farrell, A. P., and Stecyk, J. A. W. (2007). The heart as a working model to explore themes and strategies for anoxic survival in ectothermic vertebrates. *Comp. Biochem. Physiol.* **147A,** 300–312.

Faust, H. A., Gamperl, A. K., and Rodnick, K. J. (2004). All rainbow trout (*Oncorhynchus mykiss*) are not created equal: Intra-specific variation in cardiac hypoxia tolerance. *J. Exp. Biol.* **207,** 1005–1015.

Forster, M. E., Axelsson, M., Farrell, A. P., and Nilsson, S. (1991). Cardiac function and circulation in hagfishes. *Can. J. Zool.* **69,** 1985–1992.
Fritsche, R., and Nilsson, S. (1989). Cardiovascular responses to hypoxia in the Atlantic cod, *Gadus morhua. Exp. Biol.* **48,** 153–160.
Fritsche, R., and Nilsson, S. (1990). Autonomic nervous control of blood pressure and heart rate during hypoxia in the cod, *Gadus morhua. J. Comp. Physiol.* **160B,** 287–292.
Fritsche, R., Axelsson, M., Franklin, C. E., Grigg, G. G., Holmgren, S., and Nilsson, S. (1993). Respiratory and cardiovascular responses to hypoxia in the Australian lungfish. *Resp. Phyisol.* **94,** 173–187.
Fritsche, R., Schwerte, T., and Pelster, B. (2000). Nitric oxide and vascular reactivity in developing zebrafish, *Danio rerio. Am. J. Physiol.* **29,** R2200–R2207.
Furimsky, M., Cooke, S. J., Suski, C. D., Wang, Y., and Tufts, B. L. (2003). Respiratory and circulatory responses to hypoxia in largemouth bass and smallmouth bass: Implications for "live-release" angling tournaments. *Trans. Am. Fish. Soc.* **132,** 1065–1075.
Gamperl, A. K., and Farrell, A. P. (2004). Cardiac plasticity in fishes: Environmental influences and intraspecific differences. *J. Exp. Biol.* **207,** 2539–2550.
Gamperl, A. K., Axelsson, M., and Farrell, A. P. (1995). Effects of swimming and environmental hypoxia on coronary blood flow in rainbow trout. *Am. J. Physiol.* **269,** R1258–R1266.
Gamperl, A. K., Todgham, A. E., Parkhouse, W. S., Dill, R., and Farrell, A. P. (2001). Recovery of trout myocardial function following anoxia: Preconditioning in a non-mammalian model. *Am. J. Physiol.* **281,** R1755–R1763.
Gamperl, A. K., Faust, H. A., Dougher, B., and Rodnick, K. J. (2004). Hypoxia tolerance and preconditioning are not additive in the trout (*Oncorhynchus mykiss*) heart. *J. Exp. Biol.* **207,** 2497–2505.
Gamperl, A. K., Swafford, B. L., and Rodnick, K. J. (2008). The impact of elevated water temperature and zatebradine-induced bradycardia on cardiovascular function in male and female rainbow trout (*Oncorhynchus mykiss*). *Integr. Comp. Biol.* 47 (S1), e182.
Gehrke, P. C., and Fielder, D. R. (1988). Effects of temperature and dissolved oxygen on heart rate, ventilation rate and oxygen consumption of spangled perch, *Leiopotherapon unicolor* (Gunther 1859), (Percoidei, Teraponidae). *J. Comp. Physiol.* **157B,** 771–782.
Gesser, H. (2002). Mechanical performance and glycolytic requirement in trout ventricular muscle. *J. Exp. Zool.* **293,** 360–367.
Gesser, H., and Høglund, L. (1988). Action potential force and function of the sarcoplasmic reticulum in the anerobic trout heart. *Exp. Biol.* **47,** 171–176.
Gesser, H., Andresen, P., Brams, P., and Sund-Laursen, J. (1982). Inotropic effects of adrenaline on the anoxic or hypercapnic myocardium of rainbow trout and eel. *J. Comp. Physiol.* **147,** 123–128.
Godt, R. E., and Nosek, T. M. (1989). Changes of intracellular milieu with fatigue or hypoxia depress contraction of skinned rabbit skeletal and cardiac muscle. *J. Physiol.* **412,** 155–180.
Gracey, A. Y., Troll, J. V., and Somero, G. N. (2001). Hypoxia-induced gene expression profiling in the euryoxic fish *Gillichthys mirabilis. Proc. Nat. Acad. Sci. USA* **98,** 1993–1998.
Graham, J. B. (1997). "Air-Breathing Fishes: Evolution, Diversity and Adaptation," New York, Academic Press.
Gross, E. R., and Gross, G. J. (2007). Ischemic preconditioning and myocardial infarction: An update and perspective. *Drug Discov. Today Dis. Mech.* **4,** 165–174.
Hall, J. R., MacCormack, T. J., Barry, C. A., and Driedzic, W. R. (2004). Sequence and expression of a constitutive, facilitated glucose transporter (GLUT1) in Atlantic cod *Gadus morhua. J. Exp. Biol.* **207,** 4697–4706.
Hall, J. R., Richards, R. C., MacCormack, T. J., Ewart, K. V., and Driedzic, W. R. (2005). Cloning of GLUT3 cDNA from Atlantic cod (*Gadus morhua*) and expression of GLUT1 and GLUT3 in response to hypoxia. *Biochim. Biophys. Acta* **1730,** 245–252.

Hall, J. R., Short, C. E., and Driedzic, W. R. (2006). Sequence of Atlantic cod (*Gadus morhua*) GLUT 4, GLUT2, and GPDH: Developmental stage expression, tissue expression, and relationship to starvation-induced changes in blood glucose. *J. Exp. Biol.* **209,** 4990–4502.

Haman, O. H., Zwingelstein, G., and Weber, J. (1997). Effects of hypoxia and low temperature on substrate fluxes in fish: Plasma metabolite concentrations are misleading. *Am. J. Physiol.* **273,** R2046–R2054.

Han, J., Kim, N., Kim, E., Ho, W.-K., and Earm, Y. E. (2001). Modulation of ATP-sensitive potassium channels by cGMP-dependent protein kinase in rabbit ventricular myocytes. *J. Biol. Chem.* **276,** 2140–22147.

Hanson, L. M., and Farrell, A. P. (2007). The hypoxic threshold for maximum cardiac performance in rainbow trout *Oncorhynchus mykiss* (Walbaum) during simulated exercise conditions at 18°C. *J. Fish Biol.* **71,** 926–932.

Hanson, L. M., Obradovich, S., Mouniargi, J., and Farrell, A. P. (2006). The role of adrenergic stimulation in maintaining maximum cardiac performance in rainbow trout (*Oncorhynchus mykiss*) during hypoxia, hyperkalemia and acidosis at 10°C. *J. Exp. Biol.* **209,** 2442–2451.

Holeton, G. F. (1971). Respiratory and circulatory responses of rainbow trout larvae to carbon monoxide and to hypoxia. *J. Exp. Biol.* **55,** 683–694.

Holeton, G. F., and Randall, D. J. (1967). Changes in blood pressure in the rainbow trout during hypoxia. *J. Exp. Biol.* **46,** 297–305.

Holmuhamedov, E., Jovanovic, S., Dzeja, P. P., Jovanovic, A., and Terzic, A. (1998). Mitochondrial ATP-sensitive K^+ channels modulate cardiac mitochondrial function. *Am. J. Physiol.* **275,** H1567–H1576.

Hughes, G. M., Peyraud, C., Peyraud-Waitzenegger, M., and Soulier, P. (1982). Physiological evidence for the occurrence of pathways shunting blood away from the secondary lamellae of eel gills. *J. Exp. Biol.* **98,** 277–288.

Ishimatsu, A., Iwama, G. K., and Heisler, N. (1988). *In vivo* analysis of partitioning of cardiac output between systemic and central venous sinus circuits in rainbow trout – a new approach using chronic cannulation of the branchial vein. *J. Exp. Biol.* **137,** 75–88.

Jacob, E., Drexel, M., Schwerte, T., and Pelster, B. (2002). Influence of hypoxia and of hypoxemia on the development of cardiac activity in zebrafish larvae. *Am. J. Physiol.* **283,** R911–R917.

Jensen, F. B., Nikinmaa, M., and Weber, R. E. (1993). Environmental perturbations of oxygen transport in teleost fishes: Causes, consequences and compensations. *In* "Fish Ecophysiology" (Rankin, J. C., and Jensen, F.,B., Eds.), pp. 161–179. Chapman and Hall, London.

Jonz, M. G., and Nurse, C. A. (2005). Development of oxygen sensing in the gills of zebrafish. *J. Exp. Biol.* **208,** 1537–1549.

Jørgensen, J. B., and Mustafa, T. (1980). Effect of hypoxia on carbohydrate metabolism in flounder (*Platichthys flesus* L.) – I. Utilization of glycogen and accumulation of glycolytic end products in various tissues. *Comp. Biochem. Physiol.* **67B,** 243–248.

Keen, J. E., Vianzon, D.,M., Farrell, A.,P., and Tibbits, G. F. (1993). Thermal acclimation alters both adrenergic sensitivity and adrenoceptor density in cardiac tissue of rainbow trout. *J. Exp. Biol.* **181,** 24–47.

Kolář, F., and Oš'ádal, B. (2004). Molecular mechanisms of cardiac protection by adaptation to chronic hypoxia. *Physiol. Rev.* **53,** S3–S13.

Lefrançois, C., and Domenici, P. (2006). Locomotor kinematics and behaviour in the escape response of European sea bass, *Dicentrarchus labrax* L., exposed to hypoxia. *Mar. Biol.* **149,** 969–977.

Lennard, R., and Huddart, H. (1992). The effects of hypoxic stress on the fine structure of the flounder heart (*Platichthys flesus*). *Comp. Biochem. Physiol.* **101A,** 723–732.

Lewis, J. M., Costa, I., Val, A. L., Almeida-Val, V. M. F., Gamperl, A. K., and Driedzic, W. R. (2007). Responses to hypoxia and recovery: Repayment of oxygen debt is not associated with compensatory protein synthesis in the Amazonian cichlid, *Astronotus ocellatus*. *J. Exp. Biol.* **210,** 1935–1943.

Lurman, G. J., Koschnick, N., Pörtner, H., and Lucassen, M. (2007). Molecular characterization and expression of Atlantic cod (*Gadus morhua*) myoglobin from two different acclimation temperatures. *Comp. Biochem. Physiol.* **148A,** 681–689.

MacCormack, T. J., and Driedzic, W. R. (2002). Mitochondrial ATP-sensitive K^+ channels influence force development and anoxic contractility in flatfish, yellowtail flounder *Limanda feruginea*, but not Atlantic cod *Gadus morhua* heart. *J. Exp. Biol.* **205,** 1411–1418.

MacCormack, T. J., and Driedzic, W. R. (2004). Cardiorespiratory and tissue adenosine responses to hypoxia and reoxygenation in the short-horned sculpin *Myoxocephalus scorpius*. *J. Exp. Biol.* **207,** 4157–4164.

MacCormack, T. J., and Driedzic, W. R. (2007). The impact of hypoxia on *in vivo* glucose uptake in a hypoglycemic fish, *Myoxocephalus scorpius*. *Am. J. Physiol.* **292,** R1033–R1042.

MacCormack, T. J., McKinley, R. S., Roubach, R., Almedia-Val, V. M. F., Val, A. L., and Driedzic, W. R. (2003a). Changes in ventilation, metabolism, and behaviour, but not bradycardia, contribute to the hypoxia survival in two species of Amazonian armoured catfish. *Can. J. Zool.* **81,** 272–280.

MacCormack, T. J., Treberg, J. R., Almedia-Val, V. M. F., Val, A. L., and Driedzic, W. R. (2003b). Mitochondrial K_{ATP} channels and sarcoplasmic reticulum influence cardiac force development under anoxia in the Amazonian armoured catfish *Liposarcus pardalis*. *Comp. Biochem. Physiol.* **134A,** 441–448.

MacCormack, T. J., Lewis, J. M., Almeida-Val, V. M. F., Val, A. L., and Driedzic, W. R. (2006). Carbohydrate management, anaerobic metabolism, and adenosine levels in the armoured catfish, *Liposarcus pardalis* (Castelnau), during hypoxia. *J. Exp. Zool.* **305A,** 363–375.

Marques, I. J., Leito, J. T. D., Spaink, H. P., Testerlink, J., Jaspers, R. T., Witte, F., van den Berg, S., and Bagowski, C. P. (2008). Transcriptome analysis of the response to chronic constant hypoxia in zebrafish hearts. *J. Comp. Physiol.* **178B,** 77–92.

Martínez, M. L., Landry, C., Boehm, R., Manning, S., Cheek, A. O., and Rees, B. R. (2006). Effects of long-term hypoxia on enzymes of carbohydrate metabolism in the Gulf killifish, *Fundulus grandis*. *J. Exp. Biol.* **209,** 3851–3861.

Marvin, D. E., and Heath, A. G. (1968). Cardiac and respiratory responses to gradual hypoxia in three ecologically distinct species of fresh-water fish. *Comp. Biochem. Physiol.* **27,** 349–355.

McDonald, D. G., and McMahon, B.R (1977). Respiratory development in Artic char *Salvelinus alpinus* under conditions of normoxia and chronic hypoxia. *Can. J. Zool* **55,** 1461–1467.

Moore, F.B-G., Hosey, M., and Bagatto, B. (2006). Cardiovascular system in larval zebrafish responds to developmental hypoxia in a family specific manner. *Front. Zool.* **3,** 4.

Nilsson, S. (1983). "Autonomic Nerve Function in the Vertebrates." Springer-Verlag, New York.

Nilsson, S. (1984). Innervation and pharmacology of the gills. *In* "Fish Physiology Vol. 10A, Gills: Anatomy Gas Transfer and Acid-Base Regulation" (Hoar, W. S., and Randall, D. J., Eds.), pp. 185–227. Academic Press Inc., Orlando.

Oliveira, R. D., Lopes, J. M., Sanches, J. R., Kalinin, A. L., Glass, M. L., and Rantin, F. T. (2004). Cardiorespiratory responses of the facultative air-breathing fish jeju, *Hopleythrinus unitaeniatus* (Teleostei, Erythrinidae), exposed to graded ambient hypoxia. *Comp. Biochem. Physiol.* **139A,** 479–485.

Olson, K. R. (2008). Hydrogen sulfide and oxygen sensing: Implications in cardiorespiratory control. *J. Exp. Biol.* **211,** 2727–2734.

Olson, K. R., Dombkowski, R. A., Russell, M. J., Doellman, M. M., Head, S. K., Whitfield, N. L., and Madden, J. A. (2006). Hydrogen sulfide as an oxygen sensor/transducer in vertebrate hypoxic vasoconstriction and hypoxic vasodilation. *J. Exp. Biol.* **209,** 4011–4023.

Overgaard, J., and Gesser, H. (2004). Force development, energy state and ATP production of cardiac muscle from turtles and trout during normoxia and severe hypoxia. *J. Exp. Biol.* **207,** 1915–1924.

Overgaard, J., Stecyk, J. A. W., Gesser, H., Wang, T., and Farrell, A. P. (2004a). Effects of temperature and anoxia upon the performance of *in situ* perfused trout hearts. *J. Exp. Biol.* **207,** 655–665.

Overgaard, J., Stecyk, J. A. W., Gesser, H., Wang, T., Gamperl, A. K., and Farrell, A. P. (2004b). Preconditioning stimuli do not benefit the myocardium of hypoxia-tolerant rainbow trout (*Oncorhynchus mykiss*). *J. Comp. Physiol.* **174B,** 329–340.

Padilla, P. A., and Roth, M. B. (2001). Oxygen deprivation causes suspended animation in the zebrafish embryo. *Proc. Natl. Acad. Sci. USA* **98,** 7331–7335.

Paajanen, V., and Vornanen, M. (2002). The induction of an ATP-sensitive K^+ current in cardiac myocytes of air- and water-breathing vertebrates. *Pfluegers Arch.* **444,** 760–770.

Paajanen, V., and Vornanen, M. (2003). Effects of chronic hypoxia on inward rectifier K^+ current (IK1) in ventricular myocytes of crucian carp (*Carassius carassius*) heart. *J. Membr. Biol.* **194,** 119–127.

Planas, J. V., Capilla, E., and Gutiérrez, J. (2000). Molecular identification of a glucose transporter from fish muscle. *FEBS Lett.* **481,** 266–270.

Perry, S. F., and Reid, S. (1994). The effects of acclimation temperature on the dynamics of catecholamine release during acute hypoxia in the rainbow trout *Oncorhynchus mykiss*. *J. Exp. Biol.* **186,** 289–307.

Perry, S.,F., Gilmour, K.,M., McNeill, B., Chew, S.,F., and Ip, Y.,K. (2005). Circulating catecholamines and cardiorespiratory responses in hypoxic lungfish (*Protopterus dolloi*): A comparison of aquatic and aerial hypoxia. *Physiol. Biochem. Zool.* **78,** 325–334.

Pettersson, K., and Johansen, K. (1982). Hypoxic vasoconstriction and the effects of adrenaline on gas exchange efficiency in fish gills. *J. Exp. Biol.* **97,** 263–272.

Pettersson, K., and Nilsson, S. (1979). Nervous control of the branchial vascular resistance of the Atlantic cod, *Gadus morhua*. *J. Comp. Physiol.* **129,** 179–183.

Peyraud-Waitzenegger, M., and Soulier, P. (1989). Ventilatory and circulatory adjustment in the European eel (*Anguilla Anguilla* L.) exposed to short term hypoxia. *Exp. Biol.* **48,** 107–122.

Pihl, L., Baden, S. P., and Diaz, R. J. (1991). Effects of periodic hypoxia on distribution of demersal fish and crustaceans. *Mar. Biol.* **108,** 349–360.

Ramasamy, R., Hwang, Y. C., Whang, J., and Bergmann, S. R. (2001). Protection of ischemic hearts by high glucose is mediated, in part, by GLUT-4. *Am. J. Physiol.* **281,** H290–H297.

Randall, D. J., and Daxboeck, C. (1982). Cardiovascular changes in the rainbow trout (*Salmo gairdneri* Richardson) during exercise. *Can. J. Zool.* **60,** 1134–1140.

Randall, D. J., and Perry, S. F. (1992). Catecholamines. *In* "Fish Physiology, Vol. 12B, The Cardiovascular System" (Hoar, W. S., Randall, D. J., and Farrell, A. P., Eds.), pp. 255–300. Academic Press, San Diego.

Rantin, F. T., Glass, M. L., Kalinin, A. L., Verzola, R. M. M., and Fernandes, M. N. (1993). Cardio-respiratory responses in two ecologically distinct erythrinids (*Hoplias malabaricus* and *Hoplias lacerdae*) exposed to graded environmental hypoxia. *Environ. Biol. Fish.* **36,** 93–97.

Rantin, F. T., Kalinin, A. L., Guerra, C. D. R., Maricondi-Massari, M., and Verzola, R. M. M. (1995). Electrocardiographic characterization of myocardial function in normoxic and hypoxic teleosts. *Braz. J. Med. Biol. Res.* **28,** 1277–1289.

Reid, S. G., and Perry, S. F. (1994). Storage and differential release of catecholamines in rainbow trout (*Oncorhynchus mykiss*) and American eel (*Anguilla rostrata*). *Physiol. Zool.* **67,** 216–237.

Ristori, M-T., and Laurent, P. (1989). Plasma catecholamines in rainbow trout (*Salmo gairdneri*) during hypoxia. *Exp. Biol.* **48,** 285–290.

Rodnick, K. J., Bailey, J. R., West, J. L., Rideout, A., and Driedzic, W. R. (1997). Acute regulation of glucose uptake in cardiac muscle of the American eel *Anguilla rostrata*. *J. Exp. Biol.* **200,** 2871–2880.

Roesner, A., Hankeln, T., and Burmester, T. (2006). Hypoxia induces a complex response of globin expression in zebrafish (*Danio rerio*). *J. Exp. Biol.* **209,** 2129–2137.

Sanchez, A., Soncini, R., Wang, T., Koldkjaer, P., Taylor, E. W., and Glass, M. L. (2001). The differential cardio-respiratory responses to ambient hypoxia and systemic hypoxaemia in the South American lungfish, *Lepidosiren paradoxa*. *Comp. Biochem. Physiol.* **130A,** 677–687.

Sandblom, E., and Axelsson, M. (2005). Effects of hypoxia on the venous circulation in rainbow trout (*Oncorhynchus mykiss*). *Comp. Biochem. Physiol.* **140A,** 233–239.

Sandblom, E., and Axelsson, M. (2006). Adrenergic control of venous capacitance during moderate hypoxia in the rainbow trout (*Onchorhynchus mykiss*): Role of neural and circulating catecholamines. *Am. J. Physiol.* **291,** R711–R718.

Santer, R. M. (1972). Ultrastructure and histochemical studies on the innervation of the heart of the teleost, *Pleuronectes platessa*. *Z. Zellforsch. Mikrosk. Anat.* **131,** 519–528.

Schwerte, T., and Pelster, B. (2000). Digital motion analysis as a tool for analysing the shape and performance of the circulatory system in transparent animals. *J. Exp. Biol.* **203,** 1659–1669.

Schwerte, T., U"berbacher, D., and Pelster, B. (2003). Non-invasive imaging of blood cell concentration and blood distribution in zebrafish *Danio rerio* incubated in hypoxic conditions *in vivo*. *J. Exp. Biol.* **206,** 1299–1307.

Schwerte, T., Prem, C., Mairösl, A., and Pelster, B. (2006). Development of the sympatho-vagal balance in the cardiovascular system in zebrafish (*Danio rerio*) characterized by power spectrum and classical signal analysis. *J. Exp. Biol.* **209,** 1093–1100.

Shiels, H. A., Vornanen, M., and Farrell, A. P. (2002). The force-frequency relationship in fish hearts – a review. *Comp. Biochem. Physiol.* **132A,** 811–826.

Shoubridge, E. A., and Hochachka, P. W. (1983). The integration and control of metabolism in the anoxic goldfish. *Mol. Physiol.* **4,** 165–195.

Sidell, B. D., and O'Brien, K. M. (2006). When bad things happen to good fish: The loss of hemoglobin and myoglobin expression in Antarctic icefishes. *J. Exp. Biol.* **209,** 1791–1802.

Simonot, D. L., and Farrell, A. P. (2007). Cardiac remodeling in rainbow trout *Oncorhynchus mykiss* Walbaum in response to phenylhydrazine-induced anaemia. *J. Exp. Biol.* **210,** 2574–2584.

Skals, M., Skovgaard, N., Taylor, E. W., Leite, C. A. C., Abe, A. S., and Wang, T. (2006). Cardiovascular changes under normoxic and hypoxic conditions in the air-breathing teleost *Synbranchus marmoratus*: Importance of the venous system. *J. Exp. Biol.* **209,** 4167–4173.

Smatresk, N. J., and Cameron, J. N. (1982). Respiration and acid-base physiology of the spotted gar, a bimodal breather. 1. Normal values, and the response to severe hypoxia. *J. Exp. Biol.* **96,** 263–280.

Smith, D. G., Wahlqvist, I., Nilsson, S., and Eriksson, B.-M. (1985). Nervous control of blood pressure in the Atlantic cod, *Gadus morhua*. *J. Exp. Biol.* **117,** 335–347.

Smith, M. P., Russell, M. J., Wincko, J. T., and Olson, K. R. (2001). Effects of hypoxia on isolated vessels and perfused gills of rainbow trout. *Comp. Biochem. Physiol.* **130A,** 171–181.

Smith, R. W., Houlihan, D. F., Nilsson, G. E., and Brechin, J. G. (1996). Tissue-specific changes in protein synthesis rates *in vivo* during anoxia in crucian carp. *Am. J. Physiol.* **271,** R897–R904.

Southworth, R., Davey, K. A. B., Warley, A., and Garlick, P. B. (2007). A reevaluation of the roles of hexokinase I and II in the heart. *Am. J. Physiol.* **292,** H378–H386.

Stecyk, J. A. W., and Farrell, A. P. (2006). Regulation of the cardiorespiratory system of common carp (*Cyprinus carpio*) during severe hypoxia at three seasonal acclimation temperatures. *Physiol. Biochem. Zool.* **79,** 614–627.

Stecyk, J. A. W., Stenslokken, K., Nilsson, G. E., and Farrell, A. P. (2007). Adenosine does not save the heart of anoxia-tolerant vertebrates during prolonged oxygen deprivation. *Comp. Biochem. Physiol.* **147A,** 961–973.

Steffensen, J. F., and Farrell, A. P. (1998). Swimming performance, venous oxygen tension and cardiac performance of coronary-ligated rainbow trout, *Oncorhynchus mykiss*, exposed to progressive hypoxia. *Comp. Biochem. Physiol.* **119A,** 585–592.

Stensløkken, K., Sundin, L., Renshaw, G. M. C., and Nilsson, G. E. (2004). Adenosinergic and cholinergic control mechanisms during hypoxia in the epaulette shark (*Hemiscyllium ocellatum*), with emphasis on branchial circulation. *J. Exp. Biol.* **207,** 4451–4461.

Storey, K. B. (1987). Tissue-specific controls on carbohydrate catabolism during anoxia in goldfish. *Physiol. Zool.* **60,** 601–607.

Sundin, L. I. (1995). Responses of the branchial circulation to hypoxia in the Atlantic cod, *Gadus morhua. Am. J. Physiol.* **268,** R771–R778.

Sundin, L. I., and Nilsson, S. (1992). Arterio-venous branchial blood flow in the Atlantic cod, *Gadus morhua. J. Exp. Biol.* **165,** 73–84.

Sundin, L. I., and Nilsson, G. E. (1996). Branchial and systemic roles of adenosine receptors in rainbow trout: An *in vivo* microscopy study. *Am. J. Physiol.* **271,** R661–R669.

Sundin, L. I., and Nilsson, G. E. (1997). Neurochemical mechanisms behind gill microcirculatory responses to hypoxia in trout: An *in vivo* microscopy study. *Am. J. Physiol.* **272,** R576–R585.

Sundin, L. I., Nilsson, G. E., Block, M., and Lofman, C. O. (1995). Control of gill filament blood flow by serotonin in the rainbow trout, *Oncorhynchus mykiss. Am. J. Physiol.* **268,** R1224–R1229.

Sundin, L., I, Axelsson, M., Davison, W., and Forster, M. E. (1999a). Cardiovascular responses to adenosine in the Antarctic fish *Pagothenia borchgrevinki. J. Exp. Biol.* **202,** 2259–2267.

Sundin, L. I., Reid, S. G., Kalinin, A. L., Rantin, F. T., and Milso, W. K. (1999b). Cardiovascular and respiratory reflexes: The tropical fish, traira (*Hoplias malabaricus*) O_2 chemoresponses. *Resp. Physiol.* **116,** 181–199.

Sundin, L. I., Reid, S.G, Rantin, F. T., and Milsom, W. K. (2000). Branchial receptors and cardiorespiratory reflexes in a neotropical fish, the tambaqui (*Colossoma macropomum*). *J. Exp. Biol.* **203,** 1225–1239.

Sureau, D., Lagardere, J. P., and Pennec, J. P. (1989). Heart rate and its cholinergic control in the sole (*Solea vulgaris*), acclimatized to different temperatures. *Comp. Biochem. Physiol.* **92A,** 49–51.

Taylor, E. W., Short, S., and Butler, P. J. (1977). The role of the cardiac vagus in the response of the dogfish *Scyliorhinus canicula* to hypoxia. *J. Exp. Biol.* **70,** 57–75.

Teerijoki, H., Krasnov, A., Pitkänen, T. I., and Mölsä, H. (2000). Cloning and characterization of glucose transporter in teleost fish rainbow trout (*Onchorynchus mykiss*). *Biochim. Biophys. Acta* **1494,** 290–294.

Tota, B., Amelio, D., Pellegrino, D., Ip, Y. K., and Cerra, M. C. (2005). NO modulation of myocardial performance in fish hearts. *Comp. Biochem. Physiol.* **142A,** 164–177.

Treberg, J. R., MacCormack, T. J., Lewis, J. M., Almeida-Val, V. M. F., Val, A. L., and Driedzic, W. R. (2007). Intracellular glucose and binding of hexokinase and phosphofructokinase to particulate fractions increase under hypoxia in heart of the Amazonian armoured catfish (*Liposarcus pardalis*). *Physiol. Biochem. Zool.* **80,** 542–550.

Turner, J. D., and Driedzic, W. R. (1980). Mechanical and metabolic response of the perfused isolated fish heart to anoxia and acidosis. *Can. J. Zool.* **58,** 886–889.

Val, A. L., Silva, M. N. P., and Almeida-Val, V. M. F. (1998). Hypoxia adaptation in fish of the Amazon: A never ending task. *S. Afr. J. Zool.* **33,** 107–114.

Vornanen, M., and Paajanen, V. (2004). Seasonality of dihydropyridine receptor binding in the heart of an anoxia tolerant vertebrate, the crucian carp (*Carassius carassius* L.). *Am. J. Physiol.* **287,** R1263–R1269.

Vornanen, M., and Toumennoro, J. (1999). Effects of acute anoxia on heart function in crucian carp: Importance of cholinergic and purinergic control. *Am. J. Physiol.* **277,** R465–R475.

West, J. L., Bailey, J. R., Almeida-Val, V. M. F., Val, A. L., Sidell, B. D., and Driedzic, W. R. (1999). Activity levels of enzymes of energy metabolism in heart and red muscle are higher in north temperate zone than in Amazonian teleosts. *Can. J. Zool.* **77,** 690–696.

West, T. G., Arthur, P. G., Suarez, R. K., Doll, C. J., and Hochachka, P. W. (1993). *In vivo* utilization of glucose by heart and locomotory muscles of exercising rainbow trout (*Onchorynchus mykiss*). *J. Exp. Biol.* **177,** 63–79.

Wilson, J. E. (2003). Isozymes of mammalian hexokinase: Structure, subcellular localization and metabolic function. *J. Exp. Biol.* **206,** 2049–2057.

Wood, C. M., and Shelton, G. (1980). The reflex control of heart rate and cardiac output in the rainbow trout: Interactive influences of hypoxia, hemorrhage, and systemic vasomotor tone. *J. Exp. Biol.* **87,** 271–284.

Wood, S., and Trayhurn, P. (2003). Glucose transporters (GLUT and SGLT): Expanded families of sugar transport proteins. *Br. J. Nutr.* **89,** 3–9.

Wood, C. M., Pieprzak, P., and Trott, N. J. (1979). The influence of temperature and anemia on the adrenergic and cholinergic mechanisms controlling heart rate in the rainbow trout. *Can. J. Zool.* **57,** 2240–2247.

Wright, J. R., Jr., O'Hali, W., Yang, H., Han, X.-X., and Bonen, A. (1998). GLUT-4 deficiency and severe peripheral resistance to insulin in the teleost fish tilapia. *Gen. Comp. Endocrinol* **111,** 20–27.

Zuurbier, C. J., Eerbeek, O., and Meijer, A. J. (2005). Ischemic preconditioning, insulin, and morphine all cause hexokinase redistribution. *Am. J. Physiol.* **289,** H496–H499.

8

THE EFFECTS OF HYPOXIA ON GROWTH AND DIGESTION

TOBIAS WANG
SJANNIE LEFEVRE
DO THI THANH HUONG
NGUYEN VAN CONG
MARK BAYLEY

Here we review how hypoxia affects growth and digestion in fish. Thus, the growth effects of hypoxia are explained in terms of reductions of energy intake (appetite) and assimilation efficiency as well as in terms of the costs of digestion or specific dynamic action. It is clear that the most commonly documented cause of hypoxia-related growth retardation is through loss of appetite and the regulatory physiology of this effect is discussed. Finally, the

Hypoxia: Volume 27
FISH PHYSIOLOGY

DOI: 10.1016/S1546-5098(08)00008-3

effects of hypoxia on the growth of air-breathing fish are reviewed and the most promising areas for future research on oxygen's role as a limiting factor for fish growth are highlighted.

1. INTRODUCTION

The ultimate goal of an animal, in Darwinian terms, is to propagate its genes by maximising lifetime reproductive output. Growth and reproduction are tightly linked in fish as fecundity increases with body mass (Wootton, 1998). Thus, for an animal to reproduce maximally, it must maximize its "energy surplus," which is the excess of energy after having covered household costs such as heart function, ion regulation, and ongoing synthesis of proteins, etc., which can then be converted into tissue growth and reproduction. Several factors obviously influence the amount of energy available for growth in an animal. In the following discussion of the effects of hypoxia on growth, these factors are included using the framework for abiotic influences classified by Fry (1971), and for biotic factors as proposed by Brett (1979).

Biotic factors such as interactions with conspecifics (i.e., competition for food) or other species (i.e., prey–predator interactions) can affect the amount of food that the animal has access to. In fish, especially under laboratory or aquaculture conditions, food availability is of course mostly determined by researchers (or managers), and huge efforts have been put into studying the effect of food quality and stocking density, since these parameters clearly often affect growth. However, correlations between growth and quality/density can be blurred. For instance, schooling fish may well be affected in a different way by density than fish that are normally solitary. Also, food "quality" is clearly species dependent and composition must be tailored to the needs of the individual species. Food availability is another limiting biotic factor, simply because food equates to energy.

Abiotic factors such as temperature determine the amount of energy spent on maintenance, as most biological processes (for instance protein synthesis and degradation) are temperature dependent. Temperature has accordingly been classified as a controlling abiotic factor and can have both positive and negative effects on growth. Other abiotic factors such as salinity are classified as "masking" because they change the costs of specific aspects of metabolism. Oxygen availability, which is the focus of this section, is classified by Fry (1971) as a limiting factor. Oxygen is the key electron accepter in aerobic respiration and thus directly limits the amount of energy that can be metabolized by an animal.

Hypoxia occurs naturally and on a regular basis in many habitats. As an example, hypoxia can occur regularly and predictably as a result of the lack of photosynthesis at night, but occurs more unpredictably as a result of eutrophication, stagnant water, or ice cover (Nilsson and Östlund-Nilsson, 2008). Though a naturally occurring phenomenon, the frequency, abundance, and severity of hypoxic events have increased due to anthropogenic organic and inorganic nutrient loading, and also the much discussed global warming (Diaz, 2001; see Chapter 1). Hypoxia is also a widespread problem in aquaculture, where stocking density is high, requiring the expenditure of large amounts of energy in aeration. It is therefore not surprising that most studies relating hypoxia to growth performance and digestion have been conducted on commercially important species such as Atlantic cod, trout, and catfish.

2. ENERGETIC CONSIDERATIONS FOR GROWTH

2.1. Effects of Hypoxia on Metabolism

Hypoxia exerts its general influence on growth by disturbing metabolic pathways and the reallocation of energy resources. An organism's metabolism is normally divided into basal or standard metabolic rate (SMR) and routine metabolism (RMR). SMR represents the energy expenditure to maintain basic life functions, including the maintenance of ion gradients, osmoregulation, and constitutive rates of protein synthesis, and thus represent the minimum cost of living. This notion of a SMR is obviously somewhat artificial because it changes with the condition of the animal and tends, for example, to decrease progressively during food deprivation (e.g., Van Dijk *et al.*, 2002; Wang *et al.*, 2006). Furthermore, it is difficult to measure experimentally (see Steffensen, 1989). Nevertheless it serves as a useful conceptual tool to quantify environmental or physiological changes, such as those imposed by hypoxia or the animal's feeding state. Under natural conditions, as well as in aquaculture, the metabolism of an animal is considerably higher than its SMR because of the energy expended on physical activity, food digestion, or reproduction. This metabolic rate is normally referred to as the routine metabolic rate (RMR), while the maximal oxygen uptake of an animal, typically measured during strenuous exercise, is denoted VO_2max. As explained in more detail below, RMR for a given individual will be higher the more food is being digested because the metabolic cost of digestion increases proportionally with ration.

Hypoxia leads to reductions in all three levels of metabolism, but the thresholds will normally differ, so that VO_2max is the most sensitive followed by RMR, while SMR is the least sensitive. These effects of hypoxia are illustrated in Figure 8.1, which shows how VO_2max can be expected to decline (e.g., Claireaux and Lagardère, 1999) as oxygen availability is reduced. The exact manner by which VO_2max decreases is dictated by a complex interplay

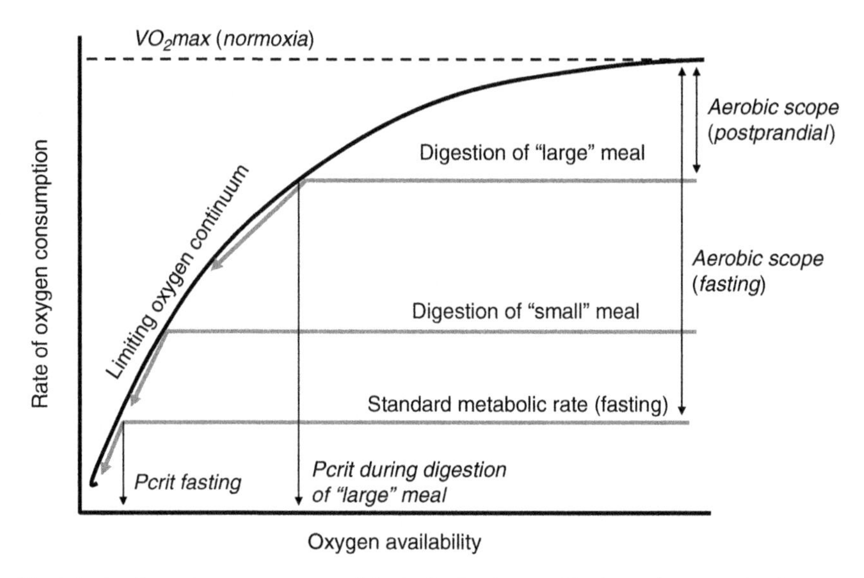

Fig. 8.1. A schematic representation of the limitation imposed by hypoxia on maximal oxygen consumption (VO_2max). Fasting fish at standard metabolic rate generally are able to sustain oxygen consumption in hypoxia until a critical level (P_{crit}), which corresponds to the interception with the line predicting VO_2max at a given oxygen availability. Digestion, by virtue of elevating the rate of oxygen consumption, increases P_{crit} and reduces the aerobic scope, which is defined as the difference between oxygen consumption at rest and exercise. [Modified from Claireaux and Lagardère (1999).]

of blood oxygen-binding characteristics and the abilities of the gills and the cardiovascular systems to transport oxygen to the metabolizing tissue (e.g., Jones *et al.*, 1970; Webb, 1994). In resting fish at SMR, the effects of hypoxia are much less pronounced, but at some level of hypoxia, oxygen delivery no longer satisfies metabolic needs and aerobic metabolism will decline (critical oxygen tension, P_{crit}). As predicted in the model represented in Figure 8.1, an elevation of metabolism is associated with an increase in P_{crit}. Thus, as metabolism rises during digestion (discussed in more detail below), the organism becomes more sensitive to hypoxia.

2.2. Basic Energy Balance, Metabolism, and Allocation to Growth

In energetic terms, the amount of energy available for growth in a non-reproducing fish is given as the difference between the energy ingested through food minus the sum of energy spent on metabolism and amount of energy that is excreted in urine and faeces:

$$E_{growth} = E_{food} - E_{metabolism} - (E_{feces} + E_{urine}) \quad (1)$$

The amount of energy that can be allocated to growth (E_{growth}) represents the difference between the energy in the food that is consumed (E_{food}) and the total amount of energy that is used for standard metabolic rate, physical activity, and digestion ($E_{metabolism}$) and the energy lost in feces and urine (E_{feces} and E_{urine}, respectively). The amount of energy lost in the feces represents food items that were not assimilated over the gut, while the energy excreted as waste products in urine (or over the gills) stems from breakdown of absorbed nutrients.

Growth is typically measured as the specific growth rate (SGR), which can be calculated as follows:

$$SGR = \frac{\ln W_t - \ln W_0}{t} \cdot 100\% \quad (2)$$

Where t is time, normally in days, and W_t is the body mass after t days. W_0 is the initial body mass.

Growth is thus expressed as the percentage of the initial body mass gained by the fish per day, and usually lies in the range of 0% day^{-1} to 4% day^{-1} (see Table 8.1), but it can also be negative in situations where food intake is insufficient to balance energy expenditure, which can occur during fasting and starvation as well as anorexia imposed by hypoxia or other challenging situations (Brett and Groves, 1979).

Another useful term when considering allocation of energy to growth is the food conversion efficiency (also called the gross conversion efficiency, K_1), which is the amount of the energy consumed that is allocated into growth:

$$K_1 = \frac{E_{growth}}{E_{food}} \cdot 100\% \quad (3)$$

The net conversion efficiency (K_2) is also defined:

$$K_2 = \frac{E_{growth}}{E_{food} - E_{RMR}} \cdot 100\% \quad (4)$$

The accuracy of this determination depends on the accuracy with which the routine metabolic rate can be measured, and as such adds little knowledge that cannot be gained from measuring the gross conversion efficiency (Brett and Groves, 1979).

A third useful term is the assimilation efficiency (AsE), which is the amount of energy consumed that is actually absorbed by the animal, i.e., it is the energy content of the feces subtracted from the energy consumed:

$$AsE = \frac{E_{food} - (E_{faeces} + E_{urine})}{E_{food}} \cdot 100\% \quad (5)$$

The assimilation efficiency influences growth in that an animal with low assimilation efficiencies will have to eat more of a particular food to absorb

Table 8.1

Specific Growth Rates (SGR) at Different Levels of Dissolved Oxygen

Common name	Latin name	Initial M_b(g)	Salinity (‰)	Feeding level	Light cycle	Temp (°C)	O_2 (%)[a]	SGR (% day^{-1})[b]	FI (g day^{-1} $fish^{-1}$)	Reference
Winter flounder	*Pseudo-pleuronectes americanus*	1.54	25.5	Ad lib.	"Natural"	20.7	87.0	2.5	–	Bejda *et al.* (1992)
							28.6	1.2	–	
Silver catfish	*Rhamdia quelen*	4.99	–	5 % 1 d^{-1}	–	22.4	77.5	1.5[c]	–	Braun *et al.* (2006)
							65.6	1.7	–	
							52.2	1.2	–	
							39.1	0.9	–	
							24.7	0.9	–	
Channel catfish	*Ictalurus punctatus*	15.0	–	1 d^{-1}	"Natural"	24.3	100	1.4[d]	–	Buentello *et al.* (2000)
							70	1.1	–	
							30	0.8	–	
Atlantic cod	*Gadus morhua*	728.2	28	3 wk^{-1}	"Natural"	10.0	93	0.8[e]	33[f]	Chabot and Dutil (1999)
							84	0.9	32	
							75	0.7	25	
							65	0.8	27	
							56	0.6	22	
							45	0.5	15	
Spotted wolffish	*Anarhichas minor*	68.5	–	1 d^{-1}	6D:18L	8.0	122.9	0.8[g]	128	Foss *et al.* (2002)
							81.4	0.9	141	
							50.8	0.7	92	
							33.9	0.4	67	
Turbot	*Scophthalmus maximus*	120	34	2 d^{-1}	8D:16L	17.0	92.3	1.0[h]	–	Pichavant *et al.* (2000)
							64.1	0.5	–	
							44.9	0.5	–	

Turbot	*Scophthalmus maximus*	66.3	34	2 d^{-1}	8D:16L	17.0	94.9	2.0[h]	–	Pichavant *et al.* (2001)
							57.7	1.1	–	
							41.0	0.7	–	
				Restricted	8D:16L	17.0	93.3	0.7	–	
European sea bass	*Dicentrarchus labrax*	60.8	34	2 d^{-1}	8D:16L	17.0	94.9	0.8[h]	–	Pichavant *et al.* (2001)
							57.7	0.5	–	
							41.0	0.3	–	
				Restricted	8D:16L	17.0	93.3	0.3	–	
Nile tilapia	*Oreochromis niloticus*	20.0	–	2 d^{-1}	12D:12L	32.3	>68.5	4.47	2.15	Tran–Duy *et al.* (2008)
							<47.9	3.55	1.54	
		140.9	–	2 d^{-1}	12D:12L	32.3	>68.5	2.19	6.08	
							<47.9	1.63	4.43	
Silver bream	*Sparus sarba*	9.4	16	3 d^{-1}	Not stated	26.6	84.7	0.61[i]	–	Chiba (1983)
							23.9	0.15	–	
Rainbow trout	*Oncorhynchus mykiss*	100	–	1 h^{-1} (light)	12D:12L	15	118.8	4.05[j]	–	Pedersen (1987)
							99.0	4.00	–	
							84.2	3.90	–	
							69.3	4.00	–	
							59.4	2.90	–	
							49.5	2.00	–	
							39.6	–0.05	–	
				1 d^{-1}	12D:12L	15	118.8	1.65	–	
							99.0	1.45	–	
							84.2	1.30	–	
							69.3	1.55	–	
							59.4	0.95	–	
							49.5	0.15	–	
							39.6	–	–	

(*continued*)

Table 8.1 *(continued)*

Common name	Latin name	Initial M_b(g)	Salinity (‰)	Feeding level	Light cycle	Temp (°C)	O_2 (%)[a]	SGR (% day^{-1})[b]	FI (g day^{-1} $fish^{-1}$)	Reference
Turbot	*Scophthalmus maximus*	54.5	34	2 d^{-1}	6D:18L	17.0	100.1	1.75	0.91[k]	Person–Le–Ruyet *et al.* (2002)
							147.4	2.00	1.00	
							223.6	2.02	1.00	
European sea bass	*Dicentrarchus labrax*	19	37	2% 1 d^{-1}	12D:12L	22	86	1.02	–	Thetmeyer *et al.* (1999)
							40	0.78	–	
White sturgeon	*Acipenser transmontanus*	<1	–	3 d^{-1}	12D:12L	15	84	1.6	–	Cech *et al.* (1984)
							58	0.6	–	
						20	84	2.6	–	
							58	2.0	–	
						25	84	2.9	–	
							58	2.3	–	
Striped bass	*Morone saxatilis*	<1		3 d^{-1}	12D:12L	15	84	1.6	–	Cech *et al.* (1984)
							58	1.4	–	
						20	84	2.2	–	
							58	2.0	–	
						25	84	3.2	–	
							58	3.0	–	
Plaice	*Pleuronectes platessa*	22.8	32	3–8% 2 d^{-1}	10D:14L	14.8	94.7	2.0[l]	–	Petersen and Pihl (1995)
							51.7	1.3	–	
							31.9	0.3	–	
Dab	*Limanda limanda*	23.5	32	3–8% 2 d^{-1}	10D:14L	14.7	94.6	1.5	–	Petersen and Pihl (1995)
							50.7	0.5	–	
							31.6	0.3	–	

Southern flounder	*Paralichthys lethostigma*	1.8	15	2 d^{-1}	10D:14L	25.0	85.9	1.9	–	Taylor and Miller (2001)
							62.6	3.0	–	
							36.9	3.5	–	
Atlantic manhaden	*Brevoortia tyrannus*	7.7	15	1 d^{-1}	10D:14L	25.0	78.9	2.5^{m}	–	McNatt and Rice (2004)
							52.6	1.7	–	
							26.3	1.5	–	
							19.7	0.9	–	
						30.0	87.0	2.6	–	
							58.0	2.7	–	
							29.0	2.5	–	
							21.7	1.1	–	
Spot	*Leiostomus xanthurus*	7.1	15	1 d^{-1}	10D:14L	25.0	78.9	2.2	–	McNatt and Rice (2004)
							52.6	2.4	–	
							26.3	2.2	–	
							19.7	1.5	–	
						30.0	87.0	1.5	–	
							58.0	2.0	–	
							29.0	2.0	–	
							21.7	0.1	–	
Sockeye salmon	*Oncorhyncus nerka*	5.6	–	4 d^{-1}	"Natural"	15.0	99.3	1.44	–	Brett and Blackburn (1981)
							69.5	0.87	–	
							29.8	0.68	–	
							20.9	–0.24	–	
Common carp	*Cyprinus carpio*	57.5	–	2% d^{-1}	12D:12L	22.0	80.1	1.4	–	Zhou *et al.* (2001)
							5.7	0.1	–	

(continued)

Table 8.1 *(continued)*

Common name	Latin name	Initial M_b(g)	Salinity (‰)	Feeding level	Light cycle	Temp (°C)	O_2 (%)[a]	SGR (% day^{-1})[b]	FI (g day^{-1} fish^{-1})	Reference
Channel catfish	*Ictalurus punctatus*	60.0	–	3% d^{-1}	–	26.6	100	5[m]		Andrews *et al.* (1973)
							60	5		
							36	3		
				Ad lib.	–	26.6	100	6		
							60	5		
							36	3		

M_b = body mass

[a]If not stated in %, the saturation was calculated as $\frac{mgO_2L^{-1}}{mgO_2L^{-1}(100\%saturation)} \cdot 100\%$assuming standard barometric pressure and accounting for temperature and salinity.

[b]$SGR = \frac{\ln W_1 - \ln W_0}{t} \cdot 100\%$

[c]Readings from figures presented in the paper.

[d]Final weight (W_1) was calculated from the weight gain increment presented in the figure from the paper. The wait gain was an average for the total period (12 weeks).

[e]Final weight was calculated from the presented change in body mass, which was averaged for the entire measurement period (84 days).

[f]Readings from figure presented in the paper.

[g]Readings from figures presented in the paper.

[h]Calculated from mean final and mean initial weight read of the figure in the paper.

[i]Averages of SGR calculated from final and initial weights, selected tanks in the data table, 6 high oxygen and 6 low.

[j]Only the growth rate for the highest and intermediate feeding ratio is presented.

[k][Food intake (% BW day^{-1}) * mean weight]/100

[l]Readings from figures presented in the paper. Only the average of the entire period is presented here.

[m]Growth rate calculated as $\frac{(W_1 - W_0)/W_{mean}}{t} \cdot 100\%$

a certain amount of energy (Jobling, 1993). Under natural conditions lower assimilation efficiency will therefore be a cost to the animal in terms of hours spent feeding.

Hypoxia may affect all components of the energy equation (1). Thus, as will be reviewed below, hypoxia inhibits appetite of fish causing E_{food} to decrease, leaving less energy available for growth. Hypoxia may also affect assimilation efficiency and hence E_{feces}. These effects are obviously interrelated and factors that influence the amount of food eaten are likely to affect the assimilation efficiency and any factor that influences the rate of the digestive processes and assimilation is likely to have an influence upon the amount of food ingested.

2.3. The Relationship between Metabolism, Aerobic Scope, and Growth

The effects of hypoxia are often interpreted in the context of "Fry's paradigm" (Fry, 1971; Kerr, 1976, 1990), where an animal's ability to perform activity is dictated by the influence of environmental factors on metabolism. In short, controlling factors such as temperature determine the rate of biochemical processes comprising metabolism and thus dictate both maximum and standard metabolic rates. Limiting factors, such as hypoxia, reduce oxygen supply and constrain aerobic metabolism (see Figure 8.1), while masking factors, such as salinity, may affect SMR by altering the energy expenditure associated with key metabolic processes.

In addition to showing how oxygen availability limits standard and maximal oxygen uptake, Figure 8.1 also illustrates how hypoxia and digestion affect the aerobic scope. Aerobic scope is defined as the proportional change in oxygen uptake between SMR and VO_2max. In a digesting animal, however, the aerobic scope is reduced because RMR becomes greatly elevated above SMR thus limiting the extent to which aerobic metabolism can be increased during physical activity. The environmental conditions that maximize the aerobic scope are often interpreted as being optimal for growth (e.g., Brett, 1979), but have only been evaluated in terms of temperature (Lefebvre *et al.*, 2001; Mallekh and Lagardère, 2002; Claireaux and Lefrançois, 2007). An example of the correlation between aerobic scope and maximal feeding rate, which presumably translates into maximal growth, is presented in Figure 8.2 for turbot (*Scophthalmus maximus*). Increased temperature leads to an elevation of both SMR and VO_2max, but because VO_2max stabilizes at the higher temperature interval, feeding rate and presumably growth is maximal at approximately 18 °C. Although temperature and hypoxia most likely affect growth and feeding behavior through different mechanisms, it has been argued that the reduced appetite and growth rates in hypoxia represent an adaptive behavioral response

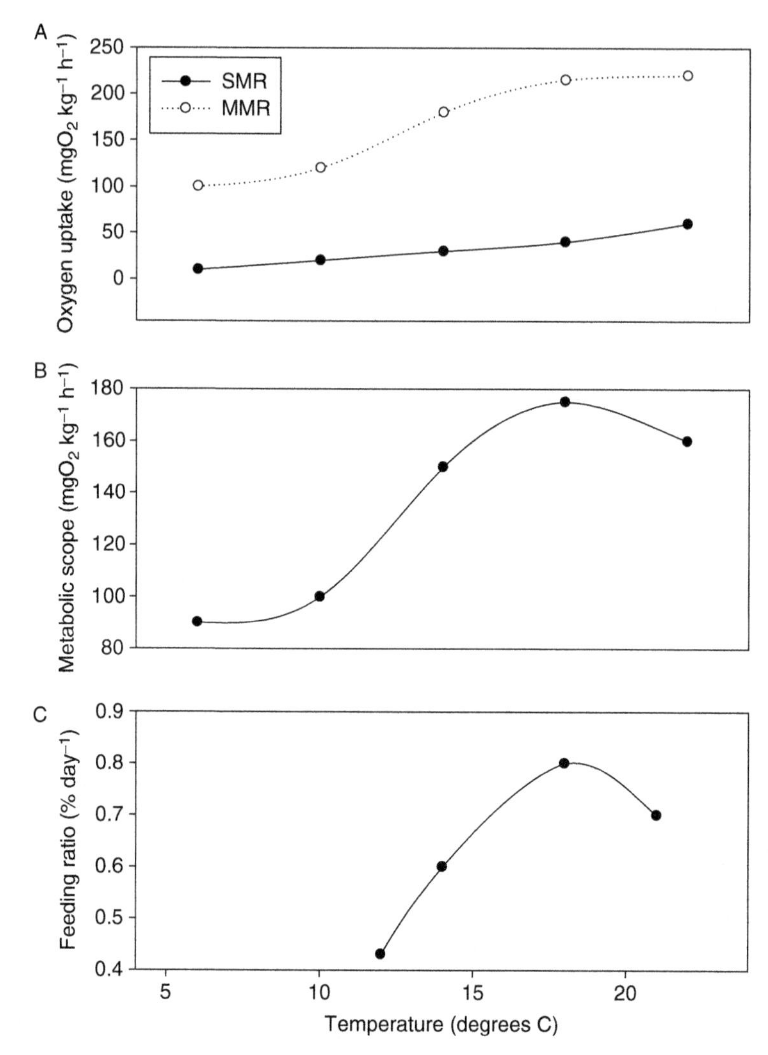

Fig. 8.2. Standard and maximal oxygen uptake (A), metabolic scope (B), and feeding ratio (C) in turbot (*Scophthalmus maximus*) at different temperatures and in normoxic water. The metabolic scope is the difference between standard metabolic rate (SMR) and maximum metabolic rate (MMR) (VO_2max), and attains its highest value at approximately 18 °C. At temperatures above 18 °C, SMR continues to increase, while VO_2max levels off, causing metabolic scope to decrease. Maximal feeding rate was observed at 18 °C where metabolic scope is highest. [Data from Mallekh and Lagardère (2002).]

to protect the aerobic scope for activity (Claireaux *et al.*, 2000; Claireaux and Lefrançois, 2007). This suggestion is certainly intuitively appealing, and while the physiological mechanisms remain to be characterized, there is certainly a compelling relation between scope for activity and overall growth performance in all species studied so far (Claireaux *et al.*, 2000; Mallekh and Lagardère, 2002; Claireaux and Lefrançois, 2007).

3. THE RISE IN METABOLISM DURING DIGESTION: SPECIFIC DYNAMIC ACTION (SDA)

Digestion causes metabolism to increase in all animals including fish. The metabolic increment is almost exclusively aerobic and has been termed "heat increment of feeding" and "calorigenic effect," but "specific dynamic action of food" (normally abbreviated as SDA) is currently the most common term (Rubner, 1902; Kleiber, 1961; Jobling, 1981). In its strictest sense, SDA only includes the metabolic costs involved with digestion, absorption, and utilization of food, whereas the "apparent SDA", measured as the change in metabolic rate throughout the postprandial period, also includes other costs associated with feeding, such as prey handling, as well as structural or functional remodeling of the digestive organs. It can be difficult to separate the individual components, and virtually all studies on energetic responses to feeding report apparent SDA responses. The SDA response is generally characterized by a rise in oxygen consumption within minutes or hours after ingestion, followed by a gradual decline to the resting level over many hours or days. The total amount of extra energy spent during digestion, i.e., the integral of the postprandial metabolism minus resting metabolic rate, is a measure of the energy expenditure associated with digestion. It can be useful to express the energy expenditure for digestion relative to the amount of ingested energy as the SDA coefficient, allowing a quantitative evaluation of the cost of digestion in relation to input. The SDA coefficient of fish normally ranges between 5% and 20% (e.g., Jobling, 1983; Eliason *et al.*, 2007), and while the SDA response may be viewed as a substantial bioenergetic "cost of growth" (Jobling, 1981), this cost is a prerequisite for assimilation and should not be regarded as a simple metabolic loss (Mallekh and Lagardère, 2002).

The contribution of the various digestive processes to the total SDA response, i.e., prey capture, muscular contraction of the stomach and gut motility, secretion of digestive juices and mucosal absorption of nutrients, and digestion, is likely to vary with food composition and meal size, and to vary among species (e.g., Jobling, 1981; Wang *et al.*, 2006; McCue, 2006). The mechanical component of digestion seems to be rather small. Thus, meals consisting of inert kaolin, which stimulates gastrointestinal motility without being attended by postabsorptive processes, produced minor changes in metabolism in plaice (*Platessa platessa*), whereas protein-rich food elicited marked and swift metabolic responses (Jobling and Davies, 1980; cf. Tandler and Beamish, 1979, 1980). Furthermore, infusion of amino acids directly into the blood stream, which induces metabolic responses similar to those elicited by feeding and inhibition of protein synthesis, completely abolished the SDA response *in vivo*. Thus, it seems that the biochemical transformation of food and *de novo* protein synthesis in the postabsorptive state are the major contributors to the SDA response (Brown and Cameron, 1991a,b; Bureau *et al.*, 2002).

Both the magnitude and the duration of the SDA response increases with meal size (e.g., Fu *et al.*, 2005; Andrade *et al.*, 2005). Large meals may elicit many-fold increases of the RMR, lasting for many days. Some fishes have a discontinuous feeding pattern, where they fast for long periods followed by the ingestion of large meals. An example is the Atlantic cod (*Gadus morhua*) where the peak VO_2 during SDA may represent up to 90% of VO_2max (Soofiani and Hawkins, 1982; Claireaux *et al.*, 2000). The SDA response also varies with food composition (Jobling and Davies, 1980), body size (Hunt von Herbing and White, 2002), and environmental factors, but both its magnitude and duration correlates with the rate and amount of food passing through the gastrointestinal system. Thus, factors that prolong the digestive processes, such as lowered body temperature, prolong the duration of the SDA response associated with lower maximal values, while the SDA coefficient generally remains unaffected (e.g., Jobling and Davies, 1980; Soofiani and Hawkins, 1982). Chabot and Claireaux (2008) note that in the common sole (*Solea solea*), which has a small stomach where maximum meal size is less than 3% of body mass, neither the peak value nor the duration of postprandial metabolism are affected until hypoxia becomes very severe (<30% saturation).

The effects of hypoxia on the SDA response was recently characterized in Atlantic cod (*Gadus morhua*) exposed to 5% O_2 or normoxia (Jordan and Steffensen, 2007; Figure 8.3). This level of hypoxia did not affect RMR, but the SDA response to a meal of approximately 5% of body mass was significantly prolonged in 5% O_2 compared to normoxia and was associated with a lower maximal rate of oxygen consumption. It was not verified whether the prolongation of the SDA response was associated with increased retention time and a slower rate of digestion, but it is likely that reduced oxygen delivery to the gastrointestinal organs and the liver delayed the digestion and assimilation. Clearly, it would be informative to perform similar studies on other species, preferably over a range of meal sizes and at different temperatures, and correlate the changes in the SDA response with temporal changes in nutrient assimilation. Also, it would be of considerable interest to establish whether the levels of hypoxia that affect the SDA response of a given species correlates with the oxygen levels that retard growth and reduce appetite.

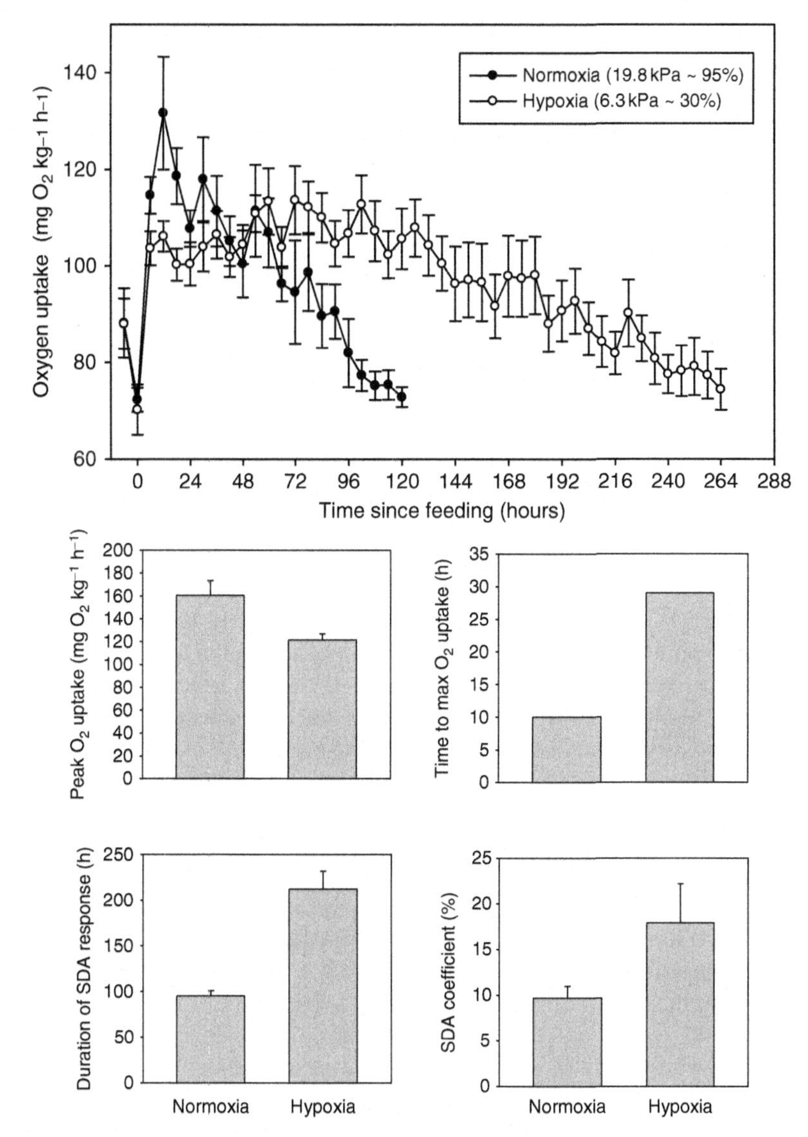

Fig. 8.3. The effects of digestion of a meal corresponding to 5% of body mass on oxygen uptake in Atlantic cod (*Gadus morhua*) maintained in normoxic or hypoxic water. Duration of SDA response, time to peak VO_2, and SDA coefficient are increased under hypoxia, while the peak VO_2 is reduced. [Data are means ± S.E. from Jordan and Steffensen (2007).]

4. GENERAL EFFECTS OF HYPOXIA ON GROWTH, APPETITE, AND ASSIMILATION

The effects of hypoxia on growth have been characterized in a number of studies in different fish, and the findings from many of these studies are collated in Table 8.1. These different growth experiments have been conducted under very different abiotic and biotic regimes (e.g., different temperatures, water composition, feeding rates, and levels of hypoxia), but it is evident that hypoxia inevitably stifles growth and that this hypoxia-related reduction in growth is primarily a result of reduced food intake (Davis, 1975; Brett, 1979). Two examples are shown in Figures 8.4 and 8.5 for silver bream (*Sparus sarba*) and the Atlantic cod (*Gadus morhua*). In both species growth is reduced under hypoxia, and while ingestion rate decreases in both species, growth is also reduced as a result of impaired food conversion efficiency in the silver bream (Chiba, 1983; Chabot and Dutil, 1999). In some species, severe hypoxia is even associated with weight loss as the reduced food intake results in a negative energy balance where basic metabolic needs are covered by internal stores. It is evident, however, that the specific level of hypoxia that retards growth varies among species and is likely to depend on the individual species' ability to compensate physiologically for the reduction in available oxygen. Thus, species with high oxygen affinities and robust cardiorespiratory responses to hypoxia are likely to be less affected than the more hypoxia-sensitive species. As extraordinary exceptions to this rule, two species of cichlids (*Astatoreochromis alluaudi* and *Haplochromis ishmaeli*) from Lake Victoria have similar growth rates in normoxia and at a PO_2 of approximately 2kPa (10%) for over a year (Rutjes *et al.*, 2007). While it is possible that these cichlids were fed on a restricted ration and that effects of hypoxia would be evident if the fish were fed to satiety, the study by Rutjes *et al.* (2007) shows that very hypoxia-tolerant fish can complete digestive processes and grow under extraordinary hypoxic conditions.

Growth effects of hypoxia also depend on the amount of available food, i.e., the food ration or feeding levels (equation 1). This is illustrated in Figure 8.6, where growth and food intake were measured in rainbow trout (*Oncorhynchus mykiss*) at 15 °C at various degrees of hypoxia and different feeding levels (Pedersen, 1987). Specific growth rate decreased with decreasing feeding levels (Figure 8.6A), and lowered food consumption seemed to explain most of the growth reduction (Figure 8.6B). Thus, the reduction in appetite was evident at all feeding levels, but growth retardation in hypoxia was most pronounced at the lower feeding levels.

All of the growth studies presented have been performed in captivity under more or less controlled conditions and given the obvious difficulties of performing growth studies in the wild, very little data is available from

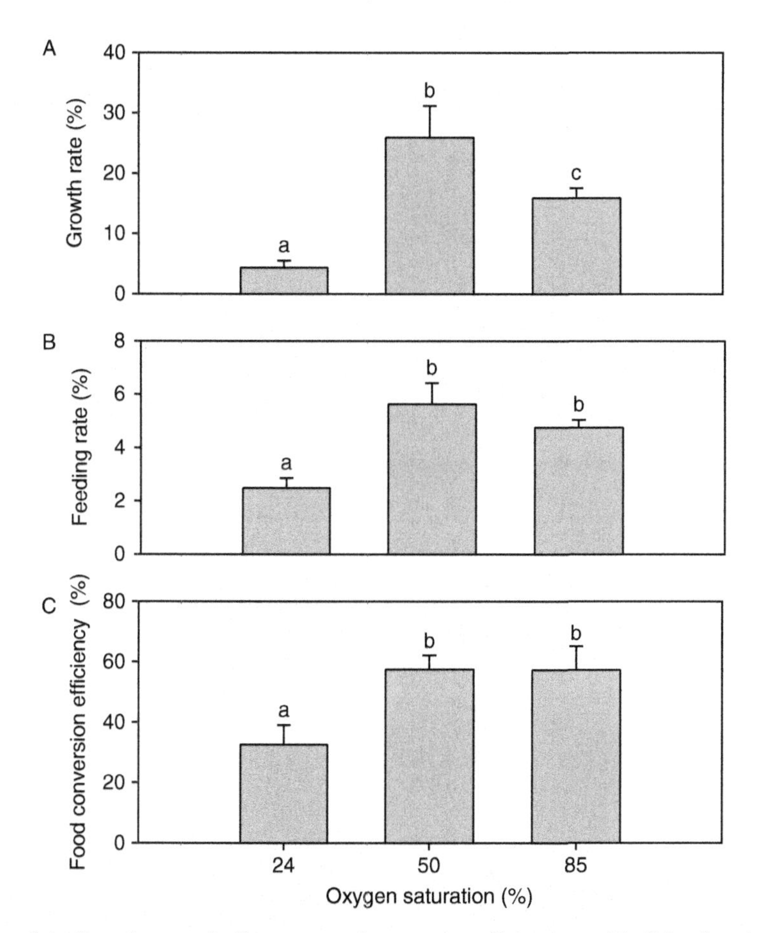

Figure 8.4. Growth rates, feeding rates, and conversion efficiencies at 27 °C in silver bream (*Sparus sarba*) at different dissolved oxygen levels. Different letters denote significantly different means. It can be seen that growth, feeding rate, and conversion efficiency are impaired at the lowest oxygen level. [Data are means ± S.E. from Chiba (1983).]

natural habitats. Hypoxic episodes in south-east Kattegat have been correlated with the abundance of smaller plaice (*Pleuronectes platessa*) and dab (*Limanda limanda*), indicating that hypoxia also limits growth under natural conditions (Petersen and Pihl, 1995). Also, a recorded decrease in growth in flathead flounder (*Hippoglossoides dubius*) was found to be correlated with an occasional decrease in dissolved oxygen levels in Funka Bay, Japan (Kimura *et al.*, 2004). Under natural conditions, hypoxia is likely to occur in combination with hypercapnia and will often be associated with elevated

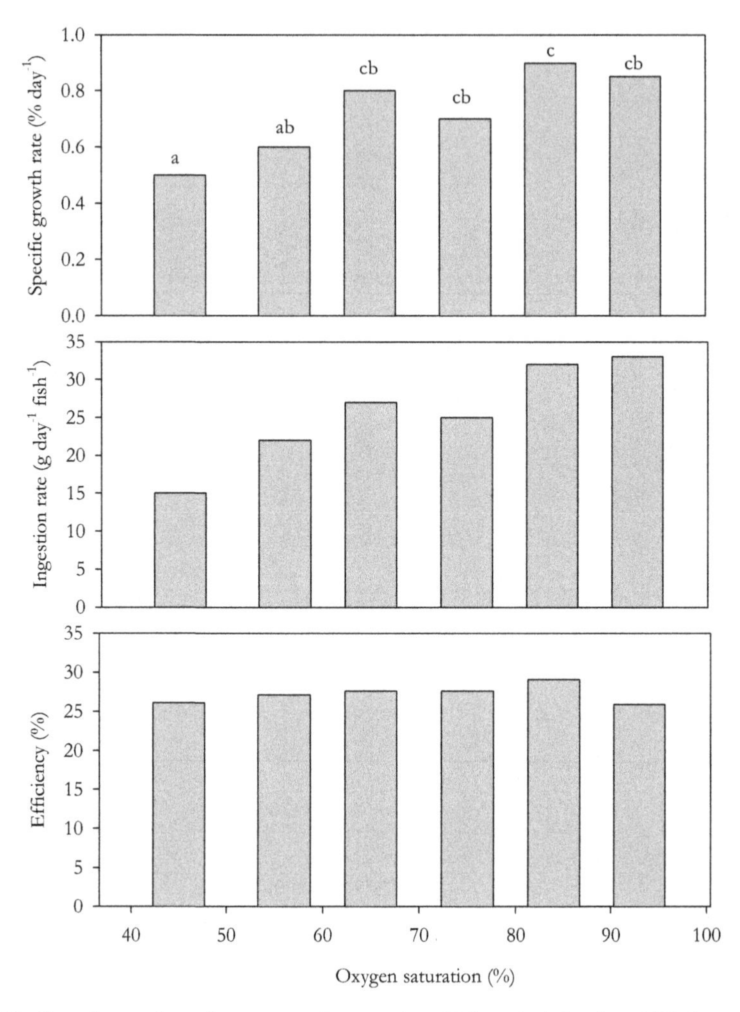

Fig. 8.5. Growth rates, ingestion rates, and conversion efficiency in Atlantic cod (*Gadus morhua*) reared at 10 °C under different oxygen levels. Letters indicates significantly different values, thus growth was less at the lowest oxygen level. There was a significant correlation between ingestion rate and oxygen level ($R^2 = 0.93$). There was a significant linear correlation between conversion efficiency and oxygen ($R^2 = 0.92$), if the efficiency at 93% saturation was excluded. [Data are means estimated from Chabot and Dutil (1999), and S.E. can therefore not be provided.]

temperatures. These additional stressors are likely to exacerbate the adverse effects of hypoxia and it would be interesting to see if future studies could assess the roles of disturbed acid–base balance and/or temperature challenge on growth and digestive performance in hypoxia.

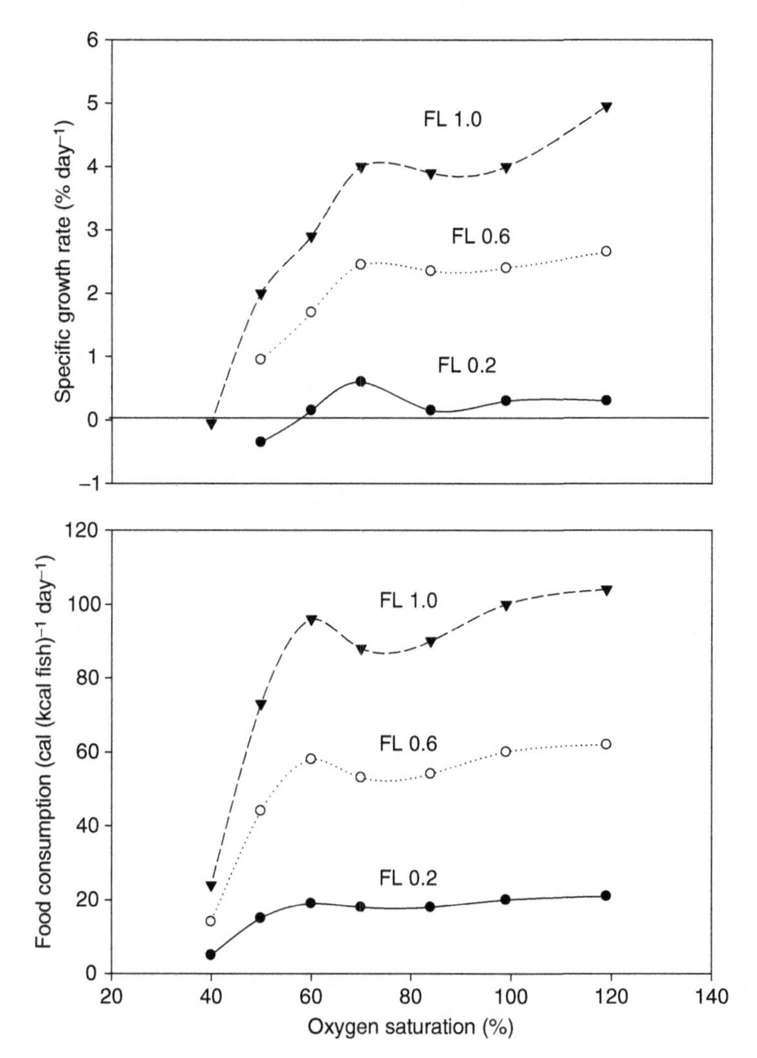

Fig. 8.6. Specific growth rates (A) and food intake (B) in rainbow trout (*Oncorhyncus mykiss*) at 15 °C under different amounts of dissolved oxygen and at different relative feeding levels (FL). [Data are means estimated from Pedersen (1987), and S.E. can therefore not be provided.]

4.1. Adaptation of Growth during Long-Term Hypoxia

While hypoxia consistently lowers growth rate, long-term adaptations to prolonged hypoxia may alleviate the negative effects of insufficient oxygen. The temporal changes in growth performance during long-term hypoxia have been studied in a few species (see Table 8.2 and Figure 8.7). In general,

Table 8.2
Specific Growth Rates at Different Durations of Hypoxia

Common name	Latin name	Initial M_b(g)	Salinity (‰)	Feeding level	lLght cycle	Temp (°C)	O_2 (%)[a]	Time (day)	SGR (% day^{-1})[b]	Reference
Spotted wolffish	*Anarhichas minor*	68.5	–	1 day^{-1}	6D:18L	8.0	34	0–32	0.35	Foss *et al.* (2002)
								33–55	0.35	
								56–76	0.63	
Dab	*Limanda limanda*	23.5	32	3–8% 2 day^{-1}	10D:14L	14.7	30	0–10	–0.1	Petersen and Pihl (1995)
								11–20	0.6	
Plaice	*Pleuronectes platessa*	22.8	32	3–8% 2 day^{-1}	10D:14L	14.8	94.7	0–10	–0.4	Petersen and Pihl (1995)
								11–20	0.8	
Southern flounder	*Paralichthys lethostigma*	1.8	15	Ad lib. 2 day^{-1}	10D:14L	25.0	36.9	0–14	1.6	Taylor and Miller (2001)
								15–21	2.2	
							62.6	0–14	2.5	
								15–21	3.5	
Turbot	*Scophthalmus maximus*	120	34	2 day^{-1}	8D:16L	17.0	45	0–15	0.3	Pichavant *et al.* (2000)
								16–30	0.5	
								31–45	0.7	
Turbot	*Scophthalmus maximus*	120	34	2 day^{-1}	8D:16L	17.0	64	0–15	0.3	Pichavant *et al.* (2000)
								16–30	0.5	
								31–45	0.7	

[a]If not stated in %, the saturation was calculated as $\frac{mgO_2L^{-1}}{mgO_2L^{-1}(100\%saturation)} \cdot 100\%$, assuming standard barometric pressure and accounting for temperature and salinity.

[b]$SGR = \frac{\ln W_1 - \ln W_0}{t} \cdot 100\%$

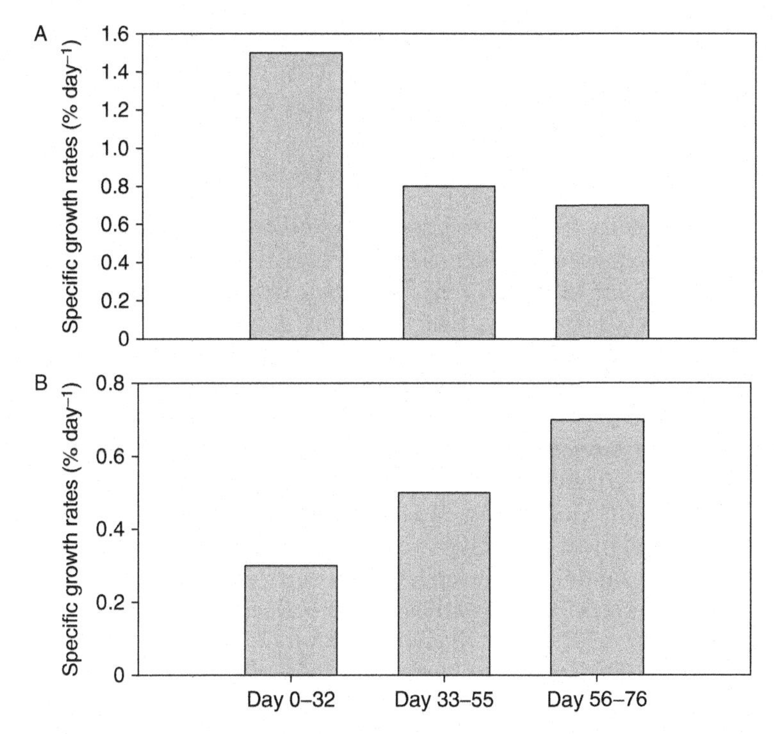

Fig. 8.7. Specific growth rates in juvenile turbot (*Scophthalmus maximus*) at (A) 95% saturation and (B) 45–65% saturation, both at 17°C. In normoxia the SGR decreases with time, while it increases in hypoxia. [Data are means estimated from Pichavant *et al.* (2000), and S.E. can therefore not be provided.]

fish adapt to the hypoxic conditions and increase growth rate progressively over time, and some species even reverse an initial weight loss to a weight gain (e.g., Petersen and Pihl, 1995). The mechanisms that underlie this adaptation to long-term hypoxia are likely to involve the common physiological responses to hypoxia, which include increased blood O_2 affinity, blood volume, and hemoglobin concentrations, as well as increased capillary density, etc. These physiological adaptations would increase oxygen transfer from the hypoxic water to the metabolizing tissue. Some studies have measured the hematocrit of animals grown in different oxygen concentrations, but with unclear results. In some experiments the hematocrit is slightly elevated in animals grown at lower oxygen levels (Taylor and Miller, 2001) while in others there is no significant difference (Andrews *et al.*, 1973). A study by Chabot and Dutil (1999) revealed no difference among groups reared at different oxygen saturations, but the hematocrit was larger in the control group at the beginning of the experiment. However, a causal relationship between the classic physiological responses and the recovery of appetite and growth remains to be established.

4.2. Effects of Dynamic Changes in Oxygen Levels on Growth

While the effects of chronic hypoxia have been characterized in some detail, the consequences of fluctuating O_2 levels for growth and digestion have been investigated more rarely. In brook trout (*Salvelinus fontinalis*), coho salmon (*Oncorhynchus kisutch*), and largemouth bass (*Micropterus salmoides*), growth seems to be more affected by fluctuations in O_2 concentrations than by exposure to constant intermediate O_2 concentrations (reviewed by Brett, 1979). Adverse effects on the growth to O_2 oscillations compared to constant concentrations [2.8–6.2 mg L^{-1} (~38–84%) and 4.74 mg L^{-1} (~65%), respectively] have recently been demonstrated in juvenile southern flounder (*Paralichthys lethostigma*) and related respiratory adjustments were described (Taylor and Miller, 2001). In comparison, growth performances of European sea bass juveniles (*Dicentrarchus labrax*) are not significantly affected by repetitive O_2 oscillations of between 6 (~85%) and 3 mg L^{-1} O_2 concentrations (~42%) compared to constant O_2 concentrations of 6 or 3 mg L^{-1}, respectively (Thetmeyer *et al.*, 1999). The effects of dynamic changes in oxygen levels are likely to differ drastically depending on the duration and severity of the hypoxic insults. Thus, as expanded on below, acute exposure to severe hypoxia can induce vomiting, and the reduction in appetite is likely to persist for many hours upon return to normoxia. Furthermore, given that some of the physiological responses to hypoxia, such as synthesis of additional red cells and angiogenesis, are accompanied by energetic costs, it is likely that dynamic changes in oxygen levels would increase RMR and reduce the amount of energy available for growth (see equation 1).

4.3. The Effect of Interactions between Temperature, Salinity, and Hypoxia on Growth Rate

The solubility of oxygen decreases as water temperature increases, while the stimulation by temperature of metabolic processes increases the need for oxygen delivery. Recently, it has been emphasized that the capacity for the cardiorespiratory system is an important determinant of temperature tolerance (e.g., Portner and Knust, 2007; see Chapter 4). Growth generally increases with temperature, as illustrated for striped bass (*Morone saxatilis*) and white sturgeon (*Acipenser transmontanus*) in Figure 8.8, which also illustrates the typical Q_{10} of 2 (Cech *et al.*,1984). Besides illustrating the profound effect of temperature on growth, this example also illustrates the

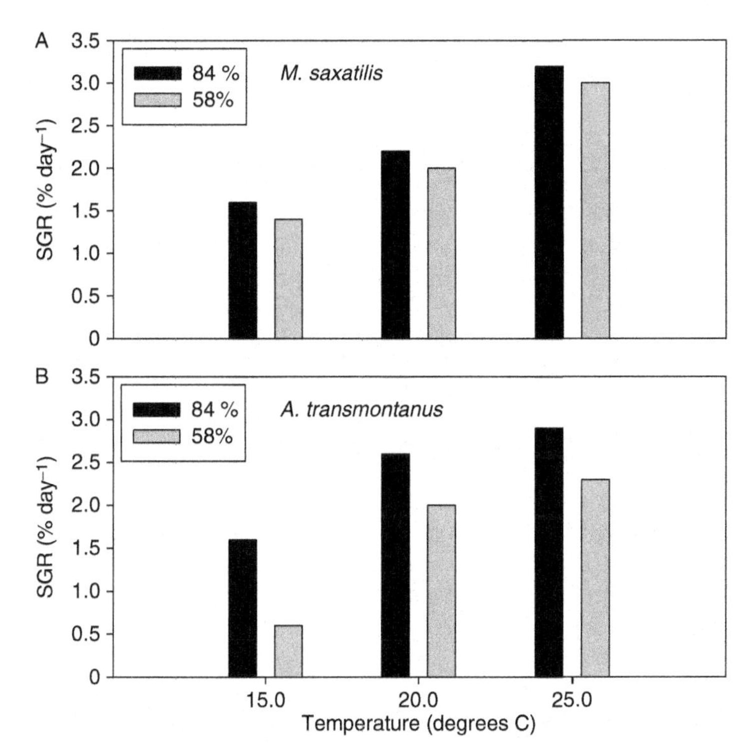

Fig. 8.8. Specific growth rates at different temperatures and dissolved oxygen levels in (A) striped bass (*Morone saxatilis*) and (B) white sturgeon (*Acipenser transmontanus*). Measurements were conducted over 30–34 days. It can be seen that SGR increases with temperature for both species at both normoxia and hypoxia, and that the SGR is lower in normoxia compared to hypoxia. [Data are means estimated from Cech *et al.* (1984) and S.E. can therefore not be provided.]

species-specific variation in sensitivity to, for instance, hypoxia. However, growth typically decreases when the optimal body temperature is exceeded. This negative effect might be directly caused by the effects of temperature on proteins, but the effects at more moderate temperatures are likely to include an inability to maintain sufficient oxygen delivery to the gastrointestinal organs during digestion. In natural systems, the incidence of hypoxia does increase with temperature as evident from the variety of air-breathing fish in the tropics. It would be interesting to characterize the effects of hypoxia at different temperatures in species from temperate and tropical regions.

In some areas, e.g., fjords and estuaries, the effect of hypoxia will, besides interactions with temperature, also be affected by varying salinity. Owing to the cost of osmoregulation in higher salinities, growth may be influenced more by hypoxia in higher salinities than at lower salinities.

5. EFFECTS OF HYPOXIA AND DIGESTIVE STATE ON OXYGEN TRANSPORT

Apart from the rise in metabolism, digestion is associated with a number of physiological changes, such as changes in acid–base status caused by gastric acid secretion, and elevated nitrogen excretion in connection with the increased protein metabolism and changes in water balance as the food items are degraded to osmotically active nutrients (e.g., Taylor *et al.*, 2007; Wood *et al.*, 2007). Along with the metabolic changes, the physiological challenge of digestion alters the manner in which fish can respond to hypoxia. Digestive state also affects the responses to other situations with elevated metabolism increases and the energetic burden imposed by digestion, for example, affects swimming ability (Blaikie and Kerr, 1996; Alsop and Wood, 1997).

Most markedly, the rise in metabolism associated with digestion and the need to increase perfusion of the gastrointestinal tract to facilitate absorption of nutrients requires cardiovascular responses that include an increase in cardiac output, through increments of heart rate and stroke volume, as well as a redistribution of blood flows to the digestive organs (Wang *et al.*, 2005). In fasting fish, blood flow to the gastrointestinal tract accounts for 10–30% of total cardiac output at rest, but this proportion increases drastically within hours after feeding, and may constitute 60–70% of cardiac output in the postprandial period (Axelsson *et al.*, 1989, 2000; Axelsson and Fritsche, 1991; Thorarensen *et al.*, 1994; Farrell *et al.*, 2001; Eliason *et al.*, 2008; Altimiras *et al.*, 2008). The rise in gastrointestinal blood flow seems to depend on meal size, rather than species differences. As in other vertebrates, blood flow to the gastrointestinal organs is normally reduced during hypoxia to prioritize oxygen-sensitive organs such as the heart and brain (e.g., Axelsson and Fritsche, 1991; Axelsson *et al.*, 2002). A reduction in blood flow to the gastrointestinal organs may compromise digestive functions and is likely to lower absorption efficiency and prolong the digestive process.

Axelsson *et al.* (2002) measured gastrointestinal blood flow during hypoxia in fasting and digesting European sea bass (*Dicentrarchus labrax*). The fasting sea bass exhibited the typical piscine cardiovascular response to hypoxia, consisting of a reduction in heart rate and a reduction in gastrointestinal blood flow (Figure 8.9). Feeding causes gastrointestinal blood flow to increase, primarily through an increased heart rate, and the proportion of cardiac output allocated to gut increases from 24% to 35%. When the sea bass was challenged by hypoxia in the postprandial period, the proportion of blood flow directed to the gastrointestinal system remained elevated although cardiac output decreased as in the fasting sea bass. As discussed by Axelsson *et al.* (2002), the maintenance of the relative gut perfusion is

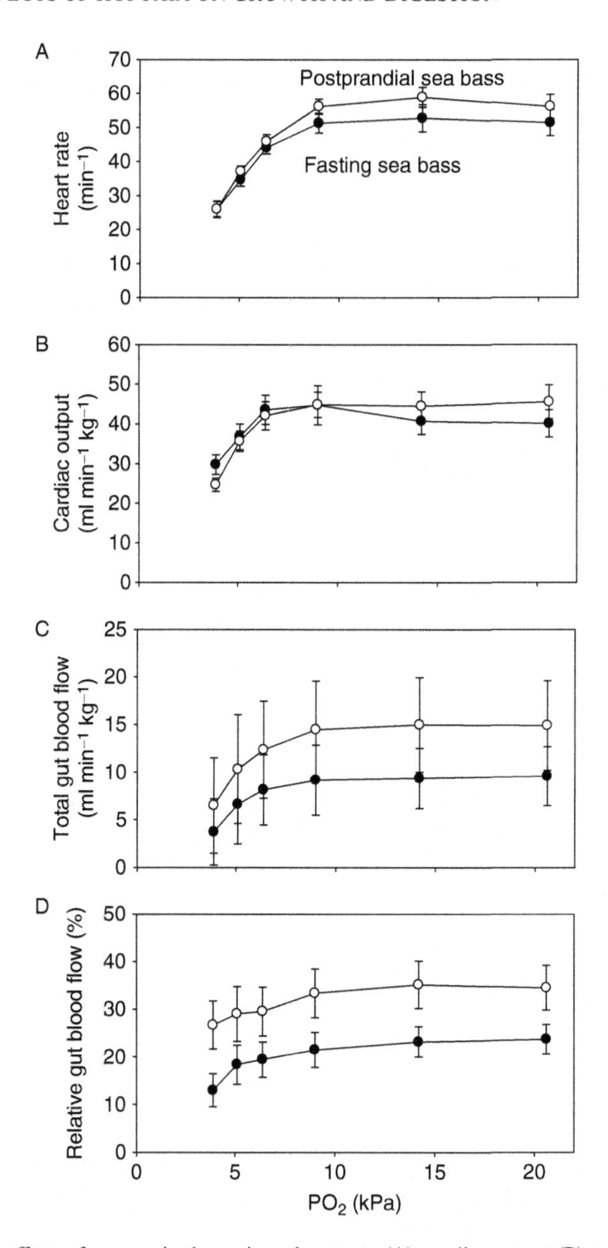

Fig. 8.9. The effects of progressive hypoxia on heart rate (A), cardiac output (B), total gut blood flow (C), and relative gut blood flow (C) in fasting and postprandial European sea bass (*Dicentrarchus labrax*). The relative gut blood flow is the total gut blood flow relative to the cardiac output. Both fasting and digesting fish respond to hypoxia with a bradycardia, but digesting fish maintain a higher gut blood flow at all hypoxic conditions. [Data from Axelsson *et al.* (2002).]

likely to stem from local release of signal transmitter substances causing relaxation of the vascular beds in the stomach and intestine associated with digestion, rather than a lack of control over gastrointestinal blood flow. Mechanical stretch of the stomach may be one of the signals causing such release of vasoactive substances, but chemical stimuli are also likely to play an important role (Seth *et al.*, 2008). Studying the same species, Altimiras *et al.* (2008) recently showed that gastrointestinal blood flow is decreased during swimming in both fasting and postprandial animals, suggesting that the regulation of the blood distribution differs between muscular exercise and hypoxia.

6. EFFECTS OF HYPOXIA ON APPETITE

Hypoxia causes significant reductions in appetite and the resulting reduction in ingested food constitutes the major part of the hypoxia-induced growth reduction (equation 1), which is illustrated by the feed intake (FI) values in Table 8.1. Acute exposure to hypoxia immediately affects the digestive processes and digesting Atlantic cod (*Gadus morhua*), for example, immediately void their stomach when exposed to hypoxia (Claireaux *et al.*, 2000). As such, the vomiting response may be viewed as part of the normal stress response to an immediate hypoxic challenge, and is likely to reflect the fact that organ systems other than those involved in digestion are prioritized during hypoxia. Using a similar teleology, it seems advantageous to reduce food intake to lessen the metabolic burden associated with digestion during long-term hypoxia, leaving more of the aerobic scope for physical activity (Figure 8.1). The mechanisms by which mild hypoxia reduces appetite over longer time scales have not been studied in fish or even in mammals, where considerable attention has been paid to understanding the mechanisms underlying anorexia and the associated weight loss at altitude (e.g., Vats *et al.*, 2007).

As in other vertebrates, long-term regulation of food intake in fish is controlled by a complex interplay of stretch and chemoreceptors within the stomach and intestine as well as hormones released either centrally or from the gastrointestinal organs (Volkoff *et al.*, 2005; Gorissen *et al.*, 2006). Most of the satiety-inducing inputs are transmitted to the nucleus lateralis tuberis within the hypothalamus, which through the nucleus preopticus is involved in regulation of appetite and the release of growth hormone. In general, appetite is stimulated by the orexigenic peptide hormone ghrelin, which is released from the fasting stomach and acts directly on the pituitary causing release of growth hormone. Satiety, on the other hand, seems more complex and involves many different hormones and signal molecules, such as cholecystokinin (CCK), as well as gastric and intestinal satiety signals induced primarily by stretch of the stomach as well as the presence of nutrients in the intestine (Volkoff *et al.*, 2005; Gorissen *et al.*, 2006; Maljaars *et al.*, 2007). More long-term effects include hormones such as leptin, which is released from adipose tissue and provides information on energy state of the organism. While many of these hormones have been characterized in fish (e.g., Volkoff *et al.*, 2005; Gorissen *et al.*, 2006; Canosa *et al.*, 2007), circulating levels of these hormones have not been measured during hypoxia and their role remains to be studied.

The effects of hypoxia on the central regulation of appetite have been studied in rainbow trout (*O. mykiss*) through measurements of forebrain corticotropin-releasing factor and urotensin (Bernier and Craig, 2005). The forebrain concentration of both peptides increased in hypoxia and after treatment with an antagonist to inhibit the receptors for corticotropin-releasing factor (Bernier and Craig, 2005). It was concluded that the hypophysiotropic factors stimulate the hypothalamic–pituitary–adrenal axis in fish and that this part of the stress response plays a role in reducing food intake during hypoxia.

The piscine stomach is innervated by stretch receptors that relay information on the presence and amount of food in the stomach, as well as chemoreceptors in the intestine providing feed-back on the presence of food in the gut (e.g., Grove *et al.*, 1978, 1985; Grove and Holmgren, 1992a,b). A significant part of the appetite reduction occurring during hypoxia may arise from prolonged stimulation of the gastric and intestinal stretch and chemoreceptors. Thus, as hypoxia prolongs the digestive processes, as indicated from the extended SDA response (Jordan and Steffensen, 2007; Figure 8.3), the stimulation of stretch receptors and chemoreceptors persists for a longer period of time extending the sensation of satiety. Certainly gut emptying time is prolonged and gut motility is inhibited by hypoxia in mammals (Yamaji *et al.*, 1996; Yoshimoto *et al.*, 2004), and similar effects are likely to be present in fish.

7. ASSIMILATION EFFICIENCY

Assimilation efficiency refers to the amount of ingested food that is assimilated over the gut (see equation 5). Assimilation efficiency is also referred to as absorption efficiency (Jobling, 1993) or digestion efficiency. In humans, assimilation seems to be reduced in severe hypoxia, but contributes only slightly to the weight loss at altitude (Westerterp *et al.*, 1994; Westerterp *et al.*, 2000). This is also the case in invertebrates where slight (3%) reductions in assimilation efficiency are only seen during very severe hypoxia (McGaw, 2008). In this study, the gut clearance time of the

Dungeness crab, *Cancer magister*, increased as PO_2 fell below 10.5 kPa, whereas assimilation efficiency was only slightly affected at the lowest PO_2 of 1.6 kPa, indicating that digestion was prolonged rather than directly impaired by hypoxia. As in these other animal groups, there are few studies that have addressed the effects of hypoxia on assimilation efficiency in fish and in this group contradicting results have been seen. Thus, in the Nile tilapia (*Oreochromis niloticus*), assimilation efficiency was reduced from ~80% in normoxia to ~54% under severe hypoxia (Tsadik and Kutty, 1987). In contrast, Pedersen (1987) found that assimilation in rainbow trout (*O. mykiss*) was unaffected in hypoxía. It is clear that this is an area for further research on a greater variety of fish species and at different oxygen levels. In particular, it would be interesting to investigate whether increased passage time is a general compensation for a less effective absorption during hypoxia.

8. EFFECTS OF HYPOXIA ON GROWTH IN AIR-BREATHING FISHES

Diurnal hypoxic events are particularly common in small and stagnant water bodies in the tropics where air-breathing fish often dominate the piscine fauna. The effects of hypoxia on growth rate, feed intake, and digestion have, however, only been studied in a few species of air-breathing fishes. This may be because obligate air-breathers, which, by virtue of their ability to obtain the vast majority of their oxygen from the air, are unlikely to be strongly affected by aquatic hypoxia (e.g., Sanchez *et al.*, 2001). Facultative air-breathers, on the other hand, are affected by aquatic hypoxia and the few existing studies point to clear effects on energetics and growth.

The obligate air-breathing fish *Ophiocephalus striatus* (now *Channa striatus*) maintains routine metabolic rate in hypoxic water by increasing surfacing and air-breathing frequency to extract oxygen from the air (Vivekanandan, 1977). Many air-breathing fish species increase the frequency of surfacing during digestion (Vivekanandan, 1977; Ponniah and Pandian, 1977; Figure 8.10A). Furthermore, Pandian and Vivekanandan (1976) observed that fed fish in unaerated water had a lower food consumption than fish in aerated water, which might indicate that the cost of elevating the air-breathing frequency at some point exceeds the gain of keeping a certain metabolic rate, leaving scope for activity and digestion. This is also apparent from starved individuals of *C. striatus* having a higher hanging frequency and duration in deeper aquariums, than more shallow aquariums Vivekanandan (1977). Vivekanandan argues that the fed fish had higher food consumption in deeper aquariums as a result of the increased surfacing costs necessary to

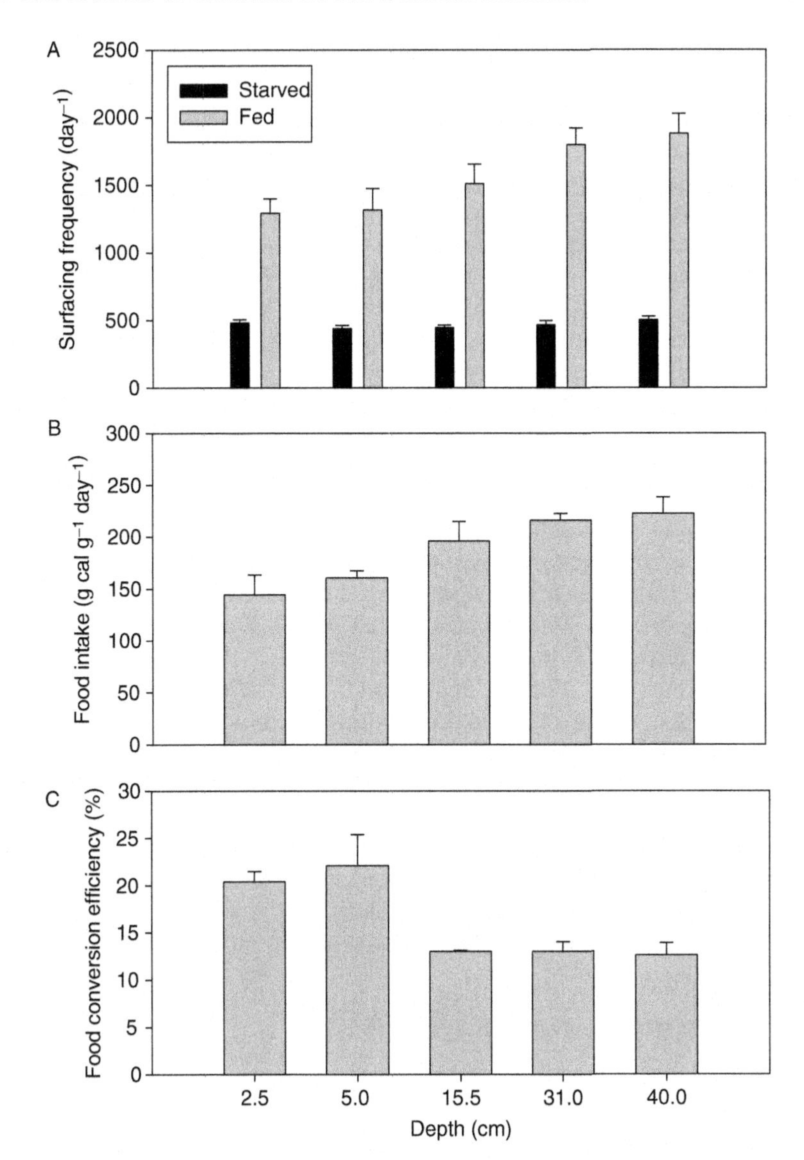

Fig. 8.10. Surfacing frequency (A), food intake (B), and food conversion efficiency (C) in Snakehead fish (*Channa striatus*) at 27°C, at different depths. The oxygen level was not controlled very accurately, resulting in higher O_2 levels in the starved group; this might explain the higher surfacing frequency in fed fish. Food intake increases with depth in fed fish. Surfacing frequency was independent of depth in starved fish. The food conversion efficiency is highest at small depths, but independent of depth below 15 cm. The fish had a body mass of 750 ± 70 mg and were measured over a period of 21 days. [Data are means ± S.D. from Vivekanandan (1977).]

obtain the necessary oxygen for their RMR (see Figure 8.10B). In the facultative air-breather *Heteropneustes fossilis*, the surfacing frequency also increased with increasing distance to the surface (Pandian and Vivekanandan, 1976). This compensation apparently only holds up to a limit where the costs of surfacing exceed the gain, and it is more profitable for the fish to rely solely on aquatic respiration, which is not possible in *C. striatus* (see also Kramer, 1987). A possible consequence of the elevated surfacing frequency during digestion and feeding is that the SDA (specific dynamic action) will be higher in air-breathing species than in non-air-breathers (Krishnan and Reddy, 1989), but further studies on this topic are needed to evaluate this hypothesis.

Juvenile air-breathing fish tend to depend on aquatic respiration until their air-breathing organ has been sufficiently developed. Therefore, growth in juvenile air-breathing fish is probably more dependent on dissolved oxygen levels than in adults. Even the giant South American obligate air-breather *Arapaima gigas*, which as an adult drowns within 10 min if prevented from access to air, is entirely water-breathing from hatching until about 9 days old after which a transition begins where the gill lamellae disappear and the air-breathing organ develops (Brauner *et al.*, 2004). It is likely that hypoxia slows down this transition as well as affecting growth itself. In support of this theory, Ebeling and Alpert (1966) found in paradise fish (*Macropodus opercularis*) that the air-breathing organ developed more slowly under hypoxic conditions than under normoxia. Finally, this ontogeny of air-breathing is probably affected by other factors such as the need to minimize predation risks in juveniles, which may be why a number of air-breathers perform synchronous surfacing behaviors (Kramer and Graham, 1976). The growth effects of hypoxia have been largely overlooked in air-breathing fish and given their fascinating position in the evolution from water to land and their increasing commercial importance, this aspect is worthy of further study.

9. CONCLUSIONS AND PERSPECTIVES

Hypoxia is common in natural areas and fish have evolved a number of physiological responses to tolerate large variations in oxygen availability. A substantial number of studies show that hypoxia limits growth primarily through a reduction in appetite, and it seems that the anorexic response and growth retardation occur at relatively mild levels of hypoxia. The overall effects of hypoxia on growth were already well established when the effects of hypoxia were reviewed by Brett for *Fish Physiology* in 1979 (see also Davis, 1975). Thus, as concluded 30 years ago, the impaired growth during hypoxia is primarily caused by a depression of food intake. More recent studies including cardiorespiratory and metabolic measurements have reinforced the link between aerobic scope and maximal growth. The causal link between scope of metabolism and appetite, however, still needs to be established. Thus future studies that provide in depth analysis of the hormonal profile along with blood flow and metabolic measurements would be very useful. An understanding of the mechanisms by which hypoxia affects growth is important to link physiology with life history traits and is also of considerable economic importance for aquaculture.

ACKNOWLEDGMENTS

The authors are supported by DANIDA (PhysCAM) and the Danish Research Council.

REFERENCES

Alsop, D., and Wood, C. (1997). The interactive effects of feeding and exercise on oxygen consumption, swimming performance and protein usage in juvenile rainbow trout (*Oncorhynchus mykiss*). *J. Exp. Biol.* **200,** 2337–2346.

Altimiras, J., Claireaux, G., Sandblom, E., Farrell, A. P., McKenzie, D. J., and Axelsson, M. (2008). Gastrointestinal blood flow and postprandial metabolism in swimming sea bass *Dicentrachus labrax*. *Physiol. Biochem. Zool.* **81,** 663–672.

Andrade, D. V., Abe, A. S., Cruz-Neto, A. P., and Wang, T. (2005). Specific dynamic action in ectothermic vertebrates: A general review on the determinants of the metabolic responses to digestion in fish, amphibians and reptiles. *In* "Adaptation in Physiological and Ecological Adaptations to Feeding in Vertebrates" (Starck, J. M., and Wang, T., Eds.), pp. 305–324. Science Publishers Inc., New Delhi.

Andrews, J. W., Murai, T., and Gibbons, G. (1973). The influence of dissolved oxygen on the growth of channel catfish. *Trans. Am. Fish. Soc.* **4,** 835–838.

Axelsson, M., and Fritsche, M. (1991). Effects of exercise, hypoxia and feeding on the gastrointestinal blood flow in the Atlantic cod *Gadus morhua*. *J. Exp. Biol.* **158,** 181–198.

Axelsson, M., Driedzic, W. R., Farrell, A. P., and Nilsson, S. (1989). Regulation of cardiac output and gut flow in the sea raven, *Hemitripterus americanus*. *Fish. Physiol. Biochem.* **6,** 315–326.

Axelsson, M., Thorarensen, M., Nilsson, S., and Farrell, A. P. (2000). Gastrointestinal blood flow in the red Irish lord, *Hemilepidotus hemilepidotus*: Long-term effects of feeding and adrenergic control. *J. Comp. Physiol. B* **170,** 145–152.

Axelsson, M., Altimiras, J., and Claireaux, G. (2002). Postprandial blood flow to the gastrointestinal tract is not compromised during hypoxia in the sea bass *Dicentrarchus labrax*. *J. Exp. Biol.* **205,** 2891–2896.

Bejda, A. J., Phelan, B. A., and Studholme, A. L. (1992). The effect of dissolved oxygen on the growth of young of the year winter flounder, *Pseudopleuronectes americanus*. *Environ. Biol. Fish.* **34,** 321–327.

Bernier, N. J., and Craig, P. M. (2005). CRF-related peptides contribute to stress response and regulation of appetite in hypoxic rainbow trout. *Am. J. Physiol.* **289,** R982–R990.

Blaikie, H. B., and Kerr, S. R. (1996). Effect of activity level on apparent heat increment in Atlantic cod, *Gadus morhua*. *Can. J. Fish. Aquat. Sci.* **53,** 2093–2099.

Braun, N., Lima de Lima, R., Moraes, B., Loro, V. L., and Baldisserotto, B. (2006). Survival, growth and biochemical parameters of silver catfish, *Rhamdia quelen* (Quoy & Gaimard, 1824), juveniles exposed to different dissolved oxygen levels. *Aquac. Res.* **37,** 1524–1531.

Brauner, C. J., Matey, V., Wilson, J. M., Bernier, N. J., and Val, A. L. (2004). Transition in organ function during the evolution of air-breathing; insights from *Arapaima gigas*, an obligate air-breathing teleost from the Amazon. *J. Exp. Biol.* **207,** 1433–1438.

Brett, J. R. (1979). Environmental factors and growth. *In* "Fish Physiology Vol. 8, Bioenergetics and Growth" (Hoar, W. S., Randall, D. J., and Brett, J. R., Eds.), pp. 599–675. Academic Press, New York.

Brett, J. R., and Blackburn, J. M. (1981). Oxygen requirements for growth of young coho (*Oncorhynchus kisutch*) and sockeye (*O. nerka*) salmon at 15 °C. *Can. J. Fish. Aquat. Sci.* **38,** 399–404.

Brett, J. R., and Groves, T. D. D. (1979). Physiological energetics. *In* "Fish Physiology Vol. 8, Bioenergetics and Growth" (Hoar, W. S., Randall, D. J., and Brett, J. R., Eds.), pp. 279–352. Academic Press, New York.

Brown, C. R., and Cameron, J. N. (1991a). The induction of specific dynamic action in channel catfish by infusion of essential amino-acids. *Physiol. Zool.* **64,** 276–297.

Brown, C. R., and Cameron, J. N. (1991b). The relationship between specific dynamic action (SDA) and protein- synthesis rates in the channel catfish. *Physiol. Zool.* **64,** 298–309.

Bureau, D. P., Kaushik, S. J., and Cho, C. Y. (2002). Bioenergetics. *In* "Fish Nutrition" (Halver, J. E., and Hardy, R. W., Eds.), pp. 2–54. Academic Press, San Diego, CA.

Buentello, J. A., Gatlin, D. M., III, and Neill, W. H. (2000). Effects of water temperature and dissolved oxygen on daily feed consumption, feed utilization and growth of channel catfish (*Ictalurus punctatus*). *Aquaculture* **182,** 339–352.

Canosa, L. F., Chang, J. P., and Peter, R. E. (2007). Neuroendocrine control of growth hormone in fish. *Gen. Comp. Endocr.* **151,** 1–26.

Cech, J. J., Jr., Mitchell, S. J., and Wragg, T. E. (1984). Comparative growth of juvenile white sturgeon and striped bass: Effects of temperature and hypoxia. *Estuaries* **7,** 12–18.

Chabot, D., and Claireaux, G. (2008). Environmental hypoxia as a metabolic constraint on fish: The case of the Atlantic cod, *Gadus morhua. Mar. Pollut. Bull* **57,** 287–294.

Chabot, D., and Dutil, J.-D. (1999). Reduced growth of Atlantic cod in non-lethal hypoxic conditions. *J. Fish Biol.* **55,** 472–491.

Chiba, K. (1983). The effect of dissolved oxygen on the growth of young Silver Bream. *B. Jpn. Soc. Sci. Fish.* **49,** 601–610.

Claireaux, G., and Lagardère, J.-P. (1999). Influence of temperature, oxygen and salinity on the metabolism of the European sea bass. *J. Sea Res.* **42,** 157–168..

Claireaux, G., and Lefrançois, C. (2007). Linking environmental variability and fish performance: Integration through the concept of scope for activity. *Philos. Trans. Roy. Soc. B.* **362,** 2031–2041.

Claireaux, G., Webber, D. M., Lagardère, J.-P., and Kerr, S. R. (2000). Influence of water temperature and oxygenation on the aerobic metabolic scope of Atlantic cod (*Gadus morhua*). *J. Sea Res.* **44,** 257–265.

Davis, J. C. (1975). Minimal dissolved-oxygen requirements of aquatic life with emphasis on Canadian species. *J. Fish. Res. Board Can.* **32,** 2295–2332.

Diaz, R. J. (2001). Overview of hypoxia around the world. *J. Environ. Qual.* **30,** 275–281.

Ebeling, A. W., and Alpert, J. S. (1966). Retarded growth of the paradisefish, *Macropodus opercularis* (L.), in low environmental oxygen. *Copeia* **1966,** 606–610.

Eliason, E. J., Higgs, D. A., and Farrell, A. P. (2007). Effect of isoenergetic diets with different protein and lipid content on the growth performance and heat increment of rainbow trout. *Aquaculture* **272,** 723–736.

Eliason, E. J., Higgs, D. A., and Farrell, A. P. (2008). Postprandial gastrointestinal blood flow, oxygen consumption and heart rate in rainbow trout (*Oncorhynchus mykiss*). *Comp. Biochem. Physiol. A.* **149,** 380–388.

Farrell, A. P., Thorarensen, H., Axelsson, M., Crocker, C. E., Gamperl, A. K., and Cech, J. J. (2001). Gut blood flow in fish during exercise and severe hypercapnia. *Comp. Biochem. Physiol.* **A128,** 551–563.

Foss, A., Evensen, T. H., and Øiestad, V. (2002). Effects of hypoxia and hyperoxia on growth and food conversion efficiency in the spotted wolffish *Anarhichas minor* (Olafsen). *Aquac. Res.* **33,** 437–444.

Fry, F. E. J. (1971). The effect of environmental factors on the physiology of fish. *In* "Fish Physiology" (Hoar, W. D., and Randall, D. J., Eds.), Vol. 6, pp. 1–98. Academic Press, New York.

Fu, S. J., Xie, X. J., and Cao, Z. D. (2005). Effect of feeding level and feeding frequency on specific dynamic action in *Silurus meridionalis*. *J. Fish Biol.* **67,** 171–181.

Gorissen, M. H. A. G., Flik, G., and Huising, M. O. (2006). Peptides and proteins regulating food intake: A comparative view. *Anim. Biol.* **56,** 447–473.

Grove, D. J., and Holmgren, S. (1992a). Intrinsic mechanisms controlling cardiac stomach volume of the rainbow trout (*Oncorhynchus mykiss*) following gastric distension. *J. Exp. Biol.* **163,** 33–48.

Grove, D. J., and Holmgren, S. (1992b). Mechanisms controlling stomach volume of the Atlantic cod (*Gadus morhua*) following gastric distension. *J. Exp. Biol.* **163,** 49–63.

Grove, D. J., Loizides, L., and Nott, J. (1978). Satiation amount, frequency of feeding and gastric emptying rate in *Salmo gairdneri*. *J. Fish Biol.* **12,** 507–516.

Grove, D. J., Moctezuma, M. A., Flett, H. J. R., Foott, J. S., Watson, T., and Flowerdew, M. W. (1985). Gastric emptying and the return of appetite in juvenile turbot, *Scophthalmus maximus* L. fed on artificial diets. *J. Fish Biol.* **26,** 339–354.

Hunt von Herbing, I., and White, L. (2002). The effects of body mass and feeding on metabolic rate in small juvenile Atlantic cod. *J. Fish Biol.* **61,** 945–958.

Jobling, M. (1981). The influences of feeding on the metabolic rates of fishes: A short review. *J. Fish Biol.* **18,** 385–400.

Jobling, M. (1983). Towards an explanation of specific dynamic action (SDA). *J. Fish Biol.* **23,** 549–555.

Jobling, M. (1993). Bioenergetics: Feed intake and energy partitioning. *In* "Fish Ecophysiology" (Rankin, J. C., and Jensen, F. B., Eds.). Chapman & HallFish and Fisheries Series 9, London, pp. 1–44.

Jobling, M., and Davies, P. S. (1980). Effects of feeding on the metabolic rate and the Specific Dynamic Action in plaice, *Pleuronectus platessa* L. *J. Fish Biol.* **16,** 629–638.

Jones, D. R., Randall, D. J., and Jarman, G. M. (1970). Graphical analysis of oxygen transfer in fish. *Resp. Physiol.* **10,** 285–300.

Jordan, A. D., and Steffensen, J. F. (2007). Effects of ration size and hypoxia on specific dynamic action in the cod. *Physiol. Biochem. Zool.* **80,** 178–185.

Kimura, M., Takahashi, T., Takatsu, T., Nakatini, T., and Maeda, T. (2004). Effetcs of hypoxia in principal prey and growth of flathead flounder *Hippoglossoides dubius* in Funka Bay, Japan. *Fisheries Sci.* **70,** 537–545.

Kleiber, M. (1961). *In* "The Fire of Life: An Introduction to Animal Energetics". John Wiley and Sons, New York and London.

Kramer, D. L., and Graham, J. B. (1976). Synchronous air-breathing, a social component of respiration in fishes. *Copeia* **1976,** 689–697.

Kramer, D. L. (1987). Dissolved oxygen and fish behavior. *Environ. Biol. Fish.* **18,** 81–92.

Kerr, S. R. (1976). Ecological analysis and the Fry paradigm. *J. Fish. Res. Board Can.* **33,** 329–335.

Kerr, S. R. (1990). The Fry paradigm: Its significance for contemporary ecology. *T. Am. Fish. Soc.* **119,** 779–785.

Krishnan, N., and Reddy, S. R. (1989). Combined effects of quality and quantity of food on growth and body composition of the air-breathing fish *Channa gachua* (Ham.). *Aquaculture* **76,** 79–96.

Lefebvre, S., Bacher, C., Meuret, A., and Hussenot, J. (2001). Modelling nitrogen cycling in a mariculture ecosystem as a tool to evaluate its outflow. *Estuar. Coast. Shelf Sci.* **52,** 305–325.

Maljaars, J., Peters, H. P. F., and Masclee, A. M. (2007). Review article: The gastrointestinal tract: Neuroendocrine regulation of satiety and food intake. *Aliment. Pharm. Therap.* **26,** 241–250.

Mallekh, R., and Lagardère, J. P. (2002). Effects of temperature and dissolved oxygen concentration on the metabolic rate of the turbot and the relationship between metabolic scope and feeding demand. *J. Fish Biol.* **60,** 1105–1115.

McCue, M. D. (2006). Specific dynamic action: A century of investigation. *Comp. Biochem. Physiol.* **A 144,** 381–394.

McGaw, I. J. (2008). Gastric processing in the Dungeness crab, *Cancer magister*, during hypoxia. *Comp. Biochem. Physiol.* **A 150,** 458–463.

McNatt, R. A., and Rice, J. A. (2004). Hypoxia-induced growth rate reduction in two juvenile estuary-dependent fishes. *J. Exp. Mar. Biol. Ecol.* **311,** 147–156.

Nilsson, G., and Östlund-Nilsson, S. (2008). Does size matter for hypoxia tolerance in fish. *Biol. Rev.* **83,** 173–189.

Pandian, T. J., and Vivekanandan, E. (1976). Effects of feeding and starvation on growth and swimming activity in an obligatory air-breathing fish. *Hydrobiologia* **49,** 33–39.

Pedersen, C. L. (1987). Energy budgets for juvenile rainbow trout at various oxygen concentrations. *Aquaculture* **62,** 289–298.

Petersen, J. K., and Pihl, L. (1995). Responses to hypoxia of plaice, *Pleuronectes platessa*, and dab, *Limanda Limanda*, in the south-east Kattegat: Distribution and growth. *Environ. Biol. Fish.* **43,** 311–321.

Person-Le Ruyet, J., Pichavant, K., Vacher, C., Le Bayon, N., Sévère, A., and Boeuf, G. (2002). Effects of oxygen supersaturation on growth and metabolism in juvenile turbot (*Scophthalmus maximus*). *Aquaculture* **205,** 373–383.

Pichavant, K., Person-Le Ruyet, J., Le Bayon, N., Sévère, A., Le Roux, A., Quéméner, L., Maxime, V., Nonnotte, G., and Boeuf, G. (2000). Effects of hypoxia on growth and metabolism of juvenile turbot. *Aquaculture* **188,** 103–114.

Pichavant, K., Person-Le Ruyet, J., Le Bayon, N., Sévère, A., Le Roux, A., and Boeuf, G. (2001). Comparative effects of long-term hypoxia on growth, feeding and oxygen consumption in juvenile turbot and European seabass. *J. Fish Biol.* **59,** 875–883.

Ponniah, A. G., and Pandian, T. J. (1977). Surfacing activity and food utilization in the air-breathing fish *Polyacanthus cupanus* exposed to constant PO_2. *Hydrobiologia* **53,** 221–227.

Portner, H.-O., and Knust, R. (2007). Climate change affects marine fishes through the oxygen limitation of thermal tolerance. *Science* **315,** 95–97.

Rubner, M. (1902). *In* "Die Gesetze des Energie verbrauchs bei der Ernährung" Frank Dauticke, Leipzig & Vienna.

Rutjes, H. A., Nieven, M. C., Weber, R. E., Witte, F., and Van den Thillart, G. E. E. J. M. (2007). Multiple strategies of Lake Victoria cichlids to cope with lifelong hypoxia include hemoglobin switching. *Am. J. Physiol.* **293,** R1376–R1383.

Sanchez, A., Soncini, R., Wang, T., Koldkjær, P., Taylor, E. W., and Glass, M. L. (2001). The differential cardio-respiratory responses to ambient hypoxia and systemic hypoxaemia in the South American lungfish, *Lepidosiren paradoxa*. *Comp. Biochem. Physiol.* **130,** 677–687.

Seth, H., Sandblom, E., Holmgren, S., and Axelsson, M. (2008). Effects of gastric extension on the cardiovascular system in rainbow trout (*Oncorhynchus mykiss*). *Am. J. Physiol.* **294,** R1648–R1656.

Soofiani, N. M., and Hawkins, A. D. (1982). Energetic costs at different levels of feeding in the juvenile cod, *Gadus morhua*. *J. Fish Biol.* **21,** 577–592.

Steffensen, J. F. (1989). Some errors in respirometry of aquatic breathers: How to avoid and correct for them. *Fish Physiol. Biochem.* **6,** 49–59.

Tandler, A., and Beamish, F. W. H. (1979). Mechanical and biochemical components of apparent specific dynamic action in largemouth bass, *Micropterus salmoides* Lacepede. *J. Fish Biol.* **14,** 343–350.

Tandler, A., and Beamish, F. W. H. (1980). Specific dynamic action and diet in largemouth bass, *Micropterus salmoides* (Lacepede). *J. Nutr.* **110,** 750–764.

Taylor, J. C., and Miller, J. M. (2001). Physiological performance of juvenile southern flounder, *Paralichthys lethostigma* (Jordan and Gilbert, 1984), in chronic and episodic hypoxia. *J. Exp. Mar. Biol. Ecol.* **258,** 195–214.

Taylor, J. R., Whittamore, J. M., Wilson, R. W., and Grosell, M. (2007). Postprandial acid-base balance and ion regulation in freshwater and seawater-acclimated European flounder, *Platichthys flesus*. *J. Comp. Physiol.* **B 177,** 597–608.

Thetmeyer, H., Waller, U., Black, K. D., Inselmann, S., and Rosenthal, H. (1999). Growth of European sea bass (*Dicentrarchus labrax* L.) under hypoxic and oscillating oxygen conditions. *Aquaculture* **174,** 355–367.

Thorarensen, H., Gallaugher, P. E., Kiessling, A. K., and Farrell, A. P. (1994). Intestinal blood flow in swimming chinook salmon *Oncorhynchus tshawytscha* and the effects of hematocrit on blood flow distribution. *J. Exp. Biol.* **179,** 115–129.

Tran-Duy, A., Schrama, J. W., van Dam, A. A., and Verreth, J. A. J. (2008). Effects of oxygen concentration and body weight on maximum feed intake, growth and haematological parameters of Nile tilapia, *Oreochromis niloticus*. *Aquaculture* **275,** 152–162.

Tsadik, G. G., and Kutty, M. N. (1987). Influence of ambient oxygen on feeding and growth of the Tilapia (*Oreochromis niloticus*) (Linnaeus). *FAO document* AC168E.

Van Dijk, P. L. M., Staaks, G., and Hardewig, I. (2002). The effect of fasting and refeeding on temperature preference, activity and growth of roach, *Rutilus rutilus*. *Oecologia* **130,** 496–504.

Vats, P., Singh, V. K., Singh, S. N., and Singh, S. B. (2007). High altitude induced anorexia: Effect of changes in leptin and oxidative stress levels. *Nutritional Neuroscience* **10,** 243–249.

Vivekanandan, E. (1977). Effects of the PO_2 on swimming activity and food utilization in *Ophiocephalus striatus*. *Hydrobiologia* **52,** 165–169.

Volkoff, H., Casona, L. F., Unniappan, J. M., Cerdá-Reverter, J. M., Bernier, N. J., Kelly, S. P., and Peter, R. E. (2005). Neuropeptides and the control of food intake in fish. *Gen. Comp. Endocrinol.* **142,** 3–19.

Wang, T., Andersen, J., and Hicks, J. W. (2005). Effects of digestion on the respiratory and cardiovascular physiology of amphibians and reptiles. *In* "Physiological and Ecological Adaptations to Feeding in Vertebrates" (Starck, J. M., and Wang, T., Eds.), pp. 279–303. Science Publishers Inc., New Delhi.

Wang, T., Hung, C. C. Y., and Randall, D. J. (2006). The comparative physiology of food deprivation: From feast to famine. *Ann. Rev. Physiol.* **68,** 223–251.

Webb, P. W. (1994). Exercise performance of fish. *Adv. Vet. Sci. Comp. Med.* **38B,** 1–49.

Westerterp, K. R., Keyser, B., Wouters, L., Le Trong, J.-L., and Richalet, J.-P. (1994). Energy balance at high altitude of 6542 m. *J. Appl. Physiol.* **77,** 862–866.
Westerterp, K. R., Meijer, E. P., Rubbens, M., Robarch, P., and Richalet, J.-P. (2000). Operation Everest III. Energy and water balance. *Pflug. Arch. Eur. J. Phy.* **439,** 483–488.
Wood, C. M., Kajimura, M., Bucking, C., and Walsh, P. J. (2007). Osmoregulation, ionoregulation and acid-base regulation by the gastrointestinal tract after feeding in the elasmobranch (*Squalus acanthias*). *J. Exp. Biol.* **210,** 1335–1349.
Wootton, J. R. (1998). *In* "Ecology of Teleost Fishes" 2nd ed, Kluwer, Dordrecht.
Yamaji, R., Sakamoto, M., Miyake, K., and Nakano, Y. (1996). Hypoxia inhibits gastric emptying and gastric acid secretion in conscious rats. *J. Nutr* **126,** 673–680.
Yoshimoto, M., Sasaki, M., Naraki, N., Morhi, M., and Miki, K. (2004). Regulation of gastric motility at simulated high altitude in conscious rats. *J. Appl. Physiol.* **97,** 599–604.
Zhou, B. S., Wu, R. S. S., Randall, D. J., and Lam, P. K. S. (2001). Bioenergetics and RNA/DNA ratios in the common carp (*Cyprinus carpio*) under hypoxia. *J. Comp. Physiol.* **B 171,** 49–57.

9

THE ANOXIA-TOLERANT CRUCIAN CARP (*CARASSIUS CARASSIUS* L.)

MATTI VORNANEN
JONATHAN A.W. STECYK
GÖRAN E. NILSSON

The crucian carp is probably the most anoxia-tolerant fish there is, surviving without oxygen for days to months depending on temperature. The anoxia tolerance has evolved in response to over-wintering in ponds and small lakes that can become anoxic for months during the winter. The exceptional anoxia tolerance of the crucian carp is based on special physiological traits that are either constitutively expressed or seasonally primed. A key to its anoxia tolerance is its constitutive ability to produce ethanol as the major anaerobic end product. The ethanol production is supported by massive stores of glycogen in various tissues, and these stores are largest in the autumn before the onset of wintertime anoxia. Metabolic depression is less pronounced than in anoxia-tolerant turtles and there is no major

Hypoxia: Volume 27
FISH PHYSIOLOGY

DOI: 10.1016/S1546-5098(08)00009-5

down-regulation of membrane permeability in brain ("channel arrest"), possibly with the exception of reduced NMDA receptor function. Increased levels of the inhibitory neurotransmitter GABA and low levels of the excitatory transmitter glutamate together with a modest activation of glycolysis probably ensure energy balance to the anoxic brain and aid to maintain normal ion gradients across neuronal membranes. The heart has been found to sustain cardiac output in anoxia, possibly to allow for substrate transport and a sufficient rate of ethanol release to the water. Like the brain, the heart also shows few signs of reduced ion permeability in anoxia. However, a lack of compensatory temperature acclimatization suggests that it is utilizing the low winter temperature to suppress its energy needs during anoxia.

1. INTRODUCTION

While the vast majority of vertebrates need a constant supply of oxygen to survive, there are a few examples of vertebrates that can survive long periods of anoxia – the complete absence of oxygen. The best studied examples of such animals are some species of North American freshwater turtles and two cyprinid fishes: the crucian carp (*Carassius carassius* L.) and the closely related goldfish (*Carassius auratus* L.). Most studies on goldfish have been made on fish obtained from the aquarium trade, where goldfish have been cultured for more than a millennium, and where the main selection pressures have been for readily breeding in captivity, and for various colors and shapes, rather than for anoxia tolerance. Thus, many of the original traits promoting anoxia tolerance may have been lost. Indeed, goldfish do not display the same extreme anoxia tolerance as the crucian carp, which survives days of anoxia at room temperature and several months of anoxia at temperatures close to 0°C (Blazka, 1958; Blazka, 1960; Piironen and Holopainen, 1986) (Figure 9.1). In this chapter we will focus on the crucian carp, although some references to studies on goldfish will be given. Additionally, when relevant, we will contrast and draw parallels between the mechanisms of anoxia tolerance that the unrelated crucian carp and freshwater turtle display.

1.1. Distribution and Habitat of Crucian Carp

The crucian carp has a wide distribution in central Asia and Europe, ranging from the Arctic Circle in Scandinavia to central France and the Black Sea in the south, and from England to the Lena River in Russia (Holopainen *et al.*, 1997b). Crucian carp exist in two distinct forms; both evolved to reduce or avoid predation (Nilsson, 1855). When crucian carp coexist with piscivorous fish in lakes, the presence of predators induces a

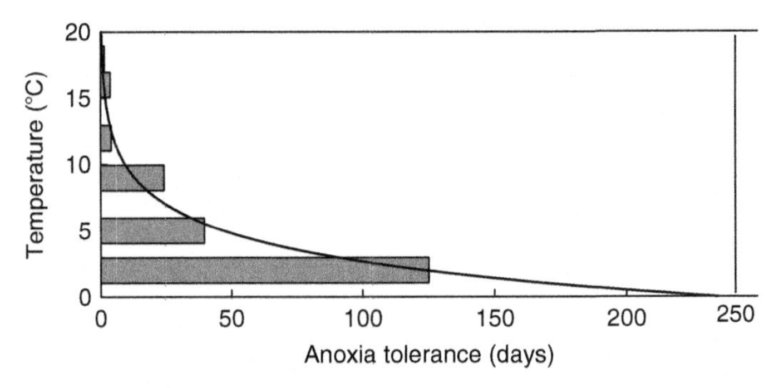

Fig. 9.1. Anoxia tolerance of crucian carp at different temperatures measured as lethal time to 50% mortality (LT_{50}). Tolerance to anoxia decreases exponentially as a function of temperature. [Data from (Piironen and Holopainen, 1986).]

change in body morphology from a shallow and long-bodied fish to a much deeper bodied form, which decreases predation efficiency of gape-limited piscivores such as pike (Brönmark and Milner, 1992; Holopainen *et al.*, 1997a), and enhances the escape locomotor performance of crucian carp (Domenici *et al.*, 2008). An alternative survival strategy of crucian carp is based on an equally prominent trait that allows it to completely avoid predatory fish: they are able to inhabit small ponds where anoxic periods exclude the presence of other fish species. In the fall and winter seasons, these ponds progressively become hypoxic because oxygen consumption by the inhabitants exceeds the photosynthetic oxygen production by plants, and ice and snow cover prevents diffusion of oxygen from the atmosphere (Nagell and Brittain, 1977). The hypoxic period lasts until the spring and frequently includes totally anoxic periods (Figure 9.2). It is the latter survival strategy that has resulted in the evolution of physiological traits that promote anoxia tolerance (Blazka, 1958; Nilsson, 2001). The ponds in which crucian carp reside also experience large and regular seasonal changes in temperature (Figure 9.2). Consequently, crucian carp are also known for their excellent eurythermicity with thermal tolerance ranging from 0 to 38°C, with a high thermal optimum of 27°C (Horoszewicz, 1973).

1.2. The Need for Oxygen in the Vertebrate Brain and Heart

An animal that survives prolonged anoxia, like the crucian carp, has to overcome a major problem: to protect its tissues from energy deficiency. The stop in oxidative phosphorylation during anoxia leaves the animal with glycolysis as the only major ATP-producing process. Glycolysis has an ATP yield that is only about 6–10% of that of the oxidation of glucose during normoxic conditions (depending on how well the mitochondria are coupled during

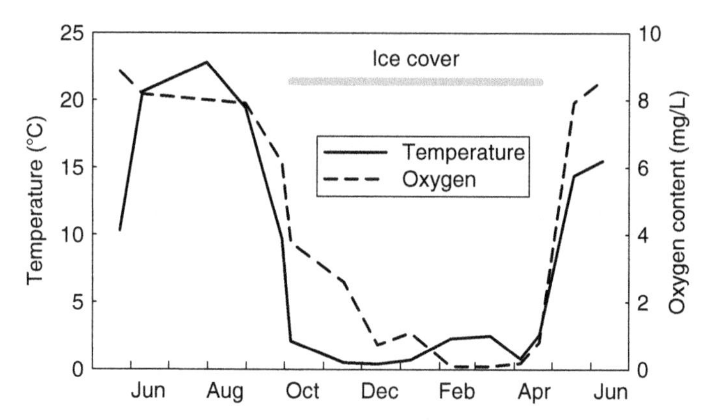

Fig. 9.2. Seasonal changes in temperature and oxygen content in a typical habitat of crucian carp in Eastern Finland at the latitude of 62°41/N. [Data from (Vornanen and Paajanen, 2004).]

normoxia; Hochachka and Somero, 2002). Consequently, it is widely accepted that the high intrinsic rates of ATP use make the vertebrate brain and heart particularly sensitive to low oxygen levels, as they will rapidly run out of ATP when the sole source of ATP is anaerobic glycolysis. In mammals, most of the ATP turnover in brain is devoted to neural signaling, particularly to the ion-pumping needed to maintain the membrane potentials of electrically active neurons (Erecinska and Silver, 1994). While fish bodies only consumes about one tenth of the energy used by mammalian bodies, fish brains do not differ much from mammalian brains when it comes to ATP use (Nilsson, 1996), and it is likely that electric activity is also the main energy consumer in the fish brain. In the majority of vertebrates, the brain will rapidly fail to function in anoxia. Key events of the anoxic brain failure in mammals include a loss of ion homeostasis, which can be detected as a rapid rise in extracellular K^+ levels, and a subsequent outflow of glutamate – the major excitatory neurotransmitter in brain. Both these events are also seen in anoxic rainbow trout (*Oncorhynchus mykiss*) brain within about half an hour in anoxia at 10°C (Nilsson *et al.*, 1993a; Hylland *et al.*, 1995). The depolarization and the glutamate out-flow triggers the opening of Ca^{2+} channels and a massive rise in intracellular Ca^{2+}, which in turn activates several degenerative pathways (Arundine and Tymianski, 2003).

Like the brain, the heart has an ATP requirement greatly exceeding that of most other tissues. For the heart to function as a muscular pump, a continual ATP supply is needed for powering the myofilament sliding by the myosin ATPase. Also, ATP is needed to power the various ATP-dependent ion-motive pumps (i.e., Na^+/K^+-ATPase and Ca^{2+}-ATPases) essential for repeated action potential generation, intracellular Ca^{2+} homeostasis, and membrane ion transport (Aho and Vornanen, 1997; Rolfe and Brown, 1997; Huss and Kelly, 2005;

Taha and Lopaschuk, 2007). For mammalian hearts, mechanical activity has been estimated to account for 75–85% of the cardiac ATP demand and the Na^+/K^+-ATPase and Ca^{2+}-ATPases for 15–25% (Schramm *et al.*, 1994; Rolfe and Brown, 1997), and it is assumed that ~2% of the cellular ATP pool is consumed in each heart beat (Balaban, 2002). Thus, like the brain, if ATP supply is not matched to ATP demand, the heart's ATP pool will quickly be depleted and a catastrophic sequence of events will occur, including failure of the ATP-dependent ion-motive pumps, disruption of cellular membrane potentials, and a loss of ionic integrity of cellular membranes. This will lead to cardiomyocyte death by necrosis, cardiac failure, and, ultimately, organismal death (Hochachka, 1986; Boutilier, 2001). The cardiac ATP budget in the carp may differ from mammals due to differences in cellular structure and Ca^{2+} management (Santer, 1985; Aho and Vornanen, 1997). Even so, the energetic cost of mechanical work likely remains the greater fraction of total cardiac ATP expenditure because myofibrillar volume density of cardiomyocytes is similar for both fish and mammals (Santer, 1985; Aho and Vornanen, 1997). An additional contributing factor to anoxic cardiac failure is the accumulation of protons from anaerobic metabolism (acidosis). Acidosis dramatically decreases the ability of the heart to pump blood by reducing contractile force and promoting fatal ventricular arrhythmias (Williamson *et al.*, 1976; Gesser and Jørgensen, 1982; Orchard and Kentish, 1990).

Obviously, the crucian carp brain and heart constitute striking exceptions to those of most vertebrates, as these organs continue to function for days to months in anoxia. Such a feat is possible because the hypoxia- and anoxia-tolerance of crucian carp depend on both (1) innate or constitutively expressed traits, e.g., ethanol production and gill remodeling, that can be recruited any time when oxygen shortage sets in (these mechanisms are described in section 2), and (2) induced traits, e.g., glycogen content of various tissues and remodeling of cardiac function, which are seasonally primed by environmental cues and set the ultimate limits for anoxia tolerance (the latter are described in Section 3).

2. MECHANISMS OF HYPOXIC AND ANOXIC SURVIVAL

2.1. Ethanol Production

The crucian carp has the exotic ability to produce ethanol as the major end-product of anaerobic metabolism. Among vertebrates, this trait has only been found in two close relatives: the goldfish (*Carassius auratus*) and the bitterling (*Rhodeus amarus*) (Shoubridge and Hochachka, 1980; Mourik *et al.*, 1982; Johnston and Bernard, 1983; Holopainen *et al.*, 1986; Nilsson, 1988;

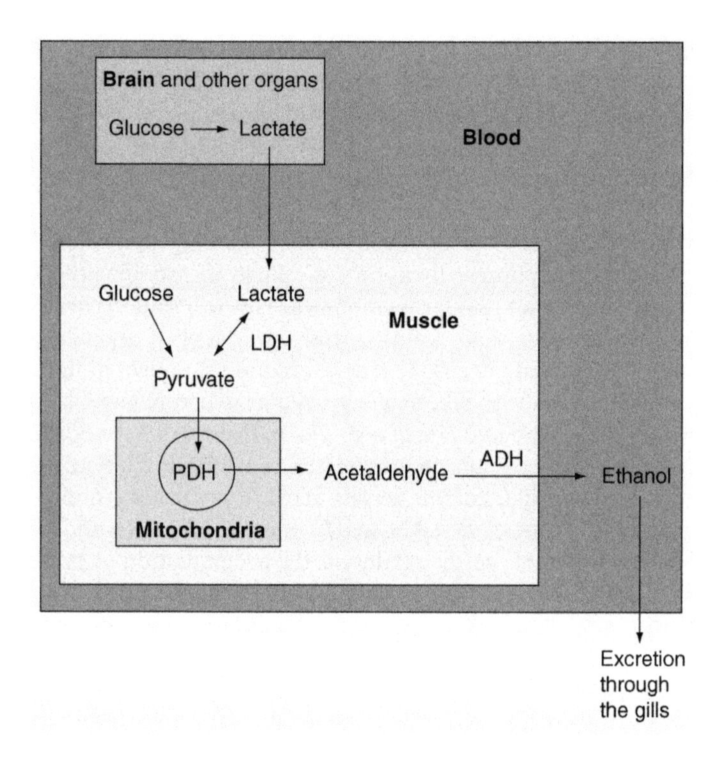

Fig. 9.3. The ethanol-producing pathway in anoxic crucian carp and goldfish. Ethanol is only produced in muscle tissue, while all other organs produce lactate during anoxia. In muscle, lactate is turned into pyruvate by lactate dehydrogenase (LDH). Pyruvate is further converted to acetaldehyde by pyruvate dehydrogenase (PDH), which in contrast to other vertebrates leaks out acetaldehyde during anoxia. The acetaldehyde is turned into ethanol by alcohol dehydrogenase (ADH), an enzyme that only occurs in muscle of crucian carp and goldfish. Ethanol readily passes biological membranes, and after it leaks out into the blood it will leave the fish over the gills. In this way, a buildup of the anaerobic end-product is avoided.

Wissing and Zebe, 1988). The enzyme system used to produce ethanol is confined to skeletal muscle (red and white) and is a three-step process. Lactate is turned into pyruvate by lactate dehydrogenase (LDH), pyruvate is turned into acetaldehyde by Enzyme 1 (also called pyruvate decarboxylase) of the pyruvate dehydrogenase (PDH) complex, and acetaldehyde is turned into ethanol by alcohol dehydrogenase (ADH) (Figure 9.3). Only the skeletal muscle of crucian carp and goldfish contains ADH, so all other tissues, including the brain and heart, will produce lactate in anoxia, which has to be transported in the blood to the muscle, where it is transformed to ethanol. The ethanol, which readily passes through cellular membranes, quickly diffuses out into the blood for transport to the gills through which it easily diffuses into the ambient water. Consequently, provided internal convection continues, the level of ethanol in blood does not rise high enough to significantly suppress nervous activity (the steady state level is below 10 mM; Van Waarde *et al.*, 1993). The unusual distribution of ADH in crucian carp (all in muscle and none in liver) is retained also during the summer, when the crucian carp is unlikely to encounter anoxia (Nilsson, 1990a).

While the ethanol production allows the crucian carp to endure long-term anoxia without suffering lactic acidosis, it has a clear energetic drawback: ethanol, a very energy-rich hydrocarbon is released to the water and forever lost. Therefore, to allow for long-term survival in anoxia, the crucian carp accumulates enormous glycogen stores prior to the winter months (see Section 3.1) and the only factor that appears to limit its anoxia endurance is the total depletion of the main glycogen store in the liver (Nilsson, 1990b). In contrast, in animals where lactate is accumulated, it can later either be used as a fuel or used to synthesize glycogen when O_2 becomes available (if the animal survives the anoxic episode).

It is likely that the ability of the crucian carp to sustain a considerable neural, cardiac, and physical activity during anoxia is closely linked to its ability to produce ethanol as the major end-product of anaerobic metabolism, allowing it to avoid self-poisoning from production of lactate and the associated H^+. Freshwater turtles, the other well-studied example of anoxia-tolerant vertebrates, cannot produce any other anaerobic end-product than lactate, which forces them to resort to a drastic depression of energy metabolism in anoxia to reduce lactate production as much as possible. Still, lactate levels as high as 200 mM may be reached in turtles, which they need to buffer with calcium carbonate from their bones and shell (Jackson, 2002). It is highly unlikely that any fish would be able to tolerate such high lactate levels as they, as water breathers, have relatively low CO_2 levels in their tissues, and therefore a low pH buffering capacity. Thus, ethanol production is probably a key prerequisite for long-term anoxic survival in crucian carp.

2.2. Gill Remodeling

Even if the crucian carp can survive without any oxygen, the steady depletion of glycogen stores and loss of the energy-rich ethanol is costly. Thus, getting access to oxygen is highly desirable from an energetic point of view. The crucian carp can reversibly adjust the morphology of its gills to match its oxygen needs (Figure 9.4) (Sollid and Nilsson, 2006; Nilsson, 2007). Sollid *et al.* (2003) showed that normoxic crucian carp kept in relatively cold water (20°C or less) have gills that lack protruding lamellae. But when the crucian carp is exposed to hypoxia, it starts remodeling the gills, resulting in a 7-fold increase in the respiratory surface area and reducing diffusion

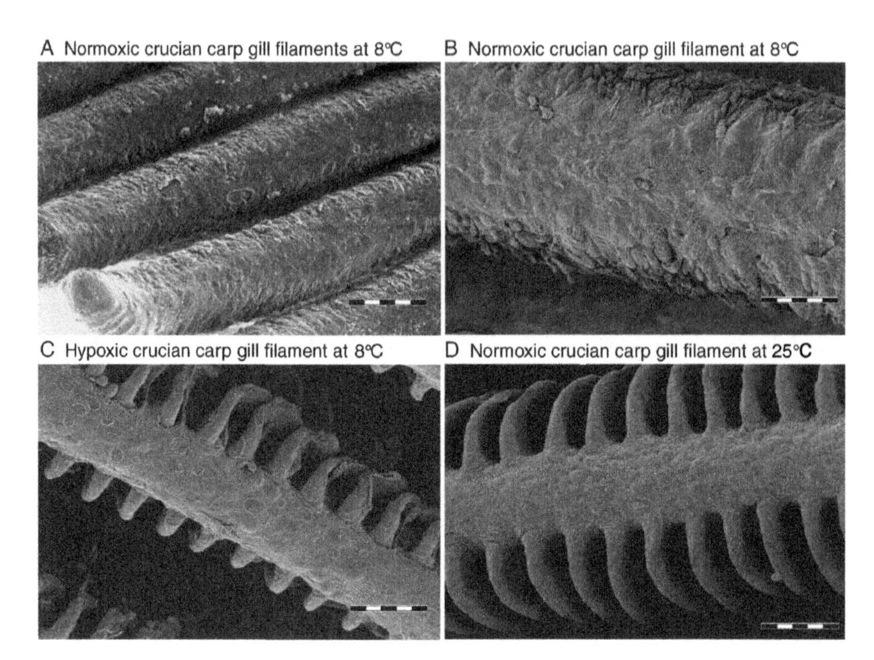

Fig. 9.4. Scanning electron micrographs of gill filaments from crucian carp kept in normoxic water at 8°C (A–B), in hypoxic water at 8°C (C), or in normoxic water at 25°C (D). Scale bars are 150 μm in A and 50 μm in B–D. [Data from (Sollid *et al.*, 2003, 2005a).]

distances, thereby boosting the ability to take up oxygen. The lamellae are actually present in normoxic crucian carp, but they are covered by a non-vascularized cell mass situated outside the gill epithelium that completely occupies the space between the gill lamellae. The mitotic and apoptotic activity in this interlamellar cell mass (ILCM) varies with ambient oxygen levels (Sollid *et al.*, 2003). Thus, mitosis dominates in normoxic water, causing ILCM to fill up the interlamellar space, while in hypoxic water, apoptosis prevails and the ILCM nearly disappears, exposing the respiratory epithelium to the water. Goldfish (*Carassius auratus*) also have the same capacity for remodeling the gills when kept at low temperature (7°C) (Sollid *et al.*, 2005a).

In addition to hypoxia, increasing the water temperature to 25°C causes gill remodeling (within hours) in crucian carp (Figure 9.4D). This indicates that an elevated oxygen demand is a key trigger for the remodeling process (Sollid *et al.*, 2005a). In contrast, a total absence of oxygen does not induce the remodeling, which makes functional sense since there is no oxygen to take up (Sollid *et al.*, 2005b). The mechanistic reason for this may be that the apoptosis needed to remove the ILCM could be oxygen dependent.

With regard to the molecular signals involved in the gill remodeling, little is known. Although hypoxia-inducible factor-1α (HIF-1α) increases in hypoxic crucian carp gills (coinciding with a reduction in the ILCM), the level of this transcription factor is also increased by a fall in temperature (which coincides with increased ILCM) (Rissanen *et al.*, 2006; Sollid *et al.*, 2006). This makes an involvement of HIF-1α less likely.

2.3. Cardiorespiratory Adjustments

In addition to remodeling the gills, crucian carp utilize other strategies to maintain oxygen transfer at the gills and tissues in the face of declining oxygen availability. Foremost, crucian carp hemoglobin has an extremely high oxygen affinity at high temperature (P_{50} = 1.8 mmHg at pH 7.7 and 20°C) that increases markedly with decreasing temperature (P_{50} = 0.8 mmHg at 10°C; Sollid *et al.*, 2005a). Further, like other fish exposed to hypoxia, carp immediately hyperventilate to maximize oxygen uptake. With the onset of anoxia, ventilation frequency (f_R) nearly doubles (Stecyk *et al.*, 2004b) (Figure 9.5). Concurrently, ventilation amplitude also increases (J. A. W. Stecyk, K- O. Stensløkken, G. E. Nilsson, unpublished observation). With prolonged anoxia exposure, f_R returns to the control normoxic rate.

The immediate cardiovascular response of most fish to oxygen deprivation is bradycardia (slowing of heart rate; f_H) (Farrell, 2007), and the crucian carp is no exception in this regard (Vornanen 1994a; Vornanen and Toumennoro 1999; Stecyk *et al.*, 2004b)(Figure 9.5). Hypoxic bradycardia may benefit oxygen uptake at the gills, although supporting evidence is equivocal in other fish species, and is believed to afford a number of direct benefits to the heart, such as increasing the residence time of blood in the heart lumen, therefore allowing a greater time for oxygen diffusion, and improving cardiac contractility through the negative force-frequency effect (see Farrell, 2007 for review). However, the bradycardia in crucian carp is transitory. By 48 h of anoxia at 8°C, cardiac output (Q), f_H, cardiac power output (PO) and cardiac stroke volume all return to control normoxic levels where they remain stable for at least 5 days (Stecyk *et al.*, 2004b). Concurrently, ventral aortic blood pressure and peripheral resistance decrease by 30% and 40%, respectively, signifying vasodilation in peripheral tissues. The ability of crucian carp to maintain cardiovascular status at normoxic levels during prolonged anoxia is unique among the vertebrates, and the normal Q during anoxia has been proposed (Stecyk *et al.*, 2004b; Farrell and Stecyk, 2007) to be perhaps essential for shuttling ethanol to the gills for excretion and rapidly distributing glucose from the crucian carp's large liver glycogen stores (Holopainen and Hyvärinen, 1985; Hyvärinen *et al.*, 1985) to metabolically active tissues. Likewise, the peripheral vasodilation during anoxia

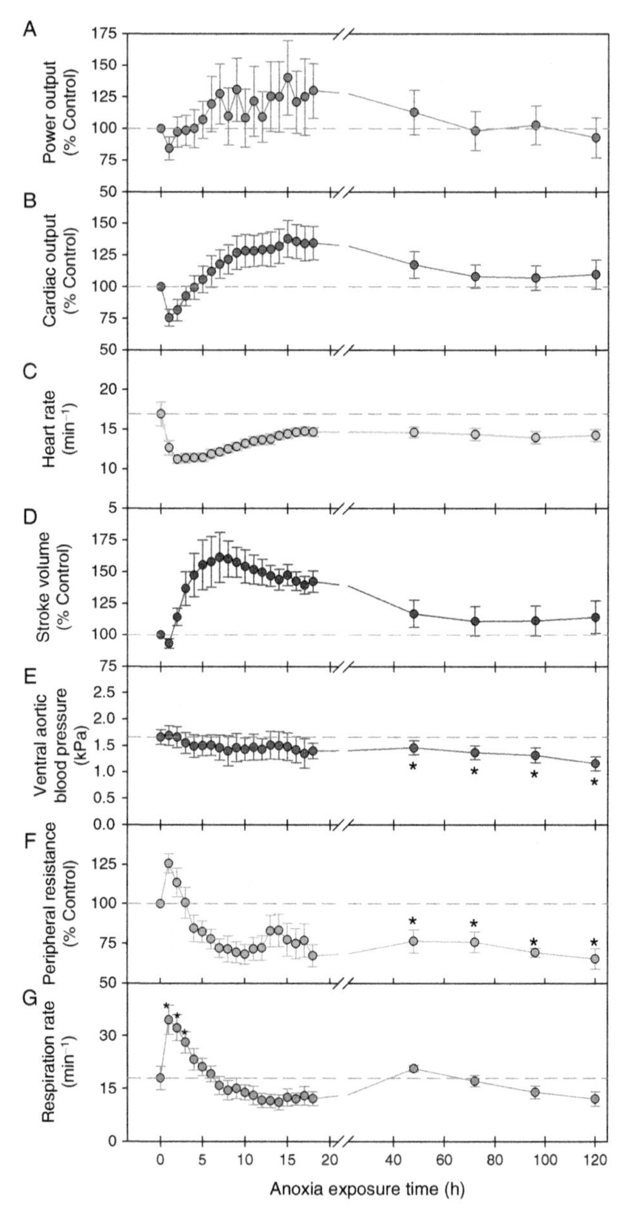

Fig. 9.5. Chronological changes of cardiorespiratory status in 8°C-acclimated crucian carp during 5 days of anoxia exposure. (A) Cardiac power output, (B) cardiac output, (C) heart rate, (D) stroke volume, (E) ventral aortic blood pressure, (F) peripheral resistance, and (G) respiration rate. Dashed lines indicate the control normoxic level for each measured parameter.

may reflect an increased perfusion of skeletal muscle where the conversion of lactate to ethanol takes place.

The cardiovascular responses of the crucian carp to anoxia are regulated by the autonomic nervous system. Cardiac inhibitory cholinergic and excitatory β-adrenergic tones, as well as a tonic α-adrenergic vasoconstriction have been revealed in anoxic crucian carp by injections of pharmacological blockers (Vornanen and Toumennoro, 1999; Stecyk *et al.*, 2004b). However, it remains to be determined if autonomic cardiovascular control persists beyond 5 days of anoxia and at colder acclimation temperatures. Thus, autonomic cardiovascular controls remain intact in the crucian carp during anoxia exposure at both warm- and cold-acclimation temperatures, consistent with its brain remaining functional (Lutz and Nilsson, 1997; Nilsson, 2001). In comparison, autonomic cardiovascular control is blunted in 5°C-acclimated anoxic turtles, which suppress brain activity in anoxia (Hicks and Farrell, 2000; Stecyk *et al.*, 2004a) (see Section 2.5).

2.4. The Heart in Anoxia

2.4.1. Balancing ATP Supply and Demand

The crucian carp's ability to maintain cardiac activity during anoxia may be possible because the cardiac ATP demand of their routine PO has been suggested to be below the maximum glycolytic potential (i.e., ATP supply) of ectothermic vertebrates (Farrell and Stecyk, 2007). In other words, during anoxia, anaerobic ATP generation is sufficient to power the heart to pump as in normoxia. This cardiac ATP conservation strategy comes about through a low arterial blood pressure compared to other teleosts, including another cyprinid fish, the common carp (*Cyprinus carpio*) (Farrell and Jones, 1992; Stecyk and Farrell, 2006). Also important for limiting cardiac activity and thus ATP demand is the low temperature during cold wintertime anoxia and inverse thermal acclimatization that prepares the heart for winter anoxia (see Section 3.2).

If it is that the crucian carp does not necessarily need to reduce cardiac ATP demands during anoxia to match energy use to supply, it is not surprising that anoxic channel arrest is not a strategy utilized by anoxic crucian carp. Channel arrest is a hypothesized energy-conserving strategy, first proposed by Lutz *et al.* (1985) and Hochachka (1986), in which the number and/or open probability of functional ion channels is reduced with either oxygen limitation or low temperature to diminish the metabolic cost of

---

For cardiac variables, significant differences (asterisks) are only indicated between normoxic control (time zero) and hours 48, 72, 96, and 120. For respiration (ventilation) rate, all significant differences from the normoxic control are indicated. Values are means ± S.E.M. from 6 to 18 fish. [Adapted from Stecyk *et al.* (2004b).] (See Color Insert.)

ion pumping to maintain ion gradients. For the heart, channel arrest could primarily be expected to involve decreases in Na^+ current and K^+ currents, which would then reduce demands by the Na^+/K^+-ATPase, and a decrease in Ca^{2+} current, which would reduce demands by Ca^{2+}-ATPases and Na^+/K^+-ATPase. Specifically, reducing Na^+ inflow during an action potential upstroke means that less Na^+ has to be extruded afterwards by Na^+/K^+-ATPase. Similarly, reduced Ca^{2+} influx means that less Na^+ is needed to extrude Ca^{2+} via the Na^+/Ca^{2+}-exchanger. K^+ channels represent a K^+ leakage pathway across the sarcolemma, allowing for continuous K^+ efflux during diastole and/or systole and placing continuous demands on the Na^+/K^+-ATPase (Roden *et al.*, 2002). Therefore, a down-regulation of K^+ channels would limit K^+ leakage and also lower ATP demand.

However, it has been discovered that crucian carp cardiac electrophysiology is largely unaffected by severe hypoxia and anoxia (reviewed by Stecyk *et al.*, 2008). Thus, the channel arrest hypothesis does not appear to be valid for the crucian carp heart. Specifically, the amplitude and kinetics (whole-cell conductance, single-channel conductance, and open probability) of ventricular inward rectifier K^+ current are unaffected after 4 weeks of severe hypoxia exposure (<0.4 mg O_2 L^{-1}) at 4°C (Paajanen and Vornanen, 2003). Even so, sarcolemmal Na^+/K^+-ATPase activity is reduced by one third within 4 days of anoxia exposure at 4°C as well as with the onset of hypoxic conditions in the natural environment (Aho and Vornanen, 1997). This change likely conserves ATP, but is at odds with the channel arrest hypothesis because it is not accompanied by a concomitant reduction of K^+ current. Further, the number of ventricular L-type Ca^{2+} channels and the density of Ca^{2+} current do not change with the seasonal decrease in water oxygen content (Vornanen and Paajanen, 2004). Therefore, cardiac down-regulation of L-type Ca^{2+} channels is not triggered by seasonal anoxia in the natural environment.

2.4.2. Cardiac Protection Against Anoxia

In the vertebrate heart, several molecular mechanisms have evolved to protect the heart against hypoxic or ischemic insults. These mechanisms include opening of the ATP-sensitive K^+ channels in sarcolemma and mitochondria of cardiac myocytes and release of adenosine, a general negative feedback regulator of cardiac function (Mubagwa and Flameng, 2001; Zingman *et al.*, 2007).

2.4.2.1. ATP-sensitive K^+ channels. ATP-sensitive K^+ channels are formed as heteromultimers of inwardly rectifying K^+ channel Kir6.2 and sulfonurea receptors SUR2A and provide cardiac protection during metabolic stress by virtue of the direct coupling between channel opening and myocyte energy balance. When cellular phosphorylation potential is high, ATP-sensitive K^+ channels are kept closed by ATP binding to the Kir6.2 proteins, whereas under metabolic stress the intrinsic ATPase of the SUR2A hydrolyses ATP and subsequent Mg-ADP binding to SUR2A overrides the inhibition of ATP on the Kir6.2 (Nichols, 2006). Binding of ATP to the Kir6.2 proteins occurs at two orders of magnitude lower ATP concentrations (Kd = 10–30 μM) than is the bulk ATP concentration (6–10 mM) of well-energized cells. Therefore, a creatine kinase and adenylate kinase systems are needed to couple energy state of the myocyte to the opening of ATP-sensitive K^+ channels in the diffusion-restricted subsarcolemmal space (Zingman *et al.*, 2007). ATP-sensitive K^+ current shortens the duration of cardiac AP, and thereby limits sarcolemmal Ca^{2+} influx and reduces cardiac contractility, which results in energy savings and possible rescue of cardiac myocytes from the anoxic cell death. ATP-sensitive K^+ channels also play a role in cardiac preconditioning, a process where short periods of ischemia provide cardiac protection against subsequent ischemic insults or reperfusion injury.

It might be expected that in anoxia-tolerant vertebrates, cardiac protection by ATP-sensitive K^+ channels would be particularly well developed in order to allow survival under severe hypoxic and anoxic stress. However, comparison of the cardiac ATP-sensitive K^+ current in several vertebrate species indicates almost the opposite. Density of the ATP-sensitive K^+ current is much larger in a mammal and a land-dwelling lizard than in an aquatic frog and two fish species (Paajanen and Vornanen, 2002). Among six vertebrate species spanning a wide range of hypoxia tolerance, crucian carp had the second smallest ATP-sensitive K^+ current, with only rainbow trout having a smaller current (Figure 9.6). The relatively small current is not, however, a limiting factor for cardiac protection, since it still clearly exceeds the densities of other outward K^+ currents and is sufficient to shrink the AP duration to almost nil (Paajanen and Vornanen, 2004). More importantly, relatively extreme measures are needed to open ATP-sensitive K^+ channels in isolated cardiac myocytes from crucian carp, especially in myocytes from cold-acclimated (4°C) fish. In warm-acclimated fish (18°C), inhibition of the aerobic metabolism with an oxygen scavenger, 0.1 mM $Na_2S_2O_3$, is sufficient to activate the ATP-sensitive K^+ current, whereas in ventricular myocytes of cold-acclimated fish, inhibition of both aerobic and anaerobic (5 mM iodoacetate) metabolism is necessary to induce the current. In excised crucian carp hearts, blocking of the aerobic metabolism with cyanide increases the duration of contraction and prolongs ventricular AP, indicating that ATP-sensitive K^+ channels are not opened in this multicellular preparation either (Vornanen and Tuomennoro, 1999). However, a small (15.1%) shortening of AP has been recorded in hypoxic ventricular myocytes of the warm-acclimated (21°C) goldfish, a close relative to crucian carp, and the opening was prevented by an inhibitor of nitric oxide synthesis, L-NAME (Cameron *et al.*, 2003; Chen *et al.*, 2005). Whether this represents a real difference in regulation of ATP-sensitive K^+ channels between closely related species or is caused by differences in experimental conditions, is not clear.

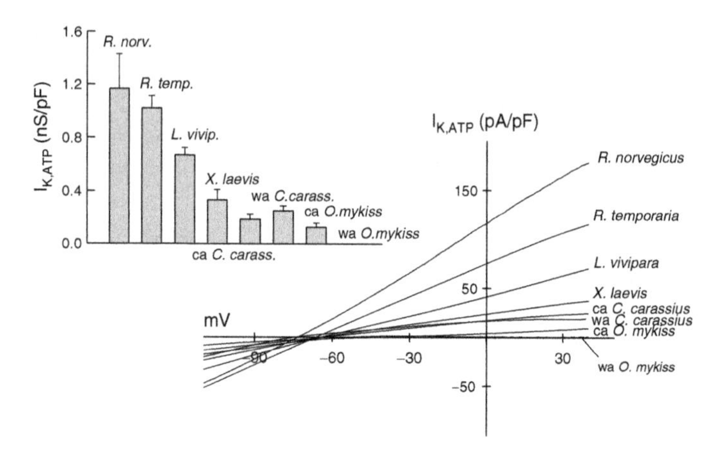

Fig. 9.6. The size of ATP-sensitive K^+ current ($I_{K,ATP}$) in ventricular myocytes of six vertebrate species. Right side of the figure shows linear current–voltage relations when metabolism of the cells was compromised with 0.1 mM $Na_2S_2O_3$ and/or 5 mM iodoacetate. Mean (± S.E.M.; $N = 4–27$) conductance of the ATP-sensitive K^+ current for the six species is shown on the left. The species are: rat (*Rattus norvegicus*), frog (*Rana temporaria*), clawed frog (*Xenopus laevis*), European common lizard (*Lacerta vivipara*), crucian carp (*Carassius carassius*) and rainbow trout (*Oncorhynchus mykiss*). The abbreviations wa and ca denote warm-acclimated and cold-acclimated, respectively. [Data from (Paajanen and Vornanen 2002).]

The high resistance of crucian carp ATP-sensitive K^+ channels to opening is even more surprising considering their regulatory properties. ATP-sensitive K^+ channels open when MgADP binding to the SUR2A overrides ATP-dependent inhibition of the Kir6.2 channels. In crucian carp and goldfish hearts, ATP-sensitive K^+ channels have a very low affinity to ATP (K_d = 1.35–1.85 mM in crucian carp), and therefore relatively small changes in cellular ATP concentration would be expected to remove ATP-dependent inhibition without SUR2A ATPase activity (Ganim *et al.*, 1998; Paajanen and Vornanen, 2004). The general ability of anoxia-tolerant organisms to maintain high $[ATP]_i$ in anoxia does not explain the resistance to opening, since intracellular perfusion of myocytes with a ATP-free solution cannot open the ATP-sensitive K^+ channels in crucian carp ventricular myocytes. These findings suggest that bulk ATP concentration alone is insufficient to signal the opening of ATP-sensitive K^+ channels even though ATP affinity of the channel is low. Current evidence suggests that the ATP-sensitive K^+ channels of the crucian carp heart are not primarily involved in anoxia protection, but rather associated with cardioprotection against severe heat stress (Ganim *et al.*, 1998; Paajanen and Vornanen, 2004).

2.4.2.2. Adenosine. In many animals, various states of energy deficiency, including hypoxia, quickly results in increased levels of adenosine as a result of a net breakdown of phosphorylated adenosines (ATP, ADP, and AMP), and it is well established that both the rise in adenosine and the fall in ATP can initiate an array of mechanisms aimed at restoring energy levels (Lipton, 1999). In mammals, adenosine is able to balance cardiac function under energy-limited conditions by reducing work load of the heart and improving glycolytic energy supply through stimulation of cellular glucose uptake (Mubagwa *et al.*, 1996). Adenosine is released in the circulation when oxygen demand of the cardiac muscle exceeds circulatory supply of oxygen, e.g., in hypoxia and anoxia. Adenosine improves blood flow in the coronary vessels by vasodilatation, reduces the velocity of impulse conduction over the heart, and weakens contractile force of atrial and ventricular myocardia (Mubagwa *et al.*, 1996). By these actions, adenosine is thought to rebalance energy demand and supply, thereby providing protection for the heart during hypoxia.

The significance of adenosinergic control of the heart under oxygen deficient conditions is still poorly documented in fish and especially the mechanisms of action are largely unexplored (Rotmensch *et al.*, 1981; Meghji and Burnstock, 1984; Lennard and Huddart, 1989; Sundin *et al.*, 1999; MacCormack and Driedzic, 2004). In the crucian carp heart, adenosine is a weak modulator of cardiac function and unlikely to have any major role in anoxia protection of the heart (Vornanen and Tuomennoro, 1999; Stecyk *et al.*, 2007). Adenosine at the concentration of 0.1 mM or lower has no effect on contractile function of isolated atrial and ventricular muscle of the crucian carp heart, and 1 mM adenosine causes a small increase of force in ventricular muscle and a small decrease of force in atrial muscle (Vornanen and Tuomennoro, 1999). Electrophysiological effects of adenosine are often mediated by activation of the ligand-gated inward rectifier K^+ channels, which generate an outward K^+ current (I_{KAdo}) that shortens the duration of cardiac AP, especially in atrial myocytes (Belardinelli *et al.*, 1995). Adenosine neither affects cardiac AP nor activates I_{KAdo} in crucian carp atrial and ventricular myocytes at a concentration of 0.1 mM, which is an effective dose in the hypoxia-sensitive trout heart (Vornanen and Tuomennoro, 1999; Aho and Vornanen, 2002). Consistent with the findings from multicellular cardiac preparations and cardiac myocytes, intra-arterial injection of aminophylline, an adenosine receptor blocker, to 5-day anoxic crucian carp did not change their cardiovascular status (Stecyk *et al.*, 2007).

2.5. The Brain in Anoxia

2.5.1. Suppression of Brain Activity

A fundamental factor in the anoxic survival strategy of both the crucian carp and the freshwater turtles is their remarkable ability to maintain brain ATP levels when exposed to anoxia (Figure 9.7) (Lutz *et al.*, 1984; Johansson

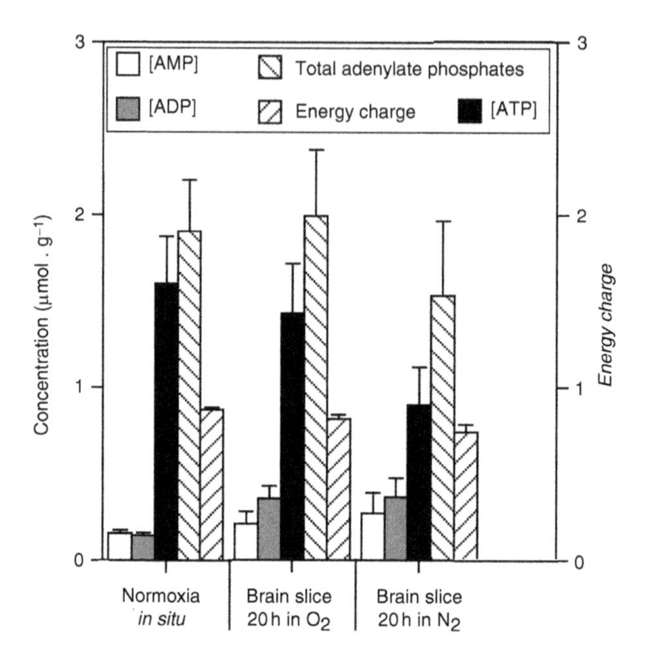

Fig. 9.7. Energy charge is maintained in the anoxic crucian carp brain. The graph shows levels of ATP, ADP, AMP, total adenylate phosphates (ATP+ADP+AMP) and energy charge (EC) in crucian carp brain and in brain slices (telencephalon) kept for 20 h in normoxic or anoxic Ringer at 12°C. EC = ([ATP] + ½ [ADP])/([ATP] + [ADP] + [AMP]). Values are means ± S.E.M. from 6 fish. [Reproduced from (Johansson *et al.*, 1995).]

and Nilsson, 1995; Johansson *et al.*, 1995), whereby many of the detrimental processes initiated by anoxia in other vertebrate brains are avoided. However, the crucian carp and the turtles seem to differ in the way they achieve this. For one, the turtles become virtually comatose in anoxia, with a brain that is nearly completely electrically silent (Fernandes *et al.*, 1997). The accompanying deep metabolic depression is thought to be the major factor enabling the turtle brain to maintain its energy charge (Nilsson and Lutz, 2004; Bickler and Buck, 2007). By contrast, the crucian carp must uphold much of its brain functions since it remains physically active in anoxia (Nilsson *et al.*, 1993b; Nilsson, 2001), although the activity level is reduced and senses like hearing and vision have been found to be suppressed in goldfish and crucian carp (Suzue *et al.*, 1987; Johansson *et al.*, 1997). Also, whole body metabolism, measured as heat production, is more suppressed in the turtle than in the crucian carp. Whole-body metabolic rate of warm-acclimated turtles (20°C–24°C) is depressed to 15–18% of the normoxic metabolic rate during prolonged anoxia exposure (Jackson, 1968; Herbert and Jackson, 1985). For cold-acclimated turtles

(3°C), the decreases in metabolic rate is even greater, reaching values less than 10% of the normoxic metabolic rate at 12 weeks of anoxia exposure (Herbert and Jackson, 1985). No such measurements have been done in crucian carp, but in the closely related goldfish there is only a 70% reduction in heat production during 3 h of anoxia at 20°C (Van Waversveld *et al.*, 1989). With regard to the crucian carp brain, microcalorimetric measurements on brain slices have indicated a modest 30% reduction in ATP turnover during anoxia, and that the anoxic crucian carp brain has to increase its glycolytic rate 3-fold to defend its ATP levels (Johansson *et al.*, 1995).

Measurements of the Na^+/K^+-ATPase activity in the brain of turtles and crucian carp also point toward a more severe energetic depression in turtles. Na^+/K^+-ATPase is the major ATP consumer in the brain and is responsible for establishing the electrochemical gradients of Na^+ and K^+ across the plasma membrane, which are necessary for negative resting membrane potential, electrical excitability, neurotransmitter uptake, and osmotic balance of neuronal cells. For example, in the mammalian brain, 50–80% of the total energy budget is devoted to the Na^+/K^+-ATPase (Erecinska *et al.*, 2004). Therefore, a depression of Na^+/K^+-ATPase activity would lead to ATP conservation. After 24 h of anoxia at 20°C, a 30% fall in the activity of Na^+/K^+-ATPase is seen in major parts of turtle brain (Figure 9.8) (Hylland *et al.*, 1997). Conversely, anoxia exposure does not decrease the number of Na^+/K^+-ATPase alpha subunits or the activity of this enzyme in the crucian carp brain (Figure 9.8) (Hylland *et al.*, 1997; Vornanen and Paajanen, 2006). In support of these findings, determinations of [^{3}H]ouabain binding and Na^+/K^+-ATPase activity in fish caught directly from the wild around the year have not indicated a depression of the Na^+/K^+-ATPase activity during the anoxic season (Vornanen and Paajanen, 2006).

Another indication of maintained brain activity in anoxia is the sustained doubling in brain blood flow, starting within the first minutes of anoxia and lasting at least over a 6 h period of anoxia at 10°C (Nilsson *et al.*, 1994). The increased brain blood flow is probably needed to shuttle glucose to, and lactate from, the brain to support the increased glycolytic activity. In turtle brain, blood flow also doubles initially in anoxia but falls back to preanoxic levels within the first hours of anoxia (Hylland *et al.*, 1994; Stecyk *et al.*, 2004a), which probably corresponds to the entrance into deep neural suppression and a reduced need for glucose supply.

2.5.2. Mechanisms of Neural Depression in Anoxia

2.5.2.1. Ion channels and protein synthesis. The differences displayed by crucian carp and turtles in physical and metabolic activity are also reflected in the mechanisms employed to suppress brain energy demands in anoxia. A major difference appears to be the utilization of channel arrest. This

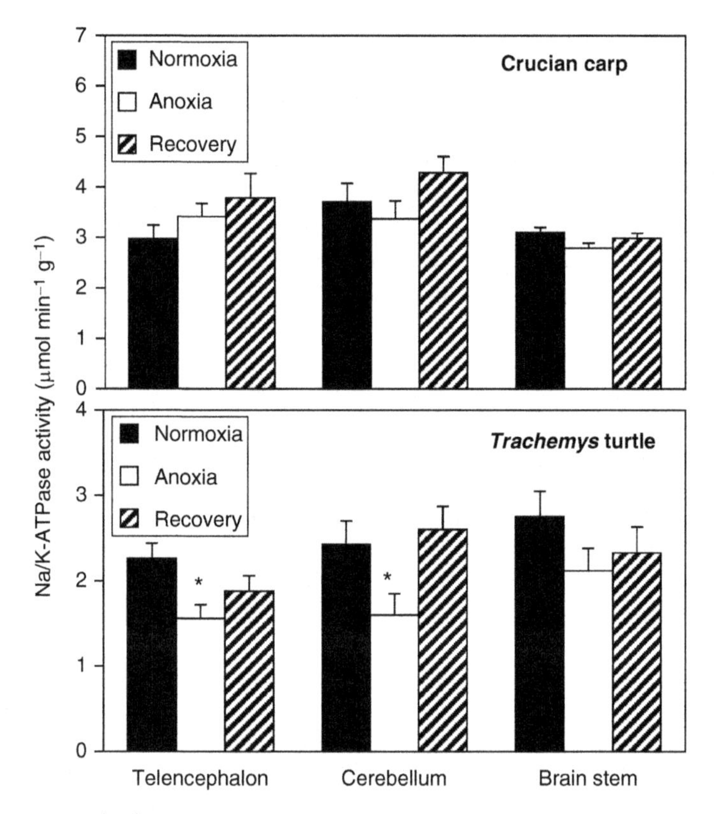

Fig. 9.8. Brain Na^+/K^+-ATPase activity is maintained in anoxic crucian carp but not turtle. Both species were exposed to 24 h of anoxia at 20°C followed by 24 h of reoxygenation. Values are means ± S.E.M. from 6 to 7 animals. [Reproduced from (Hylland *et al.*, 1997).]

mechanism appears to play a major role in suppressing neural excitability and ATP use in turtles (Bickler and Buck, 2007), but, like for the crucian carp heart (see Section 2.4), there is so far very little evidence that brain channel arrest plays any major role in the anoxia tolerance of crucian carp. In anesthetized crucian carp, brain K^+ permeability appears to be unaffected by anoxia, as measured by the efflux of K^+ to the extracellular space when Na^+/K^+-ATPase is blocked with ouabain (Johansson and Nilsson, 1995). Crucian carp brain slices also display a similar lack of change in Ca^{2+} permeability, at least during the initial hours of anoxia (Nilsson, 2001). However, there are some indications of a limited "channel arrest" in crucian carp brain: measurements of the expression of excitatory neurotransmitter receptors show that most receptors are relatively unaffected by anoxia, but that mRNA levels of certain NMDA receptor subunits, including the ubiquitous NR1 subunit, are depressed by about 50% after a week of anoxia at ~10°C (Ellefsen *et al.*, 2008). Moreover, a recent study utilizing the whole-cell patch clamp technique on telencephalic brain slices from goldfish show that acute (40 min) anoxia exposure causes a 40–50 % fall in the NMDA receptor activity (Wilkie *et al.*, 2008). The NMDA receptor is a major glutamatergic receptor with a large conductivity for Ca^{2+}, and such changes could function to reduce neural excitability. Indeed, there is good evidence for a reduced NMDA receptor function in the turtle brain (Bickler and Buck, 2007).

However, studies of ion channel gene expression in crucian carp have also revealed that mRNA levels of various subunits of voltage-gated Na^+ and Ca^{2+} channels are maintained, or even increased, in the brain of crucian carp kept in anoxia for a week (Ellefsen *et al.*, 2008). These voltage-gated channels are responsible for the generation of action potentials and neurotransmitter release and could therefore be important targets for a channel arrest strategy involving reduced gene expression, but apparently such a strategy is not utilized by the crucian carp.

Like for the heart, a possible explanation for the absence of any major channel arrest in crucian carp brain is offered by the low metabolic rate induced by the low temperature of the anoxic season (see Section 3.3), combined with the fact that the ethanol-producing pathway relieves the crucian carp of the problem of having to minimize lactate accumulation.

With regard to protein synthesis in crucian carp, measurements utilizing [^{3}H] phenylalanine incorporation have revealed maintained rates in brain, as opposed to liver (>95% suppression), muscle (50% suppression), and heart (50% suppression) (Figure 9.9) (Smith *et al.*, 1996). This unequal suppression of protein synthesis makes sense since only about 3% of the energy used by the crucian carp brain goes to protein synthesis, so not much would be saved from shutting it down. By contrast, the extreme suppression of protein synthesis in the crucian carp liver could make a significant contribution to energy savings on the whole body level, as protein synthesis can make up more than 50% of the energy use of the fish liver (Smith *et al.*, 1996). However, apparently the freshwater turtle has again turned to a more radical strategy as its rate of protein synthesis in all tissues, including brain, is virtually at a halt in anoxia (Fraser *et al.*, 2001).

2.5.2.2. Neurotransmitters and neuromodulators. Presently, most evidence points toward neurotransmitters and neuromodulators as responsible for suppressing the electric activity of the anoxic crucian carp brain. The major inhibitory neurotransmitter in the vertebrate brain is gamma-aminobutyric acid (GABA), and microdialysis studies on anesthetized crucian carp show that extracellular [GABA] rises in the brain (telencephalon) during anoxia (Hylland and Nilsson, 1999). At the same time, extracellular glutamate levels remain low (Hylland and Nilsson, 1999), which of course makes the

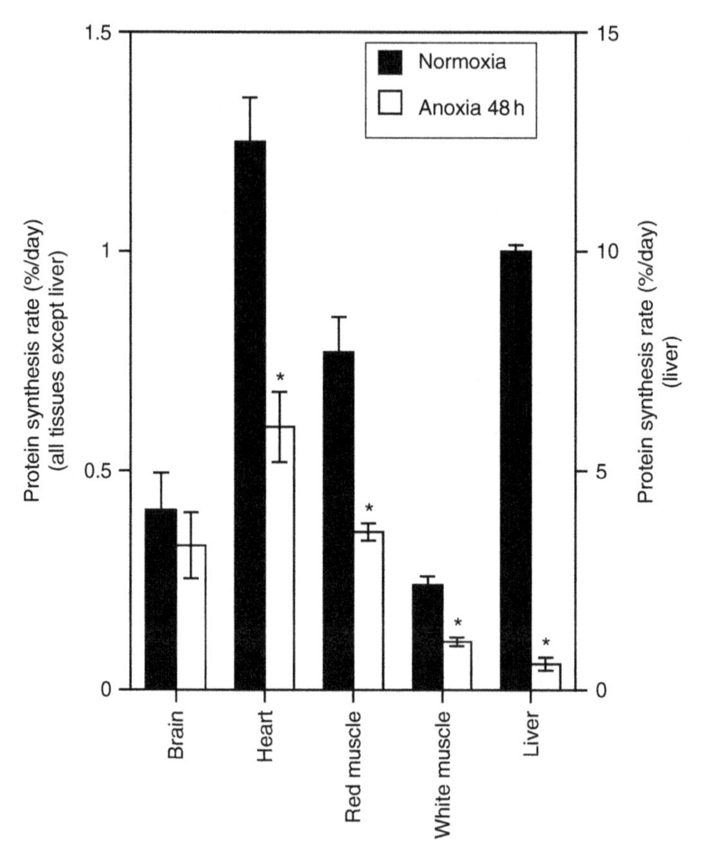

Fig. 9.9. Rates of protein synthesis *in vivo* in various tissues of crucian carp exposed to anoxia for 48 h at 9°C. Values are means ± S.E.M. from 12 to 16 fish. [Data from (Smith *et al.*, 1996).]

anoxic crucian carp brain strikingly different from anoxic mammalian brains, and also different from the anoxic rainbow trout brain (Hylland *et al.*, 1995), which all show substantial rises in extracellular glutamate. However, compared to the massive (80-fold) increase in extracellular [GABA] seen in the anoxic turtle brain (Nilsson and Lutz, 1991), the rise in extracellular [GABA] in the crucian carp brain (telencephalon) is relatively modest: on average it is doubled after 6 h of anoxia (Figure 9.10) (Hylland and Nilsson, 1999). There is also a considerable individual variation in the extracellular [GABA] rise during anoxia, varying from no change to a 6-fold increase, which indicates that the release of GABA is fine-tuned to the need for neural suppression.

The most direct evidence for a role of GABA in metabolic depression comes from studies using inhibitors of the GABA synthesizing enzyme glutamate decarboxylase (isoniazid or 3-mercaptopropionic acid), or a

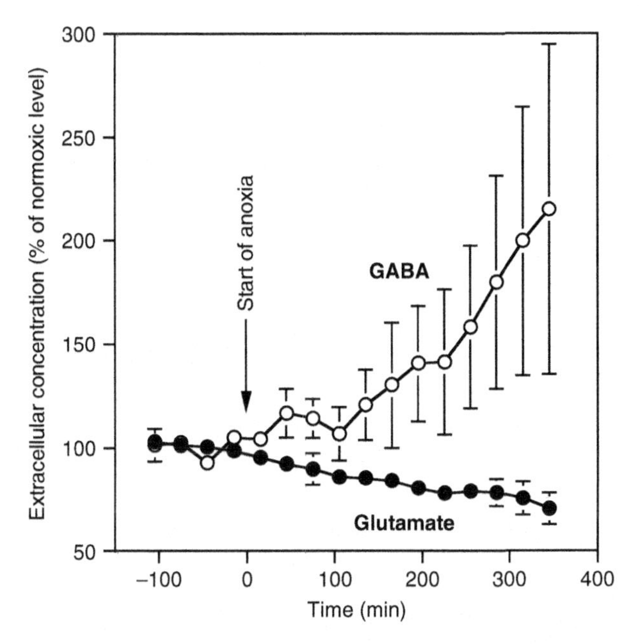

Fig. 9.10. Changes in the extracellular brain levels of GABA and glutamate in crucian carp brain, measured with a microdialysis probe placed in the telencephalon of anesthetized fish kept at 10°C. Values are means ± S.E.M. from 6 to 8 fish. [Data from (Hylland and Nilsson, 1999).]

blocker of $GABA_A$ receptors (securinine). Such manipulations make crucian carp release up to three times more ethanol to the water during anoxia (while normoxic oxygen consumption is unaffected), suggesting a profound inhibition of metabolic depression (Figure 9.11) (Nilsson, 1992).

The crucian carp may not only utilize GABA as a metabolic depressant under normal anoxic conditions, but may also employ a more massive GABA release as a last line of defense during severe neural energy deficiency. In the crucian carp brain (telencephalon), the potential for releasing GABA appears to be higher than for releasing glutamate. By running a high-[K^+] Ringer through the microdialysis probe, to depolarize the surrounding tissue, Hylland and Nilsson (1999) found a 14-fold increase in extracellular [GABA], while extracellular [glutamate] was barely doubled. Similarly, when neural ATP levels are forced to fall by superfusing the crucian carp telencephalon with the glycolytic inhibitor iodoacetate, the resultant increase in extracellular [GABA] was found to be both faster and more massive (a 10-fold increase after 30 min) than that of extracellular [glutamate] (a 3-fold increase after 2 h) (Hylland and Nilsson, 1999).

There is a close metabolic interrelation between GABA and glutamate, which is interesting from an anoxia-defense perspective. GABA is

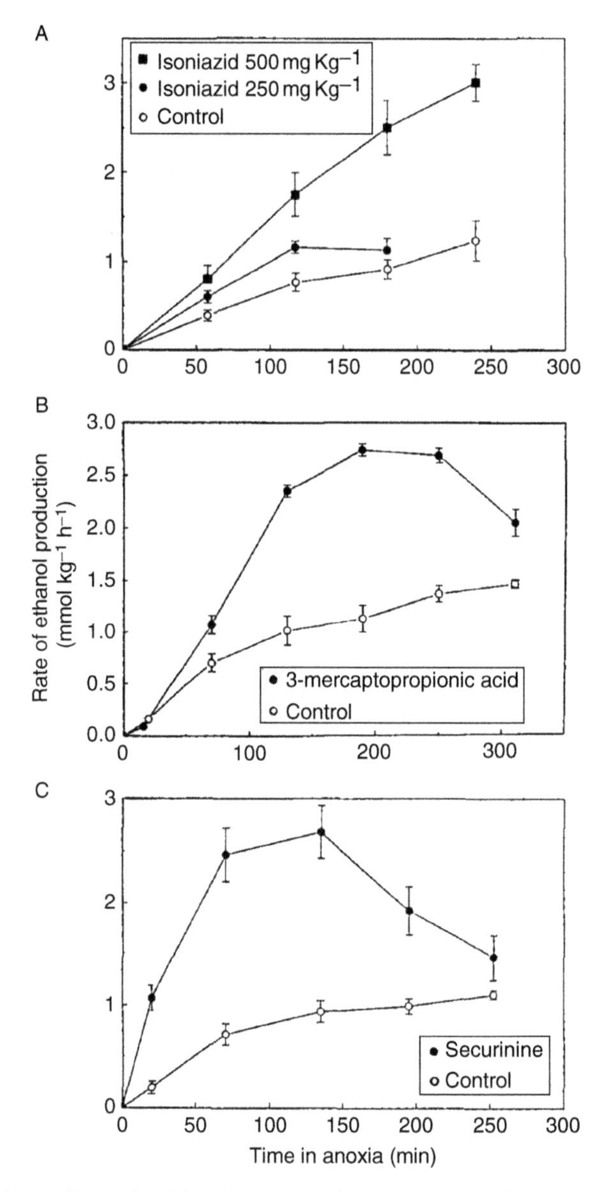

Fig. 9.11. Evidence for a role of GABA in controlling anoxic metabolic rate. The graphs show the effects of anti-GABAergic agents on the rate of ethanol release to the water by anoxic crucian carp kept at 18°C. Isoniazid (250–500 mg kg^{-1} i.p.) and 3-mercaptopropionic acid (200 mg kg^{-1} i.p.) both block the GABA-synthesizing enzyme glutamate decarboxylase. 3-mercaptopropionic acid also inhibits neuronal GABA release. Securinine (20 mg kg^{-1} i.p.) blocks $GABA_A$ receptors. Values are means ± S.E.M. from 4 to 6 fish. [Reproduced from (Nilsson, 1992).]

synthesized from glutamate in a single oxygen-independent step, catalyzed by glutamate decarboxylase. In contrast, both the synthesis of glutamate and the breakdown of GABA depend on oxygen-dependent processes that will stop in anoxia. As a result, brain tissue shows a steady increase in [GABA] and a corresponding fall in [glutamate] during anoxia (Siesjö, 1978; Nilsson and Lutz, 1993). Interestingly, GABA is the major inhibitory neurotransmitter and glutamate the major excitatory neurotransmitter in all vertebrates as well as many invertebrates (Gerschenfeld, 1973; Usherwood, 1978; Koopowitz and Keenan, 1982; McGeer and McGeer, 1989; Restifo and White, 1990). It has been hypothesized that hypoxia could be the underlying selection pressure that is responsible for maintaining GABA inhibitory and glutamate excitatory throughout animal evolution, because it provides a system where a fall in oxygen will automatically make the inhibitory neurotransmitter levels rise and the excitatory fall, thereby inducing and maintaining hypoxic metabolic depression (Nilsson and Lutz, 1993).

In the goldfish, which, as mentioned, appears to be somewhat less well adapted to anoxia than the crucian carp (a possible side effect of long domestication), elevated extracellular [glutamate] has been seen in brain during anoxia, probably as a consequence of a poorly maintained ATP level (Van Ginneken *et al.*, 1996). In this species, a release of glutamate during energy deficiency may initiate protective mechanisms mediated by one class of glutamate receptors, the group II metabotropic glutamate receptors (Poli *et al.*, 2003).

2.5.2.3. Adenosine. The role of adenosine in protecting the anoxic brain has also been investigated in anoxia-tolerant vertebrates. In the turtle (*Trachemys*) brain there is an almost immediate, substantial rise in extracellular adenosine during the initial phase of anoxia, linked to the simultaneous fall in ATP (Nilsson and Lutz, 1992). However, unlike the turtle, microdialysis experiments have so far failed to detect an increase in extracellular adenosine in the anoxic crucian carp brain (P. Hylland and G. E. Nilsson, unpublished data), but that may reflect limitations of the microdialysis method, because other evidence points at a role for adenosine in both stimulating brain blood flow and reducing metabolic rate in crucian carp during anoxia. First, the sustained doubling of cerebral blood flow in crucian carp is probably adenosine mediated since it can be blocked by superfusing the brain with the adenosine receptor blocker aminophylline (Nilsson *et al.*, 1994) (Also, the temporary elevation in cerebral blood flow seen in turtles appears to be adenosine mediated; Hylland *et al.*, 1994). Second, blocking adenosine receptors with aminophylline in anoxic crucian carp causes a 3-fold increase in the rate of ethanol release to the water (while it is without effect on normoxic oxygen consumption), suggesting that adenosine causes a significant inhibition of metabolic rate in anoxia (Nilsson, 1991). It could also be mentioned that in goldfish hepatocytes, adenosine has a powerful depressant effect on both protein synthesis and Na^+/K^+-ATPase (Krumschnabel *et al.*, 2000), and that adenosine suppresses K^+ stimulated Ca^{2+}-dependent glutamate release in goldfish cerebellar slices (Rosati *et al.*, 1995).

3. SEASONALITY OF CRUCIAN CARP PHYSIOLOGY: PREPARING FOR WINTER ANOXIA

Although anoxia tolerance of the crucian carp greatly surpasses that of most other vertebrates, it is not a fully fixed trait, but includes an inducible component that varies according to the season (Piironen and Holopainen, 1986). Experiments on seasonally acclimatized fish indicate that anoxia tolerance of adult crucian carp follows an exponential dependence on temperature and extrapolation of the curve to zero temperature suggests a theoretical maximum of 235 days for anoxia tolerance (see Figure 9.1). There are two reasons for the seasonal difference in anoxic survival time. Firstly, since metabolic rate of organisms increases with increasing temperature, anoxic survival time of ectothermic animals correlates inversely with ambient temperature. At low temperature, energy stores and essential nutrients last longer and there is less production and accumulation of toxic end-products of metabolism. Secondly, crucian carp metabolism and organ function is altered seasonally to reflect the three main phases in the year of a crucian carp: growth and multiple breeding episodes in early summer, accumulation of energy reserves for winter in late summer, and anoxic overwintering (Holopainen *et al.*, 1997). Thus, crucian carp physiology becomes primed to be beneficial for the survival of winter anoxia.

In the habitat of crucian carp, anoxia is a regular and well predictable seasonal condition that is accompanied by several environmental cues, most notably temperature (Nagell and Brittain, 1977; Vornanen and Paajanen, 2004). In fact, changes in oxygen content and temperature occur almost in parallel (Vornanen and Paajanen, 2004; see Figure 9.2). Thus, ambient temperature can function as an environmental cue that entrains the fish for winter anoxia and reoxygenation in spring.

3.1. Seasonal Changes of Glycogen Stores

In the absence of molecular oxygen, fats are unsuitable for energy production and the animal must rely on the catabolism of carbohydrate reserves in the body. Consistent with this, lipid content of the crucian carp is low (ca. 2% of wet weight) and the lipid metabolism of the tissues (e.g., liver) is active only for a short period in summer from May to September (Blazka, 1958; Piironen and Holopainen, 1986; Lind, 1992). Further, carp increase the size of glycogen stores in the body in preparation for winter anoxia.

3.1.1. Liver and Muscle Glycogen

In the vertebrate body, the liver and the skeletal muscle have the largest glycogen reserves. Glucose from liver glycogen can be released to the blood and exploited elsewhere in the body. In muscle tissue, however, glycogen is for local use since muscle lacks the glucose-6-phosphatase needed to release glucose into the circulation. The liver of crucian carp has an exceptional ability to store glycogen, which appears in enormous seasonal changes in the size of liver and glycogen concentration of the liver tissue (Figure 9.12). In winter, the liver constitutes 12–15% of the whole body mass and glycogen concentration of the liver can be 35% of the liver wet weight or 4.5% of the fish body mass (Hyvärinen *et al.*, 1985; Holopainen and Hyvärinen, 1985). In July, when glycogen stores are smallest, the liver mass and glycogen concentration of the liver are 1.5% and 2%, respectively. Thus, there is about a 15-fold seasonal variation in liver glycogen content. White myotomal muscle has similar annual glycogen dynamics as the liver, but the stores are maximally about 4% of the muscle wet weight (Hyvärinen *et al.*, 1985). In winter, liver and muscle glycogen together form about 6% of the fish body mass.

Crucian carp begin to prepare for hypoxia/anoxia in late July by accumulating glycogen deposits in the liver. The increase in liver size continues as long as the fish forage. When water temperature drops, and when the hypoxic period sets in, crucian carp begin a fast, which can last almost half a year from November to May (Penttinen and Holopainen, 1992). It is not known whether depletion of glycogen reserves causes mortality in crucian carp

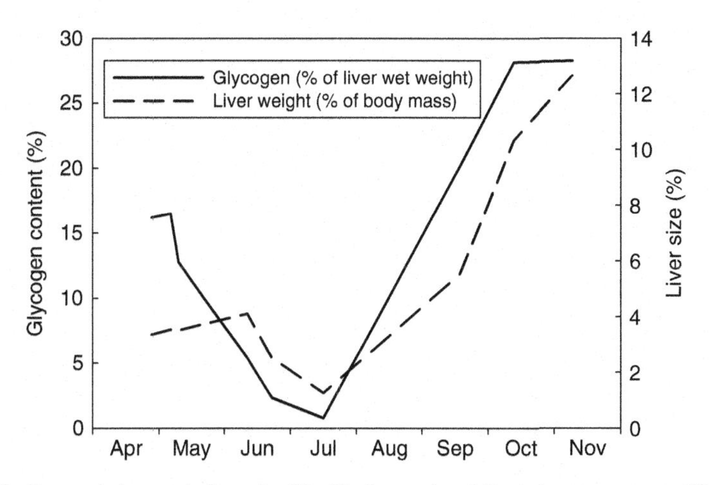

Fig. 9.12. Seasonal changes in liver size (% of body mass) and liver glycogen content (% of liver wet weight) of crucian carp. [Data taken from (Holopainen and Hyvärinen, 1985).]

populations. It seems that the total glycogen stores of the crucian carp body exceed the metabolic demands of the anoxic fish, as significant amounts of liver glycogen (about 20%) are still present at the end of April when the hypoxic/anoxic period is over. The remaining glycogen stores may represent a surplus of the safety margin, which is subsequently used in preparation for breeding. Indeed, the activity of liver glycogen phosphorylase, a glycogen hydrolyzing enzyme, peaks in May during the maturation of gonads (Holopainen and Hyvärinen, 1985). Thus, the glycogen stores can also be important for successful breeding. Still, in an experiment where crucian carp were starved for 18 days (at 8°C, before the breeding season), they did not utilize the liver glycogen store, apparently saving it for anoxic periods (Nilsson, 1990b).

3.1.2. Brain Glycogen

Glycogen stores of the vertebrate brain are usually small giving an impression that the brain tissue is unable to store significant amounts of carbohydrates. High concentrations of glycogen have been found in the lamprey (*Petromyzon marinus* L.) brain (Rovainen, 1970), indicating that brain glycogen might be an important energy source at least in some animal species. Considering the excellent anoxia tolerance of crucian carp, it is not surprising that their brains store more glycogen than mammalian brains (Schmidt and Wegener, 1988), but that the concentration of brain glycogen is comparable to the glycogen stores of the white skeletal muscle is quite impressive (Vornanen and Paajanen, 2006). The brain glycogen content of the winter-acclimatized crucian carp greatly surpasses brain glycogen stores reported for any other vertebrate species. In frogs and reptiles, including anoxia-tolerant turtles, the concentration of brain glycogen varies between 8 and 18 μmol/g, while in crucian carp the winter average is 204 μmol/g (3.3% of wet brain mass) (Table 9.1). The lamprey brain approaches the crucian carp brain with a value of 137 μmol/g (Rovainen, 1970). Like in its other tissues, the glycogen stores in the crucian carp brain are smallest in June to July and reach a maximum in December to February (Figure 9.13). The difference in brain glycogen content between summer and winter, like the liver, is about 15-fold (Vornanen and Paajanen, 2006). The large seasonal variation in the brain glycogen content suggests that glycogen is important for the anoxic survival of the brain. It is unclear, however, whether brain glycogen functions as an immediate energy source when anoxia sets in or whether it is an emergency reservoir that is recruited under prolonged anoxia if the circulation fails to provide sufficient glucose to meet the brain's demands. In this respect, it is notable that under natural conditions brain glycogen is not used during moderately hypoxic periods but only under total oxygen shortage (Vornanen and Paajanen, 2006). As the Na^+/K^+-ATPase is

Table 9.1

Glycogen concentration (glucosyl units, μmol/g wet weight) in brain, heart and liver of crucian carp in comparison with those of other vertebrates

Species	Brain	Heart	Liver	References
Crucian carp, *Carassius carassius*, in summer	13	18-86	123	Vornanen & Paajanen, 2004 and 2006; Vornanen, 1994; Hyvärinen *et al.*, 1985
Crucian carp, *Carassius carassius*, in winter	204	493	2160	Vornanen and Paajanen, 2004 and 2006; Vornanen, 1994; Hyvärinen *et al.*, 1985
Goldfish, *Carassius auratus*	13-20	142	ca 800	McDougal *et al.*, 1968; Schmidt and Wegener, 1988; Merrick, 1954
Rainbow trout, *Oncorhynchus mykiss*	0.5	25-60	110	Polakof *et al.*, 2008; Bernier *et al.*, 1996; Gesser, 2002; French *et al.*, 1981
Frog, *Rana ridibunda*	16		156	L'vova and Plotnikov, 1978
Frog, *Rana temporaria*		315	285	Donohoe and Boutilier, 1998
Turtle, *Chrysemys picta belli*	ca 8	284-413	ca 120	Packard and Packard, 2005; Daw *et al.*, 1967; Beall and Privitera, 1973
Turtle, *Trachemys (Pseudemys) scripta*	18	80-180	300-600	McDougal *et al.*, 1968; Warren *et al.*, 2006; Warren and Jackson, 2008
Rat, *Rattus norvegicus*	2-12	15	68	L'vova and Plotnikov, 1978; Støttrup *et al.*, 2006; Cruz and Dienel, 2002

References: Beall, R. J., and Privitera, C. A. (1973). Effects of cold exposure on cardiac metabolism of the turtle Pseudemys (Chrysemys) picta. *Am. J. Physiol.* **224**, 435–441; Bernier, N. J., Fuentes, J., and Randall, D. J. (1996). Adenosine receptor blockade and hypoxia-tolerance in rainbow trout and pacific hagfish. II. Effects of plasma catecholamines and erythrocytes. *J. Exp. Biol.* **199**, 497–507; Cruz, N. F., and Dienel, G. A. (2002). High glycogen levels in brains of rats with minimal environmental stimuli: Implications for metabolic contributions of working astrocytes. *J. Cereb. Blood Flow Metab.* **22**, 1476–1489; Donohoe, P. H., and Boutilier, R. G. (1998). The protective effects of metabolic rate depression in hypoxic cold submerged frogs. *Resp. Physiol.* **111**, 325–336; French, C. J., Mommsen, T. P., and Hochachka, P. W. (1981). Amino acid utilisation in isolated hepatocytes from rainbow trout. *Eur. J. Biochem.* **113**, 311–317; Gesser, H. (2002) Mechanical performance and glycolytic requirement in trout ventricular muscle. *J. Exp. Biol.* **293**, 360–367; Hyvärinen, H., Holopainen, I. J., and Piironen, J. (1985). Anaerobic wintering of crucian carp (*Carassius carassius* L.) - I. Annual dynamics of glycogen reserves in nature. *Comp. Biochem. Physiol.* **82A**, 797–803; L'vova, S. P., and Plotnikov, V. P. (1978). Content of glycogen and glucose in tissues of the frog and some reptiles. *Zh. Evol. Biokhim. Fiziol.* **14**, 400–402; Mandic, M., *et al.* (2008). Metabolic recovery in goldfish: A comparison of recovery from severe hypoxia exposure and exhaustive exercise. Comp.Biochem.Physiol.C (in press); McDougal, D. B., Jr., *et al.* (1968). The effects of anoxia upon energy sources and selected metabolic intermediates in the brains of fish, frog and turtle. *J. Neurochem.* **15**, 577–588; Merrick, A. W. (1954). Cardiac glycogen following fulminating anoxia. *Am. J. Physiol.* **176**, 83–85; Packard, M. J., and Packard, G. C. (2005). Patterns of variation in glycogen, free glucose and lactate in organs of supercooled hatchling painted turtles (*Chrysemys picta*). *J. Exp. Biol.* **208**, 3169–3176; Polakof, S., Mígues, J. M., and Soengas, J. L. (2008). Changes in food intake and glucosensing function of hypothalamus and hindbrain in rainbow trout subjected to hyperglycemic or hypoglycemic conditions. *J. Comp. Physiol. A.* **194**, 829–839; Schmidt, H., and Wegener, G. (1988). Glycogen phosphorylase in fish brain (*Carassius carassius*) during hypoxia. *Biochem. Soc. Trans.* **16**, 621–622; Støttrup, N. B., *et al.* (2006). L-glutamate and glutamate improve haemodynamic function and restore myocardial glycogen content during postischaemic reperfusion: A radioactive tracer study in the rat isolated heart. *Clin. Exp. Pharm. Physiol.* **33**, 1099–1103; Vornanen, M. (1994). Seasonal adaptation of crucian carp (*Carassius carassius* L.) heart: Glycogen stores and lactate dehydrogenase activity. *Can. J. Zool.* **72**, 433–442; Vornanen, M., and Paajanen, V. (2004). Seasonality of dihydropyridine receptor binding in the heart of an anoxia-tolerant vertebrate, the crucian carp (*Carassius carassius* L.). *Am. J Physiol.* **287**, R1263–R1269; Vornanen, M., and Paajanen, V. (2006). Seasonal changes in glycogen content and Na+-K+-ATPase activity in the brain of crucian carp. *Am. J. Physiol.* **291**, R1482–R1489; Warren, D., and Jackson, D. (2008). Lactate metabolism in anoxic turtles: An integrative review. *J. Comp. Physiol. B* **178**, 133–148; Warren, D., Reese, S., and Jackson, D. (2006). Tissue glycogen and extracellular buffering limit the survival of red eared slider turtles during anoxic submergence at 3°C. *Physiol. Biochem. Zool.* **79**, 736–744. Daw, J. C., Wenger, D. P., and Berne, R. M. (1967). Relationship between cardiac glycogen and tolerance to anoxia in the western painted turtle. Chrysemys picta bellii. *Comp. Biochem. Physiol.* **22**, 69–73.

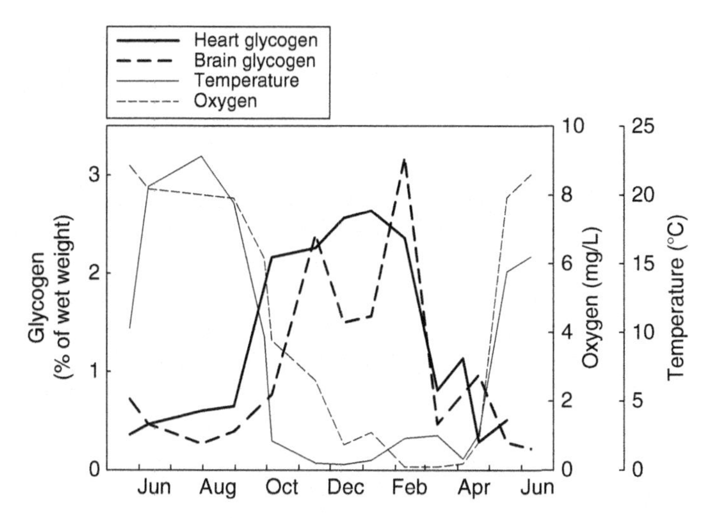

Fig. 9.13. Seasonal changes in glycogen concentration of crucian carp heart and brain. [Data from (Vornanen and Paajanen, 2004, 2006).]

preferentially fueled by ATP produced from glycolysis (see Dhar-Chowdhury *et al.*, 2007), glycogen might protect neurons against anoxic depolarization by securing ATP demand of this vital ion pump.

3.1.3. Heart Glycogen

In winter, activity of the crucian carp heart is probably fairly low due to cold temperature and absence of positive thermal compensation in contractile activity of the heart (Matikainen and Vornanen, 1992; Tiitu and Vornanen, 2001). Even so, the heart functions continuously and has a steady need for energy also during the long hypoxic/anoxic period. Functionality and anoxia tolerance of the crucian carp heart must be dependent on anaerobic glycolysis and therefore on the glycolytic capacity of the tissue, i.e., the amount of cardiac glycogen stores, glycolytic enzymes, and sarcolemmal glucose uptake for an efficient use of exogenous glucose.

Unlike mammalian cardiac muscle, crucian carp heart does not store fats, but instead has massive glycogen stores (Figure 9.13). In mid-winter, glycogen content of the heart in small fish (about 10 g) can be as high as 8% of the wet heart weight, while a minimum of 0.3% occurs in May (Vornanen, 1994a; Vornanen and Paajanen, 2004). Thus, there can be a 26-fold seasonal difference in cardiac glycogen reserves.

Glycogen phosphorylase activity of the heart is similar as in the red muscle but higher than in the white muscle (Hyvärinen *et al.*, 1985). Lactate dehydrogenase (LDH) activity of the crucian carp heart is only half of the

LDH activities measured in some other hypoxia-resistant fishes, e.g., South American lungfish (*Lepidosiren paradoxa*) and synbranchid eel (*Synbranchus marmarotus*) (Hochachka and Hulbert, 1978; Hochachka *et al.*, 1978) and there is little seasonal change in cardiac LDH activity (Vornanen, 1994a). In fact, a small depression of LDH activity is apparent toward winter (Lind, 1992; Vornanen, 1994a). It is possible that the moderate activities of glycogenolytic and glycolytic enzymes and the absence of positive thermal compensation in enzyme activities cannot sustain a fast rate of glycolytic energy production in the cold, and that the activity of the crucian carp heart has to be modest in winter. Although the glycogen stores of the crucian carp heart are large in comparison to those of other vertebrates and the cardiac energy demand is presumably rather low in winter, the heart must be largely dependent on blood glucose for energy supply, due to the small size of the organ. Uptake of exogenous glucose by the crucian carp heart has not been determined.

3.2. Seasonality of the Heart

Crucian carp stop feeding in winter (Penttinen and Holopainen, 1992) and are difficult to catch in the anoxic season, which is indicative for a relatively inactive lifestyle during anoxic winter that should reduce the demands on circulation. Therefore, it is expected that seasonal temperature acclimatization in the form of lowering temperatures could also prepare the heart for winter anoxia (Tiitu and Vornanen, 2001). Several findings indicate that the heart of crucian carp, unlike those of many other fish species, does not compensate for the depressive effects of low temperature in its function. Consequently, activity of the heart and circulation of blood will probably be depressed at the low temperatures of the anoxic winter season, although direct demonstration of this in the wild is still lacking.

3.2.1. Heart Size and Heart Rate

The rate of circulation is determined by cardiac output, which is the product of heart rate and stroke volume. The relative size of the heart is an important determinant for heart function as it directly affects the stroke volume (Graham and Farrell, 1989). The ventricle size of crucian carp heart is approximately 0.08% of the body mass, which is a quite typical value for a teleost fish, although less than in some more active fish species (Wilber *et al.*, 1961; Farrell *et al.*, 1992; Tiitu and Vornanen, 2002). Cold-induced enlargement of the heart, which is characteristic for many cold-active fish species, does not occur in seasonally acclimatized crucian carp. Instead there is marked decrease in cardiac water content in late autumn, probably due to accumulation of glycogen in the heart (Aho and Vornanen, 1997).

Considering that cold-induced hypertrophy of the heart should compensate for temperature-dependent depression of cardiac contractility and increased viscosity of the blood, the absence of this compensation in crucian carp is suggestive for temperature-dependent reduction of circulation.

Heart rate is a significant determinant of cardiac output and it is strongly modified by both acute and chronic temperature changes in fish (Farrell and Jones, 1992). In summer, normoxic heart rate of crucian carp varies from 8 beats/min at 4°C to slightly over 100 beats/min at 30°C and acclimatization to winter strongly depresses heart rates at temperatures above 10°C, while the rate at 4°C is approximately the same in summer and winter (Matikainen and Vornanen, 1992). The heart rate of winter-acclimatized crucian carp (at 4°C) is less than one third of the rate of cold-acclimated rainbow trout (*Oncorhynchus mykiss*) and burbot (*Lota lota*) hearts (about 30 beats/min). Thus, the low heart rate and absence of positive thermal compensation in beating rate typical for many temperate fish species, probably keeps cardiac ATP demands low in preparation for anoxic conditions.

3.2.2. Cardiac Contractility

Myocardial contractility describes the performance of cardiac muscle and is defined as the intrinsic ability of a cardiac muscle tissue to contract at a given sarcomere length. Adjustment of cardiac contractility to new conditions happens at the level of individual myocytes in the properties of myofilaments or in the management of intracellular Ca^{2+} concentration. The duration of ventricular twitch, especially the relaxation phase, is much longer in winter-acclimatized crucian carp than in summer-acclimatized fish (Vornanen, 1994b), suggesting seasonal differences in cardiac contractility. The rate of cardiac contraction is determined by attachment and detachment rate of cross bridges, i.e., by myosin ATPase activity and by the rate of activation induced by Ca^{2+} ions and their removal (Hoh *et al.*, 1988), and evidence is accumulating that both Ca^{2+} management and cardiac myosins are modulated by seasonal acclimatization in crucian carp. These changes probably contribute to the anoxia tolerance of the heart.

The force of cardiac contraction is directly related to the amount of free intracellular Ca^{2+}. Electrical excitation of the sarcolemma grades the size of intracellular Ca^{2+} in a process of excitation-contraction (e-c) coupling to produce adequate amounts of force and power for pumping of blood (Bers, 2002). As a part of a physiologically integrated entity, contractility and e-c coupling of the fish heart must be fine-tuned to correspond to the overall physiological demands under varying environmental conditions, and accordingly we can expect modifications at the organ, cell, and molecular level in fish exposed to prolonged anoxia.

In most fish hearts, the major part of Ca^{2+} comes from the extracellular space through L-type Ca^{2+} channels and Na^{+}–Ca^{2+} exchange (NCX), and may trigger a further release of Ca^{2+} from the sarcoplasmic reticulum (SR) via the SR Ca^{2+} release channels. Contraction ends when Ca^{2+} is returned from myofilaments back to the extracellular space and into the lumen of the SR by NCX and SR Ca^{2+}-pump, respectively (Tibbits *et al.*, 1992; Vornanen *et al.*, 2002b; Hove-Madsen *et al.*, 2003; Shiels and White, 2005). Sarcolemmal Ca^{2+} influx through both Ca^{2+} channels and NCX is critically dependent on the shape of AP, especially on plateau height and duration, and therefore any current that has influence on the shape of AP may affect voltage-dependent Ca^{2+} transport across the SL and, accordingly, e-c coupling (Edman and Johannsson, 1976). Function of the sarcolemmal K^{+} currents is especially important, since they regulate the duration of cardiac AP (Vornanen *et al.*, 2002b; Schotten *et al.*, 2007).

L-type Ca^{2+} current and NCX are the principal Ca^{2+} pathways in the crucian carp cardiac myocytes (Vornanen, 1997; Vornanen, 1999). Seasonal changes in the number of pore-forming alpha subunits of the L-type Ca^{2+} channels (DHPRs, dihydropyridine receptors) in crucian carp heart have been measured by [methyl-^{3}H]PN200-110 binding (Vornanen and Paajanen, 2004) and show that the number of Ca^{2+} channels are approximately doubled for a relatively short period of time in mid-summer (May–July), i.e., for the major part of the year the density of Ca^{2+} channels is low (Figure 9.14). Furthermore, the change in the number of Ca^{2+} channels can

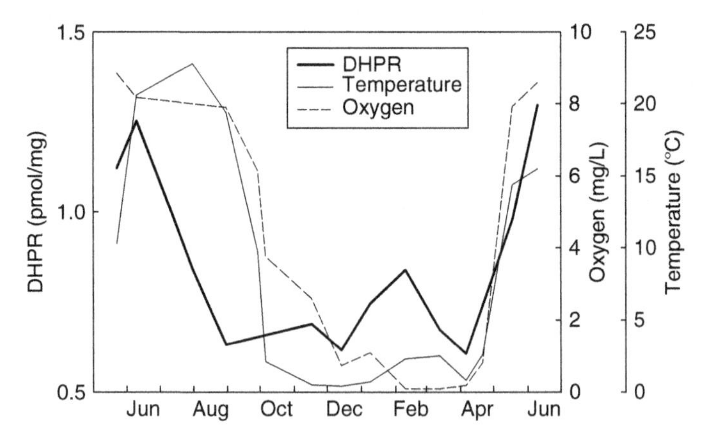

Fig. 9.14. Seasonal changes in the number of dihydropyridine receptors (DHPR; alpha subunits of the L-type Ca^{2+} channels) in the crucian carp heart. [Data from (Vornanen and Paajanen, 2004).]

be triggered by temperature acclimation (Tiitu and Vornanen, 2003). Functionally, these changes appear as 74% larger Ca^{2+} current (at 11°C) in summer-acclimatized hearts in comparison to winter-acclimatized hearts, and when measured in seasonally relevant temperatures (4°C and 18°C) the current is 6.1 times larger in summer than in winter (Vornanen and Paajanen, 2004). Even if the lengthening of ventricular AP from about 1.3 s to 2.8 s in the cold (Paajanen and Vornanen, 2004) is taken into account, sarcolemmal Ca^{2+} influx through L-type Ca^{2+} channels is at least three times larger in summer than in winter.

Although seasonal changes in e-c coupling proteins of the crucian carp heart, except myosin heavy chains and L-type Ca^{2+} channels, have not been examined yet, temperature acclimation under laboratory conditions indicate that several ion transport mechanisms are depressed by cold-acclimation. The density of Na^{+} current, which determines the rate of impulse propagation in the heart, is strongly depressed to one-fifth of that of warm-acclimated fish with cold-acclimation (Haverinen and Vornanen, 2004). Thapsigargin (a specific blocker of SR Ca^{2+}-pump)-sensitive Ca^{2+} uptake of the cardiac SR is also decreased in cold-acclimated crucian carp (Aho and Vornanen, 1998). Assuming that cold-acclimation primes the heart of crucian carp for winter, those findings suggest that several steps of the cardiac e-c coupling are down-regulated for winter and that the activity of the heart is depressed in the absence of positive thermal compensation in the cold winter waters. Indeed, tissue level experiments indicate that the kinetic properties of atrial and ventricular contraction are strongly depressed by cold-acclimation, which should appear as strongly reduced cardiac power output in the cold (Tiitu and Vornanen, 2001).

Interesting exceptions to the inverse thermal compensation of the crucian carp cardiac function are sarcolemmal K^{+} currents (Haverinen and Vornanen, 2008). Two major K^{+} currents of the fish heart are the inward rectifier K^{+} current (I_{K1}), which maintains the negative resting membrane potential and contributes to the final rate of AP repolarization, and the rapid component of the delayed rectifier K^{+} current (I_{Kr}), which is important in the regulation of plateau duration (Vornanen *et al.*, 2002a). Densities of these K^{+} currents are increased by cold-acclimation in atrial and ventricular myocytes of the crucian carp heart so that the sizes of the currents are similar in cold-acclimated fish at 4°C and in warm-acclimated fish at 18°C (Haverinen and Vornanen, 2008). Still, the duration of AP is 2.6 and 2.8 times longer at 4°C than at 18°C for ventricular and atrial muscle, respectively. Obviously positive temperature compensation in the density of K^{+} currents is not sufficient to prevent the lengthening of cardiac AP in winter.

Thus, our current knowledge of cardiac ion currents of the crucian carp indicates that the inward Na^{+} and Ca^{2+} currents are depressed and the outward K^+ currents are enhanced in the cold-acclimatized winter fish. Inward currents are excitatory in that they promote contraction, while outward currents tend to stabilize membrane to the negative equilibrium potential of K^+ ions. Therefore, opposite changes in inward and outward currents are likely to reduce excitability of the cardiac muscle.

Cardiac contractility is also affected by composition and function of myofibrillar proteins. Two myosin heavy chains have been demonstrated in crucian carp ventricle by SDS-PAGE (Vornanen, 1994b). Only one myosin heavy chain isoform is expressed in the hearts of winter-acclimatized fish and is therefore called "winter" myosin, whereas the hearts of summer-acclimatized fish express both winter and "summer" isoforms. In June and July both isoforms are almost equally represented in the ventricular muscle, but the amount of summer myosin decreases already in August and cannot be resolved any more in September. Small amounts of summer myosin appear again in May, when waters warm up (see Figure 9.2). The physiological significance of this seasonal pattern in myosin heavy chain composition probably lies in the different myosin-ATPase activities of the two isoforms: the activity is much greater in summer than in winter (Vornanen, 1994b; Tiitu and Vornanen, 2001). It is well known that the contraction of slow myosins occurs with less energy expenditure than the contraction of fast myosins (Alpert and Mulieri, 1982). Therefore, the exclusive reliance on the slow myosin in winter would improve energetic economy of the heart under conditions where energy production is severely limited by oxygen shortage. In heart function, this should appear as a reduced cardiac power output, which might be well tolerated due to reduced circulatory demands. The slow myosin may also be useful in tuning the rate of myofilament sliding to the low heart rate and the long duration of cardiac action potential in the cold. Taken together, inverse thermal compensation seems to be typical for the crucian carp, with the exception of sarcolemmal K^+ currents, which will result in temperature-dependent depression of cardiac function in cold and anoxic winter waters.

3.3. Seasonality of Brain

3.3.1. Brain Na^+/K^+-ATPase

As described above, anoxia exposure does not decrease the number of Na^+/K^+-ATPase alpha subunits or their molecular activity in the crucian carp brain (Hylland *et al.*, 1997; Vornanen and Paajanen, 2006). In contrast, determinations of [^{3}H]ouabain binding and Na^+/K^+-ATPase activity in crucian carp caught directly from the wild have revealed a strong temperature dependence of Na^+/K^+-ATPase (Vornanen and Paajanen, 2006). As a result, the activity of the sodium pump in winter is only 10–15% of its activity in summer, suggesting a considerable down-regulation of brain activity with cold acclimation. In fact, a small positive compensation of the Na^+/K^+-ATPase activity is seen in mid-winter, which might be needed to prevent the brain from depressing into a comatose state. The positive compensation in brain Na^+/K^+-ATPase activity is attained without increase in the number of pump units by a decrease in temperature dependence (Q_{10}) of the enzyme catalysis (Vornanen and Paajanen, 2006). Seasonal changes in phosphatidylethanolamine and phosphatidylserine phospholipids of the neuronal membrane might explain the reduced temperature dependence of the sodium pump (see below) (Käkelä *et al.*, 2008).

3.3.2. Brain Lipids

The brain of crucian carp retains a significant level of functionality in complete anoxia (Nilsson, 2001) and is assumed to sustain considerable nerve function under cold and hypoxic/anoxic winter conditions (see Section 2.5). Many crucial neuronal activities occur in the plasma membrane where proper function of ion channels, ion pumps, and membrane receptors is essential for electrical excitability and neurotransmission. Those molecules are embedded in the phospholipid membrane, which should provide a favorable matrix to the integral membrane proteins under all environmental conditions. Temperature has a particularly strong effect on the physical properties of biological membranes and in many ectotherms temperature acclimation strongly modifies membrane lipids to maintain the semifluid state of the plasma membrane (Sinensky, 1974). Hypoxia, ischemia, and reperfusion also affect the biochemical composition of the lipid membrane and may damage the membrane (Cao *et al.*, 2007). In the typical habitat of crucian carp, low temperature and hypoxia/anoxia occur simultaneously and therefore the brain membranes have to cope with both thermal and hypoxic stresses, which may affect brain lipids differently and which may require different adaptations. Considering the large seasonal changes in abiotic conditions of the crucian carp habitat, it is not surprising that profound seasonal changes in the composition of brain lipids occur (Käkelä *et al.*, 2008).

Comparison of membrane lipids from fish acclimated in laboratory to different temperatures and fish collected from the wild in different seasons indicates that fatty acid composition of the brain lipids display similar temperature responses in laboratory-acclimated and seasonally acclimatized crucian carp (Figure 9.15A). At low temperatures, the brains contain lower levels of saturated fatty acids, higher levels of polyunsaturated fatty acids, and the average length of the monounsaturated fatty acid chain is shorter (Käkelä *et al.*, 2008). All these changes are compatible with the model of

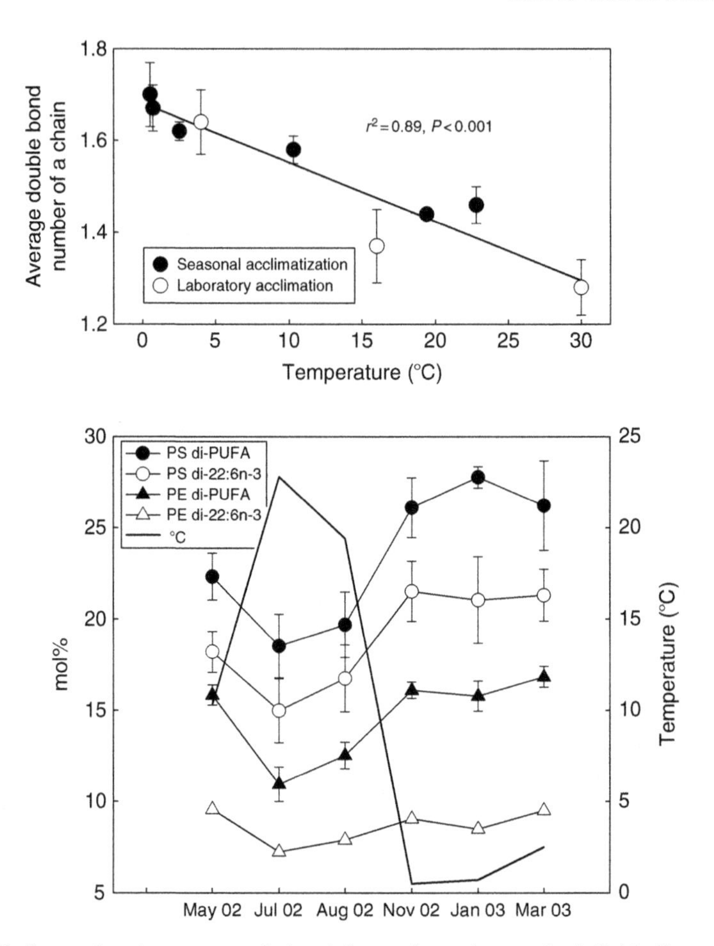

Fig. 9.15. Seasonal- and temperature-induced changes in crucian carp brain lipids. Temperature dependence of the average double bond number of the acyl and alkenyl chains in the total lipids of the crucian carp acclimated in the laboratory at three different temperatures for 4 weeks (A). Seasonal changes in the total di-PUFA and di-22:6n-3 phosphatidylserine (PS) and phosphatidylethanolamine (PE) in the brain of crucian carp collected from the wild (B). Values are means ± S.D. from 5 to 7 fish. [Data from Käkelä *et al.* (2008).]

homeoviscous adaptation of membrane fluidity (Sinensky, 1974) and suggest compensation for the direct effects of temperature to maintain the proper fluidity of the brain membranes.

Acclimation/acclimatization causes little changes in the phospholipid class composition, i.e., the relative contents of phosphatidylethanolamines, phopshatidylcholines, phosphatidylserines, phosphatidylinositols, and plasmalogens, but induces marked changes in molecular species composition (different fatty acid combinations) of the brain phospholipids (Figure 9.15B). Most notably a large increase of the di-22:6n-3 phosphatidylserine and phosphatidylethanolamine species (DHA, docosahexaenoic acid estrified to carbon-1 and carbon-2 of the phospholipids) appears in the cold. Since the increase of DHA in the total fatty acyl pool of the brain is small, the formation of di-DHA aminophospholipid species appears to be a specific molecular rearrangement for winter. Plasma membranes of eukaryotic cells are highly asymmetric, with most phosphatidylethanolamines and all phosphatidylserines residing in the inner membrane leaflet (Virtanen *et al.*, 1998), and therefore di-DHA changes should have a significant impact on this membrane compartment. Such highly unsaturated species are needed to maintain adequate membrane fluidity in the vicinity of ion transporters and other integral membrane proteins. More specifically, these changes may be behind the noticed decrease in temperature-dependence of Na^+/K^+-ATPase in winter, since DHA-containing phospholipids activate Na^+/K^+-ATPase in excitable membranes (Turner *et al.*, 2003). On the other hand, DHA-containing lipids also protect against ischemia, oxygen deprivation, and reperfusion injuries in mammals (Strokin *et al.*, 2006; Cao *et al.*, 2007). In particular, DHA-containing lipids alleviate ischemia-associated decrease in Na^+/K^+-ATPase activity and thereby reduce brain infarct size.

The brain lipidome of the crucian carp is strongly modified by seasonal acclimatization and the seasonal changes are in many respects similar, although not identical, to changes induced by temperature acclimation. This suggests that ambient temperature is the main environmental cue that primes brain lipids for winter stresses. The responses in seasonal acclimatization are sometimes smaller than in laboratory acclimation, which may represent combined effects of low temperature and hypoxia on brain lipids.

4. SUMMARY

The crucian carp employs a number of innate survival strategies to tolerate prolonged anoxia exposure. With the onset of oxygen deprivation, respiration rate is augmented and the gills remodeled. These responses, in conjunction with an extremely high hemoglobin oxygen affinity, should extend the period that oxygen can be extracted from the water when the crucian carp is faced with a steady fall in ambient oxygen levels during the early winter. A concurrent severe bradycardia likely benefits myocardial oxygen supply. When oxygen is no longer available, crucian carp produce ethanol as the major anaerobic end-product to avoid self-poisoning by lactate and H^+ ions, and increase brain blood flow to deliver fermentable fuel. Maintained cardiac activity during prolonged anoxia may be possible because the ATP demand of the heart lies below the maximum glycolytic potential for ATP production and is believed to be necessitated for transport of fermentable fuels and ethanol among tissues. Unlike the equally anoxia-tolerant freshwater turtle, the anoxic crucain carp sustains neural activity. During anoxia, there is no depression of brain Na^+/K^+-ATPase activity and K^+ and Ca^{2+} permeability remains unchanged, but some indication of limited "channel arrest" involving NMDA receptors exists.

In addition to constitutively expressed characteristics, a seasonal induction of numerous physiological traits is imperative for preparing the carp for long-term anoxic survival in the winter. In the habitat of crucian carp, anoxia is a regular and predictable seasonal condition that is accompanied by several environmental cues, most notably temperature. Changes in oxygen content and temperature of water in the lakes and ponds inhabited by crucian carp occur almost in parallel. Multiple findings indicate that temperature induces changes in physiology of crucian carp that are probably beneficial in the anoxic winter conditions, i.e., temperature prepares the body of crucian carp for winter anoxia through slow temperature acclimatization. Thermal acclimatization in crucian carp is often noncompensatory (inverse), so that the depressive effects of low temperature on metabolism and organ functions are enforced. Fittingly, positive thermal compensation is rare in crucian carp and has been documented only for the cardiac K^+ channels.

REFERENCES

Aho, E., and Vornanen, M. (1997). Seasonality of ATPase activities in crucian carp (*Carassius carassius* L.) heart. *Fish Physiol. Biochem.* **16,** 355–364.

Aho, E., and Vornanen, M. (1998). Ca-ATPase activity and Ca-uptake by sarcoplasmic reticulum in fish heart: Effects of thermal acclimation. *J. Exp. Biol.* **201,** 525–532.

Aho, E., and Vornanen, M. (2002). Effects of adenosine on the contractility of normoxic rainbow trout heart. *J. Comp. Physiol. B* **172,** 217–225.

Alpert, N. R., and Mulieri, L. A. (1982). Increased myothermal economy of isometric force generation in compensated cardiac hypertrophy induced by pulmonary artery constriction in the rabbit: A characterization of heat liberation in normal and hypertrophied ventricular papillary muscles. *Circ. Res.* **50,** 491–500.

Arundine, M., and Tymianski, M. (2003). Molecular mechanisms of calcium-dependent neurodegeneration in excitotoxicity. *Cell Calcium* **34,** 325–337.

Balaban, R. S. (2002). Cardiac energy metabolism homeostasis: Role of cytosolic calcium. *J. Mol. Cell. Cardiol.* **34,** 1259–1271.

Belardinelli, L., Shryock, J. C., Wang, D., and Srinivas, M. (1995). Ionic basis of the electrophysiological actions of adenosine on cardiomyocytes. *FASEB J.* **9,** 359–365.

Bers, D. M. (2002). Cardiac excitation-contraction coupling. *Nature* **415,** 198–205.

Bickler, P. E., and Buck, L. T. (2007). Hypoxia tolerance in reptiles, amphibians, and fishes: Life with variable oxygen availability. *Annu. Rev. Physiol.* **69,** 145–170.

Blazka, B. (1958). The anaerobic metabolism of fish. *Physiol. Zool.* **31,** 117–128.

Blazka, B. (1960). On the biology of crucian carp (*Carassius carassius* L. morpha humilis Heckel). *Zool. Zhurnal.* **39,** 1384–1389.

Brönmark, C., and Milner, J. G. (1992). Predator-induced phenotypical change in body morphology in crucian carp. *Science* **258,** 1348–1350.

Cameron, J. S., Hoffmann, K. E., Zia, C., Hemmett, H. M., Kronsteiner, A., and Lee, C. M. (2003). A role for nitric oxide in hypoxia-induced activation of cardiac K_{ATP} channels in goldfish (*Carassius auratus*). *J. Exp. Biol.* **206,** 4057–4065.

Cao, D., Yang, B., Hou, L., Xu, J., Xue, R., Sun, L., Zhou, C., and Liu, Z. (2007). Chronic daily administration of ethyl docosahexaenoate protects against gerbil brain ischemic damage through reduction of arachidonic acid liberation and accumulation. *J. Nutr. Biochem.* **18,** 297–304.

Chen, J., Zhu, J. X., Wilson, I., and Cameron, J.S (2005). Cardioprotective effects of K_{ATP} channel activation during hypoxia in goldfish *Carassius auratus. J. Exp. Biol.* **208,** 2765–2772.

Dhar-Chowdhury, P., Malester, B., Rajacic, P., and Coetzee, W. A. (2007). The regulation of ion channels and transporters by glycolytically derived ATP. *Cell. Mol. Life Sci.* **64,** 3069–3083.

Domenici, P., Turesson, H., Brodersen, J., and Bronmark, C. (2008). Predator-induced morphology enhances escape locomotion in crucian carp. *Proc. Roy. Soc. Ser. B* **275,** 195–201.

Donohoe, P. H., and Boutilier, R. G. (1998). The protective effects of metabolic rate depression in hypoxic cold submerged frogs. *Resp. Physiol.* **111,** 325–336.

Edman, K. A. P., and Johannsson, M (1976). The contractile state of rabbit papillary muscle in relation to stimulation frequency. *J. Physiol.* **254,** 565–581.

Ellefsen, S., Sandvik, G. K., Larsen, H. K., Stensløkken, K.-O., Hov, D. A.S, Kristensen, T. A., and Nilsson, G. E. (2008). Expression of genes involved in excitatory neurotransmission in anoxic crucian carp brain (Carassius carassius). *Physiol. Genomics*. **35,** 5–17.

Erecinska, M., and Silver, I. A. (1994). Ions and energy in mammalian brain. *Prog. Neurobiol* **43,** 37–71.

Erecinska, M., Cherian, S., and Silver, I. A. (2004). Energy metabolism in mammalian brain during development. *Prog. Neurobiol.* **73,** 397–445.

Farrell, A. P. (2007). Tribute to P. L. Lutz: A message from the heart – why hypoxic bradycardia in fishes. *J. Exp. Biol.* **210,** 1715–1725.

Farrell, A. P., and Jones, D. R. (1992). The Heart. *In* "Fish Physiology" Vol. 12A, (Hoar, W. S., Randall, D. J., and Farrell, A.P.), pp. 1–88. Academic Press, San Diego.

Farrell, A. P., and Stecyk, J. A. W. (2007). The heart as a working model to explore themes and strategies for anoxic survival in ectothermic vertebrates. *Comp. Biochem. Physiol. A* **147,** 300–312.

Farrell, A. P., Davie, P. S., Franklin, C. E., Johansen, J. A., and Brill, R. W. (1992). Cardiac physiology in tunas. I. *In vitro* perfused heart preparations from yellowfin and skipjack tunas. *Can. J. Zool.* **70,** 1200–1210.

Fernandes, J. A., Lutz, P. L., Tannenbaum, A., Todorov, A. T., Liebovitch, L., and Vertes, R. (1997). Electroencephalogram activity in the anoxic turtle brain. *Am. J. Physiol.* **273,** R911–R919.

Fraser, K. P. P., Houlihan, D. F., Lutz, P. L., Leone-Kabler, S., Manuel, L., and Brechin, J. G. (2001). Complete suppression of protein synthesis during anoxia with no post-anoxia protein synthesis debt in the red-eared slider turtle *Trachemys scripta elegans. J. Exp. Biol.* **204,** 4353–4360.

Ganim, R. B., Peckol, E. L., Larkin, J., Ruchhoef, M. L., and Cameron, J. S. (1998). ATP-sensitive K^+ channels in cardiac muscle from cold-acclimated goldfish: Characterization and altered response to ATP. *Comp. Biochem. Physiol. A* **119,** 395–401.

Gerschenfeld, H. M. (1973). Chemical transmission in invertebrate central nervous systems and neuromuscular junctions. *Physiol. Rev.* **53,** 1–119.
Gesser, H., and Jørgensen, E. (1982). pH_i, contractility and Ca-balance under hypercapnic acidosis in the myocardium of different vertebrate species. *J. Exp. Biol.* **96,** 405–412.
Graham, M. S., and Farrell, A. P. (1989). The effect of temperature acclimation and adrenaline on the performance of a perfused trout heart. *Physiol. Zool.* **62,** 38–61.
Haverinen, J., and Vornanen, M. (2004). Temperature acclimation modifies Na^+ current in fish cardiac myocytes. *J. Exp. Biol.* **207,** 2823–2833.
Haverinen, J., and Vornanen, M. (in press). Responses of action potential and K+ currents to chronic thermal stress in fish hearts. Phylogeny or thermal preferences? *Physiol. Biochem. Zool.*
Herbert, C. V., and Jackson, D. C. (1985). Temperature effects on the responses to prolonged submergence in the turtle *Chrysemys picta bellii*. II. Metabolic rate, blood acid-base and ionic changes and cardiovascular function in aerated and anoxic water. *Physiol. Zool.* **58,** 670–681.
Hicks, J. M. T., and Farrell, A. P. (2000). The cardiovascular responses of the red-eared slider (*Trachemys scripta*) acclimated to either 22 or 5°C. II. Effects of anoxia on adrenergic and cholinergic control. *J. Exp. Biol.* **203,** 3775–3784.
Hochachka, P. W. (1986). Defense strategies against hypoxia and hypothermia. *Science* **231,** 234–241.
Hochachka, P. W., and Hulbert, W. C. (1978). Glycogen "seas", glycogen bodies, and glycogen granules in heart and skeletal muscle of two air-breathing, burrowing fishes. *Can. J. Zool.* **56,** 774–786.
Hochachka, P. W., and Somero, G. N. (2002). "Biochemical Adaptation: Mechanism and Process in Physiological Evolution." Oxford University Press, Oxford.
Hochachka, P. W., Guppy, M., Guderley, H. E., Storey, K. B., and Hulbert, W. C. (1978). Metabolic biochemistry of water- vs. air-breathing osteoglossids: Heart enzymes and ultrastructure. *Can. J. Zool.* **56,** 759–768.
Hoh, J. F. Y., Rossmanith, G. H., Kwan, L. J., and Hamilton, A. M. (1988). Adrenaline increases the rate of cycling of crossbridges in rat cardiac muscle as measured by pseudo-random binary noise-modulated perturbation analysis. *Circ. Res.* **62,** 452–461.
Holopainen, I. J., and Hyvärinen, H (1985). Ecology and physiology of crucian carp (Carassius carassius (L.)) in small Finnish ponds with anoxic conditions in winter. *Verh. Internat. Verein. Limnol.* **22,** 2566–2570.
Holopainen, I. J., Hyvärinen, H., and Piironen, J. (1986). Anaerobic wintering of crucian carp (Carassius carassius L.) - II. Metabolic products. *Comp. Biochem. Physiol. A* **83,** 239–242.
Holopainen, I. J., Aho, J., Vornanen, M., and Huuskonen, H. (1997). Phenotypic plasticity and predator effects on morphology and physiology of crucian carp in nature and in the laboratory. *J. Fish Biol.* **50,** 781–798.
Holopainen, I. J., Tonn, W. M., and Paszkowski, C. A. (1997). Tales of two fish: The dichotomous biology of crucian carp (*Carassius carassius* (L.)) in northern Europe. *Ann. Zool. Fennici.* **28,** 1–22.
Horoszewicz, L. (1973).). Lethal and "disturbing" temperatures in some fish species from lakes with normal and artificially elevated temperature. *J. Fish Biol.* **5,** 165–181.
Hove-Madsen, L., Llach, A., Tibbits, G. F., and Tort, L. (2003). Triggering of sarcoplasmic Ca^{2+} release and contraction by reverse mode Na^+-Ca^{2+} exchange in trout atrial myocytes. *Am. J. Physiol.* **284,** R1330–R1339.
Huss, J. M., and Kelly, D. P. (2005). Mitochondrial energy metabolism in heart failure: A question of balance. *J. Clin. Invest.* **115,** 547–555.

Hylland, P., and Nilsson, G. E. (1999). Extracellular levels of amino acid neurotransmitters during anoxia and forced energy deficiency in crucian carp brain. *Brain Res.* **823,** 49–58.
Hylland, P., Nilsson, G. E., and Lutz, P. L. (1994). Time course of anoxia induced increase in cerebral blood flow rate in turtles: Evidence for a role of adenosine. *J. Cerebral Blood Flow Metab.* **14,** 877–881.
Hylland, P., Nilsson, G. E., and Johansson, D. (1995). Anoxic brain failure in an ectothermic vertebrate: Release of amino acids and K^+ in rainbow trout thalamus. *Am. J. Physiol.* **269,** R1077–R1084.
Hylland, P., Milton, S., Pek, M., Nilsson, G. E., and Lutz, P. L. (1997). Brain Na^+/K^+-ATPase activity in two anoxia tolerant vertebrates: Crucian carp and freshwater turtle. *Neurosci. Lett.* **235,** 89–92.
Hyvärinen, H., Holopainen, I. J., and Piironen, J. (1985). Anaerobic wintering of crucian carp (Carassius carassius L.) – I. Annual dynamics of glycogen reserves in nature. *Comp.Biochem. Physiol. A* **82,** 797–803.
Jackson, D. C. (1968). Metabolic depression and oxygen depletion in the diving turtle. *J. Appl. Physiol.* **24,** 503–509.
Jackson, D. C. (2002). Hibernation without oxygen: Physiological adaptations in the painted turtle. *J. Physiol.* **543,** 731–737.
Johansson, D., and Nilsson, G. E. (1995). Roles of energy status, K_{ATP} channels, and channel arrest in fish brain K^+ gradient dissipation during anoxia. *J. Exp. Biol.* **198,** 2575–2580.
Johansson, D., Nilsson, G. E., and Törnblom, E. (1995). Effects of anoxia on energy metabolism in crucian carp brain slices studied with microcalorimetry. *J. Exp. Biol.* **198,** 853–859.
Johansson, D., Nilsson, G. E., and Døving, K. B. (1997). Anoxic depression of light-evoked potentials in retina and optic tectum of crucian carp. *Neurosci. Lett.* **237,** 73–76.
Johnston, I. A., and Bernard, L. M. (1983). Utilization of the ethanol pathway in carp following exposure to anoxia. *J. Exp. Biol.* **104,** 73–78.
Käkelä, R., Mattila, M., Hermansson, M., Haimi, P., Uphoff, A., Paajanen, V., Somerharju, P., and Vornanen, M. (2008). Seasonal acclimatization of brain lipidome in a eurythermal fish (*Carassius carassius*, L.) is mainly determined by temperature. *Am. J. Physiol. Regulatory Integrative Comp. Physiol.* **294,** R1716–R1728.
Koopowitz, H., and Keenan, L. (1982). The primitive brain of platyhelminthes. *Trends Neurosci.* **5,** 77–79.
Krumschnabel, G., Biasi, C., and Wieser, W. (2000). Action of adenosine on energetics, protein synthesis and K^+ homeostasis in teleost hepatocytes. *J. Exp. Biol.* **203,** 2657–2665.
Lennard, R., and Huddart, H. (1989). Purinergic modulation of cardiac activity in the flounder during hypoxic stress. *J. Comp. Physiol. B* **159,** 105–113.
Lind, Y. (1992). Summertime and early autumn activity of some enzymes in the carbohydrate and fatty acid metabolism of the crucian carp. *Fish Physiol. Biochem.* **9,** 409–415.
Lipton, P. (1999). Ischemic cell death in brain neurons. *Physiol. Rev.* **79,** 1431–1568.
Lutz, P. L., and Nilsson, G. E. (1997). Contrasting strategies for anoxic brain survival – glycolysis up or down. *J. Exp. Biol.* **200,** 411–419.
Lutz, P. L., McMahon, P., Rosenthal, M., and Sick, T. J. (1984). Relationships between aerobic and anaerobic energy production in turtle brain in situ. *Am. J. Physiol.* **247,** R740–R744.
Lutz, P. L., Rosenthal, M., and Sick, T. (1985). Living without oxygen: Turtle brain as a model of anaerobic metabolism. *Mol. Physiol.* **8,** 411–425.
MacCormack, T. J., and Driedzic, W. R. (2004). Cardiorespiratory and tissue adenosine responses to hypoxia and reoxygenation in the short-horned sculpin Myoxocephalus scorpius. *J. Exp. Biol.* **207,** 4157–4164.
Matikainen, N., and Vornanen, M. (1992).). Effect of season and temperature acclimation on the function of crucian carp (*Carassius carassius*) heart. *J.. Exp. Biol.* **167,** 203–220.

McGeer, P. L., and McGeer, E. G. (1989). Amino acid neurotransmitters. *In* "Basic Neurochemistry" (Siegel, G. J., Agranoff, B., and Alberts, R. W., *et al.*, Eds.), pp. 311–332. Raven Press, New York.

Meghi, P., and Burnstock, G. (1984). Actions of some autonomic agents on the heart of the trout (*Salmo gairdneri*) with emphasis on the effects of adenyl compounds. *Comp. Biochem. Physiol.* **78C,** 69–75.

Mourik, J., Raeven, P., Steur, K., and Addink, A. D. F. (1982). Anaerobic metabolism of red skeletal muscle of goldfish, *Carassius auratus* (L.). *FEBS. Lett.* **137,** 111–114.

Mubagwa, K., and Flameng, W. (2001). Adenosine, adenosine receptors and myocardial protection: An updated overview. *Cardiovasc. Res.* **52,** 25–39.

Mubagwa, K., Mullane, K., and Flameng, W. (1996). Role of adenosine in the heart and circulation. *Cardiovasc. Res.* **32,** 797–813.

Nagell, B., and Brittain, J. E. (1977). Winter anoxia – a general feature of ponds in cold temperate regions. *Int. Revue Ges. Hydrobiol.* **62,** 821–824.

Nichols, C. G. (2006). K_{ATP} channels as molecular sensors of cellular metabolism. *Nature* **440,** 470–476.

Nilsson, G. E. (1988). A comparative study of aldehyde dehydrogenase and alcohol dehydrogenase activity in crucian carp and three other vertebrates: Apparent adaptations to ethanol production. *J. Comp. Physiol. B* **158,** 479–485.

Nilsson, G. E. (1990a). Distribution of aldehyde dehydrogenase and alcohol dehydrogenase in summer acclimatized crucian carp (*Carassius carassius* L.). *J. Fish Biol.* **36,** 175–179.

Nilsson, G. E. (1990b). Long term anoxia in crucian carp: Changes in the levels of amino acid and monoamine neurotransmitters in the brain, catecholamines in chromaffin tissue, and liver glycogen. *J. Exp. Biol.* **150,** 295–320.

Nilsson, G. E. (1991). The adenosine receptor blocker aminophylline increases anoxic ethanol production in crucian carp. *Am. J. Physiol.* **261,** R1057–R1060.

Nilsson, G. E. (1992). Evidence for a role of GABA in metabolic depression during anoxia in crucian carp (*Carassius carassius* L.). *J. Exp. Biol.* **164,** 243–259.

Nilsson, G. E. (1996). Brain and body oxygen requirements of *Gnathonemus petersii*, a fish with an exceptionally large brain. *J. Exp. Biol.* **199,** 603–607.

Nilsson, G. E. (2001). Surviving anoxia with the brain turned on. *News Physiol. Sci.* **16,** 218–221.

Nilsson, G. E. (2007). Gill remodeling in fish – a new fashion or an ancient secret? *J. Exp. Biol.* **210,** 2403–2409.

Nilsson, G. E., and Lutz, P. L. (1991). Release of inhibitory neurotransmitters in response to anoxia in turtle brain. *Am. J. Physiol.* **261,** R32–R37.

Nilsson, G. E., and Lutz, P. L. (1992). Adenosine release in the anoxic turtle brain: A possible mechanism for anoxic survival. *J. Exp. Biol.* **162,** 345–351.

Nilsson, G. E., and Lutz, P. L. (1993). Role of GABA in hypoxia tolerance, metabolic depression and hibernation – possible links to neurotransmitter evolution. *Comp. Biochem. Physiol. C* **105,** 329–336.

Nilsson, G. E., and Lutz, P. L. (2004). Anoxia tolerant brains. *J. Cereb. Blood Flow Metab.* **24,** 475–486.

Nilsson, G. E., Pérez-Pinzón, M., Dimberg, K., and Winberg, S. (1993a). Brain sensitivity to anoxia in fish as reflected by changes in extracellular potassium-ion activity. *Am. J. Physiol.* **264,** R250–R253.

Nilsson, G. E., Rosén, P., and Johansson, D. (1993b). Anoxic depression of spontaneous locomotor activity in crucian carp quantified by a computerized imaging technique. *J. Exp. Biol.* **180,** 153–163.

Nilsson, G. E., Hylland, P., and Löfman, C. O. (1994). Anoxia and adenosine induce increased cerebral blood flow in crucian carp. *Am. J. Physiol.* **267,** R590–R595.

Nilsson, S. (1855). Skandinavisk Fauna. Fjerde Delen Fiskarna, Berlingska Boktryckeriet.

Orchard, C. H., and Kentish, J. C. (1990). Effects of changes of pH on the contractile function of cardiac muscle. *Am. J. Physiol.* **258,** C967–C981.

Paajanen, V., and Vornanen, M. (2002). The induction of an ATP-sensitive K^+ current in cardiac myocytes of air- and water-breathing vertebrates. *Pflugers Archiv.* **444,** 760–770.

Paajanen, V., and Vornanen, M. (2003). Effects of chronic hypoxia on inward rectifier K^+ current (I_{K1}) in ventricular myocytes of crucian carp (*Carassius carassius*) heart. *J. Membr. Biol.* **194,** 119–127.

Paajanen, V., and Vornanen, M. (2004). Regulation of action potential duration under acute heat stress by $I_{K,ATP}$ and I_{K1} in fish cardiac myocytes. *Am. J. Physiol.* **286,** R405–R415.

Penttinen, O.-P., and Holopainen, I. J. (1992). Seasonal feeding activity and ontogenetic dietary shifts in crucian carp, *Carassius carassius. Environ. Biol. Fish* **33,** 215–221.

Piironen, J., and Holopainen, I. J. (1986). A note on seasonality in anoxia tolerance of crucian carp (*Carassius carassius* L.) in the laboratory. *Ann. Zool. Fennici.* **23,** 335–338.

Poli, A., Beraudi, A., Villani, L., Storto, M., Battaglia, G., Gerevini, V. D. G., Capuccio, I., Caricasole, A., D/Onofrio, M., and Nicoletti, F. (2003). Group II metabotropic glutamate receptors regulate the vulnerability to hypoxic brain damage. *J. Neurosci.* **23,** 6023–6029.

Restifo, L. L., and White, K. (1990). Molecular and genetic approaches to neurotransmitter and neuromodulator systems in *Drosophila. In* "Advances in Insect Physiology Vol. 22" (Evans, P. D., and Wigglesworth, V. B., Eds.), pp. 115–219. Academic Press, London.

Rissanen, E., Numminen, H. K., Sollid, J., Nilsson, G. E., and Nikinmaa, M. (2006). Temperature regulates hypoxia-inducible factor-1 (HIF-1) in a poikilothermic vertebrate, crucian carp (*Carassius carassius*). *J. Exp. Biol.* **209,** 994–1003.

Roden, D. M., Balser, J. R., George, A. L., Jr, and Anderson, M. E. (2002). Cardiac ion channels. *Ann. Rev. Physiol.* **64,** 431–475.

Rolfe, D. F. S., and Brown, G. C. (1997). Cellular energy utilization and molecular origin of standard metabolic rate in mammals. *Physiol. Rev.* **77,** 731–758.

Rosati, A. M., Traversa, U., Lucchi, R., and Poli, A. (1995). Biochemical and pharmacological evidence for the presence of A1 but not $A2_a$ adenosine receptors in the brain of the low vertebrate teleost *Carassius auratus* (goldfish). *Neurochem. Int.* **26,** 411–423.

Rotmensch, H. H., Cohen, S., Rubinstein, R., and Lass, Y. (1981). Effects of adenosine in the isolated carp atrium. *Isr. J. Med.* **17,** 393.

Rovainen, C. M. (1970). Glucose production by lamprey meninges. *Science* **167,** 889–890.

Santer, R. M. (1985). Morphology and innervation of the fish heart. *Adv. Anat. Embryol. Cell Biol.* **89,** 1–102.

Schmidt, H., and Wegener, G. (1988). Glycogen phosphorylase in fish brain (*Carassius carassius*) during hypoxia. *Biochem.. Soc. Trans* **16,** 621–622.

Schotten, U., de Haan, S., Verheule, S., Harks, E. G. A., Frechen, D., Bodewig, E., Greiser, M., Ram, R., Maessen, J., Kelm, M., Allessie, M., and Van Wagoner, D. R. (2007). Blockade of atrial-specific K^+-currents increases atrial but not ventricular contractility by enhancing reverse mode Na^+/Ca^{2+}-exchange. *Cardiovasc. Res.* **73,** 37–47.

Schramm, M., Klieber, H.-G., and Daut, J. (1994). The energy expenditure of actomyosin-ATPase, Ca^{2+}-ATPase and Na^+, K^+-ATPase in guinea-pig cardiac ventricular muscle. *J. Physiol.* **481,** 647–662.

Shoubridge, E. A., and Hochachka, P. W. (1980). Ethanol: Novel end product of vertebrate anaerobic metabolism. *Science* **209,** 308–309.

Shiels, H. A., and White, E. (2005). Temporal and spatial properties of cellular Ca^{2+} flux in trout ventricular myocytes. *Am. J. Physiol.* **288,** R1756–R1766.

Siesjö, B. K. (1978). Brain Energy Metabolism Chichester, Wiley.

Sinensky, M. (1974). Homeoviscous adaptation – a homeostatic process that regulates the viscosity of membrane lipids in *Escerichia coli. PNAS* **71,** 522–525.

Smith, R. W., Houlihan, D. F., Nilsson, G. E., and Brechin, J. G. (1996). Tissue specific changes in protein synthesis rates *in vivo* during anoxia in crucian carp. *Am. J. Physiol.* **271,** R897–R904.

Sollid, J., and Nilsson, G. E. (2006). Plasticity of respiratory structures – adaptive remodeling of fish gills induced by ambient oxygen and temperature. *Resp. Physiol. Neurobiol.* **154,** 241–251.

Sollid, J., De Angelis, P, Gundersen, K, and Nilsson, G. E. (2003). Hypoxia induces adaptive and reversible gross-morphological changes in crucian carp gills. *J. Exp. Biol.* **206,** 3667–3673.

Sollid, J., Weber, R. E., and Nilsson, G. E. (2005a). Temperature alters the respiratory surface area of crucian carp (*Carassius carassius*) and goldfish (*Carassius auratus*). *J. Exp. Biol.* **208,** 1109–1116.

Sollid, J., Kjernsli, A., De Angelis, P. M., Røhr, Å.K., and Nilsson, G. E. (2005b). Cell proliferation and gill morphology in anoxic crucian carp. *Am. J. Physiol.* **289,** R1196–R1201.

Sollid, J., Rissanen, E, Numminen, H, Thorstensen, T, Vuori, K. A. M., Nikinmaa, M, and Nilsson, G. E. (2006). HIF-1a and iNOS in crucian carp gills during hypoxia induced transformation. *J. Comp. Physiol. B* **176,** 359–369.

Stecyk, J. A. W., and Farrell, A. P. (2006). Regulation of the cardiorespiratory system of common carp (*Cyprinus carpio*) during severe hypoxia at three seasonal acclimation temperatures. *Physiol. Biochem. Zool.* **79,** 614–627.

Stecyk, J. A. W., Overgaard, J., Farrell, A. P., and Wang, T. (2004a). α-Adrenergic regulation of systemic peripheral resistance and blood flow distribution in the turtle (*Trachemys scripta*) during anoxic submergence at 5°C and 21°C. *J. Exp. Biol.* **207,** 269–283.

Stecyk, J. A. W., Stensløkken, K- O., Farrell, A. P., and Nilsson, G. E. (2004b). Maintained cardiac pumping in anoxic crucian carp. *Science* **306,** 77.

Stecyk, J. A. W., Stensløkken, K.-O., Nilsson, G. E., and Farrell, A. P. (2007). Adenosine does not save the heart of anoxia-tolerant vertebrates during oxygen deprivation. *Comp. Biochem. Physiol. A* **147,** 961–973.

Stecyk, J. A., Galli, G. L., Shiels, H. A., and Farrell, A. P. (2008). Cardiac survival in anoxia-toolerant vertebrates: An electrophysiological perspective. *Comp. Biochem. Physiol. C Toxicol. Pharmacol.* **148,** 339–354.

Strokin, M., Chechneva, O., Reymann, K. G., and Reiser, G. (2006). Neuroprotection of rat hippocampal slices exposed to oxygen-glucose deprivation by enrichment with docosahexaenoic acid and by inhibition of hydrolysis of docosahexaenoic acid-containing phospholipids by calcium independent phospholipase A2. *Neuroscience* **140,** 547–553.

Sundin, L., Axelsson, M., Dawison, W., and Forster, M. E. (1999). Cardiovascular responses to adenosine in the Antarctic fish *Pagothenia borchgrevinki. J. Exp. Biol.* **202,** 2259–2267.

Suzue, T., Wu, G-B., and Furukawa, T. (1987). High susceptibility to hypoxia of afferent synaptic transmission in the goldfish sacculus. *J. Neurophysiol.* **58,** 1066–1079.

Taha, M., and Lopaschuk, G. D. (2007). Alterations in energy metabolism in cardiomyopathies. *Ann. Med.* **39,** 594–607.

Tibbits, G. F., Moyes, C. D., and Hove-Madsen, L. (1992). Excitation-contraction coupling in the teleost heart. *In* "Fish Physiology Vol. 12A, The Cardiovascular System, (Hoar, W. S., Randall, D. J., and Farrell, A. P., Eds.), pp. 267–304. Academic Press, San Diego.

Tiitu, V., and Vornanen, M. (2001). Cold adaptation suppresses the contractility of both atrial and ventricular muscle of the crucian carp heart. *J. Fish Biol.* **59,** 141–156.

Tiitu, V., and Vornanen, M. (2002). Regulation of cardiac contractility in a cold stenothermal fish, the burbot *Lota lota* L. *J. Exp. Biol.* **205,** 1597–1606.

Tiitu, V., and Vornanen, M. (2003). Ryanodine and dihydropyridine receptor binding in ventricular cardiac muscle of fish with different temperature preferences. *J. Comp. Physiol. B.* **173,** 285–291.

Turner, N., Else, P., and Hulbert, A. J. (2003). Docosahexaenoic acid (DHA) content of membranes determines molecular activity of the sodium pump: Implications for disease states and metabolism. *Naturwissenschaften* **90,** 521–523.

Usherwood, P. N. R. (1978). Amino acids as neurotransmitters. *Adv. Comp. Physiol. Biochem.* **7,** 227–309.

Van Ginneken, V., Nieveen, M., Van Eersel, R., Van den Thillart, G., and Addink, A. (1996). Neurotransmitter levels and energy status in brain of fish species with and without the survival strategy of metabolic depression. *Comp. Biochem. Physiol. A* **114,** 189–196.

Van Waarde, A., Van den Thillart, G., and Verhagen, M. (1993). Ethanol formation and pH regulation in fish. *In* "Surviving Hypoxia: Mechanisms of Control and Adaptation" (Hochachka, P. W., Lutz, P. L., and Sick, T., *et al*, Eds.), pp. 157–170. CRC Press, Boca Raton.

Van Waversveld, J., Addink, A. D. F., and Van den Thillart, G. (1989). Simultaneous direct and indirect calorimetry on normoxic and anoxic goldfish. *J. Exp. Biol.* **142,** 325–335.

Virtanen, J. A., Cheng, K. H., and Somerharju, P. (1998). Phospholipid composition of the mammalian red cell membrane can be rationalized by a superlattice model. *PNAS* **95,** 4964–4969.

Vornanen, M. (1994a). Seasonal adaptation of crucian carp (*Carassius carassius* L.) heart: Glycogen stores and lactate dehydrogenase activity. *Can. J. Zool.* **72,** 433–442.

Vornanen, M. (1994b). Seasonal and temperature-induced changes in myosin heavy chain composition of crucian carp hearts. *Am. J. Physiol.* **267,** R1567–R1573.

Vornanen, M. (1997). Sarcolemmal Ca influx through L-type Ca channels in ventricular myocytes of a teleost fish. *Am. J. Physiol.* **272,** R1432–R1440.

Vornanen, M. (1999). Na^+/Ca^{2+} exchange current in ventricular myocytes of fish heart: Contribution to sarcolemmal Ca^{2+} influx. *J. Exp. Biol.* **202,** 1763–1775.

Vornanen, M., and Paajanen, V. (2004). Seasonality of dihydropyridine receptor binding in the heart of an anoxia-tolerant vertebrate, the crucian carp (*Carassius carassius* L.). *Am. J. Physiol.* **287,** R1263–1269.

Vornanen, M., and Paajanen, V. (2006). Seasonal changes in glycogen content and Na^+-K^+-ATPase activity in the brain of crucian carp. *Am. J. Physiol.* **291,** R1482–R1489.

Vornanen, M., and Tuomennoro, J. (1999). Effects of acute anoxia on heart function in crucian carp: Importance of cholinergic and adenosinergic control. *Am. J. Physiol.* **277,** R465–R475.

Vornanen, M., Ryökkynen, A., and Nurmi, A. (2002a). Temperature-dependent expression of K^+ currents in rainbow trout atrial and ventricular myocytes. *Am. J. Physiol.* **282,** R1191–R1199.

Vornanen, M., Shiels, H. A., and Farrell, A. P. (2002b). Plasticity of excitation-contraction coupling in fish cardiac myocytes. *Comp. Biochem. Physiol. A* **132,** 827–846.

Wilber, C. G., Robinson, P. F., and Hunn, J. B. (1961). Heart size and body size in fish. *Anat. Rec* **140,** 285–287.

Wilkie, M. P., Pamenter, M. E., Alkabie, S., Carapic, D., Shin, D. S., and Buck, L. T. (2008). Evidence of anoxia-induced channel arrest in the brain of the goldfish (Carassius auratus). *Comp. Biochem. Physiol. C Toxicol. Pharmacol.* **148,** 355–362.

Williamson, J. R., Safer, B., Rich, T., Schaffer, S., and Kobayashi, K. (1976). Effects of acidosis on myocardial contractility and metabolism. *ActaMedScandSuppl* **587,** 95–112.

Wissing, J., and Zebe, E. (1988). The anaerobic metabolism of the bitterling *Rhodeus amarus* (Cyprinidae, Teleostei). *Comp. Biochem. Physiol. B* **89,** 299–303.

Zingman, L. V., Alekseev, A. E., Hodgson-Zingman, D. M., and Terzic, A. (2007). ATP-sensitive potassium channels: Metabolic sensing and cardioprotection. *J. Appl. Physiol.* **103,** 1888–1893.

10

METABOLIC AND MOLECULAR RESPONSES OF FISH TO HYPOXIA

JEFFREY G. RICHARDS

Hypoxia survival requires a well-coordinated response to either secure more O_2 from the depleted environment or to defend against the metabolic consequences of too little O_2 at the mitochondria, which limits aerobic ATP production. Inhibition of aerobic ATP production during hypoxia exposure imposes a substrate-limited cap on the duration of survival because O_2-independent ATP production (anaerobic) is far less efficient than aerobic ATP production. It has long been held that hypoxia-tolerant animals are able to extend the period of survival under severely hypoxic conditions through a depression of basal metabolic rate, which limits the extent of activation of O_2-independent pathways of ATP production. This contention appears to be supported by the available literature; however, more studies measuring metabolic rate during hypoxia exposure are needed before a definitive outcome can be decided. Duration of hypoxia exposure is also an important component to consider when assessing the responses to hypoxia. Long-term hypoxia exposure (> a few hours in some cases) can result in large changes in

Hypoxia: Volume 27
FISH PHYSIOLOGY

DOI: 10.1016/S1546-5098(08)00010-1

gene expression, which underlie acclimation/acclimatization and potentially enhance hypoxic survival. Hypoxia-mediated changes in gene expression are likely regulated by the transcription factor, hypoxia inducible factor (HIF), which is well characterized in mammalian systems, but has only recently been examined in fish. Hypoxia inducible factor appears to be regulated in a similar fashion in fish as in mammals, but to date, there does not appear to be a direct link between HIF function and hypoxia tolerance in fish.

1. INTRODUCTION

Environmental hypoxia is a common, naturally occurring phenomenon in many aquatic ecosystems, the prevalence of which is increasing due to anthropogenic nutrient loading and eutrophication (reviewed in Chapter 1). In light of these O_2 fluctuations in the aquatic environment, it is perhaps not surprising that among all vertebrates, fish boast the largest number of hypoxia-tolerant species; hypoxia has clearly played an important role shaping the evolution of many unique adaptive strategies for hypoxic survival. Previous chapters in this volume have outlined a myriad of physiological and biochemical strategies that facilitate O_2 uptake under hypoxic conditions including changes in behavior, ventilation, hemoglobin-O_2 binding characteristics, and cardiovascular function. These strategies work to sustain metabolic function by maximizing O_2 extraction from the environment. Of importance to the present chapter, however, are the biochemical and molecular strategies that are responsible for defending against the metabolic consequences of O_2 levels that fall below a threshold where metabolic function is affected or cannot be maintained. Paramount to this defense strategy is a well coordinated response to maintain cellular ATP turnover, albeit at reduced levels, and the ability for hypoxic acclimation to "enhance" cellular and whole animal function under O_2 limiting conditions.

Metabolic and molecular responses to hypoxia are critical to enhance survival at O_2 levels below a species critical oxygen tension (P_{crit}). In the context of this chapter, P_{crit} is defined as the environmental O_2 tension at which an organism's O_2 consumption rate transitions from being independent of environmental O_2 to being dependent on environmental O_2 (see Figure 10.1A; Pörtner & Grieshaber 1993). As such, P_{crit} represents a whole-animal measure of O_2 extraction capacity from the environment and is considered by many researchers as an indicator of hypoxia tolerance (Chapman *et al.*, 2002). Many physiological adjustments can affect P_{crit}, and the majority of these have been outlined in previous chapters in this volume. For example, increases in O_2 extraction capacity through modifications to ventilation (see Chapter 5), O_2 transport systems (see Chapter 6),

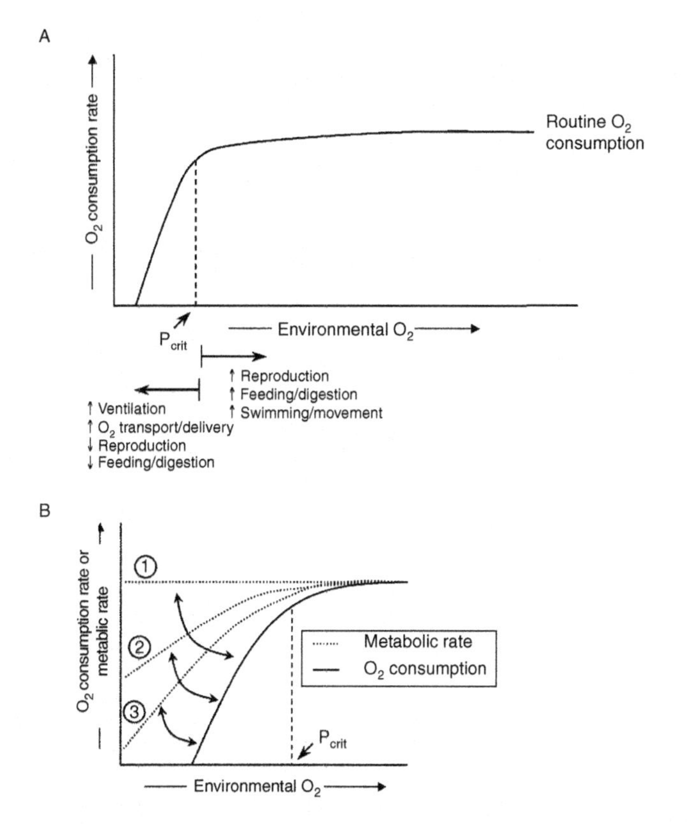

Fig. 10.1. Metabolic responses of fish to changes in environmental O_2. (A) A species' critical oxygen tension (P_{crit}) is the point at which O_2 consumption rate transitions from being independent of environmental O_2 levels (often referred to as O_2 regulation) to being dependent on environmental O_2 (often referred to as O_2 conforming). P_{crit} can be increased, detrimentally affecting hypoxia tolerance by increasing energetically expensive processes such as reproduction, growth, or digestion. P_{crit} can also be decreased, enhancing hypoxia tolerance through changes in respiration (V_E and gill perfusion), O_2 transport/delivery (changes in hemoglobin–O_2 binding affinity and cardiovascular responses), or through reductions in energetically expensive processes such as reproduction, digestion, and swimming. (B) At O_2 levels below P_{crit}, survival is dependent upon the ability of an animal to suppress basal metabolic rate to limit the extent of the activation of O_2-independent pathways of ATP production. See text for more detail.

or O_2 delivery systems (see Chapter 7) can theoretically result in a decrease in P_{crit}, and thus an enhancement of hypoxia tolerance. In contrast, increases in whole animal metabolic demands associated with, for example, gonad development and reproduction (see Chapter 3) as well as during digestion and allocation of energy to growth (see Chapter 8) can cause an increase in P_{crit}, and a decrease in hypoxia tolerance. Thus, suppression of reproduction, digestion, and growth during hypoxia exposure reduces metabolic demands and enhances hypoxia tolerance and survival (Figure 10.1A).

2. THE METABOLIC CHALLENGE OF HYPOXIA EXPOSURE

At O_2 levels below P_{crit}, the fundamental challenge is one of metabolic energy balance. Greater than 95% of the O_2 consumed by a fish in normoxia is used as the terminal electron acceptor by the mitochondrial electron transport chain for ATP production (via oxidative phosphorylation). If environmental hypoxia leads to hypoxemia (i.e., physiological mechanisms to enhance O_2 uptake are insufficient to protect the animal from its environment and blood O_2 content is reduced), then there is the potential for an O_2 limitation at the mitochondrion, which imposes limitations on the capacity for ATP production. Under these conditions, ATP can only be generated by processes such as glycolysis yielding lactate production or through direct phosphate transfer from phosphorylated intermediates such as creatine phosphate (CrP). These processes of direct phosphate transfer from a substrate to ADP forming ATP are termed substrate-level phosphorylation. Although these processes of ATP generation can occur during periods of O_2 lack, the amount of ATP produced per mole of substrate consumed is approximately 15- to 30-fold lower than if mitochondrial respiration occurs. For example, aerobic catabolism of 1 mole of glucose yields ~30 moles of ATP, while the anaerobic catabolism of glucose, involving only glycolysis and lactate production, produces 2 moles of ATP. A reduction in the ability of an organism or cell to generate sufficient ATP to meet metabolic demands presents a problem for the maintenance of cellular energy balance. Hypoxia-sensitive animals quickly succumb to hypoxia due to an inability to maintain cellular energy balance and a loss of cellular [ATP] (Boutilier, 2001). Thus, during hypoxia the inhibition of O_2-based mitochondrial ATP production imposes a potential substrate-limited cap on the duration of survival. Under these O_2 limiting conditions duration of survival is dictated by two, interrelated factors: (1) the ability to reduce metabolic demands through a controlled metabolic rate suppression; and (2) the availability of substrate for O_2-independent ATP production. Illustrated in Figure 10.1B is a conceptual framework to understand the relationship between metabolic rate suppression and capacity for O_2-independent ATP production. At environmental O_2 tensions below P_{crit}, hypoxia tolerance is likely to be dictated by the degree of metabolic rate suppression, which extends the length of time a fixed quantity of fermentable substrate can support cellular function. For example, scenario 1 would represent a severely hypoxia-sensitive fish, where at O_2 tensions below P_{crit}, the animal attempts to maintain metabolic rate, which can only be accomplished by a large activation of O_2-independent pathways of ATP production (largely glycolysis) thus utilizing fermentable fuels at a high rate (indicated by the large curved arrow). If the quantity of fermentable fuels is limited, then the animal will quickly succumb to hypoxia and die. On the other hand, scenarios 2 and 3 represent increasing levels of hypoxia tolerance, where decreases in metabolic rate limit the magnitude of the activation of O_2-independent ATP production (shorter curved arrows) and extend the period of time that can be supported by substrate-level phosphorylation. Thus, it seems reasonable to hypothesize that there should be a relationship between hypoxia tolerance, the magnitude of the hypoxia-induced metabolic rate suppression, and the availability of fermentable fuels to support O_2-independent ATP production.

At the cellular level, the precise mechanism of hypoxia-induced death is not known; however, it is clear that hypoxic death in fish is associated with catastrophic loss of substrate, failure of essential ATP consuming processes, accumulation of toxic levels of waste products (protons and lactate), and cellular necrosis. Underlying all of these potential causes of hypoxia-induced death is an inability of the animal to maintain metabolic energy balance. Boutilier and St-Pierre (2000) analyzed the available literature and proposed an elegant hypoxia-induced (and hypothermia-induced) cascade of events that yield necrotic cell death. In hypoxia-sensitive animals, hypoxia exposure leads to an inability to generate sufficient ATP to meet the metabolic demands of cellular ion regulation, protein synthesis, and other metabolic processes – a mismatch between ATP supply and demand – therefore, cellular [ATP] falls to levels that are insufficient to maintain the activity of these energy-consuming processes. Boutilier and St-Pierre (2000) pointed to cellular ion regulation to be the most critical aspect of cell survival and proposed that a loss of ATP limits the capacity of a cell to maintain trans-membrane potential resulting from net Na^+ influx and K^+ efflux. This results in depolarization of plasma and organelle membranes, Ca^{2+} accumulation in the cytosol from organelles and extracellular fluid, the activation of phospholipases and Ca^{2+}-dependent proteases, and the rupture of membranes, ultimately resulting in necrotic cell death. It has been proposed that hypoxia-tolerant animals are able to stave off these catastrophic events by initiating regulated metabolic rate suppression and stabilizing cellular [ATP].

Stable cellular [ATP] during hypoxia exposure is often accepted as the hallmark measure of a hypoxia-tolerant animal (Hochachka *et al.*, 1996; Boutilier, 2001); however, this has been demonstrated to be an over simplification. Numerous studies have shown a substantial disruption of cellular energetics during hypoxia exposure even in hypoxia-tolerant organisms (van den Thillart *et al.*, 1980,1989; Borger *et al.*, 1998; Hallman *et al.*, 2008;

Jibb and Richards, 2008; Richards *et al.*, 2008;) and changes in cellular [ATP] appear to be tissue specific. For example, in muscle, numerous studies have demonstrated that cellular [ATP] is not affected by hypoxia/anoxia exposure (van den Thillart *et al.*, 1980; Richards *et al.*, 2007, 2008); while in liver, [ATP] decreases initially upon hypoxia exposure and then stabilizes at a lower concentration (Figure 10.2A) (J. Dalla Via *et al.*, 1994; Jibb and Richards, 2008; van den Thillart *et al.*, 1980). These results are in general agreement with the results of Busk and Boutilier (2005) who showed in isolated eel hepatocytes that anoxia exposure caused an initial decrease in [ATP], followed by a stabilization at a new, lower level. In contrast, Krumschnabel *et al.* (1997) demonstrated that exposure of isolated goldfish hepatocytes to anoxia did not result in a decrease in [ATP], while the same preparation exposed to chemical anoxia, using NaCN, showed a decrease in [ATP]. This latter decrease in [ATP] was modest when compared with the large decreases of [ATP] observed in anoxia-exposed hepatocytes isolated from the hypoxia-intolerant rainbow trout (Krumschnabel *et al.*, 1997). It has been postulated that the reason for the differences in response in [ATP] between muscle and liver is related to the tissue [CrP]. Muscle [CrP] are much higher than measured in liver (20 to 50 versus <5 μmol/g wet tissue, respectively); thus, in liver, there is a lack of capacity to buffer [ATP] during the onset of hypoxia.

Whether tissue [ATP] is affected by hypoxia or not, intracellular acidosis and CrP hydrolysis result in an accumulation of [ADP_{free}] and [AMP_{free}], causing increases in [ADP_{free}]/[ATP] and [AMP_{free}]/[ATP] and substantial losses of cellular phosphorylation potential (Figure 10.2; Hallman *et al.*, 2008; Jibb and Richards, 2008; Richards *et al.*, 2008; van den Thillart *et al.*, 1989). This disruption of cellular energy status plays several important roles in the cell during hypoxia exposure. First, decreases in phosphorylation potential may affect rates of cellular ATP production and substrate oxidation. For example, hypoxia exposure was associated with a significant drop in the free energy of ATP hydrolysis ($\Delta fG'$; Figure 10.2C) (Hallman *et al.*, 2008; Jibb and Richards, 2008; Richards *et al.*, 2008). Estimates of the critical limit of $\Delta fG'$ for the maintenance of cellular function suggest that below a threshold of -52 kJ/mol, cellular processes such as ion pumping can no longer derive sufficient energy from ATP hydrolysis to be maintained (Hardewig *et al.*, 1998; Jansen *et al.*, 2003). Second, changes in cellular [ADP_{free}]/[ATP] and [AMP_{free}]/[ATP] are important signals coordinating the metabolic responses to hypoxia. For example, increases in [ADP_{free}]/[ATP] are known to allosterically activate glycolysis, increasing O_2-independent ATP production and more recent evidence indicates that increases in [AMP_{free}]/[ATP] may be vital to overall coordination of metabolic rate suppression in certain tissues of hypoxia-tolerant fish (Jibb and Richards, 2008).

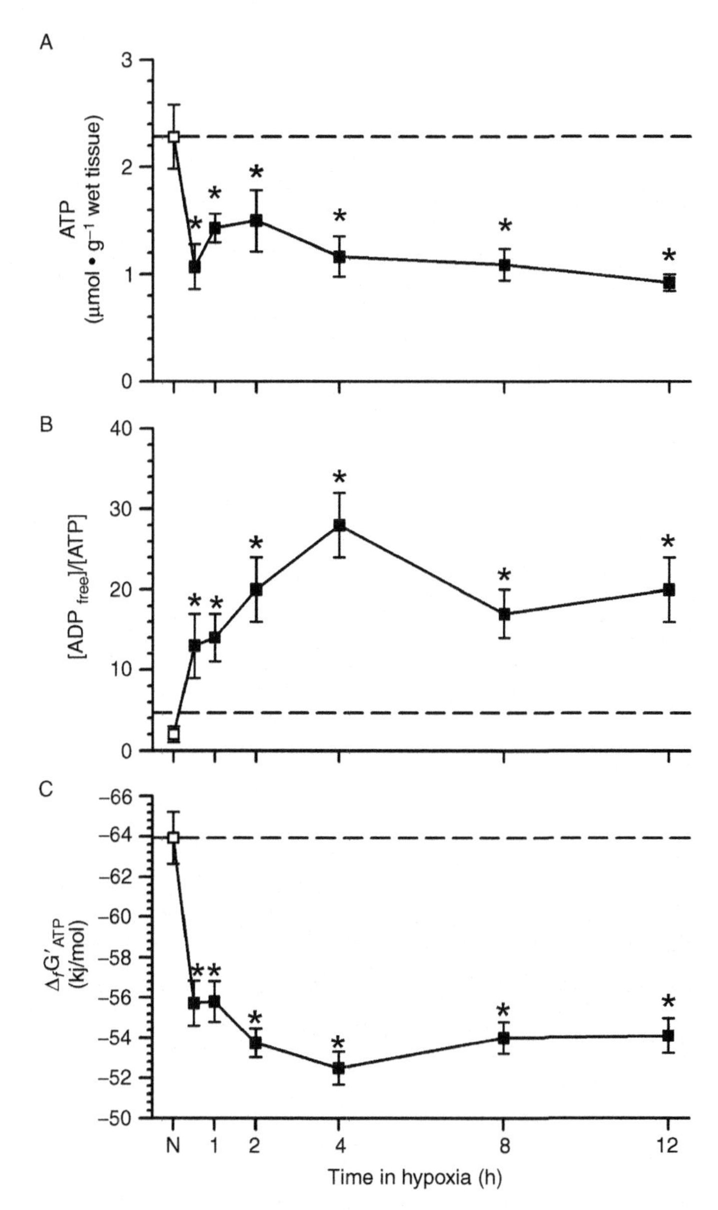

Fig. 10.2. Liver [ATP] (A), calculated [ADPfree]/[ATP] (B), Gibbs free energy of ATP hydrolysis (C) ($\Delta_f G'_{ATP}$) in goldfish exposed to normoxia and 12 h of hypoxia (<0.5% air saturation). Horizontal dashed lines through normoxia are shown as a reference. [Data from Jibb and Richards (2008) with permission.]

3. THE CONCEPT OF TIME IN THE METABOLIC RESPONSES TO HYPOXIA

When considering the physiological and biochemical responses to hypoxia, environmental O_2 levels are not the only factor to consider: the length of time spent in hypoxia can have dramatic effects on the responses to O_2 lack. Upon exposure to hypoxia, immediate survival is dependent upon the ability of the fish to quickly modify existing physiological and biochemical systems in an attempt to maintain metabolic function. If these immediate responses are sufficient for survival of the onset of hypoxia, then an animal has the opportunity to acclimate or acclimatize, which, for the most part, is thought to be of benefit in enhancing the ability of an animal to survive hypoxic exposure. At the heart of any acclimation response are changes in gene expression, which can alter the capacity of an animal to endure hypoxia. Changes in gene expression, if they are translated into functional changes in protein amount or possibly protein turnover rates, can affect hypoxia survival by either increasing or decreasing the amounts of specific proteins in a metabolic pathway. For example, large increases in the expression of the lactate dehydrogenase gene (*ldh*) and increases in LDH activity have been observed during both long- and short-term hypoxia exposure in fishes (e.g., Amazonia cichlid) (Almeida-Val *et al.*, 1995, 2006). In addition, selective changes in gene expression can result in protein isoform switching, which in some cases has been shown to enhance survival to environmental perturbation (Schulte, 2004). Environmental hypoxia is well known to affect gene expression patterns in fish with several microarray studies showing changes in the transcription of genes involved in O_2 uptake, energy turnover, growth and development, immune responses, cell signaling, and stress (Figure 10.3 and Table 10.1; Gracey *et al.*, 2001; Ton *et al.*, 2003). Thus, almost every physiological and biochemical response discussed earlier in this volume is regulated, at least in part, by changes in gene expression. Regulation and coordination of changes in gene expression in response to hypoxia exposure are mediated largely by the transcription factor hypoxia inducible factor (HIF), which has been characterized in mammalian and fish systems.

The remaining portion of this chapter is divided into two parts. The first section outlines the metabolic and molecular responses of fish to hypoxia exposure. Combining the metabolic and molecular gene expression changes is meant to emphasize that the changes observed in metabolic phenotype are also controlled to a degree by changes in gene expression, which underlie acclimation responses. The second portion of this chapter examines how these processes are coordinated at the biochemical and molecular level with emphasis on HIF as a regulator of hypoxia-induced changes in gene expression.

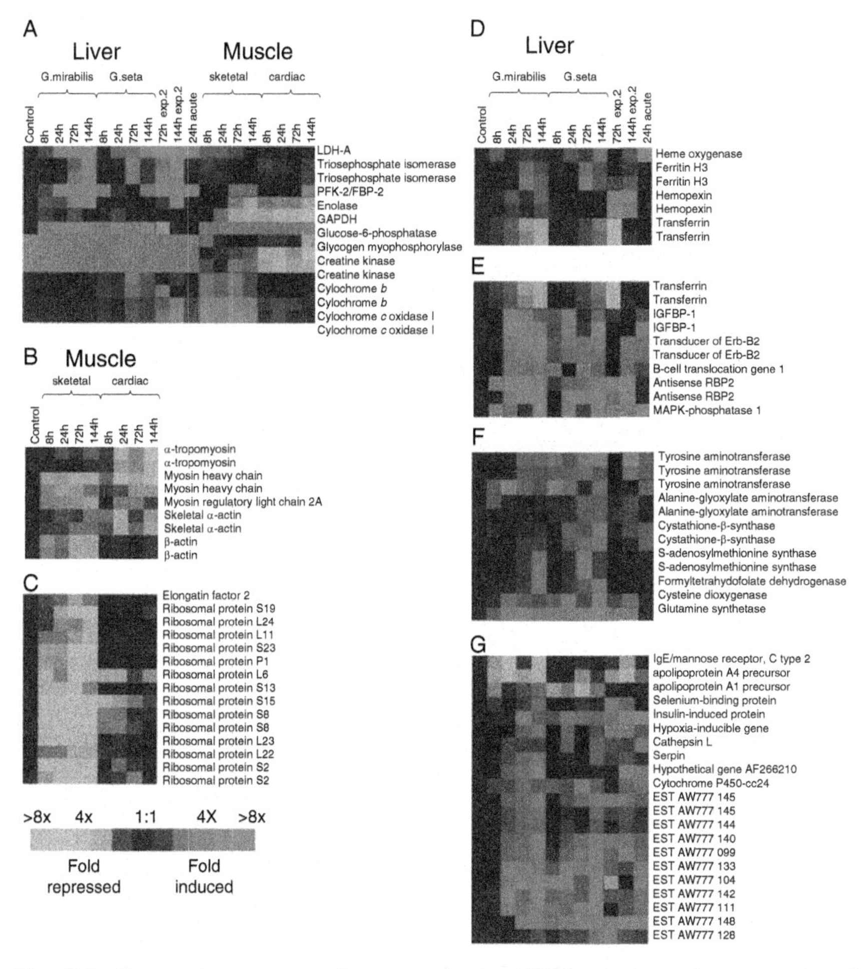

Fig. 10.3. Changes in gene expression assessed using cDNA microarray in two species of mudsucker during hypoxia exposure. Genes are categorized on the basis of their probable biological role: (A) ATP metabolism; (B) locomotion and contraction; (C) protein translation; (D) iron metabolism; (E) antigrowth and proliferation; (F) amino acid metabolism; and (G) cryptic role. [Reproduced from Gracey *et al.* (2001) with permission.] (See Color Insert.)

4. METABOLIC AND MOLECULAR RESPONSES TO HYPOXIA

Many excellent reviews have summarized the metabolic and molecular responses of fish and other lower vertebrates to hypoxia exposure (Almeida-Val *et al.*, 2006; Bickler and Buck, 2007; Nikinmaa and Rees, 2005). Many of the chapters in this book outline numerous responses to hypoxia including responses that work to increase O_2 uptake and the metabolic adjustments

Table 10.1
Molecular responses to hypoxia

Common name	Scientific name	Temp. (°C)	Hypoxia	Duration	Whole animal or tissue	Gene expression changes	Reference
Longjaw mudsucker	*Gillichthys mirabilis*	15	10% air sat.[a]	8, 24, 72, and 144 h	Liver	↑ glycolysis (7) ↑ amino acid metabolism ↑ iron and Hb metabolism (8) ↑ anti-growth & cell proliferation (10) ↓ aerobic metabolism (4)	Gracey *et al.* (2001)
					Muscle	↑ glycolysis (4) ↓ glycolysis & CK (5) ↓ aerobic metabolism (4) ↓ locomotion and contraction (9) ↓ protein synthesis (15)	
Zebrafish	*Danio rerio*	28	5% oxygen 23% air sat.	24 h	48 h post-fertilization embryos	↑ glycolysis (6) ↑ cell signaling (5) ↓ aerobic metabolism (10) ↓ creatine kinase (2) ↓ cell structure & mobility (20) ↓ ion transporting ATPases (5) ↓ protein synthesis (6) ↓ iron & Hb metabolism (5) ↓ cell division (5) ↓ cell/organism defense (5)	Ton *et al.* (2003)

Zebrafish	*Danio rerio*	28	10% air sat.	21 days[b]	Gills	↑ glycolysis (2) ↑ disease defense (12) ↑ phosphatases (6) ↑ chapterones (8) ↓ aerobic metabolism (31) ↓ protein synthesis (54) ↑ elongation factors (2) ↓ stress response (6) ↓ apoptosis (8) ↓ locomotion and contraction ↓ growth regulation (12) ↓ innume response (9) ↓ proteosome degradation (6)	van der Meer *et al.* (2005)

The numbers of genes indicated may have either increased or decreased in expression, but overall the authors concluded that the changes indicated would yield an overall change in biological outcome indicated by the arrow.

[a] P_{crit} of this species is 1.2 mg O_2/L and hypoxia exposure was 0.8 mg O_2/L

[b] Decrease in oxygen occurred gradually over 4 days.

associated with hypoxia exposure in the heart (see Chapter 7) and the metabolic aspects yielding the extraordinarily anoxia-tolerant crucian carp (*Carassius carassius*; see Chapter 9). From all of the preceding chapters it has become evident that there are three principal aspects that function to maintain cellular energy balance and these include: (1) increased O_2 uptake from the hypoxic environment to sustain a modicum of aerobic ATP production; (2) strong activation of an O_2-independent means of ATP production; and (3) a reduction in metabolic demands through regulated metabolic rate suppression, which is described in more detail below.

4.1. Increases in O_2 Transport

As outlined previously in this volume, the physiological and biochemical responses that yield increases in O_2 transport capacity are important adaptations to survive hypoxia. Indeed, recent work by Mandic *et al.* (2008) showed in a group of closely related intertidal fish species (sculpins from the family Cottidae) that approximately 75% of the variation in hypoxia tolerance (assessed as P_{crit}) could be explained by variation in physiological attributes affecting O_2 uptake (Hb–O_2 binding affinity or gill surface area) or O_2 use (routine metabolic rate). However, since other chapters have explicitly dealt with the physiological responses that increase O_2 uptake from the environment and O_2 delivery to the tissues, in this section I will solely focus on the O_2-dependent changes in gene expression that may form the foundation of possible acclimation responses. In fact, almost every microarray study performed to date has shown an effect of hypoxia exposure on mRNA levels for proteins involved in Hb metabolism and oxygen transport (Figure 10.3 and Table 10.1).

As pointed out in Chapter 6, modifications to Hb–O_2 binding affinity and blood Hb content are important responses to hypoxia. At the gene expression level, Gracey *et al.* (2001) showed dramatic changes in the expression of genes involved in heme metabolism in liver of the mudsucker in response to hypoxia exposure. Several genes involved in iron-heme catabolism and heme protein turnover were all induced by hypoxia exposure. These general changes in genes involved in iron and heme metabolism could be linked with hypoxia-induced erythropoietin (EPO) or erythropoiesis and increased demand for iron from hemoglobin synthesis. By contrast, zebrafish embryos exposed to hypoxia show a general decrease in the expression of genes involved in Hb metabolism. Specifically, Ton *et al.* (2003) showed large decreases in mRNA levels for globin, βA1, hemoglobin β chain, globin α-embryonic, globin 2 α-embryonic, and, oddly, erythropoietin. The probable explanation for these counterintuitive decreases in mRNA levels for proteins involved in blood O_2 transport is that the very small zebrafish embryos do not require blood flow for survival and O_2 uptake is mostly via diffusion.

In mammals, hypoxia is a powerful regulator of the production of erythropoietin (Semenza and Wang, 1992), which causes an increase in red blood cell production leading to increases in Ht and increases in blood O_2-carrying capacity. There is no published data directly linking hypoxia exposure and changes in EPO gene expression in fish, although injection of human EPO into goldfish stimulates red blood cell production (Taglialatela and Della Corte, 1997) demonstrating that if EPO is synthesized it could enhance red blood cell production. It must be pointed out, however, that the available data on hypoxia-induced EPO regulation in fish is not clear in its conclusion. When Fugu EPO gene and promoter region constructs (6 kb) are transfected into human carcinoma cell lines transcription is not hypoxia responsive (Chou *et al.*, 2004) and this is supported by a lack of an HIF-binding hypoxia response element (HRE) in the promoter region of Fugu. However, when these same promoter region constructs were transfected into fish cell lines, increased expression of an alternatively spliced EPO transcript was observed in cells subjected to hypoxia (Fraser *et al.*, 2006), suggesting at least some degree of hypoxia regulation of EPO in fish.

Hypoxia-induced changes in myoglobin (Mb) expression have recently received considerable attention in fish and have been discussed by Wells (see Chapter 6) and Gamperl and Driedzic (see Chapter 7). Typically, Mb is expressed at high levels in red-skeletal and cardiac muscle, but recent evidence has shown a hypoxia-induced expression of Mb in nonmuscle tissues of the hypoxia-tolerant common carp (*Cyprinus carpio*; Fraser *et al.*, 2006) and in the gills of zebrafish (van der Meer *et al.*, 2005). In the carp, increases in Mb mRNA were observed during 1–8 days of hypoxia exposure in the liver, gill, and brain. Increases in mRNA were reflected in increased protein expression determined using 2D gel electrophoresis, which suggests that the increase in Mb expression may enhance O_2 diffusion into tissues during hypoxia exposure. Enhanced expression of Mb in the gills of zebrafish (van der Meer *et al.*, 2005), suggests a potential generalized role for Mb in facilitating O_2 transport in fish tissues; however, it is interesting to reiterate that the expression of Hb genes were not affected in the same gills during hypoxia exposure. A brain-specific myoglobin was also identified in the common carp, distinct from neuroglobin, but it was not hypoxia responsive at the transcript level. Additional details on Mb and neuroglobin expression in fish can be found in Chapter 6 of this volume.

4.2. O_2-Independent ATP Production

Hypoxia exposure in fish elicits a strong activation of substrate-level phosphorylation via glycolysis and CrP hydrolysis and a decrease in aerobic metabolism. Endogenous glycogen typically serves as the carbohydrate store for glycolysis, thus the levels of tissue glycogen are indicative of the capacity of a tissue to support ATP turnover via glycolysis. Furthermore, due to the suppression of appetite and digestive function during hypoxia (see Chapter 8) endogenous stores of fermentable fuels represent the only source of substrate to support ATP production. As illustrated in Table 10.2, hypoxia-tolerant animals such as carp, goldfish, killifish, and oscar typically have higher levels of tissue glycogen relative to animals considered to be hypoxia sensitive (e.g., rainbow trout). Thus, it seems reasonable to conclude that across broad taxonomic groups of fish, those animals with more glycogen will be able to produce more ATP for longer periods of time at lower O_2 levels (see Figure 10.1B). Another striking feature illustrated by Table 10.2 is the very large glycogen stores that occur in liver compared with those observed in other tissues including the heart, brain, and skeletal muscle. Liver glycogen is thought to serve as a repository of glucose that can be used by other tissues to support glycolytic ATP production during hypoxia exposure; however, for this to occur the glucose liberated from liver glycogen must be transported between tissues. The details of glucose transport during hypoxia exposure are outlined in Chapter 7.

At the molecular level, every cDNA microarray study performed on fish has shown a typical hypoxia-induced metabolic switch, that is, a reduction in mRNA levels for proteins involved in aerobic metabolism and an increase in the mRNA levels for proteins involved in anaerobic metabolism (see Table 10.1; glycolysis, creatine kinase, and aerobic metabolism). For example, in zebrafish embryos, Ton *et al.* (2003) showed a decrease in the expression of mRNA coding for genes involved in the TCA cycle, including succinate dehydrogenase, malate dehydrogenase, and citrate synthase, and an increase in expression of genes involved in glycolysis including phosphoglycerate mutase, phosphoglycerate kinase, enolase, aldolase, and lactate dehydrogenase (details of hypoxia exposure given in Table 10.1). Similarly, in gills of zebrafish exposed to hypoxia, the levels of mRNA for proteins involved in the TCA cycle and electron transfer chain were all decreased signifying an overall decline in mitochondrial ATP production (van der Meer *et al.*, 2005). Simultaneously, increases in mRNA coding for proteins involved in glycolytic ATP production were noted in the gill during hypoxia exposure, including increases in glycogen phosphorylase and aldolase. Further, there was a general decrease in the expression of genes that code for proteins involved in fat metabolism, cellular uptake, and transport, including acyl-CoA dehydrogenase, intestinal fatty acid binding protein, and other metabolite binding proteins. Also associated with the metabolic switch from aerobic to anaerobic metabolism were highly sensitive changes in the expression of pyruvate dehydrogenase kinase, which was up-regulated in muscle of killifish (*Fundulus heteroclitus*) during hypoxia exposure, but

Table 10.2
Glycogen content in brain, liver, and muscle of fish

Common name	Scientific name	Brain	Liver	Skeletal muscle	Heart	References
Crucian carp	*Carassius carassius*	13 to 204	123 to 2160		18 to 493	Voranen *et al.* (Chapter 9)
Goldfish	*Carassius auratus*	13 to 20	800	30	142	Voranen *et al.* (Chapter 9); (Mandic *et al.* (In Press)
Rainbow trout	*Oncorhynchus mykiss*	0.5	110		25 to 60	Voranen *et al.* (Chapter 9)
Killifish	*Fundulus heteroclitus*	N/A	299 to 550	10 to 40	N/A	Fangue *et al.* (2008); Richards *et al.* (2008)
Oscar	*Astronotus ocellatus*		203 to 279	25 to 30		Chippari-Gomes *et al.* (2005); Richards *et al.* (2007)
Blue discus	*Symphysodon aequifasciatus*		100	15		Chippari-Gomes *et al.* (2005)
Tilapia	*Oreochromis mossambicus*		175 mg/g protein			Chang *et al.* (2007)
African Lungfish	*Protropterus dolloi*		98 to 180	8 to 10	50 to 60	Frick *et al.* (2008)
Pacu	*Piaractus mesopoamicus*		ca. 500	15		Moraes *et al.* (2006)
Silver catfish	*Rhamdia quelen*			21		

Glycogen content is reported in μmol glucosyl units/g wet tissue unless otherwise stated.

these changes did not yield measurable changes in PDK protein content (Richards *et al.*, 2008).

Associated with the large increases in the expression of glycolytic enzymes, increases in mRNA levels for proteins involved in amino acid catabolism have been demonstrated. In the liver of longjaw mudsucker, Gracey *et al.* (2001) noted increases in *S*-adenoylmethionine synthase and cystathione synthease, which catalyze steps in methionine degradation as well as several aminotransferases. Consistent with the induction of aminotransferases was the coexpression of glutamine synthetase, which catalyzes the major liver ammonia detoxification reaction of the synthesis of glutamine from glutamate. Catabolism of gluconeogenic amino acids, such as tyrosine and serine, yields either pyruvate or TCA cycle intermediates, both of which can serve as carbon skeletons for gluconeogenesis. Further evidence linking amino acid catabolism with hypoxia-induced gluconeogenesis is that the expression of glucose-6 phosphatase was strongly induced in response to hypoxia. Glucose-6 phosphatase catalyzes the dephosphorylation of glucose-6 phosphate to glucose, which can be transported in the circulation to other tissues to fuel glycolysis. Thus, for the longjaw mudsucker, amino acid catabolism coupled with gluconeogenesis in the liver may represent a mechanism to maintain blood glucose levels during hypoxia and may contribute to maintaining whole animal energy balance.

Changes in the mRNA levels for several metabolite transporters have also been noted in many studies. For example, mRNA for MCT4, a membrane-bound lactate/pyruvate transporter, increased in response to hypoxia exposure in zebrafish (Ton *et al.*, 2003). Furthermore, there were dramatic increases in the expression of glucose transporters (GLUT) in eye, gill, and kidney of grass carp during exposure to hypoxia (up to 170 h at ~0.6% air saturation; Zhang *et al.*, 2003). Changes in both of these transporters in response to hypoxia exposure indicate an overall increase in the movement of substrates for glycolysis and waste products (lactate).

Tissue-specific effects of hypoxia exposure have been noted in several studies suggesting that not all tissues respond similarly to hypoxia. Differential gene expression responses have been noted in the liver and muscle of the longjaw mudsucker during hypoxia exposure (Gracey *et al.*, 2001). In the liver of the mudsucker, there was an overall increase in mRNA levels for proteins involved in glycolysis, with large increases observed in mRNA for LDH-A, triosephosphate isomerase, PFK-2/FBP-2, enolase, and glucose 6-phosphatase. Smaller, yet significant increases in mRNA were also noted for cytochrome *b* and cytochrome *c* oxidase, which are proteins of the mitochondrial electron transport chain, and possibly point to an enhancement of overall capacity for the mitochondria in liver to sustain at least some level of ATP production. In muscle tissue, however only minor increases in mRNA for the glycolytic enzymes (LDH-A and PFK-2) were observed and in direct contrast to the response observed in liver, substantial decreases in mRNA levels for the glycolytic enzymes enolase, GAPDH, and glucose 6-phospate dehydrogenase, as well as creatine kinase were observed. Furthermore, unlike the effects of hypoxia on liver mRNA levels, there were substantial increases in the expression of cytochrome *b* and cytochrome *c* oxidase I in muscle.

In general, the tissue-specific responses observed in the mudsucker at the mRNA level is consistent with the tissue-specific effects of 4 weeks of hypoxia exposure (~15% air saturation) in *Fundulus grandis* (Martinez *et al.*, 2006). This study clearly demonstrated that enzyme activities of glycolysis and glycogen metabolism were strongly suppressed by hypoxia exposure in skeletal muscle, while in liver there was evidence for increases in several enzymes involved in glycolysis and carbohydrate oxidation (Figure 10.4). Fewer changes in glycolytic and glycogen enzymes were observed in the heart and brain compared with the liver and muscle and those that did change, did so with a smaller magnitude. Interestingly, among the tissues that showed general increases in enzymes within the glycolytic pathway, the enzymes that increased were not always the same enzymes. Martinez *et al.* (2006) speculated that tissue-specific differences in the responses to long-term hypoxia in *Fundulus grandis* reflect the balance of energetic demands, metabolic role, and oxygen supply to the tissues. More studies are needed to examine the tissue-specific effects of hypoxia exposure on metabolic energy supply.

4.3. Metabolic Rate Suppression

The ability to suppress cellular ATP demand to match the limited capacity for O_2-independent ATP production has emerged as the unifying adaptive strategy ensuring hypoxia survival (Hochachka *et al.*, 1996). Because ATP turnover rates cannot be measured directly *in vivo*, those interested in measuring metabolic rate must use indirect measures. The two typical indirect measures for metabolic rate is O_2 consumption and heat loss. Measurements of O_2 consumption only determine the contributions of aerobic metabolism to overall ATP turnover and therefore during periods of metabolic stress that lead to increases in substrate-level phosphorylation O_2 consumption can underestimate total ATP turnover or metabolic rate. The best indirect measure of metabolic rate (often referred to as the "direct" measure of metabolic rate to demonstrate its superiority) is the measurement of heat loss. Metabolic heat production is proportional to ATP turnover, therefore a reduction in heat loss can be directly linked with a reduction in total ATP turnover and metabolic rate suppression.

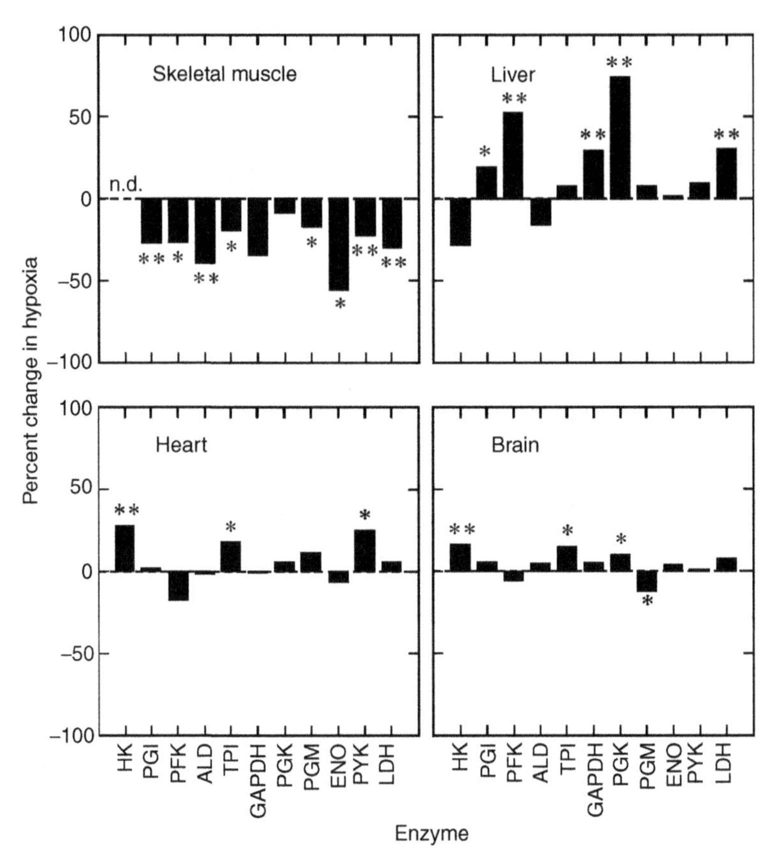

Fig. 10.4. Effects of long-term hypoxia exposure on glycolytic enzyme activities (i.u. mg^{-1} protein) in tissues of *Fundulus grandis*. The y axis represents the percentage change in the mean value for each enzyme measured from hypoxic fish relative to the normoxic value for (A) skeletal muscle, (B) liver, (C) heart, and (D) brain. HK, hexokinase; PGI, phosphoglucoisomerase; PFK, phosphofructokinase; ALD, aldolase; TPI, triose phosphate isomerase; GAPDH, glyceraldehyde-3-phosphate dehydrogenase; PGK, phosphoglycerokinase; PGM, phosphoglyceromutase; ENO, enolase; PYK, pyruvate kinase; LDH, lactate dehydrogenase. [Data from Martinez *et al.* (2006) with permission.]

4.3.1. Evidence of Metabolic Rate Suppression in Fish

Heat production in fish during hypoxia/anoxia exposure has been assessed in several species including goldfish (*Carassius auratus*; Stangl & Wegener, 1996; van Waversveld *et al.*, 1988a,b; van Ginneken *et al.*, 2004), crucian carp (*Carassius carassius*; Johansson *et al.*, 1995), tilapia

(*Oreochromis mossambiscus*; van Ginneken *et al.*, 1997, 1999), European eel (*Anquilla anquilla*; van Ginneken *et al.*, 2001), zebrafish (*Brachydanio rerio*; Stangl & Wegener, 1996), and in isolated hepatocytes from rainbow trout (*Oncorhynchus mykiss*; Rissanen *et al.*, 2006) (see Table 10.3). Interestingly, across a very broad range of fish species including those considered to be hypoxia tolerant (crucian carp, goldfish, and tilapia) and tissues from those considered to be intolerant (rainbow trout), all species show the capacity to decrease metabolic rate in response to hypoxia exposure. The most impressive reductions in metabolic rate, however, still occurred in the goldfish, tilapia, and European eel with an ~70% decrease in metabolic rate during hypoxia exposure (Table 10.3). Hepatocytes isolated from rainbow trout showed a decreased metabolic rate to a lesser degree than seen in more hypoxia-tolerant animals such as goldfish and tilapia (whole-animal measurements), but comparisons between isolated tissues and whole animals are difficult to make because of tissue-specific responses to hypoxia. Oddly, zebrafish exposed to severe hypoxia (<6% air saturation) for only 50 min showed a progressive increase in heat production indicating an overall increase in metabolic rate during hypoxia exposure (Stangl and Wegener, 1996). This increase in metabolic rate may represent increased costs associated with hypoxia-induced movement and escape behavior. Although metabolic rate suppression is clearly a response of fish to hypoxia/anoxia exposure, due to the limited number of studies available it is not possible to comment with any certainty on the direct association between the degree of metabolic rate suppression and overall hypoxia tolerance.

4.3.2. Mechanisms of Metabolic Rate Suppression

The question of how organisms are able to reduce metabolic rate below routine levels has received considerable attention over the past several decades. Original work in this field using hepatocytes isolated from the anoxia-tolerant turtle (*Chrysemys picta*), demonstrated that a 94% suppression in metabolic rate during anoxia exposure was achieved through the dramatic down-regulation of Na pumping, protein turnover, urea synthesis and gluconeogensis (Buck & Hochachka, 1993; Buck *et al.*, 1993a,b; Land *et al.*, 1993; Hochachka *et al.*, 1996; Hochachka and Lutz, 2001). It is now clear that cellular mechanisms underlying metabolic rate suppression are similar across broad taxonomic groups with metabolic rate suppression involving the controlled arrest of processes involved in membrane ion movement (Buck and Hochachka, 1993; Richards *et al.*, 2007), protein synthesis (Lewis *et al.*, 2007; Wieser and Krumschnabel, 2001), RNA transcription, urea synthesis, gluconeogensis, and other anabolic pathways (Hochachka *et al.*, 1996).

Table 10.3
Maximum recorded decreases in metabolic rate in fish during hypoxia/anoxia exposure

Common name	Scientific name	Whole animal or tissue	Temperature (°C)	Hypoxia	Duration	Metabolic Rate Suppression (% decrease from normoxia)	References
Goldfish	*Carassius auratus*	Whole animal	20	Anoxia	2 to 3h	70	van Waversveld *et al.* (1988, 1989a)
Goldfish	*C. auratus*	Whole animal	20	10% air sat.	3 h	59	van Waversveld *et al.* (1989a)
Goldfish	*C. auratus*	Whole animal	20	5% air sat.	3 h	53	van Waversveld *et al.* (1989a)
Goldfish	*C. auratus*	Whole animal	20	Anoxia	3 h	70 to 85	Stangl and Wegener (1996)
Goldfish	*C. auratus*	Whole animal	20	3% air sat. Progressive	5 h	55	van Ginneken *et al.* (2004)
Crucian carp	*Carassius carassius*	Brain slices	12	Anoxia	20 h	37	Johansson *et al.* (1995)
Tilapia	*Oreochromis mossambicus*	Whole animal	20	5%	8 h	55	van Ginneken *et al.* (1997)
Tilapia	*Oreochromis. mossambicus*	Whole animal	20	3% air sat. Progressive	1 h	64	van Ginneken *et al.* (1999)
European eel	*Anquilla anquilla*	Whole animal	20	Anoxia	1 h	70	van Ginneken *et al.* (2001)
Rainbow trout	*Oncorhynchus mykiss*	Hepatocytes	20[1a]	~4% air sat. Progressive	6 to 12 min	46	Rissanen *et al.* (2006)
Zebrafish	*Brachydanio rerio*	Whole animal	25	6% air sat.	50 min	Increase by 50[b]	Stangl and Wegener (1996)

[a]Animals were acclimated to 12°C and metabolic rate was measured in isolated hepatocytes at 20°C.
[b]Metabolic rate increased upon exposure to hypoxia.
In studies with more than one level of hypoxia shown, the degree of metabolic rate suppression for the most severe level of hypoxia is shown.

The suppression of protein synthesis has been described in both isolated hepatocytes and fish *in vivo* in species ranging from the crucian carp (Smith *et al.*, 1996; see Chapter 9), to goldfish (Jibb and Richards, 2008), to the Amazonian oscar (*Astronotus ocellatus*; Lewis *et al.*, 2007). In the oscar, severe hypoxia exposure (10% air saturation) caused tissue specific decreases in protein synthesis rates that varied from 27% decreases in protein synthesis rate in brain to 60% decreases in heart. In the crucian carp, Smith *et al.* (1996) also demonstrated substantial decreases in protein synthesis rates in heart, liver, and muscle in response to anoxia exposure and these decreases were, in part, mediated by decreases in RNA transcription rates (Smith *et al.*, 1999). In the goldfish, hypoxia exposure ($<0.5\%$ air saturation) caused a very rapid (within 0.5 h) $\sim 70\%$ decline in liver protein translation rate (assessed in cell-free isolates). These decreases in protein synthesis rates in the hypoxic goldfish were mediated through specific phosphorylation of eukaryotic elongation factor-2 (Jibb and Richards, 2008), which halts protein elongation during translation (Figure 10.5).

Few studies have examined how other ATP-consuming processes besides protein synthesis are modified during hypoxia exposure in fish, but some modifications to ion pumping have been noted. In particular, hypoxia-induced decreases in the activity of Na^+/K^+-ATPase as observed in some studies (Bogdanova *et al.*, 2005) could represent a substantial ATP saving, but results are conflicting. The crucian carp does not decrease brain Na^+/K^+-ATPase activity during anoxia exposure (reviewed in detail in Chapter 9; Hylland *et al.*, 1997) despite increases in the inhibitory neuromodulators GABA (Nilsson, 1992) and adenosine (Nilsson, 1991). This lack of an effect of anoxia/hypoxia exposure on Na^+/K^+-ATPase activity in the brain of crucian carp is unlike the response observed in turtles, which suppress the activity of Na^+/K^+-ATPase. The differential responses observed in the two champions of anoxia tolerance is probably associated with the fact that crucian carp remains active during anoxia exposure, unlike the comatose turtle (see Chapter 9). Recent work by Richards *et al.* (2007) demonstrated a substantial decrease in gill Na^+/K^+-ATPase activity in the oscar exposed to hypoxia ($\sim 5\%$ air saturation) and it was speculated that this decrease was achieved by a post-translational modification to the Na^+/K^+-ATPase protein. A similar effect of hypoxia exposure was observed in isolated trout hepatocytes, where hypoxia caused a transient down-regulation of Na^+/K^+-ATPase activity (Bogdanova *et al.*, 2005). These authors speculated that decreases in Na^+/K^+-ATPase activity in response to hypoxia may be accomplished by local changes in reactive oxygen species, but no precise mechanism was given.

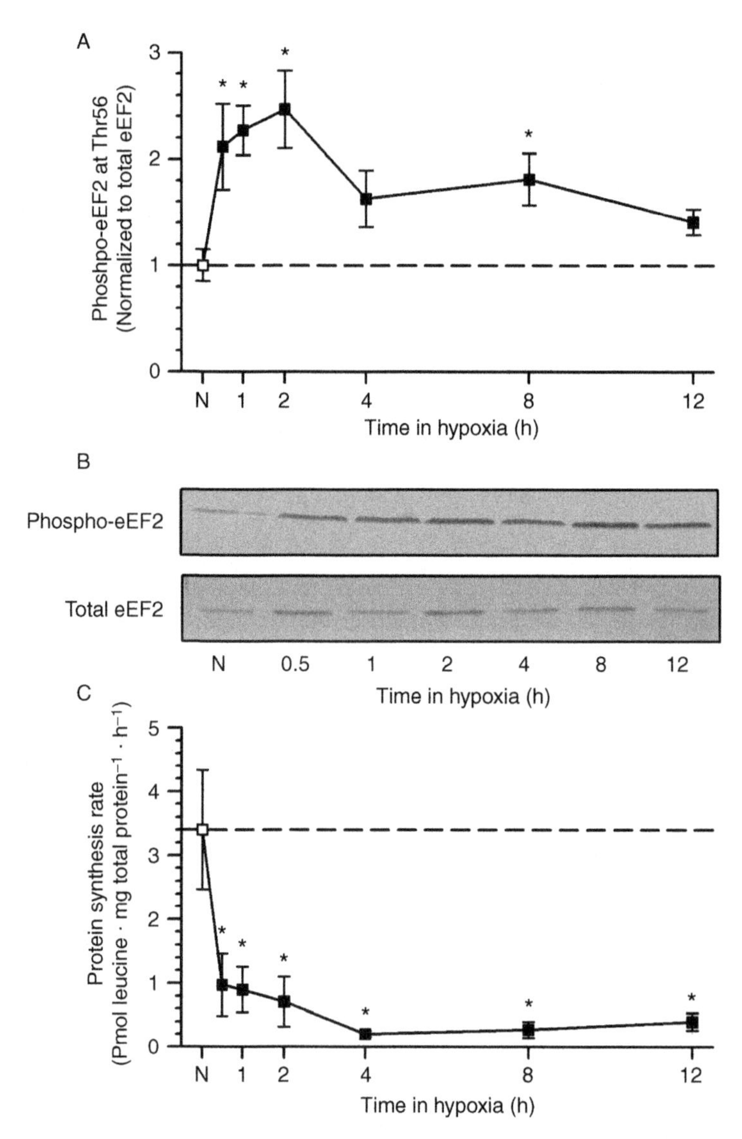

Fig. 10.5. Liver phospho-eEF2 (A), representative phosphoThr56-eEF2 and eEF2 Western blots (B), and protein synthesis rate (C) in goldfish exposed to normoxia and 12 h of hypoxia. [Data from Jibb and Richards (2008) with permission.]

4.3.3. Molecular Responses that Facilitate Metabolic Rate Suppression

As outlined above, decreases in protein synthesis rates are an important response to hypoxia exposure reducing ATP demands and facilitating whole animal metabolic rate suppression. To this end, across a number of tissues, including muscle, liver, and gills, cDNA microarray studies have demonstrated dramatic decreases in mRNA coding for proteins involved in protein synthesis. In muscle, the levels of mRNA coding for elongation factor 2 and several ribosomal proteins have all been shown to be substantially reduced in response to hypoxia exposure (Table 10.1; Gracey *et al.*, 2001). Similarly, in the gills of zebrafish exposed to hypoxia, decreases in mRNA coding for ribosomal proteins have been shown; however, the same study showed a curious accumulation of mRNA coding for elongation factors (van der Meer *et al.*, 2005).

Metabolic energy saving can also be realized through a reduction in movement (Chapter 2) and the maintenance of cellular machinery for movement. Genes involved in muscle contraction including α-tropomyosin, myosin heavy chain, myosin regulatory light chain 2A, skeletal muscle α-actin, and β-actin were, for the most part, all strongly suppressed in response to hypoxia exposure in the mudsucker (Gracey *et al.*, 2001). A similar response was also noted in zebrafish embryos exposed to hypoxia with decreases in mRNA coding for proteins involved in contraction, extracellular matrix, and cytoskeletal proteins (Ton *et al.*, 2003).

Cell growth and proliferation is generally suppressed during hypoxia exposure as a mechanism for ATP conservation. Gracey *et al.* (2001) observed mRNA increases for a number of genes involved in the suppression of cell growth and proliferation. For example, elevated levels of mRNA for insulin-like growth factor binding protein 1 (IGFBP-1), which regulates the availability of insulin-like growth factors in circulation, were observed in liver. Increases in MAP-kinase phosphates were also observed, including MKP-2, which attenuates the activity of the ERK group of MAP kinases. These kinases are phosphorylated in response to the binding of growth factor to cell-surface receptors and activate a signaling cascade that stimulates cell growth. The importance of inhibition of cell growth as an adaptive response to hypoxia exposure is best illustrated by the elegant work of Sollid and Nilsson (Sollid and Nilsson, 2006; Sollid *et al.*, 2006; see Chapter 9). Briefly, in the crucian carp, hypoxia exposure causes a dramatic increase in gill surface area, mediated primarily by a decrease in cell division and increase in apoptosis in the intralamellar space. However, hypoxia does not yield an increase in mRNA consistent with an increase in cellular apoptosis in zebrafish gills (van der Meer *et al.*, 2005).

Ton *et al.* (2003) showed repression of several genes involved cell division such as cyclin G1 and proliferating cell nuclear antigen in zebrafish embryos, which is consistent with observations that hypoxia causes these embryos to undergo developmental arrest and enter a state of suspended animation (Padilla and Roth, 2001).

5. COORDINATING THE METABOLIC AND MOLECULAR RESPONSES TO HYPOXIA

Cell survival during hypoxia exposure requires a metabolic reorganization to decrease ATP demands to match the reduced capacity for ATP production and these metabolic responses must be coordinated temporally otherwise hypoxia exposure will lead to cell death. Several signal transduction cascades have been shown to be activated in response to hypoxia exposure in mammals and other vertebrates (Storey and Storey, 2004), but considerably less work has been done in fishes. In the remaining part of this chapter, I will outline recent advances in the role of one specific signal transduction cascade, the AMP-activated protein kinase, and its role in coordinating the metabolic responses to hypoxia followed by the role of HIF in coordinating the gene expression responses described in this chapter and others.

5.1. AMP-Activated Protein Kinase as a Metabolic Coordinator

Recent evidence has suggested that AMP-activated protein kinase (AMPK) may play a critical role in coordinating the metabolic responses to hypoxia in the hypoxia-tolerant goldfish. AMPK is a heterotrimeric protein kinase comprised of a catalytic subunit (α) and two regulatory subunits, and phosphorylation of AMPK at Thr-172 on the α-subunit activates the protein (Carling, 2004). Activation of AMPK in mammals inhibits energetically expensive anabolic processes including protein synthesis (Horman *et al.*, 2002), glycogen synthesis (Nielsen *et al.*, 2002), and fatty acid synthesis (Hardie and Pan, 2002) rates. Furthermore, activation of AMPK increases skeletal muscle hexokinase activity, GLUT-4 glucose transporter expression (Holmes *et al.*, 1999), and translocation to the membrane (Kurth-Kraczek *et al.*, 1999), and increased phosphofructokinase-2 (PFK-2) activity in rat cardiomyocytes (Marsin *et al.*, 2000), all of which could enhance O_2-independent ATP production. Combined, these actions have led to AMPK being termed the cellular "energy gauge" because of its critical role in maintaining cellular energy balance.

Jibb and Richards (2008) demonstrated that AMPK was activated in the liver of goldfish exposed to severe hypoxia and that there was a close temporal change in $[AMP_{free}]/[ATP]$ and AMPK activity. Increases in AMPK activity in the liver were associated with an increase in the percent phosphorylation of a well-characterized target of AMPK, eukaryotic elongation factor-2 (eEF2), and decreases in protein synthesis rates measured in liver cell-free extracts (Figure 10.5) suggesting that a disruption of cellular energy status is important for the activation of mechanisms involved in metabolic rate suppression. AMP-activated protein kinase, however, was not activated in muscle, brain, heart, or gill during 12 h of severe hypoxia exposure in goldfish suggesting a tissue-specific regulation of AMPK and metabolic responses to hypoxia (Jibb and Richards, 2008).

5.2. Hypoxia Inducible Factor

Hypoxia-regulated gene expression was described some decades ago, but it wasn't until 1992 that the O_2-regulated transcription factor, HIF-1α, was identified as a key regulator of hypoxia-regulated gene expression (Semenza and Wang, 1992). Since its discovery, HIF has been viewed as the molecular master factor of the hypoxic response and a great deal of information is now available on the genes and gene families regulated by HIF (Semenza, 2007; Gardner & Corn, 2008). Many excellent reviews of hypoxia-regulated gene expression and HIF are present in the literature (e.g., Kenneth and Rocha, 2008) including several on fish (e.g., Nikinmaa and Rees, 2005). In the remaining part of this chapter, I will outline HIF regulation in mammals and then describe what is known of HIF function in fish using the work done in mammals as a point of reference.

Hypoxia inducible factor is a heterodimeric transcription factor composed of two subunits; an O_2-sensitive HIF α subunit and an O_2 stable HIF β (also referred to as the aryl hydrocarbon receptor nuclear translocator; ARNT). Hypoxia inducible factor α and β subunits are both members of a very large family of transcription factors known as bHLH/PAS domain-proteins, named because all members of this family contain a basic helix-loop-helix (bHLH) domain as well as one or several PAS domains (domain named after its first members Per, ARNT, and Sim). The bHLH/PAS domain-containing transcription factors constitute a superfamily of transcription factors that are capable of forming homo- and heterodimers through the bHLH and PAS domain and have been implicated in regulating the transcription of genes involved in circadian rhythm, central nervous system development, and induction of hydrocarbon metabolizing enzymes, as well as the cellular responses to hypoxia. Hypoxia inducible factor α, unlike HIF β, contains an O_2-dependent degradation domain (ODD), rendering these proteins labile in the presence of O_2.

In mammalian systems, HIF is regulated through the post-translational modifications of HIF α, which affects both protein stability and transactivation (Figure 10.6). The O_2-dependent control of HIF α is provided by the actions of two proteins, prolyl hydroxylase (PHDs) and factor inhibiting HIF (FIH; Mahon *et al.*, 2001), both of which are members of the 2-oxoglutarate-dependent dioxygenase superfamily of hydroxylases. These proteins both require iron and 2-oxoglutarate as cofactors or substrates (Schofield *et al.*, 1999), and possess many of the features of an O_2-sensitive control mechanism (Land and Hochachka, 1995). Under conditions of normal cellular O_2 tensions, HIF α and HIF β are continuously transcribed and translated; however, HIF α is rapidly hydroxylated at two conserved proline residues in the ODD by PHD. HIF α proteins containing hydroxylated

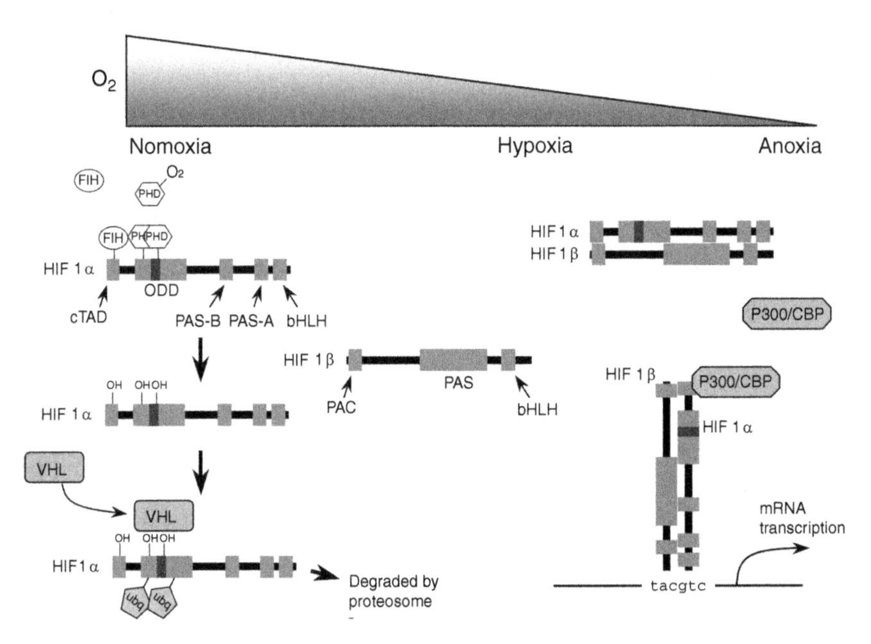

Fig. 10.6. Regulation of hypoxia inducible factor by O_2. In normoxic cells, propyl hydroxylases (PHD) and factor inhibiting HIF (FIH) enzymes use O_2 to hydroxylate key resides on the HIF α subunit in the oxygen-dependent domain (ODD). Hydroxylation of the ODD signals the von Hippel-Lindau (VHL) protein binding leading to ubiquitination and subsequent degradation by the proteosome. The stability of HIF β is not affected by O_2 levels. During periods of cellular hypoxia, PDH and FIH are inhibited resulting in the stabilization of HIF α and as HIF α accumulates it dimerizes with HIF β, recruits other co-activators (e.g. p300/CBP) and activates the transcription of genes containing hypoxia response elements in their promoter region.

proline residues are recognized by an E3 ubiquitin ligase, the von Hippel-Lindau protein (VHL), which promotes Lys48-linked ubiquitination and targets HIF α for rapid degradation by the cellular proteasome. HIF α is also hydroxylated at a conserved asparagine residue in the C-terminal-transactivation domain (cTAD) by FIH, which prevents the recruitment of the p300/CBP transcriptional coactivators leading to a reduced ability of HIF α to transactivate and an overall suppression of HIF regulated gene transcription (Linke *et al.*, 2004). Thus, under normoxic conditions, HIF α protein is continually made but prevented from accumulating or initiating transcription through the PHD-mediated ubiquitin-proteasome degradation and the FIH inhibition of transactivation. In a normoxic cell, HIF α has a half-life of approximately 5–10 min.

The onset of cellular hypoxia leads to an inactivation of PHD and FIH and a lack of HIF α hydroxylation. The lack of HIF α proyl hydroxylation prevents pVHL from recognizing HIF α and initiating the ubiquitin-regulated protein degradation. Thus, cellular hypoxia leads to an almost instantaneous stabilization and accumulation of HIF α, which migrates into the nucleus and dimerizes with HIF β. The HIF α/HIF β dimer then binds with the p300/CBP coactivator, and the complete complex binds to specific hypoxia-response elements (HRE) in the promoter regions of target genes. The absence of asparagine hydroxylation by FIH is permissive for the HIF dimer to interact with transcriptional coactivators and initiate transcription (Lando *et al.*, 2002). Numerous genes have been reported to possess HRE and associated elements in their 5′ promoter regions and their HIF regulation has been described; the known hypoxia-induced gene expression response in fish is described below.

5.2.1. HIF Isoforms

5.2.1.1. HIF α Isoforms. Three HIF α subunit isoforms have been identified in mammals (designated HIF 1α, HIF 2α, HIF 3α; Gu *et al.*, 1998) and some differences between these isoforms have been described, although the precise function of these isoforms has not been fully elucidated. Hypoxia inducible factor 1α and 2α both contain transactivation domains (cTAD domains), while HIF 3α appears to lack the cTAD and, as such, it has been proposed that HIF 3α may act as an inhibitor of HIF 1α and HIF 2α (Bardos and Ashcroft, 2005). HIF 1α and 2α have been shown to have non-redundant functions in the cell, and although HIF 1α is the best-studied isoform, recent studies in mammals have illuminated important roles for HIF 2α in cancer tumor growth (Carroll and Ashcroft, 2006; Hu *et al.*, 2006). HIF 2α has also been shown to be expressed at high levels in certain cell types such as vascular endothelial cells, kidney fibroblasts, hepatocytes, glial cells, interstitial cells of the pancreas, and epithelial cells of the intestinal lumen (Jain *et al.*, 1998).

The first fish HIF α sequence was determined for the rainbow trout by Soitamo *et al.* (2001) and since that point, a total of 38 HIF α gene sequences have been identified in fish either through direct sequencing (Powell & Hahn, 2002; Law *et al.*, 2006; Rahman and Thomas, 2007; Rojas *et al.*, 2007; Rytkönen *et al.*, 2007) or as part of genome sequencing projects (see Ensembl genome projects for Zebrafish, Fugu, Tetradon, Medaka, and Stickleback; Figure 10.7). For the most part, fish HIF α protein sequences are slightly shorter than their counterparts in tetrapods. For example, the length of HIF 1α in fish is between 699 and 778 amino acids while in tetrapods HIF 1α is between 800 and 836 amino acids long.

Phylogenetic analysis of available tetrapod, bird, and fish HIF α sequences indicate that the three major classes of HIF α sequences seen in mammals are represented in fish (Figure 10.7). For the most part, within each isoform class, fish sequences group closely together and are distinct from their tetrapod and bird counterparts. The one exception is for HIF 1α from the Russian sturgeon (*Acipenser gueldenstaedtii*), which groups more closely with the birds (chicken, *Gallus gallus*) and tetrapods. This grouping of the more pleisiomorphic sturgeon HIF 1α with tetrapods and birds suggests that the more derived teleost fish species may have a faster changing HIF 1α sequence.

Several HIF 4α isoforms have been identified in fish (Law *et al.*, 2006; also see orange-spotted grouper; *Epinephelus coioides*), however their identification at the time was based upon a lack of similarity to the scant fish and tetrapod HIF sequences. Since that time, the proliferation of available HIF α sequences and the phylogenetic analysis performed in Figure 10.7 suggests that previously identified HIF 4α sequences are in fact HIF 3α sequences.

Among the isoforms identified in fish, all the appropriate functional domains can be identified. For example, sequence analysis of deduced amino acid sequence of fish HIF 1α genes reveals the presence of four major functional domains including the bHLH domain, two PAS domains (PAS-A and PAS-B), ODD domain, and the DNA-binding domain termed cTAD. These four major functional domains are the same as those seen in tetrapod and bird HIF 1α sequences. Sliding window analysis of 11 fish HIF 1α gene sequences clearly demonstrates that the amount of amino acid sequence variability between fish species is lowest at the four major functional domains (Figure 10.8). These analyses suggest that the amino acid sequence of the important functional domains is well conserved across fish species. Sites showing a high degree of sequence variability occur in areas that have not been identified as important for HIF 1α function. Given the high degree of similarity between fish, tetrapod, and bird sequences it seems reasonable to generalize that HIF 1α functions in fish in much the same way as it does in tetrapods (Figure 10.6).

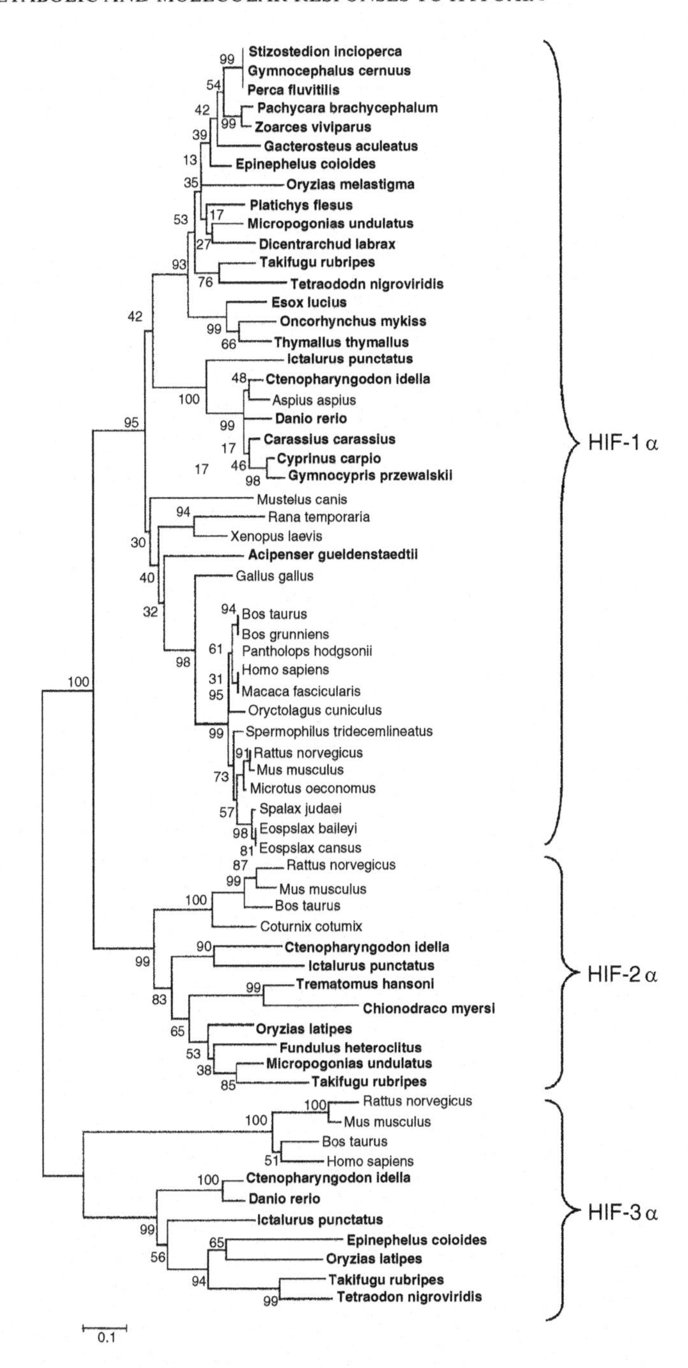

Fig. 10.7. Phylogenetic analysis of HIF α isoform amino acid sequences from fish, tetrapods, and birds. The phylogeny was created from deduced amino acid sequences from GenBank or Ensembl: *Oncorhynchus mykiss* HIF 1 (AF304864); *Thymallus thymallus* HIF 1(ABO26714);

5.2.1.2. HIF β Isoforms. At least two HIF *β* isoforms have been identified in tetrapods; HIF 1*β*, which is ubiquitously expressed in most tissues, and HIF 2*β*, which is primarily restricted to nervous system and kidneys at specific developmental stages (Hirose *et al.*, 1996). In fishes, a total of 14 HIF *β* isoforms have been identified and in general they group closely with other known HIF *β* isoforms identified in tetrapods and birds (Figure 10.9). At the sequence level, HIF *β* is similar to HIF *α* in that it is a member of the bHLH/PAS group of transcription factors and both isoforms possess bHLH and PAS domains. HIF *β* also possesses a terminal activation domain (AD). Sequence analysis of all available fish HIF *β* sequences reveals a high degree of sequence conservation in these well-identified functional domains (Figure 10.10). Overall, this high degree of sequence similarity and the conservation of important regulatory and functional domains suggest that fish HIF *β* probably functions in a similar fashion to its tetrapod and bird orthologs.

Esox lucius HIF 1(ABO26715); *Micropogonias undulates* HIF 1 (ABD32158); *Perca fluviatilis* HIF 1 (ABO26717); *Stizostedion lucioperca* HIF 1 (ABO26718); *Gymnocephalus cernuus* HIF 1 (ABO26716); *Pachycara brachycephalum* HIF 1 (AAZ52828); *Zoarces viviparous* HIF 1 (AAZ52832); *Dicentrarchus labrax* HIF 1 (AAZ95453); *Epinephelus coioides* HIF 1 (AAW29027); *Gasterosteus aculeatus* HIF 1 (ABO26719); *Oryzias latipes* HIF 1 (ENSORLT00000004404); *Rattus norvegicus* HIF 1 (NP_075578); *Tetraodon nigroviridis* HIF 1 (ENSTNIG00000017339); *Takifugu rubripes* HIF 1 (ENSTRUG00000012093); *Oryzias melastigma* HIF 1 (ABC47310); *Ctenopharyngodon idella* HIF 1 (AAR95697); *Platichthys flesus* HIF 1 (ABO26720); *Ictalurus punctatus* HIF 1 (AAZ75952); *Danio rerio* HIF 1 (AAQ91619); *Cyprinus carpio* HIF 1 (ABV59209); *Carassius carassius* HIF 1 (ABC24677); *Gymnocypris przewalskii* HIF 1 (AAW69834); *Aspius aspius* HIF 1 (ABO26713); *Acipenser gueldenstaedtii* HIF 1 (ABO26712); *Rana temporaria* HIF 1 (ABY86629); *Mustelus canis* HIF 1 (ABY86628); *Gallus gallus* HIF 1 (NP_989628); *Xenopus laevis* HIF 1 (ABF71072); *Rattus norvegicus* HIF 1 (NP_077335); *Eospalax baileyi* HIF 1 (ABB17537); *Eospalax cansus* HIF 1 (ABQ53550); *Microtus oeconomus* HIF 1 (AAY27087); *Spalax judaei* HIF 1 (CAG29396); *Spermophilus tridecemlineatus* HIF 1 (AAU14021); *Mus musculus* HIF 1 (BAA20130); *Oryctolagus cuniculus* HIF 1 (NP_001076251); *Pantholops hodgsonii* HIF 1 (AAX89137); *Bos grunniens* HIF 1 (ABH06559); *Mus musculus* HIF 1 (AAH26139); *Bos taurus* HIF 1 (NP_776764); *Homo sapiens* HIF 1 (AAF20149); *Macaca fascicularis* HIF 1 (BAE01417); *Fundulus heteroclitus* HIF 2 (AAL95711); *Micropogonias undulates* HIF 2 (ABD32159); *Takifugu rubripes* HIF 2 (ENSTRUT00000013648); *Ctenopharyngodon idella* HIF 2 (AAT76668); *Bos taurus* HIF 2 (BAA78676); *Mus musculus* HIF 2 (NP_034267); *Trematomus hansoni* HIF 2 (AAZ52830); *Ictalurus punctatus* HIF 2 (ABK27926); *Chionodraco myers* HIF 2 (AAZ52827); *Coturnix coturnix* HIF 2 (AAF21052); *Ctenopharyngodon idella* HIF 4 (AAR95698); *Danio rerio* HIF 3 (AAQ94179); *Ictalurus punctatus* HIF 3 (AAZ75953); *Epinephelus coioides* HIF 4 (AAW29028); *Rattus norvegicus* HIF 3 (NP_071973); *Mus musculus* HIF 3 (NP_058564); *Bos taurus* HIF 3 (NP_001098812); *Homo sapiens* HIF 3 (NP_690008); *Takifugu rubripes* HIF 3 (ENSTRUT00000021549); *Tetraodon nigroviridis* HIF 3 (ENSTNIT00000009762); *Oryzias latepes* HIF 3 (ENSORLT00000002500). Sequences were aligned using ClustalW and phylogenetic analysis was performed using the neighbor-joining methods with complete deletion of gaps using MEGA2 software (Kumar *et al.*, 2001). The support for each node was assessed using 500 bootstrap replicates and are presented at each branch point. Bold-face type indicates fish sequences.

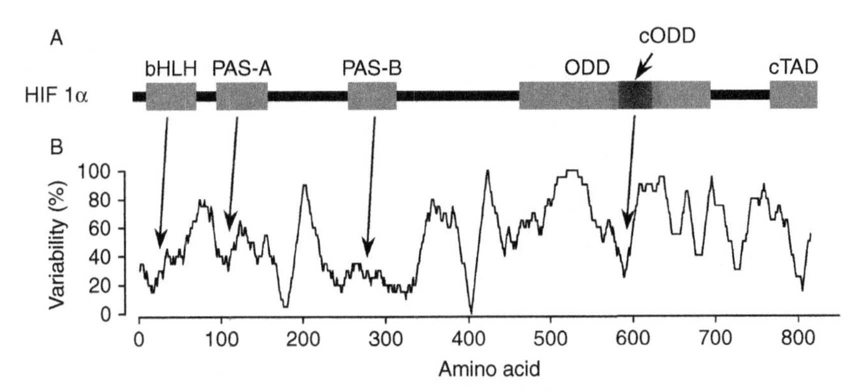

Fig. 10.8. Structural analysis of HIF 1α amino acid sequence. (A) Relative position of the four major functional domains of HIF 1α including the basic helix-loop-helix (bHLH) domain, two PAS domains (Per/ARNT/Sim), the O_2-dependent domain (ODD), and C-terminal transactivation domain (cTAD). (B) Percentage variability among 11 HIF 1α isoforms from fish as determined by sliding window analysis on predicted amino acid sequences. Sliding window analysis quantifies the variation between aligned sequences by counting the average number of differences between isoforms within overlapping windows. For the present analysis, an overlapping window of 20 amino acids was used. Sliding window analysis was performed using MEGA software (version 1.02). Arrows point to the relative sequence variability of the major functional domains. Sliding window analysis of HIF 1α was performed using sequences from the following species: *Perca fluviatilis, Stizostedion lucioperca, Gymnocephalus cernuus, Pachycara brachycephalum, Zoarces viviparous, Dicentrarchus labrax, Gasterosteus aculeatus, Oryzias latipes, Tetraodon nigroviridis, Takifugu rubripes*, and *Tetraodon nigroviridis.*

In addition to its role in hypoxic signaling, HIF 1β, or rather ARNT, is known to play an important role in regulating gene expression changes in response to toxic aryl-hydrocarbon exposure (Hahn *et al.*, 2006). A number of the gene expression responses to aryl-hydrocarbon exposure are similar to those observed in response to hypoxia exposure including increases in lactate dehydrogenase gene expression. HIF 1β regulates aryl hydrocarbon-mediated changes in gene expression through the binding of the aryl hydrocarbon to a specific receptor, the aryl hydrocarbon receptor (AHR). The AHR then binds to its partner, HIF 1β, and the AHR/HIF 1β heterodimer moves to the nucleus where it binds to xenobiotic responsive elements (XREs). Binding of the AHR/HIF 1β heterodimer to XRE regions adjacent to aryl hydrocarbon-inducible genes increases their transcription. Because HIF 1β is known to be involved in the responses to hypoxia and aryl-hydrocarbon exposure, possible interactions between responses may exist. Kraemer and Schulte (2004) demonstrated an antagonistic interaction

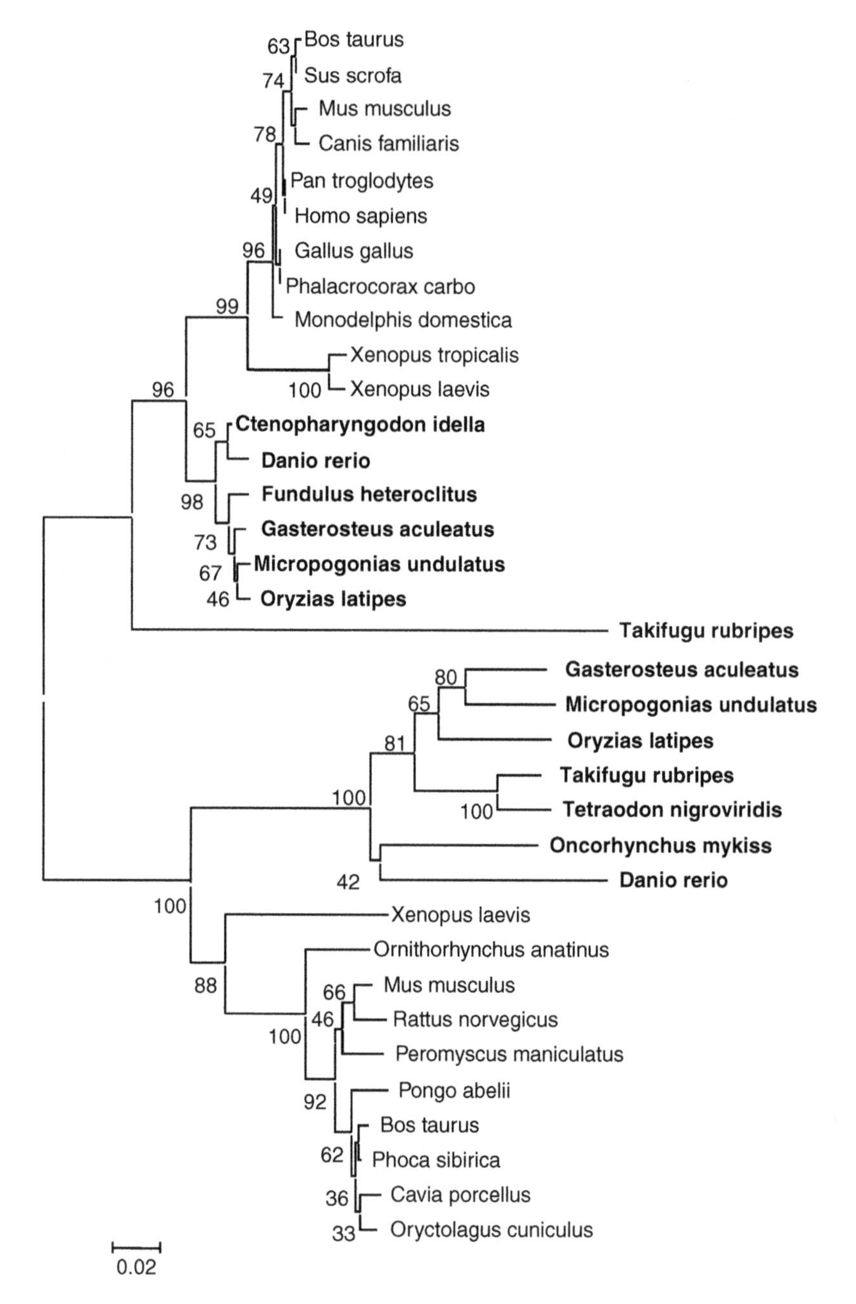

Fig. 10.9. Phylogenetic analysis of HIF β isoform amino acid sequences from fish, tetrapods, and birds. The phylogeny was created from deduced amino acid sequences from GenBank or Ensembl: *Pongo abelii* HIF 1β (NP_001125275); *Gasterosteus aculeatus* HIF 1β

between exposure to PCBs (3,3′,4,4′-tetrachlorobiphenyl) and hypoxia (~15% air saturation) in killifish and suggested that prior PCB exposure could make these fish less tolerant of environmental hypoxia.

5.2.1.3. PHD Isoforms. Four PDH isoforms have been identified in mammals, numbered PDH 1 to 4, and so far only PDH 1, 2, and 3 have been shown to hydroxylate HIF. Biochemical analysis has shown PHD 2 to have a higher affinity for HIF 1α, whereas PHD 1 and PDH 3 have higher affinity for HIF 2α (Appelhoff *et al.*, 2004). Prolyl hydroxylase sequences have been found in fish as a result of genome sequencing projects, but to date no study has explicitly characterized the sequence or function of PHD isoforms in fish. This will undoubtedly be an important and fruitful area of research in the next few years as PHDs are now considered the cellular O_2 sensors responsible for initiating the HIF response.

5.2.2. Regulation of HIF Activity in Fish

The O_2-dependent regulation of HIF in fish has received remarkably little attention since the literature was reviewed by Nikinmaa and Rees (2005). However, given the available data on sequence similarity between tetrapod and fish HIF α and β sequences it seems reasonable to speculate that the same or similar mechanisms of O_2-dependent regulation of HIF shown in Figure 10.6 are at play in fish. Specifically, as pointed out by Rahman and Thomas (2007) for Atlantic croaker (*Micropogonias undulates*) and shown in

---

(ENSGACG00000011686); *Oryzias latipes* HIF 1β (ENSORLG00000010551); *Takifugu rubripes* HIF 1β (ENSTRUG00000014504); *Ornithorhynchus anatinus* HIF 1β (XP_001517995); *Peromyscus maniculatus* HIF 1β (AAN52084); *Bos taurus* HIF 1β (ABG67008); *Mus musculus* HIF 1β (AAH12870); *Oryctolagus cuniculus* HIF 1β (NP_001075675); *Phoca sibirica* HIF 1β (BAE16957); *Cavia porcellus* HIF 1β (BAF02596); *Rattus norvegicus* HIF 1β (AAO89090); *Xenopus laevis* HIF 1β (NP_001082130); *Oncorhynchus mykiss* HIF 1β (AAC60052); *Micropogonias undulates* HIF 1β (ABD32160); *Fundulus heteroclitus* HIF 2β (AAD09750); *Micropogonias undulates* HIF 2β (ABD32161); *Ctenopharyngodon idella* HIF 2β (AAT70730); *Danio rerio* HIF 2β (NP_571749); *Mus musculus* HIF 2β (BAA09799); *Rattus norvegicus* HIF 2β (AAB05247); *Canis familiaris* HIF 2β (XP_850172); *Gallus gallus* HIF 2β (XP_413854); *Phalacrocorax carbo* HIF 2β (BAF44221); *Bos taurus* HIF 2β (XP_612854); *Sus scrofa* HIF 2β (XP_001926107); *Monodelphis domestica* HIF 2β (XP_001367955); *Pan troglodytes* HIF 2β (XP_001156233); *Homo sapiens* HIF 2β (NP_055677); *Xenopus tropicalis* HIF 2β (NP_001093686); *Xenopus laevis* HIF 2β (AAQ91608); *Danio rerio* HIF 2β (AAG25919); *Gasterosteus aculeatus* HIF 2β (ENSGACG00000013947); *Xenopus laevis HIF* 2β (NP_001083622); *Oryzias latipes* HIF 2β (ENSORLG00000019479); *Takifugu rubripes* HIF 2β (ENSTRUG00000007832); *Tetraodon nigroviridis* HIF β (ENSTNIG00000008064). Sequences were aligned using ClustalW and phylogenetic analysis was performed using the neighbor-joining methods with complete deletion of gaps using MEGA2 software (Kumar *et al.*, 2001). The support for each node was assessed using 500 bootstrap replicates and is presented at each branch point. Bold-face type indicates fish sequences.

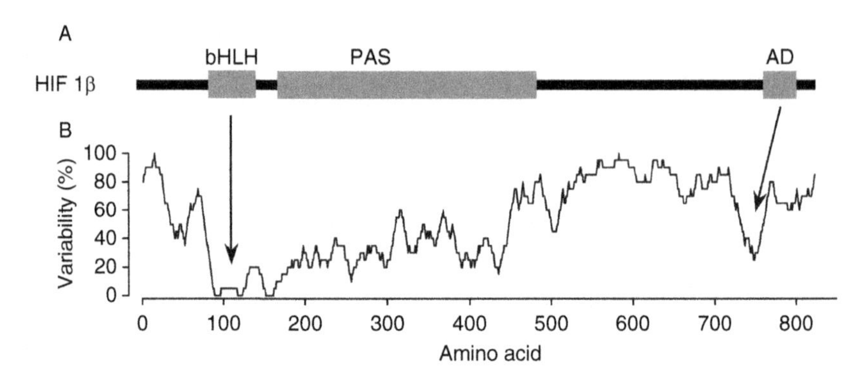

Fig. 10.10. Structural analysis of combined HIF 1*β* and 2*β* amino acid sequence. (A) Relative position of the four major functional domains of HIF *β* including the basic helix-loop-helix (bHLH) domain, a PAS domain (Per/ARNT/Sim), and the activation domain (AD). Panel B shows the percentage variability among all available HIF 1*β* and HIF 2*β* isoforms available in fish as determined by sliding window analysis on predicted amino acid sequences.

Figure 10.8, there is a high degree of sequence similarity in the core O_2-dependent degradation domain regions of fish HIF *α* sequences, suggesting a similar mechanism of HIF degradation to that in other vertebrate species.

The first and only study to address the issue of O_2-dependent regulation of HIF in fish was that of Soitamo *et al.* (2001), which demonstrated that although HIF 1*α* was present under normoxic conditions (air saturation) in rainbow trout and salmon cell lines, the levels of HIF 1*α* protein increased during hypoxia exposure. Oddly, however, the maximum levels of HIF 1*α* protein were noted in cells cultured at 5% O_2, which as the authors pointed out is similar to typical venous PO_2. These data suggest that *in vivo*, HIF 1*α* may accumulate under what should be considered as normoxic conditions in tissues. Additional research is needed to understand how HIF functions in fish cells and whether there are differences in O_2 sensitivity in HIF 1*α*-regulated gene expression among fish that vary in hypoxia tolerance.

5.2.3. Hypoxia-Regulated HIF *α* mRNA Expression

Unlike in mammals, where there appears to be little or no regulation of HIF at the mRNA level, hypoxia-induced changes in HIF *α* mRNA and protein expression have been noted in several fish species. Law *et al.* (2006) examined the mRNA and protein levels of two HIF *α* isoforms (1 and 3; note that these authors incorrectly named HIF 3*α* as HIF 4*α*) from the hypoxia-tolerant grass carp (*Ctenopharyngodon idella*) and showed substantial increases in HIF 1*α* mRNA in gill and kidney after 4 h exposure to ~7% air saturation compared with normoxia-exposed fish. In the same fish, no or few changes in HIF 1α were noted in brain, eye, gill, heart, kidney, liver, and muscle. On the other hand, substantial increases in the HIF 3α isoform (identified as HIF 4α) were observed during hypoxia exposure in all tissues examined. Similarly, Rahman and Thomas (2007) demonstrated that both HIF 1α and HIF 2α from the hypoxia-tolerant Atlantic croaker were hypoxia responsive in ovaries during short-term (3–7 days at $\sim$20% air saturation) and longer-term hypoxia exposure (3 weeks at $\sim$20–40% air saturation). There does not, however, appear to be a good relationship between hypoxia tolerance and HIF α expression, although the available data are limited. Specifically, HIF 1α mRNA levels have also been shown to increase in the liver of the hypoxia-sensitive sea bass (*Dicentrarchus labrax*; Terova *et al.*, 2008) during both acute hypoxia exposure (4 h at $\sim$20% air saturation) and 15 days of chronic hypoxia ($\sim$50% air saturation).

5.2.4. Relationship Between HIF Function and Hypoxia Tolerance in Fish

The fact that there is enormous variation in hypoxia tolerance among fish species raises the question of whether there is a relationship between HIF function and hypoxia tolerance. In fact, careful comparisons among fish species known to vary in hypoxia tolerance open the possibility of elucidating which aspects of HIF function are adaptive and thus potentially most important in dictating hypoxia tolerance. Surprisingly however, most of our current understanding of HIF regulation and function comes from mammalian models, which typically only experience hypoxia as a result of disease such as cancer (Gort *et al.*, 2008).

To begin to address the question of whether HIF structure or responsiveness to hypoxia differ among hypoxia-sensitive and hypoxia-tolerant fish species, Rytkönen *et al.* (2007) sequenced HIF 1α from nine species of fish that varied in lifestyle related to O_2 requirements (hypoxia tolerance was not quantified). Analysis of sequence variation among the available fish HIF 1α amino acid sequences showed that there was no clear protein signature associated with O_2 requirements (Rytkönen *et al.*, 2007). Further analysis of these sequences and others revealed that the overall evolutionary rate in teleost HIF 1α was approximately twice as fast as the predicted evolutionary rate in mammalian HIF 1α (Rytkönen *et al.*, 2008). Despite the faster sequence divergence, however, crucial functional domains in HIF 1α (Figure 10.6) were found to be under stringent purifying selection in all vertebrates. As a result, the faster sequence divergence occurred in the less crucial areas of sequence. Some evidence for positive selection on HIF 1α amino acid sequence was observed, but was not associated with sequence variation in the O_2 sensitive ODD, but was associated with the bHLH/PAS domains.

5.2.4. Oxygen-dependent Gene Expression

Hypoxia exposure in fish is well known to initiate a complex suite of gene and protein expression responses, many of which have been outlined above (see Table 10.1; Gracey *et al.*, 2001; Ton *et al.*, 2002, 2003; Bosworth *et al.*, 2005; van der Meer *et al.*, 2005). However, in many cases a direct link between changes in gene or protein expression and the transcriptional regulator HIF has not been directly assessed, therefore the reader is cautioned against assuming all responses described above are mediated by HIF. In reality, remarkably few studies, especially in fish, have focused on identifying functional HRE in the promoter regions of the hypoxia-responsive genes. The only definitive studies conducted in fish that have shown a direct relationship between HIF and hypoxia-regulated gene expression is for insulin-like growth factor binding protein in zebrafish (Kajimura *et al.*, 2005, 2006). In mammalian and carcinoma cell lines, however, HIF has been directly implicated in regulating the expression of genes involved in a number of physiological and biochemical responses to hypoxia (outlined above).

Patterns of gene expression in response to hypoxia exposure can vary between tissues and in some cases the differences can be dramatic. Ju *et al.* (2007) using an 8046 gene microarray showed substantial tissue-specific gene regulation and few consistent responses between tissues. In response to hypoxia exposure, 501 genes in the brain, 442 genes in the gills, and 715 genes in the liver were differentially expressed in hypoxia-exposed medaka (*Oryzias latipes*) and there were a number of pathways affected (Table 10.1). Among the up-regulated genes there were remarkably few overlapping genes with 24, 21, and 20 genes showing the same expression patterns between brain and gill, brain and liver, and gill and liver, respectively (Figure 10.11). Of the genes that were shown to be down-regulated, 65, 24, and 26 genes were common between brain and gill, brain and liver and gill and liver, respectively (Ju *et al.*, 2007). Only nine genes in total changed in a consistent fashion across all tissues examined. Of all the tissues examined, liver showed the greatest number of differentially expressed genes.

6. CONCLUSIONS AND PERSPECTIVES

Hypoxia survival requires a rapid reorganization of physiological and biochemical systems to either maximize O_2 uptake from the hypoxic environment to support the maintenance of a routine metabolic rate or cellular adjustments to function under O_2-limiting conditions. Survival under O_2-limiting conditions requires a cellular metabolic reorganization to reduce ATP consumption through a regulated metabolic rate suppression to

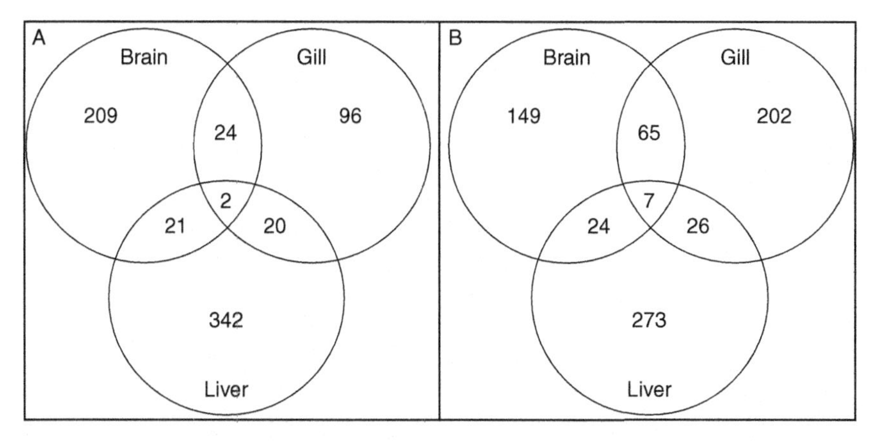

Fig. 10.11. Venn diagram showing differentially expressed genes in medake during hypoxia exposure. (A) Number of up-regulated genes in response to hypoxia exposure; (B) number of down-regulated genes in response to hypoxia exposure. [Data from Ju *et al.* (2007) with permission.]

match the limited capacity for O_2-independent ATP production. As outlined above, controlled metabolic rate suppression is essential to extend the length of time that can be supported by the limited levels of fermentable fuels. Thus, it appears reasonable to speculate that the degree of metabolic rate suppression and the quantity of stored fermentable fuel is likely strongly selected for in hypoxia-tolerant fishes. Indeed, this chapter has outlined and summarized the available information on the degree of metabolic rate suppression in a variety of fish species as well the quantity of tissue glycogen and, broadly speaking, there was a reasonable relationship between fish lifestyle (that being sluggish, hypoxia-tolerant carp species c.f. athletic, intolerant salmonid species, for example) and stored fermentable fuels, but the relationship between metabolic rate suppression and hypoxia tolerance is, however, oddly not clear. This is primarily because of the scant data available on the topic. Further still, the study of HIF in fish and hypoxia-regulated gene expression has been fruitful in demonstrating that HIF function in fish appears at least superficially similar to that observed in mammals, but the relationship between HIF function and hypoxia tolerance is still lacking. Despite the wealth of information available on the metabolic and molecular responses of a variety of fish species to hypoxia, we are still far from a unified concept of the important adaptations underlying hypoxia tolerance. However, fish provide an incredibly tractable system to understand the evolution of hypoxia tolerance because of the incredible diversity of fishes as well as their diverse O_2 habitats.

REFERENCES

Almeida-Val, V. M., Chippari-Gomes, A. R., and Lopes, N. P. (2006). Metabolic and physiological adjustments to low oxygen and high temperature in fishes of the amazon. *In* "The Physiology of Tropical Fishes Vol. 21" (Val, A. L., Almeida-Val, V. M., and Randal, D. J., Eds.), pp. 443–491. Elsevier, San Diego.

Almeida-Val, V. M. F., Farias, I. P., Paula-Silva, M. N., and Duncan, W. P. (1995). Biochemical adjustments to hypoxia by Amazon cichlids. *Braz. J. Med. Biol. Res.* **28,** 1257–1263.

Appelhoff, R. J., Tian, Y. M., Raval, R. R., Turley, H., Harris, A. L., Pugh, C. W., Ratcliffe, P. J., and Gleadle, J. M. (2004). Differential function of the prolyl hydroxylases PHD1, PHD2, and PHD3 in the regulation of hypoxia-inducible factor. *J. Biol. Chem.* **279,** 38458–38465.

Bardos, J. I., and Ashcroft, M. (2005). Negative and positive regulation of HIF-1: A complex network. *Biochim. Biophy. Acta. – Rev. Cancer* **1755,** 107–120.

Bickler, P. E., and Buck, L. T. (2007). Hypoxia tolerance in reptiles, amphibians, and fishes: Life with variable oxygen availability. *Annu. Rev. Physiol.* **69,** 145–170.

Bogdanova, A., Grenacher, B., Nikinmaa, M., and Gassmann, M. (2005). Hypoxic responses of Na^+/K^+ ATPase in trout hepatocytes. *J. Exp. Biol.* **208,** 1793–1801.

Borger, R., De Boeck, G., Van Audekerke, J., Dommisse, R., Blust, R., and Van der Linden, A. (1998). Recovery of the energy metabolism after a hypoxic challenge at different temperature conditions: A P-31 nuclear magnetic resonance spectroscopy study with common carp. *Comp. Biochem. Physiol. A* **120,** 143–150.

Bosworth, C. A. T., Chou, C. W., Cole, R. B., and Rees, B. B. (2005). Protein expression patterns in zebrafish skeletal muscle: Initial characterization and the effects of hypoxic exposure. *Proteomics* **5,** 1362–1371.

Boutilier, R. G. (2001). Mechanisms of cell survival in hypoxia and hypothermia. *J. Exp. Biol.* **204,** 3171–3181.

Boutilier, R. G., and St-Pierre, J. (2000). Surviving hypoxia without really dying. *Comp. Biochem. Physiol. A* **126,** 481–490.

Buck, L. T., and Hochachka, P. W. (1993). Anoxic suppression of Na^+/K^+-ATPase and constant membrance potential in hepatocytes: Support for channel arrest. *Am. J. Physiol.* **265,** R1020–R1025.

Buck, L. T., Hochachka, P. W., Schon, A., and Gnaiger, E. (1993a). Microcalorimetric measurement of reversible metabolic suppression induced by anoxia in isolated hepatocytes. *Am. J. Physiol.* **265,** R1014–1019.

Buck, L. T., Land, S. C., and Hochachka, P. W. (1993b). Anoxia-tolerant hepatocytes: Model system for study of reversible metabolic suppression. *Am. J. Physiol.* **265,** R49–56.

Busk, M., and Boutilier, R. G. (2005). Metabolic arrest and its regulation in anoxic eel hepatocytes. *Physiol. Biochem. Zool.* **78,** 926–936.

Carling, D. (2004). The AMP-activated protein kinase cascade – a unifying system for energy control. *Trends. Biochem. Sci.* **29,** 18–24.

Carroll, V. A., and Ashcroft, M. (2006). Role of hypoxia-inducible factor (HIF)-1 alpha-versus HIF-2 alpha in the regulation of HIF target genes in response to hypoxia, insulin-like growth factor-1, or loss of von Hippel-Lindau function: Implications for targeting the HIF pathway. *Cancer Res.* **66,** 6264–6270.

Chang, J. C. H., Wu, S. M., Tseng, Y. C., Lee, Y. C., Baba, O., and Hwang, P. P. (2007). Regulation of glycogen metabolism in gills and liver of the euryhaline tilapia (*Oreochromis mossambicus*) during acclimation to seawater. *J. Exp. Biol.* **210,** 3494–3504.

Chapman, L. J., Chapman, C. A., Nordlie, F. G., and Rosenberger, A. E. (2002). Physiological refugia: Swamps, hypoxia tolerance and maintenance of fish diversity in the Lake Victoria region. *Comp. Biochem. Physiol. A* **133,** 421–437.

Chippari-Gomes, A. R., Gomes, L. C., Lopes, N. P., Val, A. L., and Almedia-Val, V. M. (2005). Metabolic adjustments in two Amazonian cichlids exposed to hypoxia and anoxia. *Compo. Biochem. Physiol. B.* **141,** 347–355.

Chou, C. F., Tohari, S., Brenner, S., and Venkatesh, B. (2004). Erythropoietin gene from a teleost fish, *Fugu rubripes. Blood* **104,** 1498–1503.

Dalla Via, J., van den Thillart, G., Cattani, O., and de Zwaan, A. (1994). Influence of long-term hypoxia exposure on the energy metabolism of *Solea solea*. II. Intermediary metabolism in blood, liver and muscle. *Mar. Ecol. Prog. Ser.* **111,** 17–27.

Fangue, N. A., Mandie, M., Richards, J. G., and Schulte, P. M. (2008). Swimming performance and energetics as a function of temperature in killifish *Fundulus heteroclitus. Physiol. Biochem. Zool.* **81,** 389–401.

Fraser, J., de Mello, L. V., Ward, D., Rees, H. H., Williams, D. R., Fang, Y. C., Brass, A., Gracey, A. Y., and Cossins, A.R (2006). Hypoxia-inducible myoglobin expression in non-muscle tissues. *Proc. Natl. Acad. Sci. USA* **103,** 2977–2981.

Frick, N. T., Bystriansky, J. S., Ip, A. K., Chew, S. F., and Ballantyne, J. S. (2008). Carbohydrate and amino acid metabolism in fasting and aestivating African lungfish (*Protopterus dolloi*). *Compo Biochem. Physiol. A* **151,** 85–92.

Gardner, L. B., and Corn, P. G. (2008). Hypoxic regulation of mRNA expression. *Cell Cycle* **7,** 1916–1924.

Gort, E. H., Groot, A. J., van der Wall, E., van Diest, P. J., and Vooijs, M. A. (2008). Hypoxic regulation of metastasis *via* hypoxia-inducible factors. *Cur. Mol. Med.* **8,** 60–67.

Gracey, A. Y., Troll, J. V., and Somero, G. N. (2001). Hypoxia-induced gene expression profiling in the euryoxic fish *Gillichthys mirabilis. Proc. Natl. Acad. Sci. USA* **98,** 1993–1998.

Gu, Y. Z., Moran, S. M., Hogenesch, J. B., Wartman, L., and Bradfield, C. A. (1998). Molecular characterization and chromosomal localization of a third alpha-class hypoxia inducible factor subunit, HIF3alpha. *Gene. Expr.* **7,** 205–213.

Hahn, M. E., Karchner, S. I., Evans, B. R., Franks, D. G., Merson, R. R., and Lapseritis, J. M. (2006). Unexpected diversity of aryl hydrocarbon receptors in non-mammalian vertebrates: Insights from comparative genomics. *J. Exp. Zool.* **305A,** 693–706.

Hallman, T. M., Rojas-Vargas, A., Jones, D. R., and Richards, J. G. (2008). Differential recovery from exercise and hypoxia exposure measured using ^{31}P- and ^{1}H-NMR in white muscle of the common carp *Cyprinus carpio*. *J. Exp. Biol.* **211,** 3237–3248.

Hardewig, I., Van Dijk, P. L., and Portner, H. O. (1998). High-energy turnover at low temperatures: Recovery from exhaustive exercise in Antarctic and temperate eelpouts. *Am. J. Physiol.* **274,** R1789–1796.

Hardie, D. G., and Pan, D. A. (2002). Regulation of fatty acid synthesis and oxidation by the AMP-activated protein kinase. *Biochem. Soc. Trans.* **30,** 1064–1070.

Hirose, K., Morita, M., Ema, M., Mimura, J., Hamada, H., Fujii, H., Saijo, Y., Gotoh, O., Sogawa, K., and Fujii-Kuriyama, Y. (1996). cDNA cloning and tissue-specific expression of a novel basic helix-loop-helix/PAS factor (Arnt2) with close sequence similarity to the aryl hydrocarbon receptor nuclear translocator (Arnt). *Mol. Cell. Biol.* **16,** 1706–1713.

Hochachka, P. W., Buck, L. T., Doll, C. J., and Land, S. C. (1996). Unifying theory of hypoxia tolerance: Molecular/metabolic defense and rescue mechanisms for surviving oxygen lack. *Proc. Natl. Acad. Sci. USA* **93,** 9493–9498.

Hochachka, P. W., and Lutz, P. L. (2001). Mechanism, origin, and evolution of anoxia tolerance in animals. *Comp. Biochem. Physiol. B* **130,** 435–459.

Holmes, B. F., Kurth-Kraczek, E. J., and Winder, W. W. (1999). Chronic activation of 5′-AMP-activated protein kinase increases GLUT-4, hexokinase, and glycogen in muscle. *J. Appl. Physiol.* **87,** 1990–1995.

Horman, S., Browne, G., Krause, U., Patel, J., Vertommen, D., Bertrand, L., Lavoinne, A., Hue, L., Proud, C., and Rider, M. (2002). Activation of AMP-activated protein kinase leads to the phosphorylation of elongation factor 2 and an inhibition of protein synthesis. *Curr. Biol.* **12,** 1419–1423.

Hu, C. J., Iyer, S., Sataur, A., Covello, K. L., Chodosh, L. A., and Simon, M. C. (2006). Differential regulation of the transcriptional activities of hypoxia-inducible factor 1 alpha (HIF-1 alpha) and HIF-2 alpha in stem cells. *Mol. Cell. Biol.* **26,** 3514–3526.

Hylland, P., Milton, S., Pek, M., Nilsson, G. E., and Lutz, P. L. (1997). Brain Na^+/K^+-ATPase activity in two anoxia tolerant vertebrates: Crucian carp and freshwater turtle. *Neurosci. Lett.* **235,** 89–92.

Jain, S., Maltepe, E., Lu, M. M., Simon, C., and Bradfield, C. A. (1998). Expression of ARNT, ARNT2, HIF1 alpha, HIF2 alpha and Ah receptor mRNAs in the developing mouse. *Mech. Dev.* **73,** 117–123.

Jansen, M. A., Shen, H., Zhang, L., Wolkowicz, P. E., and Balschi, J. A. (2003). Energy requirements for the Na^+ gradient in the oxygenated isolated heart: Effect of changing the free energy of ATP hydrolysis. *Am. J. Physiol.* **285,** H2437–2445.

Jibb, L. A., and Richards, J. G. (2008). AMP-activated protein kinase activity during metabolic rate depression in the hypoxic goldfish, *Carassius auratus*. *J. Exp. Biol.* **211,** 3111–3122.

Johansson, D., Nilsson, G. E., and Tornblom, E. (1995). Effects of anoxia on energy metabolism in crucian carp brain slices studied with microcalorimetry. *J. Exp. Biol.* **198,** 853–859.

Ju, Z. L., Wells, M. C., Heater, S. J., and Walter, R. B. (2007). Multiple tissue gene expression analyses in Japanese medaka (*Oryzias latipes*) exposed to hypoxia. *Comp. Biochem. Physiol. C* **145,** 134–144.

Kajimura, S., Aida, K., and Duan, C. (2005). Insulin-like growth factor-binding protein-1 (IGFBP-1) mediates hypoxia-induced embryonic growth and developmental retardation. *Proc. Natl. Acad. Sci. USA* **102,** 1240–1245.

Kajimura, S., Aida, K., and Duan, C. (2006). Understanding hypoxia-induced gene expression in early development: *In vitro* and *in vivo* analysis of hypoxia-inducible factor 1-regulated zebra fish insulin-like growth factor binding protein 1 gene expression. *Mol. Cell. Biol.* **26,** 1142–1155.

Kenneth, N. S., and Rocha, S. (2008). Regulation of gene expression by hypoxia. *Biochem. J.* **414,** 19–29.

Kraemer, L. D., and Schulte, P. M. (2004). Prior PCB exposure suppresses hypoxia-induced up-regulation of glycolytic enzymes in *Fundulus heteroclitus*. *Comp. Biochem. Physiol. C* **139,** 23–29.

Krumschnabel, G., Schwarzbaum, P. J., Biasi, C., Dorigatti, M., and Wieser, W. (1997). Effects of energy limitation on Ca^{2+} and K^+ homeostasis in anoxia-tolerant and anoxia-intolerant hepatocytes. *Am. J. Physiol.* **273,** R307–316.

Kumar, S., Tamura, K., Jakobsen, I. B., and Nei, M. (2001). MEGA2: Molecular evolutionary genetics analysis software. *Bioinformatics* **17,** 1244–1245.

Kurth-Kraczek, E. J., Hirshman, M. F., Goodyear, L. J., and Winder, W. W. (1999). 5′ AMP-activated protein kinase activation causes GLUT4 translocation in skeletal muscle. *Diabetes* **48,** 1667–1671.

Land, S. C., and Hochachka, P. W. (1995). A heme-protein-based oxygen-sensing mechanism controls the expression and suppression of multiple proteins in anoxia-tolerant turtle hepatocytes. *Proc. Natl. Acad. Sci. USA* **92,** 7505–7509.

Land, S. C., Buck, L. T., and Hochachka, P. W. (1993). Response of protein synthesis to anoxia and recovery in anoxia-tolerant hepatocytes. *Am. J. Physiol.* **265,** R41–48.

Lando, D., Peet, D. J., Whelan, D. A., and Whitelaw, M. L. (2002). Asparagine hydroxylation of the HIF transactivation domain: A hypoxic switch. *Science* **295,** 858–861.

Law, S. H. W., Wu, R. S. S., Ng, P. K. S., Yu, R. M. K., and Kong, R. Y. C. (2006). Cloning and expression analysis of two distinct HIF-alpha isoforms – gcHIF-1alpha and gcHIF-4alpha – from the hypoxia-tolerant grass carp, *Ctenopharyngodon idellus*. *BMC Mol. Biol.* **7,** 5.

Lewis, J. M., Costa, I., Val, A. L., Almeida-Val, V. M., Gamperl, A. K., and Driedzic, W. R. (2007). Responses to hypoxia and recovery: Repayment of oxygen debt is not associated with compensatory protein synthesis in the Amazonian cichlid, *Astronotus ocellatus*. *J. Exp. Biol.* **210,** 1935–1943.

Linke, S., Stojkoski, C., Kewley, R. J., Booker, G. W., Whitelaw, M. L., and Peet, D. J. (2004). Substrate requirements of the oxygen-sensing asparaginyl hydroxylase factor-inhibiting hypoxia-inducible factor. *J. Biol. Chem.* **279,** 14391–14397.

Mahon, P. C., Hirota, K., and Semenza, G. L. (2001). FIH-1: A novel protein that interacts with HIF-1alpha and VHL to mediate repression of HIF-1 transcriptional activity. *Genes Dev.* **15,** 2675–2686.

Mandic, M., Todgham, A. E., and Richards, J. G. (in press). Mechanisms and evolution of hypoxia tolerance in fish. *Proc. R. Soc. B.*

Marsin, A. S., Bertrand, L., Rider, M. H., Deprez, J., Beauloye, C., Vincent, M. F., Van den Berghe, G., Carling, D., and Hue, L. (2000). Phosphorylation and activation of heart PFK-2 by AMPK has a role in the stimulation of glycolysis during ischaemia. *Curr. Biol.* **10,** 1247–1255.

Martinez, M. L., Landry, C., Boehm, R., Manning, S., Cheek, A. O., and Rees, B. B. (2006). Effects of long-term hypoxia on enzymes of carbohydrate metabolism in the Gulf killifish, *Fundulus grandis*. *J. Exp. Biol.* **209,** 3851–3861.

Moraes, G., Avilez, I. M., and Hori, T. S. F. (2006). Comparison between biochemical responses of the teleost Pacu and its hybrid Tambacu (*Piraractus mesopotamicus x Colossoma macropomum*) to short term nitrite exposure. *Braz. J. Biol.* **66,** 1103–1108.

Nielsen, J. N., Wojtaszewski, J. F., Haller, R. G., Hardie, D. G., Kemp, B. E., Richter, E. A., and Vissing, J. (2002). Role of 5′AMP-activated protein kinase in glycogen synthase activity and glucose utilization: Insights from patients with McArdle's disease. *J. Physiol.* **541,** 979–989.

Nikinmaa, M., and Rees, B. B. (2005). Oxygen-dependent gene expression in fishes. *Am. J. Physiol.* **288,** R1079–1090.

Nilsson, G. E. (1991). The adenosine receptor blocker aminophylline increases anoxic ethanol excretion in crucian carp. *Am. J. Physiol.* **261,** R1057–R1060.

Nilsson, G. E. (1992). Evidence for a role of GABA in metabolic depression during anoxia in crucian carp (*Carassius carassius*). *J. Exp. Biol.* **165,** 243–259.

Padilla, P. A., and Roth, M. B. (2001). Oxygen deprivation causes suspended animation in the zebrafish embryo. *Proc. Natl. Acad. Sci. USA* **98,** 7331–7335.

Powell, W. H., and Hahn, M. E. (2002). Identification and functional characterization of hypoxia-inducible factor 2alpha from the estuarine teleost, *Fundulus heteroclitus*: Interaction of HIF-2α with two ARNT2 splice variants. *J. Exp. Zool.* **294,** 17–29.

Pörtner, H., and Grieshaber, M. K. (1993). Critical Po_2(s) in oxyconforming and oxyregulating animals: Gas exchange, metabolic rate and the mode of energy production. *In* "The Vertebrate Gas Transport Cascade: Adaptations to Environment and Mode of Life" (Eduardo, J., and Bicudo, P. W., Eds.). CRC Press, Boca Raton.

Rahman, M. S., and Thomas, P. (2007). Molecular cloning, characterization and expression of two hypoxia-inducible factor alpha subunits, HIF-1α and HIF-2α, in a hypoxia-tolerant marine teleost, Atlantic croaker (*Micropogonias undulatus*). *Gene* **396,** 273–282.

Richards, J. G., Sardella, B. A., and Schulte, P. M. (2008). Regulation of pyruvate dehydrogenase in the common killifish, *Fundulus heteroclitus*, during hypoxia exposure. *Am. J. Physiol.* **295,** R979–R990.

Richards, J. G., Wang, Y. S., Brauner, C. J., Gonzalez, R. J., Patrick, M. L., Schulte, P. M., Choppari-Gomes, A. R., Almeida-Val, V. M., and Val, A. L. (2007). Metabolic and ionoregulatory responses of the Amazonian cichlid, *Astronotus ocellatus*, to severe hypoxia. *J. Comp. Physiol. B* **177,** 361–374.

Rissanen, E., Tranberg, H. K., and Nikinmaa, M. (2006). Oxygen availability regulates metabolism and gene expression in trout hepatocyte cultures. *Am. J. Physiol.* **291,** R1507–R1515.

Rojas, D. A., Perez-Munizaga, D. A., Centanin, L., Antonelli, M., Wappner, P., Allende, M. L., and Reyes, A. E. (2007). Cloning of hif-1α and hif-2α and mRNA expression pattern during development in zebrafish. *Gene. Exp. Pat.* **7,** 339–345.

Rytkönen, K. T., Vuori, K. A. M., Primmer, C. R., and Nikinmaa, M. (2007). Comparison of hypoxia-inducible factor-1 alpha in hypoxia-sensitive and hypoxia-tolerant fish species. *Comp. Biochem. Physiol. C* **2,** 177–186.

Rytkönen, K. T., Ryynänen, H. J., Nikinmaa, M., and Primmer, C. R. (2008). Variable patterns in the molecular evolution of the hypoxia-inducible factor-1 alpha (HIF-1α) gene in teleost fishes and mammals. *Gene* **420,** 1–10.

Schofield, C., Hsueh, L. C., Zhang, Z. H., Robinson, J. K., Clifton, I., and Harlos, K. (1999). Mechanistic studies on 2-oxoglutarate dependent oxygenases. *J. Inorg. Biochem.* **74,** 49–49.

Schulte, P. M. (2004). Changes in gene expression as biochemical adaptations to environmental change: A tribute to Peter Hochachka. *Comparative Biochemistry and Physiology, Part B* **139,** 519–529.

Semenza, G. L. (2007). Hypoxia-inducible factor 1 (HIF-1) pathway. *Science STKE*: cm8 [DOI: 10.1126/stke.4072007cm8].

Semenza, G. L., and Wang, G. L. (1992). A nuclear factor induced by hypoxia via de novo protein synthesis binds to the human erythropoietin gene enhancer at a site required for transcriptional activation. *Mol. Cell. Biol.* **12,** 5447–5454.

Smith, R. W., Houlihan, D. F., Nilsson, G. E., and Brechin, J. G. (1996). Tissue-specific changes in protein synthesis rates in vivo during anoxia in crucian carp. *Am. J. Physiol.* **271,** R897–R904.

Smith, R. W., Houlihan, D. F., Nilsson, G. E., and Alexandre, J. (1999). Tissue-specific changes in RNA synthesis in vivo during anoxia in crucian carp. *Am. J. Physiol.* **277,** R690–R697.

Soitamo, A. J., Rabergh, C. M. I., Gassmann, M., Sistonen, L., and Nikinmaa, M. (2001). Characterization of a hypoxia-inducible factor (HIF-1α) from rainbow trout – Accumulation of protein occurs at normal venous oxygen tension. *J. Biol. Chem.* **276,** 19699–19705.

Sollid, J., and Nilsson, G. E. (2006). Plasticity of respiratory structures – Adaptive remodeling of fish gills induced by ambient oxygen and temperature. *Resp. Physiol. Neurobiol.* **154,** 241–251.

Sollid, J., Rissanen, E., Tranberg, H. K., Thorstensen, T., Vuori, K. A., Nikinmaa, M., and Nilsson, G. E. (2006). HIF-1alpha and iNOS levels in crucian carp gills during hypoxia-induced transformation. *J. Comp. Physiol. B* **176,** 359–369.

Stangl, P., and Wegener, G. (1996). Calorimetric and biochemical studies on the effects of environmental hypoxia and chemicals on freshwater fish. *Thermochim. Acta* **271,** 101–113.

Storey, K. B., and Storey, J. M. (2004). Metabolic rate depression in animals: Transcriptional and translational controls. *Biol. Rev.* **79,** 207–233.

Taglialatela, R., and Della Corte, F. (1997). Human and recombinant erythropoietin stimulate erythropoiesis in the goldfish *Carassius auratus*. *Eur. J. Histochem.* **41,** 301–304.

Terova, G., Rimoldi, S., Cora, S., Bernardini, G., Gornati, R., and Saroglia, M. (2008). Acute and chronic hypoxia affects HIF-1α mRNA levels in sea bass (*Dicentrarchus labrax*). *Aquaculture* **279,** 150–159.

Ton, C., Stamatiou, D., Dzau, V. J., and Liew, C. C. (2002). Construction of a zebrafish cDNA microarray: Gene expression profiling of the zebrafish during development. *Biochem. Biophy. Res. Commun.* **296,** 1134–1142.

Ton, C., Stamatiou, D., and Liew, C. C. (2003). Gene expression profile of zebrafish exposed to hypoxia during development. *Physiol. Genomics* **13,** 97–106.

van den Thillart, G., Kesbeke, F., and Waarde, A. V. (1980). Anaerobic energy metabolism of goldfish, *Carassius auratus* (L) – influence of hypoxia and anoxia on phosphorylated compounds and glycogen. *J. Comp. Physiol.* **136,** 45–52.

van den Thillart, G., van Waarde, A., Muller, H., Erklens, C., Addink, A., and Lugtenberg, J. (1989). Fish muscle energy metabolism measured by *in vivo* ^{31}P-NMR during anoxia and recovery. *Am. J. Physiol.* **256,** R992–R929.

van der Meer, D. L. M., van den Thillart, G. E. E. J. M., Witte, F., de Bakker, M. A. G., Besser, J., Richardson, M. K., Spaink, H. P., Leito, J. T. D., and Bagowski, C. P. (2005). Gene expression profiling of the long-term adaptive response to hypoxia in the gills of adult zebrafish. *Am. J. Physiol.* **289,** R1512–R1519.

van Ginneken, V. J., van Den Thillart, G. E., Muller, H. J., van Deursen, S., Onderwater, M., Visee, J., Hopmans, V., van Vliet, G., and Nicolay, K. (1999). Phosphorylation state of red and white muscle in tilapia during graded hypoxia: An in vivo (31)P-NMR study. *Am. J. Physiol.* **277,** R1501–R1512.

van Ginneken, V. J. T., Addink, A. D. F., van den Thillart, G. E. E. J. M., Korner, F., Noldus, L., and Buma, M. (1997). Metabolic rate and level of activity determined in tilapia (*Oreochromis mossambicus* Peters) by direct and indirect calorimetry and videomonitoring. *Thermochim. Acta* **291,** 1–13.

van Ginneken, V. J. T., Onderwater, M., Olivar, O. L., and van den Thillart, G. E. E. J. M. (2001). Metabolic depression and investigation of glucose/ethanol conversion in the European eel (*Anguilla anguilla* Linnaeus 1758) during anaerobiosis. *Thermochim. Acta* **373,** 23–30.

van Ginneken, V. J. T., Snelderwaard, P., van der Linden, R., van der Reijden, N., van den Thillart, G. E. E. J. M., and Kramer, K. (2004). Coupling of heart rate with metabolic depression in fish: A radiotelemetric and calorimetric study. *Thermochim. Acta* **414,** 1–10.

van Waversveld, J., Addink, A. D. F., van den Thillart, G., and Smit, H. (1988). Direct calorimetry on free swimming goldfish at different oxygen levels. *J. Therm. Anal.* **33,** 1019–1026.

van Waversveld, J., Addink, A. D. F., and van den Thillart, G. (1989a). The anaerobic energy metabolism of goldfish determined by simultaneous direct and indirect calorimetry during anoxia and hypoxia. *J. Comp. Physiol. [B]* **159,** 263–268.

van Waversveld, J., Addink, A. D. F., and van den Thillart, G. (1989b). Simuultaneous direct and indirect calorimetry on normoxic and anoxic goldfish. *J. Exp. Biol.* **142,** 325–335.

Wieser, W., and Krumschnabel, G. (2001). Hierarchies of ATP-consuming processes: Direct compared with indirect measurements, and comparative aspects. *Biochem. J.* **355,** 389–395.

Zhang, Z. P., Wu, R. S. S., Mok, H. O. L., Wang, Y. L., Poon, W. W. L., Cheng, S. H., and Kong, R. Y. C. (2003). Isolation, characterization and expression analysis of a hypoxia-responsive glucose transporter gene from the grass carp, *Ctenopharyngodon idellus*. *Eur. J. Biochem.* **270,** 3010–3017.

11

DEFINING HYPOXIA: AN INTEGRATIVE SYNTHESIS OF THE RESPONSES OF FISH TO HYPOXIA

ANTHONY P. FARRELL
JEFFREY G. RICHARDS

This chapter attempts to synthesize the responses of fish to hypoxia presented in this *Fish Physiology* volume. The previous chapters are built on by differentiating between environmental hypoxia and functional hypoxia, and by outlining the possible compensatory mechanisms that fish use to counteract these forms of hypoxia. Environmental hypoxia is most simply defined as the water PO_2 when physiological function is compromised, thus the definition of environmental hypoxia is dependent upon the physiological system under examination. Hypoxia-induced decrements in maximal oxygen consumption and thus reduced aerobic scope occur at higher water O_2 levels than changes in routine oxygen consumption, which when compromised, is quantified as the critical oxygen tension (P_{crit}). At water O_2 levels below P_{crit}, duration of survival is dependent upon the capacity to reduce metabolic demands to match the limited supply of fermentable fuels. Functional hypoxia, on the other hand, occurs during situations where tissue O_2 demands exceed circulatory O_2 supply, which can be evident during exercise, temperature extremes, anemia, acidosis, and changes in gill structure, but the physiological strategies used to survive environmental hypoxia are not necessarily utilized to endure functional hypoxia.

Hypoxia: Volume 27
FISH PHYSIOLOGY

DOI: 10.1016/S1546-5098(08)00011-3

1. SCOPE OF THE CHAPTER

While a complete picture of the consequences of hypoxia in fishes will require much work to finalize, several central messages have emerged, which are detailed in the preceding chapters of this volume. The aim here is to synthesize these messages, where possible, and point to where future research might be most valuable.

2. DEFINING HYPOXIA

Simply put, hypoxia is a shortage of O_2. Anoxia is a complete lack of O_2. In its simplest context, regulators define aquatic hypoxia as dissolved O_2 concentrations below 2–3 mg O_2/L in marine and estuarine environments and below 5–6 mg O_2/L in freshwater environments. With these thresholds, regulators are aiming to protect the environment of the most sensitive fish species and for North American and European freshwaters this is often a salmonid. However, as pointed out by Diaz and Breitburg (see Chapter 1), this is clearly an oversimplification. Indeed, in a recent meta-analysis of toxicological literature (lethal and sublethal indicators of hypoxia), fish were found to be generally the most sensitive of marine taxa (Vaquer-Sunyer and Duarte, 2008). Furthermore, the current literature range for defining hypoxia of 0.2 to 4.0 mg O_2/L, with a mean of 2.1 mg O_2/L, fails to adequately protect sensitive species. Instead, Vaquer-Sunyer and Duarte (2008) suggest that 4.6 mg O_2/L may be more appropriate to protect the 90th percentile of the distribution of mean lethal O_2 concentrations.

Clearly, using an environmental concentration of O_2 is a poor way to describe hypoxia. Foremost, what is functionally hypoxic for one fish is most certainly not functionally hypoxic for all fish. Indeed, Vaquer-Sunyer and Duarte (2008) show environmental thresholds for sublethal responses to hypoxia in fishes ranging from 2 to 10 mg O_2/L. Furthermore, the O_2 concentration in water tells us relatively little about what is happening in the fish itself.

As pointed out in many of the preceding chapters, hypoxia develops for a variety of reasons. Consequently, hypoxia can be described as several forms. Wells (see Chapter 6) highlights and differentiates between environmental hypoxia and functional hypoxia, while Pörtner and Lannig (see Chapter 4) further expand on functional hypoxia by highlighting temperature-induced hypoxia. Temperature is a very important consideration when evaluating responses to hypoxia in fish because, not only does it have dramatic effects on the fish's O_2 demands (Q_{10} values of up to 4 to 6 have been reported for O_2 consumption), but it also affects the amount of dissolved O_2 available in the water (see Chapter 1). The temperature dependence of O_2 solubility in water is clearly another reason for not relying on O_2 concentration alone to define environmental hypoxia. Within physiological temperature ranges for most fish (0 to 40°C), there is a 10 and 20% decrease in dissolved O_2 for every 10°C increase in temperature.

Here we attempt to bring together all forms of hypoxia into an integrated framework, building upon the more detailed knowledge and citations in the individual chapters. In doing so, we rely on the partial pressure of O_2 (PO_2) in our definitions of hypoxia since this is what drives oxygen diffusion and, in part, determines O_2 concentration.

2.1. Environmental Hypoxia

In the broadest possible context, environmental hypoxia can be defined as the water PO_2 when physiological function is first compromised, i.e., a sublethal effect in toxicological terms. From a physiological perspective, environmental hypoxia can be defined as any water PO_2 that decreases the arterial blood O_2 concentration (C_aO_2), because such a decrease has the potential to decrease the arterial O_2 transfer factor [T_aO_2 = the product of cardiac output (Q) and C_aO_2]. At these water PO_2 levels, the fish is limited in its capacity to acquire O_2 from the environment and its blood is hypoxemic. Even so, hypoxemia does not mean that the tissues are hypoxic; routine O_2 needs can still be met through compensatory mechanisms.

In using the above definition of environmental hypoxia, it becomes clear how a resting fish can initially maintain TaO_2 by compensating for the decrease in water PO_2 and the arterial hypoxemia. Compensations include increasing gill ventilation (to deliver more, but O_2-depleted water to the gills if this is a short fall), increasing gill perfusion (if O_2 transfer across the gill secondary lamellae is perfusion rather than diffusion limited, which is often the case), increasing Q (to deliver more blood to tissues), increasing the blood hemoglobin (Hb) concentration (to increase the O_2-carrying capacity of blood usually through splenic red blood cell (rbc) release), or increasing tissue O_2 extraction (to remove more O_2 from the arterial blood and increase the arterial-venous O_2 difference). However, a fish's metabolic state must also be taken into consideration when assessing the relative importance of these compensatory responses. Full expression of these compensatory mechanisms may be possible only in resting fish. Indeed, if fish are exercising at maximum MO_2 before the hypoxic challenge started, all these compensatory mechanisms may have been already fully utilized to support the elevated O_2 requirement of the locomotory skeletal muscles. Thus, exercising fish are more sensitive than resting fish to decreasing water PO_2 (Figure 11.1) and hypoxemia. Postprandial fish likely lie somewhere in between these two extremes.

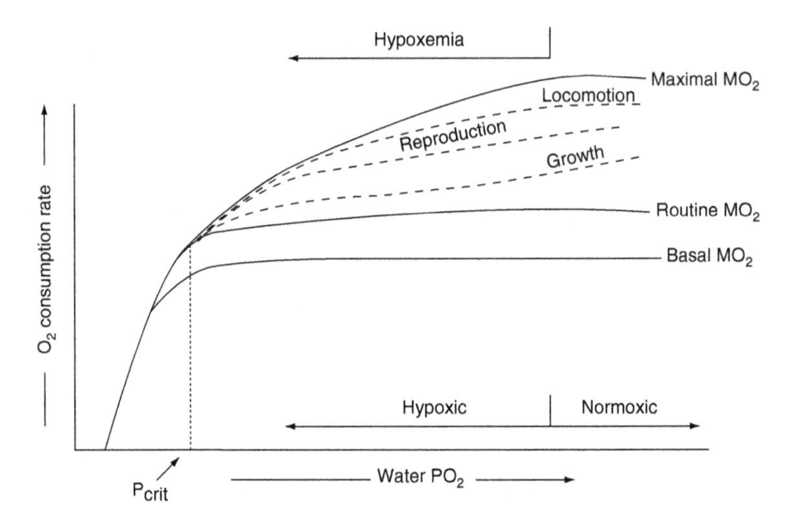

Fig. 11.1. Metabolic responses of fish to environmental hypoxia. Solid lines indicate O_2 consumption rates $\dot{M}O_2$ to support maximal, routine, and basal metabolic rates. Dashed lines show the theoretical effects of decreases in environmental O_2 tension on physiological processes. The critical O_2 tension (P_{crit}) is defined here as the point at which routine O_2 consumption transitions from being independent of environmental PO_2 to being dependent upon environmental PO_2. The difference between routine and maximal $\dot{M}O_2$ is the aerobic scope, which decreases with increasing levels of hypoxia. See text for more details.

Even when water PO_2 values are well below 100% saturation, there is an additional safety factor before hypoxemia sets. It is possible for most fishes to achieve 85–95% O_2 saturation of Hb at PO_2 values <15 kPa (see Chapter 6). Thus, fish have a zone of insensitivity to environmental hypoxia that is determined by the O_2-binding affinity of Hb. However, once water PO_2 falls below this zone of insensitivity, the zone of environmental hypoxia begins. Mechanistically, the zone of environmental hypoxia begins when arterial O_2 saturation falls below that seen in normoxia, the four compensatory mechanisms mentioned above kick in, and functional activities start to be compromised (Figure 11.1).

Clearly, the O_2 binding affinity of Hb is the principal factor setting the exact water PO_2 when integrated function begins to be lost. Consequently, since the O_2-binding affinity of Hb varies enormously among fishes, it is fairly obvious why any definition of environmental hypoxia using water Po_2 can be done only in a species-specific context. A simple index of fish's resilience to a low water PO_2 is its Hb P_{50} (the PO_2 at which arterial blood is 50% saturated); a low P_{50} value reflects a high Hb–O_2 binding affinity. Fish with a low P_{50} will curtail activities at a lower water PO_2 than fish with a higher P_{50}.

This raises the interesting point of why, in an environment where large changes in water O_2 saturation are commonplace (see Chapter 1), any fish would have a low Hb–O_2 binding affinity if a high affinity Hb can confer such an advantage? Wells (Chapter 6) makes the point that fishes adapted to environmental hypoxia have high-affinity Hbs whereas fishes adapted for functional hypoxia (exercise in this case) have low affinity Hbs. In other words, hypoxia-tolerant fishes are rarely athletic and vice versa. Low-affinity Hbs also result in a higher arterial PO_2 and this favors a faster rate of O_2 diffusion during unloading at tissues. Thus, P_{50} should vary inversely with maximum MO_2, an idea that could be easily tested among the great diversity of fish species. Unlike Hb affinity, elevated Hb and even tissue myoglobin (Mb) concentrations are adaptations common to both environmental and functional hypoxia (see Chapter 6). In both situations an elevated Hb concentration conveys the advantages of increased O_2-carrying capacity, T_aO_2, buffering capacity, and CO_2 capacity.

The zone of environmental hypoxia can also be defined as beginning at a water PO_2 when aerobic scope must begin to decrease with declining water PO_2 (Figure 11.1). As water PO_2 decreases further, a point is ultimately reached when there is no aerobic scope and only routine metabolic activities can be maintained. Above this water PO_2, a minimum routine O_2 consumption is maintained and below this water PO_2, routine O_2 consumption must conform to water PO_2 The water PO_2 at which this transition occurs is defined in the context of this chapter as the P_{crit}. Simply put, P_{crit} is the water PO_2 at which fish transitions its O_2 consumption rate (MO_2) from being independent of water PO_2 to being dependent upon water PO_2 and aerobic scope (maximal $\dot{M}O_2$-routine $\dot{M}O_2$) is zero.

In reality, however, P_{crit} can be more broadly used for as the water PO_2 when any physiological function changes as a function of water PO_2. For example, P_{crit} is widely used to indicate the initiation of the hypoxic bradycardia response common in fishes (but not other vertebrates; Farrell, 2007a). The P_{crit} for bradycardia, however, is unlikely to be analogous to the P_{crit} determined for MO_2. This is because bradycardia has variable effects on Q and hence T_aO_2 that are species-specific (see Chapter 7). For species such as trout, cod, flounder, sharks, and sturgeon, bradycardia is associated with an increase in stroke volume to maintain or slightly increase routine Q, and MO_2. Hence, the P_{crit} for bradycardia in these species lies within the zone for hypoxia but above the P_{crit} for MO_2. However, the decrease in Q, despite associated increases in stroke volume with bradycardia in sea bass, lingcod, and eels, can only indicate either a state of collapse or oxygen conformity.

Species differences in P_{crit} must then reflect interspecific differences in Hb–O_2 binding affinity and P_{50} values. Indeed, recent work by Mandic *et al.* (2008) has shown a tight relationship between P_{crit} and Hb P_{50} among several species of intertidal fishes from the family Cottidae. This then allows for a simpler definition of the zone of environmental hypoxia: a range of water

PO_2 values over which aerobic scope progressively falls to zero. Obviously, this is the zone where physiologists have focused much of their study because many fascinating behavioral, physiological, biochemical, and molecular changes occur as a fish tries to either compensate and maintain functional activities (oxyregulating), or curtail functions and reduce O_2 demand (oxyconforming). These responses might be loosely termed stress responses, although the primary stress hormones, such as epinephrine and norepinephrine are not typically released until arterial PO_2 levels fall below the blood P_{50} (Perry and Reid, 1992).

For all fishes, the zone of environmental hypoxia is characterized by a progressive loss of physiological function as the hypoxic state deepens and aerobic scope declines. The progressive loss of physiological function is likely an orderly affair. However, we know of no one who has fully characterized the exact order of these functional losses. Consequently, the order and degree of loss represented in Figure 11.1 is educated guesswork on our part. We do know that fish cannot swim as fast in hypoxia as in normoxia. But does this mean that avoidance behaviors are more likely to be powered anaerobically as P_{crit} is approached? Or is a small component of aerobic swimming capacity retained until just before P_{crit} for such purposes? Hypoxic fish also cease feeding. Mechanistically, this could be because blood flow is diverted away from the gut to more needy or O_2-sensitive tissues, as we know that absolute gut blood flow certainly decreases in hypoxic fish, even if hypoxia occurs after they have been fed. However, the physiological mechanisms integrating such a response are unknown. Hypoxia can also act as an emetic, which would be a serious concern in open sea cage aquaculture operations if hypoxic conditions were prevalent. In particular, anoxic deepwater up-wellings, eutrophication, and even elevated water temperature could dramatically lower water PO_2 and potentially impair digestion and growth well before lethal oxygen levels are reached. Reproductive development is also suspended at some point with environmental hypoxia. Reproduction may resume without serious fitness consequences for iteroparous fishes when either normoxia is restored or fish acclimate to the hypoxic conditions. However, for semelparous fishes such as salmon, which only have one opportunity to spawn, hypoxic suspension of reproductive development could have serious fitness consequences. Consideration of this potential concern may be all the more important with global warming, given that high temperature is known to halt the spawning migrations of adult salmon in the Columbia and Fraser Rivers, likely as a result of a state of functional hypoxia (more on this below).

A state of physiological anoxia exists at a water PO_2 when O_2 no longer loads onto either Hb or tissue Mb. Experimentally, this state is difficult to achieve and confirm, especially when the P_{50} values for hypoxia-tolerant fishes are <1 kPa. Therefore, to reflect this uncertainty, researchers often use the term severe hypoxia when it is clear that an animal has switched to a state of temporary anaerobic existence that is supported by glycolytic metabolism.

At O_2 levels below P_{crit}, the basic challenge is one of balancing metabolic energy demand with supply. The principal problem is that an inhibition of O_2-dependent mitochondrial ATP production imposes a substrate-limited cap on the duration of survival and as Richards points out (see Chapter 10), the duration of survival under these O_2-limiting conditions is likely dictated by two key factors: compensations, which is the ability to reduce basal metabolic demands through a controlled metabolic rate suppression; and provisions, which is the amount of substrate available for O_2-independent ATP production. A third factor, which is apparently evident only in the cyprinid family (see Chapter 9), is one of effective waste handling. In the long term, acidosis may become a more serious physiological challenge than anoxia if the fish has provisioned an extensive glycogen store and has down-regulated metabolism so that these energy stores can be metered out at a slower rate. As Vornanen and colleagues point out in Chapter 9, the conversion of lactate to ethanol and the associated consumption of protons limit the development of acidosis in the anoxia-tolerant crucian carp. An analogous situation exists in anoxic freshwater turtles that buffer protons and lactate using their shell.

2.2. Functional Hypoxia

Functional hypoxia can come in various forms, but it always reflects a situation where tissue O_2 demands exceed circulatory supply. Below we will discuss four physiological states that result in functional hypoxia: exercise, anemia, acidosis, and changes in gill structure, plus the confounding effects of temperature on the development of functional hypoxia.

2.2.1. Exercise

The most obvious example of functional hypoxia is exercise. There are numerous examples showing that as fish swim faster (increasing the O_2 demand of locomotory muscle), there comes a point when anaerobically powered locomotion takes over and lactate starts appearing in the tissues and blood. The precise mechanisms underlying this aerobic/anaerobic transition are not entirely resolved (Farrell, 2007b). Some hold that it's an O_2 supply limitation whereby the heart has reached its maximum anatomical (stroke volume) and physiological (heart rate and power development) pumping capacity, meaning internal convection can no longer deliver O_2 to the working muscle at the rate required to sustain aerobic metabolism. This internal convection limitation typically manifests itself as a plateau in Q during incremental swimming trials, as in salmonids (e.g. Steinhauser *et al.*,

2008). O_2 supply can also become limited if Hb is not fully oxygenated as it passes through the secondary lamella (as with environmental hypoxia), which could be a product of either an external convection limitation (gill ventilation), or a limitation in gill O_2 diffusion (see below). Still others view muscle capillarity as being a limiting factor for O_2 supply to working muscle (a muscle O_2 diffusion limitation).

Other research groups support the notion that the aerobic/anaerobic transition during exercise in fish is due to metabolic inertia and an energy demand that outpaces the capacity for aerobic metabolism (e.g., Richards *et al.*, 2002). Specifically, the reason lactate accumulates during intense exercise is simply that the rate of ATP production from oxidative phosphorylation cannot keep pace with the high ATP demands of contracting white muscle. As a result, pathways of ATP production shift from the slow mitochondrial pathway to the much faster glycolytic pathways and even faster PCr hydrolysis pathway. These faster metabolic pathways likely become increasingly important when tailbeat or fin sculling frequencies increase with swimming speed. The maximum contraction frequency of fish skeletal muscle is set, in part, by their ATP production pathways, as well as the kinetic properties of their contractile proteins and supporting excitation-contraction processes. Given that aerobic metabolism isn't really even an option due to the low mitochondrial content in white muscle, this means O_2 delivery to white muscle may not be an important component of why white muscle makes lactate and why measurements of white muscle mitochondrial $NADH/NAD^+$ indicate no O_2 limitation (Richards *et al.*, 2002).

What is also clear in this debate is that as fish swim faster, they ultimately "shift muscular gears," The well-capillarized, mitochondrial-rich red muscle is no longer able to power locomotion. Instead, the 20-times more abundant, and therefore mechanically more powerful, white muscle takes over. What is remarkable in this transition is that around 50% of a salmon's body mass is switched into a fully functional state with apparently little initial impact on either Q or venous PO_2 (P_vO_2). This observation is consistent with P_vO_2 reaching a lower plateau state as salmonids approach maximum prolonged swimming speed. The most parsimonious explanation for the P_vO_2 plateau is that O_2 diffusion to white muscle becomes diffusion limited under extreme exercise states. Whether red muscle also becomes functionally hypoxic due to a perfusion limitation is unknown, however.

Unravelling the "chicken and egg" conundrum of O_2 supply during exercise will be on going for some time to come since species differences undoubtedly exist. Furthermore, the potential role that body mass may play in determining which step in the O_2 cascade limits O_2 delivery in fish will require consideration beyond the recent review of Nilsson and Ostlund-Nilsson (2008). Although Nilsson and Ostlund-Nilsson (2008) generally concluded that body size had little impact on the dynamics of O_2 uptake from hypoxic environments, other anecdotal evidence suggests body size may be an important consideration in O_2 delivery. In this regard, a fascinating discovery made over 30 years ago was that Mb in the tuna ventricle increases abruptly once the fish reaches about 20 kg in body mass (Poupa *et al.*, 1981), suggesting that body mass plays a role in O_2 delivery in fishes. Myoglobin increases tissue O_2 storage and high Mb concentrations are thought to be adaptations for both environmental and functional hypoxia. However, as pointed out by Gamperl and Driedzic (see Chapter 7), the potential role of Mb in facilitating O_2 diffusion in fish hearts is not entirely resolved.

2.2.2. Anemia

This may come about as a result of accidental and experimental blood loss, pathologies (e.g., hemorrhagic septicemia), and adaptations. A low Hb concentration reduces C_aO_2 and potentially T_aO_2. Effects on C_aO_2 can be compensated for by increases in Q, which over time appear to stimulate an increase in relative ventricular mass. These compensatory responses of Q and relative ventricular mass to low Hb concentrations have been seen with experimentally induced anemia (Simonot and Farrell, 2007), as well as with adaptations to anemia. For example, some flatfishes have naturally low Hb concentrations, as do polar fishes (apparently as an adaptation to cold-induced increases in blood viscosity), and both have high Q. Nevertheless, while hemoglobin-free Antarctic icefishes have a Q and ventricular mass that rivals that of tunas, neither their maximum $\dot{M}O_2$, heart rate, nor cardiac power production come close to those of tunas (Axelsson, 2005). Consequently, the fish heart is a very plastic organ, one that responds over experimental and evolutionary time scales to ensure adequate oxygen supply. Nevertheless, there is a clear tradeoff in terms of functional ranges and aerobic scope is considerably lower in these anemic fish compared with tunas.

2.2.3. Acidosis

Many fish Hbs have a high Bohr coefficient and Root effect with some of the largest values among vertebrates (see Chapter 6). These effects potentially increase O_2 unloading from Hb as a result of CO_2 and H^+ release from tissues during capillary blood transit. Consequently, a high Bohr coefficient and Root effect are seen as adaptive for athletic fish. Added to the Bohr and Root effects is the effect of catecholamine release, which is a primary stress response during hypoxia (probably when CaO_2 falls below the Hb P_{50}) and stimulates the rbc Na^+/H^+ exchanger, elevating rbc pH and securing O_2 uptake at the gills during acidosis, but negatively affecting O_2 delivery to tissues. Thus, if fish can maintain P_aO_2 during exercise, it would seem prudent not to release catecholamines during exercise at least in terms of tissue O_2 delivery.

Collectively, a large Bohr and Root effect may cause a decrease in Hb–O_2 binding affinity, visualized as a right-shift in the Hb–O_2 dissociation curve, which increases the PO_2 gradient for unloading O_2 at tissues, enhancing the rate and amount of O_2 diffusion. Despite such benefits, the right-shifted Hb–O_2 dissociation curve reduces the PO_2 gradient driving O_2 diffusion across the gill lamellae, potentially impairing both the rate and the amount of O_2 diffusion. Of course, such impairments would be manifest only if gill O_2 diffusion was diffusion limited. Implicit to these adaptations that modulate Hb function is that a diffusion limitation of O_2 transfer at the gills is more unlikely for athletic fishes than for hypoxia-tolerant fishes. Indeed, this appears to be the case as the epithelial barrier of the secondary lamellae of eels is considerably thicker than that of tuna, for example. Species comparisons of gill diffusion capacitance might then be revealing in terms of hypoxia tolerance. A cautionary note in this regard is that such morphological changes may reflect other challenges of hypoxic environments rather than a primary modulation of gill O_2 diffusion.

2.2.4. Changes in the Gill Lamellae

The lamellae of the gills are the primary O_2 exchange sites for all water-breathing fishes heavier than about 5 g. Here, the O_2 gradient between water and blood, the lamellar surface area, and the thickness of the lamellar epithelial barrier are the primary determinants of the rate of O_2 diffusion. Various lamellae adaptations among fish have been extensively quantified (see Volume 10 in the *Fish Physiology* series). In addition, it is also clear that the lamellar morphology is extremely plastic. Functional changes in gill morphology are well documented for hypoxia, low temperature, and a variety of toxicant exposures.

Hypoxia can increase the lamellar surface area and reduce the blood to water diffusion barrier. In the scaleless carp these changes, which clearly benefit O_2 diffusion at the gills, can occur rapidly in 12–24 h (Matey *et al.*, 2008). On the other hand, low temperature exposure decreases lamellar surface area through mitotic expansion of the filament epithelial cell between adjacent lamella (see Chapter 9). These changes decrease the diffusive capacity of the gills for all molecules, presumably as a protective mechanism to limit diffusive ion movement, but at the same time decreases in lamellar surface area reduce aerobic scope. The so-called osmorespiratory compromise (Gonzalez and McDonald, 1992) highlights the potential tradeoffs of having a multifunctional gill (that being an organ involved in both ion regulation and O_2 transfer) and that modifications to the gill epithelium that limit molecule movement are beneficial for ion regulation, but detrimental for gas exchange, in some cases inducing functional hypoxia. It seems probable that O_2 conformity during hypoxia relaxes the need to optimize gill diffusing capacity perhaps decreasing ion loses (in freshwater) or gain (in seawater), decreasing toxicant uptake from water, and increasing the protective barrier for pathogen entry.

2.2.5. Temperature-induced Hypoxia

The recent discovery that temperature extremes lead to hypoxic states is generating considerable interest in ecologists, physiologists, biochemists, and molecular biologists. Pörtner and Lanning (see Chapter 4) detail the evidence for, and consequences of, this form of hypoxia. While temperature-induced hypoxia obviously reflects an environmental change, it would be misleading to characterize this as a form of environmental hypoxia, as there is good evidence that some species maintain T_aO_2 at high temperature but are still functionally hypoxic when temperature exceeds its optimum (Steinhausen *et al.*, 2008). In this case, it is the routine O_2 demand that increases as a result of the Q_{10} effect on routine metabolic rate. In fact, much like exercise, increasing temperature increases the tissue O_2 demand until it eventually outstrips the ability of the heart to deliver O_2. Once T_aO_2 is maximal, further increases in temperature may lead to cardiac collapse as revealed by a declining Q and cardiac arrhythmias (Clark *et al.*, 2008).

The mechanism(s) for cardiac collapse at high temperature is (are) still under scrutiny (see Chapter 7). One possible mechanism explaining cardiac collapse at high temperatures relates to the fact that at least 50%, if not all, of the ventricular muscle in fishes depends on the venous PO_2 as the pressure driving O_2 diffusion to cardiac mitochondria. Thus, a hypothetical temperature-induced hypoxic spiral to death might start with Q first reaching its anatomical (maximum stroke volume) and physiological (maximum heart rate and power) limits. Any increase in O_2 extraction from arterial blood to meet the increased peripheral tissue O_2 demands would ultimately lower venous PO_2, which then limits cardiac O_2 supply, restricting cardiac performance and T_aO_2. Attempts by fish to exercise at high temperature, and perhaps even the energetic requirements of avoidance behaviors, could easily make matters worse by increasing O_2 extraction by the muscle and further reducing P_vO_2. Further work into the causes of cardiac collapse is urgently needed.

Fish appear to have evolved a number of strategies to avert or delay the hypoxic cardiac spiral. Foremost, and as noted above, venous PO_2 may be maintained above a threshold level in one of several ways. One is that unloading of O_2 at peripheral tissues may eventually become diffusion limited once internal convection has reached its maximum capacity and the arterial to venous O_2 difference has been fully exploited. This may be inevitable if the metabolic demand of white muscle, which has an inherent tissue O_2 diffusion limitation, takes on an increasing portion of the overall metabolic rate as temperature increases. In addition, the temperature-induced decrease in Hb–O_2 binding affinity potentially elevates P_vO_2 with increasing temperature, enhancing tissue O_2 delivery, including that to the heart. If, however, tissue O_2 uptake becomes diffusion limited, as suggested, the temperature-induced decrease in Hb affinity would increase tissue O_2 delivery without decreasing P_vO_2. Indeed, acute warming can result in decreases or no change in P_vO_2 when T_aO_2 has been maximized (Steinhausen *et al.*, 2008). Lastly, a coronary circulation represents a more secure O_2 supply for the ventricle because it brings arterial blood directly from the gills to at least the outer part of the ventricle. The phylogenetic distribution of the coronary circulation in relation to temperature is unknown. Those teleosts that have a coronary circulation tend to be athletic or hypoxia tolerant, and so temperature tolerance may not be a primary factor determining the presence or absence of a coronary circulation in fish. Interestingly, all elasmobranchs have a coronary circulation, but most teleosts do not.

While it may seem reasonable to conclude that temperature-induced hypoxia is best categorized as a state of functional hypoxia, there are intriguing functional parallels with environmental hypoxia that need to be further explored. In characterizing the responses to environmental hypoxia (Figure 11.1), we identified a zone of independence of O_2 consumption from water PO_2, where aerobic scope was preserved. Is this zone homologous or just functionally analogous with the zone for optimal temperature described by Pörtner and Lannig (see Chapter 4)? Similarly, is the zone of hypoxia functionally homologous to the pejus temperatures and the P_{crit} functionally homologous to the critical temperatures?

2.3. Exposure Time in Defining Hypoxia

An important component of defining hypoxia is exposure time. The rate of change in water PO_2 during the development of hypoxia is crucial for both the type of physiological and biochemical responses initiated as well as animal survival. The faster the rate of change in water PO_2, the more likely dire consequences will follow. This is primarily because the capacity for responding to hypoxia over short time periods is dictated by functioning physiological and biochemical systems that are in place at the time hypoxia is imposed. If, for example, during progressive hypoxia exposure the physiological mechanisms for maintaining O_2 extraction from the environment do not maintain pace with the falling environmental levels, then the fish falls into a cascade toward eventual death. As mentioned previously in this chapter, this transition point is classically defined at P_{crit}. At water PO_2 levels below P_{crit}, the time until death will be dictated primarily by two interlinked biochemical responses: (1) reductions in basal metabolic demands to match the limited capacity for ATP production; and (2) the availability of substrate for O_2-independent ATP synthesis. There is another possible metabolic option that can be employed to extend survival at O_2 levels far below P_{crit} and that is the capacity to modify the mitochondria to function at reduced O_2 tensions, but unfortunately little work has examined this option in fish.

Acclimation or acclimatization to hypoxia exposure involves processes that are distinct from the acute response, but are likely initiated simultaneously with hypoxia exposure, and involve the restructuring of physiological and biochemical responses to enhance function and extend survival. As Richards points out in Chapter 10, all cDNA microarray studies performed to date with hypoxia-exposed fish show a suite of gene expression changes that are consistent with optimizing O_2 uptake from the environment (e.g., gill apoptotic factors causing a thinning of the gill; see also Chapter 9), O_2 distribution and circulation (e.g., changes in heme synthesis and iron metabolism; see also Chapters 5 and 6), and metabolic energy balance (e.g., up-regulation of glycolysis and suppression of energy consumption; see Chapters 7, 9, and 10). Hypoxia-induced changes in gene expression in fish are likely regulated in a similar fashion to mammals and by the hypoxia inducible factor (HIF; see Chapter 10), but remarkably, a direct linkage between HIF function and hypoxia tolerance among fish species has not been demonstrated at this time.

Overall, acclimation occurs over hours, days and weeks, and involves changes predominately in gene expression, which facilitate the reorganization of the physiological process of O_2 uptake and distribution and biochemical processes of cellular energy metabolism. While the gill secondary lamella and cardiac tissues are both extremely plastic in fishes, secondary lamellae can be altered after as little as 12 h of hypoxia, but it takes at least 2 weeks for the heart to increase in mass during anemia. Temperature may also play a role in how quickly these compensations take place, and temperature may even trigger for hypoxic acclimations. The interaction between cold acclimation and hypoxic acclimation is worth considering since cooling of the water of course precedes the progression to hypoxia in ice-covered lakes.

Although few, if any studies, have examined the effects of hypoxic acclimation on P_{crit}, it is likely that acclimation would yield a decrease in P_{crit} brought about by changes in gill structure, cardiac function, and Hb profiles. Acclimation may also shift O_2 thresholds for the display of avoidance behaviors. At water PO_2 levels below P_{crit}, the advantage of acclimation responses in up-regulating tissue-specific capacities for O_2-independent ATP production is reasonably clear, which can dramatically affect survival time. Clearly, prolonged exposure to environments depleted in O_2 can affect the hypoxic response and must be considered when deciding whether a particular fish species is hypoxic or not.

When considering the potential impacts of acclimation/acclimatization on hypoxia tolerance, not only should chronic depletions in water PO_2 be considered, but also oscillations in water PO_2. Many of the most hypoxia-tolerant fish species inhabit environments that undergo diurnal fluctuations in O_2 that are due primarily to plant respiration (see Chapter 1 for more detail). Some of the best examples are fish found in the Amazon (see Volume 21 on Tropical Fishes in the *Fish Physiology* series) and fishes found within intertidal environments. It is presently unknown if oscillations in water PO_2 achieve the same degree of acclimation as chronic hypoxia, but it is likely that some degree of environmental entrainment occurs. The differentiating effects of oscillations in water PO_2 from chronic hypoxia exposure is the need to recover from the oscillating hypoxia. Hypoxic recovery has not been extensively studied, but in general is has been shown to occur at a faster rate than recovery from exercise (Hallman *et al.*, 2008) and involves an enhancement of $\dot{M}O_2$ to facilitate an aerobic recovery from the anaerobic period. Further work into hypoxia recovery is needed.

It has long been assumed that the prevalence of environmental hypoxia in the aquatic environment has been a powerful evolutionary pressure resulting in the selection of the traits described throughout this volume. Indeed, fish species represent the ideal "model" system to understand the selection of traits underlying hypoxia tolerance because of the highly specious nature of fish and the extremely O_2 diverse environments they inhabit. Certainly, the study of diverse fish groups has led to the general consensus of the selection-driven traits associated with hypoxia tolerance, but the capacity to address questions of adaptation is far more limited. With the advent of the phylogenetic comparative method, comparative physiologists can now attempt to identify the repeated evolution of a trait correlated with one or more putatively selective variables while factoring out the possible confounding effects of shared ancestry among species (Felsenstein, 1985). The application of phylogenetically independent contrasts assists in identifying selection-driven traits but requires an understanding of the phylogenetic relationships among the species under study. A phylogenetic comparative approach was recently taken by Mandic *et al.* (2008) using a group of fish from the family Cottidae (sculpins), which are distributed along the marine intertidal zone and experience oscillating hypoxia to varying degrees. In these fish, there was a phylogenetically independent relationship between P_{crit} and routine metabolic rate, total gill surface area, and rbc Hb–O_2 binding affinity, such that variation in these components accounted for 75% of the variation in P_{crit} (Mandic *et al.*, 2008). More studies that take into account phylogenetic relationships are needed to isolate adaptation from phylogenetic signal.

Ultimately, beyond a critical PO_2 only a passive, anaerobic existence is possible, one that is rarely exploited among fishes. Carps and hagfishes may be the important exceptions in this regard. However, sessile invertebrates inhabiting the extremely variable intertidal environment use metabolic depression, anaerobic energy production, and stress protection mechanisms to provide short- to medium-term tolerance of this extremely challenging environment. Mobile intertidal fishes, on the other hand, employ a complex suite of behavioral, physiological, and biochemical strategies for long-term hypoxic survival, which are influenced to a large degree by many ecological factors (see Chapter 2) such as perceived risk of predation and availability of cover. The use of simple whole-animal measures, such as P_{crit}, will undoubtedly be of benefit to examine the relationships between environment, ecology, and hypoxia tolerance among the numerous species of small fishes that inhabit challenging environments.

3. CONSIDERATIONS FOR THE FUTURE

We are entering a golden age for comparative physiology, propelled by the development of tools to dissect the responses of fish from gene through to the ecosystem and to place these responses into an evolutionary and ecological context. Although we have accumulated a wealth of information on how many fish species respond to hypoxia (both environmental and functional) and the potential adaptations underling hypoxia tolerance, there remain several areas of study that have yet to be adequately explored. Furthermore, the ability to compare among studies is paramount, but as pointed out above, hypoxia has no simple definition and therefore mechanisms for comparisons among studies must be developed. We offer the following suggestions for consideration to anyone interested in hypoxic research.

1. Because "hypoxia" must be considered in light of the species under study, a water PO_2 that is hypoxic for one fish species could conceivably have no effects on another fish species. For example, what might be considered hypoxic for a rainbow trout may have almost no measurable effect on an extremely hypoxia-tolerant fish such as the crucian carp. Thus, as pointed out earlier in this chapter, reporting the water PO_2 at which a physiological response occurs is more useful if it is put into the context of the organism under study. Being able to standardize hypoxia exposures across species is essential to understanding processes of adaptation. Expressing hypoxia exposures relative to P_{crit}, as defined by the transition from an oxyregulator to an oxyconformer, may provide an overall framework to allow researchers to standardize responses observed in diverse groups of fish.
2. There is a need for more studies that better replicate environmental conditions. Studies of chronic hypoxic exposure are in short supply, as

are oscillations that might reflect diurnal rhythms. Synergistic responses between hypoxia and other environmental changes such as hypercarbia (elevated water CO_2), acidosis, and temperature would be worthwhile as multiple loadings are predicted to decrease aerobic scope. Furthermore, studies that integrate across levels of biological organization and put behavioral, physiological, biochemical, and molecular responses into an ecological context are essential.

3. Additional curiosity driven research should explore whether the ultimate outcomes of environmental hypoxia exposure and functional hypoxia exposure in fish are similar, particularly at the biochemical and molecular level. Some research has demonstrated that responses to environmental hypoxia and functional hypoxia induced by exercise differ, but the precise reasons for these differences are not known. Does, for example, exercise-induced hypoxia elicit the same gene expression responses in muscle as environmental hypoxia?

ACKNOWLEDGMENTS

The authors would like to thank the Natural Sciences and Engineering Council of Canada for financial support of their research.

REFERENCES

Axelsson, M. (2005). The circulatory system and its control. *In* "Fish Physiology, Polar Fishes" (Farrell, A. P., and Steffensen, J. F., Eds.), Vol. 22, pp. 239–280. Academic Press, San Diego.

Clark, T. D., Sandblom, E., Cox, G. K., Hinch, S. G., and Farrell, A. P. (2008). Circulatory limits to oxygen supply during an acute temperature increase in the Chinook salmon (*Oncorhynchus tshawytscha*). *Am. J. Physiol.* **295,** R1631–R1639.

Farrell, A. P. (2007a). Tribute to P. L. Lutz: A message from the heart – why hypoxic bradycardia in fishes? *J. Exp. Biol.* **210,** 1715–1725.

Farrell, A. P. (2007b). Cardiorespiratory performance during prolonged swimming tests with salmonids: A perspective on temperature effects and potential analytical pitfalls. *Phil. Trans. R. Soc. B.* **362,** 2017–2030.

Felsenstein, J. (1985). Phylogenies and the comparative method. *Am. Nat.* **125,** 1–15.

Gonzalez, R. J., and McDonald, D. G. (1992). The relationship between oxygen consumption and ion loss in a freshwater fish. *J. Exp. Biol.* **163,** 317–332.

Hallman, T. M., Rocha, A., Jones, D. R., and Richards, J. G. (2008). Metabolic recovery from exercise and hypoxia exposure measured using ^{31}P- and ^{1}H-NMR in the common carp, *Cyprinus carpio*. *J. Exp. Biol.* **211,** 3237–3248.

Mandic, M., Todgham, A. E., and Richards, J. G. (2008). Mechanisms and evolution of hypoxia tolerance in fish. *Proc. Roy. Soc. B.* DOI: 10.1098/rspb.2008.1235.

Matey, V., Richards, J. G., Wang, Y. X., Wood, C. M., Rogers, J., Semple, J., Murray, B. W., Chen, X.-Q., Du, J., and Brauner, C. J. (2008). The effect of hypoxia on gill morphology and ionoregulatory status in the endangered Lake Qinghai scaleless carp, *Gymnocypris przewalskii*. *J. Exp. Biol.* **211,** 1063–1074.

Nilsson, G. E., and Ostlund-Nilsson, S. (2008). Does size matter for hypoxia tolerance in fish? *Biol. Rev. Camb. Philos. Soc.* **83,** 173–189.

Poupa, O, Lindstrom, L., Maresca, A., and Tota, B. (1981). Cardiac growth, myoglobin, proteins and DNA in developing tuna (*Thunnus thynnus thynnus* L.). *Comp. Biochem. Physiol. A* **70,** 217–222.

Perry, S. F., and Reid, S. D. (1992). Relationship between blood O_2-content and catecholamine levels during hypoxia in rainbow trout and American eel. *Am. J. Physiol.* **263,** R240–R249.

Richards, J. G., Heigenhauser, G. J. F., and Wood, C. M. (2002). Glycogen phosphorylase and pyruvate dehydrogenase transformation in white muscle of trout during high-intensity exercise. *Am. J. Physiol.* **282,** R828–R836.

Simonot, D. L., and Farrell, A. P. (2007). Cardiac remodelling in rainbow trout *Oncorhynchus mykiss* Walbaum in response to phenylhydrazine-induced anaemia. *J. Exp. Biol.* **210,** 2574–2584.

Steinhausen M. F., Sandblom E., Eliasson E., Verhille C., and Farrell A. P. (2008). The effect of acute temperature increases on the cardiorespiratory performance of resting and swimming sockeye salmon *(Oncorhynchus nerka)*. *J. Exp. Biol.* **211,** 3915–3926.

Vaquer-Sunyer, R., and Duarte, C. M. (2008). Thresholds of hypoxia for marine diversity. *PNAS* **105,** 15452–15457.

INDEX

A

D

E

H

I

J

K

L

N

O

P

R

S

T

OTHER VOLUMES IN THE FISH PHYSIOLOGY SERIES

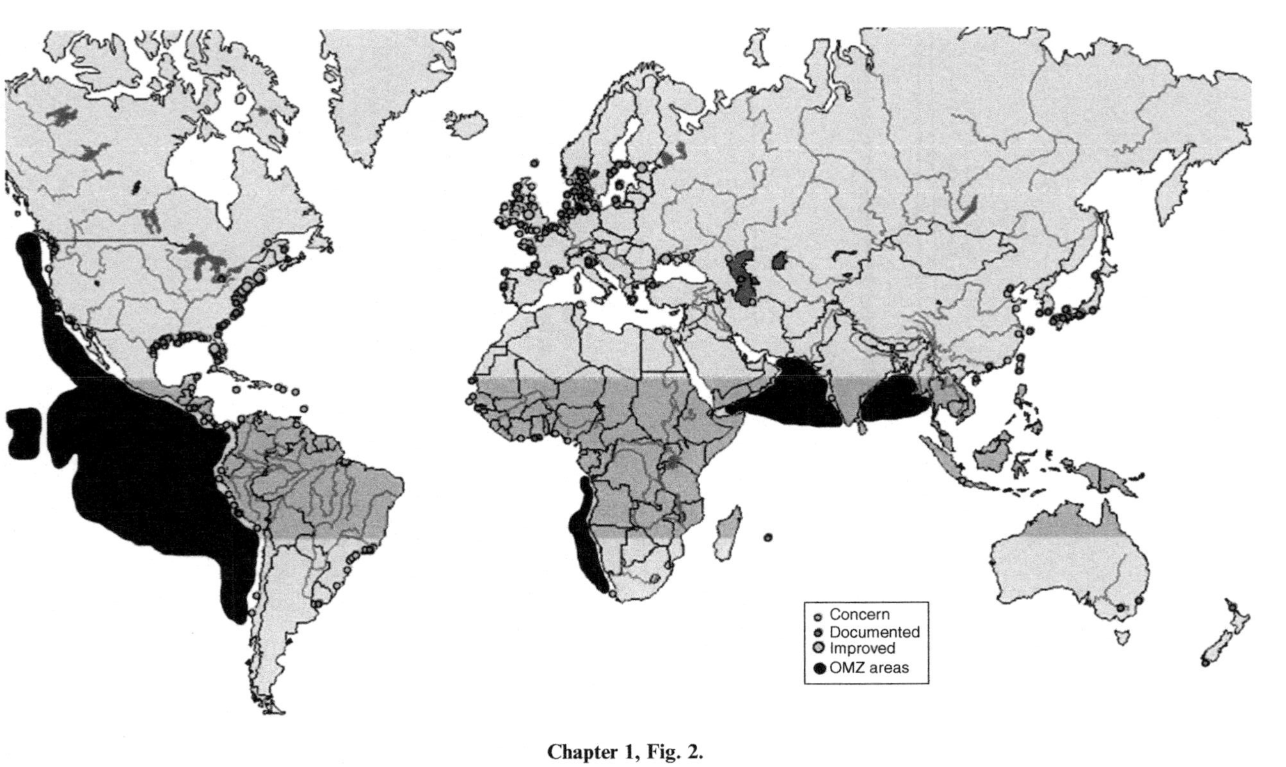

Chapter 1, Fig. 2.

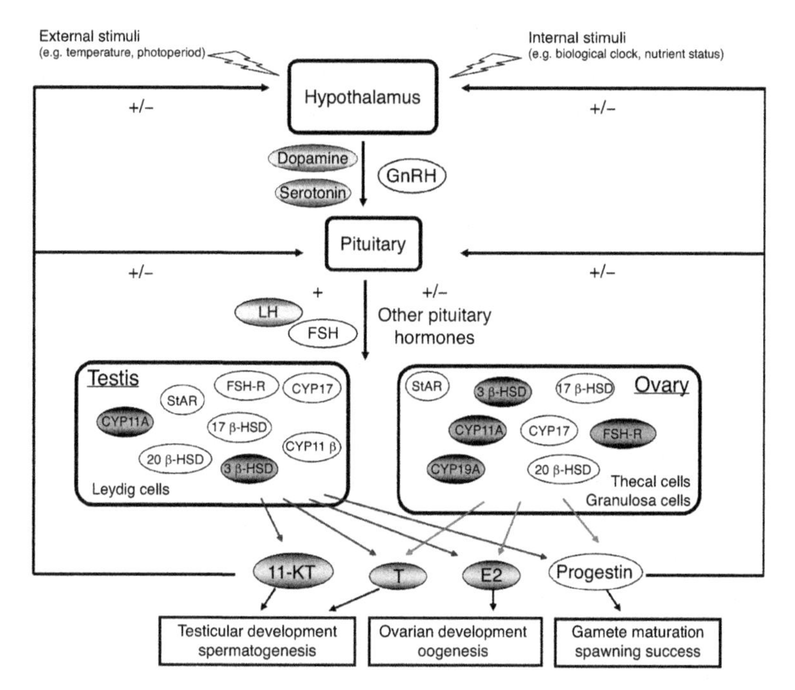

Chapter 3, Fig. 2.

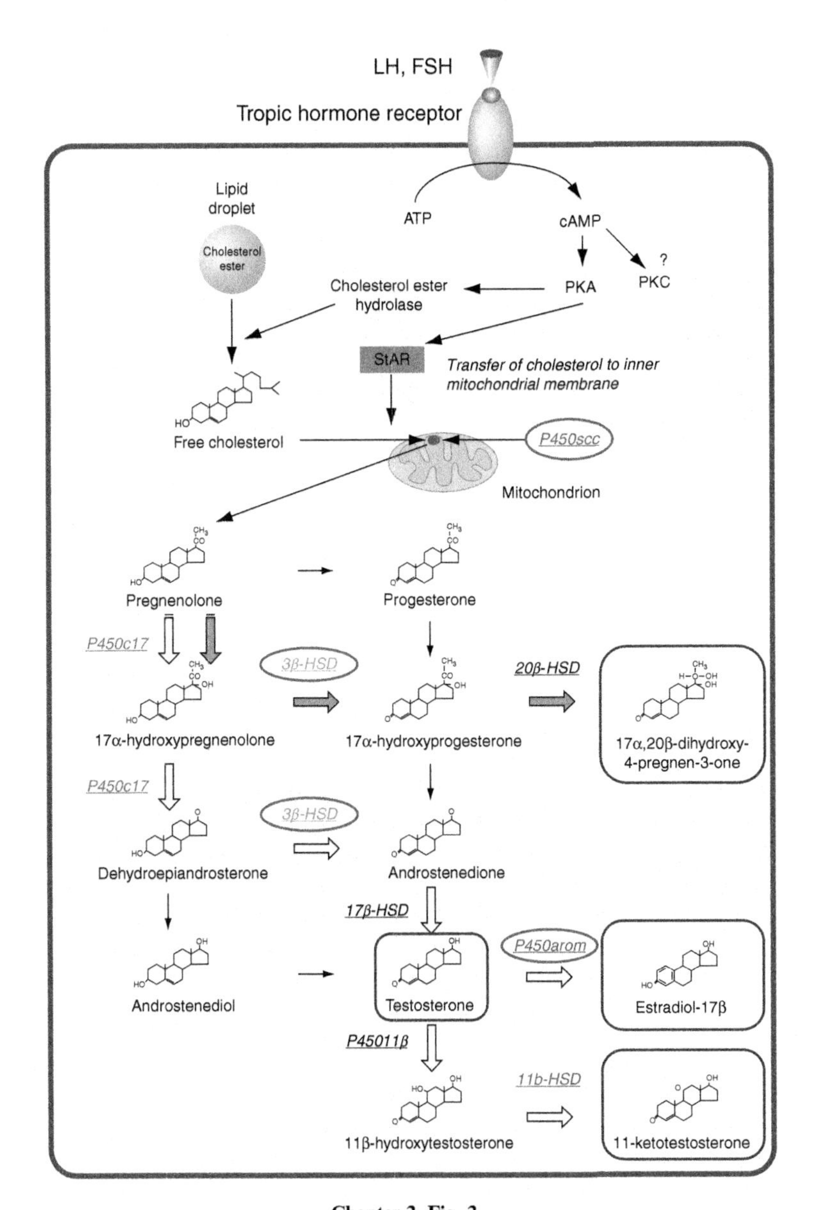

Chapter 3, Fig. 3.

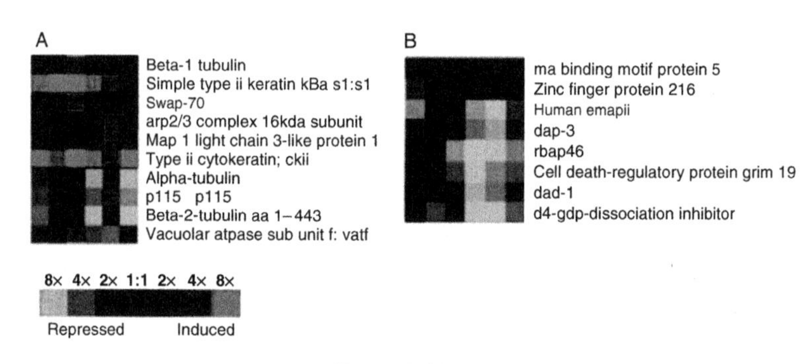

Chapter 3, Fig. 4.

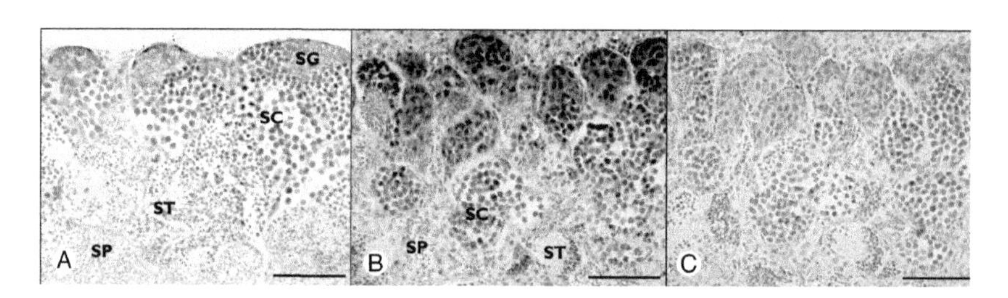

Chapter 3, Fig. 23.

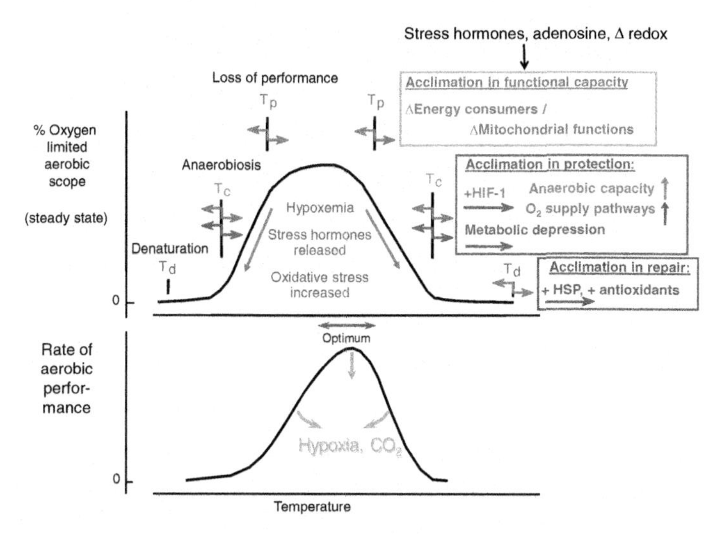

Chapter 4, Fig. 1.

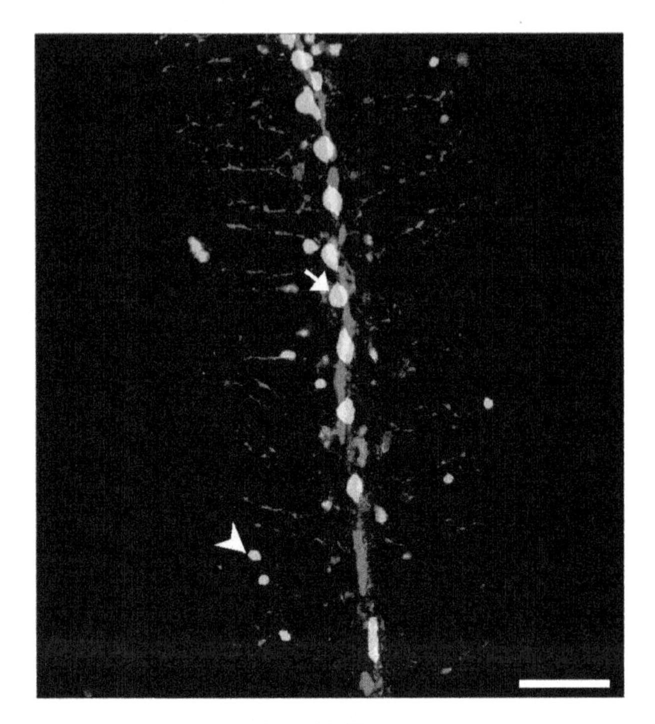

Chapter 5, Fig. 6.

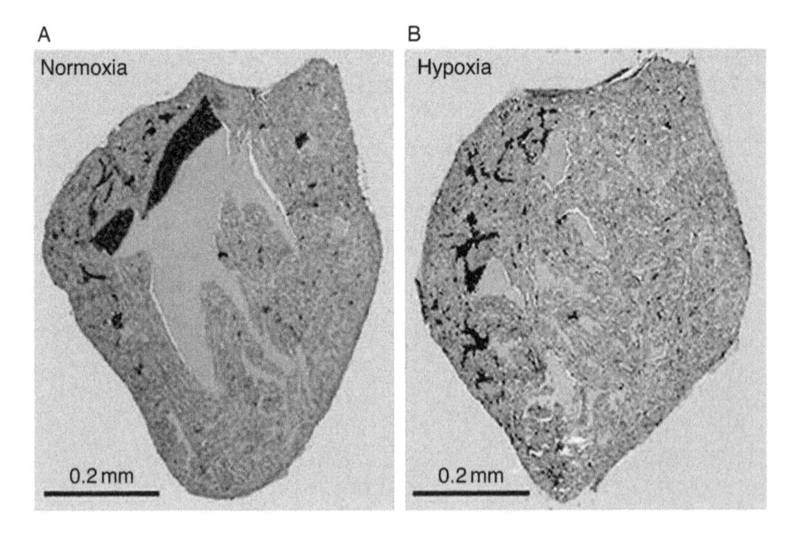

Chapter 7, Fig. 13.

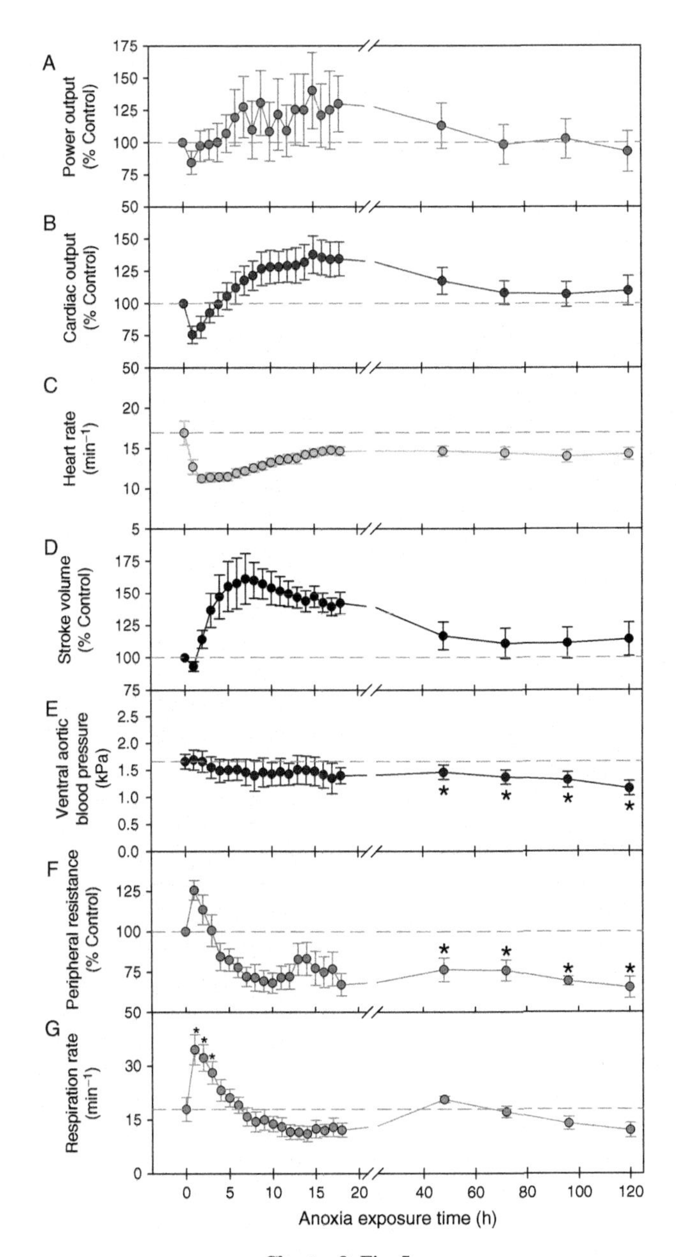

Chapter 9, Fig. 5.

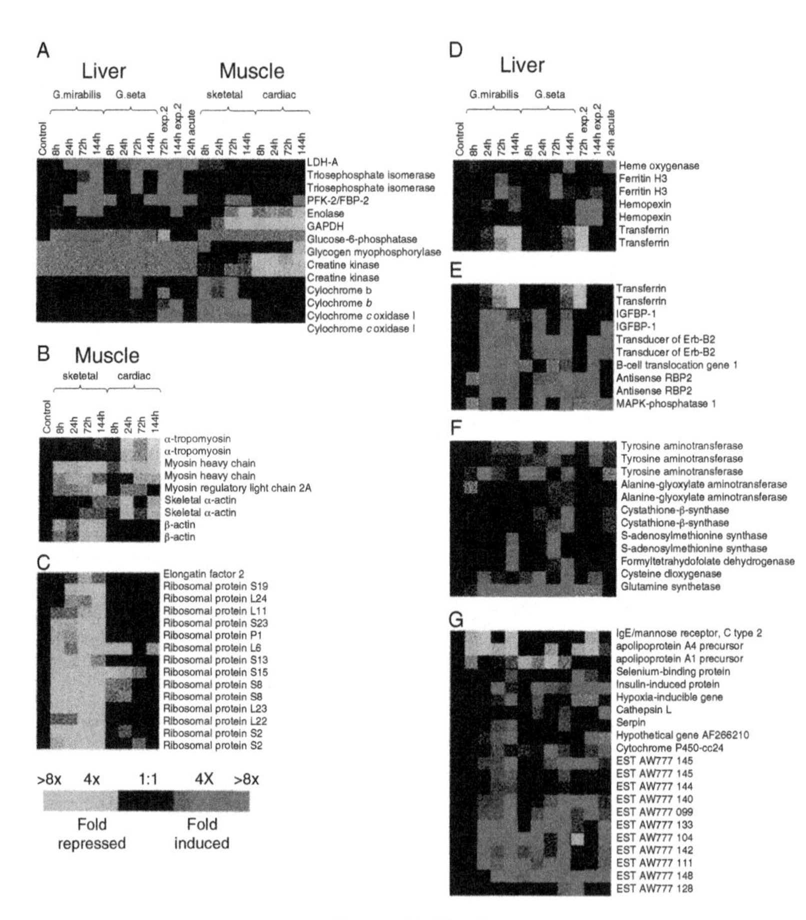

Chapter 10, Fig. 3.

Printed in the United States
By Bookmasters